SOIL-WATER-SOLUTE PROCESS CHARACTERIZATION

An Integrated Approach

SOIL-WATER-SOLUTE PROCESS CHARACTERIZATION

An Integrated Approach

Edited by
JAVIER ÁLVAREZ-BENEDÍ
RAFAEL MUÑOZ-CARPENA

CRC Press
Taylor & Francis Group
Boca Raton London New York

CRC Press is an imprint of the
Taylor & Francis Group, an **informa** business

CRC Press
Taylor & Francis Group
6000 Broken Sound Parkway NW, Suite 300
Boca Raton, FL 33487-2742

First issued in paperback 2019

ISBN-13: 978-1-56670-657-5 (hbk)
ISBN-13: 978-0-367-39335-9 (pbk)

Library of Congress Cataloging-in-Publication Data

Soil-water-solute process characterization: an integrated approach / [edited by]
Javier Álvarez Benedí and Rafael Muñoz-Carpena.
p. cm.
Includes bibliographical references and index.
ISBN 1-56670-657-2 (alk. paper)
1. Soil moisture–Mathematical models. 2. Soils–Solute movement–Mathematical
models. 3. Soil permeability–Mathematical models. 4. Groundwater flow–
Mathematical models. I. Álvarez-Benedí, Javier. II. Muñoz-Carpena, Rafael.
III. Title.

S594.S6935 2005
631.4'32'011–dc22   2004015853

Library of Congress Card Number 2004051920

Preface

The development and application of methods for monitoring and characterizing soil-water-solute processes are among the most limiting factors in understanding the soil environment. Experimental methods are a critical part of scientific papers, and their design and implementation are usually the most time-consuming tasks in research. When selecting a method to characterize a property governing a soil process, the practitioner or researcher often faces complex alternatives. In many cases these alternatives are bypassed in favor of recommendations from colleagues on well-established methods that might not be the most suitable for the specific conditions of a study.

Several factors add to the complexity of selecting the best characterization method for a particular case:

- The governing properties or parameters are referred to by similar names although in fact their actual values depend greatly on the conceptual model selected to explain the process (e.g., several empirical models of soil infiltration have different parameters associated with saturated hydraulic conductivity).
- The ultimate goal of the characterization effort, whether it be general classification of the soil, qualitative estimation of an output, exploratory modeling to gain insight on a process, quantitative modeling prediction, etc., may determine the method of choice.
- Since many of the soil characteristics are intrinsically variable (spatially and temporally), the most accurate method might not necessarily be the best choice when compared with a simpler one that can provide a larger number of samples with the same or lower investment.

An integrated approach for soil characterization is needed that combines available methods with the analysis of the conceptual model used to identify the governing property of a soil process, its intrinsic nature (variability), and the ultimate use of the values obtained. This holistic approach should be applied to the selection of methods to characterize energy and mass transfer processes in the soil (i.e., water and solute flow), sorption, transformation, and phase changes, including microbiological processes.

This book applies this integrated approach to present a comparative discussion of alternative methods, their practical application for characterization efforts, and an evaluation of strengths, weaknesses, and trade-offs. This book is not a laboratory or field handbook. The authors present the information with a critical spirit, showing benefits, limitations, and alternatives to the methods when available. Numerous references to some of the excellent handbooks and publications available are given for details on each of the methods. Some nontraditional state-of-the-art characterization methods (NMRI, x-ray tomography, fractals) and modeling techniques are also presented as alternatives or as integral components.

The book is divided into six sections. (Fig. P.1) The first section defines the basis for the integrated strategy that will be developed in the following sections (i.e., need and use, issues of spatial and temporal variability, and modeling as an integral part of the process). Sections II–IV present the critical evaluation of methods available for energy and water transfer, chemical transport, and soil microbiological processes. Different methods of characterization are presented and compared using numerous tables and diagrams to help the users identify the most suitable option for their application. Section V discusses tools and applications to account for the intrinsic temporal and spatial variability and scale of soil processes. The last section is devoted to modeling aspects including uncertainty, inverse modeling, and practical recommendations.

Section I contains three chapters. In Chapter 1, Corwin and Loague discuss the problem of subsurface non-point-source pollution and present the application of an integrated array of multidisciplinary, advanced information technologies useful in characterizing the process. This chapter introduces

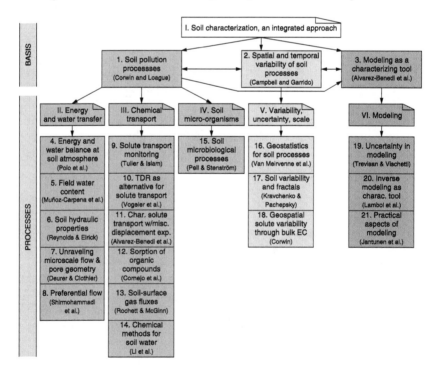

FIGURE P.1. Book structure and contents roadmap.

the methods Sections II–IV. In Chapter 2, Campbell and Garrido explore the role of deterministic and stochastic approaches to describe soil processes, and how their intrinsic temporal and spatial variability affect method selection in field studies. These issues remain the greatest challenges for field research and limit the quantitative comparison of existing field studies. This chapter provides the background for the chapters in Section V. Chapter 3, written by Alvarez-Benedí et al., proposes that modeling (conceptual and mathematical) is at the core of the characterization effort, affecting not only the method selection but also the final application of the study. A review of the conceptual building blocks needed to construct a soil-water-solute mathematical model is presented here to illustrate current assumptions and limitations when modeling soil-water-solute phenomena. This chapter leads into the final section (VI) of the book.

Section II, devoted to soil physical processes, opens with Chapter 4, in which Polo et al. offer a review of water and energy exchange processes between soil-plant and the atmosphere. A comparison of methods to account for energy and water balance, with emphasis on evapotranspiration, is presented. This chapter serves as background for the rest of the section, in which water or energy are discussed. The authors reflect on how the selection and success of the physically based methods presented require knowledge of the relevant spatial and temporal scales and a determination of the uncertainty associated with the variables of interest. In Chapter 5 Muñoz-Carpena et al. present current field methods to monitor soil water status. Soil water potential and soil water content devices are presented and compared in terms of desired moisture range to measure, soil type, accuracy, soil volume explored by the device, soil salinity levels, device maintenance and installation issues, and cost. A criterion to select the most suitable method for a given application is presented. Chapter 6, by Reynolds and Elrick, introduces current methods to characterize soil hydraulic parameters that control soil water redistribution and flow. They point out that *in situ* measurements are essential for dealing with the extreme complexity of the field, and that rather than using a single method, the correct approach seems to use a "suite" of complementary methods. They propose the infiltrometer, permeameter, and instantaneous profile methods as the core of such a suite of methods. In Chapter 7, Deurer and Clothier give an in-depth look at the complex soil topology using two state-of-the-art methods — NMRI and x-ray tomography. These complementary techniques provide a new look at the microscopic scale and topology of pore geometry and of water and solute transport. Although for most practical applications these methods are not yet cost-effective, they may be so in the near future. Chapter 8, by Shirmohammadi et al., offers a critical assessment of one of the most daunting problems encountered when describing flow and transport processes in the soil: preferential flow. Preferential flow, probably more often than not, presents a limit to our classical description of such processes. Different experimental methods to quantify the presence of preferential paths are compared. A detailed presentation of the theoretical representation of the process and alternative models follows, with emphasis on their limitations. It is concluded that our handling of the preferential flow either fails to be properly represented mathematically, or fails in the parametrization for proper representation of the system.

Within Section III, dedicated to solute processes, Tuller and Islam present in Chapter 9 an exhaustive review of field methods for characterizing solute transport. They conclude that our ability to measure and characterize spatial distribution of chemicals and preferential migration pathways is restricted due to the application of *in situ* point measurements with limited volume and geophysical techniques that only work indirectly or qualitatively. The authors present electrical methods such as time domain reflectometry (TDR), electrical resistivity tomography (ERT), and magnetic induction as the most promising for large-scale and real-time monitoring. In Chapter 10, Vogeler et al. show the modern application of the TDR technique to measure not only water content but also saline solute concentration through soil electrical conductivity (EC_a) changes. The method can be applied reliably and successfully to study nonreactive and reactive solutes. Although the estimation of EC_a with TDR is well established, the relation of that with solute concentration is soil specific, influenced by soil texture/structure and bulk density, and not yet fully understood. Two weaknesses of the method are the relatively small zone of influence and inability to discriminate between different ionic species. The method should not replace existing monitoring techniques, but rather complement them. In Chapter 11, Alvarez-Benedí et al. build on Chapters 3, 5, 9, and 10 to discuss the laboratory characterization of solute transport through miscible displacement experiments. This method is presented as the most important for characterizing solute transport at small to large lysimeter (column) scale, especially if several experiments can be performed varying hydrodynamic conditions and tracers in the same column. However, extending this methodology to the field scale is usually not feasible, and field experiments like the ones presented in Chapter 9 are preferred for validation purposes of the parameters obtained in the column studies. Chapter 12, by Cornejo et al., compares methods to determine sorption of pesticides in the soil. This process controls pesticide transport in different soils and conditions and has important environmental implications. The selection of the method is governed by the accuracy required for the intended use and regulatory environment. In Chapter 13, Rochette and McGinn take a critical look at state-of-the-art methods to quantify another controlling factor in the distribution and degradation of contaminants from the soil, volatilization. Three types of techniques are compared: soil mass balance, chambers, and micrometeorology. Because of the usually significant error associated with any one technique, the authors recommend the use of two techniques when possible to increase confidence in the gas flux estimates. Further research is recommended in all these techniques to reduce current uncertainty in measurements. Chapter 14 completes Section III. Li et al. present a critical and exhaustive look at one important aspect often overlooked by field researchers and practitioners in soil solute characterization studies, i.e., the chemical analysis of the samples. There is a multitude of available techniques to analyze any given element or compound. The selection of the appropriate method is often complex, since new methods and techniques are continuously entering the market. In addition, the intrinsic uncertainty, interferences, and method detection limit (MDL) are not always taken into account when interpreting the results, although these can vary greatly across methods. Comparison of results obtained with a standard method is the important criterion in the selection of an appropriate method. Laboratory accreditation is discussed as a growing trend that will benefit the scientist and clientele of analyses.

Section V (and Chapter 15) is devoted to the emerging area of soil microbiological processes. Pell and Stenström discuss the fact that, although soil quality is closely related to soil microbiology, the latter has received little attention. This common oversight is at the root of many of the difficulties found in measuring or predicting reactive solute transport of important contaminants such us pesticides and fertilizers. The authors describe how microbial respiration and nitrification/denitrification processes affect soil sample handling, soil reactive behavior, and how microbial parameters can be used in soil function description and assessment of special variability at different scales. The authors conclude that cooperation between soil physicists, chemists, and microbiologists is needed to advance our understanding of soil processes.

Section VI reviews available techniques that could be incorporated in methods to address the intrinsic soil variability. In Chapter 16, Van Meirvenne et al. give a comprehensive review of available geostatistical techniques and how to incorporate them in field and laboratory methods. Despite its promise, there is no single solution for all situations, and the user must understand the underlying hypotheses and limitations before embarking on a geostatistical analysis. In Chapter 17, Kravchenko and Pachepsky present the use of fractals as an innovative technique to address scaling issues in soil processes. Fractal and multifractal techniques show promise in identifying scaling laws in soil science. Although soils are not ideal fractals and because fractal scaling is only applicable within a range of scales, these models present limitations. One important advantage of fractal models of variability is their ability to better simulate "rare" occurrences in soils (i.e., large pores, preferential pathways, very high conductivities, localized bacteria habitats, etc.). These rare occurrences often define soil behavior at scales coarser than observational ones. Corwin provides in Chapter 18 an overview of the characterization of soil spatial variability using EC_a-directed soil sampling for three different landscape-scale applications: (1) solute transport modeling in the vadose zone, (2) site-specific crop management, and (3) soil quality assessment. Guidelines, methodology, and strengths and limitations are presented for characterizing spatial and temporal variation in soil physicochemical properties using EC_a-directed soil sampling. Fast geospatial EC_a measurements can be made with available mobile electrical resistivity (ER) or electromagnetic induction (EMI) equipment coupled with GPS. The author stresses that without ground-truthing with soil samples, the interpretation of measurements is questionable and is not advised.

Section VI is devoted to modeling tools. In Chapter 19, Trevisan and Vischetti present the issue of modeling uncertainty across different scales. Uncertainty analysis techniques are presented as well as sources of errors. Data availability, choice of model, parameter estimation error, error propagation in model linkages, and upscaling are presented as significant sources of error that must be controlled in the modeling application. Chapter 20, by Lambot et al., examines the utility of inverse modeling (IM) techniques to obtain parameters for characterizing a soil process. Although IM is attractive, since it can reduce the cost associated with experimental measurement of model parameters, the success of the procedure depends on the suitability of the forward model, objective function, identifiability of parameters, uniqueness and stability of the inverse solution, and robustness of the IM algorithm. A comparison of available techniques and possible pitfalls of this promising technique are presented. Finally, Chapter 21 by Jantunen et al. discusses the practical aspects of choosing a suitable model for a given purpose and how to use it correctly and also reviews recent pesticide-fate models and their practical applications.

In the words of D. Hillel (1971, *Soil and Water, Physical Principles and Processes*), "No particular book by one or even several authors is likely to suffice. The field [...] is too important, too complex and too active to be encapsulated in any one book, which necessarily represents a particular point of view." We hope that the views presented herein by the excellent group of authors will spark a critical sense in the reader when discussing methods for soil process characterization.

R. Muñoz-Carpena
J. Álvarez-Benedí

Editors

Javier Álvarez-Benedí obtained his Science and Doctoral Degrees at the University of Valladolid (Spain) in 1988 and 1992. His doctoral work was related to the characterization and modeling of soil heat flux and heat balance at the soil surface in greenhouses with energy support at a fixed soil depth. After completing his Ph.D. degree, he worked as a researcher at the Servicio de Investigación Agraria in Valladolid (Spain). In this research center, he was involved in the characterization of soil–solute processes such as sorption, transport, and volatilization of soil applied pesticides at different working scales. The focus of his research has been modeling and characterization as close-coupled topics. In 2003, Dr. Álvarez-Benedí joined Instituto Tecnológico Agrario de Castilla y León, Valladolid, Spain, where he provides technical oversight and program management. Dr. Álvarez-Benedí has been an active member of the Scientific Committee of the Spanish Vadose Zone Group "Zona no Saturada," and he was the president of the organizing committee at the biannual meeting held at Valladolid in November 2003. He is a member of the Soil Science Society of America and the International Association of Hydrogeologists.

Rafael Muñoz-Carpena is an assistant professor in hydrology and water quality at the University of Florida's IFAS/TREC and Department of Agricultural and Biological Engineering (United States), and tenured researcher on leave at the Instituto Canario de Investigaciones Agrarias (Spain). He obtained his professional engineering degree at Universidad Politécnica de Madrid (Spain) and his Ph.D. at North Carolina State University (United States), where he developed and tested a surface water quality numerical model, VFSMOD. He has taught courses internationally in hydrology, soil physics for irrigation, and instrumentation for hydrological research. Currently his work is focused in hydrological and water quality issues surrounding the Everglades restoration effort in Florida (United States), one of the most expensive and ambitious environmental projects in history. His work involves field and computer modeling activities to understand water flow and quality in the area, including solute transport in the soil. He has been an active member of the Scientific Committee of the Spanish Vadose Zone Group for the last 10 years, where soil characterization research has been a central issue. Dr. Muñoz-Carpena serves as Associate Editor for *Transactions of ASAE* and *Applied Engineering in Agriculture* and is a member of the American Society of Agricultural Engineers and the American Geophysical Union.

Contributors

Javier Álvarez-Benedí
Instituto Tecnológico Agrario de Castilla y León
Valladolid, Spain

Lars Bergström
Division of Water Quality Management
Swedish University of Agricultural Sciences
Uppsala, Sweden

S. Bolado
Departamento de Ingeniería Química
Universidad de Valladolid
Valladolid, Spain

David Bosch
Research Hydraulic Engineer
USDA-ARS, SEWRL
Tifton, Georgia

Moira Callens
Laboratory of Hydrology and Water Management
Ghent University
Gent, Belgium

Chris G. Campbell
Earth Sciences Division
Lawrence Livermore National Laboratory
Livermore, California

Ettore Capri
Istituto di Chimica Agraria ed Ambientale
Università Cattolica del Sacro Cuore
Piacenza, Italy

Rafael Celis
Instituto de Recursos Naturales y Agrobiología
de Sevilla, CSIC
Sevilla, Spain

Brent E. Clothier
Environment and Risk Management Group
HortResearch Institute
Palmerston North, New Zealand

Juan Cornejo
Instituto de Recursos Naturales y Agrobiología
de Sevilla, CSIC
Sevilla, Spain

Dennis L. Corwin
USDA-ARS
George E. Brown, Jr. Salinity Laboratory
Riverside, California

Lucía Cox
Instituto de Recursos Naturales y Agrobiología
de Sevilla, Sevilla
CSIC
Spain

Markus Deurer
Institute for Soil Science
University of Hannover
Hannover, Germany

Ahmed Douaik
Department of Soil Management and Soil Care
Ghent University
Gent, Belgium

María P. González-Dugo
Department of Soils and Irrigation
CIFA, Alameda del Obispo, IFAPA
Córdoba, Spain

David E. Elrick
Department of Land Resource Science
University of Guelph
Guelph, Ontario, Canada

Fernando Garrido
Department of Soils
Centro de Ciencias Medioambientales, CSIC
Madrid, Spain

Juan Vicente Giráldez
Department of Agronomy
University of Córdoba
Córdoba, Spain

María P. González-Dugo
Department of Soils and Irrigation
CIFA, Alameda del Obispo, IFAPA
Córdoba, Spain

Steve Green
HortResearch Institute
Palmerston North, New Zealand

Mª Carmen Hermosín
Instituto de Recursos Naturales y Agrobiología
de Sevilla CSIC, Sevilla, Spain

F. Hupet
Department of Environmental Sciences and Land Use Planning
Catholic University of Louvain
Louvain-la-Neuve, Belgium

Mohammed R. Islam
Soil and Land Resources Division
University of Idaho, Moscow, Idaho

Anna Paula Karoliina Jantunen
Department of Biology, University of Joensuu
Joensuu, Finland

M. Javaux
Department of Environmental Sciences and Land Use Planning
Catholic University of Louvain
Louvain-la-Neuve, Belgium

A. N. Kravchenko
Department of Crop and Soil Sciences
Michigan State University
East Lansing, Michigan

S. Lambot
Department of Environmental Sciences and Land Use Planning
Catholic University of Louvain
Louvain-la-Neuve, Belgium

Yuncong Li
Department of Soil and Water Sciences
Tropical Research and Education Center
University of Florida
Homestead, Florida

Keith Loague
Department of Geological and Environmental Sciences
Stanford University
Stanford, California

Sean M. McGinn
Lethbridge Research Centre
Agriculture and Agri-Food Canada
Lethbridge, Alberta, Canada

H. Montas
Biological Resources Engineering Department
University of Maryland
College Park,
Maryland

Rafael Muñoz-Carpena
Agricultural and Biological Engineering Department
IFAS/TREC
University of Florida
Homestead, Florida

Y. A. Pachepsky
USDA-ARS Environmental Microbial Safety Laboratory
Beltsville, Maryland

Mikael Pell
Swedish University of Agricultural Sciences
Department of Microbiology
Uppsala, Sweden

María José Polo
Department of Agronomy
University of Córdoba
Córdoba, Spain

Carlos M. Regalado
Departamento de Suelos y Riegos
Instituto Canario de Investigaciones Agrarias (ICIA)
La Laguna, Tenerife, Spain

W. Daniel Reynolds
Greenhouse and Processing Crops Research Centre
Agriculture and Agri-Food Canada
Harrow, Ontario
Canada

Axel Ritter
Departamento de Suelos y Riegos
Instituto Canario de Investigaciones Agrarias
La Laguna, Tenerife, Spain

Philippe Rochette
Soils and Crops Research and Development Centre
Agriculture and Agri-Food Canada
Sainte-Foy, Quebec,
Canada

Ali Sadeghi
Environmental Quality Laboratory
USDA-ARS, Beltsville, Maryland

Adel Shirmohammadi
Biological Resources Engineering Department
University of Maryland
College Park, Maryland

John Stenström
Department of Microbiology
Swedish University of Agricultural Sciences
Uppsala, Sweden

Marco Trevisan
Istituto di Chimica Agraria ed Ambientale
Università Cattolica del Sacro Cuore
Piacenza, Italy

Markus Tuller
Soil and Land Resources Division
University of Idaho
Moscow, Idaho

M. Vanclooster
Department of Environmental Sciences and Land Use Planning
Catholic University of Louvain
Louvain-la-Neuve, Belgium

Karl Vanderlinden
Department of Soils and Irrigation
CIFA, Las Torres-Tomejil, IFAPA
Sevilla, Spain

Marc Van Meirvenne
Department of Soil Management and Soil Care
Ghent University
Gent, Belgium

Niko E.C. Verhoest
Laboratory of Hydrology and Water Management
Ghent University
Gent, Belgium

Lieven Vernaillen
Department of Soil Management and Soil Care
and Laboratory of Hydrology and Water Management
Ghent University
Gent, Belgium

Costantino Vischetti
Dipartimento di Scienze Ambientali e delle Produzioni Vegetali
Università Politecnica delle Marche
Ancona, Italy

I. Vogeler
Environment and Risk Management Group
HortResearch Institute
Palmerston North, New Zealand

Jianqiang Zhao
Bureau of Pesticide
Division of Agricultural Environmental Services
Florida Department of Agriculture and Consumer Services
Tallahassee, Florida

Meifang Zhou
Water Quality Analysis Division
Environmental Monitoring & Assessment Department
South Florida Water Management District
West Palm Beach, Florida

Table of Contents

Chapter 2   Spatial and Temporal Variability of Soil Processes: Implications for Method Selection and Characterization Studies   59

Chris G. Campbell and Fernando Garrido

Chapter 3   Modeling as a Tool for the Characterization of Soil Water and Chemical Fate and Transport   87

Javier Álvarez-Benedí, Rafael Muñoz-Carpena and Marnik Vanclooster

Section II
Soil and Physical Processes: Energy and Water

               M. J. Polo, J. V. Giráldez, M. P. González-Dugo and
               K. Vanderlinden

Section III
Soil and Solutes Processes

Section IV
Soil and Microorganisms

Section V
Spatial Variability and Scale Issues

Section VI
Modeling Tools

1 Multidisciplinary Approach for Assessing Subsurface Non-Point Source Pollution

Dennis L. Corwin
USDA-ARS, George E. Brown, Jr. Salinity Laboratory, Riverside, California

Keith Loague
Stanford University, Stanford, California

CONTENTS

1-5667-0657-2/05/$0.00 + $1.50
© 2005 by CRC Press

1

1.1   INTRODUCTION

Humankind's need or inexorable desire to alter their surroundings, whether for better or for worse, has changed little over the centuries. The rich Mesopotamian marshlands known as the Fertile Crescent (i.e., the ancient region extending from the confluence of the Tigris and Euphrates Rivers around the north of the Syrian Desert to the eastern shore of the Mediterranean Sea) stands out as an example of how an ancient people ingeniously utilized the meager water resources of a region to alter the agricultural productivity and vegetative landscape of an entire region. In the absence of soil and water stewardship, the resultant adverse effects upon the Fertile Crescent became irreversible. In part because of the degradation of the region to the semi-desert conditions that now prevail, the present inhabitants have on the whole a lower standard of living than its ancient inhabitants. Novotny and Olem (1994) assert that "the history of the Middle East shows that if land stewardship is absent, the well-being of the people who misuse the land and water resources declines."

Even though the extent of the deterioration of parts of the Fertile Crescent is striking, what has clearly changed over centuries is humankind's increased ability and capacity to alter the environment. When left unchecked, humankind's current capacity to alter or even devastate vast land masses and water bodies is unparalleled as evident from the deforestation of the Brazilian Amazon and Southeast Asia, and the pollution of North America's Great Lakes and Chesapeake Bay. It is the progressive increase in this ability to alter and degrade the environment and the extent of the impact exacerbated by an ever increasing world population that has drawn world attention to global environmental issues such as climatic change, ozone layer depletion, deforestation, desertification, and non-point source (NPS) pollution.

The global awareness of complex environmental problems that are not respecters of political or physical boundaries spawned the 1992 Rio de Janeiro World Summit, which in turn set the stage for forging the Kyoto Protocol of

1997. The heightened global environmental consciousness is largely the consequence of the information revolution, which has produced ever-greater knowledge disseminated at faster speeds, and the space age, which provides constant visual reminders from space of our finite resources and the need for environmental stewardship to protect them. Corwin et al. (1999a) pointed out that

> *Ostensibly, the point has been reached where the awareness of NPS pollution is less of a concern than the voluntary and regulatory actions that must follow. However, before decision makers can formulate effective regulatory actions the ability to reliably and cost-effectively assess NPS pollutants on a real-time and predictive basis is of paramount importance.... Equipped with an arsenal of advanced information technologies, modelers of NPS pollution now stand at the threshold of aggressively breaching barriers that have thwarted their efforts. The ability to accurately assess present and future NPS-pollution impacts on ecosystems provides a powerful tool for environmental stewardship and guiding human activities.*

In response to the challenge of breaching the barriers that impede NPS pollution assessment, this chapter discusses an integrated array of multi-disciplinary, advanced-information technologies (geographic information systems [GIS], solute transport modeling, fuzzy logic, fractals, geostatistics, neural networks, transfer functions, digital terrain models, scale and hierarchy theory, wavelet analysis, remote sensing, and uncertainty analysis) useful in assessing subsurface NPS pollutants. The focus of this chapter is to describe an integrated, multidisciplinary approach for assessing pollutants in the complex, spatially heterogeneous vadose zone of soil. The approach is compatible with resident information resources and current infrastructure for capturing spatial environmental data to practically address real-world environmental issues pertaining to NPS pollution at landscape, regional, and global scales.

1.1.1 DEFINITION AND CHARACTERISTICS OF NPS POLLUTION

Non-point source pollution was not generally recognized until the mid-1960s. At first NPS pollution was associated entirely with pollution from storm water and runoff. Since that time NPS pollution has expanded to encompass all forms of diffuse pollutants. NPS pollutants are defined as "contaminants of [air and] surface and subsurface soil and water resources that are diffuse in nature and cannot be traced to a point location" (Corwin and Wagenet, 1996). Often NPS pollutants occur naturally, e.g., as salts and trace elements in soils, or are the consequence of direct application by humans (e.g., pesticides and fertilizers), but regardless of their source they are generally the direct consequence of human activities including agriculture, urban runoff, feedlots, hydromodification, and resource extraction. Specifically, NPS pollutants include: (1) excess fertilizers, herbicides, and insecticides from agricultural lands and residential areas; (2) oil, grease, and toxic chemicals from urban runoff and energy production; (3) sediment from improperly managed construction sites, crop and forest lands, and eroding stream banks; (4) naturally occurring salts and trace elements from irrigation practices; (5) acid drainage from abandoned mines; (6) pathogens (i.e., viruses and bacteria) and nutrients from livestock, and pet wastes; and (7) atmospheric deposition.

Characteristically, NPS pollutants (1) are difficult or impossible to trace to a source; (2) enter the environment over an extensive area; (3) are related, at least in part, to certain uncontrollable meteorological events, and existing geographic and geomorphologic conditions; (4) have the potential for maintaining a relatively long active presence in the global ecosystem; and (5) may result in long-term chronic effects on human health and soil–aquatic degradation.

1.1.2 THE NPS POLLUTION PROBLEM

The environmental problems stemming from NPS pollution are a consequence of the widespread nature of NPS pollutants and the resultant chronic effects on human health. The essence of the NPS pollution problem has been cogently articulated by Duda and Nawar (1996):

> *Through mankind's pursuit of unsustainable economic development policies and projects, we now have the capacity not only to injure our surrounding environment and ourselves, but also remote environments and their inhabitants (like the Arctic) and future generations of humans through the buildup of persistent toxic substances that mimic and disrupt human hormone systems. ... Not only is it environmentally unsustainable and economically inefficient to incur non-point related environmental deficit, but it is ethically and morally wrong to ask citizens and their future offspring to shoulder the human suffering, impaired health and economic burden associated with toxic, hazardous, and radioactive contamination.*

1.1.2.1 The Issue of Health

The real significance of the NPS pollution problem lies in the long-term ramifications on human health and the proper functioning of ecosystems. Throughout the world, ground and surface waters are the source of drinking and irrigation water. The protection of ground and surface water resources has become a primary global concern because of the public's apprehension over long-term health effects resulting from drinking water containing low levels of toxic chemicals. A secondary concern is the accumulation of inorganic (e.g., salinity and trace elements) and organic chemicals (e.g., pesticides) in soil that detrimentally impact agricultural productivity.

Reminiscent of the silent killer tobacco, which remained disassociated from its impact on human health for decades due to inconclusive cause-and-effect evidence, the impact of NPS pollutants on human health is just beginning to be understood. Some chemicals (e.g., As, Cd, Pb, Hg, DDT and its degradation products, methoxychlor, triazine herbicides, synthetic pyrethroids, lindane, chlordane, PCBs, dioxins, and others) are suspected to disrupt the endocrine system, causing metabolic, neurological, and immune system abnormalities

(Colborn et al., 1993). The hypothesis of Colborn et al. (1996) that certain NPS pollutants behave as endocrine disruptors, which may play a role in a range of problems from reproductive and development abnormalities to neurological and immunological defeats to cancer, is beginning to be supported by evidence (Stillman, 1982; Colborn et al., 1993; Porterfield, 1994; Jacobson and Jacobson, 1996; Lipschultz, 1996; Repetto and Baliga, 1996; Toppari et al., 1996). However, scientists are a long way from understanding at what levels of exposure these hormone-mimicking and blocking effects occur in wildlife, let alone humans (Guillette et al., 1995).

The impact of low-level contaminants is only now becoming apparent. Radionuclides, pesticides, solvents, heavy metals, and petroleum compounds are well known because of their association with high-level point source pollution, but a new generation of carcinogens, mutagens, and teratogens is found to cycle from one medium or form to another. These include various classes of polychlorinated biphenyls (PCBs), furans, dioxins, and organochlorines. Whether consumed as residues in drinking water, in fruits and vegetables, or in the fat of fish and meat, these chemicals pose potentially serious, long-term health risks.

Some common toxic chemicals and their associated maximum concentration levels (MCLs) are presented in Table 1.1. Maximum concentration levels have been determined by the U.S. Environmental Protection Agency (USEPA) to ensure protection against human carcinogenic or non-carcinogenic impacts resulting from chronic environmental chemical exposure.

1.1.2.2 Global Scope and Significance

Satisfying the ever-growing need for natural resources to meet food and living standard demands while minimizing impacts upon an environment that already shows signs of serious levels of biodegradation is among the foremost global issues (Corwin and Wagenet, 1996). There is worldwide concern for the future availability of limited natural resources such as water and productive agricultural soil. Not only the availability of finite resources, but the condition of those resources as impacted by domestic, industrial, and agricultural activities is of concern. The condition of soil and water resources is largely a consequence of the presence of pollutants, particularly NPS pollutants.

On a national scale, the USEPA (1995) reports that NPS pollution is the leading remaining cause of water quality problems in the United States. Agriculture is recognized as the single greatest contributor of NPS pollutants to surface and subsurface waters, followed by urban runoff and resource extraction (USEPA, 1995). Roughly 80% of the assessed rivers and lakes in the United States are impaired by NPS pollutants and only 20% by point sources (USEPA, 1995). This disparity is similar in other developed countries because developed countries have concentrated their effort on the control of point source pollutants due to the difficulty in regulating diffuse sources of pollution.

TABLE 1.1
Toxicological Profiles for Selected Chemicals

Class	Contaminant	MCL[a] (μg/L)	Toxic	Neurotoxic	Carcinogenic[b]	Teratogenic	Mutagenic
Metals	Lead	15	✓	✓	E	✓	✓
	Arsenic	50	✓		A	✓	✓
	Mercury	2	✓	✓	E	✓	
	Cadmium	5	✓		D	✓	
	Chromium	100			D	✓	
Petroleum hydrocarbons	Benzene (BTEX)	5	✓	✓	A	✓	✓
	Toluene	1000	✓		D	✓	✓
	Ethylene	—			E	✓	
	Xylene	10000	✓	✓	D	✓	
Chlorinated organic solvents	Hexachlorobenzene	1			B	✓	
	Trichloroethylene (TCE)	5	✓	✓	B		
	Perchloroethylene (PCE)	5	✓	✓	B		
	Methylene chloride	5	✓	✓	B		
	Chloroform	—	✓	✓	B		
	Carbon tetrachloride (CTC)	5	✓	✓	B	✓	

Category	Chemical	MCL[a]			Class[b]		
Wood-preserving chemicals	Vinyl chloride (VC)	2			A	✓	✓
	Trichlorophenol	—	✓	✓	B		✓
	Pentachlorophenol (PCP)	1	✓		B		
	Creosote	—	✓		B		
PCBs	Polychlorinated biphenyls	0.5	✓		B	✓	✓
Dioxins	Dioxins	0.00005	✓	✓	B	✓	✓
Pesticides	DDT	—		✓	B	✓	✓
	EDB	0.05			B	✓	
	Chlordane	2	✓	✓	B		
	Heptachlor	0.4		✓	B		
	Toxaphene	3	✓	✓	B		✓
	Lindane	0.2			B		

[a]Maximum concentration levels (MCLs) for selected chemicals are taken from Fetter, 1993.

[b]EPA Carcinogen Classification System: (A) human carcinogens, (B) probable human carcinogens, (C) possible human carcinogens, (D) not classsifiable, and (E) noncarcinogenic to humans.

Source: Freeze, 2000.

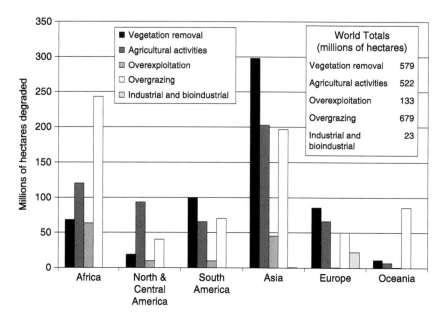

FIGURE 1.1 Human-induced soil degradation by region and by cause from 1945 to the late 1980s. (From World Resources Institute, 1998.)

The impact upon soil and water resources from NPS pollutants extends over millions of hectares of land and billions of liters of water. Throughout the world 30–50% of the earth's land is believed affected by NPS pollutant degradation from erosion, fertilizers, pesticides, organic manures, and sewage sludge (Pimental, 1993). Worldwide, NPS pollutants are recognized as *the* major contributors to surface and groundwater contamination (Duda, 1993), with agriculture as the single greatest contributor of NPS pollutants (Humenik et al., 1987). Agricultural activities result in the movement of NPS pollutants from the soil surface into rivers and streams via runoff and erosion and into subsurface soil and groundwater via leaching. Figure 1.1 reflects the worldwide extent of human-induced degradation of soil by region and by cause over the period 1945 to the late 1980s.

1.1.2.3   Common NPS Pollutants

The most common global NPS pollutants of soil and groundwater resources include biosolids and manure, persistent organic pollutants (POPs), nutrients (e.g., nitrates and phosphates), salinity, toxic heavy metals (e.g., Bi, Co, Sn, Te, Ag, Pt, Tl, Sb, Hg, As, Cd, Pb, Cr, Ni), trace elements (e.g., Se, B, Mo, Cu, Zn), and pathogens. Each type of pollutant is directly associated with agricultural activity. Table 1.2 shows the relative importance of each NPS pollutant with respect to its source.

TABLE 1.2
Relative Importance of Pollutant Concentrations in Soil–Water Systems

Non-point source	Suspended solids and sediments	BOD	Nutri-ents	Toxic metals	Trace elements	Pesticides	Patho-gens	Salinity/TDS
Urban storm runoff	M	L-M	L	H	M	L	H	M
Construction	H	N	L	N-L	N-L	N	N	N
Highway deicing	N	N	N	N	N	N	N	H
In-stream hydrologic Modification		H	N	N N-H	N-H	N	N	N-H
Nonc oal mining	H	N	N	M-H	M-H	M-H	N	M-H
Agriculture:								
Nonirrigated crop production	H	M	H	N-L	N-L	H	N-L	N
Irrigated crop production	L	L-M	H	N-L	H	M-H	N	H
Pasture and range	L-M	L-M	H	N	N-L	N	N-L	N-L
Animal production	M	H	M	N-L	N-L	N-L	L-H	N-L
Forestry:								
Growing	N	N	L	N	N	L	N-L	N
Harvesting	M-H	L-M	L-M	N	N	L	N	N
Residuals management	N-L	L-H	L-M	L-H	N-H	N	L-H	N-H
On-site sewage disposal	L	M	H	L-M	L-M	L	H	N
In-stream sludge accumulation		H	H	M-H	L-H	L-H	M-H	L
Direct precipitation	N	N	N-M	L	L	L	N-L	N
Air pollution fallout	M	L	L-M	L-H	N-M	L-M	N-L	N
"Natural" background		L-M	L-H	M N-M	N-M	N	N-L	N-H

N = negligible; L = low; M = moderate; H = high; TDS = total dissolved solids.

Source: Peirce et al., 1998.

1.1.3   JUSTIFICATION FOR ASSESSING NPS POLLUTION IN SOIL

The past century has brought unprecedented gains in many of the indicators used to gauge human progress, from increased life expectancy to increased literacy. Nevertheless, during the same time period human impact on the natural world has risen dramatically due to the scope and intensity of human activities (World Resources Institute, 1998). Environmental issues such as climatic change, ozone depletion, biodiversity, erosion, deforestation, desert-ification, and NPS pollution are global concerns. These problems are exacerbated by the trends in growing world population and consumption. World population has doubled since 1950 and is expected to range from 8 to 12 billion in 2050, while yearly consumption of natural resources by modern industrial economies is at a staggering 45–85 metric tons of material per person (World Resources Institute, 1998). Concomitantly, satisfying the ever-growing food and living standard demands of the projected world's population will be a formidable task. This poses a dilemma for agriculture's prominent role as a contributor of NPS pollutants to soil and water resources. There is growing pressure to meet the food demands of a growing world population, but in doing so the likelihood of detrimentally impacting the environment seems inescapable.

To meet the increased global food demand, substantial intensification of agriculture is essential both on land currently cropped and on land to be converted for agricultural use, especially in developing countries. Both actions have far-reaching implications for health concerns and environmental quality due to the increased levels of NPS pollution created. Health concerns related to agricultural intensification stem from (1) increased exposure to toxic substances such as pesticides, (2) a higher incidence of infectious disease associated with the expansion of irrigation systems and the use of waste or drainage water for irrigation, and (3) increased human exposure to infectious agents as ecosystems of developing countries are converted to agricultural land (World Resources Institute, 1998). From an environmental perspective, if current developing world agricultural practices continue, soil resources could be degraded through erosion, loss of fertility, and accumulation of organic and inorganic chemicals and water resources could be depleted and degraded.

Barring unexpected technological breakthroughs, sustainable agriculture is viewed as the most viable means of meeting the food demands of the projected world's population. The concept of sustainable agriculture is predicated on a delicate balance of maximizing crop productivity and maintaining economic stability while minimizing the utilization of finite natural resources and the detrimental environmental impacts of associated NPS pollutants. The ability to assess the environmental impact of NPS pollutants on soil-groundwater systems at local, regional, and global scales is a key component to achieving the sustainability of agriculture. Only through assessment can (1) the true extent of the NPS-pollution problem be established, (2) an evaluation of mitigating management practices and regulatory policies be made, and (3) a prognostication of future potential problems be made.

Assessment involves the determination of change of some constituent over time. In a spatial context, assessment is a spatio-temporal evaluation of trend. This change can be measured in either real time or predicted with a model. Real-time measurements reflect the activities of the past, whereas model predictions are a glimpse into the future. Both means of assessment are valuable, although the distinct advantage of prediction, like preventative medicine, is that it can be used to alter the occurrence of detrimental conditions before they occur. The ability to assess environmental contaminants such as NPS pollutants provides a means for humans to optimize the use of the environment by sustaining its utility without detrimental consequences and preserving its esthetic qualities to serve human's need of spirituality (Corwin and Wagenet, 1996).

Increased attention has been given to NPS pollution of subsurface soil and water. There are numerous reasons for this. The vadose zone (i.e., the zone from the soil surface to the groundwater table) is a complex physical, chemical, and biological ecosystem that regulates the passage of NPS pollutants from the soil surface or near-surface, where they have been deposited or accumulated due to agricultural activities, to groundwater. The accumulation of salinity and trace elements in soil can significantly reduce crop productivity, while loading of salts, nitrates, trace elements, and pesticides into groundwater supplies can degrade a significant source of drinking and irrigation water. Limited surface water resources and continued contamination of surface water supplies have increased the reliance upon groundwater to meet growing water demands in nearly all industrialized nations. Groundwater accounts for half of the drinking water and 40% of the irrigation water used in the United States. The degradation of groundwater, particularly by NPS pollutants, has become a growing public concern primarily because of the concern over long-term health effects. Non-point source pollutants pose a tremendous threat to soil and groundwater resources because of the areal extent of their contamination and the difficulty of effective remediation once soils and groundwater are contaminated. The scope of the contamination by NPS pollutants can be over entire basins, watersheds, and aquifers and can cross state, national, and even continental boundaries. Additionally, the assessment of subsurface NPS pollutants is no longer as formidable and intractable a problem because of multidisciplinary, advanced information technologies that have recently developed and continue to develop.

1.2 MULTIDISCIPLINARY APPROACH FOR ASSESSING SUBSURFACE NPS POLLUTANTS

Solutions to complex global environmental problems, such as assessing NPS pollutants, stand on the shoulders of technological and interdisciplinary scientific achievements (Corwin et al., 1999b). To solve the problems of climatic change, ozone layer depletion, deforestation, desertification, and NPS pollution, it is necessary to examine these issues from a multidisciplinary, systems-based approach and to look at these problems in a spatial context with an awareness of scale (Corwin et al., 1999b).

The formidable barriers to assessing NPS pollutants are the consequence of the complexities of geographic scale and position; the complexities of the physical, chemical, and biological processes of solute transport in porous media; and the spatial complexities of the soil media's heterogeneity. The knowledge, information, and technology needed to address each of these issues crosses several subdisciplinary lines, including classical and spatial statistics, remote sensing, GIS, surface and subsurface hydrology, soil science, and space science. Spatial statistics is useful in dealing with the uncertainty and variability of spatial information (e.g., Cressie, 1993); remote sensing provides measurements of physical, chemical, and biological properties needed in environmental models (e.g., Barnes et al., 2003); GIS is a means of organizing,

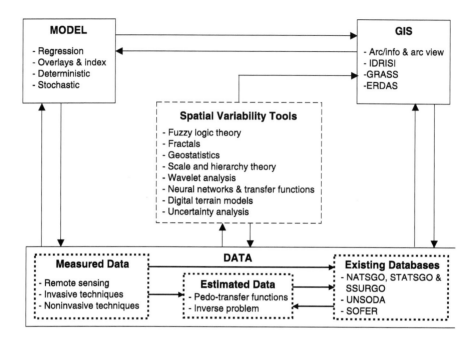

FIGURE 1.2 Integrated components of a GIS-based NPS pollutant model system. Arrows show the flow of information. (Modified from Corwin et al., 1997.)

manipulating, storing, and displaying spatial data; and water flow and solute transport models developed within soil science and hydrogeology are the tools for simulating future scenarios to assess potential temporal and spatial changes (e.g., Burrough, 1996). Precise geographic location and areal extent are captured with the space science technology of the global positioning system (GPS).

The essential components of modeling NPS pollutants consist of (1) a *model* of transport and/or accumulation, (2) input and parameter *data* for the model, and (3) a *GIS* to handle the input and output of spatial data. Figure 1.2 shows the interaction between these three basic components based on a flow of information. Figure 1.2 reflects the multidisciplinary nature of modeling NPS pollutants in the environment.

Because of the complex spatial heterogeneity of the vadose zone, a variety of sophisticated techniques are useful as tools to deal with the vicissitudes of soil (Figure 1.2). Fuzzy logic theory provides a means of handling vague and imprecise data, as either a means to characterize map units or transitional boundaries between map units (e.g., Altman, 1994; Zhu, 1999). Fractal geometry with its scale independence may offer a means of bridging a variety of gaps related to spatial variability from determining the predictability of complex spatial phenomena such as solute transport to relating difficult-to-measure soil hydraulic properties to other soil variables available from soil surveys (e.g., Crawford et al., 1999) (see Chapter 17). Geostatistics is useful in interpolating sparse spatial data and providing associated uncertainty (e.g., Bourgault et al., 1997) (see Chapter 16). Hierarchical theory establishes an organizational hierarchy of pedogenetic modeling approaches and their appropriate scale of application (e.g., O'Neill, 1988; Hoosbeek and Bryant, 1993). Wavelet analysis provides a means of determining spatial scales and the dominant processes at those scales (e.g., Lark and Webster, 1999; Lark et al., 2003). Neural networks and transfer functions provide a means of deriving complex hydraulic parameters from easily measured data (e.g., Wösten and Tamari, 1999). Digital terrain or digital elevation models (DEM) provide spatial geomorphologic information. Uncertainty analysis serves as a means of establishing the reliability of simulated model results based on model errors and data uncertainties (e.g., Loague and Corwin, 1996) (see Chapter 19).

1.2.1 DETERMINISTIC MODELING PROCESS

The classical deterministic modeling process consists of six stages (Corwin, 1996): model conceptualization, verification, sensitivity analysis, calibration, validation, and simulation (Figure 1.3). Corwin (1996) indicates that

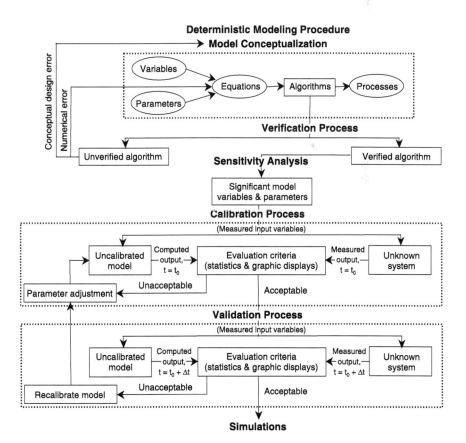

FIGURE 1.3 Schematic of modeling process. (From Corwin, 1996.)

a generic procedure for the development of deterministic models consists of (1) formulation of a simplified conceptual model consisting of integrated processes characterizing the system, (2) representation of each individual process by an algorithm consisting of mathematical expressions of variables and parameters, (3) verification of the algorithm(s) to ascertain if the conceptual model is truly represented, (4) sensitivity analysis to determine the relative importance of the variables and parameters, (5) model calibration, (6) model validation, and (7) application of the model for simulation.

Hillel (1987) points out four philosophical axioms that should guide model design: parsimony, modesty, accuracy, and testability. The principle of parsimony simply states that a model should be no more complex than necessary and thereby should have the fewest parameters and the parameters easiest to infer from the observed data. Modesty refers to scope in that a model should not pretend to do too much. Next, a model does not need to be any more accurate than the ability to measure a phenomenon. Finally, a model must be designed so that it can be validated and the validation should provide insights into the model's limitations.

Models incorporate descriptions of the key processes that determine a system's behavior with varying degrees of sophistication. However, a "good model must not only produce accurate results, but must do so for the right reasons" (Klemes, 1986). This implies knowledge of "cause-and-effect" relationships within the model that can only be known through experimentation and observation.

1.2.1.1 Model Conceptualization

Model conceptualization seeks to extract the essence from experimental data through the formulation of mathematical representations of the relevant processes constituting a real system. The conceptual design of a mathematical model requires the formulation of meaningful parameters and variables into equations that can be translated into algorithms that represent each process and then the integration of the processes to represent the real system. The algorithms are then turned into source code, and the source code is translated into object code useable by a computer. In the case of NPS pollutants, the algorithms describe the appropriate chemical, physical, and biological processes involved with solute transport within the spatial context of a GIS containing georeferenced model input variables.

Model conceptualization is governed by the intended application of the model. This presupposes that there is no definitive model, which complies with the philosophically established principle of modesty. The intended application determines the scale; the scale determines the predominant processes and the level of spatial variability that must be accounted for in the model; the predominant processes determine the type of modeling approach (i.e., functional or mechanistic, qualitative or quantitative) that should be used.

Donigian and Rao (1986) state that the primary differences between models are the level of detail at which the fundamental processes are described and the conceptual completeness in terms of the number of processes that are included in the model. This is true for both point and NPS pollutant models. Rao et al. (1982) point out that the primary differences are the consequence of the following factors: (1) current state of understanding of the system and its components, (2) how the modeler conceptualizes the system components, (3) model approach and the acceptable error limits allowable, (4) temporal scale over which the model is to be applied, and (5) spatial scale at which the model is to be applied. The inherent difference between point and NPS models relates to the consideration of spatial variations and spatial scale.

1.2.1.2 Model Parameters

Another consideration of model conceptualization is the availability and type of input data. Ideally, a model should be developed with the availability and type of input data as a primary consideration. For instance, in a laboratory setting where sophisticated instrumentation can be utilized under controlled conditions and precise measurements of rate parameters can be made on column experiments, it makes perfect sense to develop and apply a sophisticated mechanistic model. However, in field situations where the resources required just to characterize spatial variability are prohibitive in cost and labor, a reliance on easily measured and less spatially variable capacity parameters is more appropriate, making the application of a functional model more suitable.

The parameters and input variables found in functional and mechanistic models reflect the appropriate scale of the application of each type of model. Capacity parameters (e.g., bulk density, field capacity, wilting point water content, etc.) are generally associated with functional models, while rate parameters (e.g., hydraulic conductivity, infiltration rate, etc.) are generally associated with mechanistic models. Characteristically, capacity parameters are less spatially variable than rate parameters and require fewer samples to determine a representative value (Jury, 1986). Generally speaking, sampling intensity requirements favor the use of functional models at larger spatial extents.

Aside from sampling intensity, there is also the consideration of the physical size of the sample volume used to assess the value of the physical or chemical parameter. The magnitude and variability of a soil property often varies with measurement scale. Sample volume is an important scale issue with regards to measurements used to develop input data or measurements needed for model validation (Wagenet et al., 1994). As stated by Wagenet and Hutson (1996), "Until sampling and measurement approaches are consistent in scale with the models being used at that scale, our assessments of model performance will be plagued by ambiguity that arises from this lack of appreciation of scale dependency." Wagenet and Hutson (1996) conclude that sampling and measurement methods for both input parameters and observations to evaluate model performance need to be consistent with the modeling scale. They also note that models, which are developed for a specific spatial scale, are often calibrated with data from a much different scale, resulting in model parameters that have little physical significance and a modeling approach that is questionable.

1.2.1.3   Verification

The second stage of the modeling process involves verification of the model. Verification is an examination of the analytical or numerical approach in the computer code to ascertain that it is truly representative of the conceptual model. This often involves a comparison with a previously accepted model or analytical solution that has passed close scrutiny and has been judged to be valid for the restrictive assumptions upon which it was derived. Most often the test is done by simply comparing the results of a numerical model to that of an analytical model. Perfect agreement between the numerical code and the analytical model only means that the code can accurately solve the governing equations, but not that it will do so under any and all circumstances. In essence, verification demonstrates the ability of a model to solve the governing equations, thereby representing a certification that the computational program that solves the mathematical equations representing the system is mathematically correct in formulation and solution.

Ideally, each process described within the model is examined separately and then in combination with one another. However, if the numerical model introduces complexities beyond those of the analytical model, as is most often the case, then there is no way to confirm the accuracy of the code regarding these complexities. Only simple mass-balance checks can be made that offer modest assurances. Verification may or may not require experimental data since previously developed verified and validated models are generally used for comparison purposes. At times, experimental data have been relied upon to verify those components of a model that cannot be evaluated through a comparison with an accepted model. In this way, verification and validation are seemingly interchangeable. In some cases verification and validation have been erroneously used interchangeably to indicate that model predictions and experimental observations are consistent with one another. However, it is a misleading implication that verification and validation are synonymous.

Verification, as classically defined above and in its narrowest sense, is best referred to as "bench marking," which denotes a reference to an accepted standard. Verification in the classical sense limits a model's range of verification to the realm of the analytical model to which it was compared. Oreskes et al. (1994) contend that "The congruence between a numerical and an analytical solution entails nothing about the correspondence of either one to material reality . . . (and) . . . the raison d'être of numerical modeling is to go beyond the range of available analytical solution. . . . Therefore, in application, numerical models cannot be verified."

1.2.1.4   Sensitivity Analysis

Donigian and Rao (1986) state that sensitivity analysis is the "degree to which the model result is affected by changes in a selected model parameter."

Parameter as used in this context is taken from Addiscott and Tuck (1996) to be *a quantity that is constant in a particular case considered, but [potentially] varies in other cases.* In this context a parameter differs from a variable in that a variable is susceptible to variation. However, this distinction needs further clarification, because as Addiscott and Tuck (1996) point out, hydraulic conductivity, which is considered a parameter, varies significantly from point to point and as a result appears ineligible as a parameter. The key point is that hydraulic conductivity may vary in space, but varies less over time under set conditions, whereas a variable such as rainfall can vary both in space and time.

Sensitivity analysis provides a means of identifying those parameters with the greatest influence on the simulated output, thereby indicating which parameters should be more accurately measured. The robustness of a model depends upon its sensitivity to fixed variations in the input (Corwin, 1995). The sensitivity of a model to a fixed percentage change in its input parameters reflects the model effects due to conceptual design, whereas the sensitivity of a model to the standard deviation range of its inputs (i.e., plus or minus one standard deviation of the measured mean) reflects not only model effects, but also uncertainties in the measured input parameters (Corwin, 1995).

As presented by Loague and Corwin (1996), the sensitivity of a model's output to a given input parameter is the partial derivative of the dependent variable with respect to the parameter:

$$x_{ij} = \frac{\partial \hat{y}_i}{\partial a_j} \tag{1.1}$$

where, x_{ij} is the sensitivity coefficient of the model dependent variable $\hat{y}$ with respect to the jth parameter at the ith observation point. Sensitivity analysis can be extremely useful in identifying the most important (sensitive) parameters in the "trial-and-error" calibration of a hydrologic-response model (e.g., see Loague, 1992). The sensitivity coefficient in Eq. (1.1), with respect to a given parameter, can be approximated by making small perturbations in the parameter of particular focus while keeping all the other parameters constant and then dividing the change in the dependent variable by the change in the parameter (Zheng and Bennett, 1995):

$$x_{ij} = \frac{\partial \hat{y}_i}{\partial a_j} = \frac{\hat{y}_i(a_j + \Delta a_j) - \hat{y}_i(a_j)}{\Delta a_j} \tag{1.2}$$

where Δa_j is the small change (perturbation) in the parameter. Equation (1.1) can be normalized by the parameter value so that the sensitivity coefficient with respect to any parameter is the same unit as that for the dependent variable:

$$x_{ij} = \frac{\partial \hat{y}_i}{\partial a_j / a_j} \tag{1.3}$$

Based on Eq. (1.3), Eq. (1.2) can be written as:

$$x_{ij} = \frac{\partial \hat{y}_i}{\partial a_j / a_j} = \frac{\hat{y}_i(a_j + \Delta a_j) - \hat{y}_i(a_j)}{\Delta a_j / a_j} \tag{1.4}$$

Sensitivity analysis is extremely useful in determining the level of accuracy that input variables need to be measured and that adjustable parameters need to be calibrated.

Addiscott and Tuck (1996) suggest that sensitivity analysis should also be used in NPS pollutant model discrimination. Addiscott (1993) suggests that an additional form of sensitivity analysis is needed in which the model is tested not only for changes in means of its parameters, but for changes in the variances of the parameters to determine if the model is nonlinear. Nonlinear models are at greater risk because of misleading results that may occur in parameterization and in validation. This stems from the fact that nonlinear models exhibit a discrepancy between output resulting from the use of spatially averaged parameters and output resulting from the spatial averaging of the model output generated with discrete parameter values. If a model is nonlinear, the calibration and validation procedures must be standardized to avoid discrepancy.

1.2.1.5 Calibration

Calibration is a test of a model with known input and output information that is used to adjust or estimate parameters for which measured data are not available. Calibration involves "varying parameter values within reasonable ranges until the differences between observed and simulated values are minimized" (Konikow and Bredehoeft, 1992). Although trial-and-error estimates can be used in the minimization process, there are inverse and optimization techniques that are generally applied (Kauffmann et al., 1990; Medina et al., 1990). Calibration by means of automatic optimization methods is feasible on fast computers even for large grids. Both linear and nonlinear least-square fits of the adjustable parameter(s) to the known input and output data are commonly used methods in calibration. Various calibration techniques have been presented using sensitivity analysis, Monte Carlo techniques, and geostatistics (Hoeksema and Clapp, 1990; Jakeman et al., 1990; Kauffmann et al., 1990; Keidser et al., 1990; Medina et al., 1990).

Calibration does not necessarily yield a unique set of parameters; as a result, best professional judgment is often a crucial element in determining what is acceptable and what is not. Konikow and Bredehoeft (1992) cite three reasons for not obtaining a unique set of parameters: (1) conceptual model errors due to neglecting relevant processes, representing inappropriate processes or invalid scientific principles, (2) numerical errors arising in the equation-solving algorithm, and (3) errors in the known input and output information. Because of this lack of uniqueness, Beven (1989) argues that the comparison of observed data and simulated model results serves as a necessary test, but not a sufficient test in both calibration and validation stages.

One means of dealing with the problem of lack of uniqueness is to employ a method of robust calibration under different uncertainties (Klepper and Hendrix, 1994). In this case, the inverse problem results in a probabilistic representation in parameter space so that the posterior probability distribution of the parameters is based on prior information and the probability distribution of measurements. Uncertainty in parameter values is reduced using information on the actual system. The simultaneous calibration of flow and transport using concentration data has also removed the nonuniqueness problem of flow parameters found when just calibrating flow models (Kauffmann et al., 1990).

After the constraints of the model conceptualization assumptions, calibration is the next action toward setting application limits on the model. Calibration utilizes a data set over the range of conditions that ultimately establishes the site-specific limits of the model. This restricts the use of the model to the area(s) where calibration occurred. For example, when a model is calibrated at a particular location, scale, and set of conditions, then the model is restricted to a validation and future application at that location and scale. Calibration is a site-, scale-, and scenario-specific process.

1.2.1.6 Validation

Validity is classically defined as *the quality of being based on sound evidence* consequently, validation is a comparison of model results with numerical data derived from experiments or observations of the environment. As such, validation "demonstrates the ability of a site-specific model to represent cause-and-effect relations at a particular field area" (Konikow and Bredehoeft, 1992). Importantly, the experiments and data used in validation should be temporally independent from those used in calibration. A comparison of observed and simulated data with graphical and statistical methods evaluates how well a model represents the real system.

The philosophy of validation falls into two schools of thought, as pointed out in a thought provoking and cogent paper by Konikow and Bredehoeft (1992) concerning whether or not subsurface hydrologic models can be validated. One school professes that "theories are confirmed or refuted on the basis of critical experiments designed to verify the consequences of the theories" (Matalas et al., 1982). In contrast, the second school advocates that "as scientists we can never validate a hypothesis, only invalidate it" (Popper, 1959) because "any physical theory is always provisional, in the sense that it is only a hypothesis: you can never prove it" (Hawking, 1988). Most present-day scientists probably fall into the latter school of thought. Because no model is ever assured of 100% accuracy with 100% reliability, no model can be validated; consequently, models are always subject to further testing (Oreskes et al., 1994).

Validation must be regarded as an ongoing, dynamic process continually dependent upon new and more reliable observations. A growing body of scientific research has clearly shown that site-specific models that provide good representations of processes and are capable of historically matching data still do not necessarily render accurate predictions at future times (Konikow and Person, 1985; Konikow, 1986; Konikow and Swain, 1990). A good match does not prove validity, nor should it be sufficient for acceptability. To be scientifically valid or even acceptable, not only must a model fit the observed data, but it must do so for the right reasons (Klemes, 1986). In other words, the model's robustness is also a prime consideration. Validation must consider both accuracy and robustness.

Along with calibration, validation establishes the context of the model for future application or in other words sets the reasonable limits of the model's application. It is crucial, therefore, to obtain comparison data over the complete range of conditions of the model's intended application.

Even though some definite progress has been made in the attempt to validate models of the vadose zone, the instances of comprehensive validation attempts are isolated. However, a growing school of thought convincingly argues that validation, per se, is a futile objective (Konikow and Bredehoeft, 1992; Maloszewski and Zuber, 1992; Oreskes et al., 1994). Oreskes et al. (1994) argue that verification and validation of models of natural systems are impossible because natural systems are open, so model results are always nonunique. At best, models of natural systems can only be *confirmed* by demonstrating the agreement between observation and prediction, and even confirmation is inherently partial because "complete confirmation is logically precluded by the fallacy of affirming the consequent and by incomplete access to natural phenomena" (Oreskes et al., 1994).

In actuality, the philosophical argument that no model is capable of validation may be one of semantics and degree because from a practical standpoint the intention of validation is to provide the user with confidence in the reliability and utility of the model. Once confidence in a model's reliability and utility is established in the eyes of the user and that confidence is maintained by ongoing observation, sufficient validation for the user's needs has been attained. Because Newton's theory of gravity does not explain the motion of planets as accurately as Einstein's general theory of relativity, it has been shown to be invalid by comparison, but it is still practical in most situations, so it continues to be used. Though this argues strongly in favor of the status quo, it points out that observations are singular realizations and the application of a model has historically continued to be justified as long as it is in agreement with those observations within the limits of its intended applicability.

Arguably, the work that has been done by Wierenga, Rockhold, and Hills on the Las Cruces trench experiment constitutes the most complete field validation of solute transport modeling efforts for the vadose zone (Hills et al., 1989a, 1989b; Hills and Wierenga, 1991, 1994; Hills et al., 1991; Rockhold et al., 1996). The work by Rockhold et al. (1996) stands among the most successful attempts to model water flow and transport (see also Rockhold, 1999).

1.2.1.7 Simulation and Uncertainty Analysis

The final step of the modeling process is simulation. Whether for the purpose of understanding or prognostication, a simulation is not simply a matter of running the model for a set of measured or agreed-upon inputs and conditions. Knowledge of the level of uncertainty associated with each simulation is crucial. This is particularly true for simulations of NPS pollutants because of the uncertainties associated with spatial variability and scale. A map of simulated results is only of value when accompanied with an associated map of uncertainty. This is clearly pointed out in the work by Loague and colleagues, summarized by Loague and Corwin (1996) and Loague et al. (1996). (Further discussion of uncertainty assessment associated with the simulation process can be found in Chapter 19.)

Loague and Corwin (1996) point out several potential sources of uncertainty. The reliability of a model is determined by the error associated with its simulated output and the intended use of the simulated output. Error is inherent in all models, no matter how sophisticated or complex. Three sources of error are inherent to all NPS pollutant models (Loague et al., 1996): model error, input error, and parameter error. Model error results in the inability of a model to simulate the given process, even with the correct input and parameter estimates. Model error can be due to the characteristic oversimplification of the complexities of the actual processes described within the model. Input error is the result of errors in the source terms (e.g., soil-water recharge and chemical application rates). Input error can arise from measurement, juxtaposition, and/or synchronization errors. Input (or data) error is inherent not only in estimated information, but measured data as well; therefore, uncertainty is associated with all data. Parameter error has two possible connotations. For models requiring calibration, parameter error usually is the result of model parameters that are highly interdependent and nonunique. For models with physically based parameters, parameter error results from an inability to represent aerial distributions on the basis of a limited number of point measurements. The combination of input and parameter errors is reflected in the quality of the model simulations (relative to ground truth) and in the reliability of the simulations for use in making decisions. The aggregation of model error, input error, and parameter error is the simulation (or total) error. Simulation error is complicated further, for multiple-process and comprehensive models, by the propagation of error between model components.

Different methods have been used to evaluate uncertainty in NPS pollutant models. These methods fall into two distinct categories: (1) sensitivity analysis, where the primary concern is assessing the propagation of error between model components; and (2) uncertainty analysis, where the causes of simulation uncertainty are the focus of concern. Uncertainty analysis considers the inherent uncertainty in model input and parameter information and the subsequent effect this uncertainty has upon simulation results. Uncertainty analysis can be carefully designed to uncover information shortfalls and process misrepresentation. Sensitivity analysis, on the other hand, makes no use of information related to the sources or ranges of uncertainty in the model input; i.e., only considering the sensitivity of the model outputs to slight changes in an input variable/parameter.

A number of uncertainty methods have been developed and applied specifically to hydrologic and water resource problems (Dettinger and Wilson, 1981; Beck, 1987; Schanz and Salhotra, 1992; Summers et al., 1993). Loague et al. (1989, 1990) and Zhang et al. (1993) used uncertainty analysis specifically for deterministic transport models in the vadose zone.

Methods for estimating the uncertainty of model predictions from deterministic models fall into two general categories (Loague and Corwin, 1996): (1) first-order variance propagation and (2) Monte Carlo methods. Monte Carlo simulations (see Zhang et al., 1993; Bobba et al., 1995) involve the repeated sampling of the probability distribution for model parameters, boundary conditions, and initial conditions and the use of each generated set of samples in a simulation to produce a probability distribution of model predictions. Monte Carlo simulations are computationally intensive, particularly if the contaminant transport model is numerically complex. An alternative approximate technique, the Rackwitz-Fiessler method, can be used when computation times prohibit the use of Monte Carlo simulation (Veneziano et al., 1987; Schanz and Salhotra, 1992). First-order variance propagation methods such as first-order second-moment (FOSM) analysis require the calculation of a deterministic output trajectory for the model followed by the quantification of the effects of various small amplitude sources of output uncertainty about the reference trajectory (Burges and Lettenmaier, 1975; Argantesi and Olivi, 1976). The application of FOSM analysis is limited to simple models that are continuous with respect to model parameters and time. Furthermore, FOSM approximation deteriorates when the coefficient of variation is greater than 10–20%, which, of course, is common for soil properties in solute transport modeling (Zhang et al., 1993).

The intended use of model simulations determines the level of error that can be tolerated for the simulations to be of value. There are generally three regional-scale uses for GIS-based NPS pollutant models: (1) assessment of existing conditions resulting from legacies, (2) prediction of future impacts resulting from ongoing or future activities, and (3) development of concepts for the design of future experiments to improve the understanding of processes.

The major problem in applying simulated NPS vulnerability assessments to real problems is that it has not been possible to rigorously and unequivocally validate, based upon field observations, any regional-scale earth science modeling approach (Konikow and Bredehoeft, 1992; Oreskes et al., 1994). The model validation problem is directly linked to the uncertainties associated with simulation errors and can be tremendous at regional scales. It should be pointed out that performance standards have not yet been established for any of the applied problems in which NPS pollutant models are used. Future NPS simulation efforts will be greatly improved if well-defined model testing protocols, including model performance standards, are established.

Complete model evaluation requires both operational and scientific examination (Willmott et al., 1985). The operational component of model evaluation is the assessment of accuracy and precision. Accuracy is the extent to which model-predicted values approach a corresponding set of measured observations. Precision is the degree to which model-predicted values approach a linear function of measured observations. The concept of scientific evaluation is the assessment of consistency between model-predicted results and the prevailing scientific theory (Willmott et al., 1985). The concept of scientific evaluation is well suited to evaluating deterministic-conceptual and stochastic-conceptual models. However, it is not appropriate for deterministic-empirical or stochastic-empirical models. Evaluation of an NPS pollutant model's performance should include both statistical criteria and graphical displays (Loague and Green, 1991).

1.2.2 SPATIAL FACTORS TO CONSIDER WHEN MODELING NPS POLLUTANTS IN SOIL

Mulla and Addiscott (1999) cite several factors that must be considered when modeling NPS pollutant models of the vadose zone at moderate to large scales: (1) development of adequate datasets for validation, (2) development/selection of a model that adequately accounts for transport and fate processes that exist at each scale, (3) methods for parameterizing the model, including methods for evaluating goodness of fit, (4) issues of nonlinearity, (5) methods for dealing with spatial and temporal variability in input parameters, and (6) methods for dealing with uncertainty. An awareness and understanding of the influence of scale and spatial variability on these factors serves as the linchpin for optimizing and legitimizing the modeling process.

1.2.2.1 Scale

Scale, as used in soil science and hydrology, refers to the "characteristic length in the spatial domain" and to the "characteristic time interval in the temporal domain" (Baveye and Boast, 1999). So, even though space and time are continuous, there is only a discrete set of scales that is of interest based upon specific features that make them of particular use or interest (Baveye and Boast, 1999). The existence of a hierarchy of scales has been postulated to relate spatial or temporal features of systems of interest (see Figure 1.4).

Temporal and spatial scales dictate the general type of model. The consideration of scale in model development requires observed information for the real system being modeled at the spatial and temporal scales of interest. This is to say that microscopic-scale models developed in the laboratory are not appropriate for macroscopic-scale applications, and vice versa.

Models of solute transport in the vadose zone exist at all scales. A hierarchical depiction of the scales from molecular to global showing the relationship between scale and model type is depicted in Figure 1.4. An important consideration in model conceptualization is for the model to

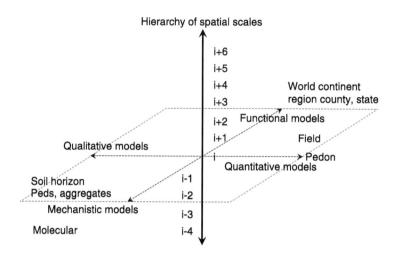

FIGURE 1.4 Organizational hierarchy of spatial scales pertinent to NPS pollutant models. (From Hoosbeek and Bryant, 1993.)

account for the predominant processes occurring at the spatial and temporal scales of interest. This complies with the guideline of parsimony. Qualitatively speaking, as spatial scale increases, the complex local patterns of solute transport are attenuated and dominated by macroscale characteristics. For this reason, mechanistic models are utilized more frequently at the (i) to $(i-4)$ scales, while functional models are more often applied to scales ranging from $i+1$ to $(i+6)$. The stochastic application of deterministic models is found at the $(i+1)$ scale, and stochastic models generally are used at $(i+1)$ and $i+2)$ scales. Statistical models are applied most often at the larger scales, $(i+3)$ to $(i+6)$.

The relevance of temporal domain is also a consideration not to be overlooked. Larger spatial scales appear more constant because the rapid dynamics of the lower scales are disregarded (O'Neill, 1988). For this reason, time steps of functional models can expand over days, such as the time between irrigation or precipitation events, while the time steps of mechanistic models characteristically extend over minutes. A complete discussion of the application of models at different spatial and temporal scales is presented by Wagenet (1996) and Wagenet and Hutson (1996). Scale and its implications on modeling are also discussed by Mulla and Addiscott (1999) and Baveye and Boast (1999). In particular, Baveye and Boast (1999) offer a unique and controversial perspective concerning the influence of scale on model conceptualization.

The integration of solute transport models of the vadose zone into a GIS provides the ability to dynamically describe NPS pollutant transport at a range of spatial scales allowing the user to rapidly scale "up" and "down." However, this introduces incompatibilities between the model and data, and raises basic questions regarding (1) the compatibility of the model with input and validation data, and (2) the relevance of the model to the applied spatial scale. Wagenet and Hutson (1996) addressed the issue of scale dependency and proposed three scale-related factors to consider when applying GIS-based solute transport models to the simulation of NPS pollutants in soils: (1) the type of model (i.e., functional or mechanistic) must consider the scale of application and the nature of the available data at that scale, (2) sampling and measurement of input and validation data must be spatially consistent with the model, and (3) measurement and monitoring methods must be relevant at the temporal domain being modeled.

1.2.2.2   Spatial Variability and Structure

The spatial variability of soil has been the focus of numerous books (Bouma and Bregt, 1989; Mausbach and Wilding, 1991; Robert et al., 1993) and review articles (Beckett and Webster, 1971; Wilding and Drees, 1978; Warrick and Nielsen, 1980; Peck, 1983; Jury, 1985, 1986), and a compendium of Pedometrics-92 Conference papers. Ever since the classic paper by Nielsen et al. (1973) concerning the variability of field-measured soil water properties, it has been well known that soil properties exhibit considerable variability (Warrick and Nielsen, 1980; Jury, 1985; White, 1988). As discussed in Chapter 2, spatial and temporal variability of soil processes have profound implications on field-scale study design and characterization.

Studies suggest that much of the field-scale variability occurs at rather small spatial scales (van Wesenbeeck and Kachanoski, 1991, 1994; Poletika et al., 1995; Ellsworth and Boast, 1996; Ellsworth et al., 1996). Ellsworth (1996) points out that assessments of NPS pollution could be obtained using a relatively smaller number of "large"-sized samples. Measurement methods should be adopted that can give an integrated value of a soil property over large areas and volumes considered homogeneous with respect to that property. This will aid NPS pollution modeling by enhancing discrimination between models and improving parameter estimation efforts.

Prior to the emergence of GIS, the incorporation of spatial variability into the modeling of solute transport in the vadose zone was accomplished with numerical techniques such as finite elements. Even with today's super-computers, the application of finite elements to regional-scale problems such as NPS pollutants is impractical primarily because of the astronomical input data requirements and the limited availability of supercomputer time. GIS is currently used as a practical tool for incorporating a spatial capability into one-dimensional models, which are more widely used and generally more easily understood than the multidimensional finite-element models.

The nature of soil variability is dependent on one's perspective, or scale of resolution. For example, when viewed from the moon, the spatial diversity of the earth's surface appears as land and water, whereas low-level aerial photography yields tonal patterns of soils, landforms, geomorphic features, erosion, and vegetative patterns. If soil is observed at still greater resolution and in vertical cross section, spatial variability is seen in three dimensions as a succession of soil horizons and features not evident at the soil surface. Therefore, spatial variability can be recognized to varying degrees within two or three dimensions at microscopic, plot, field or landscape, regional, and global scales. Spatial variation is recognized as a continuum from short-range to long-range order.

Since the inception of the Soil Survey, the users of soil maps, most notably solute transport modelers, have desired to know to what extent they could assume that all the soil mapped as one class had equal potentialities. Users want and need confidence limits, probabilities, and frequency analyses on the composition of map units and information on how inclusions within a given map unit influence interpretations and behavior. The obvious question for a soil scientist to ask is how many samples are needed to characterize soil spatial variability? The response to this question depends on the magnitude of variability within the population for the parameter in question and the probability level placed on the confidence limits (Wilding and Drees, 1978).

The following discussion utilizes the coefficient of variation as a measure to compare soil property variation. The coefficient of variation (CV%) is defined as sample standard deviation expressed as a percentage of the sample mean. Some of the input variables and parameters needed for solute transport models of the vadose zone are dominated by the bulk characteristics of the solid matrix of the soil; consequently, the spatial variability of these properties are relatively small, which reflects the uniformity of soil genesis processes (Jury, 1986). These properties include porosity, bulk density, field capacity (i.e., soil-water content after free drainage has occurred, approximately 0.3 bar), and wilting point (i.e., soil-water content when plants begin to wilt, approximately 15 bars). Characteristically, these variables/parameters are associated more often with functional, deterministic models of solute transport. Properties dominated by the bulk characteristics of the soil matrix are low to moderate in variability irrespective of field size or soil type. This is reflected in the low coefficients of variation as tabulated by Jury (1986): porosity (CV = 7–11%), bulk density (CV = 3–26%), 0.1 bar soil-water content (CV = 4–20%), and 15 bar soil-water content (CV = 14–45%). In contrast, water transport parameters including saturated hydraulic conductivity, infiltration rate, and hydraulic conductivity–water content or hydraulic conductivity–matric potential relations, are characterized by a high variability of at least 100% or greater. Finally, the calibration and validation of NPS pollutant models depend upon the comparison of predicted and measured solute concentrations. Solute transport experiments tabulated by Jury (1986) have shown coefficients of variation of 60–130% for observed and simulated solute concentrations.

Not only do many model input variables and parameters vary considerably across a field, but substantial local-scale variability can also be found. It is common to find 50% of the variation in many soil properties within a 1 m radius (Corwin et al., 2003a). Local-scale variability occurs because soils vary significantly from one location to the next in their structural properties, textural composition, and mineralogical constituents. Human influence also has considerable effect. For instance, on agricultural lands, salinity can vary significantly over short distances merely due to variations in surface topography and how water infiltrates into the soil. On soils with bed-furrow flood irrigation, the salinity within the bed can be an order of magnitude higher than the salinity below the furrow, which is just a few centimeters away. The increased salinity is due to the lateral and upward flow of irrigation water into the bed from the furrow that causes the accumulation of salts in the bed, while the salts directly below the furrow are continuously leached downward.

The local-scale structure is a feature that must be considered in relation to its influence upon the overall scale of interest. In other words, are local-scale influences in relation to the dominant processes of the "big picture" inconsequential, or must they be taken into account? This is a question useful in determining whether a sophisticated mechanistic or a simple functional model should be applied to a given NPS pollutant problem. Qualitatively it is recognized that as the spatial scale increases, the complex local patterns of solute transport are attenuated and are dominated by macroscale characteristics. Furthermore, knowledge of the local-scale structure is not only needed for model discrimination, but is of value in estimating the minimum volume of the soil sample necessary to represent a property at a given location. This will allow an estimation of the minimum spatial scale at which the field-scale parameters dominate solute transport behavior.

The scale of the averaging process becomes very important. Replicated measurements of representative variables/parameters where large field areas are involved must be substantial enough so that their mean values give a representative average. The type of model — functional or mechanistic — can be a factor in determining the scale of the averaging process. Table 1.3, originally presented by Jury (1986), shows sample sizes necessary to have at least a 95% probability of detecting a relative change of 20, 40, and 100% in the value of the mean of various field-scale solute transport model parameters when using a one-sample two-tailed t-test with a probability of Type 1 error set at $\alpha = 5\%$. The last column, which has been added to show whether the parameter is a capacity or rate parameter, indicates the type of model — functional or mechanistic — with which the parameter is associated. Table 1.3 clearly shows that when using functional models of solute transport, the number of samples needed to represent the parameters is significantly less than for mechanistic models. Furthermore, the uncertainty of the measurement as indicated by the sample variance is as important as the mean value because it indicates the precision of the mean and uniformity of the measurement. Because most models of NPS pollutants in the vadose zone are one-dimensional, uniformity is particularly significant to illustrate the extent of validity of the assumption of one-dimensionality for a defined volume of soil.

There is a need to quantify soil variability and to determine the scale or scales of its occurrence. Such information is increasingly needed for modeling of water flow and contaminant transport in GIS applications and for

TABLE 1.3

Sample Sizes Required for a 95% Probability of Detecting a Change of 20, 40 and 100% in the Mean of Solute Transport Parameters Using a *t*-Test with α = 5%

Parameter	Number of Studies	Number of Samples			Average CV SD	Capacity or Rate Parameter
		20%	40%	100%		
Bulk density or porosity	13	6	*	*	10 ± 6	Capacity
Percent sand or clay	10	28	9	*	28 ± 18	Capacity
0.1 bar soil-water content	4	9	*	*	14 ± 7	Capacity
15 bar soil-water content	5	23	7	*	25 ± 14	Capacity
K_{sat}[a]	13	502	127	22	124 ± 71	Rate
Infiltration rate content	8	135	36	8	64 ± 26	Rate
$K(\theta)$[b]	4	997	251	442	25 ± 14	Rate
Ponded solute velocity	1	1225	308	51	194	Rate
Unsaturated solute velocity	5	127	33	7	62 ± 9	Rate

*Sample size estimates are less than 5 and should not be used.
K_{sat} is the saturated hydraulic conductivity.
$K(\theta)$ is the unsaturated hydraulic conductivity as a function of water content θ.

Source: Jury, 1986.

environmental impact assessment. Different approaches have been proposed for quantifying variability in soil map unit delineations. Traditionally, map unit composition has been quantified by transecting selected delineations of the map unit and determining at each point on a transect whether or not the soil is the same as, or similar to, the selected series. Confidence intervals were calculated using either the Student's *t*-distribution or a binomial method (Brubaker and Hallmark, 1991). The major advantage of the *t*-statistic for calculating map composition is that it allows an estimate of the amount of variability within delineations, provided that more than one set of samples is taken for each delineation, as well as the amount of variability between delineations. The primary disadvantage of this approach is that it can result in biased estimates if care is not taken to account for differences in the size of the delineations and the associated difference is the number of samples taken within each delineation. Care should be taken to ensure that the sampling density is the same for all delineations.

Currently, several techniques are useful in the quantification and delineation of spatial variability, including electromagnetic induction (EMI) (see Chapter 18), time domain reflectometry (TDR), ground-penetrating radar (GPR), aerial photography, and multi- and hyperspectral imagery. However, none of these methods has been as extensively studied as EMI for characterizing spatial variability (Corwin and Lesch, 2005a). Spatial domains or map units of "homogeneous" water flow characteristics, referred to as stream tubes, are promising and potentially well adapted to GIS applications.

The stream-tube model is discussed in detail by Jury and Roth (1990) and Jury (1996). Examples of this approach for simulating field-scale solute transport are given by Bresler and Dagan (1981), Destouni and Cvetkovic (1991), and Toride and Leij (1996a, 1996b). However, considerably less effort has been directed to a realistic evaluation of this approach or to the actual delineation of a stream tube. Corwin et al. (1998) proposed a potential means of delineating stream tubes in the field using EMI. Subsequently, the stream-tube approach was used by Corwin et al. (1999c) in the spatial modeling of salt loading to tile drains over a 5-year period using a GIS-linked solute transport model in a 2400 ha study area. The stream tubes were spatially defined using an EC_a-directed soil sampling design, which minimized the number of soil samples needed to characterize the spatial variability of soil-related inputs for the functional solute transport model. The stream tubes were simply defined as Thiessen polygons formulated around the soil sample sites, which served as the centroids for the polygons. In addition to delineating stream tubes, EMI has been used to map the spatial variability of physicochemical properties for applications in soil quality assessment (Corwin et al., 2003a, 2005) and precision agriculture (Corwin et al., 2003b).

The implications of soil and climatic variability on broad-scale modeling of NPS pollutants has been studied by Jury and Gruber (1989), Foussereau et al. (1993), and Wilson et al. (1996). Jury and Gruber (1989) showed that soil and climatic variability can introduce a small probability that some mass of even relatively immobile NPS pollutant will migrate below the soil surface even when the projected mass is negligible, as determined from models neglecting variability by using average values for soil and climatic properties. This is significant in lieu of the fact that some regulatory decisions have established a compliance surface below which pesticides may not migrate (State of California Legislature, 1985). Foussereau et al. (1993) demonstrated a means of replicating soil variability by using bootstrapping to generate pseudo-profiles of soils from pedon characterization data. Their approach permitted an assessment of the uncertainty associated with model output due to the variability of soil input data. Wilson et al. (1996) explored a means of capturing real-world soil variability through the use of existing databases (i.e., the USDA-NRCS State Soil Geographic Database [STATSGO], the county-level Soil Survey Geographic Database [SSURGO], and the Montana Agricultural Potential System [MAPS]. Their findings revealed that the higher resolution of the SSURGO database was needed to identify those areas where potential chemical applications are likely to contaminate groundwater.

Though not as extensively studied as the spatial variability of soil, the aspect of temporal variability, particularly of soil hydraulic properties, is of concern. Temporal variation is attributed to both intrinsic factors (i.e., natural processes), such as freezing and thawing, root growth and exudates, wetting and drying cycles, carbon turnover and biological activity, and extrinsic factors (i.e., human-related activities), such as tillage operations. Temporal changes have been demonstrated to occur for total porosity (Cassel, 1983; Scott et al., 1994), bulk density (Cassel, 1983; Scott et al., 1994), water retention

(Cassel, 1983; Gantzer and Blake, 1978; Anderson et al., 1990), saturated hydraulic conductivity (Scott et al., 1994), macroporosity (Skidmore et al., 1975; Cassel, 1983; Carter, 1988), and infiltration (Starr, 1990; van Es et al., 1991; van Es, 1993). Tillage affects both the magnitude and variability of soil properties because it physically disrupts the structure of the soil and causes changes in water and solute flow patterns, which may change again with time as soil settles and continuous macropores develop through active soil biota and/or physical processes of nature (e.g., freezing and thawing, wetting and drying). To handle temporal data within existing soil survey databases, Grossman and Pringle (1987) provided a description of a record to join together the use and time invariant information from soil survey documentation with use-dependent temporal quantities. From the GIS standpoint, Langran (1989) reviewed temporal research in information processing, contrasted various proposed temporal designs, and summarized the problem of adapting it to GIS requirements.

1.2.3   MODELING NPS POLLUTANTS IN SOIL

Crucial to the success of sustainable agriculture is an ability to assess detrimental environmental impacts with both real-time measurements and model predictions. Real-time measurements reflect the activities of the past and provide an inventory of the problem, whereas model predictions are glimpses into the future based upon a simplified set of preeminently valid assumptions. These "what-if" scenarios can be used to alter the occurrence of detrimental conditions before they develop.

The ability to model NPS pollutants provides a means to optimize the use of the environment by sustaining its utility without detrimental consequences while preserving its esthetic qualities. The basic reasons for developing models of NPS pollutants in the vadose zone are (1) to increase the level of understanding of the cause-and-effect relationships of the processes occurring in soil systems and (2) to provide a cost-effective means of synthesizing the current level of knowledge into a useable form for making decisions in the environmental policy arena both spatially and temporally (Beven, 1989; Grayson et al., 1992).

Modeling the fate and movement of NPS pollutants in the vadose zone is a spatial problem well suited for the integration of a deterministic solute transport model with a GIS. A GIS characteristically provides a means of representing the real world through integrated layers of constituent spatial information. To model NPS pollution within the context of a GIS, each transport parameter or variable of the deterministic transport model is represented by a three-dimensional layer of spatial information. The three-dimensional spatial distribution of each transport parameter/variable must be simulated (Journel, 1996), measured, or estimated. This creates a tremendous volume of spatial information due to the complex spatial heterogeneity exhibited by the numerous physical, chemical, and biological processes involved in solute transport through the vadose zone. GIS serves as the tool for organizing, manipulating and visually displaying this information efficiently.

Some of the greatest interest in the use of GIS for environmental problem solving is to apply the technology to translate the results of models into environmental policy. Specifically, GIS-based models of NPS pollutants provide diagnostic and predictive outputs that can be combined with socio-economic data for assessing local, regional, and global environmental risk or natural resource management issues (Steyaert, 1993).

In their simplest form, GIS-based environmental models are comprised of three basic components (Burrough, 1996): data, GIS, and environmental model. An understanding of the application of GIS to the modeling of NPS pollutants in the vadose zone requires a cursory understanding of each component and the interrelationship between these components.

1.2.3.1   Data

All models require input data from which simulated output is generated. The single greatest challenge to modeling NPS pollutants is to obtain sufficient data to characterize the temporal and spatial distribution with knowledge of its uncertainty. Over a decade ago, Maidment (1993) insightfully pointed out that the most limiting factor to hydrologic modeling is the ability not to mathematically characterize the processes, but to accurately specify the values of model parameters and input data. This is reiterated even today by Beven (2002), who stated that it has become clear that the real constraint on predictability by environmental models is not the detail of the model structures, but defining the characteristics of individual places.

The effectiveness of a model to simulate a practical application is highly dependent on how well model inputs and model parameters are identified. Basically, there are three sources of input and parameter data for NPS pollutant models (Corwin et al., 1997): measured data, estimated data, and existing data. Each source of data carries distinct advantages and limitations.

1.2.3.1.1   Measured Data

Measured data are the most desirable, but often the least readily available and the most expensive to obtain. A review of current physical measurements to determine flow-related properties of subsurface porous media and soil physical properties is provided by Dane and Molz (1991) and Topp et al. (1992), respectively. The measurement of variables and parameters related to solute transport along with characterization of initial and boundary conditions necessary for model simulation, calibration, and validation constitutes a considerable investment of time and labor because of the tremendous volume of data required. Although direct measurement of transport parameters and variables is the most reliable means of obtaining accurate information for modeling purposes, it is also the most labor intensive and costly. A quick and easy means of obtaining these measurements is crucial to the cost-effective modeling of NPS pollutants.

Remote sensing and noninvasive measurement techniques have the greatest potential for meeting the thirst for measured spatial data by NPS pollutant modelers, but so far these techniques have not reached their potential (Corwin, 1996). In most cases remote sensing provides measurements of only the top few centimeters; consequently, it suffers from a lack of depth information needed in modeling the vadose zone. Noninvasive techniques such as EMI can provide information down to several meters in depth, but these techniques generally require measurements taken at or near the soil surface, and their measurement volume is limited to tens of cubic meters or less. Nevertheless, geospatial measurements of apparent soil electrical conductivity (EC_a) with EMI is currently the most widespread and reliable means of characterizing the spatial variability of a variety of physicochemical properties in the vadose zone including salinity, texture, water content, cation exchange capacity (CEC), organic matter (OM), and bulk density (Corwin and Lesch, 2003, 2005a, 2005b; Corwin et al., 2005) (see Chapter 18).

Corwin (1996) provided a cursory review of some of the instrumental techniques developed for the remote and noninvasive measurement of variables and parameters found in transport models for the vadose zone. The review covers geophysical resistivity methods, aerial photography, x-ray tomography, ground-penetrating radar, magnetic resonance imaging (MRI), microwaves, multispectral imagery, thermal infrared imagery, and advanced very high-resolution radiometry (AVHRR). Barnes et al. (2003) provided a more recent review of remote- and ground-based sensor technology for mapping soil properties. Even though considerable progress has been made over the past decade in the area of remote sensing, Corwin (1996) concluded that "the array of instrumentation needed to measure all the parameters and variables in even the simplest of transport models for the vadose zone is not available and in most cases is not even on the drawing board"; consequently, "the greatest progress [in the modeling of NPS pollutants] needs to be made in the area of instrumentation." Aside from the fact that remote sensing/noninvasive methods are still in their infancy, in most cases the parameters measured are often not directly applicable to solute transport models. For instance, the use of EMI to measure soil salinity is not a direct measure of salinity in the soil solution, but rather measures EC_a, which includes the conductivity of both the solid and liquid phases, thereby requiring ground-truth soil samples for calibration.

1.2.3.1.2 Estimated Data

The extreme spatio-temporal variability and the nonlinearity of many soil processes make parameterization of landscape-scale models a particularly daunting task. The inability of remote measurement techniques to meet the demand for spatial and temporal parameter and input data has resulted in the development of transport parameter estimation techniques that estimate parameters by fitting data or are based upon the formulation of transfer functions. Inverse modeling (discussed in detail in Chapter 20) is a powerful and practical means of estimating flow and transport parameters for landscape-scale solute transport models using advanced optimization algorithms. Transfer functions relate readily-available and easy-to-measure soil properties to more complex transport variables/parameters needed for simulation.

Corwin et al. (1997) provide a referenced list of the estimation methods for many of the commonly used parameters in solute transport models of the vadose zone. The most common of the transfer functions, the pedo-transfer function (PTF), uses particle-size distribution, bulk density, and soil organic-carbon content to yield soil-water retention or unsaturated hydraulic conductivity functions (Bouma and van Lanen, 1987). Rawls et al. (1991) provides a review of soil-water retention estimation methods. Reviews of methods of estimating soil hydraulic parameters for unsaturated soils have been written by van Genuchten and Nielsen (1985), van Genuchten et al. (1992), and Timlin et al. (1996).

Pedo-transfer functions have been developed to predict the hydraulic characteristics of a textural class using more easily measured soil data. However, pedo-transfer functions are limited in accuracy. For example, an evaluation of PTFs has shown that greater than 90% of the variability of simulations for a map unit was due to the variability in the estimated hydraulic parameters with the PTFs, which brings the value of PTFs into question (Vereecken et al., 1992).

Although estimation methods are cheap and ease to use, their limited accuracy makes them less desirable than directly measured data. Nonetheless, if measured parameter data are not available, then estimations using transfer functions or derived from inverse modeling are usually the next best alternative.

1.2.3.1.3   Existing Data

In most instances, limited resources do not permit the measurement or even estimation of needed input or parameter data. In these instances, the use of existing data is crucial. SSURGO (Soil Survey Geographic Database), STATSGO (State Soil Geographic Database), and NATSGO (National Soil Geographic Database) are existing soil databases for the USA, SSURGO (Soil Survey Geographic Database; map scale ranges from 1 : 12,000 to 1 : 63,360) is a county level database and it is the most detailed GIS database available from NRCS. STATSGO (State Soil Geographic Database; map scale 1 : 250,000) is the state-level database designed for state, large watershed, and small river basin purposes. NATSGO (National Soil Geographic Database; map scale 1 : 7,500,000) is the national soil database whose map units are defined by major land resource area (MLRA) and land resource region (LRR) boundaries. Even though considerable data are available through existing databases, most soil databases do not meet minimum data requirements for many of the distributed-parameter models used for NPS pollutants in the vadose zone, nor do they provide useful statistical information concerning the uncertainty of the soil property data (Wagenet et al., 1991); consequently, there is a need for a reevaluation of the types of information collected in soil surveys to meet the quantitative requirements of environmental and agricultural management models (Bouma, 1989). An excellent example of the use of existing data sources in a GIS-based solute transport modeling application is the work of Wilson et al. (1996). Table 1.4 provides a list of some of the existing databases.

The typical parameter surface maps (e.g., soil survey maps) used in NPS pollutant models are most often based upon point value measurement averages that are extrapolated to large unsampled regions without consideration for the variability (uncertainty) in the measured data. By not considering the variability in the data, there is obviously tremendous opportunity for error propagation (Heuvelink et al., 1989).

The problem with the use of generalized rather than measured data has been the associated uncertainties. Loague et al. (1996) extensively reviewed the uncertainty associated with the use of an existing database for non-point source groundwater vulnerability and concluded that assessments based on this type of data are relegated to guiding data-collection strategies rather than their intended purpose of groundwater vulnerability assessment. Measured input data that capture natural variability both in space and time are essential for diminishing uncertainty in simulations.

1.2.3.2   GIS

A GIS is defined by Goodchild (1993) as a "general-purpose technology for handling geographic data in digital form with the following capabilities: (1) the ability to preprocess data from large stores into a form suitable for analysis (reformatting, change of projection, resampling, and generalization), (2) direct support for analysis and modeling, and (3) postprocessing of results (reformatting, tabulation, report generation, and mapping)." In the context of NPS pollutant modeling, a GIS is a tool used to characterize the full information content of the spatially variable data required by solute transport models. GIS is characterized by its capability to integrate layers of spatially oriented information. The advantages of GIS in its application to general spatial problems include "the ease of data retrieval; ability to discover and display information gained by testing interactions between phenomena; ability to synthesize large amounts of data for spatial examination; ability to make scale and projection changes, remove distortions, and perform coordinate rotation and translation; and the capability to discover and display spatial relationships through the application of empirical and statistical models" (Walsh, 1988).

The use of GIS in environmental modeling has proliferated over the past 25 years. In its infancy GIS was primarily used to create inventories of natural resources. However, over the past 15 years modeling and analysis applications with GIS have become more prevalent, especially in the environmental-assessment arena. The principal benefit of coupling GIS to environmental

TABLE 1.4
Some of the Existing Databases for Use in Modeling NPS Pollutants with GIS

Database	Source	Description
Soil databases:		
SOTER	ISRIC[a]	World Soils and Terrain Digital Database: global-scale database of soils, terrain, climate, vegetation and land use data; scale of 1 : 1,000,000.[b,c]
NATSGO	USDA-NRCS[b,c]	National Soil Geographic database: national-level soils database of USA; scale of 1 : 7,500,000. Application: national, regional, and multi- state resource appraisal, planning and monitoring. Linked to Soil Information Record (SIR) database for soil property data. Soil mapping percentage of the map unit having the queried properties.
STATSGO	USDA-NRCS[b,d]	State Soil Geographic database: state-level soils database of USA scale of 1 : 250,000. Application: state and regional studies of large watersheds, small river basins. Linked to SIR. Map units consist of 1 to 21 components with each component consisting of up to 25 physical and chemical properties.
SSURGO	USDA-NRCS[b,e]	Soil Survey Geographic database: county-level (most detailed) soils database of USA; scale 1 : 12,000 to 1 : 63,360. Duplicate of original soil survey maps. Application: resource planning and management of private property, townships, and counties. Linked to Map Unit.
Meteorologic databases:		
	NOAA	Weather station data comprised of daily rainfall, daily min/max temperature, daily average temperature, relative humidity, etc.
SNOTEL	USDA-NRCS[f]	Daily snow and precipitation amounts at specific locations within specified states and regions.
Miscellaneous databases:		
CIMIS	California Dept.	Seasonal crop evapotranspiration estimates of Water Resources[g]
UNSODA	USDA-ARS[h]	Database of measured unsaturated hydraulic properties (water retention, hydraulic conductivity, and soil water diffusivity) and basic soil properties (particle-size distribution, bulk density, organic matter, etc.).

Internation Soil Reference and Information Center, P.O. Box 353, 6700 AJ Wageningen, The Netherlands.

Technical Information: National Soil Survey Center; USDA-ARS; Federal Bldg., Room 152, 100 Centennial Mall, North; Lincoln, NE 68508-3866; Phone: 402-437-4149. Data Source: USDA-NRCS; National Cartography and Geospatial Center; 501 Felix St., Bldg. 23; P.O. Mail 6567; Fort Worth, TX 76115; Phone: 800-672-5559.

Web site:http://www.ncg.nrcs.usda.gov/natsgo.html

Web site:http://www.ncg.nrcs.usda.gov/statsgo.html

Web site:http://www.ncg.nrcs.usda.gov/ssurgo.html

Web site:http://www.ncg.nrcs.usda.gov/water.html

California Department of Water Resources, Office of Water Conservation, P.O. Box 942836, Sacramento, CA 94236-0001.

USDA-ARS, George E. Brown Jr. Salinity Laboratory, 450 West Big Springs Road, Riverside, CA 92507-4617.

Source: Corwin et al., 1997.

models is to enable the models to deal with large volumes of spatial data that geographically anchor many environmental processes. This is especially true of surface and subsurface hydrologic processes. GIS applications to hydrologic modeling have been used in the past most widely and effectively by surface hydrologists and to a lesser extent by soil scientists and groundwater hydrologists for NPS pollutant applications. Only within the past decade have soil scientists begun to utilize GIS as a tool in data organization and spatial visualization of NPS pollution model simulation in the vadose zone.

1.2.3.3 Models

By definition, mathematical models integrate existing knowledge into a framework of rules, equations, and relationships for the purpose of quantifying how a system behaves (Moore and Gallant, 1991). As long as models are applied over the range of conditions from which they were initially developed, they serve as a useful tool for prognostication. Models can range in complexity from the simplest empirical equation to complex sets of partial differential equations that are only solvable with numerical approximation techniques.

Since the 1950s numerous models have been developed to simulate the one-, two-, and three-dimensional movement of solutes through the vadose zone. Addiscott and Wagenet (1985) discussed a categorization of these models based upon conceptual approach. Their categorization distinguished between (1) deterministic and stochastic and (2) mechanistic and functional. According to Addiscott and Wagenet (1985), the key distinction between deterministic and stochastic models is that deterministic models "presume that a system or process operates such that the occurrence of a given set of events leads to a uniquely definable outcome," while stochastic models "presuppose the outcome to be uncertain." Deterministic models are based on conservation of mass, momentum, and energy. Theoretically, if all essential parameters and variables of the predominant transport processes are known for every point in a soil system, then a mechanistic model of solute transport can be applied with confidence. Practically speaking, this is unlikely and has spawned an interest in stochastic models of solute transport. Stochastic models consider the statistical credibility of both input conditions and model predictions, whereas deterministic models ignore any uncertainties in their formulation. The second level of model distinction is between mechanistic and functional models. As stated by Addiscott and Wagenet (1985), "mechanistic is taken to imply that the model incorporates the most fundamental mechanisms of the process, as presently understood," whereas the term functional is used for "models that incorporate simplified treatments of solute and water flow and make no claim to fundamentality but do thereby require less input data and computer expertise for use."

Because of lateral and vertical variation of soil, it is not reasonable to expect that three-dimensional models capable of describing point-to-point variability could be calibrated by any conceivable combination of measurements at field scales (i.e., hundreds or thousands of hectares); consequently, field-scale models of processes in which large surface and subsurface areas are treated relatively uniformly will need to be one-dimensional (Jury, 1986). For this reason the vast majority of models that have been coupled to a GIS to simulate NPS pollutants in the vadose zone have been one-dimensional.

A comprehensive review of GIS-based NPS pollutant modeling in the vadose zone has been presented by Corwin et al. (1997). To date, the vast majority of models of NPS pollutants in the vadose zone have utilized deterministic models of solute transport coupled to a GIS (Corwin, 1996; Corwin et al., 1997). However, there is a growing recognition that stochastic approaches may offer the most viable means of modeling NPS pollution (Jury, 1996; Corwin et al., 1999a; Loague et al. 1999). The number of stochastic models of solute transport coupled to a GIS is gradually increasing, such as the stochastic application of GLEAMS by Wu et al. (1996) and the conditional simulations incorporating geostatistical methods by Rogowski (1993).

1.2.3.3.1 GIS-Based Deterministic Models

A detailed conceptualized development of the soil processes involved in water flow and solute transport is presented in Chapter 3. The methods for measuring, characterizing, and evaluating the spectrum of soil physicochemical processes (e.g., water flow, mass transport, volatilization, sorption, transformations, etc.) that comprise a deterministic approach to modeling the fate and movement of solutes in soil are discussed in detail in Chapters 5–15. These chapters provide the fundamental concepts pertaining to most GIS-based landscape-scale models of solute transport.

Corwin (1996) provided a review of GIS applications of one-dimensional, deterministic solute transport models to field-, basin-, and regional-scale assessment of NPS pollutants in the vadose zone. The use of deterministic transport models with GIS has been justified on practical grounds based upon availability, usability, widespread acceptance, and the assumption that a heterogeneous medium macroscopically behaves like a homogeneous medium with properly determined parameters and variables. The philosophy of modeling NPS pollutants in the vadose zone with a one-dimensional deterministic model of solute transport is based upon the coupling of a GIS to the simulation model. The physical, chemical, and biological properties influencing transport in the vadose zone are represented using a distributed parameter structure. The validity of the assumption that a heterogeneous medium macroscopically behaves like a homogeneous medium depends upon whether spatial domains can be defined and characterized that behave as stream tubes [see Jury and Roth (1990) for a discussion of stream tubes] or "representative element volumes" [for a discussion of REVs, see Mayer et al. (1999) and Chapter 2].

Three categories of deterministic models have been coupled to GIS to simulate NPS pollution in the vadose zone: regression models, overlay and index models, and transient-state solute transport models. Regression models have generally used multiple linear regression techniques to relate various causative factors to the presence of a NPS pollutant. These causative factors have included soil properties or conditions related to groundwater vulnerability or to the accumulation of a solute in the soil root zone (Corwin et al., 1989; Skop, 1995; Wang et al., 1995; Teso et al., 1997). For instance, Corwin et al. (1989) related soil salinization factors (i.e., soil permeability, irrigation efficiency, and groundwater quality) to the development of salinity in the root zone for the entire Wellton-Mohawk irrigation district (170 mi.^{2}). More recently, logistic regression techniques have been utilized to identify areas of groundwater vulnerability to pesticides (Teso et al., 1997) and the development of soil salinity (Wang et al., 1995). Overlay and index models refer to those models that compute an index of NPS pollutant mobility from either a simple functional model of steady-state solute transport (Merchant et al., 1987; Corwin et al., 1988; Corwin and Rhoades, 1988; Khan and Liang, 1989; Evans and Myers, 1990; Regan, 1990; Halliday and Wolfe, 1991; Rundquist et al., 1991) or a steady-state mechanistic model (Wylie et al., 1994). Two types of overlay and index models have been developed: property-based and process-based. Property-based index models are established upon hydrogeologic setting (e.g., DRASTIC) or NPS pollutant properties (e.g., GUS). Process-based index models are founded upon the characterization of transport processes (e.g., Rao's attenuation factor model). Overlay and index models have been used largely to assess groundwater pollution vulnerability to pesticides and nitrates. Transient-state, process-based solute transport models include deterministic models capable of handling the movement of a pollutant in a dynamic flow system. Transient-state, process-based models describe some or all of the processes involved in solute transport in the vadose zone: water flow, solute transport, chemical reactions (adsorption-desorption, exchange, dissolution, precipitation, etc.), root growth, plant-water uptake, vapor phase flow, degradation, and dispersion/diffusion. The most recent progress has occurred in the coupling of transient-state solute transport models to GIS (Bleecker et al., 1990; Petach et al., 1991; Corwin et al., 1993a, 1993b; Wilson et al., 1993; Tiktak et al., 1996).

1.2.3.3.2 GIS-Based Stochastic Models

Jury (1986) pointed out that the difficulty of constructing a three-dimensional model of chemical transport as a consequence of field variability has two significant implications: (1) any hope of attempting to estimate a continuous spatial pattern of chemical transport must be abandoned, and (2) there exists a possibility of extreme deviations from average movement so that significant concentrations of chemical may flow within relatively small fractions of the total cross-sectional area, which may be nearly impossible to detect from point measurements. The latter implication has fostered the development of stochastic solute transport models for the vadose zone as opposed to deterministic models.

Two distinct stochastic approaches are currently in use for dealing with the spatial variability encountered in modeling NPS pollutants in the vadose zone: geometric scaling and regionalized variables. Jury (1986) indicates that geometric scaling uses specific "standardized variables to scale the differential equations describing transport and relates the standardized variables to some measurable or definable property of each local site of a heterogeneous field." Once the variables are defined, the onerous task of characterizing the variability is reduced to determining the statistical and spatial distribution of these scaling parameters. In contrast, Jury (1986) explains that the regionalized variable approach regards the "various parameters relevant to a field-wide description of transport as random variables characterized by a mean value and a randomly fluctuating stochastic component."

In comparison to deterministic models, the coupling of a stochastic solute transport model to GIS is relatively unexplored. In a paper discussing the potential compatibility of stochastic transport models with GIS, Jury (1996) suggested that stochastic-convective stream tube modeling seems the most compatible with GIS because it "utilizes a relatively simple local process driven by parameters that might be associated with soil morphological features, and could be integrated up to a large scale by simple arithmetic averaging over the local sites." A stochastic stream tube model is made up of parallel, noninteracting one-dimensional soil columns whose properties are locally homogeneous, but vary from one soil column to the next. The collection of all stream tubes constitutes the field-, basin-, or regional-scale area being represented. This approach is in essence the same approach that has been undertaken in the past where deterministic piston-flow local transport models have been coupled to soil survey information, only now there is an associated stochastic component of information. Jury (1996) warns that the challenge of this approach will be "to develop a reasonable local-scale model whose parameters can be related to identifiable local-scale features."

1.2.4 ROLE OF GEOSTATISTICS AND FUZZY SET THEORY

Geostatistics can provide support for the modeling process by: (1) determination of the suitability of a particular data set for use as input data for a model; (2) estimation of data values required by models at locations where actual measured values are not available; and (3) postprocessing of computed results. The GIS data pool provides data for both geostatistical analysis and modeling. Results of either geostatistical analyses or modeling are also returned to the data pool. A more detailed discussion of geostatistical procedures for characterizing soil processes is found in Chapter 16.

Tools for preliminary statistical examination of spatial data sets include *h*-scatterplots for analyzing actual data values and indicator maps as grayscale plots of standardized ranks of the data (Bourgault et al., 1997). The GIS can support these operations by providing data selection and sorting in addition to map making. These preliminary analyses assist in discovering features of the data such as spatial trends, spatial correlation, and specific data values that appear outside the general grouping of the data and may be questionable (termed erratic).

Among the linear geostatistical methods of analyzing spatial data, a common tool is the experimental semivariogram (from now on referred to simply as semivariogram), which represents the effect of distance between sampling points on variability (Journel and Huijbregts, 1978). There are several related measures of spatial correlation such as plots of spatial covariance, correlograms, and indicator semivariograms (Deutsch and Journel, 1992). A semivariogram indicates how the mean-squared variation between pairs of data values changes with spacing between the two measurement locations. Ideally, a semivariogram will show zero variability at a sample spacing of zero and semivariance increasing with increasing spacing up to a point where it levels off at a sill value. Real data seldom have such ideal behavior; normally there is a finite, but difficult-to-quantify, semivariance at a spacing of zero known as the nugget effect. Also, the sill value is often not constant because, at continually greater spacing, new soil types are encountered, causing further increases in the variation between pairs of measured values.

Modeling of the spatial covariance is necessary for estimation of data values at unsampled points using any of the kriging-type techniques. The spatial covariances are obtained from modeling semivariograms using, for example, the spherical, exponential, or power models. Anisotropy can also be represented by defining an azimuth angle for rotation of the principal axes from north in addition to providing model parameters for each of the two principal directions (Deutsch and Journel, 1992).

Deterministic water flow and solute transport models require a complete data set at every site where simulations will be performed. However, sampling for measurement of data required by the model often cannot be carried out so extensively. Thus, flow and transport modeling in a GIS context usually requires local estimation of at least some of the input data. One approach is through pedo-transfer functions as previously described. Alternatively, geostatistical techniques such as kriging can be employed in situations where the density of existing data points is sufficient to make spatial interpolation a practical method for estimating a parameter. Advantages of estimation by kriging techniques are: (1) the spatial structure of the data, as represented by semivariogram modeling, will be integrated into the estimation; (2) for locations where data points are known, kriging estimates the data value exactly; and (3) the data provide the starting point for determining the weighting as compared to other methods such as inverse-distance-squared weighting, which rely on an assumption (Wackernagel, 1995). Kriging estimates demand that the data conform to the requirements of second-order stationarity, meaning the semivariogram is independent of location, and there is no regional trend in the data (Journel and Huijbregts, 1978; Henley, 1981). For estimation at unknown points based on a normal distribution of data for a single parameter, the ordinary point kriging method is appropriate. If two or more parameters are cross-correlated, then cokriging may make a more accurate estimation.

Another possibility for generating data at unsampled locations is stochastic simulation. A simulation generates data values by drawing them randomly from a Gaussian distribution. Methods include turning bands (Journel, 1974; Mantoglou, 1987), spectral methods (Robin et al., 1993), sequential Gaussian simulation (Deutsch and Journel, 1992), and matrix decomposition (Davis, 1987). A conditional simulation ensures that those values are statistically consistent with a set of measured data (the conditioning data). Non-Gaussian distributions are common for many types of soil data, but transformation of conditioning data from such distributions into Gaussian form allows implementation of the simulation methods (Gotway and Rutherford, 1994). Unlike regression or kriging, which produce a single estimation at unsampled points, stochastic simulation can generate many data sets known as realizations. Deutsch and Journel (1992) advise: "Inasmuch as a simulated realization honors the data deemed important, it can be used as an interpolated map for those applications where reproduction of spatial features is more important than local accuracy." For water flow and transport modeling, stochastic simulation of input data is useful for studying the propagation of uncertainty within the models. However, computational demands of these models may preclude running such a calculation for a large number of realizations. Some discretion is necessary because stochastic simulation is a rapidly evolving field and some of the simulation methods have not yet been extensively tested with real data.

Finally, geostatistics can assist in postprocessing model results. Maps are especially useful for representation of postprocessed data because nonspecialists are often familiar with interpreting maps, whereas graphs and histograms frequently require more detailed knowledge. A map could simply represent computed values; for example, the TIN module of ARC/INFO* can drape a surface representing computed values over an irregularly spaced set of locations (Corwin et al., 1993a), or kriging the computed data can estimate the results at locations on a grid for mapmaking by raster GIS methods.

The proposed use of geostatistical methods (Rogowski and Wolf, 1994) and fuzzy set theory (Burrough, 1989; Odeh et al., 1990, 1992; Burrough et al., 1992; Altman, 1994) to incorporate variability and imprecision, respectively, into soil map unit delineations has gained recognition and favor. The obvious advantage of using geostatistics is that it provides not only an estimate of a value of a property at a given point or over a given area, but also an estimate of the error associated with that estimated value. However, there are disadvantages that have precluded its routine application: (1) the method is sample intensive, requiring a large number of samples within the area being

---

*ARC/INFO was designed by ESRI, 380 New York, Redlands, CA 92373. Trade and company names are provided for the benefit of the reader and do not imply any endorsement by the U.S. Department of Agriculture.

described to accurately estimate the semivariogram, and (2) the method is site specific, so the results have limited use outside of the sampled area. Geostatistics is useful in characterizing variability that exists due to random processes within the spatial system, and hence statistical and probabilistic models are appropriate.

However, there are certain aspects associated with variability (i.e., imprecision or vagueness in data) that cannot be attributed to randomness whether due to complexity, missing information, imprecision, and/or the use of natural language. Because soil variation is more continuous than discrete and consequently calls for a continuous approach, fuzzy set theory (Zadeh, 1965) offers an appropriate means of modeling the imprecision or vagueness in a continuous system by allowing the matching of membership on a continuous scale rather than on a Boolean binary or an integer scale. Fuzzy set theory is a generalization of classical Boolean algebra to situations where zones of gradual transition divide classes rather than conventional crisp boundaries. Fuzzy sets are especially useful when insufficient data exist to characterize variability using standard statistical measures (e.g., mean standard deviation, and distribution type). The central concept of fuzzy set theory is the membership function. The membership function is a mathematical relationship that defines the grade of membership with 1 representing full membership, 0 representing nonmembership, and a suitable function defining the flexible membership grades between 0 and 1. Aside from the representation of imprecision occurring within a map unit, fuzzy set theory also has been applied to represent the imprecision of boundary location and the gradual changes that actually occur between map unit boundaries on thematic maps (Wang and Hall, 1996).

1.3 CASE STUDY

1.3.1 SAN JOAQUIN VALLEY GROUNDWATER VULNERABILITY STUDY

The use of agrochemicals in California's San Joaquin Valley (see Figure 1.5a) has resulted in widespread groundwater contamination. Assessing the impacts of pesticide legacies (such as EDB) and currently used agrochemicals (such as atrazine) is of considerable interest to those in the decision-management arena. The example given here, distilled from the work of Blanke (1999) and Mills (2003), is a regional-scale NPS groundwater vulnerability assessment driven by (in a GIS format) a relatively simple leaching index with consideration for the uncertainty in the soil, recharge, and chemical data sets. The approach by Blanke (1999) and Mills (2003) spins off of the work of Loague and co-workers for the islands of Hawaii (see Loague et al., 1989, 1990; Loague, 1991, 1994; Kleveno et al., 1992; Giambelluca et al., 1996; Bernknopf et al., 1999) and Tenerife (see Diaz-Diaz et al., 1998, 1999; Diaz-Diaz and Loague, 2000a, 2000b, 2001).

The index approach used by Blanke (1999) and Mills (2003) to assess groundwater vulnerability for the San Joaquin Valley (SJV) is the

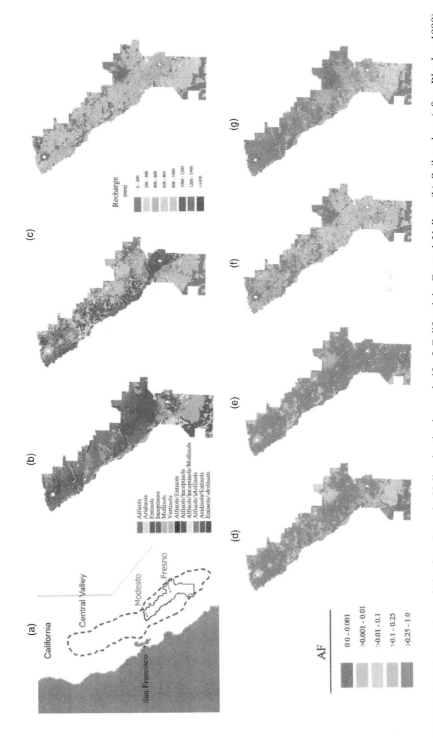

FIGURE 1.5 (a) Location of the San Joaquin Valley in the lower half of California's Central Valley. (b) Soil orders (after Blanke, 1999). (c) Recharge estimates (after Mills, 2003): annual (left), monthly (right). (d) AF estimates for EDB based on the annual recharge estimates (after Mills, 2003). (e) $AF + S_{AF}$ estimates for EDB based upon annual recharge estimates (after Mills, 2003). (f) AF estimates for EDB based on the

dimensionless attenuation factor (AF), proposed by Rao et al. (1985), defined as:

$$AF = \exp\left(\frac{-0.69dRF\theta_{FC}}{qt_{1/2}}\right) \tag{1.5}$$

where d is the distance to groundwater (or some compliance depth) from the surface [L], RF $\left[RF = 1 + \left(\rho_b f_{oc} K_{oc}/\theta_{FC}\right)\right]$ is the retardation factor [dimensionless], θ_{FC} is the soil-water content at field capacity [volume fraction], q is the net groundwater recharge [LT^{-1}], $t_{1/2}$ is the pesticide half-life [T], ρ_b is the soil bulk density [ML^{-3}], f_{oc} is the soil organic carbon (mass fraction), and K_{oc} is the pesticide sorption coefficient [L^{3}M^{-1}]. The range of values for AF is between 0.0 and 1.0; the larger the value, the more likely it is that the chemical will leach. The assumptions (Kleveno et al., 1992) underlying AF are too restrictive to allow the index to be used for quantitative prediction of pesticide leaching. Therefore, the AF index should only be used as a screening tool for comparing the approximate leaching potential of various chemicals at a given location or for ranking the vulnerability to leaching of different locations with variable soil properties.

Blanke (1999) and Mills (2003) both employed first-order analysis to characterize the uncertainty in their assessments. For the special case where the parameters are uncorrelated, first-order analysis reduces to first-order error propagation. The total uncertainty in an AF estimate is given by:

$$S_{AF} =^1 \left(\sum_{i=1}^{n} C_i^2\right)^{1/2} \tag{1.6}$$

where $=^1$ means equal in the first-order sense, C_i is the component uncertainty contributed by the ith parameter, n is the number of parameters, and S_{AF} is the standard deviation of the AF estimate. The equations for estimating the AF component uncertainties in Eq. (1.6) are given by Loague et al., (1990). It should be pointed out that Ugalde (2000) considered correlated soil parameters in uncertainty analyses for AF assessments of NPS groundwater vulnerability and found the impact to be relatively small.

Figure 1.5b shows the distribution of the six soil orders within the SJV. Blanke (1999) developed the AF soil database for the SJV. Blanke used a single average recharge estimate (gleaned from Loague et al., 1998a, 1998b) for the entire SJV. Mills (2003) considered spatially variable recharge across the SJV. Figure 1.5c shows recharge estimates based upon annual and monthly water balances. Blanke (1999) and Mills (2003) both use the chemical database prepared by Diaz-Diaz and Loague (2000a). The conservative (risk averse) compliance depth used by Blanke (1999) and Mills (2003) is 1 m.

Ranked AF estimates for 32 agrochemicals for the six soil orders in the SJV for an average recharge rate are given in Table 1.5. It is worth pointing out, for

TABLE 1.5
Ranking and Classification, Based on Leaching Estimates with the AF Index, for 32 Agrochemicals and the Six Soil Orders in the San Joaquin Valley

Chemical	Soil order					
	Alfisols	Aridisols	Entisols	Inceptisols	Mollisols	Vertisols
1,2-Dichloropropane	1	1	1	1	1	1
Tebuthiuron	2	2	2	2	2	2
Prometon	3	3	3	3	3	3
DBCP	4	4	4	4	4	4
EDB	5	5	5	5	5	5
Dalapon	6	6	6	6	6	6
Bromacil	7	8	8	8	8	8
Carbofuran	8	7	7	7	7	7
Propazine	9	9	9	9	10	9
Aldicarb	10	10	10	10	11	10
Dinoseb	11	11	11	11	12	11
Atrazine	12	13	13	13	14	13
Dicamba	13	12	12	12	9	12
Metribuzin	14	14	14	14	13	14
Metolachlor	15	15	15	15	15	15
Simazine	16	16	16	16	16	16
Methomyl	17	17	17	17	17	17
2,4-D	18	18	18	18	18	18
Diuron	19	19	19	19	19	19
Prometryn	20	20	20	20	20	20
Diphenamid	21	21	21	21	21	21
Alachlor	22	23	23	23	23	23
Oxamyl	23	22	22	22	22	22
Cyanazine	24	24	24	24	24	24
Disulfoton	25	25	25	25	25	25
Diazinon	26	26	26	26	26	26
Carbaryl	27	27	27	27	27	27
Chlorothalonil	28	28	28	28	28 (tie)	28
DCPA	29	29	29	29	28 (tie)	29
Carboxin	30	30 (tie)	30	30	28 (tie)	30 (tie)
Trifluralin	31 (tie)	30 (tie)	31 (tie)	31 (tie)	28 (tie)	30 (tie)
2,4-DP	31 (tie)	30 (tie)	31 (tie)	31 (tie)	28 (tie)	30 (tie)

Scale (Khan et al., 1986)   AF estimate

	Very Likely?	≥ 0.25 and ≤ 1.0
	Likely	≥ 0.1 and < 0.25
	Moderately Likely	≥ 0.01 and < 0.1
	Unlikely	≥ 0.0001 and < 0.01
	Very Unlikely	≥ 0.0 and < 0.0001

Source: Adapted from Blanke, 1999.

example, that EDB, a banned legacy known to have contaminated ground-water in the SJV, is near the top of the list in Table 1.5. Blanke (1999) and Mills (2003) provide rankings similar to the one in Table 1.5 showing the impacts of recharge variability and data uncertainties. Figures 1.5d through 1.5g are regional-scale (San Joaquin Valley) NPS assessments of groundwater vulner-ability for EDB. Figures 1.5d and 1.5e show, respectively, AF and $AF + S_{AF}$ estimates for the annual recharge estimates (see Figure 1.5c, left). Figures 1.5f and 1.5g show, respectively, AF and $AF + S_{AF}$ estimates for the monthly recharge estimates (see Figure 1.5c, right). It is obvious that Figures 1.5d through 5g do not show the same overall assessment. Therefore, it is reasonable to conclude that (1) careful characterization of recharge is important (also see Table 1.5) and (2) there is considerable uncertainty in the soil, recharge, and chemical data sets. There is no question that the spatial variations in the soils and the recharge (shown in Figures 1.5b and 1.5c) are propagated to the vulnerability assessments. Comparative inspection of Figures 1.5c, 1.5d, and 1.5f shows that recharge estimates are (on average) higher for the monthly water balance, which results in lower *AF* estimates. Blanke (1999) and Mills (2003) both present vulnerability maps (AF and $AF + S_{AF}$) for the SJV for each of the 32 chemicals listed in Table 1.5.

REFERENCES

Addiscott, T.M., Simulation modeling and soil behavior, *Geoderma*, 60, 15–40, 1993.

Addiscott, T.M. and Tuck, G., Sensitivity analysis for regional-scale solute transport modeling, in *Applications of GIS to the Modeling of Non–point Source Pollutants in the Vadose Zone*, Corwin, D.L. and Loague, K., eds., SSSA Special Publication No. 48, Soil Science Society of America, Madison, WI, 1996, 153–162.

Addiscott, T.M. and Wagenet, R.J., Concepts of solute leaching in soils: a review of modelling approaches, *J. Soil Sci.*, 36, 411–424, 1985.

Altman, D., Fuzzy set theoretic approaches for handling imprecision in spatial analysis, *Int. J. Geogr. Inf. Syst.*, 8 (3), 271–289, 1994.

Anderson, S.H., Gantzer, C.J., and Brown, J.R., Soil physical properties after 100 years of continuous cultivation, *J. Soil Water Conserv.*, 45, 117–121, 1990.

Argantesi, F. and Olivi, L., Statistical sensitivity analysis of a simulation model for biomass-nutrient dynamics in aquatic ecosystems, in *Proceedings of the 4th Summer Computer Simulation Conference*, Simulation Council, LaJolla, CA, 1976, 389–393.

Barnes, E.M., Sudduth, K.A., Hummel, J.W., Lesch, S.M., Corwin, D.L., Yang, C., Daughtry, S.T., and Bausch, W.C., Remote- and ground-based sensor techniques to map soil properties, *Photogramm. Eng. Remote Sens.*, 69 (6), 619–630, 2003.

Baveye, P. and Boast, C.W., Physical scales and spatial predictability of transport processes in the environment, in *Assessment of Non–Point Source Pollutants in the Vadose Zone*, Corwin, D.L., Loague, K., and Ellsworth, T.R., eds., Geophysical Monograph 108, American Geophysical Union, Washington, DC, 1999, 261–280.

Beck, M.B., Water quality modeling: a review of the analysis of certainty, *Water Resour. Res.*, 23, 1393–1442, 1987.

Beckett, P.H.T. and Webster, R., Soil variability: a review, *Soils Fertilizers*, 34 (1), 1–15, 1971.

Bernknopf, R.L., Lenkeit, K.A., Dinitz, L.B., and Loague, K., Estimating a societal value of earth science information in the assessment of non-point source pollutants, in *Assessment of Non–Point Source Pollution in the Vadose Zone* Corwin, D.L., Loague, K., and Ellsworth, T.R., eds., Geophysical Monograph 108, AGU Press, Washington, DC, 1999, 291–308.

Beven, K.J., Changing ideas in hydrology — the case of physically-based models, *J. Hydrol.*, 105, 157–172, 1989.

Beven, K.J., Towards an alternative blueprint for a physically-based digitally simulated hydrologic response modeling system, *Hydrol. Process.*, 16 (2), 189–206, 2002.

Blanke, J.S., Vulnerability to groundwater contamination due to agrochemical use in the San Joaquin Valley, California, M.S. thesis, Department of Geological and Environmental Sciences, Stanford University, Stanford, CA, 1999.

Bleecker, M., Hutson, J.L., and Waltman, S.W., Mapping groundwater contamination potential using integrated simulation modeling and GIS, in *Proceedings of the Application of Geographic Systems, Simulation Models, and Knowledge-Based Systems for Landuse Management*, Virginia Polytechnic Institute and State University, Blacksburg, VA, November 12–14, 1990, 319–328.

Bobba, A.G., Singh, V.P., and Bengtsson, L., Application of uncertainty analysis to groundwater pollution modeling, *Environ. Geol.*, 26, 89–96, 1995.

Bouma, J., Using soil survey data for quantitative land evaluation, *Adv. Soil Sci.*, 9, 177–213, 1989.

Bouma, J. and Bregt, A.K., eds., *Land Qualities in Space and Time*, Pudoc, Wageningen, the Netherlands, 1989.

Bouma, J. and van Lanen, H.A.J., Transfer functions and threshold values: from soil characteristics to land qualities, in *Proceedings of the International Workshop on Quantified Land Evaluation Procedures*, Beck, K.J., Burrough, P.A., and McCormack, D.E., eds., ITC Publication No. 6, Washington, DC, 1987, 106–110.

Bourgault, G., Journel, A.G., Rhoades, J.D., Corwin, D.L., and Lesch, S.M., Geostatistical analysis of a soil salinity data set, *Adv. Agron.*, 58, 241–292, 1997.

Bresler, E. and Dagan, G., Convective and pore scale dispersive solute transport in heterogeneous fields, *Water Resour. Res.*, 17, 1683–1693, 1981.

Brubaker, S.C. and Hallmark, C.T., A comparison of statistical methods for evaluating map unit composition, in *Spatial Variabilities of Soils and Landforms*, Mausbach, M.J. and Wilding, L.P., eds., SSSA Special Publication No. 28, Soil Science Society of America, Madison, WI, 1991, 73–88.

Burges, S.J. and Lettenmaier, D.P., Probabilistic methods in stream quality management, *Water Resour. Bull.*, 11, 115–130, 1975.

Burrough, P.A., Fuzzy mathematical methods for soil survey and land evaluation, *J. Soil Sci.*, 40, 477–492, 1989.

Burrough, P.A., Opportunities and limitations of GIS-based modeling of solute transport at the regional scale, in *Applications of GIS to the Modeling of Non–Point Source Pollutants in the Vadose Zone*, Corwin, D.L. and Loague K., eds., SSSA Special Publication No. 48, Soil Science Society of America, Madison, WI, 1996, 19–38.

Burrough, P.A., MacMillan, R.A., and van Deursen, W., Fuzzy classification methods for determining land suitability from soil profile observations and topography, *J. Soil Sci.*, 43, 193–210, 1992.

Carter, M.R., Temporal variability of soil macroporosity in a fine sandy loam under mouldboard ploughing and direct drilling, *Soil Tillage Res.*, 12, 37–51, 1988.

Cassel, D.K., Spatial and temporal variability of soil physical properties following tillage of Norfolk loamy sand, *Soil Sci. Soc. Am. J.*, 47 (2), 196–201, 1983.

Colborn, T., Dumanoski, D., and Myers, J.P., *Our Stolen Future*, Dutton, New York, 1996.

Colborn, T., Vom Saal, F.S., and Soto, A.M., Developmental effects of endocrine disrupting chemicals in wildlife and humans, *Environ. Health Perspect.*, 101, 5, 379, 1993.

Corwin, D.L., GIS applications of deterministic solute transport models for regional-scale assessment of non-point source pollutants in the vadose zone, in *Applications of GIS to the Modeling of Non–Point Source Pollutants in the Vadose Zone*, Corwin, D.L. and Loague, K., eds., SSSA Special Publication No. 48, Soil Science Society of America, Madison, WI, 1996, 69–100.

Corwin, D.L., Sensitivity analysis of a simple layer-equilibrium model for the one-dimensional leaching of solutes, *J. Environ. Sci. Health*, A30 (1), 201–238, 1995.

Corwin, D.L. and Lesch, S.M., Apparent soil electrical conductivity measurements in agriculture, *Comput. Electron. Agric.*, 2005a (in press).

Corwin, D.L. and Lesch, S.M., Application of soil electrical conductivity to precision agriculture: theory, principles, and guidelines, *Agron. J.*, 95, 455–471, 2003.

Corwin, D.L. and Lesch, S.M., Survey protocols for characterizing soil spatial variability with apparent soil electrical conductivity, *Comput. Electron. Agric.*, 2005b (in press).

Corwin, D.L. and Rhoades, J.D., The use of computer-assisted mapping techniques to delineate potential areas of salinity development in soils: II. Field verification of the threshold model approach, *Hilgardia*, 56 (2), 18–32, 1988.

Corwin, D.L. and Wagenet, R.J., Applications of GIS to the modeling of non-point source pollutants in the vadose zone: a conference overview, *J. Environ. Qual.*, 25, 403–411, 1996.

Corwin, D.L., Carrillo, M.L.K., Vaughan, P.J., Rhoades, J.D., and Cone, D.G., Evaluation of a GIS-linked model of salt loading to groundwater, *J. Environ. Qual.*, 28, 471–480, 1999c.

Corwin, D.L., Kaffka, S.R., Hopmans, J.W., Mori, Y., van Groenigen, J.W., van Kessel, C., Lesch, S.M., and Oster, J.D., Assessment and field-scale mapping of soil quality properties of a saline-sodic soil, *Geoderma*, 114, 231–259, 2003a.

Corwin, D.L., Lesch, S.M., Oster, J.D., and Kaffka, S.R., Characterizing spatio-temporal variability with soil sampling directed by apparent soil electrical conductivity, *Geoderma*, 2005 (in press).

Corwin, D.L., Lesch, S.M., Shouse, P.J., Soppe, R., and Ayars, J.E., Identifying soil properties that influence cotton yield using soil sampling directed by apparent soil electrical conductivity, *Agron. J.*, 95, 352-364, 2003b.

Corwin, D.L., Loague, K., and Ellsworth, T.R., Advanced information technologies for assessing nonpoint source pollution in the vadose zone: conference overview, *J. Environ. Qual.*, 28, 357–365, 1999b.

Corwin, D.L., Loague, K., and Ellsworth, T.R., GIS-based modeling of nonpoint source pollutants in the vadose zone, *J. Soil Water Conserv.*, 53 (1), 34–38, 1998.

Corwin, D.L., Loague, K., and Ellsworth, T.R., Introduction: assessing non-point source pollution in the vadose zone with advanced information technologies, in *Assessment of Non–Point Source Pollutants in the Vadose Zone*, Corwin, D.L., Loague, K., and Ellsworth, T.R., eds., American Geophysical Union, Washington, DC, 1999a, 1–20.

Corwin, D.L., Sorensen, M., and Rhoades, J.D., Field-testing of models which identify soils susceptible to salinity development, *Geoderma*, 45, 31–64, 1989.

Corwin, D.L., Vaughan, P.J., and Loague, K., Modeling nonpoint source pollutants in the vadose zone with GIS, *Environ. Sci. Technol.*, 31, 8, 2157–2175, 1997.

Corwin, D.L., Vaughan, P.J., Wang, H., Rhoades, J.D., and Cone, D.G., Coupling a solute transport model to a GIS to predict solute loading to the groundwater for a non-point source pollutant, in *Proceedings of the ASAE Application of Advanced Information Technologies: Effective Management of Natural Resources*, Spokane, WA, June 18–19, 1993, ASAE, St. Joseph, MI, 1993a, 485–492.

Corwin, D.L., Vaughan, P.J., Wang, H., Rhoades, J.D., and Cone, D.G., Predicting areal distributions of salt-loading to the groundwater, Paper No. 932566, 1993 ASAE Winter Meeting, Chicago, Illinois, December, 1993, American Society of Agricultural Engineers, St. Joseph, MI, 1993b.

Corwin, D.L., Werle, J.W., and Rhoades, J.D., The use of computer-assisted mapping techniques to delineate potential areas of salinity development in soils: I. A conceptual introduction, *Hilgardia*, 56 (2), 1–17, 1988.

Crawford, J.W., Baveye, P., Grindrod, P., and Rappoldt, C., Application of fractals to soil properties, landscape patterns, and solute transport in porous media, in *Assessment of Non–Point Source Pollutants in the Vadose Zone*, Corwin, D.L., Loague, K., and Ellsworth, T.R., eds., American Geophysical Union, Washington, DC, 1999, 151–164.

Cressie, N.A.C., *Statistics for Spatial Data*, Wiley-Interscience, New York, NY, 1993.

Dane, J.H. and Molz, F.J., Physical measurements in subsurface hydrology, in *Reviews of Geophysics*, Supplement, U.S. National Report to International Union of Geodesy and Geophysics 1987–1990, American Geophysical Union, Washington, DC, 1991, 270–279.

Davis, M.W., Production of conditional simulations via the LU triangular decomposition of the covariance matrix, *Math. Geol.*, 19, 91–98, 1987.

Destouni, G. and Cvetkovic, V., Field scale mass arrival of sorptive solute into the groundwater, *Water Resour. Res.*, 27, 1315–1325, 1991.

Dettinger, M.D., Wilson, J.L., First order analysis of uncertainty in numerical methods of groundwater flow, Part 1. Mathematical development, *Water Resour. Res.* 17 (1), 149–161, 1981.

Deutsch, C.V. and Journel, A.G., *GSLIB: Geostatistical Software Library and User's Guide*, Oxford University Press, London, 1992.

Diaz-Diaz, R. and Loague, K., Comparison of two pesticide leaching indices, *J. Am. Water Resour. Assoc.*, 36, 823–832, 2000a.

Diaz-Diaz, R. and Loague, K., Regional-scale leaching assessments for Tenerife: impact of data uncertainties, *J. Environ. Qual.*, 29, 835–847, 2000b.

Diaz- Diaz, R. and Loague, K., Assessing the potential for pesticide leaching for the pine forest areas of Tenerife, *Environ. Toxicol. Chem.*, 20, 1958–1967, 2001.

Diaz-Diaz, R., Garcia-Hernandez, J.E., and Loague, K., Leaching potentials of four pesticides used for bananas in the Canary Islands, *J. Environ. Qual.*, 27, 562–572, 1998.

Diaz-Diaz, R., Loague, K., and Notario, J.S., An assessment of agrochemical leaching potentials for Tenerife, *J. Contam. Hydrol.*, 36, 1–30, 1999.

Donigian, A.S. and Rao, P.S.C., Overview of terrestrial processes and modeling, in *Vadose Zone Modeling of Organic Pollutants*, Hern, S.C. and Melancon, S.M., eds., Lewis Publishers, Chelsea, MI, 1986, 3–36.

Duda, A.M., Addressing nonpoint sources of water pollution must become an international priority, *Water Sci. Technol.*, 28, 1–11, 1993.

Duda, A. and Nawar, M., Implementing the World Bank's water resources management policy: a priority on toxic substances from nonpoint sources, *Water Sci. Technol.*, 33, 4–5, 45–51, 1996.

Ellsworth, T.R., Influence of transport variability structure on parameter estimation and model discrimination, in *Application of GIS to the Modeling of Non–Point Source Pollutants in the Vadose Zone*, Corwin, D.L. and Loague, K., eds., SSSA Special Publication No. 48, Soil Science Society of America, Madison, WI, 1996, 101–130.

Ellsworth, T.R., Shouse, P.J., Skaggs, T.H., Jobes, J.A., and Fargerlund, J., Solute transport in an unsaturated field soil: experimental design, parameter estimation and model discrimination, *Soil Sci. Soc. Am. J.*, 60 (2), 397–407, 1996.

Ellsworth, T.R. and Boast, C.W., The spatial structure of solute transport variability in an unsaturated field soil, *Soil Sci. Soc. Am. J.*, 60 (5), 1355–1367, 1996.

Evans, B.M. and Myers, W.L., A GIS-based approach to evaluating regional groundwater pollution potential with DRASTIC, *J. Soil Water Conserv.* 45 (2), 242–245, 1990.

Fetter, C.W., *Contaminent Hydrogeology*. Prentice-Hall, Englewood Cliffs, NJ, 1993.

Foussereau, X., Hornsby, A.A., and Brown, R.B., Accountiing for variability within map units when linking a pesticide fate model to soil survey, *Geoderma*, 60, 257–276, 1993.

Freeze, A., *The Environmental Pendulum: A Quest for the Truth About Toxic Chemicals, Human Health, and Environmental Protection*, University of California Press, Berkeley, CA, 2000.

Gantzer, C.J. and Blake, G.R., Physical characteristics of Le Sueur clay loam soil following no till and conventional tillage, *Agron. J.*, 70, 853–857, 1978.

Giambelluca, T.W., Loague, K., Green, R.E., and Nullet, M.A., Uncertainty in recharge estimation: impact on groundwater vulnerability assessments for the Pearl Harbor Basin, Oahu, Hawaii, *J. Contam. Hydrol.*, 23, 85–112, 1996.

Goodchild, M.F., The state of GIS for environmental problem-solving, in *Environmental Modeling with GIS*, Goodchild, M.F., Parks, B.O., and Steyaert, L.T., eds., Oxford University Press, New York, 1993, 8–15.

Gotway, C.A. and Rutherford, B.M., Stochastic simulation for imaging spatial uncertainty: comparison and evaluation of available algorithms, in *Geostatistical Simulations*, Armstrong, M. and Dowd, P.A., eds., Geostatistical Simulation Workshop, Fontainebleau, France, May 27–28, 1993, Kluwer Academic Publishers, Dordrecht, 1994, 1–21.

Grayson, R.B., Moore, I.D., and McMahon, T.A., Physically based hydrologic modeling — II. Is the concept realistic?, *Water Resour. Res.*, 26, 2659–2666, 1992.

Grossman, R.B. and Pringle, F.B., Describing surface soil properties — their seasonal changes and implications for management, in *Soil Survey Techniques*, SSSA Special Publication No. 20, Soil Science Society of America, Madison, WI, 1987, 57–75.

Guillette, Jr., L., Crane, D.A., Rooney, A., and Pickford, D., Organization versus activation: the role of endocrine-disrupting contaminants during embryonic development in wildlife, *Environ. Health Perspectives*, 103, 7, 161, 1995.

Halliday, S.L. and Wolfe, M.L., Assessing ground water pollution potential from nitrogen fertilizer using a geographic information system, *Water Resour. Bull.* 27 (2), 237–245, 1991.

Hawking, S.W., *A Brief History of Time: From the Big Bang to Black Holes*, Bantam Books, New York, 1988.

Henley, S., *Nonparametric Geostatistics*, Applied Science Publishers, London, 1981.

Heuvelink, G.B.M., Burrough, P., and Stein, A., Propagation of errors in spatial modeling with GIS, *Int. J. Geogr. Inf. Syst.*, 3, 303–322, 1989.

Hillel, D., Modeling in soil physics: a critical review, in *Future Developments in Soil Science Research (A Collection of Soil Science Society of America Golden Anniversary Contributions)*, 1986 ASA-CSSA-SSSA Annual Meeting, New Orleans, Nov. 30–Dec. 5, 1986, Soil Science Society of America, Madison, WI, 1987, 35–42.

Hills, R.G. and Wierenga, P.J., Model validation at the Las Cruces Trench site, U.S. Nuclear Regulatory Commission Rpt. NUREG/CR-5716, U.S. Nuclear Regulatory Commission, Washington, DC, 1991.

Hills, R.G. and Wierenga, P.J., INTRAVAL Phase II Model Testing at the Las Cruces Trench Site, U.S. Nuclear Regulatory Commission Rpt. NUREG/CR-6063, U.S. Nuclear Regulatory Commission, Washington, DC, 1994.

Hills, R.G., Porro, I., Hudson, D.D.B., and Wierenga, P.J., Modeling one-dimensional infiltration into very dry soils 1. Model development and evaluation, *Water Resour. Res.*, 25 (6), 1259–1269, 1989a.

Hills, R.G., Hudson, D.B., Porro, I., and Wierenga, P.J., Modeling one-dimensional infiltration into very dry soils 2. Estimation of the soil water parameters and model predictions, *Water Resour. Res.*, 25 (6), 1271–1282, 1989b.

Hills, R.G., Wierenga, P.J., Hudson, D.B., and Kirkland, M.R., The second Las Cruces Trench experiment: experimental results and two-dimensional flow predictions, *Water Resour. Res.*, 27 (10), 2707–2718, 1991.

Hoeksema, R.J. and Clapp, R.B., Calibration of groundwater flow models using Monte Carlo simulations and geostatistics, in *ModelCARE90: Calibration and Reliability in Groundwater Modeling*, Kovar, K., ed., IAHS Publ. No. 195, IAHS Press, Wallingford, UK, 1990, 33–42.

Hoosbeek, M.R. and Bryant, R.B., Towards the quantitative modeling of pedogenesis: a review, *Geoderma*, 55, 183–210, 1993.

Humenik, F.J., Smolen, M.D., and Dressing, S.A., Pollution from nonpoint sources: where we are and where we should go, *Environ. Sci. Technol.*, 21, 8, 737–742, 1987.

Jacobson, J. and Jacobson, S., Intellectual impairment in children exposed to polychlorinated biphenyls *in utero*, *N. Engl. J. Med.*, 335, 11, 783, 1996.

Jakeman, A.J., Ghassemi, F., Dietrich, C.R., Musgrove, T.J., and Whitehead, P.G., Calibration and reliability of an aquifer system model using generalized sensitivity analysis, in *ModelCARE90: Calibration and Reliability in Groundwater Modeling*, Kovar, K., ed., IAHS Publ. No. 195, IAHS Press, Wallingford, UK, 1990, 43–51.

Journel, A.G., Geostatistics for conditional simulation of ore bodies, *Econ. Geol.*, 69, 673–680, 1974.

Journel, A.G., Modelling uncertainty and spatial dependence: stochastic imaging, *Int. J. Geogr. Inf. Syst.*, 10 (5), 517–522, 1996.

Journel, A.G. and Huijbregts, C.J., *Mining Geostatistics*, Academic Press, San Diego, CA, 1978.

Jury, W.A., Spatial variability of soil physical parameters in solute migration: a critical literature review, in *Electrical Power Research Institute (EPRI) Report EA-4228*, EPRI, Palo Alto, CA, 1985.

Jury, W.A., Spatial variability of soil properties, in *Vadose Zone Modeling of Organic Pollutants*, Hern, S.C. and Melancon, S.M., eds., Lewis Publishers, Chelsea, MI, 1986, 245–269.

Jury, W.A., Stochastic solute transport modeling trends and their potential compatibility with GIS, in *Application of GIS to the Modeling of Non–point Source Pollutants in the Vadose Zone*, Corwin, D.L. and Loague, K., eds., SSSA Special Publication No. 48, Soil Science Society of America, Madison, WI, 1996, 57–67.

Jury, W.A., and Gruber, J., A stochastic analysis of the influence of soil and climatic variability on the estimate of pesticide groundwater pollution potential, *Water Resour. Res.*, 25 (12), 2465–2474, 1989.

Jury, W.A. and Roth, K., Stochastic stream tube modeling, in *Transfer Functions and Solute Movement Through Soil*, Birkhauser Verlag, Basel, 1990, 63–84.

Kauffmann, C., Kinzelbach, W., and Fried, J.J., Simultaneous calibration of flow and transport models and optimization of remediation measures, in *ModelCARE90: calibration and Reliability in Groundwater Modeling*, Kovar, K., ed., IAHS Publ. No. 195, IAHS Press, Wallingford, UK, 1990, 159–170.

Keidser, A., Rosbjerg, D., Høgh Jensen, K., and Bitsch, K., A joint kriging and zonation approach to inverse groundwater modeling, in *ModelCARE90: calibration and Reliability in Groundwater Modeling*, Kovar, K., ed., IAHS Publ. No. 195, IAHS Press, Wallingford, UK, 1990, 171–183.

Khan, M.A. and Liang, T., Mapping pesticide contamination potential, *Environ. Manage.*, 13 (2), 233–242, 1989.

Klemes, V., Dilettantism in hydrology: transition or destiny?, *Water Resour. Res.*, 22, 177S–188S, 1986.

Klepper, O. and Hendrix, E.M.T., A method of robust calibration of ecological models under different types of uncertainty, *Ecol. Modell.*, 74, 161–182, 1994.

Kleveno, J.J., Loague, K., and Green, R.E., An evaluation of a pesticide mobility index: impact of recharge variation and soil profile heterogeneity, *J. Contam. Hydrol.* 11, 83–99, 1992.

Konikow, L.F., Predictive accuracy of a ground-water model — lessons from a postaudit, *Ground Water*, 24 (2), 173–184, 1986.

Konikow, L.F. and Bredehoeft, J.D., Ground-water models cannot be validated, *Adv. Water Resour.*, 15, 75–83, 1992.

Konikow, L.F. and Person, M.A., Assessment of long-term salinity changes in an irrigated stream-aquifer system, *Water Resour. Res.*, 21 (11), 1611–1624, 1985.

Konikow, L.F. and Swain, L.A., Assessment of predictive accuracy of a model of artificial recharge effects in the upper Coachella Valley, in *Selected Papers on Hydrogeology from the 28th International Geological Congress*, Vol. 1, Simpson, E.S. and Sharp, Jr., J.M., eds., International Association of Hydrogeologists, Verlag Heinz Heise, Hannover, Germany, 1990.

Langran, G., A review of temporal database research and its use in GIS applications, *Int. J. Geogr. Inf. Syst.* 3 (3), 215–232, 1989.

Lark, R.M. and Webster, R., Analysis and eleucidation of soil variation using wavelets, *Eur. J. Soil Sci.*, 50, 185–206, 1999.

Lark, R.M., Kaffka, S.R., and Corwin, D.L., Multiresolution analysis of data on electrical conductivity of soil using wavelets, *J. Hydrol.*, 272, 276–290, 2003.

Lipschultz, L., The debate continues — the continuing debate over the possible decline in semen quality, *Fertility and Sterility*, 65, 5, 910, 1996.

Loague, K., The impact of landuse on estimates of pesticide leaching potential: assessments and uncertainties, *J. Contam. Hydrol.*, 8, 157–175, 1991.

Loague, K., Regional scale ground water vulnerability estimates: impact of reducing data uncertainties for assessments in Hawaii, *Ground Water*, 32, 605–616, 1994.

Loague, K., Simulation of organic chemical movement in Hawaii soils with PRZM: 3. Calibration, *Pacific Sci.*, 46 (3), 353–373, 1992.

Loague, K., Abrams, R.H., Davis, S.N., Nguyen, A., and Stewart, I.T., A case study simulation of DBCP groundwater contamination in Fresno County, California: 2. Transport in the saturated subsurface, *J. Contam. Hydrol.*, 29, 137–163, 1998b.

Loague, K. and Corwin, D.L., Uncertainty in regional-scale assessments of non-point source pollutants, in *Applications of GIS to the Modeling of Non–point Source Pollutants in the Vadose Zone*; Corwin, D.L. and Loague, K., eds., SSSA Special Publication No. 48, Soil Science Society of America, Madison, WI, 1996, 131–152.

Loague, K. and Green, R.E., Statistical and graphical methods for evaluating solute transport models: overview and application, *J. Contam. Hydrol.*, 7, 51–73, 1991.

Loague, K., Bernknopf, R.L., Green, R.E., and Giambelluca, T.W., Uncertainty of groundwater vulnerability assessments for agricultural regions in Hawaii: a review, *J. Environ. Qual.*, 25 (3), 475–490, 1996.

Loague, K., Corwin, D.L., and Ellsworth, T.R., Are advanced information technologies the solution to non-point source pollution problems?, in *Assessment of Non–Point Source Pollutants in the Vadose Zone*, Corwin, D.L., Loague, K., and Ellsworth, T.R., eds., American Geophysical Union, Washington, DC, 1999, 363–369.

Loague, K., Green, R.E., Giambelluca, T.W., Liang, T.C., and Yost, R.S., Impact of uncertainty in soil, climatic, and chemical information in a pesticide leaching assessment, *J. Contam. Hydrol.*, 5, 171–194, 405, 1990.

Loague, K., Lloyd, D., Nguyen, A., Davis, S.N., and Abrams, R.H., A case study simulation of DBCP groundwater contamination in Fresno County, California: 1. Leaching through the unsaturated subsurface, *J. Contam. Hydrol.*, 29, 109–136, 1998a.

Loague, K.M., Yost, R.S., Green, R.E., and Liang, T.C., Uncertainty in a pesticide leaching assessment for Hawaii, *J. Contam. Hydrol.*, 4, 139–161, 1989.

Maidment, D.R., GIS and hydrologic modeling, in *Environmental Modeling with GIS* Goodchild, M.F., Parks, B.O., and Steyaert, L.Y., eds., Oxford University Press, New York, 1993, 147–167.

Maloszewski, P. and Zuber, A., On calibration and validation of mathematical models for the interpretation of tracer experiments in groundwater, *Adv. Water Resour.* 15, 47–62, 1992.

Mantoglou, A., Digital simulation of multivariate two- and three-dimensional stochastic processes with a spectral turning bands method, *Math. Geol.*, 19, 129–149, 1987.

Matalas, N.C., Landwehr, J.M., and Wolman, M.G., Prediction in water management, in *Scientific Basis of Water Management*, National Research Council, National Academy Press, Washington, DC, 1982.

Mausbach, M.J. and Wilding, L.P., eds., *Spatial Variabilities of Soils and Landforms* SSSA Special Publication No. 28, Soil Science Society of America, Madison, WI, 1991.

Mayer, S., Ellsworth, T.R., Corwin, D.L., and Loague, K., Identifying effective parameters for solute transport models in heterogeneous environments, in *Assessment of Non–Point Source Pollutants in the Vadose Zone*, Corwin, D.L., Loague, K., and Ellsworth, T.R., eds., American Geophysical Union, Washington, DC, 1999, 119–133.

Medina, A., Carrera, J., and Galarza, G., Calibration inverse modeling of coupled flow and solute transport problems, in *ModelCARE90: calibration and Reliability in Groundwater Modeling*, Kovar, K., ed., IAHS Publ. No. 195, IAHS Press, Wallingford, UK, 1990, 185–194.

Merchant, J.W., Whittemore, D.O., Whistler, J.L., McElwee, C.D., and Woods, J.J., Groundwater pollution hazard assessment: a GIS approach, in *Proceedings of the International GIS Symposium* (Vol. III), Arlington, Virginia, Nov. 15–18, 1987, Assoc. Am. Geographers, Washington, DC, 1987, 103–115.

Mills, M.B., Groundwater vulnerability due to pesticide use in the San Joaquin Valley, California, M.S. thesis, Department of Geological and Environmental Sciences, Stanford University, Stanford, CA, 2003.

Moore, I.D., and Gallant, J.C., Overview of hydrologic and water quality modeling, in *Modeling the Fate of Chemicals in the Environment*, Moore, I.D., ed., Center for Resource and Environmental Studies, Canberra, The Australian National University, 1991, 1–8.

Mulla, D.J. and Addiscott, T.M., Validation approaches for field-, basin-, and regional-scale water quality models, in *Assessment of Non–Point Source Pollutants in the Vadose Zone*, Corwin, D.L., Loague, K., and Ellsworth, T.R., eds., Geophyscial Monograph 108, American Geophysical Union, Washington, DC, 1999, 63–78.

Nielsen, D.R., Biggar, J.W., and Erh, K.T., Spatial variability of field-measured soil-water properties, *Hilgardia*, 42 (7), 215–259, 1973.

Novotny, V. and Olem, H., *Water Quality — Prevention, Identification, and Management of Diffuse Pollution*, Van Nostrand Reinhold, New York, 1994.

Odeh, I.O.A., McBratney, A.B., and Chittleborough, D.J., Design of optimal sampling spacings for mapping soil using fuzzy-k-means and regionalized variable theory, *Geoderma*, 47, 93–122, 1990.

Odeh, I.O.A., McBratney, A.B., and Chittleborough, D J., Soil pattern recognition with fuzzy-c-means: applications to classification and soil-landform inter-relationships, *Soil Sci. Soc. Am. J.*, 56, 505–516, 1992.

O'Neill, R.V., Hierarchy theory and global change, in *Scales and Global Change* Rosswall et al., ed., Scientific Committee on Problems of the Environment (SCOPE), John Wiley & Sons, New York, 1988, 29–45.

Oreskes, N., Shrader-Frechette, K., and Belitz, K., Verification, validation and confirmation of numerical models in the earth sciences, *Science*, 263, 641–646, 1994.

Peck, A.J., Field variability of soil physical properties, in *Advances in Irrigation*, Vol. 2, Hillel, D., ed., Academic Press, New York, 1983, 189–221.

Peirce, J.J., Weiner, R.F., and Vesilind, P.A., *Environmental Pollution and Control*, 4th ed., Butterworth-Heinemann, Boston, 1998.

Petach, M.C., Wagenet, R.J., and DeGloria, S. D., Regional water flow and pesticide leaching using simulations with spatially distributed data, *Geoderma*, 48, 245–269, 1991.

Pimental, D., Ed., *World Soil Erosion and Conservation*, Cambridge University Press, Cambridge, England, 1993.

Poletika, N.N., Jury, W.A., and Yates, M.V., Transport of bromide, simazine, and MS-2 coliphage in a lysimeter containing undisturbed, unsaturated soil, *Water Resour. Res.*, 31, 801–810, 1995.

Popper, Sir Karl, *The Logic of Scientific Discovery*, Harper & Row, New York, 1959.

Porterfield, S.P., Vulnerability of the developing brain to thyroid abnormalities: environmental insults to the thyroid system, *Environ. Health Perspect.*, 102, 125–130, 1994.

Rao, P.S.C., Hornsby, A.G., and Jessup, R.E. Indices for ranking the potential for pesticide contamination of groundwater, *Soil Crop Sci. Soc. Florida Proc.*, 44, 1–8, 1985.

Rao, P.S.C., Jessup, R.E., and Hornsby, A.G., Simulation of nitrogen in agroecosystems: criteria for model selection and use, *Plant Soil*, 67, 35–43, 1982.

Rawls, W.J., Gish, T.J., and Brakensiek, D.L., Estimating soil water retention from soil physical properties and characteristics, *Adv. Soil Sci.*, 16, 213–234, 1991.

Regan, J.J., DRASTIC: ground water pollution potential mapping in Arizona counties using a PC-based GIS, in *Protecting Natural Resources with Remote Sensing: The Third Forest Service Remote Sensing Applications Conference*, Tucson, AZ, April 9–13, 1990, 232–240.

Repetto, R. and Baliga, S., *Pesticides and the Immune System: The Public Health Risks* World Resources Institute, Washington, DC, 1996.

Robert, P.C., Rust, R.H., and Larson, W.E., eds., *Proceedings of Soil Specific Crop Management*, Soil Science Society of America, Madison, WI, 1993.

Robin, M.J.L., Gutjahr, A.L., Sudicky, E.A., and Wilson, J.L., Cross-correlated random field generation with the direct Fourier transform method, *Water Resour. Res.*, 29, 2385–2397, 1993.

Rockhold, M.L., Parameterizing flow and transport models for field-scale applications in heterogeneous, unsaturated soils, in *Assessment of Non–point Source Pollutants in the Vadose Zone*, Corwin, D.L., Loague, K., and Ellsworth, T.R., eds., Geophyscial Monograph 108, American Geophysical Union, Washington, DC, 1999, 243–260.

Rockhold, M.L., Ross, R.E., and Hills, R.G., Application of similar media scaling and conditional simulation for modeling water flow and tritium transport at the Las Cruces trench site, *Water Resour. Res.*, 32 (3), 595–609, 1996.

Rogowski, A.S., Conditional simulation of percolate flux below a rootzone, in *Proceedings of the Second International Conference/Workshop on Integrating Geographic Information Systems and Environmental Modeling*, Brackenridge, Colorado, Sept. 26–30, 1993, NCGIA, University of California, Santa Barbara, CA, 1993.

Rogowski, A.S. and Wolf, J.K., Incorporating variability into soil map unit delineations, *Soil Sci. Soc. Am. J.*, 58 (1), 163–174, 1994.

Rundquist, DC, Rodekohr, D.A., Peters, A.J., Ehrman, R.L., Di, L., and Murray, G., Statewide groundwater-vulnerability assessment in Nebraska using the DRASTIC/GIS model, *Geocarto Int.*, 2, 51–58, 1991.

Schanz, R.W. and Salhotra, A., Evaluation of the Rackwitz-Fiessler uncertainty analysis method for environmental fate and transport models, *Water Resour. Res.*, 28 (4), 1071–1079, 1992.

Scott, H.D., Mauromoustakos, A., Handayani, I.P., and Miller, D.M., Temporal variability of selected properties of loessial soil as affected by cropping, *Soil Sci. Soc. Am. J.*, 58 (5), 1531–1538, 1994.

Skidmore, E.L., Cartenson, W.A., and Banbury, E.E., Soil changes resulting from cropping, *Soil Sci. Soc. Am. J.*, 39 (5), 964–967, 1975.

Skop, E., GIS mapping of nitrate leaching in Denmark: regional scale calculations, in *Proceedings of the 1995 Bouyoucos Conference*, Corwin, D.L., ed., U.S. Salinity Laboratory, Riverside, CA, 1995, 210–220.

Starr, J.L., Spatial and temporal variation of ponded infiltration, *Soil Sci. Soc. Am. J.* 54 (3), 629–636, 1990.

State of California Legislature, The Organic Chemical Contamination Prevention Act, Assembly Bill 2021, Sacramento, CA, 1985.

Steyaert, L.T., A perspective on the state of environmental simulation modeling, in *Environmental Modeling with GIS*, Goodchild, M.F., Parks, B.O., and Steyaert, L.T., eds., Oxford University Press, New York, 1993, 16–30.

Stillman, R.J., Inutero exposure to diethylstilbestrol: adverse effects on the reproductive tract and reproductive performance in male and female offspring, *Am. J. Obstet. Gynecol.*, 142, 905–921, 1982.

Summers, J.K., Wilson, H.T., and Kou, J., A method for quantifying the prediction uncertainties associated with water quality models, *Ecol. Modell.*, 65, 161–176, 1993.

Teso, R.R., Poe, M.P., Younglove, T., and McCool, P.M., Use of logistic regression and GIS modeling to predict groundwater vulnerability to pesticides, *J. Environ. Qual.*, 25 (3), 425–432, 1997.

Tiktak, A., van der Linden, A.M.A., and Leine, I., Application of GIS to the modeling of pesticide leaching on a regional scale in the Netherlands, in *Applications of GIS to the Modeling of Non–Point Source Pollutants in the Vadose Zone* Corwin, D.L. and Loague, K., eds., SSSA Special Publication No. 48, Soil Science Society of America, Madison, WI, 1996, 259–281.

Timlin, D.J., Ahuja, L.R., and Williams, R.D., Methods to estimate soil hydraulic parameters for regional-scale applications of mechanistic models, in *Applications of GIS to the Modeling of Non–Point Source Pollutants in the Vadose Zone* Corwin, D.L. and Loague, K., eds., SSSA Special Publication No. 48, SSSA, Madison, WI, 1996, 185–203.

Topp, G.C., Reynolds, W.D., and Green, R.E., eds., *Advances in Measurement of Soil Physical Properties: Bringing Theory into Practice*, SSSA Special Publication No. 30, Soil Science Society of America, Madison, WI, 1992.

Toppari, J., Larsen, J.C., Christiansen, P., Giwercman, A., Grandjean, P., Guillette, L.J., Jegou, B., Jensen, T.K., Jouannet, P., Keiding, N., Leffers, H., Mclachlan, J.A., Meyer, O., Muller, J., Rajpertdemeyts, E., Scheike, T., Sharpe, R., Sumpter, J., and Skakkebaek, N.E., Male reproductive health and environmental xenoestrogens, *Environ. Health Perspect.*, 104, 4, 753–754, 1996.

Toride, N. and Leij, F.J., Convective-dispersive stream tube model for field-scale solute transport: I. Moment analysis, *Soil Sci. Soc. Am. J.*, 60 (2), 342–352, 1996a.

Toride, N. and Leij, F.J., Convective-dispersive stream tube model for field-scale solute transport: II. Examples and calibration, *Soil Sci. Soc. Am. J.*, 60 (2), 352–361, 1996b.

Ugalde, L.I., First-order variance characterization of uncertainty in a pesticide leaching index: the impact of correlation between soil input variables, M.S. thesis, Department of Geological and Environmental Sciences, Stanford University, Stanford, CA, 2000.

U.S. Environmental Protection Agency (USEPA), *National Water Quality Inventory* 1994 Report to Congress, EPA 841-R-95-005, Office of Water, U.S. Government. Printing Office, Washington, DC, 1995.

van Es, H.M., Verheijden, S.L.M., and Pleasant, J., in *Soil Structure Research in E. Canada*, Stone, J.A., et al., eds., Herald Press, Windsor, Ontario, 1991, 197–210.

van Es, H.M., Evaluation of temporal, spatial, and tillage-induced variability for parameterization of soil infiltration, *Geoderma*, 60, 187–199, 1993.

van Genuchten, M.Th. and Nielsen, D.R., On describing and predicting the hydraulic properties of unsaturated soils, *Ann. Geophys.*, 3 (5), 615–628, 1985.

van Genuchten, M.Th., Leij, F.J., and Lund, L.J., eds., *Indirect Methods for Estimating the Hydraulic Properties of Unsaturated Soils*, Proc. Int'l. Workshop, Riverside, California, Oct. 11–13, 1989, University of California: Riverside, CA, 1992.

van Wesenbeeck, I.J. and Kachanoski, R.G., Spatial scale dependence of *in situ* solute transport, *Soil Sci. Soc. Am. J.*, 55, 3–7, 1991.

van Wesenbeeck, I.J. and Kachanoski, R.G., Effect of variable horizon thickness on solute transport, *Soil Sci. Soc. Am. J.*, 58, 1307–1316, 1994.

Veneziano, D., Kulkarni, R.K., Luster, G., Rao, G., and Salhotra, A., Improving the efficiency of Monte Carlo simulation for groundwater transport models, in *Proceedings of the Conference on Geostatistical, Sensitivity, and Uncertainty Methods for Groundwater Flow and Radionuclide Transport Modeling*, Battelle Press, Columbus, Ohio, 1987.

Vereecken, H., Diels, J., Van Orshoven, J., Feyen, J., and Bouma, J., Functional evaluation of pedotransfer functions for the estimation of soil hydraulic properties, *Soil Sci. Soc. Am. J.*, 56, 1371–1378, 1992.

Wackernagel, H., *Multivariate Geostatistics*, Springer Verlag, Berlin, 1995.

Wagenet, R.J., Description of soil processes and mass fluxes at the regional scale, in *The Environmental Fate of Xenobiotics*, Del Re, A.A.M., et al., eds., Proceedings of the X Symposium on Pesticide Chemistry, Piacenza, Italy, Sept. 30–Oct. 2, 1996, Castelnuovo Fogliani, La Goliardica Pavese, Pavia, Italy, 1996, 1–18.

Wagenet, R.J. and Hutson, J.L., Scale-dependency of solute transport modeling/GIS applications, *J. Environ. Qual.*, 25 (3), 499–510, 1996.

Wagenet, R.J., Bouma, J., and Grossman, R.B., Minimum data sets for use of soil survey information in soil interpretive models, in *Spatial Variabilities of Soils and Landforms*, SSSA Special Publication No. 28, Soil Science Society of America, Madison, WI, 1991, 161–182.

Wagenet, R.J., Bouma, J., and Hutson, J.L., Modeling water and chemical fluxes as driving forces of pedogenesis, in *Quantitative Modeling of Soil Forming Processes*, Bryant, R.B. and Arnold, R.W., eds., SSSA Special Publication 39, Soil Science Society of America, Madison, WI, 1994, 17–35.

Walsh, S.J., Geographic information systems: an instructional tool for earth science educators, *J. Geogr.*, 87 (1), 17–25, 1988.

Wang, F. and Hall, G.B., Fuzzy representations of geographical boundaries in GIS, *Int. J. Geogr. Inf. Syst.*, 10 (5), 573–590, 1996.

Wang, H., Corwin, D.L., Lund, L.J., Rhoades, J.D., Vaughan, P.J., and Cone, D.G., Applications of GIS and logistic regression in evaluating salinity development in irrigated lands, in *Proceedings of the 1995 Bouyoucos Conference*, Corwin, D.L., ed., U.S. Salinity Laboratory, Riverside, CA, 1995, 341–355.

Warrick, A.W. and Nielsen, D R, Spatial variability of soil physical properties in the field, in *Applications of Soil Physics*, Hillel, D., ed., Academic Press, New York, 1980, 319–344.

White, I., Measurement of soil physical properties in the field, in *Flow and Transport in the Natural Environment: Advances and Applications*, Steffen, W.L. and Denmead, O.T., eds., Springer-Verlag, New York, 1988, 59–85.

Wilding, L.P. and Drees, L.R., Spatial variability: a pedologist's viewpoint, in *Diversity of Soils in the Tropics*; Drosdoff, M., Daniels, R.B., and Nicholaides, J.J., eds., Soil Science Society of America, Madison, WI, 1978, 1–12.

Willmott, C.J. et al., Statistics for the evaluation and comparison of models, *J. Geophys. Res.*, 90, 8995–9005, 1985.

Wilson, J.P., Inskeep, W.P., Rubright, P.R., Cooksey, D., Jacobsen, J.S., and Snyder, R.D., Coupling geographic information systems and models for weed control and groundwater protection, *Weed Technol.*, 7 (1), 255–264, 1993.

Wilson, J.P., Inskeep, W.P., Wraith, J.M., and Snyder, R.D., GIS-based solute transport modeling applications: scale effects of soil and climate data input, *J. Environ. Qual.*, 25 (3), 445–453, 1996.

Wösten, J.H.M. and Tamari, S., Application of artificial neural networks for developing pedotransfer functions of soil hydraulic parameters, in *Assessment of Non–Point Source Pollutants in the Vadose Zone*, Corwin, D.L., Loague, K., and Ellsworth, T.R., eds., American Geophysical Union, Washington, DC, 1999, 235–242.

World Resources Institute, *World Resources 1998–99: A Guide to the Global Environment*, Oxford University Press, New York, 1998.

Wu, Q.J., Ward, A.D., and Workman, S.R., Using GIS in simulation of nitrate leaching from heterogeneous unsaturated soils, *J. Environ. Qual.*, 25 (3), 526–534, 1996.

Wylie, B.K., Shaffer, M.J., Brodahl, M.K., Dubois, D., and Wagner, D.G., Predicting spatial distributions of nitrate leaching in northeast Colorado, *J. Soil Water Conserv.*, 49 (3), 288–293, 1994.

Zadeh, L.A., Fuzzy sets, *Information and Control*, 8, 338–353, 1965.

Zhang, H., Haan, C.T., and Nofziger, D.L., An approach to estimating uncertainties in modeling transport of solutes through soils, *J. Contam. Hydrol.*, 12, 35–50, 1993.

Zheng, C. and Bennett, G.D., *Applied Contaminant Transport Modeling: Theory and Practice*, Van Nostrand Reinhold, New York, 1995.

Zhu, A.-X., Fuzzy inference of soil patterns: implications for watershed modeling, in *Assessment of Non–Point Source Pollutants in the Vadose Zone*, Corwin, D.L., Loague, K., and Ellsworth, T.R., eds., American Geophysical Union, Washington, DC, 1999, 135–150.

2 Spatial and Temporal Variability of Soil Processes: Implications for Method Selection and Characterization Studies

Chris G. Campbell
Lawrence Berkeley National Laboratory, Berkeley,
California, U.S.A.

Fernando Garrido
Centro de Ciencias Medioambientales, CSIC, Madrid, Spain

CONTENTS

1-5667-0657-2/05/$0.00 + $1.50
© 2005 by CRC Press

2.1   INTRODUCTION

The soil is a three-dimensional continuum, a temporally dynamic and spatially heterogeneous anisotropic medium comprising the outer layer of the solid earth in which liquids, gases, and solids interact over an extreme range of space and time scales. This life-sustaining open system evolves through weathering processes driven by soil formation factors: parent material, climate, organisms, and topography acting over time. The variation in these factors across different locations provides the soil with its inherent heterogeneity from one site to another (even for close locations). As a result, characterizing soil processes requires a capacity to consider both the mechanisms and the magnitudes of spatial and temporal variability in soil features. The magnitude of these scales may range from ion exchange and sorption reactions on the surface of kaolinite and iron oxide clay minerals, occurring at nanosecond time and microscopic scales, to the development of well-structured soil illuvial horizons and the formation of toposequences over thousands of years at the watershed scale.

Common to all these processes, and a major controlling mechanism, is the movement and redistribution of water in the soil profile. As stated by Jenny (1941), "as long as water passes through the *solum*, substances are dissolved, translocated, precipitated, and flocculated, and the soil is not in a state of rest." However, soil hydraulic properties are spatially and temporally variable. This fact is of special importance when considering the soil functional view within the environment: regulating water and solute flow and filtering organic and inorganic substances. In this regard, understanding the physical, chemical, and biological processes acting on pollution during transport through soils is necessary to avoid further degradation of human-influenced and natural ecosystems. However, accurate predictions of pollution transport are dependent on the quality of the data and information available about the soil. Even when high-quality field data are available, the applicability of models strictly depends on assumptions (simplifications) as well as the field study design, techniques used to gather data, and, perhaps most importantly, the question being asked. Thus, the translation of laboratory results or model simulations into explanations of processes in natural systems remains a major challenge in soil science.

2.1.1   NEED FOR FIELD STUDIES

Although an academic division between those who collect field data and those who develop models often exists, it is an unnecessary partition. These two groups are interdependent, as without models to compare to field data (and vice versa), the ability to improve our scientific understanding would be severely limited. While relevant studies can be performed in the laboratory or on computers, field investigations remain necessary until we can completely explain the main processes in our environment.

At the same time, field investigations are difficult, uncontrolled, time-consuming, expensive, and potentially dangerous. Furthermore, the results of field investigations often create more questions than answers. This is both the exciting and extraordinarily frustrating activity of "field work" in soil science. It is also why it may be easier, or more appropriate in many cases, to satisfy a funding agency or client with modeling results.

As field studies are difficult, they require sound design to (1) meet the objectives of the study (answer the question), (2) minimize external (outside the study) influences or bias in the results, and (3) maintain a scope that falls within the limitations of space, time, and funding. It is the goal of this chapter to provide some basic considerations to meet many of these needs in field investigation design.

2.1.2 PRELIMINARY ISSUES

In designing field studies, three preliminary issues deserve explicit attention: questions, assumptions, and philosophical-scientific approach. Each of these will be discussed in greater detail in the sections that follow.

Before any research can begin, defining a clear question provides the basis to consider the study design, including the operational scale, possible modeling approaches, and assumptions. The exercise of defining the study question should be clear whether (1) specific hypotheses will be tested or (2) the study is an observational investigation. The difference between hypothesis testing and observation can be blurred and is often mixed in environmental sciences. However, both have advantages and disadvantages in study design.

Hypothesis testing may be used to isolate and examine environmental processes, providing clear results or seeking concrete answers. In this approach all the factors involved in a process are controlled, except the variable under examination, identifying the effect of that variable on the process. However, it is possible that interactions between variables are unaccounted for and that the control of other variables biases the experimental results. It is also possible (and even common) that results from field work disprove a hypothesis or show that it cannot be tested.

Observational studies include monitoring system response under ambient conditions and in response to perturbations (e.g., water or tracer additions, soil disturbance, or other alterations). These studies are excellent for determining how soil processes may change under different conditions. However, they are limited by a reduced ability to identify cause and effect in the processes under observation.

In designing either a hypothesis test or an observational study, the assumptions involved must have the researchers' attention. Many fundamental assumptions of both study and model design provided by academic advisors or scientific dogma may be unconsciously incorporated into research. Assumptions in science are most often found in the form of simplifications. For example, when describing the infiltration of water through soil, it is unnecessary to describe the movement of every water molecule. Instead, we describe a macro-scale process of wetting front advance. However, this simplification overlooks the importance of micro-scale interactions between water and the surfaces of solid particles such as soil aggregates. While these smaller-scale processes may not be important near soil saturation, during drainage surface tension exerts greater control on the process. Therefore, an examination of the assumptions should consider what processes are being simplified and if that is justified for the question being asked.

A major scientific consideration for study design is whether the soil processes under examination are deterministic, stochastic, or complex and chaotic. This important scientific issue, while not directly involved in a specific project, will inevitably constitute the theoretical framework influencing study design, measurement methods, as well as model selection. It could be argued that this is a physical, as opposed to philosophical, question. However the impossibility of proving such an argument requires a conceptual leap of faith.

A process is deterministic when there is a defined relationship between inputs and outputs of a system (or equation) for which there is a unique solution, as in newtonian physics. Stochasticity occurs when there is random variability in one of those inputs that results in a probability distribution of possible outcomes as opposed to a unique solution. We will define complex systems as those in which the multitude of factors in a process produce a response that may appear random, but is actually not truly random. Chaotic systems are defined as processes in which the outcome is extremely sensitive to initial conditions, and sometimes they are considered a subset of complexity. The starting point may be a single number for deterministic chaos or a distribution for stochastic chaos.

A reasonable case could likely be made for one or many of these physical processes operating in any given situation. For example, water flow in porous media is represented as deterministic in the Richards equation, stochastic in the Bresler-Dagan model (Dagan and Bresler, 1983), and chaotic by Faybishenko (2002), all of which will be discussed in the sections that follow. However, the researcher must select methods that assume one of these scientific approaches.

Jury and Scotter (1994) identified three theoretical methodologies (excluding chaos) to represent solute transport in soils: deterministic, stochastic-continuum, and stochastic-convective. Chaotic and complex processes are usually considered unpredictable over long time scales, but this does not mean that the observations of complex and chaotic behavior are without interest. Such observations may be significant in identifying what we may be able to predict within defined space, time, and error limits and what is unpredictable with reasonable certainty.

Selection of a theoretical methodology should then be a statement of how the researcher believes nature works. A clear statement of theory for each study is directly applicable, as it is a statement of assumptions often unspoken in soil science. It is also useful to consider whether a research-oriented or a practice-oriented model is necessary for the question being answered (Kutilek and Neilsen, 1994). In some cases simpler models have been found to adequately describe complex processes, such as the transport of nitrate (Reineger et al., 1990) or pesticide occurrence in U.S. groundwaters (Worral, 2001). The selection of a practice-oriented operational model could be an attempt to avoid selecting a theoretical framework, but, pragmatism might also be considered a valid philosophy.

2.1.2.1 Determinism in Soil Processes

Based in newtonian physics, deterministic systems are those with a predictable (usually unique) outcome assuming that all the factors influencing a process are included in a descriptive equation. This has been the dominant method historically in soil science including Darcy's law and its extension for unsaturated conditions, the Richards equation. Most models of water and solute flow in porous media are based on Darcy's law when saturated:

$$J_w = -K_s \frac{dH}{dz} \tag{2.1}$$

and its derivative for unsaturated conditions, the Richards equation (Tindall and Kunkel, 1999):

$$\frac{\partial \theta}{\partial t} = \frac{\partial}{\partial z}\left[D(\theta)\frac{\partial \theta}{\partial z}\right] + \frac{\partial K(\theta)}{\partial z} \tag{2.2}$$

In these equations J_w is the water flux, H the hydraulic head, t the time, and K the hydraulic conductivity of the soil, with the subscript s for saturated conditions, and (θ) is the volumetric water content at depth z. $D(\theta)$ is defined as the diffusivity or $K(\theta)\partial h/\partial \theta$, where h is the pressure head (Tindall and Kunkel, 1999). Further analysis of this approach is developed in Chapter 3. These equations of water flow provide the basis for contaminant transport analysis as the equations describing the reactions of contaminants within soil are coupled to these equations based on the law of mass conservation (Jury et al., 1991). Solutions for Eq. (2.2) are found analytically for various boundary conditions or numerically using finite element and finite difference schemes (van Genuchten and Sudicky, 1999; Jury et al., 1991).

2.1.2.2 Stochasticity in Soil Processes

Although it generally requires more data collection, the application of stochastic methods is becoming more common to construct spatial relation-ships and process models. Stochastic-continuum models use the existing deterministic equations with the hydraulic conductivity as space random functions with associated spatial relationships provided by experimental data (Rubin, 2003; Dagan, 1986; Dagan and Bresler, 1983). Stochastic character-ization and hydrology will receive more attention later in this chapter and in Chapters 16 through 18. While defining a spatial correlation function for a specific physical or model variable can be difficult, this method is quite effective at describing heterogeneous media. However, it should also be recognized that while this approach transforms the deterministic Richards equation into a stochastic model by defining input parameters as random variables, the functional equation is still based on a deterministic model. This reliance on the Richards equation may be a problem if transport conditions are actually chaotic (Faybishenko, 2002; Sposito, 1999).

Perhaps the simplest method to represent solute transport is stochastic-convective theory, or transfer function models (Jury and Scotter, 1994). This approach uses measured breakthrough curves (solute concentration measured over time under steady-state conditions) as a probability distribution of solute travel times through isolated stream tubes in a given transport volume (Jury and Scotter, 1994; Jury and Roth, 1990; Sposito and Jury, 1988). This so-called black box methodology relies less on a mechanistic understanding of the physical processes occurring, and more on characterizing solute transport data collected at a site using transport parameters (or the 1st and 2nd statistical moments). In this case, the breakthrough curve measured under steady-state conditions is a continuous function, or the transfer function (TF) (Jury and Sposito, 1985; Jury, 1982). The parameterization of a transfer function, model then requires measuring a consistent transfer function at an appropriate scale and precision.

Whether the study of water flow is approached under a deterministic or stochastic viewpoints can have implications for describing contaminant transport. How contaminants migrate through soil relates to the chemical, physical, and biological processes acting on solutes during transport. If solute only flows through a small percentage of the soil, not only will the velocity be faster, but the retardation and degradation of a labile (degradable) solute will also be different. Therefore, the chemical reactions that a solute may participate in during flow through the soil cannot be correctly described unless the models describe transport pathways.

2.2   ON SPATIAL VARIABILITY

A major challenge in accurately predicting the transport of water and solutes through soils has been spatial variability or soil heterogeneity resulting from irregularities in the soil physical structure. Among these, macropores, vertical and horizontal anisotropy (layering), and other structural soil characteristics have a major impact (White, 1985; Beven and Germann, 1982). Preferential flow of solutes that results from the soil heterogeneity has been demonstrated in numerous field studies (Flury et al., 1994; Edwards et al., 1993; Ghodrati and Jury, 1992, Andreini and Steenhuis, 1992; Ghodrati and Jury, 1990).

Spatial variability has been a particular focus of soil science since before the classic field scale transport studies by Biggar and Nielsen (1976). Although the term spatial variability has been applied to explain data in many situations, there are at least five independent sources for variability observed in field studies: (1) physical heterogeneity, (2) scale differences, (3) measurement error, (4) changing boundary conditions, and (5) nonstationarity in the observed process. The latter two may contribute to field observations of spatial variability, but are more correctly classified as temporal variability.

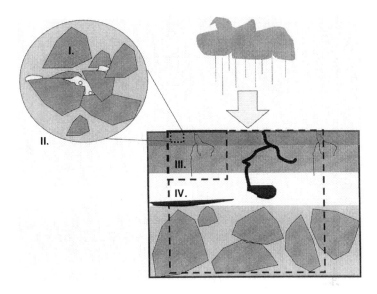

FIGURE 2.1 Sketch of a heterogeneous field soil, I. Aggregate, II. Matrix, III. Pedon, and IV. Polypedon or field-scale classifications.

Physical heterogeneity refers to the nonuniform structure of the soil. Soil is not a homogeneous porous media, it is comprised of different soil layers (horizons), irregularly shaped aggregates, rock fragments, cracks, plant roots, micro- and macro-invertebrate burrows, hydrophobic lenses, and much more (Figure 2.1). Given this starting point, field studies also encounter variability resulting from measurements collected with different sampling volumes (Garrido et al., 2001). As if this were not enough, measurement devices add in their own error, and it is quite difficult to maintain constant boundary conditions (e.g., irrigation rate, temperature, pressure conditions, water content, tracer concentrations, etc.) during data collection periods. Finally, the dominant process may change through either deformation (changes in the physical structure) of the soil during a study or a switch in process over time. This final source of variability is nonstationarity or the lack of a consistent process response at a specific location under the same boundary conditions.

In a critical review of variability in soil processes, Jury (1985) reported that static soil properties such as soil porosity, bulk density, and particle size fraction each had a coefficient of variation (standard deviation as a percent of the mean) ranging from 9 to 55%. Dynamic or process-related soil characteristics such as water and solute transport were found to have a much greater range in the coefficients of variation of 72 to 124% (Jury, 1985).

The spatial dependence of solute transport processes in the soil (scale related) has long been recognized and has been examined in numerous studies (Radcliffe et al., 1998; Shouse et al., 1994; Van Weesenbeck and Kachanoski, 1991; Kachanoski et al., 1985; Biggar and Neilsen, 1976). More recently, *in situ*) measurement techniques, for example time domain reflectometry (TDR), have been used to gather spatially distributed solute transport data at appropriate scales (Garrido et al., 2001; Jacques et al., 1998; Radcliffe et al., 1998; Rudolph et al., 1996).

Model development, the distribution of model parameters, and estimates of parameter values must have a clear relation to the distribution of field properties (Mayer et al., 1999). Advances toward including spatial heterogeneity found in soils into process characterization and models have been established through stochastic techniques such as Monte Carlo simulations (Mallants et al., 1996) and geostatistical methodologies (Goovaerts, 1999). These methods may be incorporated into models as suggested by: (1) correlating spatial associations between soil attributes and landscape factors (i.e., soil type, slope, aspect, vegetation) and using these landscape factors to incorporate the dominant processes, (2) defining a variable with a population distribution assuming some spatial independence between soil attributes to perform simulations, (3) normalizing the variability using scaling theory, (4) characterizing the spatial association between the specific soil attributes using geostatistical tools (e.g., semivariogram), and (5) applying some sort of hybrid of the previous approaches (McBratney et al., 2002; Park and Vlek, 2002; Jury, 1985).

The first of these four approaches has its greatest application in agricultural study design and statistical analyses. Through carefully designed studies that control many (if not all) environmental factors, treatments may be applied to determine how soil attributes, processes, or both change. Results can be analyzed with analysis-of-variance techniques or statistical regression analysis, the natural (background) variability may be defined (assuming a distribution — often the normal distribution), and relationships between a location and variables like soil type, depth of water table, vegetation, or the like may be established. If this characterization occurs at the appropriate scale, the idea then is that processes in adjacent fields, or modeling cells, may simply be summed to a total result. This type of correlation analysis in soil science has been called the pedotransfer function method (Pachepsky and Rawls, 1999). A discussion of the advantages and weaknesses of this pedotransfer functions is beyond the scope of this chapter; however, problems occur in the assumptions defining variable distributions, spatial independence, and scale.

Recently an example of the pedotransfer function method has been applied to study pesticide transport in the Central Valley of California. In this case, based on the transfer function model proposed by Jury (1982), different transfer functions are established for each soil type using the methods defined by Stewart and Loague (1999, 2003). Then large-scale models that incorporate spatial variability at the scale of soil types are used to predict pesticide leaching. Applications of the pedotransfer functions like this one have also been successful in a number of other cases (McBratney et al., 2002; Gonçalves et al., 2001; Woster et al., 2001; Pachepsky and Rawls, 1999).

The second approach involves using scaling techniques to normalize or coalesce the variability into a single characteristic or characteristic function

(Warrick et al., 1977). Scaling of soil properties has been based on the concept of similar media. In this approach, scale-invariant relationships exists for soil water properties based on the structure of the porous media (Neilsen et al., 1998). A review of this type of scaling methods and applications may be found in Neilsen et al. (1998). Fractal scaling methods will be discussed in greater detail in Chapter 17.

The third approach to defining the distribution of the input variables into the models and performing a diverse range of simulations has been named Monte Carlo after the famous city of chance and gambling. Amoozegar-Fard et al. (1982) applied this Monte Carlo method to simulate soil solute transport, obtaining comparable results to field data. A systematic examination of the relative contribution of each parameter to the total model variability using Monte Carlo techniques is called sensitivity analysis. This technique examines how model response changes with each specific variable (De Roo et al., 1992). The Bresler-Dagan model of water and solute transport in soils is another classic example of a stochastic model that has been tested and applied in field conditions (Dagan and Bresler, 1983). In this model the parameters for solutions of the Richards equation are defined as space random variables with statistical means and distributions defined by field measurements.

The fourth approach is essentially geostatistics, which has recently been an area of active research. More on geostatistics and its applications in soil science can be found in Chapters 16 and 18.

Finally, combinations of these approaches exist across the literature. For example, Sobieraj et al. (2002) recently presented a study using geostatistical techniques (semivariograms) to examine relationships between soil hydraulic conductivity (K_s) and landscape features (soil type and topography). It is interesting to note that in this paper, the authors found no relationship between the spatial structure (semivariogram) observed for K_s and either soil type or topography. They suggest that the process may be dominated by macropores and preferential flow operating on a different scale or under an alternative spatial organization. For an interesting application of most of the mentioned approaches to a single location and data set, the authors recommend the papers by Keith Loague and his colleagues on the R-5 catchment (Loague and VanderKwaak, 2002; VanderKwaak and Loague, 2001; Loague and Kyriakidis, 1997). These studies focused on combining a large number of infiltration measurements collected throughout a watershed into a predictive model of runoff generation.

2.3.   ON TEMPORAL VARIABILITY

Jenny (1941) stated in his fundamental equation of soil-forming factors that the magnitude of any soil property is related to time. However, relative to the consideration given to spatial variability, temporal variability in soil processes has received limited attention. In the case of solute transport, there are three major sources of temporal variability: (1) changes in the soil (or porous media) structure, (2) inconsistency in the major transport processes, and (3) changes in boundary conditions or external factors. Messing and Jarvis (1993) recognized that temporal variability in soil hydraulic conductivity could result from nonequilibrium macropore flow, soil shrinking and swelling, and macropore formation and flow pathway clogging. In addition, alteration of soil macropore structure due to earthworm activity, root growth, and other bioturbation have been demonstrated to cause changes in solute transport (Li and Ghodrati, 1995, 1994; Edwards et al., 1993).

Inconsistency in the major transport processes refers to a shift in the physical process controlling the flow of water and chemicals in soil. There are several distinct physical processes affecting solute transport through soil. One of these is the process of diffusion, or mass transport along a concentration gradient (Fickian diffusion), and the others are processes of advection, or mass transport along pressure (heat and elevation) gradients (Jury et al., 1991). These advective processes include surface film flow (Tokanaga et al., 2000), matrix flow, and preferential flow (Jury et al., 1991). In addition to the inconsistencies of water transport, chemical nonequilibrium has the potential to produce temporal changes as chemically active sites are exhausted and molecules degrade and recombine into other molecules (Sparks, 1995).

Finally, the third and likely most important source of temporal variability in solute transport relates to changes in the site boundary conditions, including water content and distribution, drainage change (perched water table or zero flux boundary), rainfall or irrigation intensity, and other external factors (e.g., temperature and relative humidity).

The following examples are combinations of all three of the sources of temporal variability. The temporal variation in soil water content has perhaps been the most studied soil characteristic, probably because of the recent development of advanced measurement technologies capable of monitoring water content at high temporal and spatial resolutions (Heimovaara et al., 1995; Comegna and Basile, 1994; Van Weesenbeck and Kachanoski, 1988; Vachaud et al., 1985) (also see Chapter 5). Van Weesenbeck and Kachanoski (1988) demonstrated that the water profile in the first 20 cm of a cultivated soil was not stable in time. However, Vachaud et al. (1985) found patterns in soil water content to have temporally stable relationships to other points in the same field. They defined time stability as the time-invariant association between spatial locations and statistical parametric values of soil deterministic properties such as texture or topography (also called temporal persistence). In this regard, Kachanoski and de Jong (1988) demonstrated the scale-dependent nature of time stability of soil water stored in the profile along a 720 m long soil transect. During the drying period of measurements, they observed a temporal stability across all spatial scales. However, during the wet periods, temporal stability of soil wetness was only observed for scales larger than 40 m and it was always related to the spatial pattern of the soil surface curvature across the transect. A similar finding by Comegna and Basile (1994) adds support to the observation that spatial relationships (patterns) in soil water content are temporally stable.

On the time scale of a steady-state miscible displacement, studies on temporal variability have been performed in the laboratory (Lennartz and Kamra, 1998), on an intact soil monolith (Buchter et al., 1995), and in a tile drained field (Lennartz et al., 1999) to compare displacement of consecutive tracer applications to the soil surface. Although Lennartz and Kamra (1998) suggested acting with prudence, these studies found temporal similarity in consecutive solute breakthrough curves. However, Campbell et al., (2001) found significant variability in solute transport with time in a single plot with measurements collected at the point and plot scales. At point scale of a 3 mm diameter fiberoptic miniprobes, measurements of tracer transport were consistent at some probes and wildly erratic at others. While the temporal variability at the larger measurement scale of 20 cm long TDR probes was less than that observed by the fiberoptic probes, the temporal difference in measured tracer transport was still evident. The different observations by Campbell et al. (2001) and Lennartz et al. (1999) may be the result of different measurement scales, or they may be because Lennartz's studies were performed in a tile drained field or simply that the results could relate to site-specific processes.

In another recent study, Jaynes et al. (2001) examined the consistency in the transport of pesticides and a bromide tracer in a tile drained field within a single leaching study. Consecutive pulses of bromide and pesticides were added through an irrigation system to a field scale plot (24.4×42.7 m) tile drained at 1.2 m depth. Similar to Campbell et al. (2001), the study found that the preferential flow was not a uniform process during a constant leaching event, as tracers applied later in the study had faster breakthrough times (Jaynes et al., 2001).

2.4   ISSUES IN FIELD STUDY DESIGN

2.4.1   ISSUES OF SCALE

The topic of scale has received much attention in hydrology. Although an exhaustive review is not necessary for this chapter, a brief discussion is required. For further information on this topic, the readers are referred to the excellent chapters in Sposito (1998) and Pachepsky et al. (2003). Many practical issues relating to the study of soil science involve scale, whether consciously or unconsciously included in the study design. There have been a number of attempts to define scales in hierarchical schemes based on physical features of processes (Wagenet, 1998). It has been suggested that characteristic scales may exist for specific processes.

Consider, for example, soil porosity (volume fraction void space) in a level grassland. Let us assume that we have a magic measurement system that can be applied to collect precise porosity measurements at a defined length scale (or "support," as defined in Chapter 17) collected over the entire surface of the grassland soil. In our experiment we start looking at the variability between individual porosity measurements for just a few adjacent measurements and

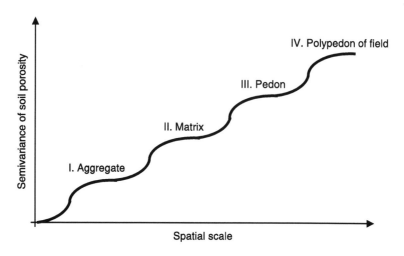

FIGURE 2.2 Hypothetical representation of the semivariance of soil porosity with increasing scale in a hierarchical system. (Adapted from Rubin, 2003.)

then a few more, increasing the combined scale of measurement. One hypothesis is that the observed variability in porosity measurements will increase until a characteristic length scale is reached, as illustrated in Figure 2.2. Then, when the characteristic length scale is exceeded (moving up a scale in Figure 2.1), the variability (in this case represented as semi-variance) will again increase until it reaches the next characteristic scale, and so on. In a hierarchical model the plateaus in Figure 2.2 would correspond to the numbered units (I–IV) in Figure 2.1. From aggregate to field scale, hierarchical characteristic length scales may apply, as demonstrated in the schematic diagram in Figure 2.2.

While there is some evidence for the previous discussion for relatively *static* soil properties like bulk density (Gajem et al., 1981), unfortunately, of universal scales at which a multitude of *processes* in nature occur have yet to be defined or may not exist. As a consequence, soil scientists have been left to heuristically defined comparable scales such as the pedon, plot, and field scales. The legacy of the limited ability of environmental scientist to standardize scales is that many studies are not quantitatively comparable.

A more reasonable set of operationally defined scales used in environmental studies have been discussed by Baveye and Boast (1999): natural, theoretical, arbitrary or system, computational, and measurement scales. These five scale classifications more or less cover all the study designs in natural sciences. Although qualitative, the value in a classification system such as this one is that scale is at least an explicit part of a study, considered in the design, even if the physical meaning of the scale classification is unknown. Each of these scale classifications will be briefly described below.

Natural scales are those based on intrinsic system properties or physical boundaries. These scales are commonly applied in the environmental sciences, for example, in the classic ecological hierarchy: individual, population, community, ecosystem, and landscape. The example in Figure 2.2 would be an attempt to define natural or physically based scales. The natural scale is also the basis for defining characteristic length scales known as representative elementary volume (REV), which will be discussed in greater detail below.

Theoretical scales refer to the application of defined scales (i.e, meter length) in the development and applications of models. These are scales set by definition in a modeling grid. They differ from natural scales in that they are exactly defined and applied to systems with a predefined physical relationship to process. These predefined relationships have emerged from the derivation and applications of theories to the study of the environment.

Arbitrary or *system* scales are dimensions or time frames that apply to studies that are unrelated to the objective of the studies itself. These include political and property boundaries as well as regulatory time frames. While these scales may be very real constraints or research boundaries, they may in fact have little to do with the process under investigation.

Computational scales are again related to modeling as it refers to the discretization of time and space within models. Computational scales are often limited by computational capacity and model runtime available for an examination.

Measurement scales are perhaps the most commonly encountered and important scales to field experimental scientists. These scales are physically real boundaries of the measurement tools and the time required to collect that measurement. It is a necessity for field scientists to know the measurement scales in their studies and to report them. Measurement scales are as close to a standardization of scale as possible in the environmental sciences. This issue will be discussed further in the following section.

2.4.2 CHARACTERIZING SCALE OF STUDY

All of the different scale classes mentioned above should be considered in a properly designed field study, but two in particular deserve specific attention: natural and measurement scales. The following discussion considers the main techniques to characterize and apply scale analysis in environmental research.

The natural scale is most often found in soil science as the representative elementary volume (REV), or the minimum soil volume where the variance does not greatly increase for an average value over many soil units of this volume (Bear, 1972; Hubbert, 1956). In this sense the REV is the minimum variance volume, where observations at greater soil volumes (increasing scale) would not result in an increase in total variance. This means that the variability would rise to a plateau, as in Figure 2.2, but thereafter remains flat. This idea

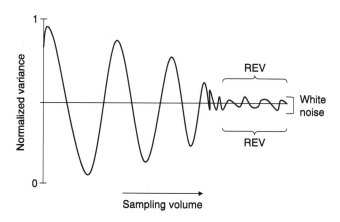

FIGURE 2.3 Hypothetical representation of normalized variance in a physical soil parameter up to the scale soil volume of a representative elementary volume (REV). (Adapted from Baveye and Sposito, 1984.)

was originally developed to allow aggregation of minimum units in modeling. A conceptualization of the change in the normalized variance in a soil property with increasing volume is illustrated in Figure 2.3 (Brown et al., 2000; Baveye and Sposito, 1984). Two experiments confirming the REV concept have been published (Baveye et al., 2002; Brown et al., 2000), but due to the difficulty involved, field validation of this concept has not yet been possible. Also, since each soil property has its own REV, a common value of REV for all kinds of observations may not exist, and a specific soil property may have different REV values in different soils (Kutilek and Neilsen, 1994).

A concept related to REV called the representative elemental area (REA) exists in hillslope hydrology. The REA is the minimum watershed area necessary to be able to aggregate into larger watershed models (Wood et al., 1988). It has also been argued by Baveye and Sposito (1984) that the REV concept is unnecessary for modeling transport in porous media. Instead these authors propose a relativistic continuum approach that incorporates measurement volume into macroscopic physical variables (Baveye and Sposito, 1984).

A more operationally defined REV may be based on the measurement devices used and is called the minimum measurement variance volume (MMVV) (Baveye and Boast, 1999). A MMVV is the number of measurements using a specific technology (or the total volume of the combined measurements) that reduces the total variability to the smallest possible values. This is akin to the old-fashioned statistical estimate of the total number of samples necessary to adequately represent the mean and standard deviation (within a defined probability) for a variable of the statistical population under examination. The MMVV differs from the REV in that it is based on a combined variability of the measurement error and inherent measurement scale, whereas a true REV is based only on physical properties of the media.

Although not always explicitly stated, the MMVV methodology is commonly used to justify length scales selected for a study. For example,

Campbell et al. (2002) used MMVV to determine the number of 20 cm time domain reflectometry (TDR) probes needed to characterize the vertical transport of a conservative tracer in a sloping soil. A total of 16 TDR probes were vertically inserted into a meter-squared plot on a hillslope, and a tracer solution of calcium chloride was applied to the entire plot and then displaced. The temporal moments (see Appendix 1 and Chapter 11) were then estimated from the breakthrough curves. The number of TDR probes required to produce a mean within a 95% probability (i.e., within 10% of the "true" mean) was estimated to be 2 probes for the first moment (μ), 8 probes for the second moment (μ^2), and 22 probes for mass recovery (Campbell et al., 2002). In this case, the MMVV for μ would be 2 TDR probes for a plot of 1×1 m surface area and 0.2 m depth and at a confidence limit of 10%.

This example illustrates one of the major problems in experimental characterization of MMVV scales, that is, different processes are likely to have different MMVVs. The first temporal moment is mostly an advective process, the second moment includes both advection and dispersion, and the mass recovery is sensitive to advection, dispersion, and bypass flow. As a result, the estimated MMVV for each moment of the solute breakthrough curve is different.

2.4.3 IRRIGATION, SOLUTE DELIVERY, AND THREE-DIMENSIONAL FLOW

In addition to the scale of study and measurement method (and scale or support as defined in Chapter 17), another major factor in field study design is the number of dimensions over which observations will be collected. For solute transport, dimensionality means that the solute breakthrough curves will be collected in one dimension or that all three dimensions of the solute plume will be imaged. Such considerations will then influence the decision as to what measurement techniques and experimental set-up, as well as models, will be most appropriate to quantify and represent the study.

The selection of the number of dimensions to include in an examination depends on the question being asked and the amount of resources available to answer that question. Not designing the dimensions (both space and time) under examination can reduce the overall effectiveness of the data. For example, if the objective of the study is to observe a transient process like stormwater runoff, erosion, or soil moisture, then temporal resolution may be of greater importance than gathering extra spatial data.

However, observational problems may occur when using too few spatial dimensions. For example, consider a lysimeter study where a solute tracer is applied at the soil surface and measured in the drainage collected in the lysimeter. The lysimeter may only recover a small amount of the total tracer applied at the soil surface. This low recovery of the solute mass has three possible explanations: (1) the tracer is somehow chemically active (sorbing or labile), (2) the irrigation applied was not enough to leach the solute to the depth of the lysimeter, or (3) the tracer bypassed the lysimeter by moving laterally in the soil. It should be recognized that the third result has been documented in many studies and has the potential to be common in soil solute transport (Garrido et al., 2001; Forrer et al., 1999; Fluehler et al., 1996). In this example a one dimensional (1-D) study design would not allow the researcher to rule out explanations (2) and (3). However, additional monitoring to capture lateral flow in the second dimension would provide the researcher with information to decide if the low tracer recovery in the lysimeter is the result of reactive tracer transport or bypass of the lysimeter.

It is possible in many instances to control the boundary conditions in a field plot to minimize the number for dimensions necessary to measure. The method of irrigation in solute transport studies is a common example. Irrigation methods include flooding, various types of sprinkler irrigation, and reverse inundation (bottom to top) from some depth in the soil to the surface. The use of each of these methods will again depend on study objectives. Common to all these irrigation methods is the inclusion of a boundary layer to avoid artificial influences of the surrounding soil not characteristic of the designed boundary conditions. Various suggestions exist in the literature as to the required dimensions of a boundary area, including half the plot width. Beven et al. (1993) suggested that in laboratory column studies the column diameter should be twice the length to avoid the column boundaries from influencing transport processes. Of course, such a requirement might depend upon the characteristic length scale, so no standard rule may be applied.

Examples of solute delivery as a point source, line source, or sheet source are illustrated in Figure 2.4. The careful application of these solute delivery methods can limit a study to 3-D, 2-D, and 1-D analysis, respectively. For example, Campbell et al. (2004) used these different application methods as well as instrumentation designs to isolate the influence of the leaf litter layer on vertical and lateral transport in hillslope plots in a oak woodland. Without

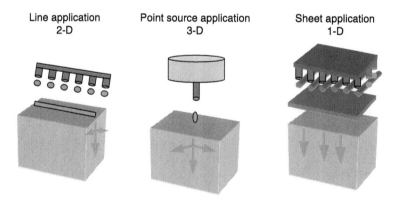

FIGURE 2.4 Examples of different solute application methods to create a 1-D, 2-D, or 3-D flow plume for studies.

the perspectives provided by the different solute application methods and instrument designs, it would not have been possible to identify how the litter layer influences the transport processes.

Field study designs and plot characteristics for a number of investigations over the past 10–15 years are summarized in Table 2.1. Notice the differences in measurement techniques, plot sizes, and solute delivery for these few studies with relatively similar objectives. It is the nature of science that each study is done differently to remain unique and likely more publishable. In addition, each investigator brings his or her own tools and set of skills to the study design. The result, however, is that quantitative comparisons of the data from each of these studies would not be scientifically justified. Those studies using the statistical moments to characterize solute break-through curves have adopted a stochastic theoretical framework, while others, using parameters of the convective-dispersion equation, have used a deterministic approach. Some studies include 2-D and 3-D flow, while others have limited the conditions to 1-D. Finally, the time scales of the study differ from days to months, and the measurement length scales range from a few to hundreds of centimeters. Any one of these differences could bias the observations for comparison with other studies but were the appropriate designs for the original question posed by the investigator. This suggests that the greatest return can be gained from qualitative comparisons of the soil processes.

2.5 SUMMARY AND CONCLUSIONS

It was the objective of this chapter to discuss a portion of the basic knowledge necessary for appropriate field study design in soil science. Focusing on soil processes, in particular solute transport, the importance of defining a clear study question, identifying the assumptions, and selecting a scientific approach were discussed. While numerous assumptions underlie our experimental and modeling methodologies in soil science, it is our opinion that this issue has not received the necessary attention. The alternative to selecting a particular theoretical framework is the unconscious acceptance of one provided by mentors or others in the field.

Increasing our understanding of the spatial and temporal variability of soil characteristics and processes and its application to regional and global environmental issues remains one of the greatest challenges for future research. In particular, the description of chemical transport across scales ranging from laboratory to field or watershed is far from being reasonably successful. The interaction and inseparability of physical heterogeneity, unstable boundary conditions, scale relationships, and measurement error have limited the quantitative comparison of field studies. However, it is not clear how to overcome these matters. As a result, scientific emphasis must remain on qualitative comparison of larger-scale process studies, along with detailed investigations of the physics of dynamic soil processes.

TABLE 2.1
Selected Examples of Field Solute Transport Studies over the Past 10–15 Years Summarizing Study Design Issues

Ref.	Measurement scale	Study length (time)	Plot dimensions	Soil type	Irrigation technique	Solute delivery	Measurement objectives	Boundary area
Butters et al. (1989)	Suction lysimeters	Months	80 × 80 m to 3.05 m depth	Typic Xeropsamment	Sprinkler	Same	Tracer transport and Mass recovery	Outer 60 m of plots
Campbell et al. (2002)	20 cm long TDR probes	Days	3 Plots A) 1.5 × 1.5 m to 0.50 m depth B) 2.0 × 3.0 m to 0.50 m depth C) 1.0 × 9.0 m to 0.50 m depth	Botella Clay Loam (Pachic Argixerol)	Sprinkler	A) same as irrigation B) 1-D line source 2.0 m upslope C) 1-D line source 0.50 m upslope	Tracer transport (μ, σ, Mr) inverse est.	Greater than outer 0.50 m of plot
Ellsworth et al. (1991)	6.35 cm dia. soil cores	Months	2.0 × 2.0 to 5.0 m depth	Typic Xeropsamment	Drip irrigation	Resident concentration and sprinkler application	Tracer transport (μ, σ, Mr) inverse est.	Greater than outer 2.0 m of plot
Feyen et al. (1999)	2 m drainage troughs	Week	13 m^2 to 0.60 m depth	Umbric and mollic Gleysols	Mobile spray bar	1-D line source surface and subsurface 3.3 m upslope	Tracer transport	Unknown
Flury et al. (1995)	Lateral soil cores 10 cm long, 5.6 cm dia.	Days	6, 1.4 × 1.4 m plots to 1.0 m depth	Loamy (Typic Hydraquent) and sandy (Mollic/Aquic Udifluvent) soils	Mobile spray bar	Same as irrigation	Tracer and pesticide transport	Outer 0.50 m of plots

Reference	Method	Time scale	Dimensions	Soil type	Water application	Flow pattern	Measurement/analysis	Sampled area
Garrido et al. (2001)	3 mm fiberoptic probes 5 cm TDR probes 20 cm TDR	Days	0.60 × 0.60 m to 0.20 m	Botella Clay Loam (Pachic Argixerol)	Sprinkler	Same as irrigation	Tracer transport (μ, σ, Mr) inverse est.	Outer 0.23 m of plot
Ghodrati and Jury (1992)	7.5 cm dia., 0.5 m long cores	Week	64, 1.5 × 1.5 m to 1.0 m deep plots	Tujunga loamy sand	Flood & sprinkler	2-D sheet front	Pesticide mass recovery	Outer 0.25 m of each plot
Jacques et al. (1998)	120, 50 cm long TDR probes	Weeks	2 plots, 2.5 × 12.0 m to 0.90 m depth	Eutric Regosol and Stagnic Podzoluvisol	Sprinkler	Same	Tracer transport (μ, σ, Mr) inverse est.	Outer 1.25 m of each plot
Jaynes and Rice (1993)	50 mm dia. suction lysimeters at 7 depths down to 3 m	Months	37 m² to 3 m depth	Avondale clay loam	Flood and drip irrigation	2-D sheet front	Tracer transport (v, D, Mr) inverse est. CXTFIT	24 m²
Kachanoski et al. (1992)	20 cm long TDR probes and 25 mm dia. Suction lysimeters	Days	2.0 × 12.0 m to 0.2 m depth	Loamy sand (Typic Hpludalf)	Drip	Same as irrigation	Tracer transport	Unknown
Kung (1990)	Dye pattern every 25 cm²	Weeks	3.0 × 3.6 m to approx. 6.6 m depth	Sandy soil	Precipitation and furrow flood irrigation	Furrow flood irrigation	Dye tracer transport	Unknown
Mertens et al. (2002)	Single ring pressure infiltrometer (9.5 cm dia.)	Days	20 × 80 m at surface	Luvisol sandy-loam	None	None	Hydraulic conductivity	Does not apply

(Continued)

TABLE 2.1
Continued

Ref.	Measurement scale	Study length (time)	Plot dimensions	Soil type	Irrigation technique	Solute delivery	Measurement objectives	Boundary area
Parkin et al. (1995)	0.2, 0.3, 0.4 m long TDR probes	Days	2.0×2.0 m	Brunisolic Grey Brown Luvison (Typic Haplualf)	Sprinkler	Same	Soil moisture and hydraulic conductivity	Unknown
Radcliffe et al. (1996) and Radcliffe et al. (1998)	20 cm long TDR probes, Tile drains, and soil cores (6.0–8.5 cm dia.)	Weeks	2 plots, 12.5×3 0.5 m to 1.5 m depth	Typic Kanhapludult	Sprinkler	Same	Tracer transport (v, D, Mr)	Unknown
Rudolph et al. (1996)	112 suction lysimeters (2.5 cm dia.) and 168 TDR (25–200 cm)	Weeks	2.0×10.0 m to 2.0 m depth	Sand	Sprinkler	Same	Infiltration and tracer transport	Greater than outer 0.50 m of plot
Simmonds and Nortcliff (1998)	6×6 cm lysimeter	Days	1.22×1.22 m to 1.0 m depth	Sandy loam	108 drip irrigation points	Same as irrigation	Tracer transport (μ, σ, Mr)	Outer 0.37 m of plot

ACKNOWLEDGMENTS

The authors wish to thank to Dr. García-González, Dr. Macías, and Dr. Persoff for their review and suggestions to this chapter. A portion of this work was performed under the auspices of the U.S. Department of Energy by the University of California, Lawrence Berkeley National Laboratory.

APPENDIX 1: BREAKTHROUGH CURVE DATA ANALYSIS

The tracer mass is commonly presented as the mass measured versus the mass applied, or mass recovery (M/M_o). The equation is:

$$\frac{M}{M_o} = \sum \left[\left(\frac{C_{(z,t)}}{C_o} \right) V_w(t) \right]$$

(2.3)

where $C_{(z,t)}/C_o$ is calculated for Eq. (2.2) and $V_w(t)$ is the volume of water moving past the probe during each sampling period (Campbell et al., 1999).

MOMENT ANALYSIS

Moment analysis is a method to quantitatively represent a statistical probability distribution using characteristic values (Valocchi, 1990). Although an infinite number of moments may be calculated, the first two moments are usually adequate to describe statistical distributions. Using the breakthrough curves as probability distributions of solute travel time in the soil, the first temporal moment (μ^1) is:

$$\mu^1 = \int_0^\infty t \cdot f(t) dt$$

(2.4)

where t is time on the x-axis of the breakthrough curve. This temporal moment characterizes the mean displacement time of the solute. The second temporal moment (μ^2) represents the spreading of the distribution along the x-axis. This value is defined as:

$$\sigma^2 = \int_0^\infty (t - \mu^1)^2 \cdot f(t) dt$$

(2.5)

where μ^1 is the first temporal moment and t is again time on the x-axis (Mallants et al., 1994). Therefore, within stochastic-convection theory, the breakthrough curves represent the frequency of stream tubes transporting solute at a given travel time during the miscible displacement. The moments may therefore be thought of as representing the probability distribution of these travel times over the time axis. Chapter 11 develops the application of moments to the analysis of breakthrough curves.

TEMPORAL ANALYSIS

The Spearman's rank test may be used to examine the temporal stability in the ranked responses measured between two breakthrough curves (Campbell et al., 2001; Vachaud et al., 1985). This test compares the ranked order of a data series for two different times or reproductions of the series. In this case the test examines if the order of μ^1 and μ^2 for all the probes in the first breakthrough curve is the same order measured in the second breakthrough curve. The more similar the ranked values of the two breakthrough curves are, the closer to 1 the Spearman's coefficient will be.

Specifically, Spearman's test uses a rank R_{ij} of the measured variable (in this case μ_{ij} and σ^2_{ij}) and $R_{ij'}$ the rank of the same variable at the same location i) at a different time (j'). The Spearman's rank correlation coefficient is:

$$r_s = 1 - \frac{6 \sum_{i=1}^{n} (R_{ij} - R_{ij'})^2}{n(n^2 - 1)} \tag{2.6}$$

where n is the sample size. If r_s is close to 1, the variable is temporally stable. Critical r_s values, below which the difference observed with time is considered significant, may be found in standard statistics texts (Daniel, 1995).

REFERENCES

Amoozegard-Fard, A., D.R. Nielsen, and A.W., Warrick. 1982. Soil solute concentration distributions for spatially varying pore water velocities and apparent diffusion coefficients. *Soil Sci. Soc. Am. J.*, 46: 3.

Andreini, M.S. and T.S. Steenhuis. 1992. Preferential paths under conventional and conservation tillage. *Geoderma*, 46: 85.

Baveye, P. and C. Boast. 1999. Physical scales and spatial predictability of transport processes in the environment, in *Assessment of Non–Point Source Pollution in the Vadose Zone*, Corwin, D.L., Loague, K., and Ellsworth, T.R., eds. Geophysical Monongraph 108. American Geophysical Union. Washington, DC, 261.

Baveye, P. and G. Sposito. 1984. The operational significance of the continuum hypothesis in the theory of water movement through soils and aquifers. *Water Resour. Res.*, 20: 521.

Baveye, P., H. Rogasik, O. Wendroth, I. Onsasch, and J.W. Crawford. 2002. Effect of sampling volume on the measurement of soil physical properties: simulation with x-ray tomography data. *Measurement Environ. Sci. Technol.*, 13: 775.

Bear, J. 1972. *Dynamics of Fluids in Porous Media*. Elsevier Science, New York, 15–26.

Beven, K.J., D.E. Henderson, and A.D. Reeves. 1993. Dispersion parameters for undisturbed partially saturated soil. *J. Hydrol.*, 143: 19.

Beven, K. and P. Germann. 1982. Macropores and water flow in soils. *Water Resour. Res.*, 18: 1311.

Biggar, J.W. and D.R. Nielsen. 1976. Spatial variability of the leaching characteristics of a field soil. *Water Resour. Res.*, 12: 78.

Brown, G.O., H.T. Hsieh, and D.A. Lucero. 2000. Evaluation of laboratory dolomite core sample size using representative elementary volume concepts. *Water Resour. Res.*, 36: 1199.

Buchter, B., C. Hinz, M. Flury, and H. Fluehler. 1995. Heterogeneous flow solute transport in an unsaturated stony soil monolith. *Soil Sci Soc. Am. J.*, 59: 14.

Butters, G.L., W.A. Jury, and F.F. Ernst. 1989. Field scale transport of bromide in an unsaturated soil 1. Experimental methodology and results. *Water Resour. Res.*, 25: 1575.

Campbell, C.G., Garrido, F., and Ghodrati, M. 2004. Role of leaf litter in initiating solute transport pathways in a woodland hillslope soil. *Soil Sci.*, 169(2): 100.

Campbell, C.G., M. Ghodrati, and F. Garrido. 2002. Using TDR to characterize shallow solute transport in an oak woodland hillslope in northern California, USA. *Hydrol. Proc.*, 16: 2921.

Campbell, C.G., M. Ghodrati, and F. Garrido. 2001. Temporal consistency of solute transport in a heterogeneous field plot. *Soil Sci.*, 166: 491.

Campbell, C.G., Ghodrati, M., and Garrido, F. 1999. Comparison of time domain reflectometry, fiber optic mini-probes, and solution samplers for real time measurement of solute transport in soil. *Soil Sci.*, 164: 156.

Comegna, V. and A. Basile. 1994. Temporal stability of spatial patterns of soil water storage in a cultivated Vesuvian soil. *Geoderma*, 62: 299.

Dagan, G. 1986. Statistical theory of groundwater flow and transport: pore to laboratory, laboratory to formation, and formation to regional scale. *Water Resour. Res.*, 22: 120S.

Dagan, G. and E. Bresler. 1983. Unsaturated flow in spatially variable fields II. Application of water flow models to various fields. *Water Resour. Res.*, 19: 421.

Daniel, W.W. 1995. *Biostatistics: A Foundation For Analysis in Health Sciences.* 6th ed., John Wiley and Sons, Inc., New York, 613.

De Roo, A.P.J., L. Hazelhoff, and G.B.M. Heuvelink. 1992. Estimating the effects of spatial variability of infiltration on the output of a distributed runoff and soil erosion model using Monte Carlo methods. *Hydrol. Process*, 6: 127.

Edwards, W.M., M.J. Shipitalo, L.B. Owens, and W.A. Dick. 1993. Factors affecting preferential flow of water and atrazine through earthworm burrows under continuous no-till corn. *J. Environ. Qual.*, 22: 453.

Ellsworth, T.R. and W.A. Jury., 1991. A three-dimensional field study of solute transport through unsaturated, layered, porous media 2. Characterization of vertical dispersion. *Water Resour. Res.*, 27: 967.

Faybishenko, B. 2002. Chaotic dynamics in flow through unsaturated fractured media. *Adv. Water Resour.*, 25: 793.

Feyen, H., H. Wunderli, H. Wydler, and A. Papritz. 1999. A tracer experiment to study flow paths of water in a forest soil. *J. Hydrol.*, 225: 155.

Fluehler, H., W. Durner, and M. Flury. 1996. Lateral solute mixing processes—a key for understanding field-scale transport of water and solutes. *Geoderma* 70: 165.

Flury, M., J. Leuenberger, B. Studer, and H. Fluehler. 1995. Transport of anions and herbicides in a loam and sandy field soil. *Water Resour. Res.*, 31: 823.

Flury, M., H. Fluehler, W.A. Jury, and J. Leuenberger. 1994. Susceptibility of soils to preferential flow of water: a field study. *Water Resour. Res.*, 30: 1945.

Forrer, I., R. Kasteel, M. Flury, and H. Fluhler. 1999. Longitudinal and lateral dispersion in an unsaturated field soil. *Water Resour. Res.*, 35: 3049.

Gajem, Y.M., A.W. Warrick, and D.E. Myers. 1981. Spatial dependence of physical properties of a typic torrifluvent soil. *Soil Sci Soc. Am. J.*, 45: 709.

Garrido, F., M. Ghodrati, C.G. Campbell, and M. Chendorain. 2001. Detailed characterization of solute transport processes in a heterogeneous field soil. *J. Environ. Qual.*, 30: 573.

Ghodrati, M. and W.A. Jury. 1992. A field study of the effects of soil structure and irrigation method on preferential flow of pesticides in unsaturated soil. *J. Contam. Hydro.*, 11: 101.

Ghodrati, M. and W.A. Jury. 1990. A field study using dyes to characterize preferential flow of water. *Soil Sci. Soc. Am. J.*, 54: 1558.

Gonçalves, M.C., F.J. Leij, and M.G. Schaap. 2001. Pedotransfer functions for solute transport parameters of Portuguese soils. *Eur. J. Soil Sci.*, 52: 563.

Goovaerts, P. 1999. Geostatistics in soil science: state of the art and perspectives, *Geoderma*, 89: 1.

Heimovaara, T.J., A.G. Focke, W. Bouten, and J.M. Verstraten. 1995. Assessing temporal variations in soil water composition with time domain reflectometry. *Soil Sci. Soc. Am. J.*, 59: 689.

Hubbert, M.K. 1956. Darcy's law and the field equations of the flow of underground fluids. *J. Petrol. Technol.*, 8: 222.

Jacques, D., D.J. Kim, J. Diels, J. Vanderborght, H. Vereecken, and J. Feyen. 1998. Analysis of steady state chloride transport through two heterogeneous field soils. *Water Resour. Res.*, 34: 2539.

Jaynes, D.B., S.I. Ahmed, K.-J.S. Kung, and R.S. Kanwar. 2001. Temporal dynamics of preferential flow to a subsurface drain. *Soil Sci. Soc. Am. J.* 65: 1368.

Jaynes, D.B. and Rice, R.C. 1993. Transport of solutes as affected by irrigation method. *Soil Sci. Soc. Am. J.*, 57: 1348.

Jenny, H., 1941. *Factors of Soil Formation. A System of Quantitative Pedology* McGraw-Hill, Inc., New York, Chap. 3, 281 pp.

Jury, W.A. and D.R. Scotter. 1994. A unified approach to stochastic-convective transport problems. *Soil Sci. Soc. Am. J.*, 58: 1327.

Jury, W.A., W.R. Gardner, and W.H. Gardner, 1991. *Soil Physics.* 5th ed. Wiley & Sons, Inc., New York, 328 pp.

Jury, W.A. and K. Roth. 1990. *Transfer Functions and Solute Movement Through Soil* Birkhäuser Verlag, Basel.

Jury, W.A. 1985. Spatial variability of soil physical parameters in solute migration: a critical literature review. *Report to the Electrical Power Research Institute* EPRI EA-4228, 3–1.

Jury, W.A. and G. Sposito. 1985. Field calibration and validation of solute transport models for the unsaturated zone. *Soil Sci. Soc. Am. J.*, 49: 1331.

Jury, W.A., 1982. Simulation of solute transport using a transfer function. *Water Resour. Res.*, 18: 363.

Kachanoski, R.G., Pringle, E., and Ward, A. 1992. Field measurement of solute transport travel times using time domain reflectometry. *Soil Sci. Soc. Am. J.* 56: 47.

Kachanoski, R.G. and E. de Jong. 1988. Scale dependence and the temporal persistence of spatial patterns of soil water storage. *Water Resour. Res.*, 24: 85.

Kachanoski, R.G., D.E. Rolston, and E. de Jong. 1985. Spatial variability of a cultivated soil as affected by past and present microtopography. *Soil Sci. Soc. Am. J.*, 49: 1082.

Kung, K.-J.S. 1990. Preferential flow in a sandy vadose zone: 1. Field observation. *Geoderma*, 46: 51.

Kutilek, M. and D.R. Nielsen. 1994. *Soil Hydrology*. Catena Verlag, Cremlingen-Destedt, 370 pp.

Lennartz, B., J. Michaelsen, W. Wichtmann, and P. Widmoser. 1999. Time variance analysis of preferential solute movement at a tile-drained field site. *Soil Sci. Soc. Am. J.*, 63: 39.

Lennartz, B. and S.K. Kamra. 1998. Temporal variability of solute transport under vadose zone conditions. *Hydrol. Proc.*, 12: 1939.

Li, Y.M. and M. Ghodrati. 1994. Preferential transport of nitrate through soil columns containing root channels. *Soil Sci. Soc. Am. J.*, 58: 653.

Li, Y.M. and M. Ghodrati. 1995. Transport of nitrate in soils as affected by earthworm activities. *J. Environ. Qual.*, 24: 432.

Loague, K. and J.E. VanderKwaak. 2002. Simulating hydrological response for the R-5 catchment: comparison of two models and the impact of the roads. *Hydrol. Proc.*, 16: 1015.

Loague, K. and P.C. Kyriakidis. 1997. Spatial and temporal variability in the R-5 infiltration data set—déjà vu and rainfall-runoff simulations. *Water Resour. Res.*, 33: 2883.

Mallants, D., D. Jacques, M. Vanclooster, J. Diels, and J. Feyen. 1996. A stochastic approach to simulate water flow in a macroporous soil. *Geoderma*, 70: 299.

Mallants, D., M. Vanclooster, M. Meddahi, and J. Feyen. 1994. Estimating solute transport in undisturbed soil columns using time-domain reflectometry. *J. Contam. Hydrol.*, 17: 91.

Mayer, S., T.R. Ellsworth, D.L. Corwin, and K. Loague. 1999. Identifying effective parameters for solute transport models in heterogeneous environments, in *Assessment of Non–Point Source Pollution in the Vadose Zone*, Corwin, D.L., Loague, K., and Ellsworth, T.R., eds. Geophysical Monongraph 108. American Geophysical Union. Washington, DC, pg. 119.

McBratney, A.B., B. Minasny, S.R. Cattle, and R.W. Vervoort. 2002. From pedotransfer functions to soil inference systems. *Geoderma*, 109: 41.

Mertens, J., D. Jacques, J. Vanderborght, and J. Feyen. 2002. Characterization of the field-saturated hydraulic conductivity on a hillslope: *in situ* single ring pressure infiltrometer measurements. *J. Hydrol.*, 263: 217–229.

Messing, I. and N.J. Jarvis. 1993. Temporal variation in the hydraulic conductivity of a tilled clay soil as measured by tension infiltrometers. *J. Soil Sci.*, 44: 11.

Neilsen, D.R., J.W. Hopmans, and K. Reichardt. 1998. An emerging technology of scaling field soil-water behavior, in *Scale Dependence and Scale Invariance in Hydrology*, G. Sposito, ed. Cambridge University Press, New York, 1998, pg. 136–166.

Pachepsky, Y.A., D.E. Radcliffe, and H.M. Selim. 2003. *Scaling in Soil Physics*, CRC Press, Boca Raton, FL, 456 pp.

Pachepsky, Y.A. and W.J. Rawls. 1999. Accuracy and reliability of pedotransfer functions as affected by grouping soils. *Soil Sci. Soc. Am. J.*, 63: 1748.

Park, S.J. and P.L.G. Vlek. 2002. Environmental correlation of three-dimensional soil spatial variability: a comparison of three adaptive techniques. *Geoderma* 109: 117.

Parkin, G.W., R.G. Kachanoski, D.E. Elrick, and R.G. Gibson. 1995. Unsaturated hydraulic conductivity measured by time domain reflectometry under a rainfall simulator. *Water Resour. Res.*, 31: 447.

Radcliffe D.E., S.M. Gupta, and J.E. Box Jr. 1998. Solute transport at the pedon and polypedon scales. *Nutr. Cycl. Agroecosys.*, 50: 77.

Reiniger, P., J. Hutson, H. Jansen, J. Kragt, H. Piehler, M. Swerts and H.Vereecken, 1990. Evaluation and testing of models describing nitrogen transport and transformations in soils: A European project. Proceedings of the 14th International Congress of Soil Science, Kyoto, Japan, Vol I: 56–61.

Rubin, Y. 2003. *Applied Stochastic Hydrogeology.* Oxford University Press, New York, 391 pp.

Rudolph, D.L., R.G. Kachanoski, M.A. Celia, D.R. LeBlanc, and J.H. Stevens. 1996. Infiltration and solute transport experiments in unsaturated sand and gravel, Cap Cod, Massachusetts: experimental design and overview of results. *Water Resour. Res.*, 32: 519.

Shouse, P.J., T.R. Ellsworth, and J.A. Jobes. 1994. Steady-state infiltration as a function of measurement scale. *Soil Sci.*, 157: 129.

Simmons, L.P. and Nortcliff S. 1998. Small scale variability in the flow of water and solutes, and implications for lysimeter studies of solute leaching. *Nutr. Cycl. Agroecosys.*, 50: 65.

Sobieraj, J.A., H. Elsenbeer, R.M. Coelho, and B. Newton. 2002. Spatial variability of soil hydraulic conductivity along a tropical rainforest catena. *Geoderma.* 108: 79.

Sparks, D.L. 1995. *Environmental Soil Chemistry.* Academic Press, San Diego, 267 pp.

Sposito, G. 1999. On chaotic flows of water in the vadose zone. American Geophysical Union Annual Meeting. Abstract published. Eos, Trans of the AGU. 80(46): F403.

Sposito, G. 1998. *Scale Dependence and Scale Invariance in Hydrolog.* Cambridge University Press, New York, 423 pp.

Sposito, G. and W.A. Jury. 1988. The lifetime probability density function for solute movement in the subsurface zone. *J. Hydro.*, 102: 503.

Stewart, I.T. and K. Loague. 2003. Development of type transfer functions for regional-scale nonpoint source groundwater vulnerability assessments. *Water Resour. Res.* 39(12): 135.

Stewart, I.T. and K. Loague. 1999. A type transfer function approach for regional-scale pesticide leaching assessments. *J. Environ. Qual.*, 28: 378.

Tindall, J.A. and J.R. Kunkel. 1999. *Unsaturated Zone Hydrology for Scientists and Engineers*, Prentice Hall, Upper Saddle River, NJ, 624 pp.

Tokunaga, T.K., J. Wan. and S.R. Sutton. 2000. Transient film flow on rough fracture surfaces. *Water Resour. Res.*, 3: 1737.

Vachaud, G., A. Passerat De Silans, P. Balabanis. and M. Vauclin. 1985. Temporal stability of spatially measured soil water probability density function. *Soil Sci. Soc. Am. J.*, 49: 822.

Valocchi, A.J. 1990. Use of temporal moment analysis to study reactive solute transport in aggregated porous media. *Geoderma*, 46: 233.

VanderKwaak, J.E. and K. Loague. 2001. Hydrologic-response simulations for the R-5 catchment with a comprehensive physics-based model. *Water Resour. Res.*, 37: 999.

van Genuchten, M. and E.A. Sudicky. 1999. Recent Advances in Vadose Zone Flow and Transport Modeling. Chapter 6 in *Vadose Zone Hydrology* M.B. Parlange and J.W. Hopmans, eds. Oxford University Press, New York, 155–193.

Van Weesenbeck, I.J. and R.G. Kachanoski. 1991. Spatial scale dependence in *in situ* solute transport. *Soil Sci. Soc. Am. J.*, 55: 3.

Van Weesenbeck, I.J. and R.G. Kachanoski. 1988. Spatial and temporal distribution of soil water in the tilled layer under a corn crop. *Soil Sci. Soc. Am. J.*, 52: 363.

Wagenet, R.J. 1998. Scale issues in agroecological research chains. *Nutr. Cycl. Agroecosys.*, 50: 23.

Warrick, S.R. and D.R. and Nielsen. 1980. Spatial variability of soil physical properties in the field, in *Applications of Soil Physics*, Hillel, D., ed., Academic Press, New York, pg. 319.

Warrick, A.W., G.J. Mullen, and D.R. Nielsen. 1977. Scaling field-measured soil hydraulic properties using a similar media concept. *Water Resour. Res.*, 13: 355.

White, R.E. 1985. The influence of macropores on the transport of dissolved and suspended matter through soil. *Adv. Soil Sci.*, 3: 95.

Wood, E.F., K. Beven, M. Sivapalan, and L. Band. 1988. Effects of spatial variability and scale with implication to hydrology modeling. *J. Hydrol.*, 102: 29.

Worral, F. 2001. A molecular topology approach to predicting pesticide pollution of groundwater. *Environ. Sci. Technol.*, 35: 2282.

Wosten, J.H.M., Y.A. Pachepsky, and W.J. Rawls. 2001. Pedotransfer functions: bridging the gap between available basic soil data and missing soil hydraulic characteristics. *J. Hydrol.*, 251: 123.

3 Modeling as a Tool for the Characterization of Soil Water and Chemical Fate and Transport

Javier Álvarez-Benedí
Instituto Tecnológico Agrario de Castilla y León,
Valladolid, Spain

Rafael Muñoz-Carpena
IFAS/TREC, University of Florida, Homestead,
Florida, USA

Marnik Vanclooster
Catholic University of Louvain, Louvain-la-Neuve, Belgium

CONTENTS

1-5667-0657-2/05/$0.00+$1.50
© 2005 by CRC Press

3.1   INTRODUCTION

Modeling is at the core of any experimental characterization, and hence also the characterization of soil water and chemical fate and transport processes. The use of a conceptual model (rather than a numerical model) is necessary before undertaking any process characterization. Thus, from this point of view, modeling can be considered as the starting point of any soil characterization process. It is important to recognize that the end result of most of the characterization efforts is to describe how the processes will evolve in space and time through a mathematical model. Hence, while the characterization process as such is based on modeling, soil process simulation modeling also builds on the results of the soil characterization. Conceptual models, for instance, are often translated into mathematical models as a basis for the interpretation of soil observations, thereby supporting the process characterization. Modeling can further be used to optimize the characterization effort in terms of data collection quantity and quality (frequency of data needed, parameters to monitor, need to assess variability, etc.).

Chapter 2 established the need for field studies and outlined a number of preliminary issues, including the selection of a theoretical methodology that takes into account the purpose of characterization (research, consulting, etc.). This is, in fact, the core of the integrated approach for process characterization, as it requires the coupling of a model with a monitoring methodology, both designed as a trade-off between the desired degree of detail of process description and cost. No general guidelines can be given for shifting from a more generalized to a more detailed degree of process description in modeling processes. But as stated in Chapter 2, this will depend very much on the modeling objective, or, as stated in Chapter 1, it will be dictated by the spatial-temporal scale of process or the process description. Bouma (1997) illustrates this question with three examples:

1. Soil chemical phase transfer and transformation processes are often characterized under equilibrium conditions in the laboratory, which are quite different from the field conditions, where equilibria often do not occur. These experiments could provide a basic understanding of process characterization and approximate independent estimations of certain parameters, but will yield soil parameters that may be ineffective in predictive simulation modeling. Hence, it should be questioned when soil properties determined under equilibrium conditions in the laboratory should no longer be used in predictive modeling and when

characterization should shift to procedures based on larger *in situ* experiments performed under transient conditions.

2. Most simulation models for solute transport in soil implicitly assume soils to be homogeneous and isotropic, although field soils are rarely homogeneous and isotropic. More general approaches for representing heterogeneous systems include stochastic, stratified, and two-domain or dual-porosity approaches. Therefore, one should also question in which conditions more detailed and complicated solute transport models should be preferred.

3. Complex, deterministic models are often used to model solute fluxes within landscapes. However, taking into account the lack of data to feed part of these models, simpler modeling approaches could be considered in some conditions in a predictive simulation context.

Hence, the complexity of the modeling-based soil characterization procedure will depend on the modeling objectives and the accuracy required. For the description of nitrate fate and transport in soil, for example, the characterization of the hydrodynamic dispersion process will not be significant, given the low sensitivity of the simulated nitrate concentration profiles and fluxes to this parameter. On the other hand, when dealing with fate and transport at low concentration levels, as is the case when predicting pesticide transport at the ppm level, then the consideration of the hydrodynamic dispersion in the flow and transport model is essential. This illustrates clearly that the complexity of the model to be considered in the soil characterization process depends very much on the characterization and modeling objective.

Taking into account the characterization-modeling tandem as the basis for an integrated approach to process characterization, a previous modeling perspective must be given before introducing characterization strategies and methods of measurement. Thus, the purpose of this chapter is to develop a modeling framework for soil process description that will serve as the basis for characterization methods described in the next sections of this book. This will be done in three steps. First, a general classification of soil processes based on their conceptual and mathematical description will be presented. This classification includes equilibrium, reversible and irreversible rate processes, and transport processes. The starting points will be simple conceptualizations, and the discussion will be directed towards higher description levels, taking into account the constraints imposed by scale and variability. Second, as the transport mechanisms of the solvent and solute are different in nature, driving forces for water flow and solute movement will be described using different conceptual and mathematical models. For this reason, soil-water and soil-solute processes (including biological transformations) will be analyzed separately. Finally, the development of complex models is presented by integrating the description of several individual processes. Inverse modeling is introduced as a characterization tool that will be further developed in Chapter 20.

3.2 GENERAL CONCEPTUALIZATION OF SOIL PROCESSES

Jury and Flühler (1992) proposed five elements in a general soil chemical fate and transport model: (1) division of chemical mass into appropriate phases requiring separate description (e.g., the gaseous, the dissolved, the adsorbed, and the nonaqueous liquid phase); (2) interphase mass transfer; (3) intraphase mass transformation describing the rate of appearance or disappearance of mass per unit volume from the system; (4) mass conservation; and (5) mass transport according to a given flux law for each mobile phase. Considering the corresponding mathematical formulation of the above elements and taking into account the rate (temporal scale) and nature (mass transfer, thermodynamics, etc.), a general classification of individual soil process description for characterization purposes could be:

1. Local phase transfer and transformation process description: (a) *instantaneous equilibrium soil process description*: at a particular time scale, the rate of the process is fast enough to be considered instantaneous and reversible (e.g., instantaneous sorption of a chemical released in the soil solution at the liquid solid inter-phase); (b) *irreversible soil process description*: nonreversible rate-limited processes (e.g., hydrolysis of a chemical); (c) *reversible soil process description* reversible rate-limited processes (e.g., some sorption-desorption processes).
2. Transport process description: processes where spatial coordinates are required for the description. This group includes the description of several transport mechanisms that are different for solvent (water) and solute. These will be described separately in more detail in Sections 3.3 and 3.4.

It is interesting to note that a particular process can be considered instantaneous, rate limited, or stationary depending on the model time scale, and thus, the temporal scale will dictate the appropriate description for rated processes. Figure 3.1 shows different examples of rate processes simultaneously occurring at different time scales. When modeling the process represented by the central block (e.g., sorption), the process occurring at a lesser time scale can be considered as "instantaneous," while the process taking place at a higher time scale can be considered as "stationary."

Thus, a general chemical transport and transformation model will use these basic process descriptions as "building stones" to construct each of the five elements. For example, the inter-phase mass transfers can be described using equilibrium, kinetic, or mass transfer descriptions, while the reaction term can be described through rate equations (equilibrium, reversible, and irreversible kinetics). Finally, transport processes will combine mass conservation and flux laws for each phase. Each of these building stones will next be described in detail.

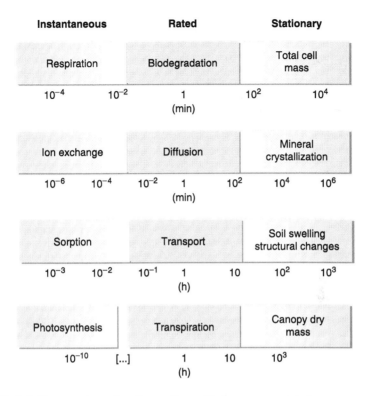

FIGURE 3.1 Time scales of soil reactions. Each row represents an example of three simultaneous processes occurring at different time scales. The central process can be considered as rate limited, the processes at smaller time scales (left) can be considered instantaneous when observed at the "central" time scale, and those occurring at larger time scales (right) can be considered as "stationary." Time scales are approximated.

3.2.1 INSTANTANEOUS EQUILIBRIUM

Equilibrium is rarely achieved in natural environments. However, this simple conceptualization sometimes gives reasonable approximations when characterizing soil processes. For example, a simple linear equilibrium model describing the partitioning of a solute between two phases is given by:

$$c_i = K_D c_{ii} \tag{3.1}$$

where c_i [e.g., M M^{-1}] and c_{ii} [e.g., M L^{-3}] are the concentrations in the two phases and K_D [e.g., L^3 M^{-1}] is the distribution or partition coefficient. This conceptualization can be used for describing liquid-solid, liquid-liquid, or liquid-gas partitioning. The above linear relationship for describing equilibrium is a simplified conceptualization that is usually valid only for a limited range of concentrations and temperatures. Other soil properties such as the soil pH or the ionic strength of the solution can limit the validity of the linear equilibrium for a soil chemical process. For example, Henry's law, which is the linear equilibrium model [Eq. (3.1)] for the liquid-gas phase transfer, is only valid for diluted solutions. In addition, the partition of a solute between two immiscible liquids can only be described successfully with a linear equilibrium model for a small range of concentrations. More complex equilibrium models can be obtained by using nonlinear models (such as the Freundlich sorption model described in Sec. 3.4) or the use of conceptual models based on thermodynamic theory, which consider activity coefficients, description of ionic interaction, or computation of the free Gibbs energy.

3.2.2   IRREVERSIBLE KINETICS

Soil processes are often rate limited due to the heterogeneity of the porous medium and the simultaneous occurrence of several coupled mechanisms. The kinetics of slow reactions act together with scale-dependent transport mechanisms (e.g., diffusion processes) that take place simultaneously, and thus the applicability of a particular model is restricted to the scale of use (e.g., batch, column, field, landscape, region, etc.).

This is the general case for batch methods, which can be used for characterizing process kinetics and equilibrium (or pseudo-equilibrium if equilibrium is not fully achieved). Batch methods for obtaining kinetic data present several advantages, including low-cost equipment, the elimination of several mechanisms by mixing such as film diffusion and sometimes particle diffusion, achievement and control of a constant soil-to-solution ratio, and control of reaction conditions (O_2, pH, etc.). However, these experiments also have a number of limitations: the values of several experimental variables can be far from those found in natural scenarios, the products are not removed and are allowed to accumulate in the closed system, etc. (Amacher, 1991).

In a general process A→B, rates can be expressed as a decrease in the concentration of the source A as a function of time, or an increase in the product formation as a function of time:

$$\text{Rate} = -\frac{dc_A}{dt} = \frac{dc_B}{dt} \qquad (3.2$$

where c_A and c_B are the concentration of reactant and products [$M\,L^{-3}$]; and *Rate* is the reaction rate [$M\,L^{-3}\,T^{-1}$]. The most frequently used kinetics are, by far, *first-order kinetics*, given by:

$$\frac{dc}{dt} = -k_1 c \qquad (3.3$$

where c [M L^{-3}] is the concentration and k_1 [T^{-1}] is the first-order kinetics constant. The corresponding integrated form is:

$$c = c_0 e^{-k_1 t} \tag{3.4}$$

where c_0 [M L^{-3}] is the initial concentration. The *half-life* $(t_{1/2})$ represents the time corresponding to a concentration $c = c_0/2$. Substituting this concentration in Eq. (3.4), the half-life for first-order kinetics is obtained:

$$t_{1/2} = \frac{\ln(2)}{k_1} \tag{3.5}$$

Also, the *zero-order kinetics* can represent a stationary source or sink of a chemical of concentration c:

$$\frac{dc}{dt} = \pm k_0 \tag{3.6}$$

where k_0 [M L^{-3} T^{-1}] is the zero-order rate parameter.

Table 3.1 summarizes the most frequently used kinetics equations used in soil science. The integrated forms are given for solute sinks (solute

---

TABLE 3.1
Equations for Extended Kinetics Models

Order	Differential	Integrated sinks, $-k$	Half-life	Integrated form sources, $+k$
		(initial condition: $t-0, C=C_0$)	(sinks)	(initial condition: $t=0, C=0$)
Zero-order	$\frac{dC}{dt} = \pm k$	$C = C_0 - kt$	$t_{1/2} = \frac{1}{2k}$	$C = kt$
First-order	$\frac{dC}{dt} = \pm kC$	$\frac{C}{C_0} = e^{(-kt)}$	$t_{1/2} = \frac{\ln(2)}{k}$	$\frac{C}{C_0} = 1 - e^{(-kt)}$
-Order	$\frac{dC}{dt} = \pm kC^n$	$\frac{1}{C^{n-1}} = \frac{1}{C_0^{n-1}} + (n-1)kt$	$t_{1/2} = \frac{(2^{n-1} - 1)}{C_0^{n-1}(n-1)k}$	
Elovich	$\frac{dC}{dt} = \alpha^{-\beta C}$			$C = \frac{1}{\beta}\ln(1 + \alpha\beta t)$
Elovich (simplified)*	$\frac{dC}{dt} = \alpha^{-\beta C}$			$C_t = a + b\ln t$
Power		$C = C_0 - at^b$	$t_{1/2} = \left(\frac{C_0}{2a}\right)^{1/b}$	$C_t = at^b$
Parabolic diffusion law		$C = C_0 - at^{1/2}$	$t_{1/2} = \sqrt{\frac{C_0}{2a}}$	$C = at^{1/2}$

* Assuming $\alpha\beta t \gg 1$, $b = 1/\beta$, $a = b\ln(\alpha\beta)$.

disappearance considering as initial condition $c = c_0$) and sources (considering $c = 0$ as initial condition). In the case of solute sinks, the variation of solute concentration c with time is a negative value $(-k)$. On the other hand, solute production from sources is represented mathematically by a positive variation of c with time $(+k)$. Power and parabolic diffusion law kinetics are generally enunciated directly in its integrated form. In fact, parabolic diffusion law can be considered as a singular case of the more generalized power kinetics (with $b = 1/2$). In addition, Elovich and simplified Elovich kinetics are reflected in Table 3.1.

In the simplest cases, the kinetics parameters are considered as constants. However, in many soil chemical fate and transport models, rate parameters are considered as variables. Variable rates have been applied to the case when several time scales are characterized simultaneously. For example, Bloom and Nater (1991) described the rate of mineral dissolution as weathering reactions proceed, with rates decreasing several orders of magnitude from the first day to several years. Another example is the reduction of the rate parameters for biologically mediated processes when soil environmental conditions (e.g., soil temperature, soil moisture, soil pH) will limit the rate of certain processes. Soil temperature dependency of biologically mediated rate parameters, for instance, can easily be encoded by multiplying the reference rate constant with a Q_{10} reduction function (e.g., Reduction $= Q_{10}^{[(T - T_{ref})/10]}$ in Vanclooster et al., 1996). Another popular concept is the use of the trapezoidal soil moisture–dependent reduction function, allowing reaction rate parameters to reduce when soil becomes too dry or too wet.

3.2.3 REVERSIBLE KINETICS

The above equations refer to irreversible processes, although they can be considered as a simplified form of the more generalized model of reversible kinetics. The rate of variation of the concentration for a given compound C will be proportional to the kinetics of solute loss plus the corresponding kinetics of solute increase. For example, in a reversible first-order kinetics sorption process, the rate of variation of the concentration of the dissolved phase (c) is given by:

$$\frac{\partial c}{\partial t} = -k_{ads}c + k_{des}\frac{\rho}{\theta}s \qquad (3.7$$

where c [M L^{-3}] is the concentration in the dissolved phase, s is the concentration in the sorbed phase [M M^{-1}], ρ is the bulk density of soil [M L^{-3}], θ is the soil water content [L^3 L^{-3}], and k_{ads} [T^{-1}] and k_{des} [T^{-1}] are the rate parameters for the sorption and desorption reactions, respectively.

Another example is the partitioning processes between two liquids, two regions, or two phases, which can be described by means of a mass transfer coefficient, assuming that the rate of exchange is proportional to the concentration difference between the two liquid regions:

$$\frac{dc}{dt} = \alpha(c_{II} - c_I) \tag{3.8}$$

where c_I and c_{II} are the concentrations in phases or regions I and II, respectively, and α is the mass transfer coefficient $[T^{-1}]$. This conceptual description has been used to describe mass transfer between water in different soil domains in physical nonequilibrium models (Sec. 3.4.2) and transfer of gases from soil to atmosphere (Sec. 3.4.5).

3.2.4 TRANSPORT

Mass flow in the liquid phase requires a different treatment for solvent (usually water) and solute, because solvent and solute transport in soil obey different driving forces. This holds also for mass flow in the gaseous phase.

Water flow in soil is mainly governed by capillarity (Ψ), gravity (z), pressure Ψ_p), and osmotic forces (Ψ_o), which are the dominant components of the total hydraulic head, H [L]. The total hydraulic head is then used for describing the liquid water flux, J_w $[LT^{-1}]$, which for one-dimensional flow yields:

$$J_w = -K\frac{\partial H}{\partial x} \tag{3.9}$$

where K is the unsaturated hydraulic conductivity function $[LT^{-1}]$, which also strongly depends on the water potential and properties of the porous media.

In the case of chemical transport in the soil liquid phase, the two dominant transport mechanisms are advection (transport of dissolved species due to a water flow) and dispersion (a macroscopic effect that accounts for microscopic variability in the flow field). Then, the total mass flow in the liquid phase, J_S $[M\ L^{-2}\ T^{-1}]$, can be described through the sum of each mass flow component:

$$J_S = J_c + J_D \tag{3.10}$$

with J_c $[M\ L^{-2}\ T^{-1}]$ the convective mass flux ($J_c = J_w\ C$), and J_D $[M\ L^{-2}\ T^{-1}$ the dispersive mass flux. Note that the total solute (J_S) and dispersive flux J_D) are expressed in terms of solute mass per cross-sectional area of soil and per unit of time $[M\ L^{-2}\ T^{-1}]$, whereas the water flux $[J_w$ in Eq. (3.9)] corresponds to the Darcian velocity $[LT^{-1}]$. The description of the chemical flux term J_S requires accounting for each possible transport mechanism. The dispersive flux per unit area of soil can be written as:

$$J_D = -\theta D\frac{\partial C}{\partial x} \tag{3.11}$$

where D is an effective dispersion coefficient (see Sec. 3.4.3). A description of other possible mechanisms can be found in Scanlon et al. (2002).

Within a continuum approach, the flux laws are further combined with the mass conservation equation to yield the governing soil flow and transport equations. The above descriptions of water and solute transport are the simplest classical conceptualizations, which consider the soil as a continuous homogeneous system. Sections 3.3 and 3.4 will discuss the main constraints of classical approaches and alternatives for description of soil-water and soil-solute.

3.3 SOIL-WATER TRANSPORT PROCESSES

3.3.1 CLASSICAL DESCRIPTION OF WATER MOVEMENT

Water movement in soils is generally described by assuming the soil to be a homogeneous (and rigid) medium. Assuming laminar one-dimensional flow and neglecting the inertial terms of the momentum conservation equation, the macroscopic water flux is given by Darcy's flow equation [Eq. (3.9)], which can be written for one-dimensional vertical flow in terms of the hydraulic head:

$$q = J_w = -K\frac{\partial H}{\partial z}$$

(3.12

where H [L] is the hydraulic head, and z [L] is the depth. In rigid unsaturated soils, the hydraulic head consists of the gravimetric potential or piezometric head z [L] and the pressure or suction head ψ [L], often referred to as the matric potential, which result from capillary forces. Equation (3.12) indicates that the water flow through the soil is in the direction of and proportional to the hydraulic gradient, which is the driving force, and proportional to the hydraulic conductivity, which is an intrinsic property of the medium. Darcy's law is valid only for laminar flow, since the linearity of flux versus hydraulic gradient fails at high flow velocities (where inertial forces are not negligible). In addition, another assumption of this equation is that the osmotic potential is not a significant mass transfer factor, which is generally acceptable at the macroscopic scale when the soil solution is diluted. Water movement in rigid, homogeneous, isotropic, variably saturated soil is described by combining mass conservation with Darcy's flow equation in an equation proposed by Richards (1931):

$$\frac{\partial \theta}{\partial t} = \frac{\partial}{\partial z}\left[K(\psi)\left(\frac{\partial \psi}{\partial z}+1\right)\right]$$

(3.13

where $\theta = \theta(\psi)$ is the volumetric water content [$L^3 L^{-3}$], t is time [T], and $K(\psi$ the unsaturated hydraulic conductivity [$L\ T^{-1}$]. Both $\theta(\psi)$ and $K(\psi)$ are functions of the matric potential head ψ. Equation (3.13) is known as the *mixed form*, as it contains two dependent variables θ and ψ. By defining the *hydraulic diffusivity* function $\zeta(\theta)$ $[L^2\, T^{-1}]$ as,

$$\zeta(\theta) = K(\theta)\frac{\partial\psi}{\partial\theta} \tag{3.14}$$

the *diffusive* or θ-*form* of Richards' equation is obtained:

$$\frac{\partial\theta}{\partial t} = \frac{\partial}{\partial z}\left[\zeta(\theta)\frac{\partial\theta}{\partial z} + K(\theta)\right] \tag{3.15}$$

In addition to the simplification of the equation with just one variable, $\zeta(\theta)$ varies less over the range of θ than $K(\theta)$, which improves the stability of the numerical solution of Richards' equation. The drawback of this form is that $\zeta(\theta)$ is only defined for unsaturated conditions, which makes the equation unusable for general conditions where the soil is saturated. An alternative formulation can be obtained by defining the *hydraulic capacity* function, $C(\psi)$ $[L^{-1}]$ as:

$$\frac{\partial\theta}{\partial t} = \frac{\partial\theta}{\partial\psi}\frac{\partial\psi}{\partial t} = C(\psi)\frac{\partial\psi}{\partial t} \tag{3.16}$$

yielding the *capacitive* or ψ-*form* of Richards' equation:

$$C(\psi)\frac{\partial\psi}{\partial t} = \frac{\partial}{\partial z}\left[K(\psi)\left(\frac{\partial\psi}{\partial z} + 1\right)\right] \tag{3.17}$$

Although this is a simpler formulation in terms of a single variable, it contains two rapidly varying functions in ψ (C and K), which complicates its solution and usually translates to poor numerical mass balance.

Celia et al. (1990) showed that the *mixed form* of the Richards' equation (Eq. 3.13), although somewhat more complex to solve numerically, presents the advantage of being valid for the entire soil moisture range (i.e., saturated and unsaturated conditions) and improves the mass balance of the numerical solution as compared to the other formulations.

Although Richards' equation provides a complete description of water flow in soils, there are important limitations. Although this was originally defined for homogeneous isotropic media, soil heterogeneity can be partially handled through different numerical methods by assigning different properties to the nodes of the numerical grid. The equation in its basic form will not be valid for nonrigid media (i.e., swelling soils). However, by introducing an overburden component in the total hydraulic head, flow in the matrix of nonrigid media can be described with a modified version of the Richards' equation (Kim et al., 1993). The equation will also not be valid when specific forms of preferential flow occur, such as macropore flow in well structured soils, by-pass flow in hydrophobic soils (highly organic or certain sandy soils), or funneled flow in sloping stratified soils. Another problem or limitation in applying the Richard's equation is the difficulty of correctly describing the intrinsic soil properties such as the $\theta(\psi)$ relationship (soil moisture retention characteristic) and the unsaturated hydraulic conductivity, $K(\psi)$. These two important soil hydraulic functions are presented in the next section.

3.3.2 CHARACTERIZATION OF WATER CONTENT–PRESSURE HEAD AND HYDRAULIC CONDUCTIVITY–PRESSURE HEAD RELATIONSHIPS

Soil volumetric moisture content θ [$L^3 L^{-3}$] and suction head ψ [L] are related by the soil water characteristic (soil moisture retention) curve, $\theta = \theta(\psi)$. Since suction is related to capillarity forces and these to pore diameter, soils with coarse structure (larger pore size in general) release moisture readily even at low suction values, where for the same suction values fine soils (small pores) retain most of the moisture. Thus, the shape of the nonlinear relationship depends largely on the soil pore size distribution and hence on soil texture (Figure 3.2). A similar effect, although more pronounced, can also be seen in the unsaturated conductivity function, $K(\psi)$ (Figure 3.2). However, for this latter relationship, soil structural properties related to the pore water connectivity and tortuosity will additionally determine the shape of the curve.

Several analytical expressions for $\theta(\psi)$ and $K(\psi)$ have been proposed in the literature (Table 3.2). The analytical description of these functions typically exhibits limitations close to saturation and dry soil conditions. Close

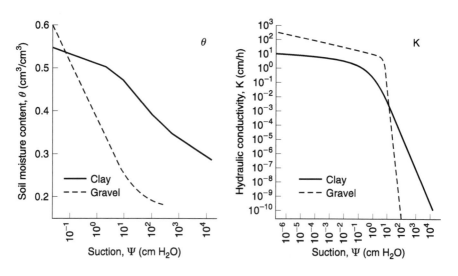

FIGURE 3.2 Effect of soil texture on soil water characteristics and unsaturated hydraulic conductivity functions.

TABLE 3.2
Soil Moisture Characteristic and Hydraulic Conductivity Equations

Type	Equation*	Parameters	Range (frequent)	Comments
Soil water characteristic function				
Hyperbolic	$\lvert\psi\rvert = a\theta^b$	a, b	—	General shape but does not fit well values close to saturation.
Brooks and Corey (1964)	$S_e = \left(\dfrac{\psi_{\neq}}{\psi}\right)^\lambda$	$\psi_{\neq}$ λ	Point at $d\theta/d\psi = 0$ 2–5 — (measured/fitted)	High values of λ indicate uniform pore distribution and low values wide range of pores sizes. Depending on soil type, does not fit well values close to saturation.
van Genuchten (1980)	$S_e = [1+(\alpha\psi)^n]^{-m}$ for $\psi \geq 0$	α n m (θ_r, θ_s)	0–1 (0.005–0.05 cm^{-1}) >1 (1.2–4) 0–1 (1 − 1/n) — (measured)	More parameters required but fits values close to saturation well. On the Mualem version ($m = 1 - 1/n$) and if θ_r and θ_s are measured, only 2 parameters.
Hydraulic conductivity function				
Leibenzon (1947)	$K_r = S_e^{n_i}$	n_i (θ_r, θ_s)	1–4 (3) $3+\frac{2}{\lambda}$ (Brooks Corey) — (measured/fitted)	Derived from Kozeny's principle. Works well for low permeable media (compacted clay). The Brooks and Corey exponent is more general and lends some physical significance to the parameter, although inherits some of its limitations.
Brooks and Corey	$K_r = \left(\dfrac{\psi_A}{\psi}\right)^m$	ψ_A m	Point at $d\theta/d\psi = 0$ 3–11	Frequently used. Derived from studies of capillarity in sands.
Gardner (1958)	$K_r = e^{\alpha\psi}$	α	0.1–0.01	α relates to soil texture and structure, Simple exponential formulation is easy to integrate and use in inverse problems. Usually only valid in the wet range up to a certain limit value.
Mualem (1976), van Genuchten (1980)	$K_r = S_e^{1/2}\left[1 - (1 - S_e^{1/m})^m\right]^2$	m n (θ_r, θ_s)	0–1 (1 − 1/n) >2 — (measured)	Works well with a wide range of soils. Parameters can be estimated graphically from Brooks and Corey values or by non-linear fitting.

*with $S_e = (\theta - \theta_r)/(\theta_s - \theta_r)$; and $K_r = K(\psi)/K_s$

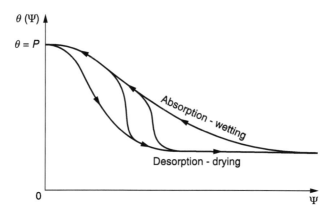

FIGURE 3.3 Hysteresis in the soil water characteristic curve.

to saturation, θ_s, some soils display inflexion points in the $\theta(\psi)$ and $K(\psi)$ relationships that are better described by alternative models (Table 3.2). The main problem in the dry range of the curve is related to the determination of the residual moisture content, θ_r [L^3 L^{-3}]. In some instances, if θ is estimated by fitting experimental soil moisture or hydraulic conductivity data, the values cannot be interpreted physically. A simple method for independently estimating θ_r has not been tested on a broad scale (Kutilek et al., 1994).

Another complicating factor is the occurrence of hysteresis in the soil moisture retention characteristic (Figure 3.3). The irregular pore geometry of natural soils and in particular solid-liquid interphase mechanisms on a microscopic scale results in different θ-ψ relationships for the draining and moistening phase of soils. Different conceptual models have been proposed to describe soil water hysteresis. Viaene et al. (1994), for example, reviewed different hysteresis models of the soil hydraulic functions. Among these, Mualem's model (Mualem, 1976) was found to be one of the most accurate. This is a two-branch conceptual model, where the soil hydraulic status is defined as a function of the two principal drying and wetting retention curves, each described by a set of Mualem–van Genuchten parameters (Table 3.2).

Critical comparisons of field methods to measure the $\theta(\psi)$ and $K(\psi$ relationships are presented in Chapter 5 and Chapter 6, respectively.

3.3.3 DUAL POROSITY MODELS

Soil homogeneity was one of the assumptions made by the Richards' equation in the classical description of water flow. A more realistic representation of soil water flow is given by a dual-porosity model, which accounts for two flow domains. The first domain comprises unsaturated flow, which is described by Richards' equation. The second flow domain accounts for the flow through macropores. When the soil is (nearly) saturated, vertical water flow in structured soils will be dominated by macropore flow, as these pores drain at low suctions. Although macropores represent a small part of the porosity, they induce preferential flow paths, allowing fast transfer of an important fraction of the flow (Bouma and Wosten, 1979; Luxmoore, 1981; Beven and Germann, 1982). The generation of robust methodologies for the *in situ* direct characterization of the macroporous flow domain of soils remains a challenging task. Indirect methods are often proposed where $K(\psi)$ or $K(\theta)$ is estimated from more easily measured soil properties, such as the retention curve (van Genuchten and Leij, 1992). Additional approaches have been developed for bi-modal porosity models exhibiting macropore flow (Othmer et al., 1991; Ross and Smettem, 1993; Durner, 1994) but remain subject to much uncertainty. Field methods for *in situ* measurement of soil hydraulic properties are discussed in Chapter 6, and preferential flow characterization is described in Chapter 8.

3.4   SOIL-SOLUTE TRANSPORT PROCESSES

The four major processes that control the movement of contaminants in porous media are advection (or convection), dispersion, interphase mass transfer, and reaction or transformation (Brusseau, 1994). Advection refers to the movement of a solute with the flowing water and is described by the water flux and dissolved solute concentration ($J_w C$). Dispersion represents the spreading of solute about a mean position, such as the center of mass and is a consequence of several transport mechanisms depending on the water regime. Phase changes include sorption, volatilization, and partitioning (e.g., two immiscible liquids such as octanol-water). Reaction can be conceptualized as a sink or source of the solute, expressed in terms of a rate equation.

3.4.1   CLASSICAL DESCRIPTION OF SOLUTE MOVEMENT

The simplest and classical approach describing "ideal" solute transport assumes that soil is homogeneous and rates of interphase mass transfer are fast enough to be considered as instantaneous (*local equilibrium assumption*). The expression of the one-dimensional advection-dispersion equation (ADE) for solute transport in a homogeneous soil is [from Eqs. (3.9) and (3.10), considering $J_S = J_w C + J_D$, and mass conservation: $\partial \theta c / \partial t = -\nabla J_S \pm \theta \Gamma$]:

$$\frac{\partial \theta c}{\partial t} = -\frac{\partial}{\partial x}\left(J_w c - \theta D \frac{\partial c}{\partial x}\right) + \theta \Gamma \qquad (3.18)$$

where θ is the water content [$L^3 L^{-3}$], c is the solute resident concentration [$M L^{-3}$], t is time [T], x is the distance [L], J_w is the water flux [$L T^{-1}$], D is the dispersion coefficient [$L^2 T^{-1}$], and $\theta \Gamma$ is the solute sink/source term

[M L^{-3} T^{-1}]. Equation (3.18) is also designated as the convective-dispersive equation (CDE).

When the water content is constant with time and space (e.g., under saturated conditions), the equation can be simplified as:

$$\frac{\partial c}{\partial t} = D\frac{\partial^2 c}{\partial x^2} - v\frac{\partial c}{\partial x} + \Gamma \qquad (3.19$$

where $v = J_w/\theta$ [L T^{-1}] is the average linear velocity of the fluid in the pores of the medium (*pore water velocity*). This equation is a second-order partial differential equation (classified as a parabolic equation). Analytical solutions of Eq. (3.19) for specific boundary conditions are described by van Genuchten and Alves (1982) and Leij and van Genuchten (2002). When the solute is sorbed to soil constituents, the mass conservation principle should take into account the total mass of solute $c_T = \theta c + \rho s$. Following the same mathematical derivation, the transport ADE (or CDE) equation for sorbed solutes is given by:

$$R\frac{\partial c}{\partial t} = D\frac{\partial^2 c}{\partial x^2} - v\frac{\partial c}{\partial x} + \Gamma \qquad (3.20$$

where R is the retardation factor ($R = 1 + \rho(\partial s/\partial c)/\theta$). Mathematical expressions of the retardation factor for linear and Freundlich isotherms are developed in Chapter 11.

3.4.2 NONEQUILIBRIUM MODELS

The governing transport model for solute transport in porous media (i.e., ADE) was based on assumptions that the porous medium was homogeneous and that the interphase mass transfers were linear and essentially instantaneous. Brusseau (1998) reviewed several nonideal transport analyses of reactive solutes in porous media. Four major factors were identified as responsible for nonideal transport:

1. Physical nonequilibrium: Most solute transport models consider that all soil water contributes to solute transport, while this is often not the case in structured or heterogeneous soils. Soil domains with minimal flow and advection originate in soil regions with smaller hydraulic conductivity.

2. Rate-limited sorption: Most field-scale solute transport models include the assumption that equilibrium for mass-transfer processes is attained instantaneously, while experimental evidence shows that sorption-desorption of many organic compounds can be significantly rate limited.

3. Nonideal sorption: Most solute transport models include the assumption of a linear equilibrium sorption isotherm. Yet many chemicals, especially in field conditions, do not obey this simplified isotherm relationship.
4. Field-scale heterogeneity: The influence of spatially variable hydraulic conductivity on water flow and solute transport at the field scale generates apparent values of longitudinal dispersivity which are much larger than those observed in soil columns.

The four above-mentioned factors can be described with different modeling approaches. The simplest conceptualization of physical nonequilibrium consists of dividing the soil flow domain into mobile (i.e., advective flow) and immobile soil regions. Thus, the total volumetric water content of the soil is divided into mobile and immobile (stagnant) fractions, and solute concentration is considered different for each region. This mobile-immobile concept was proposed by Deans (1963) and Coats and Smith (1964) and was later developed and applied in soil science by Skop and Warrick (1974), van Genuchten and Wierenga (1976), and Vanclooster et al. (1991), among others. Alternatively, the simplest conceptualization of solute nonequilibrium sorption models is the kinetic adsorption (one-site) model introduced by van Genuchten et al. (1974). Cameron and Klute (1977) applied a two–adsorption site model based on the different affinities of soil components to solutes. More sophisticated conceptualisations have been published, which include multi-reaction retention models and nonequilibrium based on maximum adsorption capacity (second-order models). These models were reviewed by Ma and Selim (1998).

The two-site and two-region nonequilibrium models have equivalent dimensionless formulations according to Nkedi-Kizza et al. (1984) and van Genuchten and Wagenet (1989):

$$\beta R \frac{\partial C_I}{\partial t} + (1 - \beta)R \frac{\partial C_{II}}{\partial T} = \frac{1}{P} \frac{\partial^2 C_I}{\partial X^2} - \frac{\partial C_I}{\partial X} - \xi C_I$$

$$(1 - \beta)R \frac{\partial C_{II}}{\partial T} = \omega(C_I - C_{II}) - \eta C_{II} \qquad (3.21)$$

where C_I and C_{II} are dimensionless equilibrium and nonequilibrium concentrations; T is dimensionless time, X is dimensionless distance, P is the Peclet number, ω is a dimensionless mass transfer coefficient, and the subscripts I and II refer to the equilibrium and nonequilibrium phases, respectively. R is the retardation factor. For the two-region model:

$$X = x/L, T = vt/L, P = vL/D, R = 1 + \rho_b K_D/\theta$$

$$\beta = \frac{\theta_m + f\rho_b K_D}{\theta + \rho_b K_D}, \omega = \frac{\alpha L}{\theta v}, C_I = \frac{c_m}{c_0}, C_{II} = \frac{c_{im}}{c_0} \qquad (3.22)$$

where x is distance [L]; v is the pore water velocity [L T^{-1}]; ρ_b and K_D are the soil bulk density [M L^{-3}] and distribution coefficient for linear sorption [L^3 M^{-1}]; f is the dimensionless fraction of sorption sites in equilibrium with the fluid of the mobile region; θ is the volumetric water content [L L^{-3}]; α is a mass transfer coefficient between the two regions [T^{-1}]; c_m, c_{im} are the resident concentration [M L^{-3}], respectively, in the mobile and the immobile soil region; L is the characteristic length [L]; and c_0 is the arbitrary characteristic concentration. The subscripts m and im denote mobile and immobile regions, respectively. ξ and η are the dimensionless degradation rate parameters:

$$\xi = \frac{L}{J_w}\left(\theta_m\mu_{lm} + f\rho K_D\mu_{sm}\right), \ \eta = \frac{L}{J_w}\left(\theta_{im}\mu_{lim} + (1-f)\rho K_D\mu_{sim}\right) \qquad (3.23$$

where μ is a first-order rate degradation coefficient [T^{-1}], lm and sm denote liquid and solid mobile phases, and lim and sim denote liquid and solid immobile phases, respectively.

3.4.3 Solute Dispersion

Soil (and heterogeneous porous media in general) is a complex and irregular distribution of voids and solids. Fried and Combarnous (1971) described three levels in the study of dispersion phenomena, which could be applied to the characterization of any physical property or process: a *local level*, where the physical quantity is described in an "infinitely small" volume element; a *fluid volume* level, which considers means of the corresponding local parameters over a pore or a set of pores; and a *macroscopic level*. The macroscopic level is only used in porous media, when a solid matrix exists, to define an equivalent continuum. Hence, assumptions of a geometrical structure of the porous media are not required. The parameters at this level are the averages of the corresponding local parameters taken over a finite volume of the medium.

Dispersion is due to a combination of both, a purely mechanical phenomenon and a physico-chemical phenomenon. The first phenomenon is denoted as *mechanical dispersion*, originated by boundary and geometrical effects when a fluid flows through a porous medium. Mechanical dispersion increases with the fluid velocity. The physical-chemical dispersion is referred to as *molecular diffusion*, which results from a chemical potential gradient. This mechanism is not dependent on fluid velocity (and therefore it exists even when there is no flow). However, the diffusion coefficient may depend on concentration if the viscosity of the mixture varies with concentration or if the mixture is not ideal (Harned and Owen, 1963). In addition, a rise in temperature will increase molecular agitation with consequent change of diffusion coefficients. In practice, diffusion in porous media has been described by the general diffusion equation with the introduction of an *effective diffusion coefficient*, which depends on the texture of the medium. This equation, based on Fick's first law of diffusion, is used for convenience to describe macroscopic solute flux, despite the conceptual differences between diffusion and dispersion.

Since it is impractical, if not impossible, to define flow at the microscopic scale, averaging of the water flow over a representative volume is necessary. This averaging introduces some uncertainty in the velocity, which implies an uncertainty as to the dispersion coefficient. The averaging, or the scale which is chosen to define the flow, can result in a scale and flow rate dependency of the mechanical dispersion coefficient or dispersivity (Pickens and Grisak, 1981; Morel-Seytoux and Nachabe, 1992; Logan, 1996; Vanderborght et al., 1997; Javaux and Vanclooster, 2002). These authors developed expressions for dispersivity as a function of the mean travel depth and flow rate. van Wesenbeeck and Kachanoski (1994) also reported a spatial scale dependence of dispersion measured in terms of variance of solute travel time.

When representing the dispersion coefficient versus mean velocity, several velocity ranges can be recognized:

1. Pure molecular diffusion: This regime occurs when the mean velocity is very small or in a fluid at rest. The porous media slows down the diffusion processes so that the effective dispersion coefficient is always lower than the molecular diffusion ($D/D_0 < 1$):

$$D = \frac{D_0}{T_p} \tag{3.24}$$

2. Superposition: The contribution of mechanical dispersion becomes of the same order as the molecular diffusion.
3. Major mechanical dispersion: The contribution of mechanical dispersion is predominant, but molecular diffusion cannot be neglected and reduces the effects of mechanical dispersion.

$$D = \frac{D_0}{T_p} + \lambda v \tag{3.25}$$

4. Pure mechanical dispersion: The influence of molecular diffusion becomes negligible.
5. Mechanical dispersion: the flow regime is out of the domain of Darcy's law.

For a one-dimensional system, the macroscopic (effective) dispersion coefficient is usually defined by means of Eq. (3.25) (Freeze and Cherry, 1979; Brusseau, 1993). In this equation, T_p is the dimensionless tortuosity factor (> 1), defined as $T_p = (L_{dif}/L)^2$ with L_{dif} [L] and L [L] being the actual and the shortest path lengths for diffusion. And λ is the (longitudinal) dispersivity [L]. It should be noticed that the above equation can be found as:

$$D = \tau D_0 + \lambda v \tag{3.26}$$

where τ is the *apparent tortuosity factor* (< 1) defined as $\tau = (L/L_{dif})^2$, and thus, terms such as *tortuosity* (defined as L_{dif}/L), *tortuosity factor*, and *apparent tortuosity factor* have not been consistently used in the literature (Leij and van Genuchten, 2002).

A second approach for modeling the effect of multiple sources of dispersion to the *lumped* coefficient consists of coupling additional coefficients to Eq. (3.24). Brusseau (1993) presented a modified version of the equation of Horvath and Lin (1976), written in terms of the lumped dispersion coefficient D^*:

$$D^* = \frac{D_0}{T_p} + \frac{\lambda v_m}{1 + [(6(1 - \theta_m)\delta)/(\theta_m d_n)]} + \frac{(d_n v_m)^2 F^2 \varepsilon \theta_i}{36 D_0 (1 - \theta_m)[1 + F(\theta_i)/(\theta_m)]^2}$$
$$+ \frac{(d_n v_m)^2 F \theta_i T_i}{60 D_0 \theta_m [1 + F(\theta_i)/(\theta_m)]^2} \tag{3.27}$$

where the four terms account for axial diffusion, hydrodynamic dispersion, film diffusion, and intra-particle diffusion, respectively, and where δ is the thickness of the stagnant water film surrounding the particles [L], ε is the porosity of the particles, d_n [L] is the nominal particle diameter, T_p is the tortuosity factor (interparticle), T_i is the intraparticle tortuosity factor, θ_m [L^3 L^{-3}] is the volumetric water content of the mobile domain, θ_i [L^3 L^{-3}] is the volumetric water content of the immobile domain (intraparticle), F is the fraction of the intraparticle porosity accessible by the solute, and v_m is the pore water velocity (J_w/θ_m).

Also, a *lumped* or apparent dispersivity (λ^*) could be defined as:

$$D^* = \lambda^* v_m \tag{3.28}$$

Dependence of the axial diffusion [i.e., the first term of Eq. (3.27)] with soil volumetric water content can be described through a dependence of the tortuosity factors on water contents (Bear, 1972; Šimůnek et al., 1999):

$$T_p = \frac{\theta^{7/3}}{\theta_s^2} \tag{3.29}$$

where θ and θ_s are the actual and the saturated volumetric water contents, respectively. Alternatively, Vanclooster et al., (1996) used an equation previously introduced by Kemper and Van Schaik (1966) to estimate the effective diffusion coefficient from the chemical diffusion coefficient [first term in Eq. (3.27)] and the following tortuosity factor:

$$T_p = \frac{\theta_m}{ae^{b\theta_m}} \tag{3.30}$$

However, Beven et al. (1993) reported differences in dispersivities and effective dispersion coefficients of more than four orders of magnitude, which reveals a tremendous difficulty in characterizing dispersion as a macroscopic (effective) process. It is not surprising, then, that several authors have suggested omitting the modeling of physical dispersion when using numerical methods with numerical dispersion in the solution. Bresler et al. (1982), when discussing the salt dynamics and distribution in fallow soils, observed that salt distribution profiles computed using both physical and numerical dispersion terms were similar to the profiles obtained by the numerical solution from a expression in which the dispersion terms were omitted. Analogous observations were made by García-Delgado et al. (1997). The role of dispersion in stochastic models has also been the focus of much discussion. For example, Isabel and Villeneuve (1991) observed that the dispersion coefficient had little influence on a stochastic convection-dispersion model. They found, through a set of numerical simulations, that the stochastic convection model produced numerically equivalent results under most soil conditions. Thus, the stochastic convection model was proposed to replace the stochastic convection-dispersion conceptualization in the modeling of solute transport at the field scale. Zhang et al. (1996) compared stochastic and deterministic models to simulate solute transport through the vadose zone at the field scale. Comparison of the simulation results with field data showed that the models described the mean concentration reasonably well without considering the pore scale dispersivity. However, the pore scale dispersivity had a significant impact on the estimation of the concentration variance. This will therefore have important consequences when predicting leaching fluxes, especially for trace elements. Indeed, at low concentration levels, fluxes will be extremely sensitive to variance of the resident concentration profile and therefore to the dispersion parameters. This is a serious point of concern when predicting, for instance, pesticide leaching in soils at the ppm level (Boesten, 2004).

3.4.4 SORPTION

The degree of accuracy required in the characterization of the sorption processes is dependent upon its interaction with other soil–solute processes. Among them, the most important processes are solute transformations (e.g., degradation) and volatilization (or other distribution between physical phases). The characterization of sorption is discussed in Chapter 12. Conceptual description of sorption requires accounting for kinetics and equilibrium, although the most simplistic assumption assumes only a conceptual model for equilibrium, e.g., the Freundlich model (Table 3.3).

A useful simplifying assumption of the Freundlich equation is given by isotherm linearity ($N_f = 1$, linear isotherm). The first concern, as stated before, is the role of kinetics in the scenario in which sorption processes take place, as equilibrium rarely, if ever, occurs in soil-solute experiments. A more accurate

TABLE 3.3
Equilibrium Isotherms Used for Characterizing Sorption Processes in Soil and Corresponding Description of Hysteresis for the Desorption Process*

Isotherm	Equation	Parameters	Hysteresis
Linear	$C_s = K_D C_e$	K_D	
Langmuir	$C_s = \dfrac{c_m K_l C_e}{1 + K_l C_e}$	c_m, K_l	
Freundlich	$C_s = K_f C_e^{N_f}$	K_f, N_f	$\dfrac{N_{f,ads}}{N_{f,des}} = a + b C_{s,max}^c$ (van Genuchten et al., 1974)
Modified Langmuir model	$C_s = \dfrac{c_m K_g e^{-2bC_s} C_e}{K_l e^{-2bC_s} C_e + (C_e/C_{me})^h}$	K_l, c_m, b, h	C_{me}, h

(Gu et al., 1994; Cartón et al., 1997)

*C_s: solute concentration in the sorbed phase; C_e: Solute concentration at equilibrium in the dissolved phase; K_D, K_f, K_l, K_g: equilibrium constants (for a given temperature); N_f: nonlinearity parameter of the Freundlich isotherm; $C_{s,max}$: maximum adsorbed concentration; C_{me}: equilibrium adsorbate concentration after adsorption. h: hysteresis coefficient ($h = 0$ for complete reversibility), c_m, a, b, c: isotherm parameters. The subscripts $_{ads}$ and $_{des}$ represent adsorption and desorption, respectively.

description of sorption would include a hysteresis effect, as represented in Figure 3.4.

Rambow and Lennartz (1994) evaluated the effect of different degrees of complexity for describing herbicide sorption in the estimation of leaching. Total atrazine discharge calculated with the linear isotherm was approximately three times larger than for the Freundlich and for the hysteresis version. These results demonstrate that careful consideration of the applied sorption-desorption assumptions must be taken when assessing the transport of chemicals in soils. Koskinen et al. (1979) and Brusseau and Rao (1989) suggested several possible mechanisms responsible for hysteresis: chemical precipitation, variation of the binding mechanism with time, incorporation of the solute into the soil matrix, and physical trapping. Also, several experimental artifacts can generate an "apparent hysteresis effect."

A higher degree of complexity for the description of sorption is given by considering temperature effects on sorption isotherm. Sorption is an exothermic process, and, depending on the magnitude of the sorption enthalpies, a decrease in sorption with temperature occurs. For example, Koskinen and Cheng (1983) and Cartón et al. (1997) found that low-magnitude values of the isosteric heats of adsorption revealed a slight increase in adsorption as temperature decreased. Finally, the effect of competitive solutes (co-solutes) in scenarios where competitive sorption could occur could be included in more complex descriptions of sorption.

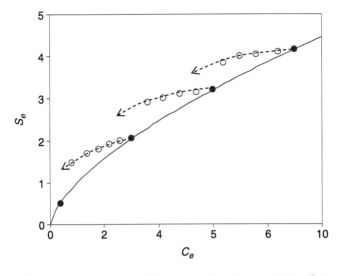

FIGURE 3.4 Sorption-desorption equilibrium for the dissolved (C_e) and sorbed (S_e solute concentrations in the presence of a hysteresis effect. Solid circles represent the sorption equilibrium points. Empty circles represent the desorption equilibrium starting from different initial concentrations at equilibrium.

3.4.5 VOLATILIZATION AND GAS SOLUBILITY

Volatilization and gas solubility can be conceptualized as solute exchanges from the liquid to the gas phase (volatilization) or vice versa (gas solubility), which take place in the bulk soil. This phenomenon should be taken into account when the process of solute transport in the gaseous phase is considered. Henry's law describes the distribution of a chemical between gaseous and liquid phases at equilibrium:

$$K_H = \frac{c_a}{c_w} \tag{3.31}$$

where K_H is the distribution coefficient (Henry's constant) and c_a and c_w are the concentrations of the solute in air and water, respectively. A similar expression could be used to represent solid and gas. Henry's law can be expressed in terms of partial pressure of the chemical in the gaseous phase (p_a):

$$K_H^p = \frac{p_a}{c_w} \tag{3.32}$$

The equivalence between the two expressions is given by:

$$K_H = \frac{K_H^p}{RT} \tag{3.33}$$

where R is the gas constant [8.3 kJ/mol °K] and T [K] is temperature. Care must be taken, because sometimes Henry's law is expressed in terms of gas solubility instead of solute volatility, and thus Eq. (3.33) is inverted (i.e., expressed as a ratio of water concentrations divided by air concentrations or partial pressures). Henry's law assumes equilibrium and is therefore applicable only at small scales under laboratory-controlled conditions. K_H increases with increasing temperature (Bamford et al., 1999). A simple description of temperature dependence is given by Sander (1999):

$$K_H = K_H^{ref}\left[\exp\left(\frac{\Delta H_{soln}}{R}\left(\frac{1}{T}-\frac{1}{T^{ref}}\right)\right)\right]^{-1} \tag{3.34}$$

where ΔH_{soln} is the enthalpy of solution.

Volatilization processes can be studied in chambers at laboratory scale under controlled conditions of temperature, wind speed, and air humidity. In these scenarios, volatilization and sorption processes are two dominating processes that can be described with a conceptual coupled model (Álvarez-Benedí et al., 1999). Although sorption-desorption processes and volatilization can be described with equilibrium equations as given by the sorption isotherm and Henry's law, kinetics or mass transfer equations should be used to account for the limited availability of the desorbed chemical as well as the volatilization rate. Thus, the boundary condition at the soil surface can be described by a mass-transfer equation:

$$J_o = -k_{atm}\left(c_{atm} - c_g\right)\big|_{z=0} \tag{3.35}$$

where J_o [M L^{-2} T^{-1}] is the solute flux to the atmosphere, k_{atm} is a mass transfer coefficient analogous to the one used in water evaporation (Brussaert, 1975), and c_{atm} is the solute concentration in the atmosphere (which can be assumed negligible in certain experimental conditions, such as wind tunnels with an external input air stream). The mass balance for a sorbed solute in soil under volatilization, considering $c_{atm}=0$, could be written as:

$$\theta\frac{\partial c}{\partial t} + \rho\frac{\partial s}{\partial t} = -k_{atm}\theta c \tag{3.36}$$

A major disadvantage of this formulation is the strong dependence of the kinetics constant k_{atm} with experimental conditions. The above conceptualization is very useful for characterizing the sorption-volatilization coupled processes at the laboratory scale, as a tool to study the kinetics and reversibility of sorption-desorption processes, and also to quantify the relative importance of volatilization under varying conditions of soil and air humidity and temperature. However, at larger scales and considering field environments, there are several experimental variables with simultaneous influence on volatilization that further complicate the process.

Most techniques for characterizing the gas exchange at the soil surface can be categorized using a soil mass balance, chambers, or micrometeorological methods. Chapter 13 studies the application of an appropriate methodology taking into account the availability of equipment, sample analysis capacity, resolution of sensors over the sampling period, treatment plot size or sampling interval. Micrometeorological techniques employ a combination of atmospheric turbulence theory and measurement to estimate gas flux to or from a surface. These techniques allow near-continuous flux estimates; some techniques also allow temporal averaging through sampling accumulation. Micrometeorological techniques can accommodate a wide range of plot sizes. The aerodynamic technique consists of the measurement of gas concentration and wind speed at two (or more) different heights. Other approaches reviewed in Chapter 13 are the Bowen ratio–energy balance technique (which does not require a wind speed profile) and other available alternatives based on different conceptual models and measurement strategies.

3.4.6 TRANSFORMATION

Despite the enormous variety of different possible solute transformations, the strategy of characterization follows the same methodological objectives, which consist of the estimation of the reaction pathways and rates of transformation (kinetics) for each solute studied. First, reaction pathways must be known. Second, data on solute concentrations for each chemical species must be determined with time in order to finally define a kinetics conceptual model. Concentration-time relationships can be obtained in incubation studies by chemical analysis or using radiolabeled compounds. For example, pesticide transformations can be represented as a simple linear pathway (Wagenet and Hutson, 1987):

$$
\begin{array}{ccccc}
\textit{Parent} & & \textit{Daughter} & & \textit{Daughter} \\
& \rightarrow & & \rightarrow & \\
\textit{Pesticide} & & \textit{Product}1 & & \textit{Product}2
\end{array}
\qquad (3.37)
$$

Note that each of the above chemical species can be involved in sorption and volatilization processes. Thus, it is important to simplify the transformation scheme as much as possible when working with field scenarios. Again, a trade-off between the degree of the complexity of description and the needs for data must be achieved in order to finally select a reasonable conceptual model and the corresponding sampling strategy.

Most of the chemical and physically mediated transformation processes can be well described by zero- or first-order kinetics. Rate equations for these processes have been summarized in Table 3.1. Microbial mediated processes, however, present a more complex component in which microbial biomass and activity can require a higher degree of description complexity. In this case, the rate constants are a function of microbial biomass and activity, and these in turn are a function of several additional environmental factors such as soil water content, temperature, aeration, pH, substrate availability, etc. Hence, it is important to have a conceptual model for microbial growth, maintenance, and decay (Starr, 1983). If a population of microorganisms uses a compound as its source of energy (and C), its population will be dependent on substrate concentration. At low concentrations the growth rate of the microorganisms would be slow because it is limited by the availability of substrate. Conversely, the growth rate will increase with concentration until a maximum level of growth is achieved. This conceptualization was formulated by Monod (1949):

$$\mu_g = \frac{\mu_{max} c}{K_C + c} \tag{3.38}$$

where μ_g is the specific growth rate [M T^{-1}] of the bacterium, μ_{max} is the maximum growth rate [M T^{-1}], c is the substrate concentration [M L^{-3}], and K_C [M L^{-3}] is a constant representing the substrate concentration at $\mu = \mu_{max}/2$.

In general, the rate that a given substrate c is utilized by microorganisms in soil can be described as (McLaren, 1973):

$$\frac{\partial c}{\partial t} = -A \frac{dm}{dt} - \alpha_b m - \beta_b m \tag{3.39}$$

where m is the biomass of the microbes [M] and A [L^{-3}], α_b [$L^{-3} T^{-1}$], and β_b [$L^{-3} T^{-1}$] are coefficients related to growth, maintenance and waste, respectively. If a steady state is assumed and the biomass m is constant, the expression may reduce to the Michaelis-Menten rate:

$$\frac{\partial c}{\partial t} = -\frac{kmc}{K + c} \tag{3.40}$$

where k and K are the Michaelis-Menten rate constants. Note that $K \gg C$ gives a first-order kinetics and $K \ll c$ is equivalent to a zero-order kinetics. As stated by Starr (1983), a major factor limiting the use of such conceptual models in the field is the lack of input data.

Some examples of the above-mentioned concepts are given below. Estrella et al. (1993) applied a nonlinear regression analysis of the simultaneous solution of the Monod equation for growth of cell mass (X), substrate utilization, and CO_2 production:

$$\frac{\partial X}{\partial t} = \mu_m \cdot \frac{XC}{K_{hs} + C} - k_{dr} X$$

$$\frac{\partial C}{\partial t} = -\frac{1}{Y} \cdot \mu_m \cdot \frac{XC}{K_{hs} + C} \tag{3.41}$$

$$\frac{\partial CO_2}{\partial t} = Y_{CO_{2substrate}} \left(-\frac{\partial C}{\partial t} \right) + Y_{CO_{2endogenous}} k_{dr} X$$

where X is the cell mass concentration, t is time, C is the solution phase concentration of substrate, μ_m is the maximum specific growth rate, K_{hs} is the half-saturation constant, k_{dr} is the cell death rate coefficient, Y is the cell mass yield from substrate degradation, $Y_{CO_{2substrate}}$ is the CO_2 mass yield from substrate degradation, and $Y_{CO_{2endogenous}}$ is the CO_2 mass yield from endogenous decay of cells.

The complexity level of the selected transformation model will strongly depend on the desired level of explanation. For describing nitrogen turnover in soil for instance, a simple first-order decay model might give a good description of the mineralization processes shortly after the addition of organic matter. However, this model might fail to describe the long-term mineralization process involving decomposition of the organic matter fractions (Vanclooster et al., 1996). On the other hand, a more detailed description reduces the applicability for extrapolation purposes and requires a greater experimental effort for characterization. A long-term approach for nitrogen transformations and resulting mineralization of organic carbon was developed by Johnsson et al. (1987). The conceptualization is based on a three-pool concept, which consists of (1) organic matter/microbial biomass complex (soil litter pool), (2) receiving fresh organic matter, and (3) slow cycling pool of stabilized decomposed products (soil humus pool). This representation was used by Vereecken et al. (1990, 1991) and was further included in the WAVE model (Vanclooster et al. 1996).

In conclusion, each microbiological transformation scenario should be characterized with a particular degree of description complexity. The fundamental aspects related to the characterization of soil microbiological processes are reviewed in Chapter 15, including methods for soil microbial characterization, sampling, soil handling, and choice of methods.

3.5 MODELING SOIL PROCESSES

3.5.1 Building Soil Process Models

The construction of more complete models is generally based on the above equations, which can be considered "pieces" for more sophisticated conceptualization. The simplest case would be an equilibrium description for solute sorption, in which the solute concentrations at equilibrium in the dissolved (C_e) and sorbed (S_e) phases are related through a sorption isotherm. This model can be represented by two boxes for C_e and S_e, respectively, and an equilibrium relationship between them (Figure 3.5). The equilibrium could be given by any of the equations presented in Table 3.3. Similarly, a rate-limited reversible sorption is described by a kinetic model considering the rate of approach towards equilibrium between C_e and S_e (e.g., Hornsby and Davidson, 1973). In this case, a double arrow in Figure 3.5 represents the reversible kinetics process.

The two–region model considers mass transfer between two flow domains (mobile with a solute concentration C_m, and immobile with a solute

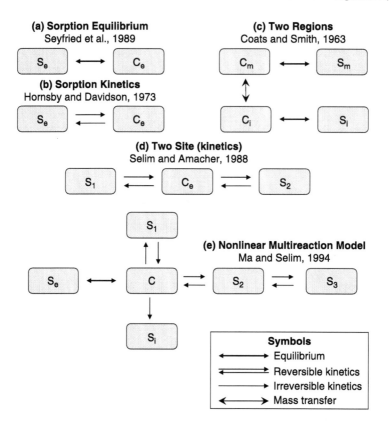

FIGURE 3.5 Schematic of some conceptual models for sorption processes: (a) sorption equilibrium (Seyfried et al., 1989); (b) sorption kinetics (Hornsby and Davidson, 1973), (c) two region (Coats and Smith, 1963), (d) two site (kinetics) (Selim and Amacher, 1988), (e) nonlinear multireaction model (Ma and Selim, 1994).

concentration C_{im}), where both domains are assumed to have the same retention mechanism (e.g., Coats and Smith, 1963). This conceptualization requires four boxes: mobile and immobile concentrations of sorbed and dissolved solutes, a mass transfer relationship between dissolved mobile and immobile solute, and sorption equilibrium for relating sorbed and dissolved solute amounts (Figure 3.5c). A two-site kinetic sorption model considering different rate-limited sorption with soil sites of type I and II (Selim and Amacher, 1988) is also depicted in Figure 3.5d. Brusseau et al. (1989) developed a transport model including two reversible kinetics, two irreversible kinetics, and a mass transfer between six different domains. An alternative generalized multireaction model was also described by Ma and Selim (1998), which considers five possible processes (Figure 3.5e). Depending on the purpose of a specific study and/or the data available, the appropriate process can be selected.

3.5.2 INVERSE CHARACTERIZATION OF SOIL PROCESSES

If a particular soil process can be described by a particular forward simulation model, then inverse methods can be used to characterize the soil process properties. Inverse methods are now becoming popular tools to identify the soil flow and transport properties. For example, Dane and Hruska (1983) used a numerical solution of Richards' equation assuming the equations of van Genuchten (1980) for $\theta(\psi)$ and $K(\psi)$. The solution was optimized to find α and n in these equations. This procedure was also applied in layered soils by Lambot et al. (2002), and Ritter et al. (2003, 2004) comparing several methodological approaches for characterizing hydraulic parameters in a layered soil.

The required premises to apply the inverse estimation are identifiability and uniqueness. Different parameter sets must lead to different solutions. If not, the parameters are unidentifiable. Additionally, nonuniqueness appears when one parameter set results in more than one solution. Care must also be taken with the simultaneous optimization of several model parameters of a physically based model as the curve can be fitted by a set of parameters that do not correspond with the true set (global minimum). In such a case, the model parameters are no longer representing the physics of the soil system. For example, the use of single parameter model to systems affected by at least more than one parameter will yield a lumped empirical single parameter. The single parameter in this case will only be valid for that specific situation. For instance, this is the case when fitting the governing ADE solute transport model with local equilibrium assumption to data from transport experiments affected by nonequilibrium flow. In this case the dispersion parameter will be a lumped parameter encompassing some of the terms of Eq. (3.26). The apparent hydrodynamic dispersivity inferred from such an experiment is not necessarily a measure of the microscopic variability of the flow field in the soil matrix.

Additional problems occur when parameters are correlated. For example, the parameters of the nonequilibrium two-region model are often strongly interdependent. This strong correlation between parameters implies that changes in one parameter can be compensated by corresponding changes in the correlated parameter. Thus, a simultaneous optimization of all of the model parameters will result in "shape parameter values" without a physical plausibility (Koch and Flühler, 1993).

Chapter 20 presents a general discussion on the use of a inverse estimation methodology. The application of inverse techniques to estimate hydraulic properties are discussed in Chapter 5, and Chapter 11 shows the application to estimate solute transport properties.

ACKNOWLEDGMENTS

This chapter was supported by the Florida Agricultural Experiment Station and approved for publication as Journal Series No. R-10129.

NOTATION

c	Concentration (dissolved phase) $[M\ L^{-3}]$
c_A, c_B	Concentration of reactant and products $[M\ L^{-3}]$
c_i, c_{ii}	Concentration of a partitioning solute in phases I and II $[M\ L^{-3}$
c_a, c_w	Concentration of a volatile solute in air and water $[M\ L^{-3}]$
c_I, c_{II}	Concentrations in phases or region I and II, respectively $[M\ L^{-3}$
C_I, C_{II}	Dimensionless concentration in regions I and II $(C_I = c_I/c_0$ $C_{II} = c_{II}/c_0)$ $[-]$
c_m, c_{im}	Resident concentration in the mobile and immobile regions $[M\ L^{-3}]$
c_0	Characteristic concentration $[M\ L^{-3}]$
c_T	Total mass of solute $[ML^{-3}]$
$C(\psi)$	Hydraulic capacity function $[L^{-1}]$
D	Effective dispersion coefficient $[L^2\ T^{-1}]$
H	Hydraulic head $[L]$
F	Fraction of the intraparticle porosity accessible by the solute $[-]$
J_c	Convective mass flux $(J_c = J_w\ C)$ $[M\ L^{-2}\ T^{-1}]$
J_D	Dispersive mass flux $[M\ L^{-2}\ T^{-1}]$.
J_o	Solute flux of a volatile compound to the atmosphere $[M\ L^{-2}\ T^{-1}]$
J_w	Water flux $[LT^{-1}]$
J_S	Total mass flow in the liquid phase $[M\ L^{-2}\ T^{-1}]$
k_0	Zero-order rate parameter $[M\ L^{-3}\ T^{-1}]$
k_1	First-order kinetics constant $[T^{-1}]$
k_{ads}	Rate parameters for the sorption reaction $[T^{-1}]$
k_{des}	Rate parameters for the desorption reaction $[T^{-1}]$
K	Unsaturated hydraulic conductivity $[LT^{-1}]$
K_D	Distribution or partition coefficient
K_H	Henry's distribution coefficient
L	Characteristic length $[L]$
P	Peclet (vL/D) $[-]$
R	Retardation factor $[-]$
s	Concentration in the sorbed phase $[M\ M^{-1}]$
	Time $[T]$
T	Number of pore volumes (vt/L)
T_p	Tortuosity factor $[-]$
$_{1/2}$	Half-life $[T]$
x	Distance $[L]$
X	Dimensionless distance $(X = x/L)$ $[-]$
z	Depth $[L]$

GREEK

Γ	Solute sink/source term $[M\ L^{-3}\ T^{-1}]$
α	Mass transfer coefficient $[T^{-1}]$

	Thickness of the stagnant water film surrounding (soil) particles [L]
ε	Porosity of the particles $[-]$
λ	Dispersivity [L]
μ	First-order rate degradation coefficient $[T^{-1}]$
v_m	Pore water velocity of the mobile region $(v_m = J_w/\theta_m)$ $[L\ T^{-1}]$
v	Average linear velocity $(v = J_w/\theta)$ $[L\ T^{-1}]$
θ	Soil volumetric water content $[L^3\ L^{-3}]$
θ_m	Volumetric water content of the mobile domain $[L^3\ L^{-3}]$
θ_i	Volumetric water content of the immobile domain $[L^3\ L^{-3}]$
ρ	Bulk density of soil $[M\ L^{-3}]$
	Apparent tortuosity factor $\tau = (L/L_{dif})^2$ $[-]$
ψ	Pressure or suction head [L]
$\zeta(\theta)$	Hydraulic diffusivity function $[L^2\ T^{-1}]$
$[-]$	Dimensionless

REFERENCES

Álvarez-Benedí, J., Tabernero, M.T., and S. Bolado. 1999. A coupled model representing volatilisation and sorption of soil incoporated herbicides. *Chemosphere*, 38,7,1583–1593.

Amacher, M.C. 1991. Methods of obtaining and analyzing kinetic data. In Sparks, D.L. and Suarez D.L. (eds.), Rates of Soil Chemical Processes. SSSA Special Publication Number 27, 19–59, Soil Science Society of America, Madison, WI.

Bamford, H.A., Porter, D.L., and J.E. Baker. 1999. Temperature dependence of Henry's law constants of thirteen polycyclic aromatic hydrocarbons between 4°C and 31°C. *Environ. Toxicol. Chem.*, 18, 1905–1912.

Bear, J., 1972. Dynamics of Fluid in Porous Media. Elsevier, New York.

Beven, K., and P. Germann, 1982. Macropores and water flow in soils. *Water Resour. Res.*, 18, 1311–1325.

Beven, K.J., Henderson, D.E., and A.D. Reeves. 1993. Dispersion parameters for undisturbed partially saturated soil, *J. Hydrol.*, 143, 19–43.

Bloom P.R, and E.A. Nater. 1991. Kinetics of dissolution of oxide and primary silicate minerals, in: Rates of Soil Chemical Processes, Sparks, D.L. and D.L. Suarez (*Eds.*), SSSA Special Publication Number 27, Soil Science Society of America, Madison, WI, 151–189.

Boesten J., 2004. Influence of dispersion length on leaching calculated with PEARL, PELMO and PRZM for FOCUS groundwater scenarios. *Pest Managm. Sci* 60(10):971–980.

Bouma J. 1997. Soil environmental quality: a European perspective. *J. Environ. Qual.* 26, 26–31.

Jury W.A., and H. Flühler. 1992. Transport of chemiclas through soil: mechanisms, models, and field applications. *Adv. Agron.*, 47.

Bouma, J., and J.H.M. Wosten. 1979. Flow patterns during extended saturated flow in two undisturbed swelling clay soils with different macropores. *Soil Sci. Soc. Am. J.*, 43, 16–22.

Bressler, E., McNeal, N.L., and D.L. Carter. 1982. Saline and Sodic Soils, Principles-Dynamics-Modeling. Springer-Verlag, New York.

Brooks, R.H. and A.T. Corey. 1964. Hydraulic properties of porous media. Hydrology Paper no. 3, Civil Engineering Dept. Colorado State Univ., Fort Collins, COL.

Brusseau, M.L. 1993. The influence of solute size, pore water velocity, and intraparticle porosity on solute dispersion and transport in soil. *Water Resour.Res.* 29, 1071–1080.

Brusseau, M.L. 1994. Transport of reactive contaminants in heterogeneous porous media. *Rev. Geophys.,* 32,3,285–313.

Brusseau, M.L. 1998. Multiprocess nonequilibrium and nonideal transport of solutes in porous media. In: Selim H.M., and Ma, L. (eds.), Physical Nonequilibrium in Soils, Modeling and Application. Ann Arbor Press, Chelsea, pp. 83–115.

Brusseau, M.L., Jessup, R.E., and P.S.C. Rao, 1989. Modeling the transport of solutes influenced by multiprocess Nonequilibrium. *Water Resour. Res.,* 25, 1971–1988.

Brusseau M.L., and P.S.C. Rao. 1989. Sorption nonideality during organic contaminant transport in porous media. *CRC Crit. Rev. Environ. Cont.,* 19, 33–99.

Brutsaert, W. 1975. A theory for local evaporation (or heat transfer) from rough and smooth surfaces at ground level. *Water Resour. Res.,* 4, 543–550.

Cameron, D.R., and A. Klute. 1977. Convective-dispersive solute transport with a combined equilibrium and kinetic adsorption model. *Water. Resour. Res.,* 13, 183–188.

Cartón, A., Isla, T., and J. Álvarez-Benedí. 1997. Sorption-desorption of imazamethabenz on three Spanish soils. *J. Agric. Food Chem.,* 45, 1454–1458.

Celia, M.A., Bouloutas, E.T. and R.L. Zabra. 1990. A general mass-conservative numerical solution for the unsaturated flow equation. *Water Resour. Res.,* 26, 1483–1496.

Coats K.H., and B.D. Smith. 1964. Dead-end pore volume and dispersion in porous media. *Soc. Pet. Eng. J.,* 4, 73–84.

Deans, H.H. 1963. A mathematical model for dispersión in the direction of flor in porous media. *Soc. Pet. Eng. J.,* 3, 49–52.

Durner, W. 1994. Hydraulic conductivity estimation for soils with heterogeneous pore structure. *Water Resour. Res.,* 30, 211–223.

Estrella M.R., Brusseau, M.L., Maier, R.S., Pepper I.L., Wierenga, P.J., and R.M. Millar, 1993. Biodegradation, sorption, and transport of 2,4-dichlorophenoxyacetic acid in saturated and unsaturated soils. *Appl. Environ. Microbiol.,* 59, 4266–4273.

Freeze, R.A., and J.A. Cherry. 1979. Groundwater. Prentice-Hall, Englewood, NJ.

Fried J.J., and M.A. Combarnous, 1971. Dispersion in porous media, in Chow V.T., ed., Advances in Hydroscience. Academic Press, New York, pp. 169–282.

García-Delgado, R., and A.D. Koussis, 1997. Grownd-water solute transport with hydrogeochemical reactions. *Ground Water,* 35, 243–249.

Gardner, W.R. 1958. Some steady-state solutions of the unsaturated moisture flow equation with application to evaporation from a water table. *Soil Sci.,* 85:228–232.

Gu, B., Schmitt, J., Chen, Z., Llang, L., and J.F. McCarty. 1994. Adsorption and desorption of natural organic matter on iron oxide: mechanisms and models. *Environ. Sci. Technol.,* 28, 38–46.

Harned, H.S., and B.B. Owen. 1963. The Physical Chemistry of Electrolytic Solutions. Reinhold, New York.

Hornsby, A.G., and J.M. Davidson. 1973. Solution and adsorbed fluometuron concentration distribution in a water-saturated soil: experimental and predicted evaluation. *Soil Sci. Soc. Am. Proc.* 37, 983–828.

Horvath, C., and H. Lin. 1976. Movement and band spreading of unsorbed solutes in liquid chromatography. *J. Chromatogr.*, 126, 401–420.

Isabel, D., and J.P. Villeneuve. 1991. Significance of the dispersion coefficient in the stochastic modeling of pesticides transport in the unsaturated zone. *Ecol. Modelling*, 59, 1–10.

Javaux M. and M. Vanclooster. 2003. Scale and rate dependent solute transport in an unsaturated sandy soil monolith. *Soil Sci. Soc. Am. J.*, 67, 1334–1343.

Johnson, D.S., Bergstrom, P., Jansson, P., and K. Paustiaen. 1987. Simulating nitrogen dynamics and losses in a layered agricultural soil. *Agric. Ecosystems Environ.* 18, 333–356.

Kemper, W.D., and J.C. Van Schaik. 1966. Diffusion of salts in clay water systems. *Soil Sci. Soc. Am. Proc.*, 30, 534–540.

Kim D.J., Vereecken, H., Feyen, J., Vanclooster, M., and L. Stroosnijder. 1993. A numerical model of water movement and soil deformation in a ripening clay soil. *Modelling Geo-Biosphere Processes*, 1:185–203.

Koch, S., and H. Flühler, 1993. Solute transport in aggregated porous media: comparing model independent and dependent parameter estimation. *Water Air Soil Poll.*, 68, 275–289.

Koskinen, W.C, O'Connor, G.A., and Cheng, H.H. 1979. Characterization of hysteresis in the desorption of 2,5,6-T from soils. *Soil Sci. Soc. Am. J.*, 43, 871–874.

Koskinen, W.C., and Cheng, H.H. 1983. Effects of experimental variables on 2,4,5-T adsorption-desorption in soil. *J. Environ. Qual.*, 12, 325–330.

Kutílek, M., and D.R. Nielsen. 1994. Soil Hydrology. Cremlingen-Destedt: Catena Verlag.

Lambot, S., Javaux, M., Hupet, F., and M. Vanclooster. 2002. A global multilevel coordinate search procedure for estimating the unsaturated soil hydraulic properties. *Water Resour. Res.*, 6, 1– 15.

Leibenzon, L.S. 1947. Flow of Natural Liquids and Gases in Porous Medium (in Russian). Gostekhizdat, Moscow.

Leij, F.J., and M. Th. van Genuchten. 2002. Solute transport, in Warrick, A.W. (ed.), Soil Physics Companion. CRC Press, Boca Raton, FL, 189–248.

Leistra M. 1986. Modelling the behaviour of organic chemicals in soil and ground water. *Pestic. Sci.*, 17, 256 254.

Logan, J.D. 1996. Solute transport in porous media with scale-dependent dispersion and periodic boundary conditions, *J. Hydrol.*, 184, 261–276.

Luxmoore, R.J. 1981. Micro-, meso-, and macroporosity of soil. *Soil. Sci. Soc. Am. J.*, 45, 671–672.

Ma, L., and H.M. Selim. 1994. Predicting the transport of atrazine in soils: second-order and multireaction approaches. *Water Resour. Res.*, 30, 3489–3498.

Ma, L., and H.M. Selim. 1998. Coupling of retention approaches to physical non-equilibrium models, in H.M. Selim, and Ma, L. (eds.), Physical Nonequilibrium in Soils, Modeling and Application. Ann Arbor Press, Chelsea, pp. 83–115.

Mclaren, A.D. 1973. A need for counting microorganisms in soil mineral cycles. *Environ. Lett.*, 5:143–154.

Monod, J. 1949. The growth of bacterial cultures. *Annu. Rev. Microbiol.*, 3, 371–394.

Morel-Seytoux, H. and M. Nachabe. 1992. An effective scale-dependent dispersivity deduced from a purely convective flow field. *Hydrol. Sci. J.*, 37, 2, 4, 93–104.

Mualem, Y. 1976. A new model for predicting the hydraulic conductivity of unsaturated porous media. *Water Resour. Res.*, 10, 514–520.

Nassar, I.N., and R. Horton. 1997. Heat, water, and solute transfer in unsaturated porous media: I-theory development and transport coefficient evaluation. *Transport Porous Media*, 27, 17–38.

Nkedi-Kizza, P., Biggar, J.W., Selim, H.M, van Genuchten, M.Th., Wierenga, P.J., Davidson, J.M., and D.R. Nielsen. 1984. On the equivalence of two conceptual models for describing ion exchange during transport through an aggregated Oxisol. *Water Resour. Res.*, 20, 1123–1130.

Othmer, H., Diekkrüger, B., and M. Kutilek. 1991. Bimodal porosity and unsaturated hydraulic conductivity. *Soil Sci.*, 152, 139–150.

Pickens J.F., and Grisak G. 1981. Scale-dependent dispersion in a stratified granular aquifer. *Water Resour. Res.*, 17, 1191,1211.

Rambow, J., and B. Lennartz. 1994. Simulation of the migration behaviour of herbicides in unsaturated soils with a modified LEACHP-version. *Ecol. Modelling* 75/76, 523–526.

Richards, L.A. 1931. Capillary conduction of liquids through porous media. *Physics*, 1, 318–333.

Ritter, A., F. Hupet, R. Muñoz-Carpena, S. Lambot and M. Vanclooster. 2003. Using inverse methods for estimating soil hydraulic properties from field data as an alternative to direct methods. *Agr. Water Management*, 59(2):77–96.

Ritter, A., R. Muñoz-Carpena, C.M. Regalado, S. Lambot and M. Vanclooster. 2004. Analysis of alternative measurement strategies for the inverse optimization of the hydraulic properties of a volcanic soil. *J. Hydrol.*, 295:124–139 .

Ross, P.J., and K.R.J. Smettem, 1993. Describing soil hydraulic properties with sums of simple functions. *Soil Sci. Soc. Am. J.*, 57, 26–29.

Sander, R. 1999. Compilation of Henry's Law Constants for Inorganic and Organic Species of Potential Importance in Environmental Chemistry. Air Chemistry Department, Max Plank Institute of Chemistry, Mainz, Germany, http://www.mpch-mainz.mpg.de/~sander/res/henry.html.

Scanlon, B.R., Nicot, J.P., and J.W. Massmann. 2002. Soil gas movement in unsaturated systems, in Warrick, A.W. (ed.), Soil Physics Companion. CRC Press, Boca Raton, FL, pp. 297–341.

Selim, H.M., and M.C. Amacher. 1988. A second order kinetic approach for modelling solute retention and transport in soils. *Water Resour. Res.*, 24, 2061–2075.

Seyfried, M.S., Sparks. D.L., Bar-Tal, A. and S. Feigenbaum. 1989. Kinetics of Ca-Mg exchange on soil using a stirred-flow reaction chamber. *Soil Sci. Soc. Am. J.*, 53:406–410.

Skopp, J. and A.W. Warrick. 1974. A two-phase model for the miscible displacement of reactive solutes in soils. *Soil Sci. Soc. Am. J.*, 38, 545–550.

Starr J.L. 1983. Assessing nitrogen movement in the field, in Nelson D.W., Elrick, D.E., Tanji, K.K. (eds.), Chemical Mobility and Reactivity in Soil Systems. S.S.S.A. Special Publication Number 11, 79–92.

Šimůnek, J., Šejna, M., and M. Th. van Genuchten. 1999. The Hydrus 2D software package for simulating the two-dimensional movement of water, heat, and multiple solutes in variably saturated media. V. 2.0, U.S. Salinity Laboratory, Riverside, CA.

van Genuchten, M.Th. 1980. A closed-form equation for predicting the hydraulic conductivity of unsaturated soils. *Soil Sci. Soc. Am. J.*, 44:892–989.

van Genuchten, M.T. and F.J. Leij. 1992. On estimating the hydraulic properties of unsaturated soils, in *Proc. of the Intern. Workshop, Indirect Methods for*

Estimating the Hydraulic Properties of Unstaurated Soils, University of California, Riverside, CA, pp. 1–14.

van Genuchten, M.Th., and W.J. Alves. 1982. Analytical solutions of the one-dimensional convective-dispersive solute transport equation. USDA Technical Bulletin 1661.

van Genuchten, M.Th, Davidson, J.M., and P.J. Wierenga. 1974. An evaluation of kinetic and equilibrium equations for the prediction of pesticide movement through porous media. *Soil Sci. Soc. Am. Proc.*, 38, 29–35.

van Genuchten M.Th, and R.J. Wagenet. 1989. Two-site/two-region models for pesticida transport and degradation: theoretical development and analytical solutions. *Soil. Sci. Soc. Am. J.*, 53(5), 1303–1310.

van Genuchten, M.Th., and P.J. Wierenga. 1977. Mass transfer studies in sorbing porous media: I. Analytical solutions. *Soil Sci. Soc. Am. J.*, 41, 272–277.

van Wesenbeeck, I.J., and R.G. Kachanoski. 1994. Spatial scale dependent of *in situ* solute transport. *Soil Sci. Soc. Am. J.*, 55, 3–7.

Vanclooster M., Vereecken, H., Diels, J., Huysmans, F., Verstraete W., and J. Feyen. 1992. Effect of mobile and immobile water in predicting nitrogen leaching from cropped soils. Modelling Geo-Biosphere Processes, 1, 23–40.

Vanclooster M., Viaene, P., Cristianes, K., and S. Ducheyne. 1996. WAVE: Water and Agrochemicals in Soil and Vadose Environment, Rel. 2.1, Institute for Land and Water Management, Katholieke Universiteit Leuven.

Vanderborght J., Gonzalez, C., Vanclooster, M., Mallants D., and J. Feyen. 1997. Effects of soil type and water flux on solute transport. *Soil Sci. Soc. Am. J.*, 61, 372–389.

Vereecken, H.M., Vanclooster M., and M. Swerts. 1990. A simulation model dor the estimation of nitrogen leaching with regional applicability, in Mercks, R., Vereecken H. (eds.), Fertilization and the Environment. Leuven, Academic Press, Belgium, pp. 250–263.

Vereecken, H.M., Vanclooster, M., Swerts M., and J. Diels. 1991. Simulating water and nitrogen behavior in soil cropped with winter wheat. *Fert. Res.*, 27, 233–243.

Viaene, P., Vereecken, H., Diels, J., and J. Feyen. 1994. A statistical analysis of six hysteresis models for moisture retention characteristic. *Soil Sci.*, 157(6), 345–353.

Wagenet R.J., and J.L. Hutson. 1987. LEACHM: Leaching Estimation and Chemistry Model, a process based model of water and solute movement, transformations, plant uptake and chemical reactions in the unsaturated zone, *Continuum 2* Dep. of Agronomy, Cornell University, Ithaca, NY.

Zhang, R., Yang, J., and Z. Ye. 1996. Transport through the vadose zone: a field study and stochastic analyses. *Soil Sci.*, 161, 5, 270–277.

4 Techniques for Characterizing Water and Energy Balance at the Soil-Plant-Atmosphere Interface

M. J. Polo and J. V. Giráldez
Department of Agronomy, University of Córdoba, Spain

M. P. González-Dugo
Alameda del Obispo, IFAPA, Córdoba, Spain

K. Vanderlinden
Rivers and Reservoirs Group, University of Granada, Spain

CONTENTS

1-5667-0657-2/05/$0.00 + $1.50
© 2005 by CRC Press

4.1   THE COMPONENTS OF WATER AND ENERGY BALANCES: DESCRIPTION AND NATURE OF PROCESSES

The partitioning of the precipitation at the soil surface into its components (infiltration, evapotranspiration, and runoff) during a period of time is called the water balance. This water exchange at the soil surface is conditioned by soil physical properties, vegetation characteristics, and the climate pattern shown by the dry and wet period distribution. Their respective influence on the processes involved is not independent, due to the close relationship between the physical characteristics of the soil surface and the vegetation cover distribution and conditions, which, in turn, are related to climatic and meteorological parameters. Since vapor fluxes return water to the atmosphere, water and energy balances are closely linked at the soil surface. In a similar fashion, the incoming energy at the soil surface, net radiation, is distributed into soil and air heat storage, and evaporation into the atmosphere. The resulting balance equations can be written in terms of a physical description of all these water and energy fluxes, which add up to zero because the surfaces have no storage capacity. The balance equations at the soil surface are:

$$P = I + R + E \tag{4.1}$$

$$R_{ng} = G + \lambda E + H \tag{4.2}$$

where P, I, R, and E are precipitation, infiltration, runoff, and evaporation fluxes, respectively, in the water balance equation, R_{ng}, G, E, and H are net radiation, soil heat flux, latent heat flux, and sensible heat flux, respectively, in the energy balance equation and λ is the latent heat of vaporization for water.

Moreover, it is not always realistic to regard the soil surface as a "surface," that is, a two-dimensional boundary between the atmosphere and soil. The existence of a vegetation cover constitutes an additional source of variability to the intrinsically heterogeneous nature of soil, and the structural characteristics of the vegetation cover define several characteristic length scales for each process, such as root-depth, canopy height, surface coverage, etc. (Niklas, 1994). Thus, vegetation changes the soil surface from a two-dimensional limit into a three-dimensional transition zone between soil and atmosphere, which is called the soil-plant-atmosphere interface. The surface concept is thus substituted by a control volume with water/energy storage capacity.

4.1.1 DESCRIPTION AND NATURE OF PROCESSES AND ASSOCIATED UNCERTAINTY

The dynamic coupling between soil, vegetation, and atmosphere through the physical processes that produce the transport of water mass and thermal energy results in a variation of the water and heat stored in the control volume. The balance equations (4.1) and (4.2) can then be rewritten as:

$$\frac{dS}{dt} = P - (I_d + R + E) \tag{4.3}$$

$$\frac{dQ}{dt} = R_{ng} - (G_d + \lambda E + H) \tag{4.4}$$

where dS/dt is the time rate of change of water stored in the control volume, S, the infiltration flux I is substituted by I_d, the deep infiltration flux (referring to the water that leaves the control volume and reaches deeper layers of soil), dQ/dt is the time rate of change of heat stored in the control volume, Q, and G_d is the deep soil heat flux leaving the control volume.

For a given rainfall event, the fraction of the incoming precipitation accepted by the soil depends on its initial water content and other physical characteristics such as structure, porosity, permeability, and depth, together with rainfall intensity and duration. Between rainfall events, some of this previously stored water is pumped back to the surface in response to the evaporative demand (atmospheric conditions and transpiration flux of vegetation), while the remaining water eventually reaches the water table. Thus, E depends on the matching between the external demand applied and the water previously stored in the soil. There is also a biological control on the water consumed, with a growing relevance as soil moisture decreases. The influence of the actual E value on the energy balance can be derived from Eq. (4.4), and, with it, the buffering effects of the vegetation can easily be outlined.

In practice it is difficult to distinguish between the amount of water evaporated directly from bare soil and the amount transpired by the vegetation cover in a portion of land surface, the former also being affected by the structure of the vegetation (i.e., micrometeorological conditions at the soil surface are affected by shadowing, humidity of air, etc.). As a result, the two processes are usually combined and referred to as evapotranspiration, or ET. A thorough discussion of the physics behind the evapotranspiration process can be found in Brutsaert (1982) and Eagleson (2002). The relevant aspects of its quantification are included in following sections of this chapter.

Besides the direct use of water, the interception of precipitation water by the vegetation canopy results in a decrease in the total fraction of water that actually reaches the soil surface. The vegetation itself constitutes an intermediate sink in the path followed by water from the atmosphere to the soil. On the one hand, vegetation stores a certain amount of water during a rainfall event, which will be later partially evaporated from its canopy during the period between events; on the other hand, the morphology of the canopy structure determines the flux of water leaking from leaves and stems, which changes the effective value of I at the soil surface under the canopy. Other aspects, such as raindrop size and velocity, are also affected and determine the erosion characteristics of the precipitation.

The importance of the interception fraction in the water balance depends on the final interaction of several factors: duration/intensity and frequency of rainfall events, temperature range in the interstorm period, spatial coverage of vegetation, and canopy structure. In Mediterranean environments the uneven occurrence of rains leads to a significant annual interception amount, which must be included in the water balance quantification. The main approaches followed are discussed later in this chapter. In general, the interception component is regarded as being a decrease in the precipitation fraction, the final equation being:

$$\frac{dS}{dt} = P_e - (I_d + R + E) \qquad (4.5$$

where P_e, the effective precipitation flux, is the difference between precipitation, P, and interception fluxes.

Through the quantification of each component of the balances, different parameters must be determined in the control volume adopted, together with the boundary and initial conditions. Natural systems undergo spatially complex processes, this complexity being the result of the interaction between all the factors involved. Putting aside the question of any observational error, uncertainty plays an important role in nature. Any approach to characterizing natural processes must include the probability distributions of both the state descriptors and the pursued results chosen for each individual process. This uncertainty lies in the structure given to the physical system modeled, mainly the definition of the parameters and input, state and output variables, along with the observations required and the quality of the available records. For example, precipitation and runoff are usually considered input and output variables, respectively, in water balance models, with a certain local/regional pattern of the spatiotemporal variability of precipitation. With regard to the uncertainty associated with precipitation occurrence, this uncertainty is transferred to the resulting runoff values depending on the way the physical processes are described (i.e., its mathematical expression). The intrinsic spatial variability of soil characteristics is very likely the second most important factor when quantifying result uncertainty. This uncertainty determines the validity of the results when soil properties are spatially averaged. The spatial pattern of vegetation distribution and its variability through the year should also be included as an uncertainty source. A good insight into the physical basis for uncertainty in the components of the annual water balance is provided by Eagleson (1978).

4.1.2 DIFFERENT APPROACHES AND SPATIOTEMPORAL SCALES

Water and energy balance equations are valid over all scales, but the parameterization of each component is scale-dependent. Once the spatial boundaries of the system are defined, the balance equations can be used either as a distributed model or a lumped model. A distributed model accounts for spatial variability in inputs, processes, and parameters. The modeled system is divided into a control volume consisting of a number of elementary units, where the values for the variables and parameters within each elementary unit are assumed constant. Examples of the inclusion of explicit spatial variability in processes are the European hydrologic model SHE (Abbot et al., 1986a,b), and the three-dimensional finite-element catchment model of Paniconi and Wood (1993); a statistical representation of the variability patterns can be found in TOPMODEL (Beven and Kirkby, 1979) and later variants (Moore et al., 1984; Famiglietti and Wood, 1994; Liang et al., 1994; Beven, 2000). A lumped model, on the other hand, regards the modeled system as the control volume, i.e., spatially homogeneous in relation to inputs and parameters. The unit hydrograph model for the runoff generation in a watershed, and its variants, are representative of this approach.

Scale problems arise when the parameterization adopted for a distributed/ lumped model at a certain scale does not apply to either smaller or larger scales. Blöschl and Sivapalan (1995) define scaling as the transfer of information between different spatiotemporal scales. Most physically based analytical models were developed from descriptions of processes at the point scale. The low level of agreement often found between the predicted and measured output from these models when describing the water or energy balance of larger systems is due mainly to these scale incompatibilities. Moreover, water and energy balances involve highly nonlinear processes, with both strong interaction between components and vertical–horizontal spatial correlation with soil properties. One of the earlier approaches to the scaling problem was provided by Dooge (1982), and a good overview of progress in this topic can be found in Gupta et al. (1986) and, more recently, in Kalma and Sivapalan (1995).

The development of geographic information systems (GIS) as a tool for storing spatially distributed information has facilitated the combination of analytical models with the spatial heterogeneity exhibited by surface characteristics to better describe the behavior of complex systems along medium or large spatial and temporal scales, from watershed to regional scale. The results are quasi-distributed models, since it is the combination of a lumped model and its application to a grid of cells that represents the land surface. This scale problem then consists of finding the link between the parameterization of point-scale processes (known) and the parameterization of watershed-scale processes (unknown), and it is commonly designated as "upscaling." "Downscaling," on the other hand, represents the obtainment of smaller scales from large-scale observations. In upscaling, the problem is usually posed as finding the effective value of the parameters at a watershed scale that produce a result equivalent to the aggregation of the spatially distributed results obtained with the original spatially distributed parameter values (what has been called "flux matching"). Sivapalan and Wood (1986) and Wood et al. (1986) showed that a strong bias was introduced into mean infiltration fluxes when only mean values of soil properties and rainfall intensities were used. More recently, Wood (1997) evaluated the effect of the spatial aggregation of soil moisture on the modeling of the evaporative fraction. These are only some examples of scaling approaches used when modeling water and energy balances at the land surface.

4.1.3 REMOTE SENSING: POTENTIAL AS A GLOBAL DATA SOURCE

The parallel and growing development of remotely sensed sources of information, automated on ground sensors, and spatial data storing and processing by GIS has gathered the largest observation records in the history of science. Data acquired by remote systems provide both the scientific and technical communities with spatially distributed observations with a high temporal frequency when compared with on-ground field data. However, the relations between remote sensing observations and water and energy flux variables and parameters are usually indirect; what is more, the various measurements apply to different scales and different noise levels (Stewart et al.,

1996). The final user must therefore be in a position to process and analyze large amounts of information, which requires a comprehensive knowledge of how data assimilation techniques and the scale effect influence the results. Finally, the need to account for uncertainty is very important when combining different sources of information, in general, but it is absolutely essential when data derived from remote sensing observations are merged with on-ground field observations. In his integrated approach to hydrologic data assimilation, McLaughlin (2002) concludes that the effectiveness of data assimilation techniques relies on the success in realistically describing the uncertainty associated with the processes modeled.

4.2 MODELING OF THE WATER AND ENERGY BALANCE AT THE SOIL-PLANT-ATMOSPHERE INTERFACE AND SCALE EFFECTS

4.2.1 THE USE OF MODELS FOR THE DESCRIPTION OF SOIL-PLANT-ATMOSPHERE EXCHANGE PROCESSES

There are a large number of water and energy exchange processes at the soil, plant, and lower atmosphere interface, depending on soil conditions, plant types, and atmospheric conditions, although most have common character- istics. Therefore, models are developed to describe the essential part of the processes and are then inserted into a general frame. As information on the processes increases, the model can be improved to better represent the behavior of the interface.

There are several models for describing energy and water exchange in the lower atmosphere. In this chapter some of them will be analyzed from the point of view of determining the best method(s) for measuring the most important processes.

The refinement of general circulation models (GCM) has renewed the interest of hydrologists and climatologists in the role played by soil as a regulator of many large-scale hydrologic processes. This interaction of land in atmospheric processes is a key process for the understanding of soil water balance.

4.2.1.1 A Simple Water and Energy Balance Model: The Interaction Between Land and Atmosphere

One simple way to represent the interaction between the land and the atmosphere was proposed by Brubaker and Entekhabi (1995), who divided the soil-plant-atmosphere interface into two different control volumes or slabs, which consist of the soil-plant domain and the lower portion of the atmosphere. The two control volumes interchange precipitation and evaporation fluxes through their common boundary. The basic water balance equation (4.1) is rewritten for both systems, as Eqs. (4.6) and (4.7), respectively:

$$\frac{ds}{dt} = \frac{1}{\rho_w n z_h}(P - R - E) \tag{4.6}$$

$$\frac{dq_m}{dt} = \frac{g}{p_s - p_h}(\Delta Q + E - P) \tag{4.7}$$

where s is the soil moisture volumetric content, set in relative terms, or degree of saturation; ρ_w, the density of water; n, the average soil porosity; and z_h, the soil depth affected by water changes. P and E constitute a moisture source and loss, respectively. In the atmospheric slab, q_m is the air specific humidity (the ratio between the masses of water and moist air); the pressure at the soil surface and the mean air pressure are p_s and p_h, respectively; the moisture contribution from the atmosphere is $\Delta Q = Q_{in} - Q_{out} - Q_{top}$ (the water flux from the windward side minus the losses leeward and upward, represented, respectively, by the three terms in this expression); and g stands for the acceleration of gravity. In this case, E and P are an atmospheric moisture source and loss, respectively.

Similarly, the energy balance equation (4.2) can be expressed as Eq. (4.8) in the soil system, and Eq. (4.9) in the atmospheric domain:

$$\frac{dT_g}{dt} = \frac{1}{c_{soil} z_t}\left[R_{ng} - H - \lambda E\right] \tag{4.8}$$

$$\frac{d\theta_m}{dt} = \frac{g}{c_p(p_s - p_h)}(R_{na} + \Delta H) \tag{4.9}$$

Equation (4.8) describes the temporal change of the ground temperature, T_g, averaged along the soil depth affected by energy exchanges, z_t, where c_{soil} is the volumetric heat capacity of the soil. The sensible and latent heat fluxes, H and λE, constitute the energy losses from the soil system. The net radiation flux, R_{ng}, is the result of the long- and shortwave radiation component balance at the ground surface, given by Eq. (4.10), where RS is the shortwave radiation flux that arrives at the surface, and α stands for the surface albedo, i.e., the reflected shortwave radiation fraction; RL_{ad} is the longwave radiation flux from the overlying atmosphere, and ε_{col} stands for the air emissivity/absorptivity, i.e., the longwave radiation fraction absorbed by the air, while RL_{sd} and RL_{gu} are the longwave fluxes emitted by the air downward and by the ground upward, respectively:

$$R_{ng} = RS(1 - \alpha) + RL_{ad}(1 - \varepsilon_{col}) + RL_{sd} - RL_{gu} \tag{4.10}$$

The energy balance in the air slab due to the net radiation and sensible heat fluxes, R_{na} and ΔH, is expressed in Eq. (4.9) in terms of the time variation of θ_m, the potential temperature of the air, i.e., the adiabatic equivalent for the surface pressure, and c_p is the specific heat of the air at constant pressure. The long- and shortwave radiation component balance in Eq. (4.11) determines R_{na} as the result of the absorbed fractions (ε_{col}) of the longwave radiation fluxes emitted by the atmosphere above the slab, RL_{ad}, and of the soil beneath, RL_{gu}, minus the loss fluxes upward, RL_{su}, and downward, RL_{sd}:

$$R_{na} = \left(RL_{ad} + RL_{gu}\right)\varepsilon_{col} - RL_{sd} - RL_{su} \qquad (4.11)$$

The net flux of sensible heat in the air slab, ΔH, is obtained from the flux balance given by Eq. (4.12) as the result of the contribution from the ground, H, plus the windward supply, H_{in}, minus the export leeward, H_{out}, and upward, H_{top}:

$$\Delta H = H + H_{in} - H_{out} - H_{top} \qquad (4.12)$$

With this scheme, Brubaker and Entekhabi (1995) showed how the conditions for equilibrium between the atmospheric and the land domains are influenced by external factors. In particular the authors described the diurnal cycles of the water and energy fluxes, in good agreement with published data, even at a lower, hourly, scale. More important than the latter descriptions is the usefulness of the model as a tool for analyzing the responses to the forcing parameters, whose random behavior was considered in a companion paper (Entekhabi and Brubaker, 1995).

4.2.1.2　The Force Restore Approach

The surface fluxes in the ISBA (interactions soil-biosphere-atmosphere) model of Noilhan and Mahfouf (1996) are formulated in a similar way, but adopting the force restore model proposed first by Deardorff (1978), which evaluates the water and energy balance in soil considering two different control volumes, the soil surface, and the deeper horizons of the soil profile as a whole. The water balance equation at the surface soil layer gives the time change of soil water content as

$$\frac{\partial s_s}{\partial t} = \frac{c_1}{\rho_w n z_s}(P - E) - \frac{c_2}{\tau}(s_s - s_m) \qquad (4.13)$$

with s_s as the average volumetric moisture content in the surface soil layer, degree of saturation, with a depth z_s, s_m is the average volumetric moisture content of the soil profile, c_1 and c_2 being two coefficients, and τ the time constant, equal to one day. Similarly, the average volumetric moisture content of the soil profile is s_m, over a depth z_m, and its time change is given by

$$\frac{\partial s_m}{\partial t} = \frac{1}{\rho_w n z_m}(P - E) - \frac{c_3}{\tau}\max\left[0, \left(s_m - s_{fc}\right)\right] \tag{4.14}$$

where c_3 is another coefficient and s_{fc} is the degree of saturation at field capacity as a reference state for the soil profile.

The time change of the average soil surface temperature, T_s, is derived from the energy balance equation as

$$\frac{\partial T_s}{\partial t} = \frac{1}{c_T z_t}\left(R_{ng} - H - \lambda E\right) - \frac{2\pi}{\tau}(T_s - T_m) \tag{4.15}$$

where T_m is the deeper soil profile temperature, c_T is the surface soil-vegetation volumetric heat capacity. For the deeper soil profile, the thermal variation is

$$\frac{\partial T_m}{\partial t} = \frac{T_s - T_m}{\tau} \tag{4.16}$$

Milly (1986) adapted this method in his model for the evolution of soil moisture. More recently, Jacobs et al. (2000) used this scheme for the study of the water and temperature changes in a dry desert system.

4.2.1.3 Dynamics of Soil Moisture Using a Simple Water Balance

In a closer approach to the soil-plant-atmosphere interface, Rodríguez-Iturbe et al. (2001) explored the plant role with another simple model based on the water balance in the soil profile with a partial vegetation cover at the surface. They described the water balance in the control volume defined by the soil depth explored by the plant roots, z_r, expressing the temporal change of the depth-averaged soil moisture content, degree of saturation,

$$\frac{ds}{dt} = \frac{1}{\rho_w n z_r}(P - I_n - R - E - I_d) = \frac{1}{\rho_w n z_r}(\varphi - \chi) \tag{4.17}$$

where I_n represents the interception, R the runoff, and I_d the percolation fluxes, respectively. The φ and χ functions express the rate of water infiltration into the soil and soil water losses, respectively, being $\varphi = P - I_n - R$, $\chi = E + I_d$. As Laio et al. (2001) indicated, the function φ is the stochastic component forcing the system to change. In their development of the model, the authors adopted a simple probability distribution function for the characterization of the rainfall process, based on two properties: the time period between events and the depth of individual events. The interception is known to be dependent on several environmental factors (e.g., Zeng et al., 2000), leading to the adoption of a Poisson process for the estimation of the net rainfall.

$$P - I_n = \sum_i h'_i \delta(t - t'_i) \tag{4.18}$$

where h'_i is the depth of net rainfall, δ the Dirac delta function, and t'_i the arrival time of the events.

The evapotranspiration flux, E, was approached with a two-step linear expression for several intervals of the saturation range, written as:

$$E = \begin{cases} E_w\,(s - s_h)/(s_w - s_h) & s_h < s \le s_w \\ E_w + (E_{max} - E_w)\,(s - s_w)/(s_* - s_w) & s_w < s \le s_* \\ E_{max} & s_* < s \le n \end{cases} \tag{4.19}$$

with s_h as the hygroscopic water, i.e., the moisture content from which evapotranspiration starts at the lowest rate, linearly increasing with water content up to E_w at the wilting point, s_w. As the moisture increases, so does the evapotranspiration flux, with a higher trend towards its maximum value, E_{max}, at the degree of saturation, s_*. In the moisture range (s_*, n) the plant loses water without any stomatal regulation. At the bottom of the profile, water percolates towards deeper horizons following the gravity gradient; therefore, the deep infiltration unit flow, i_d, is equal to the hydraulic conductivity of the soil, k:

$$i_d(s) = k(s) = k_s \frac{\exp(\beta_k s) - \exp(\beta_k s_{fc})}{\exp(\beta_k n) - \exp(\beta_k s_{fc})} \tag{4.20}$$

where k_s is the saturated hydraulic conductivity of soil, and β_k a coefficient. The subscript fc stands for field capacity conditions in the soil.

Based on this water balance model, Laio et al. (2001) estimated the probability density function of the moisture content, which showed the effect of different environmental influences, such as soil pore type, soil depth, and wet or dry climates. Another interesting result of the model is the evaluation of the contribution of different hydrologic processes to water balance, as shown in Figure 4.1. This estimation could act as a guide for the selection of the measurement method.

4.2.1.4 Exploration of Optimal Conditions for Vegetation Through a Water Balance Model

A closer view of the water and energy balance was formulated by Eagleson (2002, Chapters 6 and 7), in a thorough study linking plant and soil physical aspects in a common framework. Eagleson (2002) posed the energy balance for a control volume at the surface of the land in a similar way to the Eq. (4.2), discarding some terms. Specifically, the biological rate of energy consumption, the transfer of sensible heat from the canopy to the soil, or substrate, and the

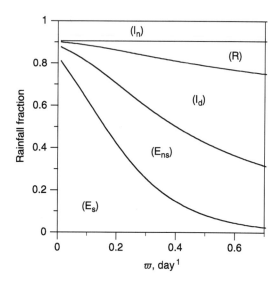

FIGURE 4.1 Contribution of the different processes of the soil water balance according to the model of Laio et al. (2001) as a function of the inverse of the average time between rainfall events, ω. (I_n, interception; R, runoff; I_d, deep infiltration; E_{ns}, non-stressed evapotranspiration; E_s, stressed evapotranspiration.)

horizontal flux divergence of sensible and latent heat were considered to be very small. As a result, the energy balance equation was simplified to:

$$R_{ng} = \lambda E + H \qquad (4.21$$

The water balance proposed by Eagleson (2002) was elaborated from earlier work on the physical processes of soil water (Eagleson, 1978), aiming to characterize the role of water in the development of the vegetation, with simple approximated analytical solutions of the equations describing the processes. The simplest form of the water balance is

$$P_A = E_{TA} + R_{sA} + R_{gA} \qquad (4.22$$

where the average annual precipitation, P_A, is equal to the sum of the average annual transpiration, E_{TA}, average annual surface runoff, R_{sA}, and average annual groundwater runoff, R_{gA}. The average values were determined by the integration of detailed processes. The average rainfall is simplified to the mean storm arrival rate multiplied by the mean water depth of each storm. The expected value of surface runoff, $E[R_{sj}]$, equivalent to the expected value of storm rainfall excess, $E[R^*_{sj}]$, minus the expected value of the storm rainfall excess, $E[E_r]$, can be expressed in terms of soil and climate parameters:

$$E[R_{sj}] = \frac{\varsigma(\sigma_*)}{\omega\xi e^{G_*}} \approx \left[\omega\xi\exp(G_* + 2\sigma_*^{3/2})\right]^{-1} \qquad (4.23)$$

where ω and ξ are the respective inverse of the average storm intensity and duration; therefore, the inverse of the product of ω and ξ is the mean storm depth, G_*, which is also given by

$$G_* = \omega k_s\left(\frac{1 + s_0^{c_{BC}}}{2} - \frac{\omega}{k_s}\right) \qquad (4.24)$$

where s_0 is the initial value of soil moisture, and c_{BC} is the pore permeability index, or pore size distribution index of the soil for the Brooks and Corey hydraulic conductivity function (e.g., Kutilek and Nielsen, 1994, § 4.3.2), and σ_* a parameter dependent on soil and climate properties. Finally, $\varsigma(\sigma_*)$ is what Eagleson (1978) called a surface runoff function.

The evapotranspiration consists of three parts: unstressed or potential transpiration, moisture-constrained transpiration, and bare soil evaporation. Soil water evaporation is expressed as

$$E[E_{sj}] = \beta_s\xi^{-1}E_{ps} \qquad (4.25)$$

where β_s is the bare soil evaporation efficiency,

$$\beta_s = 1 - \left(1 + 2^{1/2}E_*\right)e^{-E_*} + (2E_*)^{1/2}\Gamma(3/2, E_*) \qquad (4.26)$$

where the parameter E_* is the bare soil evaporation effectiveness of Eagleson (1978), dependent as well on soil properties. Equation (4.25) is the equation for natural surface evaporation discussed by Milly (1994), among other authors.

Plant transpiration at a potential rate is E_v:

$$E_v = k_v^*\xi^{-1}E_{ps} \qquad (4.27)$$

using the unstressed canopy conductance, k_v^*. The moisture limited transpiration is expressed in a similar way as soil water evaporation:

$$E[E_{vj}] = \beta_v\xi^{-1}k_v^*E_{ps} \qquad (4.28)$$

where the canopy transpiration efficiency, β_v, is

$$\beta_v \approx 1 - e^{-\xi t'_s} \left\{ \left[1 + \frac{\xi}{\lambda_0 E_{pv}} \right]^{-\kappa_0} \frac{\gamma \left[\kappa_0, \left(\lambda_0 + \xi E_{pv}^{-1} \right) \right]}{\Gamma(\kappa_0)} + e^{-\xi h_0 / E_{pv}} \left[1 - \frac{\gamma(\kappa_0, \lambda_0 h_0)}{\Gamma(\kappa_0)} \right] \right\}$$

(4.29

in which t'_s is the transpiration period of the plant, h_0 is the depth of surface water retention, κ_0 and λ_0 are parameters of the rain depth distribution, and $\Gamma(a_1)$ and $\gamma(a_1, a_2)$ are, respectively, the complete and the incomplete gamma function.

With these contributions, the total annual evapotranspiration is

$$E_{TA} = \frac{m_v E_{ps}}{\xi} \left[(1 - M)\beta_s + M k_v^* \beta_v \right]$$

(4.30

for a fraction of surface M covered by the canopy.

Finally the normalized seasonal water balance is written for the case of very deep water table as

$$1 - e^{-G_* - 2\sigma_*^{3/2}} = \frac{\lambda_0 E_{ps}}{\xi \kappa_0} \left[(1 - M)\beta_s + M k_v^* \beta_v \right] + \frac{m_\tau k_s}{P_\tau} s_0^{cBC}$$

(4.31

with the mean length of the growing season, m_τ, and the precipitation for such a period, P_τ. The balance of water stresses the importance of the vegetation variables, the cover fraction, M, and the unstressed canopy conductance, k_v^*. With these two factors Eagleson (1982) compared the traits of contrasting climates. The inclusion of a critical moisture state, at which the canopy conductance falls abruptly from its potential value due to stomatal closure, adds another useful perspective to the vegetation state space plots of Eagleson (2002, § 6.K). Ridolfi et al. (2000) proposed the use of the first-passage times for processes driven by shot noises for the study of the occurrence of critical periods in vegetation, as an alternative to the Eagleson approach.

4.2.1.5   Strengths and Weaknesses

The models described are efficient frameworks for arranging the information that is being gathered from the observations, measurements, and experiments under different conditions. The combination of theory and experiments has been very useful to interpret new findings or even to suggest new hypotheses such as the well-known proposal of Philip of the three stages of evaporation, later corroborated in field studies, although with some conditions (e.g., Kutilek and Nielsen, 1994, § 6.4.2). The strength of the models is their basic physical foundation, developed with mathematical tools. At the same time, their weaknesses include the difficult coverage of the changing environmental conditions in a simple, manageable set of equations.

4.2.2 INTERACTION OF MODEL DEVELOPMENT AND TEMPORAL AND SPATIAL SCALES

The usefulness of simple water balance models depends on the scale of application, as Jothityangkoon et al. (2001) have demonstrated in a medium-size watershed. Simple schemes are useful, but their utility is conditioned by the processes described. Jothityangkoon et al. (2001) started with the water balance similar to that of Laio et al. (2001), Eq. (4.17), for the annual data:

$$\frac{dw}{dt} = P - Q_{se} - E \tag{4.32}$$

where w is the water storage in the soil profile and Q_{se} is the saturation excess runoff rate expressed as a function of the difference between the current value and the storage capacity of the soil, W_b,

$$Q_{se} = \max[(w - W_b)/\Delta t, 0] \tag{4.33}$$

and the evaporation rate as a fraction of the potential value, $E = (wE_p)/W_b$.

At this level the influence of the spatial variability of rainfall and soil properties was incorporated through the use of a multiple-bucket model to represent the watershed. Reducing the time scale to monthly periods, the authors added a term for the subsurface runoff, Q_{ss}, and separated the contributions of bare soil water evaporation, E_b, and plant transpiration, E_v,

$$\frac{dw}{dt} = P - Q_{se} - Q_{ss} - E_b - E_v \tag{4.34}$$

with

$$Q_{ss} = \max\left[(w - W_{fc})/t_c, 0\right] \tag{4.35}$$

where W_{fc} is the soil water storage at field capacity and t_c a characteristic time for the catchment response in the groundwater flow. Bare soil water evaporation and transpiration rates depended on two respective characteristic times, t_e and t_g,

$$E_b = \frac{w(1 - M)E_p}{W_b} = \frac{w}{t_e} \tag{4.36}$$

while plant transpiration flux is

$$E_v = \begin{cases} Mk_v E_p & w > W_{fc} \\ wMk_v E_p / W_{fc} = w/t_g & w \leq W_{fc} \end{cases} \tag{4.37}$$

where k_v is analog to the canopy conductance of Eagleson.

Finally, for the daily periods, the model incorporated a nonlinear dependency on the subsurface flow rate and the water storage,

$$Q_{ss} = \begin{cases} \left[(w - W_{fc})/a_J \right]^{1/b_J} & w > W_{fc} \\ 0 & w \leq W_{fc} \end{cases} \tag{4.38}$$

where a_J and b_J are two constants, and another deep groundwater store, whose content, W_g, is nonlinearly related to the deep groundwater flow rate, Q_{sg}, by

$$Q_{sg} = \left(\frac{W_g}{a_J} \right)^{1/b_J} \tag{4.39}$$

with the following final expression for the runoff rate

$$Q_{total} = (1 - \lambda_g)Q_{ss} + Q_{sg} + Q_{se} \tag{4.40}$$

Note that only a part, λ_g, of the deep store contributes to the final runoff from the watershed.

4.2.3   Hydrologic Data Assimilation

Improved techniques for assimilation of hydrologic data are being developed because of the large amount of data collected by the new satellite-bound and ground-based sensors, and because of the advent of more complete hydrological models. The new techniques are all referred to as assimilation methods. As McLaughlin (2002) indicated in a recent review, there are several problems that the new methods have to solve, including high dimensionality of the data, the need for efficiency, optimality, and robustness, and the need to estimate uncertainty. The main methods may be grouped as interpolation, smoothing, and filtering.

Interpolation methods compute a value, or state vector at a given time, from the state and input variables of a system. State variables are those that completely describe the system, while input variables are those that induce the system changes. Thus, the system is at steady state and the solutions to the state equation can be formulated as $x = A_0(\alpha)$, and, for the input variables $z = M[x, \omega]$, in the notation of McLaughlin (2002). The most common interpolation technique has been the least-squares estimate of x and α although, as McLaughlin (2002) clearly states, this method is only acceptable when the functions A_0 and M are linear and the errors are independent with zero mean and Gaussian random distribution. McLaughlin (2002) described an interpolation method based on a tree structure, a multiplicative cascade model to assimilate radar and satellite microwave rainfall data, where the linear interpolation methods may be applied (Gorenburg et al., 2001).

Smoothing techniques are, as McLaughlin (2002) indicates, a dynamic extension of the interpolation techniques. Therefore, in this case time is a variable, and the problem is to estimate the state variables $x(t)$ from a set of input and output measurements of the time interval (t_0, t'_{nz}) for $t_0 \leq t < t'_{nz}$ The least-squares method is used assuming that the errors are additive. The variational approach employed is based on the hypothesis that all errors generated at different times are independent and time invariant, so that it is known as the batch estimation method. McLaughlin (2002) suggested the use of the representer method, adopting a linear form for the state and measurement equations, as Reichle et al. (2001) did in an estimation of soil moisture in an 80×160 km area in Oklahoma based on radio brightness measurements. Nevertheless, as McLaughlin (2002) accepted, more research is needed to find simpler methods for the practical application of iterative least-squares methods to nonlinear problems.

Filtering techniques are adopted for the estimation of state variables at the most recent time, or in the above expression the characterization of $x(t'_{nz})$ from the input and output measurements of the time interval (t_0, t'_{nz}). Being an extension of smoothing techniques, the method could be based on variational methods at each measurement time, but in order to increase its computational efficiency different methods may be used. A sequential estimation algorithm consists of two steps: the propagation step computes the changes in state estimates between measurement times, and the update step adapts the changes to the recently acquired information. Another possibility is the use of ensemble filtering that generates random replicates of the state equation, which are later updated with the new measurements. Margulis et al. (2002) used an ensemble Kalman filter method for the derivation of soil moisture estimates from airborne L-band microwave observations and ground-based measurements of soil and air temperature, soil texture, and vegetation type. Although there are implicit assumptions, such as the Gaussian character of the system for the update steps, the ensemble filter is a good method for filtering purposes. Similar conclusions about the good performance of the Kalman filter were reached by Walker et al. (2001) in their study of soil moisture and temperature retrieval. They compared the updating steps of direct insertion, replacing the state variables obtained with the TOPLATS model, with observed data, where available, and Kalman filtering. The results, based on synthetic data, indicated superior performance of extended Kalman filtering over the insertion algorithm. Hoeben and Troch (2000) also found the Kalman filter method to be superior in the assimilation of active microwave data to estimate the soil moisture profile.

4.3   THE VEGETATION COMPONENTS: MEASUREMENT METHODS

4.3.1   INTERCEPTION

In order to determine the effective precipitation flux, P_e, in the water balance equation, the canopy water interception flux, I_n, must be quantified. In this regard, the water balance equation applied to a control volume constituted by an elementary unit of the canopy, whose upper and lower limits are the atmosphere and soil surface, respectively, leads to:

$$I_n = \frac{dS_{can}}{dt} = P - P_c - P_{th} - E \qquad (4.41$$

where P_c and P_{th} are, respectively, the flux of precipitation water that leaks from the canopy (stems and leaves) and the flux of throughfall (precipitation water that passes through the canopy structure without being intercepted), both fluxes finally reaching the soil surface. P and E are the precipitation and evaporation fluxes, respectively. The amount of water stored in the canopy, S_{can}, can be expressed as:

$$S_{can} = S_{can0} + \int_{S_{can0}}^{S_{can}} dS = S_{can0} + \int_{t_0}^{t} (P - P_c - P_{th} - E)dt \qquad (4.42$$

where S_{can0} is the initial water content of the canopy. The maximum storage capacity of the canopy is represented by $S_{can\,max}$, and once the canopy becomes "saturated" the interception flux falls to zero.

The early empirical relationships (static models) between the total water intercepted by the canopy and the total precipitation water for an event, with linear or exponential expressions (e.g., Bras, 1990, § 5.5), were surpassed by the dynamic models after the proposal of Rutter et al. (1971), which explicitly account for each water balance component in the canopy, and thus describe the temporal change in the water stored in the canopy. Expressing P_{th} as a fraction f of P, and P_c and E as linear relations with the degree of saturation of the canopy, i.e., $S_{can}/S_{can\,max}$, the water balance in the canopy can be written as,

$$\frac{dS_{can}}{dt} = (1 - f)P - (k_{can} + E_0)\frac{S_{can}}{S_{can\,max}} \qquad (4.43$$

where k_{can} and E_0 (potential evapotranspiration flux) are the maximum values of P_c and E, respectively. Watanabe and Mizatuni (1996) found that the Rutter et al. (1971) model results agreed with measurements made in vegetated zones with intermediate leaf area index (LAI) values, with overestimation and underestimation, respectively, in the case of larger and smaller values of LAI.

Similar approaches are given by Gash (1979) and Massman (1983) and their revised versions. All these dynamic models require the use of automated meteorological data stations, with rainfall measurements under the canopy (throughfall plus stemflow) and the calculation of $S_{can\,max}$ for each canopy category from field trials. Interception is determined as the difference between measured P and $(P_{th} + P_c)$.

4.3.1.1 Methods of Estimation of Interception

Interception modeling requires extensive meteorological data and a good definition of vegetation structure to characterize the maximum storage capacity of the canopy, since this stands as an essential parameter in these models (Teklehaimanot and Jarvis, 1991). Canopy parameters obtained from the literature can provide good results, but the degree of uncertainty in the selection is usually too large, and there is little information from irregular and sparse vegetation. This led to modifications in the original Gash model (Gash et al., 1995; Valente et al., 1997) for sparse forests, which achieved better agreement with field measurements. It is very common to derive the canopy parameter values from experimental rainfall-throughfall data, although the possibilities for extrapolation are very limited (Klaassen et al., 1998). Direct methods mainly consist of the artificial wetting of vegetation surfaces (e.g., Liu, 1998); in practice, the difficulties in the upscaling of results are a great constraint.

Precipitation measurements above and below the canopy are usually taken by bucket gauges (e.g., Lloyd and Marques, 1988; Gash et al., 1995). In areas with short vegetation, lysimeters have been used to estimate interception loss from the temporal increase of lysimeter weight (Calder, 1990). All these methods are limited by the sampling size and the spatial representativeness of the sample. Attempts to overcome these limitations include use of plastic-sheet net-rainfall gauges (Calder and Rosier, 1976) in combination with lysimeters and determination of canopy density by the γ-ray attenuation (Olszyczka, 1979), from which evaporation and interception can be inferred, or canopy water content estimation by microwave transmission measurements (Bouten et al., 1991).

4.3.1.2 Strengths and Weaknesses

Different interception component measurements are indicated when applied to small area studies due to the great effort involved in thorough characterization of vegetation structure. For a catchment scale the most relevant interception parameter is the maximum storage capacity of the canopy cover. An adequate parameterization of the interception fraction for different temporal scales is essential for the proper adjustment of hydrologic models.

4.3.2 EVAPOTRANSPIRATION

Knowledge of evapotranspiration on a regional or local scale enables hydrologists to perform water balance calculations and understand the hydrological cycle. It also allows agronomists to assess crop water requirements. Therefore, adequate estimations or measurement techniques are required. Special attention should be paid to the spatiotemporal scale (ranging spatially from the global scale to the individual plant and temporally from minutes to one or more years) of the (agro)hydrological problem under study and the associated scale of the evapotranspiration process. It is therefore of fundamental interest to select the most appropriate measurement or estimation technique for each case, in accordance with the required spatiotemporal scale. Here we shall address the principal estimation and measurement techniques for actual evapotranspiration (ET) and reference crop evapotranspiration (ET$_0$). ET can be directly measured or estimated from meteorological observations. The value obtained will be representative of the spatiotemporal scale relevant for the method used.

Rana and Katerji (2000) proposed a classification of ET measurement and estimation techniques. The measurement methods are further subdivided into hydrological, micrometeorological, and plant physiological approaches. Among the estimation methods they make a distinction between analytical, empirical, and water balance approaches. Wilson et al. (2001) compared four different techniques for measuring ET within a mixed deciduous forest in the southeastern United States. In general, methods can be classified according to the following features: direct and indirect methods, measurement and estimation methodology, and the spatio-temporal scale at which the measurements or estimates are made or for which they are representative. The methods may be grouped as follows:

Methods based on the principles of the conservation of mass (water balance equation)
Methods based on the conservation of energy (energy balance equation)
Plant physiological methods
Methods based on ET modeling

Table 4.1 summarizes the different methods considered here and classifies them according to different criteria.

4.3.2.1 Conservation of Mass Approach

The water balance at the plot scale for a time period, T, can be written in terms of the change in root zone soil water storage, ΔS_T:

$$\Delta S_T = P_T - R_T - E_T \qquad (4.44$$

TABLE 4.1
Methods for Estimating Evapotranspiration

Method	Classification	D/I	Application	Time scale	Spatial scale
Soil water balance	Conserv. mass	I	Irrigation researcher adviser	15 min weekly	$10 \, \text{m}^2 - 10 \, \text{km}^2$
Weighing lysimeter	Conserv. mass	D	Researcher adviser	1 h–24 h	$1 - 10 \, \text{m}^2$
Microlysimeter	Conserv. mass	D	Researcher adviser	1 h–1 week	10 cm
Bowen ratio energy balance	Conserv. energy	I	Researcher adviser	30 min–1 h	Fetch dependent
Eddy correlation	Conserv. energy	D	Researcher	30 min–1 h	Fetch dependent, 100 m upwind
Agrodynamic	Conserv. energy	D		30 min–1 h	Fetch dependent
Sap flow	Physiological	I	Researcher	24 h	Individual plant
Plant chamber	Physiological	I	Researcher	30 min–24 h	Individual plants, $\pm 1 \, \text{m}^2$
Penman-Monteith	Modeling	I	Irrigation researcher adviser	30 min–24 h	100 m–10 km
Empirical models	Modeling	I	Irrigation researcher adviser	24 h monthly	100 m–10 km

D/I, direct/indirect method.

where P_T is the precipitation amount (including irrigation) for T, R_T is the total runoff, and E_T is the amount of evapotranspiration. It is clear from Eq. (4.44) that E_T equals ΔS_T when no precipitation or runoff occurs, but in general P_T and R_T must be measured or estimated.

Since it is often very difficult to accurately measure all the terms of Eq. (4.44), simplifications are introduced by considering the annual water balance. On an annual basis, both soil water storage, ΔS_A, and subterranean fluxes are negligible when considering a large area (e.g., a catchment), and this allows Eq. (4.44) to be written as

$$P_A = R_A + E_A \qquad (4.45$$

Equation (4.45) is essential to the assessment of the average annual water balance of a region and has received noticeable attention (Budyko, 1974; Milly, 1994; Eagleson, 2002).

E_T can be indirectly obtained from Eq. (4.44) when ΔS_T is either measured (gravimetrically or using one of the techniques described in Chapter 5) or assessed numerically, and when the other terms are estimated, calculated, or assumed to be equal to zero. The main gravimetric technique is lysimetry, on either a medium or small scale (microlysimetry). A lysimeter (Allen et al., 1991) consists of an isolated vegetated soil column, which is provided with a weighing mechanism to monitor changes in soil moisture storage and for which water input and output are measured. Marek et al. (1988) supply a detailed discussion of the design and construction of a lysimeter. Lysimeters are considered to give the most accurate measurements of actual evapotranspiration, and they are often used as a standard to compare with results obtained by other methods (Jensen et al., 1990).

A microlysimeter consists of a soil core enclosed within a plastic tube with typical dimensions of 10 cm in diameter and a soil depth of 10–40 cm. Measurement is accomplished by extracting the lysimeter from the soil and consecutive weighing (Lascano and Hatfield, 1992). See Evett et al. (1995) for the operational details of the method.

4.3.2.2 Conservation of Energy Approach

The energy balance equation at the land surface can be written as Eq. (4.2). Transport of vapor from the vegetated land surface towards the atmosphere occurs by both diffusion and convection. Diffusion occurs in the laminar air layer close to the plant and soil surface, where the air movement is parallel to the surface, and convection occurs in the turbulent layer, situated above the laminar layer. The vapor flux is basically convective so that it principally consists of moving air parcels or eddies in the turbulent layer. Wind speed and surface roughness increase this process and decrease the thickness of the laminar layer. In the absence of wind, eddies are not caused by turbulent flow, but by free convection, due to the differential warming-up of air parcels. Generally, as air moves over a crop, a constant flux layer develops in which the vertical fluxes of heat and vapor are constant. This layer is of special transcendence for the measurement of relevant atmospheric variables, which are required by the different measurement techniques. Monteith and Unsworth (1990) provide a full discussion of the subject, with an emphasis on soil and plant boundary layers.

Through the measurement of the rest of Eq. (4.2) components, the latent heat flux, λE, can be obtained. R_{ng} can be measured with a net radiometer, G using heat flux plates, and H can be directly measured with the eddy correlation method or estimated with a straightforward resistance equation.

The Bowen ratio is defined as the ratio of sensible to latent heat flux, $\beta = H/\lambda E$, so Eq. (4.2) can be written as:

$$\lambda E = \frac{R_{ng} - G}{1 + \beta} \tag{4.46}$$

Where β can be determined experimentally from the temperature, T, and vapor pressure, e, differences between two different heights (subindexes 1 and 2) within the constant flux layer:

$$\beta = \gamma \frac{(T_2 - T_1)}{(e_2 - e_1)} \tag{4.47}$$

where γ is the psychrometric constant. The Bowen ratio energy balance (BREB) method (Baker and Norman, 2002, §3.7.3.1) is an indirect method that has proven to be very accurate under normal atmospheric conditions (Dugas et al., 1991). It greatly depends on upwind atmospheric conditions and therefore requires a fetch of several hundreds of meters. The main advantage is that it does not require any wind velocity measurements. Under certain circumstances (e.g., just after sunrise and before sunset) the temperature and humidity differences, as well as R_{ng}, may be very small, leading to an imprecise estimation of λE with Eq. (4.46). A critical issue is still the adequate measurement of air humidity, especially in arid regions where the difference between e_1 and e_2 can be very small. The sensors required are two temperature and relative humidity probes, mounted at two heights (eventually, one probe on a moving arm in order to overcome differences in instrument calibration), a net radiometer to measure R_{ng} and a set of soil-heat flux plates and thermocouples to determine G (see also Sec. 13.4.2).

A different approach is given by the eddy correlations method. This direct λE measurement method is based on the fact that the vertical vapor flux within the turbulent layer occurs in eddy forms. Fluctuations from the time-average observations of vertical velocity, u', and vapor density, q', are simultaneously measured at very high frequencies (e.g., 20 Hz) using fast-response sensors. Under conditions of upward vapor transport, the time-averaged product (from 15 min to 1 h) of both terms (covariance), $\overline{u'q'}$, gives λE:

$$\lambda E = \lambda \overline{u'q'} \tag{4.48}$$

The eddy correlation method is very accurate and, due to its important instrumental needs, is generally only used for research purposes (e.g., Moncrieff et al., 1997; Villalobos, 1997). However, it depends heavily on upwind atmospheric conditions, and, in the absence of an adequate fetch, it can be difficult to interpret its results. The system consists basically of a (three-dimensional) sonic anemometer and a fast-response krypton hygrometer. The system is usually equipped with additional sensors in order to evaluate the system performance with the aid of Eq. (4.2) variants: a fine-wire thermocouple to compute H, a net radiometer to measure R_{ng}, a set of soil-heat flux plates and thermocouples to determine G, and eventually a temperature and relative humidity probe. Rochette and McGinn (see Chapter 13) discuss the topic in more detail.

4.3.2.3   Plant Physiology

Since the measurement of evapotranspiration fluxes is scale-dependent, the determination of transpiration flux from either individual plants or certain vegetative parts of the plant requires specific methods, such as those based on sap flow measurements and ET measurements made in plant chambers.

The earliest method developed for sap flow measurement, the heat pulse method, has exhibited accuracy and calibration problems and will not be further discussed here. The alternative heat balance method was proposed by Cermak et al. (1976) for use in trees and was used by Sakuratini (1981) in herbaceous plants. Basically, this method consists of the heating of a section of the plant stem and the simultaneous upward, downward, and radial monitoring of heat conduction. The difference between the supplied and conducted heat is assumed to be directly related to the sap flow and to plant transpiration. Kjelgaard et al. (1997) deal with improvements in micro-measurement equipment, and Smith and Allen (1996) give a detailed discussion of sap flow methods.

Sap flow measurements can be highly inaccurate and depend greatly on operational details. They are generally not appropriate for field (agro) hydrological purposes and do not take into account bare soil evaporation.

The use of plant chambers is a direct measurement method to evaluate the evapotranspiration of individual plants or small communities by measuring the increased water content of the surrounding air. This can be done by installing a closed polyethylene chamber over the plant(s) and monitoring vapor pressure within the chamber (Reicosky, 1990) or by using an open chamber where the difference between the vapor pressure of incoming and outgoing air is recorded (García et al., 1990). This method should be used with caution because the air temperature within the polyethylene chamber is artificially increased and triggers the biological control mechanism of the leaves so that transpiration is reduced. This method also takes into account direct evaporation from the soil below the plant(s).

4.3.2.4   ET Modeling

Numerous improvements of the original Penman (1948) equation for estimating the latent heat flux from free water surfaces have led to the development of the widely used Penman-Monteith combination equation:

$$\lambda E = \frac{\Delta(R_{ng} - G) + \rho_a c_p (e_s - e_a)/r_a}{\Delta + \gamma(1 + r_s/r_a)} \tag{4.49}$$

where Δ is the slope of the saturation water vapor pressure-temperature curve, ρ_a is the air density, c_p is the specific heat of air, γ the psychrometric constant, e_s–e_a the saturation vapor pressure deficit, r_a is the bulk aerodynamic resistance, and r_s is the bulk surface resistance. In order to calculate λE, R_{ng} G, and e_s–e_a (e_s can be calculated from air temperature) must be calculated. The parameters r_a and r_s must be modeled (see Allen et al., 1998).

This well-known method models evapotranspiration flux, recognizing that the latent heat flux is due on the one hand, to net radiation, and, on the other, to "mass transfer." The method is seldom used for direct estimation of λE due to difficulties in modeling r_a and r_s. However, under well-watered conditions the model has been seen to reproduce lysimeter measurements well (Jensen et al., 1990) and has become a standard method in agriculture and environmental applications for calculating reference crop evapotranspiration (ET$_0$) (see Allen et al., 1998, for a thorough discussion).

The empirical methods used to calculate crop evapotranspiration (ET$_c$) basically consist of characterizing the crop coefficient, K_c, and its evolution through the crop cycle, which transforms ET$_0$ into ET$_c$. Allen et al. (1998) compiled the crop coefficients for different conditions and management practices.

The FAO Penman-Monteith method is recommended as being the sole method for determining ET$_0$, except when observations are available from an evaporation Class A pan (Allen et al., 1998). When there are no wind or humidity measurements, one has to rely on empirical temperature based methods (e.g., Hargreaves and Allen, 2003). Jensen et al. (1990) provide an extensive comparison of empirical methods for estimating ET$_0$.

4.3.2.5   Strengths and Weaknesses

Evapotranspiration methods are based on sound physical principles yielding robust models like the Penman combination equation. The development of new techniques is improving the accuracy of measurements. On the other hand, the quick diffusion of water vapor into the atmosphere and the inherent variability of the air movement hinder the estimation of water losses from the soil-plant boundary. The problem is greater in large areas where simplifying hypotheses are used, like the complementary relationship of Bouchet (Brutsaert, 1982, § 10.3b), which require additional field contrast (e.g., Parlange and Katul, 1992).

4.3.3   RECHARGE AND TEMPORAL SOIL WATER
CONTENT VARIATIONS

The replenishment of the soil profile with water, or recharge, is characterized by different methods, based on the water balance equations described above. Nevertheless, more refined equations like the Richards equation for soil water flow (e.g., Kutilek and Nielsen, 1994, § 5.3.3) can be used when estimation of the equation various input parameters is simplified. Scott et al. (2000) proposed a method to characterize the soil moisture recharge below the depth of 0.3 m in the semiarid rangeland of Arizona, using the one-dimensional soil water movement model HYDRUS (Šimunek et al., 1997). The model parameters were estimated via soil profile description and the use of pedotransfer functions (Schaap et al., 1998). The authors compared this direct method with an inverse method based on the shuffled through the complex evolution algorithm of Duan et al. (1992). The model parameters were those of the van Genuchten soil water retention equation (Kosugi et al., 2002, § 3.3.4.4 and 3.3.4.8): residual and saturated moisture contents, respectively θ_r and θ_s, the coefficient α and the exponent n, and the hydraulic conductivity at saturation, k_s, used in the Mualem equation. The results indicated that the automatic calibration of model parameters is an useful tool, although a better knowledge of parameter variability at heterogeneous field sites is needed to improve the fit of computed to observed data.

The estimation of soil water recharge is part of the larger groundwater recharge problem commonly found in arid and semiarid areas. Here a distinction is often made between the local and localized recharge. Local recharge is the amount of water that reaches the water table by percolation through the unsaturated or vadose zone, while the localized recharge is more regional (Allison, 1988). There are several methods to estimate the local recharge, which can be loosely grouped into physical and chemical methods.

Physical methods range from the simple soil water balance to the formal statement and solution of the Richards equation. Another method is the zero-flux plane technique. As pointed out by Kutilek and Nielsen (1994, § 6.4.2), the zero flux plane marks the position of zero water flux density. Above the zero-flux plane water moves upward to an evaporation boundary at the soil surface, and below the plane water drains downward toward the water table. The recharge is estimated by the addition of all the water content changes, which can be detected with a tensiometer or an electromagnetic device. Lysimetry would be a good method, as recognized by Gee and Hillel (1988), although it has many limitations associated with installation and maintenance.

Chemical methods for recharge estimations are based on the coupled solute and water movement in the soil. Natural tracer methods (e.g., deuterium, tritium, oxygen-18, cesium) are based on the position of the maximum tracer concentration in the soil profile and assume that the water above this depth is the amount of recharge that has occurred after the natural tracers event has ceased (e.g., fallout from nuclear bomb testing).

Another method simply estimates the amount of tracer in the soil profile, e.g., the chloride balance method (CBM), which Wood and Sanford (1995) used to obtain reliable estimates of groundwater recharge (11–77 mm/y) in the southern high plains of Texas using chloride and tritium methods that were also noted for their economy and ease of use. Unfortunately, tracer movement through soil is often too complex to be analyzed adequately using the simplified approaches. Melayah et al. (1996) proposed a complete model for the coupled transport of water and solutes, including transport. Their results generally agreed well with measured deuterium and ^{18}O profiles, although some differences were found that the authors attributed to the initial state of the profile and model sensitivity to small changes in the convective transport in the liquid phase. Scanlon and Milly (1994) stressed the importance of temperature-induced fluctuations in water potential for recharge study in the Chihuahan Desert of Texas; they attributed extensive variability in the predominantly upward liquid fluxes to differences in water retention functions and to the initial conditions that affected the hydraulic conductivity values. Walker et al. (1991) proposed a simplification of the Melayah et al. (1996) method integrating water and chloride balances to estimate water recharge, although spatial variability caused some interpretation problems.

A new approach to tracer studies was proposed by Ginn and Murphy (1997), based on neglect of the diffusive and dispersive solute transport processes. The tracer mass balance equation is

$$\frac{\partial}{\partial t}(c\theta) + \frac{\partial}{\partial x}(cq) = 0 \qquad (4.50)$$

where c is solute (tracer) concentration, the volumetric water content is θ the flow rate is q, t is time, and x is the vertical coordinate. The mass balance for the water content is

$$\frac{\partial \theta}{\partial t} + \frac{\partial q}{\partial x} = -q_{ex} \qquad (4.51)$$

where q_{ex} is the evaporative removal specific flow of water in the root zone. Rearranging both equations leads to a kinematic water-type equation:

$$\theta \frac{\partial c}{\partial t} + q \frac{\partial c}{\partial x} = cq_{ex} \qquad (4.52)$$

The solution of this equation subjected to the initial condition, $c(x,0) = c_i(x) = 0$, and to the boundary conditions of flux, the annual precipitation, $p(t)$, and tracer concentration, $c_0(t_0) = m_0(t_0)/p(t_0)$, with the annual deposition of chloride, $m_0(t_0)$, is,

$$t_0(x, t) = P^{-1}[P(t) - \tau_0(x)] \qquad (4.53)$$

for the time of entry into the soil column of an element of water located at depth x and time t, $P(t)$ is the cumulative precipitation up to time t, P^{-1} is its inverse function, and $\tau_0(x)$ the travel time to depth x of the tracer for $p(t) = 1$. The concentration for the same variables is

$$c(x, t) = \frac{m_0[t_0(x, t)]}{q_0(x)p[t_0(x, t)]} \qquad (4.54$$

with $q_0(x)$ as the dimensionless water flow rate at depth x for $p(t) = 1$. These equations are useful for the estimation of the age of the water in the profile. The authors also derived an inverse solution for interpreting recharge data.

4.4   THE REMOTE SENSING PERSPECTIVE

Remotely sensed electromagnetic spectra acquired by airborne and satellite systems can inform about important surface parameters and processes. Physical quantities cannot be determined directly by these instruments, but they can be characterized by using electromagnetic properties. These spatially distributed data with a high temporal frequency and low cost permit the extension of energy and water balance calculations to large areas when combined with meteorological and ancillary data.

There is a wide variety of sensors providing a global coverage of the earth at different spatial and temporal resolutions. As they are sensitive to part of the electromagnetic spectrum, it is necessary to distinguish whether reflected or emitted radiation is being measured from the land surface. Reflected radiation is usually reflective solar, in the visible and near infrared (VIS-NIR) region of the electromagnetic spectrum, and is commonly applied to derive land cover maps and to determine the properties and classifications of vegetation, soil, and snow cover. Emitted radiation can be divided, in a growing wavelength sense, into thermal infrared (TIR) and microwave. The surface temperature, obtained from the thermal region, has been related to soil moisture content, plant transpiration, and plant water stress. Passive sensors measure microwave radiation, and active radar systems generate and transmit a signal towards the target and measure the echo produced after its interaction with it. The information from these passive and active sensors has been widely used to determine soil moisture, vegetation content, and snow properties.

The spatial resolution of available remote sensors varies from 1 m to several km, allowing the monitoring of processes from local to global scales. A mesoscale ranging from 30 m to 1 km is most frequently used in hydrological applications. The spatial variability of the different variables measured, along with the regular temporal repetition, is the main strength of remote sensing data.

TABLE 4.2
Variables Provided by Remote Sensors at the Soil-Plant-Atmosphere Interface

VIS-NIR	TIR	Microwave
Vegetation cover	Surface temperature	Soil moisture
Soil class		Vegetation cover
Snow delimitation and classification		Snow cover
Land use maps		

4.4.1 RELATIONS BETWEEN SPECTRAL MEASUREMENTS AND BIOPHYSICAL PROPERTIES

The properties of the terrestrial surface that can be measured by remote sensing are shown in Table 4.2, classified by the spectral region of the signals. The surface processes partially or completely controlled by these variables include, among others, soil evaporation and plant transpiration, rain interception, soil erosion, and snowmelt runoff.

4.4.1.1 VIS-NIR

Solar radiation reflecting properties of the upper layer of land surface, mostly soil, vegetation, or snow, depend upon its composition and state. Factors controlling that reflection are different in every material, and the information that can be inferred from it also varies.

Soils intercept much of the solar energy incident in the land surface; even in densely vegetated areas a portion of the radiation finally reaches the underlying soil. This interaction plays an important part in soil-forming processes and, more specifically, in the heat balance of soil. The reflecting properties of soils are more influenced by the moisture conditions, texture, and structural arrangement of the constituents than by their chemical composition (Mulders, 1987, §3). Therefore, the strength of remote sensing in soil studies is more related to an integration of those aspects over relatively large areas rather than detailed information on individual constituents.

Aerial photographs have been used for decades as an aid to soil mapping and a cartographic base for published soil maps (Irons et al., 1987), and the computer-aided analysis of satellite data assists the soil scientist in the preparation of soil-association maps, the delineation of parent-material boundaries, the refinement of soil-map unit boundaries, and the preparation of single-features maps, such as drainage and organic matter content (Weismiller and Kamisnsky, 1978).

In the case of vegetation, the characteristics that can be obtained by remote sensing include (Goel, 1989): (1) identity of species or families, (2) growth or development stage, (3) level of stress, (4) amount of green vegetation, expressed as the biomass or the area indices of vegetation elements (leaves, branches, stems, etc.), and (5) the architecture of the canopy (the spatial distribution of vegetated and nonvegetated areas on the ground, and the density and orientation of vegetation elements within the vegetated areas).

For individual leaves, the processes controlling their spectral properties are absorption, reflection, and transmission through the leaf, the fraction of every flux depending on leaf structure, water content, and concentration of bio-chemicals (Gates et al., 1965; Wessman, 1990; Fourty et al., 1996), and the wavelength and incidence angle of the incoming radiation (Goel, 1987; Myneni and Asrar, 1993; Meyer et al., 1995). For canopies under field conditions, single and multiple scattering from leaves and soil make the deciphering of the spectral response more complex and dependent on a number of factors such as canopy architecture, soil reflectance, and atmospheric influence.

The response of a drainage basin to the input of water and its water and energy budgets are significantly modified with the occurrence of snow, hence the quantification and monitoring of snowmelt is critically important. Snow hydrology was one of the first disciplines to make an operative use of remote sensing data. The separability of snow spectral response from other surfaces, a relative uniformity, and enough extension for satellite systems are the main reasons for this development. Most of the applications correspond to the VIS-NIR region, where the snow spectral answer depends on the impurity content, size and shape of snow needles, snow depth, liquid water content, surface roughness and density (Wiscombe and Warren, 1980). The very high reflectance of recent snow ($\alpha > 0.9$) can even decrease to 0.4 for old and dirty snow (Bohren, 1986). Snow maps, obtained from remote images with spatial resolutions of between 30 m and 1 km and a variable frequency of observations (12 h to 16 days), have shown themselves to be useful for monitoring the build-up and depletion of snow (Schmugge et al., 2002). The snow runoff model (SRM) (Martinec et al., 1982), based on the degree day approach to melting the snow cover in a basin, was designed to assimilate satellite data for snow cover measurements and has been successfully employed in 25 countries worldwide. In Spain, using NOAA-AVHRR, SRM has been used to monitor snowmelt to assist in planning hydropower production in the Pyrenees (Gomez-Landesa and Rango, 1998) and for irrigation applications in Sierra Nevada (Torres, 1998).

4.4.1.2 Thermal Infrared

Surface temperature can be obtained from radiation measurements in the thermal infrared, as stated by Planck's law. The measurement of the sensor in remote systems includes the effect of the radiation reflected by the atmosphere. The correction for the atmospheric and surface emissivity effects is the main difficulty for an accurate measurement of the surface temperature, with a deviation that can be as much as several degrees (Sobrino et al., 2000).

Transpiration is the dominant mechanism controlling plant temperature at a canopy level (Idso and Baker, 1967), as transpired water evaporates and cools the leaves below the temperature of the surrounding air. When water becomes limited, transpiration is reduced and the leaf temperature increases (Jackson, 1982). From energy balance considerations, Monteith and Szeicz (1962) derived an expression relating the difference between canopy and air temperatures to net radiation, wind speed, vapor pressure gradient, and the aerodynamic and canopy resistances. They suggested that radiometrically determined surface temperatures could be used to evaluate the effective stomatal resistance of field crops. On this basis, a crop water stress index (CWSI) was developed by Idso and Jackson (Jackson et al., 1981) as the ratio of actual to potential latent heat flux, this index being commonly used for the detection of plant stress [see review by Moran and Jackson (1991) and Norman et al. (1995)]. The same concept applied to partially vegetated fields led to the definition by Moran et al. (1994) of the water deficit index (WDI), which combined surface temperature with vegetation indices to discriminate between the vegetation fraction and its water stress. Another measurement of relative evaporation is the surface energy balance index (SEBI) by Menenti and Choudhury (1993); this algorithm scales the temperature observed in a theoretical range of differences between surface and air temperatures. The implementation of this concept in the surface energy balance system (SEBS) (Su and Jacobs, 2001; Su, 2002) has provided relevant information for water management applications (Menenti et al., 2003).

A different approach is the use of radiometric surface temperature to calculate sensible heat flux (H) and then obtain latent heat flux (λE) or ET, in water balance terms, as the residual of the energy balance equation (Choudhury et al., 1986). Kustas and Norman (1996) review the methods for obtaining the components of this equation using remote sensing data. The largest uncertainty is associated with the computation of H, while $R_{ng} - G$ can be estimated within a 10% error when reliable estimates of solar radiation are available (Schmugge et al., 2002). Another problem is the integration of the instantaneous remote observation, a solution based on the relation between λE and solar radiation being frequently applied.

4.4.1.3 Microwave

According to their mode of operation, sensors in the microwave region can be divided into two groups: active and passive sensors. Active sensors include a transmitter that sends a pulse to the target, and a receiver that measures the reflected echo, in terms of a scattering coefficient, $\sigma°$. The scattering coefficient is the dimensionless average radar cross section per unit area $\sigma° = \langle\sigma\rangle/A_0$ σ being the cross section of an object observed in a given direction, defined as the cross section of an *equivalent isotropic scatterer* that generates the *same* scattered power density as the object. With this common reference, the scattering strengths of different objects can be compared (see Fung, 1994, §1, and Ulaby et al., 1986, §13, for a detailed description), and σ can be written as

$$\sigma = \frac{4\pi R^2 |E^S|^2}{|E^i|^2} \tag{4.55}$$

where R is the distance between the target and the radar receiver, E^i the incident field, and E^S the scattered field along the direction under consideration.

Passive sensors are simple logwave radiation receivers that measure the natural radiation emission or brightness temperature of the terrain surface. Both active and passive sensors have the advantage over optical sensors of penetrating clouds, thereby providing all-weather coverage and allowing observations even at night, as the measurements are independent of solar illumination.

Remote passive sensors have recently been used to retrieve atmospheric and oceanic parameters, such as atmospheric temperature and water vapor density over oceans. Over land, with the exception of snow monitoring (Chang et al., 1987; Rango et al., 1989; Goodison and Walker, 1995), problems related to spatial resolution (approximately tens of kilometers) and the complex mechanisms responsible for microwave emission have limited the application of remote passive sensors, as reported by Ulaby et al. (1986, §19).

Snow can be considered a multilayer media, where the microwave emission from the ground surface is scattered by ice needles, resulting in a lower emission at the top of the snow. Information on snow extent and depth, snow water equivalent, and presence of liquid water can be derived from the brightness temperature measured by the microwave sensor (Fung, 1994), with the drawback that its coarse resolution limits the application to large drainage basins or regional studies.

Active sensors have a smaller pixel size, ranging from $1\,m^2$ to several hm^2, a fact that, along with the increased control over the signal, improves their suitability for land applications. The scattering behavior of a surface depends on terrain parameters like geometry, surface roughness, moisture content, and radar instrument parameters (wavelength, depression angle, polarization, etc.) (Jensen, 2000).

Remote radar systems constitute the most promising technique to measure distributed soil water content across a range of spatial scales. The basis for soil moisture estimation is the largest contrast between the value of the dielectric constant in water (80) and the value found in dry soils (~2), as the target dielectric constant is directly related to the radar backscattering coefficient σ^o). Since the early 1970s, extensive research has been conducted in this direction (Fung, 1994; Schmugge and Jackson, 1994; Ulaby et al., 1996, Verhoest et al., 1998), concluding that soil moisture content of the top 5 cm of the soil can be determined. In practice, other factors, including vegetation density, topography, and surface roughness, influence the response (Moran et al., 2000), which reduces the applicability of the microwave approach in a number of situations, such as forested or mountainous areas.

4.4.1.4   Strengths and Weaknesses

Remote sensing opens a new perspective in the study of terrestrial surface processes. The scope of the working area and the frequency of measurements allow a detailed study of the soil-plant-atmosphere interface. New techniques are being incorporated successfully every day. Nevertheless, there are still some obscure points that deserve more attention, such as the geometric and atmospheric interferences and the representativeness of the magnitudes, according to the inference of the profile characteristics from near-surface measurements of soil moisture and surface temperature of Walker et al. (2001).

4.5   RECOMMENDATIONS AND FUTURE RESEARCH

Water and energy are exchanged at the soil-plant-atmosphere interface through a great variety of processes, most of them described succinctly here. Selection of the best methods to measure these processes require knowledge of the relevant spatial and temporal scales and a determination of the uncertainty associated with the variables of interest.

There are good, reliable methods for the study of transfer processes, consisting of simple measurements based on physical principles. These methods must be a first approach to the use of more detailed instrumental methods, which require thorough calibration.

Future research may be addressed to the development of assimilation techniques, based on physical models which may enhance the efficiency of field measurements.

NOTATION

Symbol	Description	Eq.
P	precipitation flux $[ML^{-2}t^{-1}]$	(4.1)
I	infiltration flux $[ML^{-2}t^{-1}]$	(4.1)
R	runoff flux $[ML^{-2}t^{-1}]$	(4.1)
E	evaporation flux $[ML^{-2}t^{-1}]$	(4.1)
R_{ng}	net radiation flux $[Mt^{-3}]$	(4.2)
G	soil heat flux $[Mt^{-3}]$	(4.2)
λ	latent heat of vaporization of water $[L^{2}t^{-2}]$	(4.2)
H	sensible heat flux $[Mt^{-3}]$	(4.2)
I_d	deep infiltration flux $[ML^{-2}t^{-1}]$	(4.3)
S	water storage in the control volume of soil $[ML^{-2}]$	(4.3)
Q	heat storage in the control volume of soil $[Mt^{-2}]$	(4.4)
G_d	deep soil heat flux $[Mt^{-3}]$	(4.4)
P_e	effective precipitation flux $[ML^{-2}t^{-1}]$	(4.5)
I_n	interception flux $[ML^{-2}t^{-1}]$	(4.5)
	soil moisture volumetric content $[LL^{-1}]$	(4.6)
ρ_w	density of the water $[ML^{-3}]$	(4.6)
n	soil porosity $[L^{3}L^{-3}]$	(4.6)
z_h	soil depth affected by water changes $[L]$	(4.6)
q_m	specific humidity of air $[MM^{-1}]$	(4.7)
g	acceleration of the gravity $[Lt^{-2}]$	(4.7)

Symbol	Description	Eq.
p_s	atmosphere pressure at the soil surface $[ML^{-1}t^{-2}]$	(4.7)
p_h	mean atmosphere pressure of the air slab $[ML^{-1}t^{-2}]$	(4.7)
ΔQ	moisture increment in the air slab $[ML^{-2}t^{-1}]$	(4.7)
T_g	ground temperature $[T]$	(4.8)
c_{soil}	volumetric heat capacity of the soil $[ML^{-1}t^{-2}T^{-1}]$	(4.8)
c_p	specific heat of air at constant pressure $[L^2 t^{-2}T^{-1}]$	(4.9)
z_t	soil depth affected by energy changes $[L]$	(4.8)
θ_m	potential temperature of the air slab $[T]$	(4.9)
R_{na}	net radiation flux in the atmosphere slab $[Mt^{-3}]$	(4.9)
ΔH	energy increment in the air slab $[Mt^{-3}]$	(4.9)
RS	shortwave radiation flux $[Mt^{-3}]$	(4.10)
RL_{ad}	longwave radiation flux emitted from the atmosphere descending on the slab $[Mt^{-3}]$	(4.10)
RL_{sd}	longwave radiation flux descending on the slab $[Mt^{-3}]$	(4.10)
RL_{su}	longwave radiation flux ascending on the slab $[Mt^{-3}]$	(4.11)
RL_{gu}	longwave radiation flux descending on the slab $[Mt^{-3}]$	(4.10)
α	soil albedo	(4.10)
ε_{col}	air emissivity/absorptivity	(4.10)
H_{in}	windward sensible heat supply in the slab $[Mt^{-3}]$	(4.12)
H_{out}	leeward sensible heat supply in the slab $[Mt^{-3}]$	(4.12)
H_{top}	sensible heat supply at the top of the slab $[Mt^{-3}]$	(4.12)
s_s	average moisture at the surface soil layer $[L^3 L^{-3}]$	(4.13)
s_m	average moisture in the soil profile $[L^3 L^{-3}]$	(4.13)
z_s	soil depth in the water change of the force-restore model $[L]$	(4.13)
c_1	constant of the force-restore model	(4.13)
c_2	constant of the force-restore model	(4.13)
τ	time constant of the force-restore model $[t]$	(4.13)
s_{fc}	average moisture in the soil profile at soil capacity $[L^3 L^{-3}]$	(4.14)
z_m	soil depth in the water change of the force-restore model $[L]$	(4.14)
c_3	constant of the force-restore model	(4.14)
c_t	constant of the force-restore model	(4.15)
z_t	soil depth in the temperature change of the force-restore model $[L]$	(4.15)
T_m	average soil profile temperature $[T]$	(4.15)
T_s	average soil surface temperature $[T]$	(4.15)
$\varphi = P - I_n - R$	rate of water infiltration into the soil $[Lt^{-1}]$	(4.17)
$\chi = E + I_d$	rate of water loss from the soil $[Lt^{-1}]$	(4.17)
h'_i	net rainfall depth $[L]$	(4.18)
δ	function delta of Dirac	(4.18)
$'_i$	arrival time of the events $[t]$	(4.18)

Symbol	Description	Eq.
h	hygroscopic soil water content $[L^3 L^{-3}]$	(4.19)
w	water content at the wilting point $[L^3 L^{-3}]$	(4.19)
$*$	water content at full stomatal opening $[L^3 L^{-3}]$	(4.19)
E_w	lower soil water evaporation rate $[Lt^{-1}]$	(4.19)
E_{max}	higher soil water evaporation rate $[Lt^{-1}]$	(4.19)
k_s	saturated hydraulic conductivity of soil $[Lt^{-1}]$	(4.20)
β_k	a coefficient of the hydraulic conductivity of the soil	(4.20)
P_A	average annual precipitation $[L]$	(4.22)
E_{TA}	average annual transpiration $[L]$	(4.22)
R_{sA}	average annual surface runoff $[L]$	(4.22)
R_{gA}	average annual groundwater runoff $[L]$	(4.22)
ω	inverse of the average storm intensity $[L^{-1}t]$	(4.23)
ζ	inverse of the average storm duration $[t^{-1}]$	(4.23)
G_*	auxiliary function of the Eagleson model	(4.23)
0	initial value of soil moisture $[L^3 L^{-3}]$	(4.24)
c_{BC}	pore permeability index of the soil according to Brooks and Corey	(4.24)
σ_*	parameter dependent on soil and climate properties	(4.23)
$\zeta(\sigma_*)$	surface runoff function	(4.23)
β_s	bare soil evaporation efficiency	(4.25)
E_*	bare soil evaporation effectiveness	(4.26)
E_v	potential plant transpiration rate	(4.27)
k_v^*	unstressed canopy conductance	(4.27)
β_v	canopy transpiration efficiency	(4.28)
h_0	depth of surface water retention $[L]$	(4.29)
$\gamma(\cdot,\cdot)$	incomplete gamma function	(4.29)
M	fraction of surface covered by the canopy	(4.30)
w	water storage in the soil profile $[L]$	(4.33)
Q_{se}	saturation excess runoff rate $[Lt^{-1}]$	(4.33)
W_b	storage capacity of the soil $[L]$	(4.33)
Q_{ss}	subsurface runoff rate $[Lt^{-1}]$	(4.34)
E_b	bare soil evaporation rate $[Lt^{-1}]$	(4.34)
E_v	plant transpiration rate $[Lt^{-1}]$	(4.34)
W_{fc}	soil water storage at field capacity $[L]$	(4.35)
c	characteristic time of the catchment response in the groundwater flow $[t]$	(4.35)
e	characteristic time of soil evaporation process $[t]$	(4.36)
g	characteristic time of plant transpiration process $[t]$	(4.37)
a	parameter of the groundwater flow $[L]$	(4.38)
b	parameter of the groundwater flow	(4.38)
W_g	water content of the deep groundwater store $[L]$	(4.39)
Q_{sg}	deep groundwater flow rate $[Lt^{-1}]$	(4.39)
λ_g	fraction of the deep groundwater store	(4.40)
P_c	canopy drip flux $[ML^{-2}t^{-1}]$	(4.41)

Symbol	Description	Eq.
P_{th}	canopy throughfall flux $[ML^{-2}t^{-1}]$	(4.41)
S_{can}	water storage in the canopy $[ML^{-2}]$	(4.41)
S_{can0}	initial water content of the canopy $[ML^{-2}]$	(4.41)
$S_{can\,max}$	maximum water content of the canopy $[ML^{-2}]$	(4.41)
k_{can}	maximum canopy drainage flux $[ML^{-2}t^{-1}]$	(4.43)
P_T	cumulative precipitation depth $[L]$	(4.44)
β	Bowen ratio	(4.46)
γ	psychrometric constant $[ML^{-1}t^{-2}T^{-1}]$	(4.47)
$\overline{u'q'}$	covariance of the air and water upward flux $[ML^{-2}t^{-1}]$	(4.48)
Δ	slope of the saturation water vapor pressure-temperature curve $[ML^{-1}t^{-2}T^{-1}]$	(4.49)
ρ_a	air density $[ML^{-3}]$	(4.49)
e_s-e_a	saturation vapor pressure deficit $[ML^{-1}t^{-2}]$	(4.49)
r_a	bulk aerodynamic resistance $[L^{-1}t]$	(4.49)
r_s	bulk surface resistance $[L^{-1}t]$	(4.49)
c	solute concentration $[ML^{-3}]$	(4.50)
θ	water content $[L^3L^{-3}]$	(4.50)
q	solution flow rate $[Lt^{-1}]$	(4.50)
q_{ex}	evaporative removal specific flow of water in the root zone $[t^{-1}]$	(4.51)
$m_0(t_0)$	annual deposition of chloride $[ML^{-2}t^{-1}]$	(4.54)
$\tau_0(x)$	travel time to depth x $[t]$	(4.54)
σ°	scattering coefficient	
σ	cross section of an equivalent isotropic scatterer	
R	range between the target and the radar receiver	
E^i	incident field	
E^S	scattered field along the direction under consideration	

REFERENCES

Abbot, M.B., Bathurst, J.C., Cunge, J.C., O'Connell, P.E., and Rasmussen. J., An introduction to the European hydrological system — système hydrologique Européen SHE. 1. History and philosophy of a physically-based distributed modeling system, *J. Hydrol.*, 87, 45–59, 1986a.

Abbot, M.B., Bathurst, J.C., Cunge, J.C., O'Connell, P.E., and Rasmussen. J., An introduction to the European hydrological system — système hydrologique Européen SHE, 2. Structure of a physically-based distributed modeling system, *J. Hydrol.*, 87, 61–77, 1986b.

Allen, R.G., Jensen, M.E., Walter, I.A., Pruitt, W.O., and Howell, T.A., *Lysimeters for Evapotranspiration and Environmental Measurements*, Proceedings of the International Symposium on Lysimetry, July 23–25, 1991, Honolulu, Hawaii, ASCE, New York, 1991.

Allen, R.G., Pereira, L.S., Raes, D., and Smith, M., *Crop Evapotranspiration*, FAO Irrigation and Drainage Paper 56, Rome, 1998.

Allison, G.B., A review of some of the physical, chemical, and isotopic techniques for estimating groundwater recharge, in Simmers, I., ed., *Estimation of Natural Groundwater Recharge*, Reidel, Dordrecht, 49–72, 1988.

Baker, J.M., and Norman, J.M., Evaporation from natural surfaces, in Dane, J.H. and Topp, G.C., eds., *Methods of Soil Analysis, Part 4, Physical Methods*, Soil Sci. Soc. Am. Book Ser. No. 5, Soil Science Society of America, Madison, WI, 1047–1074, 2002.

Beven, K.J., *Rainfall-Runoff Modeling. The Primer*, Wiley, Chichester, 2000.

Beven, K., and Kirkby, M.J., A physically based, variable contributing area model of basin hydrology, *Hydrol. Sci. Bull.*, 24, 43–69, 1979.

Blöschl, G., and Sivapalan, M., Scale issues in hydrological modeling: a review, in *Scale Issues in Hydrological Modeling*, Kalma, J.D., and Sivapalan, M., eds., Wiley, New York, pp. 9–48, 1995.

Bohren, C.F., Applicability of effective-medium theories to problems of scattering and absorption by non-homogeneous atmospheric particles, *J. Atmos. Sci.*, 43, 463–475, 1986.

Bouten, W., Swart, P.J.F., and de Water, E., Microwave transmission, a new tool in forest hydrological research, *J. Hydrol.*, 124, 119–130, 1991.

Bras, R.L., *Hydrology*, McGraw-Hill, New York, 1990.

Brubaker, K.L., and Entekhabi, D., An analytic approach to modeling land-atmosphere interaction, 1. Construct and equilibrium behavior, *Water Resour. Res.*, 31, 619–632, 1995.

Brutsaert, W., *Evaporation into the Atmosphere. Theory, History and Applications* Reidel, Dordrecht, 1982.

Budyko, M.I., *Climate and Life*, Academic Press, New York, 1974.

Calder, I.R., *Evaporation in the Uplands*, Wiley, Chichester, 1990.

Calder, I.R., and Rosier, P.T.W., The design of large plastic-sheet net-rainfall gauges, *J. Hydrol.*, 30, 403–405, 1976.

Cermak, J., Kucera, J., and Penka, M., Improvement of the method of sap flow rate determination in full grown trees based on heat balance with direct heating of xylem, *Biol. Plant.*, 15, 171–178, 1976.

Chang, A.T.C., Foster, J.L., and Hall, D.K., Nimbus-7 SMMR derived global snow cover parameters, *Ann. Glaciol.* 9, 39–44, 1987.

Choudhury, B.J., Reginato, R.J., and Idso, S.B., An analysis of infrared temperature observations over wheat and calculation of the latent heat flux, *Agric. Forest Meteorol.*, 37, 75–88, 1986.

Deardorff, J.W., Efficient prediction of ground surface temperature and moisture with inclusion of a layer of vegetation, *J. Geophys. Res.*, 83, 1899–1903, 1978.

Dooge, J.C.I., Parameterization of hydrological processes, in *Land Surface Processes in Atmospheric General Circulation Models*, Eagleson, P.S., ed., Cambridge University Press, Cambridge, 243–288, 1982.

Duan, Q., Sorooshian, S., and Gupta, V.K., Effective and efficient global optimization for conceptual rainfall-runoff models, *Water Resour. Res.*, 28, 1163–1172, 1992.

Dugas, W.A., Fritchen, L.J., Gay, L.W., Held, A.A., Matthias, A.D., Reicosky, D.C., Steduto, P., and Steiner, J.L., Bowen ratio, eddy correlation, and portable chamber measurement of sensible and latent heat flux over irrigated spring wheat, *Agric. Forest Meteorol.*, 56, 1–20, 1991.

Eagleson, P.S., Climate, soil, and vegetation, *Water Resour. Res.*, 14, 705–776, 1978.

Eagleson, P.S., Ecological optimality in water-limited natural soil-vegetation systems. 1. Theory and hypothesis, *Water Resour. Res.*, 18, 325–340, 1982.

Eagleson, P.S., *Ecohydrology. Darwinian expressions for vegetation form and function* Cambridge University Press, Cambridge, 2002.

Entekhabi, D., and Brubaker, K.L., An analytic approach to modeling land-atmosphere interaction, 2. Stochastic formulation, *Water Resour. Res.*, 31, 633–643, 1995.

Evett, S.R., Warrick, A.W., and Matthias, A.D., Wall materials and capping effects on microlysimeter temperatures and evaporation, *Soil Sci. Soc. Am. J.*, 59, 329–336, 1995.

Famiglietti, J.S., and Wood, E.F., Multiscale modeling of spatially variable water and energy balance processes, *Water Resour. Res.*, 30, 3061–3078, 1994.

Fourty, Th., Baret, F., Jacquemond, S., Schmuck, G., and Verdebout, J., Leaf optical properties with explicit description of its biochemical composition: direct and inverse problems, *Rem. Sens. Environ.*, 56, 104–117, 1996.

Fung, A.K., *Microwave Scattering and Emission Models and Their Applications*, Artech House, Norwood, 1994.

García, R.L., Norman, J.M., and McDermitt, D.K., Measurements of canopy gas exchange using an open chamber system, *Rem. Sens. Rev.*, 5, 141–162, 1990.

Gash, J.H.C., An analytical model of rainfall interception by forests, *Quart. J. R. Met. Soc.*, 105, 43–55, 1979.

Gash, J.H.C., Lloyd, C.R., and Lachaud, G., Estimating sparse forest rainfall interception with an analytical model, *J. Hydrol.*, 170, 79–86, 1995.

Gates, D.M., Keegan, H.J., Schleter, J.C., and Wiedner, V.R., Spectral properties of plants. *Appl. Opt.* 4, 11–20, 1965.

Gee, G.W., and Hillel, D., Groundwater recharge in arid regions: review and critique of estimation methods, *Hydrol. Proc.*, 2, 255–266, 1988.

Ginn, T.R., and Murphy, E.M., A transient flux model for convective infiltration: forward and inverse solutions for chloride and mass balance studies, *Water Resour. Res.*, 33, 2065–2079, 1997.

Goel, N.S., Models of vegetation canopy reflectance and their use in estimation of biophysical parameters from reflectance data, *Rem. Sens. Rev.* 4, 1–212, 1987.

Goel, N.S., Inversion of canopy reflectance models for estimation of biophysical parameters from reflectance data, in *Theory and Applications of Optical Remote Sensing*, Asrar, G., ed., Wiley, New York, 205–251, 1989.

Gomez-Landesa, E., and Rango, A., Snow cover remote sensing and snowmelt runoff forecast in the Spanish Pyrenees using SRM model, 12 pp., in Proc. Fourth Int. Workshop Appl. Rem. Sens. in Hydr. Santa Fe, 1998.

Goodison, B.E., and Walker, A.E., Canadian development and use of snow cover information from passive microwave satellite data, in Choudhury, B.J., Kerr, Y.N., Njoku, E.G., Pampaloni P., eds. *Passive Microwave Remote Sensing of Land-Atmosphere Interactions*, Utrecht, VSP. 245–262, 1995.

Gorenburg, I.P., McLaughlin, D., and Entekhabi, D., Scale-recursive assimilation of precipitation data, *Adv. Water Resour.*, 24, 941–953, 2001.

Gupta, V.K., Rodríguez-Iturbe, I., and Wood, E.F., eds., *Scale Problems in Hydrology* Reidel, Dordrecht, 1986.

Hargreaves, G.H., and Allen, R.G., History and evaluation of Hargreaves evapotranspiration equation, *J. Irrig. Drain. Eng.*, 129, 53–63, 2003.

Hoeben, R., and Troch, P.A., Assimilation of active microwave observation data for soil moisture profile estimation, *Water Resour. Res.*, 36, 2805–2819, 2000.

Idso, S.B., and Baker, D.G., Relative importance of reradiation, convection and transpiration in heat transfer from plants, *Plant Physiol.*, 42, 631–640, 1967.

Irons, J.R., Weismiller, R.A., and Petersen, G.W., Soil reflectance, in *Theory and Applications of Optical Remote Sensing*, Asrar, G., ed., Wiley, New York, 66–106, 1987.

Jackson, R.D., Canopy temperature and crop water stress, *Adv. Irrig.*, 1, 43–85, 1982.

Jackson, R.D., Idso, S.B., Reginato, R.J., and Pinter, Jr., P.J., Canopy temperature as a crop water stress indicator, *Water Resour. Res.*, 17, 1133–1138, 1981.

Jacobs, A.F.G., Heusinkveld, B.G., and Bercowicz, S.M., Force-restore technique for ground surface temperature and moisture content in a dry desert system, *Water Resour. Res.*, 36, 1261–1268, 2000.

Jensen, M.E., Burman, R.D., and Allen, R.G., eds., Evapotranspiration and Irrigation Water Requirements. ASCE Manuals and Reports on Engineering Practices No. 70, ASCE, New York, 1990.

Jensen, J.R. *Remote Sensing of the Environment: An Earth Resource Perspective* Prentice-Hall, Upper Saddle River, NJ, 2000.

Jothityangkoon, C., Sivapalan, M., and Farmer, D.L., Process controls of water balance variability in large semi-arid catchment: downward approach to hydrological model development, *J. Hydrol.*, 254, 174–198, 2001.

Kalma, J.D., and Sivapalan, M., eds., *Scale Issues in Hydrological Modeling*, Wiley, New York, 1995.

Kjelgaard, J.F., Stöckle, C.O., Black, R.A., and Campbell, G.S., Measuring sap flow with the heat balance approach using constant and variable heat inputs, *Agric. Forest. Meteorol.*, 85, 239–250, 1997.

Klaassen, W., Bosveld, F., and de Water, E., Water storage and evaporation as constituents of rainfall interception, *J. Hydrol.*, 213, 36–50, 1998.

Kosugi, K., Hopmans, J.W., and Dane, J.H., Parametric models, in Dane, J.H., and Topp, G.C., eds., *Methods of Soil Analysis, Part 4, Physical Methods*, Soil Sci. Soc. Am. Book Ser. No. 5, Soil Science Society of America, Madison, WI, 739–757, 2002.

Kustas, W.P., and Norman, J.M., Use of remote sensing for evapotranspiration monitoring over land surfaces, *Hydrol. Sci. J.*, 41, 495–515, 1996.

Kutilek, M., and Nielsen, D.R., *Soil Hydrology*, Catena-Verlag, Cremlingen-Destedt, 1994.

Laio, F., Porporato, A., Ridolfi, L, and Rodríguez-Iturbe, I., Plants in water-controlled ecosystems: active role in hydrologic proceses and response to water stress. II. Probabilistic soil moisture dynamics, *Adv. Water Resour.*, 24, 707–723, 2001.

Lascano, R.J., and Hatfield, J.L., Spatial variability of evapotranspiration along two transects of a bare soil, *Soil Sci. Soc. Am. J.*, 56, 341–346, 1992.

Liang, X., Lettenmaier, D.P., Wood, E.F., and Burges, S.J., A simple hydrologically-based model of land surface water and energy fluxes for general circulation models, *J. Geophys. Res.*, 99, 14415–14428, 1994.

Liu, S., Estimation of rainfall storage capacity in the canopies of cypress wetlands and slash pine uplands in north-central Florida, *J. Hydrol.*, 207, 32–41, 1998.

Lloyd, C.R., and Marques, A. de O., Spatial variability of throughfall and stemflow measurements in Amazonian rainforest, *Agric. Forest. Meteorol.*, 42, 63–73, 1988.

Marek, T.H., Schneider, A.D., Howell, T.A., and Ebeling, L.L., Design and construction of large weighing monolithic lysimeters, *Trans. ASAE*, 31, 477–484, 1988.

Margulis, S.A., McLaughlin, D., Entekhabi, D., and Dunne, S., Land data assimilation and estimation of soil moisture using measurements from the Southern Great

Plains 1997 field experiment, *Water Resour. Res.*, 38(12), 1299, doi:10.1029/2001WR001114, 2002.

Martinec, J., Rango, A., and Major, E., *The Snowmelt-Runoff Model (SRM) User's Manual*, NASA Ref. Publ. 1100, Washington, DC, 1982.

Massman, W.J., The derivation and validation of a new model for the interception of rainfall by forests, *Agric. Meteorol.*, 28, 261–286, 1983.

McLaughlin, D., An integrated approach to hydrologic data assimilation: interpolation, smoothing and filtering, *Adv. Water Resour.*, 25, 1275–1286, 2002.

Melayah, A., Bruckler, L., and Bariac, T., Modeling the transport of water stable isotopes in unsaturated soils under natural conditions, 1. and 2., *Water Resour. Res.*, 32, 2047–2065, 1996.

Menenti, M., and Choudhury, B.J., Parameterization of land surface evapotranspiration using a location-dependent potential evapotranspiration and surface temperature range, in *Exchange Processes at the Land Surface for a Range of Space and Time Scales*, Bolle, H.J. ed., IAHS Publ. No. 212, 561–568, 1993.

Menenti, M., Jia, L., and Su, Z., On SEBI-SEBS validation in France, Italy, Spain, USA and China, 10pp., ICID Workshop on Remote Sensing of ET for Large Region, Montpellier, 2003.

Meyer, D., Verstraete, M., and Pinty, B., The effect of surface anisotropy and viewing geometry on the estimation of NDVI from AVHRR, *Rem. Sens. Rev.*, 12, 3–27, 1995.

Milly, P.C.D., An event-based simulation model of moisture and energy fluxes at a bare soil surface, *Water Resour. Res.*, 22, 1680–1692, 1986.

Milly, P.C.D., Climate, interseasonal storage of soil water, and the annual water balance, *Adv. Water Resour.*, 17, 19–24, 1994.

Moncrieff, J.B., Massheder, J.M., De Bruin, H.A.R., Elbers, J., Friborg, T., Heusinkveld B., Kabat, P., Scott, S., Soegaard, H., and Verhoef, A., A system to measure surface fluxes of momentum, sensible heat, water vapour and carbon dioxide, *J. Hydrol.*, 189, 589–611, 1997.

Monteith, J.L., and Szeicz, G., Radiative temperature in the heat balance of natural surfaces, *Q. J. R. Meteorol Soc.*. 88, 496–507, 1962.

Monteith, J.L., and Unsworth, M.H., *Principles of Environmental Physics*, 2nd ed., Edward Arnold, London, 1990.

Moore, I.D., O'Laughlin, E.M., and Burch, G.J., A contour-based topographic model for hydrological and ecological applications, *Earth Surf. Proc. Landforms*, 13, 305–320, 1984.

Moran, M.S., Clark, T.R., Inoue, Y., and Vidal, A., Estimating crop water deficit using the relation between surface-air temperature and spectral vegetation index, *Rem. Sens. Environ.*, 49, 246–263, 1994.

Moran, M.S., and Jackson, R.D., Assessing the spatial distribution of evaporation using remotely sensed inputs, *J. Environ. Qual.*, 20, 725–737, 1991.

Moran, M. S., Hymer, D.C., Qi, J., and Sano, E.E., Soil moisture evaluation using multi-temporal synthetic aperture radar (SAR) in semiarid rangeland. *Agr. Forest. Met.*, 105, 69–80, 2000.

Mulders, M. A., *Remote Sensing in Soil Science*, Elsevier, Amsterdam, 1987.

Myneni, R.B., and Asrar, G., Radiative transfer in three-dimensional atmosphere-vegetation media, *J. Quant. Spectrosc. Radiat. Transfer*, 49, 585–598, 1993.

Niklas, K.J., *Plant Allometry. The Scaling of Form and Processes*, University of Chicago Press, Chicago, 1994.

Noilhan, J., and Mahfouf, J.F., The ISBA land surface parameterization scheme, *Global Plan. Change*, 13, 145–159, 1996.

Norman, J.M., Divakarla, M., and Goel, N.S., Algorithms for extracting information from remote thermal-IR observations of the Earth surface. *Rem. Sens. Env.*, 51, 157–168, 1995.

Olszyczka, B., Gamma-ray determinations of surface water storage and stem water content for coniferous forests, Ph.D. dissertation, Department of Applied Physics, University of Strahclyde, Strahclyde, 1979.

Paniconi, C., and Wood, E.F., A detailed model for simulation of catchment scale subsurface hydrologic processes, *Water Resour. Res.*, 29, 1601–1620, 1993.

Parlange, M.B., and Katul, G.G, An advection-aridity evaporation model, *Water Resour. Res.*, 28, 127–132, 1992.

Penman, H.L., Natural evaporation from open water, bare soil, and grass, *Proc. Royal Soc. London, A*, 194, 120–145, 1948.

Rana, G., and Katerji, N., Measurement and estimation of actual evapotranspiration in the field under Mediterranean climate: a review, Eur. J. Agron., 13, 125–153 2000.

Rango, A., Martinec, J., Chang, A.T.C., Foster, J.L., and van Katwijk, V., Average areal water equivalent of snow in a mountain basin using microwave and visible satellite data, *IEEE Trans. Geosci. Rem. Sens.*, 27, 740–745, 1989.

Reichle, R.H., Entekhabi, D., and McLaughlin, D., Downscaling of radiobrightness measurements for soil moisture estimation: a four-dimensional variational data assimilation approach, *Water Resour. Res.*, 37, 2353–2364, 2001.

Reicosky, D.C., Canopy gas exchange in the field: closed chambers, *Remote Sens. Rev.* 5, 163–178, 1990.

Ridolfi, L., D'Odorico, P., Porporato, A., and Rodríguez-Iturbe, I., Duration and frequency of water stress in vegetation: an analytical model, *Water Resour. Res.* 36, 2297–2307, 2000.

Rodríguez-Iturbe, I., Porporato, A., Laio, F., and Ridolfi, L., Plants in water-controlled ecosystems: active role in hydrological processes and response to water stress. I. Scope and general outline, *Adv. Water Resour.*, 24, 695–705, 2001.

Rutter, A.J., Kershaw, K.A., Robins, P.C., and Morton, A.J., A predictive model of rainfall interception in forests. I. Derivation of the model from observations in a stand of Corsican pine, *Agric. Meteorol.*, 9, 367–384, 1971.

Sakuratini, T., A heat balance for measuring water flux in the stem of intact plants, *Agric. Meteorol.*, 37, 9–17, 1981

Scanlon, B.R., and Milly, P.C.D., Water and heat flux in desert soils. 2. Numerical simulations, *Water Resour. Res.*, 30, 721–733, 1994.

Schaap, M.G., Leij, F.J., and van Genuchten, M.T., Neural network analysis for hierarchical prediction for soil hydraulic properties, *Soil Sci. Soc. Am. J.*, 62, 847–855, 1998.

Schmugge, T.J., and Jackson, T.J., Mapping surface soil moisture with microwave radiometers, *Meteorol. Atmos. Phys.*, 54, 213–223, 1994.

Schmugge, T.J., Kustas, W.P., Ritchie, J.C., Jackson, T.J., and Rango, A., Remote sensing in hydrology, *Adv. Water Resour.* 25, 1367–1385, 2002.

Scott, R.L., Shuttlewoth, W.J., Keefer, T.O., and Warrick, A.W., Modeling multiyear observations of soil moisture recharge in the semiarid American southwest, *Water Resour. Res.*, 36, 2233–2247, 2000.

Simunek, J., Huang, M.K., Sejna, M., and van Genuchten, M.T., *The HYDRUS-1D Software Package for simulating the movement of water, heat, and multiple solutes in variably saturated media*, version 1.0, U.S. Salinity Lab., USDA, Riverside, 1997.

Sivapalan, M., and E.F. Wood. Spatial heterogeneity and scale in the infiltration response of catchments, in *Scale Problems in Hydrology*, Gupta, V.K., Rodríguez-Iturbe, I., and Wood, E.F., eds., Reidel, Dordrecht, 81–106, 1986.

Smith, D.M., and Allen, S.J., Measurement of sap flow in plant stems, *J. Exp. Bot.*, 47, 1833–1844, 1996.

Sobrino, J.A., N. Raissouni, Y. Kerr, A. Olioso, M.J. López-García, A. Belaid, M.H. El Kharraz, J. Cuenca, and L. Dampere, Teledetección, Sobrino, J.A., ed. Universidad de Valencia, Valencia, 2000.

Stewart, J.B., Engman, E.T., Feddes, R.A., and Kerr, Y., eds., *Scaling up in Hydrology Using Remote Sensing*, Wiley, New York, 1996.

Su, Z., and Jacobs, C., eds., *Advanced Earth Observation-Land Surface Climate*, Publ. Nat. Rem. Sens. Board (BCRS), USP-2, 01-02, 91-108, 2001.

Su, Z., The Surface Energy Balance System (SEBS) for estimation of turbulent heat fluxes, *Hydrol. Earth Syst. Sci.*, 6, 85–99, 2002.

Teklehaimanot, Z., and Jarvis, P.G., Direct measurement of evaporation of intercepted water from forest canopies, *J. Appl. Ecol.*, 28, 603–618, 1991.

Torres, M.A., *Desarrollo de una metodología para la cuantificación de la capa de nieve mediante el sensor NOAA-AVHRR. Aplicación a la previsión del aporte al río Genil*, M. Sc. diss. University of Cordoba, Cordoba, 1998.

Ulaby, F.T., Dubois, P.C., and van Zyl, J., Radar mapping of surface soil moisture, *J. Hydrol.*, 184, 57–84, 1996.

Ulaby, F.T., Moore, R.K., and Fung, A.K., *Microwave Remote Sensing. Active and Passive. Vol. III: From Theory to Applications*, Artech House Pub., Norwood, 1986.

Valente, S., David, J.S., and Gash, J.H.C., Modeling interception loss for two sparse eucalyptus and pine forests in central Portugal using Rutter and Gash analytical models, *J. Hydrol.*, 190, 141–162, 1997.

Verhoest, N.E.C., Troch, P.A., Paniconi, C., and De Troch, F.P., Mapping basin scale variable source areas from multitemporal remotely sensed observations of soil moisture behavior, *Water Resour. Res.*, 34, 3235–3244, 1998.

Villalobos, F.J., Correction of eddy covariance water vapor flux using additional measurements of temperature, *Agric. Forest. Meteorol.*, 88, 77–83, 1997.

Walker, G.R., Jolly, I.D., and Cook, P.G., A new chloride leaching approach to the estimation of diffuse recharge following a change in land use, *J. Hydrol.*, 128, 49–67, 1991.

Walker, J.P., Willgoose, G.R., and Kalma, J.D., One-dimensional soil moisture profile retrieval by assimilation of near-surface observations: a comparison of retrieval algorithms, *Adv. Water Resour.*, 24, 631–650, 2001.

Watanabe, T., and Mizatuni, K., Model study on micrometeorological aspects of rainfall interception over an evergreen broad-leaved forest, *Agric. Forest. Meteorol.*, 80, 195–214, 1996.

Wessman, C.A., Evaluation of canopy biochemistry, in *Remote Sensing of Biosphere Functioning*, Hobbs, R.J., and Mooney, H.A., eds., Springer-Verlag, New York, 135–156, 1990.

Weismiller, R.A., and Kamisnsky, S.A., Applications of remote-sensing technology to soil survey research, *J. Soil Water Conserv.*, 33, 287–289, 1978.

Wilson, K.B., Hanson, P.J, Mulholland, P.J., Baldocchi, D.D., and Wullschleger, S. D. A., A comparison of methods for determining forest evapotranspiration and its components: sap-flow, soil water budget, eddy covariance and catchment water balance, *Agric. Forest. Meteorol.*, 106, 153–168, 2001.

Wiscombe, W.J., and Warren, S.G., A model for spectral albedo of snow. I. Pure snow. *J. Atmos. Sci.*, 37, 2712–2733, 1980.

Wood, E.F., Effect of soil moisture aggregation on surface evaporative fluxes, *J. Hydrol.*, 190, 397–412, 1997.

Wood, E.F., Sivapalan, M., and Beven, K., Scale effects in infiltration and runoff production, in *Conjunctive Water Use*, IAHS publication, 156, Wallingford, 375–387, 1986.

Wood, W.W., and Sanford, W.E., Chemical and isotopic methods for quantifying ground-water recharge in a regional, semiarid environment, *Ground Water*, 33, 458–468, 1995.

Zeng, N., Shuttleworth, J.W., and Gash, J.H.C., Influence of temporal variability of rainfall on interception loss. Part I. Point analysis, *J. Hydrol.*, 228, 228–241, 2000.

5 Field Methods for Monitoring Soil Water Status

Rafael Muñoz-Carpena
University of Florida, IFAS/TREC, Homestead, FL, U.S.A.

Axel Ritter
Instituto Canario de Investigaciones Agrarias — ICIA, La Laguna, Tenerife, Spain, @icia.es

David Bosch
USDA-ARS, SEWRL, Tifton, Georgia, U.S.A.

CONTENTS

1-5667-0657-2/05/S0.00 + $1.50
© 2005 by CRC Press

5.1   INTRODUCTION

In the context of soil research and characterization, often one of the first steps in any study is to quantify the amount of water in the soil in both time and space (see Chapters 3, 4, 6). In the context of water management for irrigation, measuring and monitoring soil water status is an essential component of best management practices (BMPs) to improve the sustainability of agriculture.

Water content in the soil can be *directly* determined using the difference in weight before and after drying a soil sample. This direct technique is usually referred to as the *thermo-gravimetric method* when expressing water content as gravimetric soil moisture θ_m [M^3 M^{-3}], i.e., the ratio of the mass of water present in a sample to the mass of the soil sample after it has been oven-dried (105°C) to a constant mass. On the other hand, the *thermo-volumetric method* gives the volumetric soil moisture, θ_v or simply θ [L^3 L^{-3}], i.e., volume of water related to the volume of an oven-dried undisturbed sample (soil core). Although these direct methods are accurate $\pm 0.01\,cm^3cm^{-3}$) and inexpensive, they are destructive, slow (2 days minimum), and do not allow for making repetitions in the same location. Alternatively, many *indirect* methods are available for monitoring soil water content. These methods estimate soil moisture by a calibrated relationship with some other measurable variable. The suitability of each method depends on several issues such as cost, accuracy, response time, installation, management, and durability.

Depending on the quantity measured, indirect techniques are first classified into *volumetric* and *tensiometric methods* (Figure 5.1). While the former gives volumetric soil moisture, the latter yields soil suction or matric potential (i.e., tension exerted by capillarity and adsorptive surface forces). Both quantities are related through the soil water characteristic curve (SWC).

It is important to remember that each soil type (texture/structure) has a different curve; therefore, they cannot be related to each other the same way for all soil types (Figure 5.2). Several mathematical models have been proposed to describe the SWC (see Chapter 3). In addition, this relationship might not be unique and may exhibit hysteresis along drying and wetting cycles, especially in finer soils.

Depending on the soil physical properties and goal of the soil moisture measurement, some devices are more effective than others. First, it must be considered that although volumetric moisture is a more intuitive quantity, in fine-textured soils water is strongly retained by solid particles and therefore may not be available for plant absorption and other processes

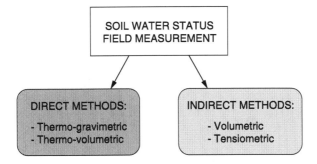

FIGURE 5.1 Methods for measuring soil moisture.

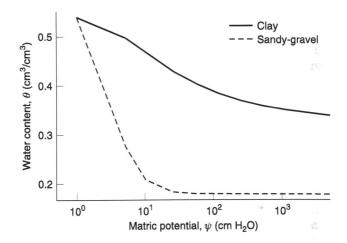

FIGURE 5.2 Measured soil water characteristic curves for two different soil types.

like flow and solute transport. In the case of plant-soil studies, soil suction may be a more useful quantity since it relates to the energy that plants have to invest to extract the water from the soil, and hence it is a more meaningful measure of plant water stress. Second, the desired sampling frequency is an important factor since different sensors' response times vary over a wide range, i.e., some devices require soil moisture to equilibrate with the sensor matrix. Third, soil physical properties (texture, shrinking/swelling) may influence the suitability of the selected method, because some require good soil-instrument contact. On the other hand, depending on soil type and hydrologic conditions (precipitation and evapotranspiration), some instruments might have higher maintenance requirements than others.

Irrigation management is a practical application of monitoring soil moisture that is becoming widespread among agricultural growers. Soil moisture–based optimized irrigation consists of keeping the soil within a target moisture range by replenishing the plant water uptake with irrigation. This practice avoids the potential for soil water excess and leaching of agrochemicals present in the soil but requires a suitable method for soil moisture estimation (Muñoz-Carpena et al., 2002, 2003). However, to calculate irrigation requirements, matric potential values from tensiometric methods need to be converted to soil moisture through the SWC.

5.2   METHODS OF CHARACTERIZATION: TRADE-OFFS. COMPARATIVE STUDY

Most practical techniques for soil water monitoring are indirect (Yoder et al., 1998; Robinson et al., 1999). A review of available techniques is given below, focusing on working principles, advantages, and drawbacks.

5.2.1   Volumetric Field Methods

Methods under this definition estimate the volume of water per volume of soil θ [L^3 L^{-3}]. This quantity is useful for determining how saturated the soil is, i.e., fraction of total soil volume filled with the soil aqueous solution. When it is expressed in dimensions of depth or equivalent depth of wetting, d_e, i.e., volume of water in soil down to a given depth over a unit surface area (m), or

$$d_e = \theta d \tag{5.1}$$

where d is the depth increment in meter. d_e can be compared with other hydrological variables such as precipitation, evaporation, transpiration, deep drainage, etc.

5.2.1.1   Neutron Moderation

Working principle: Fast neutrons are emitted from a decaying source ^{241}Am/^{9}Be), and when they collide with particles having the same mass as a neutron (i.e. protons, H^+), they slow down dramatically, building a "cloud" of thermalized (slowed-down) neutrons. Since water is the main source of hydrogen in most soils, the density of thermalized neutrons formed around the probe is nearly proportional to the volume fraction of water present in the soil:

$$CR = m\theta + b \quad \text{and} \quad CR/CS = y\theta \tag{5.2}$$

where CR is the slow neutron count rate in wet soil, CS is count rate in water in a standard absorber (i.e., shield of probe), y is a constant, and m and b are the slope and intercept of the calibration line for the probe. Many

FIGURE 5.3 Neutron probe.

researchers establish calibration curves like the one above in terms of gravimetric water content. This is more convenient because the volume of influence of the neutron probe changes with water content. It is larger when the soil is dry and decreases with increasing water content. In general the radius (r) of the sphere of influence of the probe can be estimated by r (cm) $= 15 \, (\theta)^{-1/3}$.

Description: The probe configuration is in the form of a long and narrow cylinder, containing a source and detector. Measurements are made by introducing the probe into an access tube (previously installed into the soil). Therefore, it is possible to determine soil moisture at different depths (Figure 5.3).

Advantages:
- Robust and accurate ($\pm 0.005 \, \text{cm}^3\text{cm}^{-3}$).
- Inexpensive per location, i.e., a large number of measurements can be made at different points with the same instrument.
- One probe allows for measuring at different soil depths.
- Large soil sensing volume (sphere of influence with 10–40 cm radius, depending on moisture content).
- Not affected by salinity or air gaps.
- Stable soil-specific calibration.

Drawbacks:

- Safety hazard, since it implies working with radiation. Even at 40 cm depth, radiation losses through soil surface have been detected.
- Requires certified personnel.
- Requires soil-specific calibration.
- Heavy, cumbersome instrument.
- Takes a relatively long time for each reading.
- Readings close to the soil surface are difficult.
- Manual readings; cannot be automated due to hazard.
- Expensive to buy.
- The sphere of influence may vary according to the following reasons:

  a) It increases as the soil dries because the hydrogen concentration decreases, so that the probability of collision is smaller and therefore fast neutrons can travel further from the source.
  b) It is smaller in fine texture soils because they can hold more water, and thus the probability of collision is higher.
  c) If there are layers with large differences in water content due to changes in soil physical properties, the sphere of influence can have a distorted shape.

5.2.1.2   Dielectric Methods

The next set of volumetric methods are known as *dielectric techniques* because all of them estimate soil water content by measuring the soil bulk permittivity (or dielectric constant), ε_b (Inoue, 1998; Hilhorst et al., 2001). This parameter states that the dielectric of a medium is the ratio squared of electromagnetic wave propagation velocity in a vacuum, relative to that of the medium (i.e., ε_b determines the velocity of an electromagnetic pulse through that medium). In the soil the value of this composite property is mainly governed by the presence of liquid water, because the dielectric constant of the other soil constituents is much smaller (e.g., $\varepsilon_s = 2$–5 for soil minerals, 3.2 for frozen or bounded water, and 1 for air) than that of liquid water ($\varepsilon_w = 81$).

A common approach to establish the relationship between ε_b and volumetric moisture (θ) is the empirical equation of Topp et al., (1980):

$$\theta = -5.3 \cdot 10^{-2} + 2.29 \cdot 10^{-2} \varepsilon_b - 5.5 \cdot 10^{-4} \varepsilon_b^2 + 4.3 \cdot 10^{-6} \varepsilon_b^3 \qquad (5.3$$

This third-order polynomial provides an adequate θ-ε_b relationship for most mineral soils (independent of soil composition and texture) and for $\theta < 0.5 \, cm^3 cm^{-3}$. For larger water content, organic or volcanic soils, a specific calibration is required. In this latter case, this is explained in terms of low bulk densities and large surface areas that result in a greater fraction of bounded water (Regalado et al., 2003). Moreover, high contents of aluminium and iron hydroxides may result in higher permittivity of the solid phase (Dirksen, 1999). It is worth noticing that the θ-ε_b relationship depends on the electromagnetic wave frequency. Thus, at low frequencies ($< 100\,\text{MHz}$) it is more soil-specific.

An alternative relationship to the empirical equation is the three-phase mixing model (Roth et al., 1990):

$$\theta = \frac{\varepsilon_b^\beta - (1 - \eta)\varepsilon_s^\beta - \eta\varepsilon_a^\beta}{\varepsilon_w^\beta - \varepsilon_a^\beta} \tag{5.4}$$

where η is the soil porosity; ε_w, ε_a, and ε_s are the permittivity of liquid, gaseous, and solid phase, respectively, and β describes the geometry of the medium in relation to the axial direction of the transmission line and, in general, it is considered as a fitted parameter ($-1 \leq \beta \leq 1$). When soil moisture is divided into a mobile and an immobile region, the four-phase mixing model is recommended (Dobson et al., 1985):

$$\theta = \frac{\varepsilon_b^\beta - (1 - \eta)\varepsilon_s^\beta - \eta\varepsilon_a^\beta + \theta_{bw}^\beta\left(\varepsilon_{bw}^\beta - \varepsilon_w^\beta\right)}{\varepsilon_w^\beta - \varepsilon_a^\beta} \tag{5.5}$$

where θ_{bw} is the fraction of soil water that is immobile and has a permittivity ε_{bw}. According to Dirksen and Dasberg (1993), it can be obtained from:

$$\theta_{bw} = n\delta\rho_b S_e \tag{5.6}$$

where n is the number of molecular water layers adsorbed at soil particles, δ is the thickness of a monomolecular water layer ($3\cdot10^{-10}\,\text{m}$), ρ_b is the bulk density, and S_e is the specific surface area. Alternatively, the Maxwell–De Loor equation (De Loor, 1964) is based on a theoretical model and contains only physical parameters, although it requires measurement of a significant number of other soil properties.

In addition to these approaches, some of the dielectric methods described below use empirical calibrated relationships between θ and the sensor output signal (time, frequency, impedance, wave phase). These techniques are becoming widely adopted because they have good response time (almost instantaneous measurements), do not require maintenance, and can provide continuous readings through automation.

5.2.1.2.1   Time Domain Reflectometry (TDR)

Working principle: The soil bulk dielectric constant (ε_b) is determined by measuring the time it takes for an electromagnetic pulse to propagate along a transmission line (TL) that is surrounded by the soil. Since the propagation velocity (v) is a function of ε_b, the latter is therefore proportional to the square of the transit time (t) down and back along the TL:

$$\varepsilon_b = \left(\frac{c}{v}\right)^2 = \left(\frac{ct}{2L}\right)^2 \tag{5.7}$$

where c is the velocity of electromagnetic waves in a vacuum ($3 \cdot 10^8$ m/s) and L is the length of the TL embedded in the soil.

Description: A TDR instrument requires a device capable of producing a series of precisely timed electrical pulses with a wide range of frequencies used by different devices (e.g., 0.02–3 GHz), which travel along a TL that is built with a coaxial cable and a probe. The TDR probe usually consists of two or three parallel metal rods that are inserted into the soil acting as waveguides in a similar way as an antenna used for television reception. At the same time, the TDR instrument uses a device for measuring and digitizing the energy (voltage) level of the TL at intervals down to around 100 ps. When the electromagnetic pulse traveling along the TL finds a discontinuity (i.e., probe-waveguides surrounded by soil), part of the pulse is reflected. This produces a change in the energy level of the TL. Thereby the travel time (t) is determined by analyzing the digitized energy levels (Figure 5.4).

Soil salinity or highly conductive heavy clay contents may affect TDR, since it contributes to attenuation of the reflected pulses. In other words, TDR is relatively insensitive to salinity as long as a useful pulse is reflected (i.e., as long as it can be analyzed). In soils with highly saline conditions, using epoxy-coated probe rods can solve the problem. However, this implies loss of sensitivity and change in calibration. On the other hand, pulse attenuation due to soil salinity allows using TDR for quantifying ionic solutes in the soil environment (i.e., electrical conductivity).

In combination with a neutron probe or other technique, which detects total soil moisture, TDR can be used to determine bound or frozen water, because these have much lower permittivity than liquid water.

Advantages:
- Accurate (± 0.01 cm^{3}cm^{-3}).
- Soil specific-calibration is usually not required.
- Easily expanded by multiplexing.
- Wide variety of probes configuration.

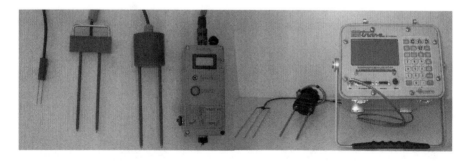

FIGURE 5.4 TDR equipment.

- Minimal soil disturbance.
- Relatively insensitive to normal salinity levels.
- Can provide simultaneous measurements of soil electrical conductivity (see Chapter 10).

Drawbacks:
- Relatively expensive equipment due to complex electronics.
- Potentially limited applicability under highly saline conditions or in highly conductive heavy clay soils.
- Soil-specific calibration required for soils having large amounts of bound water (i.e., volcanic soils) or high organic matter content.
- Relative small sensing volume (about 3 cm radius around length of waveguides).

5.2.1.2.2 Frequency Domain: Capacitance and FDR

Working principle: The electrical capacitance of a capacitor that uses the soil as a dielectric, depends on the soil water content. When connecting this capacitor (made of metal plates or rods imbedded in the soil) together with an oscillator to form a tuned electrical circuit, changes in soil moisture can be detected by changes in the circuit operating frequency. This is the basis of the frequency domain (FD) technique used in capacitance and frequency domain reflectometry (FDR) sensors. In capacitance sensors the dielectric permittivity of a medium is determined by measuring the charge time of a capacitor in that medium. In FDR the oscillator frequency is swept under control within a certain frequency range to find the resonant frequency (at which the amplitude is greatest), which is a measure of water content in the soil.

Description: Probes usually consist of two or more electrodes (i.e., parallel plates, rods, or metal rings along a cylinder) that are inserted into the soil. On the ring configuration the probe is introduced into a access tube installed in the field. Thus, when an electrical field is applied, the soil in contact with the electrodes (or around the PVC tube) forms the dielectric of the capacitor that completes the oscillating circuit. The use of an access tube allows for multiple sensors to take measurements at different depths (Figure 5.5).

A soil-specific calibration is recommended because the operating frequency of these devices is generally below 100 MHz. At these low frequencies the bulk permittivity soil minerals may be changed by soil minerals, and the estimation is more affected by temperature, salinity, bulk density, and clay content. On the other hand, using low frequencies allows for detecting bound water.

Advantages:
- Accurate after soil-specific calibration ($\pm 0.01 \, cm^3 cm^{-3}$).

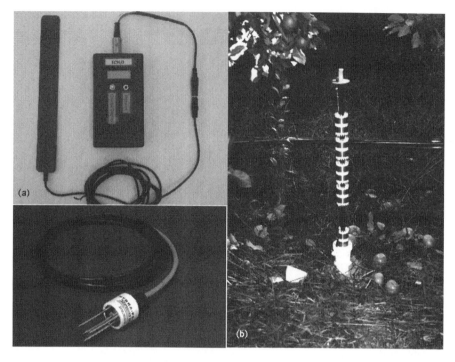

FIGURE 5.5 FD probes: (a) capacitance; (b) FDR.

- Can read at high salinity levels, where TDR fails.
- Better resolution than TDR (avoids the noise that is implied in the waveform analysis performed by TDRs).
- Can be connected to conventional loggers (DC output signal).
- Flexibility in probe design (more than TDR).
- Some devices are relatively inexpensive compared to TDR due to use of low-frequency standard circuitry.

Drawbacks:
- The sensing sphere of influence is relatively small (about 4 cm).
- For reliable measurements, it is extremely critical to have good contact between the sensor (or tube) and soil.
- Tends to have larger sensitivity to temperature, bulk density, clay content, and air gaps than TDR. Careful installation is necessary to avoid air gaps.
- Needs soil-specific calibration.

5.2.1.2.3 Amplitude Domain Reflectometry (ADR): Impedance

Working principle: When an electromagnetic wave (energy) traveling along a transmission line (TL) reaches a section with different impedance (which has two components: electrical conductivity and dielectric constant), part of the energy transmitted is reflected back into the transmitter. The reflected wave interacts with the incident wave, producing a voltage standing wave along the TL, i.e., change of wave amplitude along the length of the TL. If the soil/probe combination is the cause for the impedance change in the TL, measuring the amplitude difference will give the impedance of the probe (Gaskin and Miller, 1996; Nakashima et al., 1998). The influence of the soil electrical conductivity is minimized by choosing a signal frequency so that the soil water content can be estimated from the soil/probe impedance.

Description: Impedance sensors use an oscillator to generate a sinusoidal signal (electromagnetic wave at a fixed frequency, e.g., 100 MHz), which is applied to a coaxial TL that extends into the soil through an array of parallel metal rods, the outer of which forms an electrical shield around the central signal rod. This rod arrangement acts as an additional section of the TL, having impedance that depends on the dielectric constant of the soil between the rods (Figure 5.6).

Advantages:
- Accurate with soil-specific calibration ($\pm 0.01\,cm^3\,cm^{-3}$; $\pm 0.05\,cm^3\,cm^{-3}$ without it).

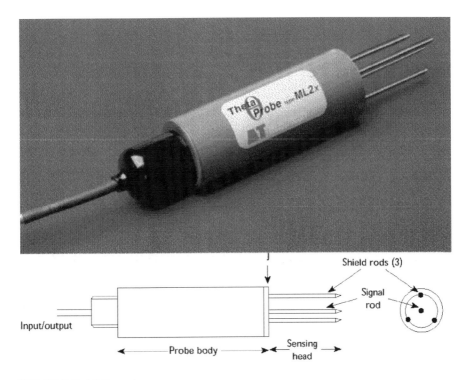

FIGURE 5.6 ADR probe.

- Allows measurements in highly saline conditions (up to 20 dS/m).
- Minimal soil disturbance.
- Can be connected to conventional loggers (DC output signal).
- Inexpensive due to standard circuitry.
- Not affected by temperature.
- *In situ* estimation of soil bulk density possible (Wijaya et al., 2002).

Drawbacks:
- Soil-specific calibration recommended for reliable measurements.
- Measurement affected by air gaps, stones or channelling water directly onto the probe rods.
- Small sensing volume (4.3 cm^3).

5.2.1.2.4 Phase Transmission (Virrib)

Working principle: After having traveled a fixed distance, a sinusoidal wave will show a phase shift relative to the phase at the origin. This phase shift depends on the length of travel, the frequency, and the velocity of propagation. Since velocity of propagation is related to soil moisture content, for a fixed frequency and length of travel, soil water content can be determined by this phase shift

Description: The probe uses a particular waveguide design (two concentric metal open rings), so that phase-measuring electronics can be applied at the beginning and ending of the waveguides (Figure 5.7).

FIGURE 5.7 Phase transmission probe.

Advantages:
- Accurate with soil-specific calibration ($\pm 0.01 \, \text{cm}^3\text{cm}^{-3}$).
- Large sensing soil volume (15–20 L).
- Can be connected to conventional loggers (DC output signal).
- Inexpensive.

Drawbacks:
- Significant soil disturbance during installation due to concentric ring sensor configuration.
- Requires soil-specific calibration.
- Sensitive to salinity levels > 3 dS/m.
- Reduced precision, because the generated pulse becomes distorted during transmission.
- Needs to be permanently installed in the field.

5.2.1.2.5   Time Domain Transmission (TDT)

Working principle: This method measures the one-way time for an electro-magnetic pulse to propagate along a transmission line (TL). Thus, it is similar to TDR, but requires an electrical connection at the beginning and ending of the TL. Notwithstanding, the circuit is simple compared with TDR instruments. Recently Harlow et al. (2003) demonstrated two different TDT approaches, showing results comparable with those of FD and TDR techniques.

Description: The probe has a waveguide design (bent metal rods), so that the beginning and ending of the transmission line are inserted into the electronic block. Alternatively, the sensor consists of a long band ($\sim$3 m), having an electronic block at both ends (Figure 5.8).

Advantages:
- Accurate ($\pm 0.01\text{–}0.02 \, \text{cm}^3\text{cm}^{-3}$).
- Large sensing soil volume (0.8–6 L).
- Can be connected to conventional loggers (DC output signal).
- Inexpensive due to standard circuitry.

Drawbacks:
- Reduced precision, because the generated pulse gets distorted during transmission.
- Soil disturbance during installation.
- Needs to be permanently installed in the field.

5.2.1.3   Other Volumetric Field Methods

Another interesting technique is ground penetrating radar (GPR). This technique is based on the same principle as TDR, but does not require direct contact between the sensor and the soil. When mounted on a vehicle or

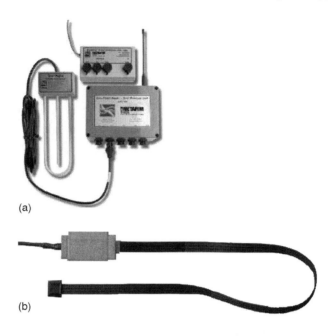

(a)

(b)

FIGURE 5.8 TDT devices.

trolley close to the soil surface, it has the potential of providing rapid, nondisturbing, soil moisture measurements over relatively large areas. Although it has been applied successfully to many field situations, GPR has not been widely used because the methodology and instrumentation are still only in the research and development phase (Davis and Annan, 2002). It is, however, likely that small, compact, and inexpensive GPR systems will be available in the near future for routine field studies.

It should be mentioned that there are several new remote sensing methods, usually mounted on airplanes or satellites, specially suited for soil moisture monitoring over large areas. Among these methods the active and passive microwave and electromagnetic induction (EMI) methods have been found useful in different applications (Dane and Topp, 2002) and are the subject of much current research. The active and EMI methods use two antennae to transmit and receive electromagnetic signals that are reflected by the soil, whereas the passive microwave just receives signals naturally emitted by the soil surface. In the microwave methods, the signal typically relates to some shallow depth below the ground surface (< 4 in.) so that only a measure of the soil moisture and electrical conductivity of the near-surface soil can be achieved. EMI does not measure water content directly, but rather soil electrical conductivity, and a known calibration relationship between the two is required. Unfortunately, this relationship is site-specific and cannot be assumed.

Other modern non-field techniques for estimating soil moisture and flow [x-ray tomography and nuclear magnetic resonance-(NMR)] are discussed in Chapter 7.

5.2.2 TENSIOMETRIC FIELD METHODS

Tensiometric methods estimate the soil water matric potential that includes both adsorption and capillary effects of the soil solid phase. The matric potential is one of the components of the total soil water potential that also includes gravitational, osmotic, gas pressure or pneumatic, and overburden components. The sum of matric and gravitational (position with respect to a reference elevation plane) potentials, i.e., hydraulic potential or total head, is the main driving force for water movement in soils and other porous media (see Chapters 3, 4, and 6).

All available tensiometric instruments have a porous material in contact with the soil through which water can move. Thereby, water is drawn out of the porous medium in a dry soil and from the soil into the medium in a wet soil. It is worth noticing that in general these instruments do not need soil-specific calibration, but in most cases they have to be permanently installed in the field, or a sufficiently long time must be allowed for equilibration between the device and the soil before making a reading.

5.2.2.1 Tensiometer

Working principle: When a sealed water-filled tube is placed in contact with the soil through a permeable and saturated porous material, water (inside the tube) comes into equilibrium with the soil solution, i.e., it is at the same pressure potential as the water held in the soil matrix. Hence, the soil water matric potential is equivalent to the vacuum or suction created inside the tube.

Description: The tensiometer consists of a sealed water-filled plastic tube with a ceramic cup at one end and a negative pressure gauge (vacuometer) at the other. Its shape and size can be variable, and the accuracy depends on the gauge or transducer used (about 0.01 bar). Typically the measurement range is 0–0.80 bar, although low-tension versions (0–0.40 bar) have been designed for coarse soils (Figure 5.9).

Advantages:
- Direct reading.
- Up to 10 cm measurement sphere radius.
- Continuous reading possible when using pressure transducer.
- Requires intimate contact with soil around the ceramic cup for consistent readings and to avoid frequent discharge (breaking of water column inside).
- Electronics and power consumption avoidable.
- Well suited for high frequency sampling or irrigation schedules.

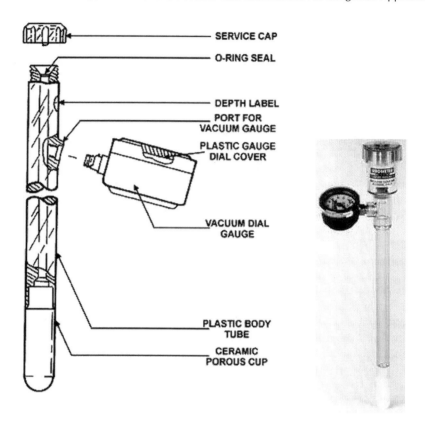

FIGURE 5.9 Tensiometers. (From Smajstrla and Harrison, 1998).

- Minimal skill required for maintenance.
- Not affected by soil salinity, because salts can move freely into and out across the porous ceramic cup.
- Inexpensive.

Drawbacks:
- Limited soil suction range (< 1 bar).
- Relatively slow response time.
- Especially in swelling or coarse soils, the ceramic cup can loose contact with soil, thus requiring reinstallation.
- Requires frequent maintenance (refilling) to keep the tube full of water, epecially in hot dry weather.

5.2.2.2   Resistance Blocks

Working principle: The electrical resistance between electrodes embedded in a porous medium (block) is proportional to its water content, which is related to the soil water matric potential of the surrounding soil. Electrical resistance reduces as the soil, hence the block, dries.

5.2.2.2.1 Gypsum (Bouyoucos) Block

Description: A gypsum block sensor constitutes an electrochemical cell with a saturated solution of calcium sulfate as electrolyte (Bouyoucos and Mick, 1940). The resistance between the block-embedded electrodes is determined by applying a small AC voltage (to prevent block polarization) using a Wheatstone bridge. Since changes to the electrical conductivity of the soil would affect readings, gypsum is used as a buffer against soil salinity changes (up to a certain level). The inherent problem is that the block dissolves and degrades over time (especially in saline soils), losing its calibration properties. It is recommended that the block pore size distribution match the texture of the surrounding soil. The readings are temperature dependent (up to 3% change/°C), and field-measured resistance should be corrected for differences between calibration and field temperatures. Some reading devices contain manual or self-compensating features for temperature or the manufacture provides correction charts or equations. Measurement range is 0.3–2.0 bar (Figure 5.10).

Advantages:
- Up to 10 cm measurement sphere radius.
- No maintenance needed.
- Simple and inexpensive.

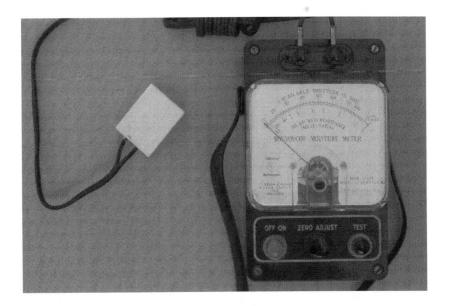

FIGURE 5.10 Gypsum (Bouyoucos) resistance block and reader.

- Salinity effects buffered up to 6 dS/m.
- Well suited for irrigation where only "full" and "refill" points are required.
- Suited to regulated-deficit irrigation.

Drawbacks:
- Low resolution, limited use in research.
- Block cannot be used for measurements around saturation (0–0.3 bar).
- Block properties change with time, because of clay deposition and gypsum dissolution. Degradation speed depends on soil type, amount of rainfall and irrigation, and also the type of gypsum block used.
- Very slow reaction time. It does not work well in sandy soils, where water drains more quickly than the instrument can equilibrate.
- Not suitable for swelling soils.
- Inaccurate readings due to the block hysteresis (at a fixed soil water potential, the sensor displays different resistance when wetting than when drying).
- Temperature dependent. If connected to a logging system, another variable and sensor for temperature must be added to the system.

5.2.2.2.2   Granular Matrix Sensors (GMS)

Description: The sensor consists of electrodes embedded in a granular quartz material, surrounded by a synthetic membrane and a protective stainless steel mesh. Inside, gypsum is used to buffer against salinity effects. This kind of porous medium allows for measuring in wetter soil conditions and lasts longer than the gypsum blocks. However, even with good sensor-soil contact, GMSs have rewetting problems after they have been dried to very dry levels. This is because of the reduced ability of water films to reenter the coarse medium of the GMS from a fine soil. The GMS material allows for measurements closer to saturation. Measurement range is 0.10–2.0 bar (Figure 5.11).

Advantages:
- Reduces the problems inherent to gypsum blocks (i.e., loss of contact with the soil by dissolving, and inconsistent pore size distribution).
- Up to 10 cm measurement sphere radius.
- No maintenance needed.
- Simple and inexpensive.
- Salinity effects buffered up to 6 dS/m.
- Suited to regulated-deficit irrigation.

Drawbacks:
- Low resolution, limited use in research.
- Slow reaction time. It does not work well in sandy soils, where water drains more quickly than the instrument can equilibrate.

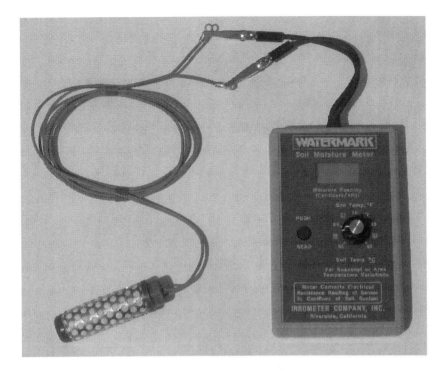

FIGURE 5.11 GMS resistance block and reader.

- Not suitable for swelling soils.
- If the soil becomes too dry, the sensor must be pulled out, resaturated and installed again.
- Temperature dependence. If connected to a logging system, another variable and sensor for temperature must be added to the system.

5.2.2.3 Heat Dissipation

Working principle: The rate of heat dissipation in a porous medium is dependent on the medium's specific heat capacity, thermal conductivity, and density. The heat capacity and thermal conductivity of a porous matrix is affected by its water content. Heat dissipation sensors contain heating elements in line or point source configurations embedded in a rigid porous matrix with fixed pore space. The measurement is based on application of a heat pulse by applying a constant current through the heating element for specified time period and analysis of the temperature response measured by a thermocouple placed at a certain distance from the heating source.

Description: A thermal heat probe consists of a porous block containing a heat source and an accurate temperature sensor. The block temperature is measured before and after the heater is powered for a few seconds. Thereby, block moisture is obtained from the temperature variation. Since the porous

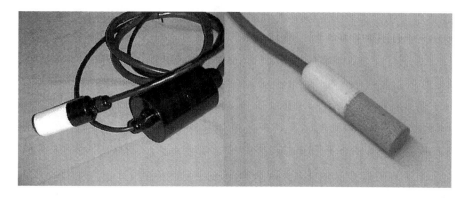

FIGURE 5.12 Heat dissipation sensors.

block, placed in contact with the soil, is equilibrated with the soil water, its SWC will give the soil water potential. Hence, the sensor must be provided with the calibrated relationship between the measured change in temperature and soil water potential. Measurement range is 0.1–30 bar (less accurate for 10–30 bar range) (Figure 5.12).

Advantages:
- Wide measurement range.
- No maintenance required.
- Up to 10 cm measurement sphere radius.
- Continuous reading possible.
- Not affected by salinity because measurements are based on thermal conductivity.

Drawbacks:
- Needs a sophisticated controller/logger to control heating and measurement operations.
- Slow reaction time. It does not work well in sandy soils, where water drains more quickly than the instrument can equilibrate.
- Fairly large power consumption for frequent readings.

5.2.2.4 Soil Psychrometer

Working principle: Under vapor equilibrium conditions, water potential of a porous material is directly related to the vapour pressure of the air surrounding the porous medium according to Kelvin's equation:

$$\phi_m = \frac{RT_a}{M_w} \ln RH \qquad (5.8$$

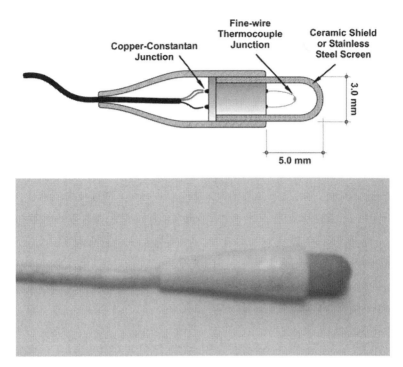

FIGRUE 5.13 Soil psychrometer.

where ϕ_m is the water potential, R is the gas constant (8.31 J mol^{-1}K^{-1}); T_a is the Kelvin temperature of the air, M_w is the molecular weight of water, and RH is the relative humidity. Thereby, the soil water potential is determined by measuring the RH of a chamber inside a porous cup equilibrated with the soil solution (Campbell and Gardner, 1971).

Description: A soil psychrometer consists of a ceramic shield or screen building an air chamber, where a thermocouple is located. The screen type is recommended for high-salinity environments, Thereby, RH in the air chamber is calculated from the "wet bulb" versus "dry bulb" temperature difference. Measurement range is 0.5–30 bar (less accurate for 10–30 bar range) (Figure 5.13).

Advantages:

- High sensitivity.
- Scientifically rigorous readings (except in wetter soil conditions).
- Suitable where typical moisture conditions are very dry.

Drawbacks:

- Not recommended at shallow soil depths, due to high susceptibility to thermal gradient.
- Small sensing volume.

- Very slow reaction time, because reaching vapour equilibrium takes time.
- Low accuracy in the wet range.
- Specialized equipment is required for the sensor's excitation and reading.

5.3 RECOMMENDATIONS AND FUTURE RESEARCH

As described above, a wide range of methods is available for measuring and monitoring soil water content. Often the selection of a technique is not simple, because all present advantages and disadvantages, that can be important in the particular situation. The selection of a suited method should take into consideration several issues:

- Soil properties (texture, organic matter content, swelling, heterogeneity)
- Application (research, monitoring, irrigation scheduling)
- Plant type (if present)
- Accuracy and moisture range needed
- Cost (capital and annual cost)
- Skill level required for operation
- Maintenance

Tables 5.1 and 5.2 display a comparison of the methods presented to provide the reader with a quick reference.

Charlesworth (2000) presented a method suggested by Cape (1997) to decide which soil moisture–measuring technique is most applicable to a particular situation. This procedure consists of answering a number of questions (Yes = 1, No = 0 (Table 5.3). The relative importance of each question is quantified with appropriate weights, and a total relative importance (T) of each sensor for a specific application is obtained by adding the individual scores from all questions and multiplying it by the score for the "effective range of measurement" criterion. This multiplication factor (0 or 1) is a modification of the original method proposed here. This implies that no sensor will be valid for an application if the field-measuring range does not match the sensor specifications. The total estimated life cost of the sensor (Cost) is estimated from capital, installation, running, and maintenance costs for the expected life of the sensor (L). The annual cost (A) of the sensor is obtained by A = Cost/L. The final sensor value for the application (V) is obtained by dividing T/A. The device with the highest value V is more suited to the needs and budget considered. An illustration example is included in Table 5.3 where the neutron probe is compared with an FDR sensor. Both alternatives include measuring moisture at one point with 10 depths. The FDR equipment includes a logger and software for graphical display of information as standard and the neutron probe a built-in display where the moisture values can be read after the site-specific calibration has been input, in addition to the count number. For the example application, both devices satisfy the criteria

TABLE 5.1
Evaluation Criteria for Volumetric Soil Water–Monitoring Methods

	Neutron moderation	TDR	FD (capacitance and FDR)	ADR	Phase transmission	TDT
Reading range	0–0.60 cm³ cm⁻³	0.05–0.50 cm³ cm⁻³ 0.05–Saturation (with soil-specific calibration)	0–saturation	0–saturation	0.05–0.50 cm³ cm⁻³	0.05–0.50 cm³ cm⁻³ 0–0.70 cm³ cm⁻³ Depending on instrument
Accuracy (with soil-specific calibration)	±0.005 cm³ cm⁻³	±0.01 cm³ cm⁻³	±0.01 cm³ cm⁻³	±0.01–0.05 cm³ cm⁻³	±0.01 cm³ cm⁻³	±0.05 cm³ cm⁻³
Measurement volume	Sphere (15–40 cm radius)	About 3 cm radius around length of waveguides	Sphere (about 4 cm effective radius)	Cylinder (about 4 cm³ cm⁻³)	Cylinder (15–20 L)	Cylinder (0.8–6 L) of 50 mm radius
Installation method	Access tube	Permanently buried in situ or inserted for manual readings	Permanently buried in situ or PVC access tube	Permanently buried in situ or inserted for manual readings	Permanently buried in situ	Permanently buried in situ
Logging capability	No	Depending on instrument	Yes	Yes	Yes	Yes
Affected by salinity	No	High levels	Minimal	No	>3 dS/m	At high levels

(Continued)

TABLE 5.1
Continued

	Neutron moderation	TDR	FD (capacitance and FDR)	ADR	Phase transmission	TDT
Soil types not recommended	None	Organic, dense, salt or high clay soils	None	None	None	Organic, dense, salt or high clay soils (depending on instrument)
Field maintenance	No	No	No	No	No	No
Safety hazard	Yes	No	No	No	No	No
Application	Irrigation researcher consultants	Irrigation researcher consultants	Irrigation researcher	Irrigation researcher	Irrigation	Irrigation
Cost (includes reader/logger/ interface if required)	$10,000–15,000	$400–23,000	$100–3,500	$500–700	$200–400	$400–1,300

TABLE 5.2
Evaluation Criteria for Tensiometric Soil Water–Monitoring Methods

	Tensiometer	Gypsum block	GMS	Heat dissipation	Soil psychrometer
Reading range	0–0.80 bar	0.3–2.0 bar	0.1–2.0 bar	0.1–10 bar	0.5–30 bar
Accuracy (with soil-specific calibration)	±0.01 bar	±0.01 bar	±0.01 bar	7% absolute deviation	±0.2 bar
Measurement volume	Sphere (> 10 cm radius)	Sphere (> 10 cm radius)	Sphere (about 2 cm radius)		Sphere (> 10 cm radius)
Installation method	Permanently inserted into augered hole	Permanently inserted into augered hole	Permanently inserted into augered hole	Permanently inserted into augered hole	Permanently inserted into augered hole
Logging capability	Only when using transducers	Yes	Yes	Yes	Yes
Affected by salinity	No	> 6 dS/m	> 6 dS/m	No	Yes, for ceramic cup type (use screen type)
Soil types not recommended	Sandy or coarse soils	Sandy or coarse soils, avoid swelling soils	Sandy or coarse soils, avoid swelling soils	Coarse	Sandy or coarse soils, avoid swelling soils
Field maintenance	Yes	No	Medium	No	No
Safety hazard	No	No	No	No	No
Application	Irrigation research	Irrigation	Irrigation	Irrigation research	Research
Cost (includes reader/logger/interface if required)	$75–250	$400–700	$200–500	$300–500	$500–1000

TABLE 5.3

Example of Evaluation Procedure for Choosing Between Alternative Soil Moisture Sensors

Attributes	Weight (A)	Neutron probe Point (B)	Neutron probe Score (C)	FDR Point (B)	FDR Score (C)
Effective range of measurement (Point: Yes = 1; No = 0 sensor is not recommended for application and total score T = 0)	—				
Is the device able to measure all ranges of soil water of interest to you?		1		1	
Accuracy (Point: Yes = 1; No = 0)	14				
Is the sensor accurate enough for your purpose?		1	14	1	14
Soil types (for use with range of soils) (Point: Yes = 0; No = 1)	11				
Is the sensor's accuracy affected by the soil type?		1	11	0	0
Reliability (Point: Yes = 1; No = 0)	13				
Do you have any personal, other users; or literature-based idea of the reliability of the sensor, and is the failure rate satisfactory to you?		1	13	1	13
Frequency/soil disturbance (Point: Yes = 1; No = 0)	8				
Can the sensor provide quick or frequent readings in undisturbed soil?		0	0	1	8
Data handling (Point: Yes = 0; No = 1)	8				
Will you have difficulty reading or interpreting data?		1	8	1	8
Communication (for remote data manipulation) (Point: Yes = 1; No = 0)	10				
Does the sensor provide data logging and downloading capabilities and friendly software for analyzing and interpreting the data?		0	0	1	10
Operation and maintenance (Point: 1/4 for each Yes answer; No = 0)	10				
Is the sensor calibration universal?		0	0	0	0
Does the sws have a long life (>5 years)?		0.25	2.5	0.25	2.5
Is the sensor maintenance free?		0	0	0.25	2.5
Is the sensor easy to install?		0.25	2.5	0.25	2.5
Safety (Point: Yes = 0; No = 1)	8				
Does use of the sensor entail any danger?		0	0	1	8
Total (T)			51		68.5
Cost (Cost) (in $)			15000		7500
Life (L) (in years)			10		10
Annual cost of sensor (A = Cost/L)			1500		750
Value of sensor (V = T/A)			0.034		0.091

Source: Adapted from Cape, 1997.

(score $= 1$) of range of measurement, accuracy, reliability, and data handling. On the other hand (score $= 0$), the FDR calibration is strongly dependent on soil type, whereas the neutron probe does not allow for quick/frequent readings, does not provide datalogging since it cannot be left unattended in the field, and needs an strict maintenance program as a radioactive device. Although the cost of installation is similar (both are tubes in the ground), the total cost of the neutron probe is higher, as is the data-collection labor (requires certified personnel). The expected life for both devices is 10 years. The value selection method indicates that FDR is a superior option for this application.

In the context of soil water monitoring, because of the soil's natural and artificially induced variability, the location and number of instruments may be crucial. Several factors can affect soil moisture reading variability: soil type and intrinsic heterogeneity, plant growth variation, rainfall interception, reduced irrigation application efficiency and uniformity, etc. Hence, in general, it is recommended to identify the average (representative) conditions in terms of soil type, depth, plant distribution, and sources of water (if irrigation) and place the instruments in each representative zone.

Since the pressure to manage water more prudently and efficiently is increasing, it is expected that research on soil water measurement will continue to produce reliable and low-cost solutions. Future research should focus on developing new techniques or improving the available actual methods to overcome the main limitation of requiring a soil-specific calibration. From a research perspective, a combined device that provides both volumetric and tensiometric *in situ* readings would be desirable, since these two state variables are often needed in mass transport studies (see Chapter 3). Further refinement of noncontact and remote sensing techniques shows promise to evaluate soil moisture distribution and variation across large scales.

ACKNOWLEDGMENT

This chapter was supported by the Florida Agricultural Experiment Station and approved for publication as Journal Series No. R-10128.

REFERENCES

Bouyoucos, G.J. and A.H. Mick, 1940. An electrical resistance method for the continuous measurement of soil moisture under field conditions. Michigan Agric. Exp. Stn. Tech. Bull. 172, East Lansing.

Campbell, G.S. and W.H. Gardner, 1971. Psychrometric measurement of soil water potential: temperature and bulk density effects. Soil Sci. Soc. Am. Proc. 35: 8–12.

Cape, J., 1997. A value selection method for choosing between alternative soil moisture sensors. Project No. AIT2, Land and Water Resources Research and Development Corporation Report.

Charlesworth, P., 2000. Soil water monitoring. Irrigation insights no. 1. National Program for Irrigation Research and Development. Land and water Resource Research and Development Corporation, Canberra, Australia, 96 pp.

Dane, J.H. and G.C. Topp (eds.), 2002. The soil solution phase, Chapter 3. Methods of Soil Analysis, Part 4.: Physical Methods. Soil Science Soc. of America, Inc., Madison, WI.

Davis, J.L. and A.P. Annan, 2002. Ground penetrating radar to measure soil water content. In: J.H. Dane and G.C. Topp (eds.) Methods of Soil Analysis, Part 4.: Physical Methods, pp.: 446–463. Soil Science Soc. of America, Inc., Madison, WI.

De Loor, G.P., 1964. Dielectric properties of heterogeneous mixtures. Appl. Sci. Res. B3: 479–482.

Dirksen, C., 1999. Soil Physics Measurements. Catena Verlag GmbH Reiskirchen, Germany.

Dirksen, C. and S. Dasberg, 1993. Improved calibration of time domain reflectometry soil water content measurements. Soil Sci. Soc. Am. J. 57: 660–667.

Dobson, M.C., F.T. Ulaby, M.T. Hallikainen, and M.A. El-rayes, 1985. Microwave dielectric behavior of wet soil. Part II: dielectric mixing models. EE Trans. Geosci. Remote Sensing GE 23: 35–46.

Gaskin, G.D. and J.D. Miller, 1996. Measurement of soil water content using simplified impedance measuring technique. J. Agric. Eng. Res. 63: 153–160.

Harlow, R.C., E.J. Burke, T.P.A. Ferre, J.C. Bennet, and W.J. Shuttleworth, 2003. Measuring spectral dielectric properties using gated time domain transmission measurements. Vadose Zone J., 2:424–432.

Hilhorst, M.A., C. Dirksen, F.W.H. Kampers, and R.A. Feddes, 2001. Dielectric relaxation of bound water versus soil matric pressure. Soil Sci. Soc. Am. J. 65: 311–314.

Inoue, M., 1998. Evaluation of measuring precision of field-type dielectric soil moisture probes using salty sand. J. Jpn. Soc. Hydrol. Water Resour. 11: 555–564.

Muñoz-Carpena, R., Y. Li, and T. Olczyk, 2002. Alternatives for low cost soil moisture monitoring devices for vegetable production in the south Miami-Dade County agricultural area. Fact Sheet ABE 333, one of a series of the Agricultural and Biological Engineering Department, Florida Cooperative Extension Service, Institute of Food and Agricultural Sciences, University of Florida, EDIS http://edis.ifas.ufl.edu.

Muñoz-Carpena, R., H. Bryan, W. Klassen, T.T. Dispenza, and M.D. Dukes, 2003. Evaluation of an automatic soil moisture-based drip irrigation system for row tomatoes. ASAE Paper No. 032093. ASAE, St. Joseph, MI.

Nakashima, M., M. Inoue, K. Sawada, and C. Nicholl, 1998. Measurement of soil water content by amplitude domain reflectometry (ADR). Method and its calibrations. Chikasui Gakkaishi 40: 509–519.

Regalado, C.M., R. Muñoz-Carpena, A.R. Socorro, and J.M. Hernández-Moreno, 2003. Time domain reflectometry models as a tool to understand the dielectric response of volcanic soils. Geoderma 117 (3–4): 313–330.

Robinson, D.A., C.M.K. Gardner, and J.D. Cooper, 1999. Measurement of relative permittivity in sandy soils using TDR, capacitance and theta probe: comparison, including the effect of bulk soil electrical conductivity. J. Hydrol. 223: 198–211.

Roth, K., R. Schulin, H. Flüher, and W. Attinger, 1990. Calibration of time domain reflectometry for water content measurement using a composite dielectric approach. Water Resour. Res. 26: 2267–2273.

Smajstrla, A.G., and D.S. Harrison, 1998. Tensiometers for soil moisture measurement and irrigation scheduling. CIRCULAR 487, Florida Cooperative Extension Service, Institute of Food and Agricultural Sciences, University of Florida, Gainesville.

Topp, G.C., Davis, J.L., and Annan, A.P., 1980. Electromagnetic determination of soil water content: measurements in coaxial transmission lines. Water Resour. Res. 16, 574–582.

Wijaya, K., T. Nishimura, and K. Makoto, 2002. Estimation of bulk density of soil by using amplitude domain reflectometry (ADR) probe. 17[th]WCSS. Thailand. Paper no. 385.

Yoder, R.E., D.L. Johnson, J.B. Wilkerson, and D.C. Yoder, 1998. Soil water sensor performance. Appl. Eng. Agric. 14(2): 121–133.

6 Measurement and Characterization of Soil Hydraulic Properties

W. D. Reynolds
Agriculture & Agri-Food Canada, Harrow, Ontario, Canada

D. E. Elrick
University of Guelph, Guelph, Ontario, Canada

CONTENTS

1-5667-0657-2/05/$0.00 + $1.50
© 2005 by CRC Press

6.1 INTRODUCTION

Soil hydraulic properties influence the entry, movement, and removal of water in the unsaturated or vadose zone. More specifically, they control the rate of water entry into the soil during the process of infiltration, the rate of water translocation within the soil during the process of redistribution, and the rate of water removal from the soil during the processes of drainage, evaporation, and plant uptake (transpiration). Through their control of water movement, soil hydraulic properties also exert a strong influence on many other soil processes, such as soil water storage (Chapters 4, 5), soil gas accumulation and flux (Chapter 13), surface water runoff, soil erosion, groundwater recharge, leaching of solutes (e.g., crop nutrients, pesticides, contaminants) (Chapters 9, 10, 12, 14), soil pedogenesis, plant growth, soil biota (Chapter 15), and the accumulation or loss of soil organic matter. Consequently, soil hydraulic properties are critically important to natural processes associated with the hydrologic cycle, and to a vast range of human activities associated with agriculture and soil–water management.

Soil hydraulic properties include hydraulic conductivity, flux potential, sorptivity, and the sorptive number, or its inverse known as the macroscopic capillary length. These parameters exist for all soil water contents, but we will focus here on their field-saturated and near-saturated values, as they exert a far greater influence on soil and hydrologic processes than the unsaturated values. We will also focus on field or "*in situ*" methods for measuring soil hydraulic properties, as opposed to laboratory methods (e.g., soil cores) or indirect methods (e.g., pedotransfer functions). It is generally felt that field methods provide more realistic measures of soil hydraulic properties than many other approaches because they sample a more representative soil volume, they generally cause less disturbance to the soil (i.e., the sampled soil volume remains largely intact), and they maintain both water and air contact between the sampled soil volume and the surrounding soil (Bouma, 1982). Methods discussed in detail include ring infiltrometers, well permeameters, and tension infiltrometers. A brief overview of the instantaneous profile method is also given.

6.2 PRINCIPLES OF SOIL WATER FLOW AND PARAMETER DEFINITIONS

Water movement in rigid, homogeneous, isotropic, variably saturated soil is most completely described by Richards' (1931) equation, which for one-dimensional vertical flow is given by

$$\frac{\partial \theta}{\partial t} = \frac{\partial}{\partial z}\left[K(\psi)\frac{\partial h}{\partial z}\right]; \quad h = \psi + z \tag{6.1}$$

where θ [$L^3 L^{-3}$] is volumetric soil water content, t [T] is time, $K(\psi)$ [LT^{-1}] is the hydraulic conductivity (K) versus pore water pressure head (ψ) relationship, h [L] is hydraulic head, and z [L] is elevation or gravitational head above an arbitrary datum (positive upward). Equation (6.1) indicates that soil water flows in the direction of decreasing hydraulic head, h, and that the rate of flow is determined by the magnitude of the hydraulic head gradient, $\partial h/\partial z$, and the hydraulic conductivity function, $K(\psi)$. The $K(\psi)$ term is the soil's water transmission relationship, and it gives the permeability of the soil to water as a function of the pore water pressure head, ψ.

The $K(\psi)$ term depends strongly on the magnitude and shape of the soil water sorption-desorption relationship, $\theta(\psi)$ [$L^3 L^{-3}$], which itself describes the change in soil water content with changing pore water pressure head. As a result, $K(\psi)$ is not a single value, but a relationship where K decreases as ψ decreases. Through its connection with the $\theta(\psi)$ relationship, $K(\psi)$ depends on the number and size distribution of the soil pores, which in turn depend on soil porosity, structure, texture, organic matter content, and clay mineralogy. Unlike $\theta(\psi)$, however, $K(\psi)$ also depends on pore morphology parameters such as tortuosity, roughness, connectivity, and continuity. These additional dependencies cause $K(\psi)$ to change by a million-fold or more over the ψ range applicable to plant growth (i.e., $\approx -150\,\mathrm{m} \leq \psi \leq 0$).

Due to the extreme sensitivity of K to pore size and pore morphology, the magnitude and shape of the $K(\psi)$ relationship changes substantially with soil texture and soil structure. Schematic examples of the $K(\psi)$ relationship for a range of soil textures and structures are given in Figure 6.1, where it is seen that near-saturated K can change very rapidly with ψ in a structured soil, and that

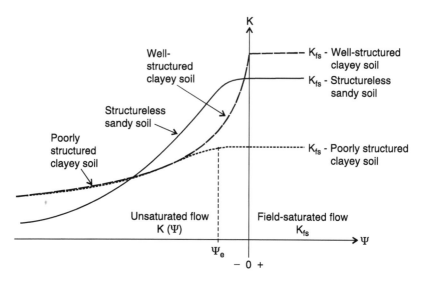

FIGURE 6.1 Schematic examples of representative hydraulic conductivity (K) vs. pore water pressure head (ψ) relationships for a structureless sandy soil, an unstructured clayey soil, and a structured clayey soil. The ψ_e parameter represents either the air-entry pressure head (ψ_a) for drainage from field saturation, or the water-entry pressure head ψ_w) for wetting up to field-saturation. The K_{fs} parameter is the field-saturated hydraulic conductivity. (Adapted from Reynolds et al., 2002.)

K can be several orders of magnitude greater when the soil is structured compared to when it is not structured. Because of this, it is not uncommon for structured loamy and clayey soils to have saturated hydraulic conductivity values (i.e., K at $\psi \geq 0$) that are comparable to, or even greater than, the saturated hydraulic conductivities of sandy soils (Figure 6.1).

When the soil is saturated ($\psi \geq 0$),

$$K(\psi) = \text{constant} = K_s \qquad (6.2$$

where K_s [LT^{-1}] is known as the saturated hydraulic conductivity. As with $K(\psi)$, the K_s value is highly sensitive to soil texture and structure, and its value ranges from as high as 10^{-2}–$10^{-4}\,\mathrm{ms}^{-1}$ in coarse-textured and/or highly structured or cracked soils, to as low as 10^{-8}–$10^{-10}\,\mathrm{ms}^{-1}$ in compacted, structureless clay soils and landfill liners. When K_s is measured via infiltration into initially unsaturated soil, it is often referred to as the *field-saturated* hydraulic conductivity, K_{fs} (Reynolds et al., 1983). This terminology is used because air entrapment during the infiltration and redistribution processes (especially downward infiltration under ponded conditions) prevents complete saturation of the soil, which in turn causes the measured soil hydraulic conductivity (i.e., K_{fs}) to be less than the truly saturated soil hydraulic conductivity (i.e., K_s). The K_{fs} value is often on the order of 0.5 K_s (Bouwer, 1966; Stephens et al., 1987; Constantz et al., 1988), although this can vary substantially depending on the nature and condition of the soil. For many unsaturated/vadose zone applications, K_{fs} is considered more appropriate than K_s because most natural and man-made infiltration processes (e.g., rainfall, irrigation, rapid rise of the water table, waste water disposal via leach fields) cause substantial air entrapment within the initially unsaturated soil (Bouwer, 1978).

Because the $K(\psi)$ relationship (including K_{fs}) determines the permeability of the soil to water, it is the most fundamental of the soil hydraulic properties. In fact, the other soil hydraulic properties (i.e., flux potential, sorptive number, sorptivity) either derive directly from, or are closely related to, the $K(\psi)$ function. This is illustrated below.

A useful empirical representation of the $K(\psi)$ relationship is given by (Gardner, 1958):

$$K(\psi) = K_{fs} \exp[\alpha(\psi - \psi_e)]; \quad 0 < \alpha < +\infty; \quad \psi < \psi_e \leq 0 \leq H \tag{6.3a}$$

$$K(\psi) = K_{fs}; \quad \psi \geq \psi_e \tag{6.3b}$$

where α [L^{-1}] is a shape parameter (i.e., slope of $\ln K$ vs. ψ) that depends primarily on soil texture and structure (Philip, 1968), ψ_e [L] is an "entry" pressure head (Figure 6.1), which represents the air-entry value (ψ_a) for drainage from field-saturation and the water-entry value (ψ_w) for wetting up to field saturation (Bouwer, 1978), and H [L] is the depth of water ponding on the soil surface. Integrating Eq. (6.3) between $\psi = \psi_i$ and $\psi = H$ produces (Figure 6.2),

$$\phi_T = \int_{\psi_i}^{H} K(u)\, du = \phi_V + \phi_S; \quad \psi_i \leq 0 \leq H \tag{6.4a}$$

$$\phi_V = \int_{\psi_e}^{H} K(u)\, du = (K_{fs} - K_i) \int_{\psi_e}^{H} du = (K_{fs} - K_i)(H - \psi_e); \quad \psi_e \leq 0 \tag{6.4b}$$

$$\phi_S = \int_{\psi_i}^{\psi_e} K(u)\, du = K_{fs} \int_{\psi_i}^{\psi_e} \exp[\alpha(\psi - \psi_e)]d\psi = [(K_{fs} - K_i)/\alpha]; \quad -\infty < \psi_i \leq \psi_e \leq 0 \tag{6.4c}$$

where $\phi_T[L^2T^{-1}]$ is the total flux potential, $\phi_V[L^2T^{-1}]$ is the velocity (flux) potential (Kirkham and Powers, 1972), ϕ_S [L^2T^{-1}] is the matric flux potential

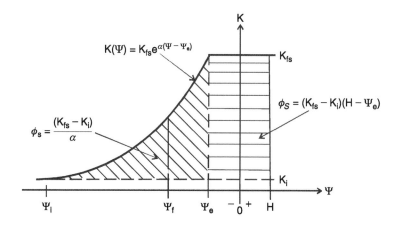

FIGURE 6.2 Schematic of the relationships among selected water transmission parameters (Adapted from Reynolds et al., 2002.)

(Gardner, 1958), ψ_i [L] is the antecedent pore water pressure head in the soil, and $K_i = K_{fs} \exp[\alpha(\psi_i - \psi_e)]$ [LT^{-1}] is the antecedent soil hydraulic conductivity corresponding to ψ_i. Mathematically, ϕ_V represents the area under the $K(\psi$ curve between $\psi = H$ and $\psi = \psi_e$, and ϕ_S represents the area under the $K(\psi$ curve between $\psi = \psi_e$ and $\psi = \psi_i$ (Figure 6.2). Physically, ϕ_V quantifies infiltration due to the hydraulic "push" of the ponded water on the soil surface, while ϕ_S quantifies infiltration or redistribution due to the absorptive "pull" or "capillarity" exerted by the unsaturated soil.

As is evident from Eq. (6.4c) and Figure 6.2, the magnitude of ϕ_S is determined by both the shape and magnitude of the $K(\psi \le \psi_e)$ relationship (which is in turn controlled by soil texture and structure), and by the antecedent pore water pressure head, ψ_i (which affects the magnitude of K_i). Consequently, coarse-textured, structured or wet soils tend to have lower capillarity [i.e., smaller area under the $K(\psi \le \psi_e)$ curve and thereby smaller ϕ_S value] than fine-textured, structureless, or dry soils. Furthermore, all soils (regardless of texture or structure) have zero capillarity ($\phi_S = 0$) when saturated or field-saturated because under that condition $\psi_i = \psi_e$, which produces $K_i = K_{fs}$ in Eq. (6.4c).

In most field environments, $\psi_e \approx 0$ and $K_i \ll K_{fs}$ when the soil water content is at or below the field capacity value (i.e., θ at $\psi_i \le -1$ m). This allows Eq. (6.4c) to be simplified to

$$\alpha \approx \alpha^* \equiv (K_{fs}/\phi_S) = \lambda^{-1} \tag{6.5}$$

where α^* [L^{-1}] is the sorptive number and λ[L] is the macroscopic capillary length. The sorptive number (or macroscopic capillary length) represents the ratio of gravity to capillarity forces during infiltration or drainage (Raats, 1976). Large α^* (or small λ) indicates dominance of the gravity force over capillarity, while small α^* (or large λ) indicates the reverse. Large α^* occurs primarily in coarse-textured and/or highly structured soils, while small α^* occurs primarily in fine-textured and/or structureless soils. Although K_{fs} and ϕ_S can range over many orders of magnitude in soil, the range of α^* is usually about 1–50 m^{-1} (White and Sully, 1987; Reynolds et al., 1992) because of the direct and partially compensatory relationship between K_{fs} and ϕ_S [Eqs. (6.4) and (6.5)]. The reduced variability of α^* and its connection with soil texture and structure make it a very useful parameter in simplified analyses for determination of K_{fs} and ϕ_S (discussed further below).

The third main soil hydraulic property is sorptivity, S, which is a measure of the ability of an unsaturated soil to absorb or store water as a result of capillarity. The S and ϕ_S parameters are related by (Philip, 1957)

$$S = [\gamma(\theta_{fs} - \theta_i)\phi_S]^{1/2} = \left[\gamma(\theta_{fs} - \theta_i)\frac{K_{fs}}{\alpha^*}\right]^{1/2} \tag{6.6}$$

where γ is a dimensionless constant (White and Sully, 1987) related to the shape of the wetting (or drainage) front ($\gamma \approx 1.818$ for wetting, but may be smaller for drainage), θ_{fs} [L^{3}L^{-3}] is the field-saturated volumetric soil water content, and θ_i [L^{3}L^{-3}] is the initial or antecedent volumetric soil water content. Note that θ_{fs} is analogous to K_{fs} in that θ_{fs} is generally less than the completely saturated soil water content, θ_s, because of air entrapment in the soil during infiltration and redistribution.

6.3 FIELD METHODS FOR *IN SITU* MEASUREMENT OF SOIL HYDRAULIC PROPERTIES

The physical characteristics of soil pores (e.g., size distribution, shape, roughness, connectivity) generally exhibit extensive spatial and temporal variation due to changes in soil texture/structure/horizonation, root growth, faunal burrowing, freeze-thaw action, tillage, and other natural and anthropogenic processes (see also Chapters 2, 7). This in turn causes field-saturated and near-saturated soil hydraulic properties to be highly variable, with coefficients of variation up to 400% or more, and statistical distributions that are highly skewed (e.g., Warrick and Nielsen, 1980; Lee et al., 1985). As a result, extensive replication of soil hydraulic property measurements in space (both laterally and vertically) and time is usually required for adequate hydrologic characterization of even small areas (see also Chapter 16). For example, K_{fs} measurements in plot-sized areas often require 10–20 or more replications for adequate identification of mean value and statistical distribution. An important consequence of the need for extensive lateral-vertical-temporal replication is that methods for measuring field-saturated and near-saturated soil hydraulic properties must be accurate and rapid, and must also employ equipment that is simple, reliable, and easily portable. Field methods that best meet these criteria include ring infiltrometers, well or borehole permeameters, and tension or disc infiltrometers.

6.3.1　RING INFILTROMETERS

Ring infiltrometers are thin-walled, opened-ended metal or plastic cylinders with one end sharpened (outside-bevel) to assist insertion into the soil. Most ring infiltrometers are 10–50 cm in diameter by 5–20 cm long, although much smaller and much larger ring diameters have been used for special-purpose applications (e.g., Youngs et al., 1996; Leeds-Harrison and Youngs, 1997). Bouma (1985) indicated that representative measures of soil hydraulic properties can be obtained from a ring infiltrometer only if the ring volume is large enough to include more than 20 soil structural units (e.g., peds delineated by a polygonal cracking pattern). Limited field data in Youngs (1987), Lauren et al., (1988), Richards (1987), and Bouma (1985) suggest that ring diameters need to be ≥ 0.05–0.1 m for single-grain sands and uniform structureless materials (e.g., compacted landfill liners); ≥ 0.3 m for stony/ heterogeneous sands, structured sandy loams, and structured silty loams; and ≥ 0.5 m for structured clays and clay loams. Ring infiltrometers are operated by inserting one or more rings into the soil (usually to a depth of 3–10 cm), ponding a known head of water inside the ring(s), and measuring the rate of water flow out of the ring(s) and into the soil. Ring infiltrometers consequently measure vertical soil water transmission parameters, which usually differ from the corresponding horizontal parameters because of vertical-horizontal anisotropy induced by macropores (cracks, wormholes, root channels), layering, and soil horizonation. Various ring infiltrometer arrangements are possible, including a single or solitary ring (single-ring and pressure infiltrometers) (Figure 6.3a), an inner measuring ring centered inside an outer buffer ring (double- or concentric-ring infiltrometer,) (Figure 6.3b), two adjacent rings (twin- or dual-ring infiltrometer) (Figure 6.7a), and three or more adjacent rings (multiple-ring infiltrometer) (Figure 6.7b). Analyses based on steady state or transient infiltration theory are available for particular ring arrangements.

6.3.1.1　Ring Infiltration Theory

6.3.1.1.1　Steady-State Infiltration

Steady state infiltration through ring infiltrometers can be described using the Reynolds and Elrick (1990) relationship, written in the form:

$$q_s = \frac{Q_s}{(\pi a^2)} = \frac{HK_{fs}}{(C_1 d + C_2 a)} + \frac{K_{fs}}{[\alpha^*(C_1 d + C_2 a)]} + K_{fs} \qquad (6.7$$

where q_s [LT^{-1}] is steady (or quasi-steady) infiltration rate, Q_s [L^{3}T^{-1}] is the corresponding steady flow rate, a [L] is ring radius, H [L] is the steady depth of

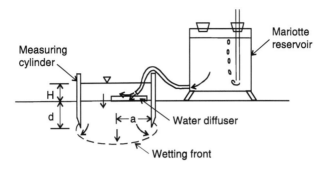

Single ring infiltrometer (cross section)

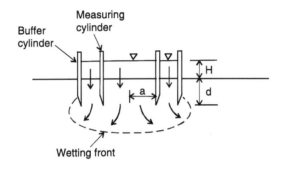

Double or concentric ring infiltrometer (cross section)

FIGURE 6.3 Schematic of the single-ring and double-ring infiltrometer. (From: Reynolds et al., 2002.)

ponded water in the ring, d [L] is the depth of ring insertion in the soil, and $C_1 = 0.316\pi$ and $C_2 = 0.184\pi$ are dimensionless quasi-empirical constants that apply for $d \geq 3$ cm and $H \geq 5$ cm (Reynolds and Elrick, 1990; Youngs et al., 1993). This relationship indicates that steady ponded infiltration from the ring (i.e., q_s or Q_s) is controlled by the field-saturated hydraulic conductivity of the soil (K_{fs}), water ponding depth (H), ring insertion depth (d), ring radius (a), and the soil sorptive number (α^*). It is also evident that there are three main components of steady flow from ring infiltrometers: flow due to the hydrostatic pressure of the ponded water in the ring [first term on the right of Eq. (6.7)], flow due to the capillarity (capillary suction) of the unsaturated soil under and adjacent to the ring [second term on the right of Eq. (6.7)], and flow due to gravity [third term on the right of Eq. (6.7)]. In addition, lateral divergence of the flow field below the ring as a result of hydrostatic pressure and capillarity is accounted for implicitly through the $(C_1 d + C_2 a)$ term. For the special case of $H = d = 0$, Eq. (6.7) collapses to the Wooding (1968) solution for steady infiltration from a shallow circular pond:

$$q_s = \frac{Q_s}{\pi a^2} = \frac{K_{fs}}{(\alpha^* C_3 a)} + K_{fs} \tag{6.8}$$

where $C_3 = 0.25\pi$. Note that for this special case there are only capillarity and gravity components of flow [first and second terms on the right of Eq. (6.8), respectively], and that lateral flow divergence due to soil capillarity is accounted for by the $(C_3 a)$ term. All ring infiltrometer analyses based on steady infiltration can be derived from Eq. (6.7) or Eq. (6.8).

6.3.1.1.2 Transient Infiltration

The Green–Ampt model for one-dimensional vertical infiltration (Green and Ampt, 1911) appears to provide the most realistic analytic description of transient infiltration through a ponded ring. Ponding encourages the sharp (step function) wetting front required by the Green–Ampt model, and the ring wall confines flow to one dimension as long as the wetting front remains within the ring.

Perhaps the most useful application of transient flow analyses is for hydraulic property characterization of low-permeability materials, such as unstructured silt and clay soils and the compacted earthen liners of landfills and waste water lagoons. The slow flow through these materials can make steady infiltration analyses (Sec. 6.3.1.1.1) impractical due to very long equilibration times, while transient analyses can be done relatively quickly. In addition, slow flow allows the use of large ponded heads to further reduce measurement time and to further encourage a steep Green–Ampt type wetting front, while still allowing the wetting front to be retained within the infiltrometer ring so that one-dimensional infiltration is ensured. Transient analyses are available for both constant head and falling head infiltration and for gravity taken into account or not taken into account (Elrick et al., 2002). Only falling head analyses including gravity will be presented here, however, as they appear to be the most useful for analyzing infiltration into low-permeability materials.

Transient Green–Ampt infiltration can be described by combining the continuity relationship with Darcy's law:

$$\Delta\theta \frac{dz_f}{dt} = K_{fs}\left(\frac{H_t - \psi_f}{z_f}\right) + K_{fs} \tag{6.9}$$

where $\Delta\theta = \theta_{fs} - \theta_i = \text{constant}$ $[L^3 L^{-3}]$ is the difference between the field-saturated soil water content (θ_{fs}) and the antecedent soil water content (θ_i), t [T] is time, z_f [L] is the time-dependent depth to the wetting front, H_t [L] is the time-dependent ponding height (pressure head) at the infiltration surface, and ψ_f [L] is the effective pore water pressure head at the wetting front (Mein and Farrell, 1974). The Green–Ampt model assumes a steep (step-function) wetting front that is stable and downward migrating. Above the front, the soil is at the field-saturated water content, θ_{fs}, and has the hydraulic conductivity, K_{fs} Below the front, the soil is at the antecedent water content, θ_i, and has the hydraulic conductivity, K_i. Integration of Eq. (6.9) for flow through a ponded ring of uniform diameter produces (Philip, 1992; Guyonnet et al., 2000; Elrick et al., 2002),

$$t = \frac{\Delta\theta}{BK_{fs}}\left[\frac{I_t}{\Delta\theta} - \frac{(H_0 - \psi_f)}{B}\ln\left(1 + \frac{BI_t}{\Delta\theta(H_0 - \psi_f)}\right)\right]; \quad 0 \leq I_t \leq H_0 \quad (6.10)$$

where t [T] is time since initiation of ponded infiltration, $B = 1 - \Delta\theta =$ constant, I_t [LT^{-1}] is cumulative infiltration into the soil at time, t, and H_0 [L] is the initial ponding height on the infiltration surface (at $t = 0$). Note that Eq. (6.10) is both implicit and nonlinear with respect to I_t and ψ_f.

6.3.1.2 Single-Ring and Double-Ring Infiltrometer Methods

The single-ring infiltrometer is typically a single open-ended measuring cylinder that is 0.1–0.5 m in diameter and 0.1–0.2 m long, although diameters as large as 1 m are occasionally used (Figure 6.3a). The double- or concentric-ring infiltrometer usually consists of a 0.1–0.3 m diameter by 0.1–0.2 m long measuring cylinder placed concentrically inside an outer buffer cylinder that is 0.2–0.6 m in diameter and the same length as the measuring cylinder (Figure 6.3b). The ASTM standard double-ring infiltrometer specifies diameters of 0.3 m and 0.6 m for the measuring cylinder and buffer cylinder, respectively (Lukens, 1981). Details on the installation and operation of single and double ring infiltrometers can be found in Reynolds et al. (2002), Clothier (2000), Reynolds (1993a), Lukens (1981), and elsewhere.

6.3.1.2.1 Traditional Steady Flow Analyses

The traditional single- and double-ring infiltrometer analysis assumes

$$q_s = K_{fs} \quad (6.11)$$

Comparison to Eq. (6.7) reveals that Eq. (6.11) accounts only for the gravity component of flow out of the ring, i.e., hydrostatic pressure, capillarity, and flow divergence are neglected. The consequence of this is that Eq. (6.11) overestimates K_{fs} to varying degrees, depending on the magnitudes of H, d, a and α^* (Reynolds et al., 2002). Setting H and d to 0.05 m (commonly used values) and α^* to 12 m^{-1} (preferred value for most agricultural soils; see Table 6.1), Eq. (6.7) predicts that a measuring cylinder diameter of 5.2 m would be required in order for Eq. (6.11) to estimate K_{fs} within 5% of the actual value. A cylinder this large does not meet the necessary criteria for a viable field method (see above), as it would be difficult to install, would be highly

TABLE 6.1

Soil Texture-Structure Categories for Site estimation of α*

Soil texture and structure category	α* (m^{-1})
Compacted, structureless, clayey, or silty materials (e.g., marine or lacustrine sediments, landfill caps and liners, liners of earthen lagoons)	1
Soils that are both fine textured (clayey or silty) and unstructured; fine sands	4
Most structured soils from clays through loams; unstructured coarse and medium sands; category most frequently applicable to agricultural soils	12
Coarse and gravely sands; highly structured or aggregated soils; soils with large and/or numerous macropores (cracks, root channels, worm burrows)	36

Source: Adapted from Elrick et al., 1989.

consumptive of water, and would require an excessively long time period to achieve steady flow.

Inclusion of the outer buffer ring in the double ring infiltrometer was an early attempt to eliminate the error in applying Eq. (6.11) by reducing flow divergence effects on the inner measuring ring (Bouwer, 1986). However, both laboratory sand tank studies (Swartzendruber and Olsen, 1961) and numerical simulation studies (Wu et al., 1997; Smettem and Smith, 2002) show that the double-ring system still systematically overestimates K_{fs} when Eq. (6.11) is applied. This occurs because infiltration from the inner ring is still affected by flow divergence and because the hydrostatic pressure and capillarity components of ring infiltration (Eq. 6.7) are not accounted for.

6.3.1.2.2   Updated Steady Flow Analyses

More recent steady flow analyses for the single- and double-ring infiltrometer methods make direct use of Eq. (6.7). This relationship presents a challenge, however, as it contains two unknowns (i.e., K_{fs} and α*) and thus cannot be solved without additional information.

One approach is to solve Eq. (6.7) for K_{fs}:

$$K_{fs} = \frac{q_s}{[H/(C_1 d + C_2 a)] + \{1/[\alpha^*(C_1 d + C_2 a)]\} + 1} \qquad (6.12$$

and obtain α* by independent measurement or estimation. Elrick et al., (1989) showed that the strong correlation between the magnitude of α* and soil texture and structure makes it possible to estimate α* from a series of fairly broad soil texture and structure categories (Table 6.1). Hence, a visual determination is made of the soil texture and structure at the measurement site, the appropriate α* value is selected from Table 6.1 and substituted into Eq. (6.12), and then K_{fs} is determined from Eq. (6.12) using the measured q_s

H, d and a values. The error introduced by estimating α^* from preset categories (i.e., Table 6.1) can be reduced by minimizing the importance of the α^* term in Eq. (6.12) relative to the other terms. This is achieved by making the ring radius (a) and water ponding depth (H) as large as practicable. The maximum potential error introduced in K_{fs} by estimating α^* can be determined by recalculating Eq. (6.12) using the two α^* categories immediately above and below the selected category. The categories in Table 6.1 are broad enough that site estimation of the appropriate category will not likely be in error by more than ±one category. If the estimated maximum potential error in K_{fs} is unacceptably large, then ring radius and/or water ponding depth should be increased (to further reduce the importance of the α^* term), or an alternative method for measuring K_{fs} should be used.

Another approach to solving Eq. (6.7) is to successively pond two or more heads in the measuring ring to produce a system of two or more simultaneous equations:

$$q_i = \frac{K_{fs}H_i}{(C_1d + C_2a)} + \frac{K_{fs}}{\alpha^*(C_1d + C_2a)} + K_{fs}; \quad i = 1, 2, 3, \ldots n; \quad n \geq 2 \quad (6.13)$$

where q_i is the steady infiltration rate corresponding to ponded head, H_i, and the heads are ponded in ascending order without allowing intervening drainage (exposure) of the infiltration surface (Figure 6.4). If two heads are ponded, K_{fs}

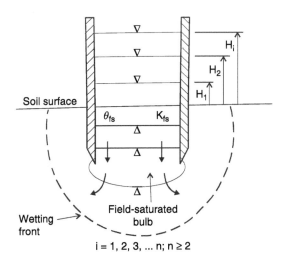

FIGURE 6.4 Schematic of the single-ring constant head infiltrometer with successive ponded heads (pressure infiltrometer).

and α^* are determined using the two simultaneous equations:

$$K_{fs} = \frac{T(q_2 - q_1)}{(H_2 - H_1)} \tag{6.14a}$$

$$\alpha^* = \frac{(q_2 - q_1)}{[q_1(H_2 + T) - q_2(H_1 + T)]} \tag{6.14b}$$

$$T = C_1 d + C_2 a \tag{6.14c}$$

where $q_1 < q_2$ and $H_1 < H_2$. If three or more heads are ponded, q_s is linearly related to H via the relationship:

$$q_i = \left(\frac{K_{fs}}{T}\right) H_i + K_{fs}\left[\left(\frac{1}{\alpha^* T}\right) + 1\right]; \quad T = C_1 d + C_2 a \tag{6.15a}$$

and thus least-squares fitting procedures can be used to obtain K_{fs} from the slope:

$$K_{fs} = T \times \text{slope} \tag{6.15b}$$

and α^* from the intercept;

$$\alpha^* = K_{fs}/[T(\text{intercept} - K_{fs})] \tag{6.15c}$$

of the q_i vs. H_i data points. Note that Eqs. (6.14) and (6.15) are different forms of the relationships developed by Reynolds and Elrick (1990) for their single-ring infiltrometer method known as the "pressure infiltrometer" (Reynolds, 1993a; Reynolds et al., 2002).

6.3.1.2.3 Traditional Transient Flow Analysis

The traditional transient analysis is a direct application of Darcy's law:

$$K_{fs} = \frac{q_a}{\left[\frac{(H - \psi_f)}{z_f} + 1\right]}; \quad \psi_f \le 0 \le H \tag{6.16}$$

where q_a [LT^{-1}] is average infiltration rate, z_f [L] is the depth from the infiltration surface to the Green–Ampt wetting front, H [L] is the pressure head (ponding height) on the infiltration surface, and ψ_f [L] is the effective pore water pressure head at the wetting front (Figure 6.2). Perhaps the most rigorous application of Eq. (6.16) is the "modified" air-entry permeameter

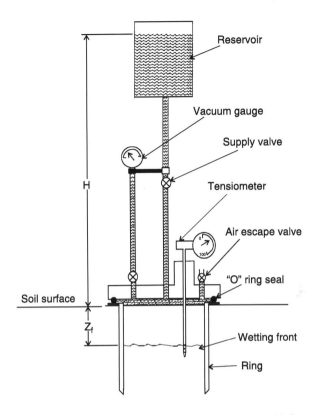

FIGURE 6.5 Schematic of the modified air-entry permeameter. (Adapted from Topp and Binns, 1976.)

method of Topp and Binns (1976) (Figure 6.5). Here, a single ring is inserted into the soil and a small, rapid-responding tensiometer is pushed through the infiltration surface to the predetermined depth, z_f, where z_f is less than or equal to the ring depth, d (z_f is typically $\approx 0.05\,\text{m}$ in clayey soils and $0.15\,\text{m}$ in sandy soils; Topp and Binns, 1976). Next, cumulative infiltration, I [L], is measured under constant head, H, and the tensiometer is monitored to determine the time, t_f [T], required for the wetting front to migrate from the soil surface to depth z_f (signified by a sudden drop in tensiometer reading to zero or near-zero). When the tensiometer responds (at t_f), infiltration is stopped, the ring is sealed off from the atmosphere, and the air-entry pressure head, ψ_a [L], of the wetted soil within the ring is measured using a Bourdon-type vacuum gauge (Topp and Binns, 1976). For many soils (Bouwer, 1966)

$$\alpha^* = -\psi_f^{-1} \approx -2(\psi_a^{-1}); \quad \psi_f \leq 0, \psi_a \leq 0 \qquad (6.17)$$

and thus ψ_a provides an estimate of ψ_f. Equation (6.16) can then be solved for K_{fs}, given that H is known and $q_a = I/t_f$. Note that an estimate of α^* can also be obtained via Eq. (6.17). Difficulties with this method occur when a diffuse and/or irregular wetting front causes an ambiguous tensiometer response (e.g., no increase or a slow increase in tensiometer reading rather than a rapid increase to zero) and when the relationship between ψ_f and ψ_a is not well estimated by Eq. (6.17). Note also that for the air-entry permeameter design of Topp and Binns (1976), H is actually the average ponding height over the infiltration period $(0 \leq t \leq t_f)$ (Figure 6.5).

6.3.1.2.4 Updated Transient Flow Analyses

More recent transient flow analyses are adaptations of Eq. (6.10). For a ring fitted with a standpipe reservoir (Figure 6.6), Eq. (6.10) combined with Eq. (6.17) takes the form (Elrick et al., 2002):

$$t = \frac{\Delta\theta}{CK_{fs}} \left[\frac{R(H_0 - H_t)}{\Delta\theta} - \frac{(H_0 + \frac{1}{\alpha^*})}{C} \ln\left(1 + \frac{CR(H_0 - H_t)}{\Delta\theta(H_0 + \frac{1}{\alpha^*})}\right) \right] \tag{6.18}$$

where $C = 1 - (\Delta\theta/R) = $ constant, H_t [L] is the time-dependent ponding height on the infiltration surface (measured via the standpipe), and the other parameters are as defined in Eq. (6.10). The R parameter is the cross-sectional area of the standpipe reservoir $(A_s = \pi r_s^2)$ divided by the cross-sectional area of the ring $(A_r = \pi r_s^2)$ (i.e., $R = A_s/A_r$; note that $R = 1$ if water is ponded directly in the ring without using a standpipe reservoir). Equation (6.18) thus describes

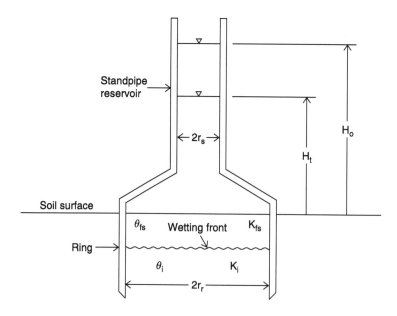

FIGURE 6.6 Schematic of the single-ring falling head infiltrometer with attached standpipe reservoir.

the time-dependent decline in standpipe water level (H_t vs. t) as water infiltrates through the ring and into the soil. As Eq. (6.18) is both implicit and nonlinear in H_t, the K_{fs} and α^* parameters are best determined by curve-fitting the equation to a sequence H_t versus t data points using numerical optimization procedures. The $\Delta\theta$, R and H_0 parameters in Eq. (6.18) must be measured independently.

If only K_{fs} is of interest, Eq. (6.18) and its solution can be simplified substantially. Equation (6.18) is solved for K_{fs} (Bagarello et al., 2003):

$$K_{fs} = \frac{\Delta\theta}{Ct_a}\left[\frac{R(H_0 - H_a)}{\Delta\theta} - \frac{\left(H_0 + \frac{1}{\alpha^*}\right)}{C}\ln\left(1 + \frac{CR(H_0 - H_a)}{\Delta\theta\left(H_0 + \frac{1}{\alpha^*}\right)}\right)\right] \qquad (6.19)$$

where H_a [L] is the measured standpipe water level at time t_a [T]. The initial standpipe water level at zero time, H_0, is preset by adding a calibrated volume of water to the reservoir. The α^* parameter is site-estimated using the soil texture-structure categories in Table 6.1. These simplifications reduce the number of required measurements to only three (i.e., H_a, t_a and $\Delta\theta$) and also allow K_{fs} to be obtained directly from Eq. (6.19) without the use of numerical optimization. What's more, even greater simplification can be achieved when it is feasible to allow the standpipe water level to fall all the way to the soil surface to produce $H_a = 0$ at $t = t_a$. Equation (6.19) can then be solved for K_{fs} using only two measurements, t_a and $\Delta\theta$, because H_a is now 0. The procedures and implications associated with site estimation of α^* are the same as those for the single-head steady flow analysis [Eq. (6.12)].

6.3.1.3 Twin-Ring and Multiple-Ring Infiltrometer Methods

These methods employ adjacent measuring rings with the same depth of water ponding (H) but different ring diameters. Two adjacent rings are used in the case of the twin- or dual-ring infiltrometer, and three or more adjacent rings are used in the case of the multiple-ring infiltrometer (Figure 6.7). The rings are typically 5–50 cm in diameter by 5–20 cm long and are installed individually (not concentrically as in the double-ring infiltrometer) with just enough separation to prevent the wetting fronts from merging before steady infiltration is achieved (Figure 6.7). Multiple rings (three or more) should be clustered, rather than placed in a line or circle (Figure 6.7), to minimize the impacts of lateral soil variability (discussed further below). To obtain sufficient measurement sensitivity, the ring diameters should differ from each other by a least a factor of 2; for the twin-ring method, Scotter et al. (1982) suggested that diameters of about 1.0 m and 0.05 m should be used.

For two or more adjacent rings with constant H, Eq. (6.7) produces the system of simultaneous equations:

$$q_i = \frac{K_{fs}H}{(C_1 d_i + C_2 a_i)} + \frac{K_{fs}}{\alpha^*(C_1 d_i + C_2 a_i)} + K_{fs}; \quad i = 1, 2, 3, \ldots n; \quad n \geq 2 \quad (6.20)$$

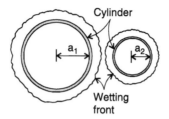

Twin ring infiltrometer ($n = 2$)

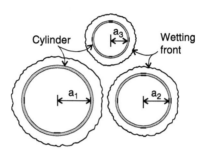

Multiple ring infiltrometer ($n \geq 2$)

FIGURE 6.7 Schematic (plan view) of the twin-ring and multiple-ring infiltrometers. (From Reynolds et al., 2002.)

where q_i is the steady infiltration rate corresponding to ring insertion depth, d_i and ring radius, a_i. If two different insertion depths and/or ring radii are used (constant H), K_{fs} and α^* are determined using the two simultaneous equations,

$$K_{fs} = \frac{(T_1 q_1 - T_2 q_2)}{(T_1 - T_2)} \qquad (6.21a$$

$$\alpha^* = \frac{(T_2 q_2 - T_1 q_1)}{[q_1(HT_1 + T_1 T_2) - q_2(HT_2 + T_1 T_2)]} \qquad (6.21b$$

where

$$T_1 = C_1 d_1 + C_2 a_1 \qquad (6.21c$$

$$T_2 = C_1 d_2 + C_2 a_2 \qquad (6.21d$$

q_1 is the measured q_s for ring 1 (insertion depth d_1; radius, a_1), and q_2 is the measured q_s for ring 2 (insertion depth, d_2; radius, a_2). If three or more ring radii and/or insertion depths are used (constant H), q_s is linearly related to $1/T$ via the relationship:

$$q_i = \frac{K_{fs}(\alpha^* H + 1)}{\alpha^*} \left(\frac{1}{T_i}\right) + K_{fs}; \quad T_i = C_1 d_i + C_2 a_i \quad (6.22a)$$

and least-squares regression can be used to obtain K_{fs} from the intercept:

$$K_{fs} = \text{intercept} \quad (6.22b)$$

and α^* from the slope:

$$\alpha^* = \frac{K_{fs}}{(\text{slope} - K_{fs}H)} \quad (6.22c)$$

of Eq. (6.22a) fitted to q_i vs. $1/T_i$ data points.

Because q_s is often highly variable, averages of replicate q_1 and q_2 values are often used in Eq. (6.21), rather than individual measurements (Scotter at al., 1982). It should also be noted that Eqs. (6.21) and (6.22) apply for any combination of ring insertion depths (d) and radii (a). Hence, K_{fs} and α^* can be obtained from these equations using two or more rings with: (1) different radius but same insertion depth; (2) same radius but different insertion depths; or (3) different radius and different insertion depths. However, using different ring insertion depths will generally not be as efficient or practical as using different ring radii. This is because q_s is not as sensitive to changing insertion depth as it is to changing radius (see Reynolds et al., 2002), because changing insertion depth over an appreciable range increases installation effort substantially, and because large ring insertion depths can cause excessive soil disturbance and impractical increases in the time required to achieve steady infiltration (equilibrium time). Nevertheless, judicious changes in both ring radius and ring insertion depth may improve the accuracy and stability of Eqs. (6.21) and (6.22).

6.3.1.4 Generalized Steady Flow Analysis for Ring Infiltrometers

Steady infiltration from all of the above ring infiltrometer methods (i.e., single, twin, multiple) can be analyzed for K_{fs} and α^* using a generalized least-squares regression relationship of the form (Reynolds et al., 2002):

$$K_{fs} = \frac{\sum_{i=1}^{n}[(H_i + T_i)/T_i]q_i \sum_{i=1}^{n}(1/T_i^2) - \sum_{i=1}^{n}[(H_i + T_i)/T_i^2]\sum_{i=1}^{n}(q_i/T_i)}{\sum_{i=1}^{n}[(H_i + T_i)/T_i]^2 \sum_{i=1}^{n}(1/T_i^2) - \left\{\sum_{i=1}^{n}[(H_i + T_i)/T_i^2]\right\}^2} \quad (6.23)$$

$$\alpha^* = \frac{\sum_{i=1}^{n}[(H_i + T_i)/T_i]q_i \sum_{i=1}^{n}(1/T_i^2) - \sum_{i=1}^{n}[(H_i + T_i)/T_i^2]\sum_{i=1}^{n}(q_i/T_i)}{\sum_{i=1}^{n}[(H_i + T_i)/T_i]^2 \sum_{i=1}^{n}(q_i/T_i) - \sum_{i=1}^{n}[(H_i + T_i)/T_i^2]\sum_{i=1}^{n}[(H_i + T_i)/T_i]q_i}$$

(6.24

$$T_i = C_1 d_i + C_2 a_i$$

(6.25

where the subscript i ($i = 1, 2, 3, \ldots n; n \geq 2$) denotes two or more sets of H, a, d and q_s values. Equations (6.23)–(6.25) apply for any combination of H, a d, and q_s values obtained from a single ring (two or more H, constant d and a), twin rings (any combination of two H, a, or d), or multiple rings (any combination of more than two H, a, or d). Although Eqs. (6.23)–(6.25) appear somewhat complicated, they are readily solved using a computer spreadsheet or simple computer program, and they may be preferred over Eqs. (6.11)–(6.15) and Eqs. (6.20)–(6.22) because of their universal applicability.

6.3.1.5   Calculation of Matric Flux Potential, Sorptivity, and Wetting Front Pressure Head

Once K_{fs} and α^* are calculated, matric flux potential (ϕ_S), sorptivity (S), and pore water pressure head at the Green–Ampt wetting front (ψ_f) can be determined using

$$\phi_S = K_{fs}/\alpha^*$$

(6.26

$$S = [\gamma(\theta_{fs} - \theta_i)\phi_S]^{1/2}$$

(6.27

$$\psi_f = -\alpha^{*-1}$$

(6.28

where γ, θ_{fs}, and θ_i are defined in Eq. (6.6).

6.3.1.6   Strengths and Weaknesses of Ring Infiltrometer Methods

The overall strengths of ring infiltrometer methods include accurate measures of vertical K_{fs}, simple and robust equipment and procedures, relatively easy and rapid spatial/temporal replication of measurements, the ability to measure water transmission parameters at the soil surface, and widespread acceptance by the science and engineering communities because of long-term usage in a vast range of porous media. The general weaknesses of ring methods include difficult use in stony soils (rings difficult to insert), potential disturbance/ alteration of the measured soil volume during the ring insertion process, inconvenience for subsurface measurements (relatively large access pits have to be dug), measurement of only the vertical water transmission properties (rings impose one-dimensional downward flow), and potentially reduced accuracy for determining the soil capillarity parameters (ϕ_S, S, α^*) because ponded infiltration maximizes the hydrostatic pressure and gravity components of flow at the expense of the capillarity component (Reynolds and Elrick, 1990; Reynolds et al., 2002).

The steady-state and transient ring infiltrometer analyses also have strengths and weaknesses. Strengths of the steady analyses include reasonably accurate and robust determination of vertical K_{fs}, extensive field testing, relatively simple measurements (q_s, H, d, a, α^*), and relatively large sample volume (often comparable to ring volume). Weaknesses of the steady analyses include potentially long equilibration times and/or extensive water consumption for large rings or highly structured soils, and high sensitivity of the twin- and multiple-ring methods to small-scale soil heterogeneity because of differing ring sizes and locations. Exacerbating these weaknesses is the fact that ring diameters may need to be ≥ 0.1 m in single-grain sands and uniform structureless materials, ≥ 0.3 m in heterogeneous sands and structured loams, and ≥ 0.5 m in structured clays and clay loams to obtain truly representative K_{fs} and α^* results. The main strengths of the transient analyses (relative to steady analyses) include reduced measurement time (especially in low-permeability soils) and potentially simpler equipment. Weaknesses of the transient analyses include potentially small sample volume (depends on rate of wetting front migration and measurement duration), the need to measure $\Delta\theta$, which requires separate collection of soil samples or use of expensive *in situ* techniques (e.g., TDR) (see Chapter 5), questionable validity of the assumed Green–Ampt (step function) wetting front in structured soils, and a general lack of field testing or comparison to more established methods.

The main strengths and weaknesses of ring infiltrometer methods are compared to other field methods in Table 6.2.

6.3.2 WELL OR BOREHOLE PERMEAMETERS

The well or borehole permeameter method (also known in the engineering literature as the shallow well pump-in method) involves augering an uncased cylindrical well (borehole) into unsaturated soil, ponding water in the well, and measuring the rate at which the water flows out of the well and into the soil. The most common form of this method uses wells that are 0.04–0.1 m in diameter by 0.1–1 m deep (although greater well diameters and depths are possible), plus simple Mariotte bottle apparatus for ponding a constant water head and measuring flow into the soil (Figure 6.8). Under constant ponded head conditions, the flow rate out of the well declines through an initial early-time transient and approaches steady state within finite time. If well diameter and ponding head are kept small (e.g., 0.04–0.1 m well diameter; 0.05–0.5 m head), steady flow can usually be reached quickly (e.g., within a few minutes to 2–3 hours), and water consumption per measurement is generally limited to a

Summary of Method Strengths and Weaknesses

	Ring infiltrometer methods	Well/Borehole permeameter methods	Tension/Disc infiltrometer methods	Instantaneous profile methods
A. Strengths	Measures vertical field-saturated and capillarity flow parameters (K_{fs}, Φ_s, S, α^*) Simple and robust equipment and simple procedures Can measure flow parameters at the soil surface Relatively easy and rapid spatial/temporal replication of measurements Data analysis options available for both steady flow and transient flow Widespread acceptance by the scientific and engineering communities	Measures 3-D field-saturated and capillarity flow parameters (K_{fs}, Φ_s, S, α^*) important for many subsurface flow problems Simple and robust equipment and procedures Generally rapid measurement with low water consumption Relatively easy and rapid depth profiling and spatial/temporal replication of measurements Usable in a wide range of soil textures including moderately stony soils Data analysis options available for steady flow, transient flow, single constant head, multiple constant head and falling head Widespread acceptance by the scientific and engineering communities	Can determine for near-saturation ($-0.2\,\mathrm{m} \leq \psi \leq 0$) a number of important 3-D flow parameters [$K(\psi)$, $S(\psi)$, $\alpha^*(\psi)$, $R(\psi)$], as well as mobile-immobile water content, solute transport characteristics, and solute sorption isotherms Simple, robust, easily portable equipment Can measure flow parameters at the soil surface Relatively easy and rapid spatial/temporal replication of measurements Causes virtually no soil disturbance, and can thus provide good flow parameter estimates of fragile aggregates and macropores Data analysis options available for steady flow, transient flow, inverse modeling, single constant head, and multiple constant head (some analyses well tested and verified)	Can yield simultaneous estimates of $K(\theta)$, $K(\psi)$, and $\theta(\psi)$ for a range of depths and horizons in the soil profile; no other method can do this When method carefully and appropriately applied, its results are considered the "benchmark" against which most other methods are evaluated In uniform soil profiles, a streamlined approach [Eqs. (6.89)–(6.92)] can be used that avoids tensiometers and allows use of simplified analysis procedures that apply from the soil surface to the maximum measurement depth

surface (requires access pit)
Measures only vertical flow
parameters (K_{fs}, Φ_S, S, α^*)
Capillarity parameter estimates
(Φ_S, S, α^*) may be of reduced
accuracy because ponded
infiltration is used
Ring insertion can cause
soil disturbance
Accuracy of some analysis
options is not yet well
established due to
limited testing

borehole wall, and by borehole
siltation by suspended silt and clay
Limited interception of discrete
biopores (esp. vertical worm
and root holes) may result in
unrepresentative parameter
estimates
Potentially long equilibration
times in low-permeability
materials
Capillarity parameter estimates
(Φ_S, S, α^*) may be of reduced
accuracy because ponded
infiltration is used
Accuracy and feasibility of some
analysis options are not yet
well established due to
limited testing

contact between infiltrometer
disc and soil surface
Results limited to the
near-saturated pressure
head range
($\approx -0.2\,\text{m} \leq \psi \leq 0$)
Difficult to use below soil
surface (requires access pit)
Accuracy of some analysis
options not yet well established
due to limited testing and
imprecise knowledge of
some proportionality
parameters

installed at various depths
below the soil surface
Can require both long
measurement times (up to
several months) and large
volumes of water for profile
saturation and drainage
Measurements difficult
to replicate due to extensive
equipment, time and water
requirements
Can determine only the "wet
end" of the $K(\theta)$ and $\theta(\psi)$
relationships ($\approx -8\,\text{m} \leq \psi \leq 0$)
Primarily subsurface method,
i.e., cannot determine $K(\theta)$
within about 0.2–0.3 m of the
soil surface unless simplified
approach used (see above)
May not yield realistic results in
highly structured soils due to
rapid "bypass" flow or in low
permeability soils due to long
times required for profile
saturation and drainage
The streamlined approach
for uniform soil profiles
[Eqs. (6.89)–(6.92)] does not
provide $\theta(\psi)$ data and
assumes validity of specific

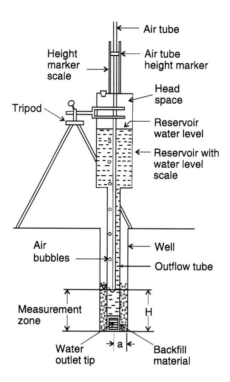

FIGURE 6.8 Schematic of the constant head well permeameter. (From Reynolds, 1993.)

few liters. Details on well permeameter equipment and procedures can be found in Topp et al. (1992), Reynolds et al. (2002), Amoozegar and Wilson (1999), and elsewhere.

Because the well is uncased (or lined with a perforated screen in the measurement zone), water flow into the soil is three-dimensional (i.e., radial and vertical). The water transmission parameters obtained from well permeameters are thus relevant to three-dimensional flow, although the horizontal-vertical weighting of the parameters changes with the depth of ponding in the well (i.e., the greater the ponding depth, the more the transmission parameters are weighted toward the horizontal direction). Most well permeameter analyses are based on steady flow theory, although approximate transient analyses exist for some special cases.

6.3.2.1 Well Permeameter Flow Theory

Steady flow out of a well can be described using the Reynolds et al. (1985) relationship, written as

$$Q_s = \frac{2\pi H^2 K_{fs}}{C_w} + \frac{2\pi H K_{fs}}{\alpha^* C_w} + \pi a^2 K_{fs} \qquad (6.29$$

where Q_s [L^{3}T^{-1}] is the steady flow rate out of the well, H [L] is the steady water ponding depth in the well, a [L] is well radius, and C_w is a dimensionless shape factor which depends on the H/a ratio and α^*. Thus, steady flow out of a ponded well is controlled by the hydraulic properties of the soil (K_{fs} and α^*), depth of water ponding in the well (H), and well radius (a). As with ring infiltration [Eq. (6.7)], the first term on the right of Eq. (6.29) represents flow due to the hydrostatic pressure of the ponded water in the well, the second term represents flow due to the capillarity of the unsaturated soil adjacent to the well, and the third term represents flow due to gravity.

The C_w shape factor is given by (Zhang et al., 1998; D. E. Elrick, personal communication):

$$C_w = \left(\frac{H/a}{2.074 + 0.093(H/a)} \right)^{0.754} \quad \text{for } \alpha^* \geq 9 \, \text{m}^{-1} \qquad (6.30a)$$

$$C_w = \left(\frac{H/a}{1.992 + 0.091(H/a)} \right)^{0.683} \quad \text{for } \alpha^* = 4 \, \text{m}^{-1} \qquad (6.30b)$$

$$C_w = \left(\frac{H/a}{2.081 + 0.121(H/a)} \right)^{0.672} \quad \text{for } \alpha^* = 1 \, \text{m}^{-1} \qquad (6.30c)$$

which are empirical relationships that have been least-squares fitted to discrete data points derived from numerical solutions of the Richards equation for steady, three-dimensional, saturated-unsaturated flow out of a ponded well (Reynolds and Elrick, 1987). Equations (6.30a,b,c) are most accurate for 0.01 m $\leq a \leq 0.05$ m, 0.005 m $\leq H \leq 0.2$ m, $0.25 \leq H/a \leq 20$ (Reynolds and Elrick, 1987). Note also that a single C_w vs. H/a curve applies for all $\alpha^* \geq 9 \, \text{m}^-$ [Eq. (6.30a)]. This occurs because the impact of capillarity on flow out of the well decreases as α^* increases.

6.3.2.2 Original Well Permeameter Analysis

The original well permeameter analysis was based on the Glover relationship (in Zangar, 1953):

$$Q_s = \frac{2\pi H^2 K_{fs}}{C_G} \qquad (6.31)$$

where the shape factor, C_G, is given by

$$C_G = \sinh^{-1}(H/a) - [(a/H)^2 + 1]^{1/2} + (a/H) \qquad (6.32)$$

Note that this relationship neglects the capillarity and gravity components of flow out of the well [i.e., the second and third terms on the right of Eq. (6.29)].

In addition, the C_G relationship is based on the Laplace equation for saturated flow around the well, rather than the Richards equation for variably saturated flow (Reynolds et al., 1983). As a result, Eqs. (6.31) and (6.32) can overestimate K_{fs} by an order of magnitude or more in dry, fine-textured soils (Philip, 1985; Reynolds et al., 1985), although the degree of overestimation can be reduced somewhat by making the H/a ratio as large as practicable. For improved accuracy of the Glover analysis, Amoozegar and Wilson (1999) recommend that H/a be ≥ 5, while Zangar (1953) recommends H/a be ≥ 10.

6.3.2.3   Updated Well Permeameter Analyses

Improved well permeameter analyses for both steady flow and transient flow have been developed. The improved steady flow analyses follow the same pattern as for the ring infiltrometer (Sec. 6.3.1), i.e., single-, double-, and multiple-head options exist. These analyses are based on Eqs. (6.29) and (6.30) and thus do not systematically overestimate K_{fs} (as in the traditional Glover analysis) because all components of flow are taken into account.

6.3.2.3.1   Improved Steady Flow Analyses

The single-head analysis for steady flow uses the relationship (Elrick et al., 1989):

$$K_{fs} = \frac{C_w Q_s}{[2\pi H^2 + C_w \pi a^2 + (2\pi H/\alpha^*)]} \qquad (6.33$$

where C_w is obtained from Eq. (6.30), and the α^* parameter must be selected from the soil texture — structure categories in Table 6.1 or determined from independent measurements. Reynolds et al. (1992) showed that setting $\alpha^* = 12\,\mathrm{m}^{-1}$ for a wide range of agricultural soil textures and structures produced K_{fs} estimates that were usually accurate within a factor of 2 and often accurate within $\pm 25\%$. They further showed that the accuracy of single-head analysis could be improved by making H as large as practicable, which decreases the sensitivity of Eq. (6.33) to error in the estimated value for α^*.

　　The double-head (or two-head) and multiple-head analyses for steady flow are based on Eq. (6.29) written as,

$$Q_i = \frac{2\pi H_i^2 K_{fs}}{C_{wi}} + \frac{2\pi H_i K_{fs}}{\alpha^* C_{wi}} + \pi a^2 K_{fs}; \quad i = 1, 2, 3, \ldots n; \quad n \geq 2 \qquad (6.34$$

where Q_i is the steady flow rate corresponding to ponded head, H_i, and C_{wi} is the shape factor [Eq. (6.30)] corresponding to H_i/a. The heads are ponded in ascending order, with $H_1 < H_2 < H_3 < \cdots$, and the water level in the well is not allowed to fall when switching from one head to the next higher head (Figure 6.9). If two heads are ponded (double-head analysis), K_{fs} and α^* can be

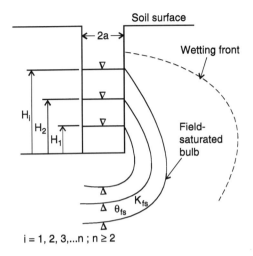

FIGURE 6.9 Schematic of the constant head well permeameter with successive ponded heads.

determined using the two simultaneous equations:

$$K_{fs} = \frac{H_1 C_2 Q_2 - H_2 C_1 Q_1}{\pi[H_1(2H_2^2 + C_2 a^2) - H_2(2H_1^2 + C_1 a^2)]} \quad (6.35a)$$

$$\alpha^* = \frac{2(Q_2 H_1 C_2 - Q_1 H_2 C_1}{Q(2C_1 H_2^2 + C_2 a^2) - Q_2(2C_2 H_1^2 + C_1 a^2)} \quad (6.35b)$$

where $Q_2 > Q_1$, $H_2 > H_1$, $C_2 > C_1$, and C_1 and C_2 are obtained using Eq. (6.30). Note that Eqs. (35) are different forms of the equations given in Reynolds et al. (1985, 2002). If two or more heads are ponded (multiple-head analysis), Eq. (6.34) can be written as (Reynolds and Elrick, 1986):

$$C_i Q_i = P_1 H_i^2 + P_2 H_i + P_3; \quad i = 1, 2, 3, \ldots n; \quad n \geq 2 \quad (6.36a)$$

and least-squares fitted to CQ vs. H data, where Q_i is the steady flow rate corresponding to steady ponding height, H_i, C_i is the C-value corresponding to H_i/a (Eq. 6.30), and

$$P_1 = 2\pi K_{fs}; \quad P_2 = 2\pi \frac{K_{fs}}{\alpha^*}; \quad P_3 = C_i \pi a^2 K_{fs} \quad (6.36b)$$

Equation (6.36) can be solved for K_{fs} and α^* using a computer program (see Reynolds and Elrick, 1986, for working equations) or the regression function of a computer spreadsheet. Once K_{fs} and α^* are determined from either the double-head or multiple-head methods, ϕ_s, S, and ψ_f can be obtained from Eqs. (6.26), (6.27), and (6.28), respectively. The double-head and multiple-head approaches give the same results when two ponded heads (H_1, H_2) are used.

6.3.2.3.2 Transient Flow Analyses

Exact analysis of transient flow from a ponded well is not possible because of the complexity of the three-dimensional, saturated-unsaturated flow process. Some approximate analyses are available, however.

Elrick and Reynolds (1986) developed an approximate falling head analysis by rewriting the constant head, steady flow relationship [Eq. (6.29)] in the form:

$$-\frac{dH_t}{dt} = \frac{Q_t}{\pi a^2} = \frac{2H_t^2 K_{fs}}{a^2 C_{wt}} + \frac{2H_t K_{fs}}{a^2 C_{wt}\alpha^*} + K_{fs} \tag{6.37}$$

where $-dH_t/dt$ describes the time rate of fall of the water level in the well, H_t, and C_{wt} is provided by Eq. (6.30). In effect, Eq. (6.37) assumes that transient, falling head flow out of the well can be approximated as a series of steady flows at declining heads, H_t. Integration of Eq. (6.37) produces

$$t = a^2 \int_H^{H_0} \frac{C_{wt}dH_t}{K_{fs}[2H_t^2 + (2H_t/\alpha^*) + a^2 C_{wt}]} \tag{6.38}$$

where t [T] is the time required for the water level in the well to fall from H (at $t = 0$) to H. No analytical solution to Eq. (6.38) has been found; however, it can be readily solved for K_{fs} and α^* by least-squares curve-fitting to H vs. t data (Elrick and Reynolds, 1986). Because Eqs. (6.37) and (6.38) are based on a constant head, steady flow relationship [i.e., Eq. (6.29)], steady flow at $H = H_0$ must be attained before the falling head phase (i.e., H_t vs. t is started. An important potential application of Eq. (6.38) is improved analysis of the standard "perc test" for site-selection and design of on-site waste water treatment facilities, such as septic tank leach fields (Elrick and Reynolds, 1986; Reynolds et al., 2002).

Assuming a Green-Ampt soil, Philip (1993) developed an approximate falling head analysis for downward outflow from a tightly cased auger hole open only at the bottom (Figure 6.10):

$$t = \frac{\pi^2 a}{8K_{fs}}\left[\left(1 + \frac{1}{2A}\right)\ln\left(\frac{A^3 - 1}{A^3 - \rho_t^3}\right) - \frac{3}{2A}\ln\left(\frac{A - 1}{A - \rho_t}\right) \right. $$
$$\left. + \frac{\sqrt{3}}{A}\left[\tan^{-1}\left(\frac{A + 2\rho_t}{\sqrt{3}A}\right) - \tan^{-1}\left(\frac{A + 2}{\sqrt{3}A}\right)\right]\right] \tag{6.39a}$$

where

$$A^3 = \frac{3\left(H_0 + \frac{1}{\alpha^*} + \frac{\pi^2 a}{8}\right)}{a(\Delta\theta)} + 1 \tag{6.39b}$$

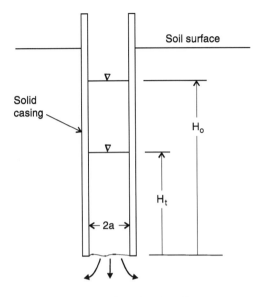

FIGURE 6.10 Schematic of the lined falling-head well permeameter.

$$\rho_t^3 = \frac{3(H_0 - H_t)}{a(\Delta\theta)} + 1 \qquad (6.39c)$$

[T] is the time since initiation of flow out of the well, $\Delta\theta = \theta_{fs} - \theta_i$, H_0 is the initial water depth in the well (at $t = 0$), and H_t [L] is the depth of water in the well at time, t. Equation (6.39) thus describes the time-dependent decline in water level (H_t vs. t) as water flows out through the bottom of the cased well and into the soil. For simultaneous determination of K_{fs} and α^* via Eq. (6.39), a minimum of two H vs. t data points are required, which Philip (1993) and Muñoz-Carpena et al. (2002) chose as t at $H = H_0/2$ and t at $H = 0$. An alternative (and perhaps more robust) approach would be to numerically curve-fit Eq. (6.39) to a sequence of H_t vs. t data points. The $\Delta\theta$ and H_0 parameters must be independently measured.

If only K_{fs} is of interest, Eqs. (6.39a) and (6.39c) can be simplified to (Elrick and Reynolds, 2002);

$$K_{fs} = \frac{\pi^2 a}{8 t_b} \left[\begin{array}{l} \left(1 + \dfrac{1}{2A}\right) \ln\left(\dfrac{A^3 - 1}{A^3 - \rho_b^3}\right) - \dfrac{3}{2A} \ln\left(\dfrac{A - 1}{A - \rho_b}\right) \\ + \dfrac{\sqrt{3}}{A} \left[\tan^{-1}\left(\dfrac{A + 2\rho_b}{\sqrt{3}A}\right) - \tan^{-1}\left(\dfrac{A + 2}{\sqrt{3}A}\right) \right] \end{array} \right] \qquad (6.40a)$$

$$\rho_b^3 = \frac{3(H_0 - H_b)}{a(\Delta\theta)} + 1 \qquad (6.40b)$$

where A is defined by Eq. (6.39b) and H_b [L] is the measured water level in the well at time, t_b [T]. In Eqs. (6.40a) and (6.40b), the initial water level at zero time (H_0) is preset by adding a calibrated volume of water to the well, and the α^* parameter is site-estimated using the soil texture-structure categories in Table 6.1. These simplifications reduce the number of required measurements to three (H_b, t_b, $\Delta\theta$) if H_b is set at some value greater than zero or to just two measurements (t_b, $\Delta\theta$) if H_b is set at zero, in which case t_b is the time required for the well to empty. The simplifications also allow K_{fs} to be obtained directly from Eq. (6.40a) without resorting to simultaneous equation or numerical curve-fitting methods. At this point in time, the accuracies of Eqs. (6.39) and (6.40) are not well established due to limited field testing.

6.3.2.4   Strengths and Weaknesses of Well Permeameter Methods

Overall strengths of the well permeameter method include the ability to measure three-dimensional K_{fs} and α^* (weighted toward the horizontal), which are relevant for many subsurface flow problems such as tile drainage and leach field infiltration; simple and robust equipment and procedures (e.g., easily portable equipment, no requirement for electronics or specialized materials); relatively rapid measurements with generally low water consumption (e.g., measurements completed within minutes to hours using only a few litres of water); easy and rapid depth profiling and spatial-temporal replication of measurements (see Chapter 2); ability to be used in a wide range of soil textures including moderately stony soils; and general acceptance in the science and engineering communities because of long-term usage. Perhaps the primary weaknesses of the well permeameter method include smearing and compaction of the soil along the well wall during the augering process and progressive siltation of the soil along the well wall by suspended silt and clay during the flow measurements. Smearing, compaction, and siltation reduce flow by plugging soil pores, which can in turn produce unrepresentatively low estimates of K_{fs} and/or α^* (e.g., smearing/compaction can cause K_{fs} to be underestimated by an order of magnitude or more in wet, fine-textured soils). Hence, proper well construction/preparation and control of siltation are paramount in the well permeameter method (see Reynolds, 1993a and Reynolds et al., 2002, for details). Other weaknesses of the well permeameter method include potential difficulty in providing representative measures of K_{fs} and α^* in soils where vertical flow is controlled primarily by isolated wormholes and root channels (due to limited interception of the biopores by the well) (see also Chapter 8), potentially slow response and long equilibration times (up to several hours) in low permeability soils (practical K_{fs} range is on the order of 10^{-4}–$10^{-8}\,\mathrm{ms^{-1}}$ although this range can be extended somewhat by adjusting the well and/or permeameter sizes), and potentially reduced accuracy for determining the soil capillarity parameters (ϕ_S, S, α^*) because ponded infiltration maximizes the hydrostatic pressure and gravity components of flow at the expense of the capillarity component (Reynolds and Elrick, 1990; Reynolds et al., 2002).

The various well permeameter analyses also have specific strengths and weaknesses. Single-head analysis [Eq. (6.33)] has the advantages of being robust, simple, generally rapid, and unperturbed by small-scale soil heterogeneity near the well. However, it also has the disadvantages of providing only K_{fs}, requiring site estimation or independent measurement of α^*, which may introduce error, and potentially needing long equilibration times in low-permeability soils. Double-head and multiple-head analyses [Eqs. (6.35) and (6.36)] have the strength of providing measures of both K_{fs} and α^*, but also the weakness that heterogeneity in the soil adjacent to the well (especially vertical heterogeneity such as layering) can cause invalid results in the form of negative K_{fs} or α^* values or α^* values outside the plausible range of $1 \, \mathrm{m}^{-1} \leq \alpha^* \leq 100 \, \mathrm{m}^{-1}$ (Elrick et al., 1989). The falling head transient analysis of Elrick and Reynolds (1986) [Eq. (6.38)] has the disadvantages of requiring the use of numerical curve-fitting to obtain K_{fs} and α^* and the need to attain steady, constant head flow before the falling head measurements are started. Some recent unpublished numerical simulation studies suggest, however, that Eq. (6.38) may be both accurate and robust. The main disadvantages of the falling head methods of Philip (1993), Muñoz-Carpena et al. (2002), and Elrick and Reynolds (2002) [Eqs. (6.39) and (6.40)] are that they require a water-tight seal between the well liner and the well wall (which can be difficult or impossible in some soils); they assume a step-function (Green–Ampt) wetting front, which is of dubious validity for three-dimensional infiltration (Philip, 1993); and they have so far received only limited testing and comparison to more established analyses. On the other hand, potential advantages of Eqs. (6.39) and (6.40) include greatly reduced measurement times in low-permeability materials and the ability to measure lower K_{fs} values than is practical with the steady flow approaches.

The main strengths and weaknesses of well permeameter methods are compared to other field methods in Table 6.2.

6.3.3 TENSION OR DISC INFILTROMETERS

The tension or disc infiltrometer method (also referred to as the disc permeameter) provides measures of both field-saturated and near-saturated soil hydraulic properties for pore water pressure heads ranging between about $-0.2 \, \mathrm{m}$ and $+0.02 \, \mathrm{m}$. The infiltrometer usually consists of a 0.05–$0.30 \, \mathrm{m}$ diameter "infiltration disc" containing a hydrophilic porous plate (e.g., ceramic, plastic) or membrane (e.g., nylon mesh, sieve screen) connected to a water reservoir and a Mariotte-type bubble tower (Figure 6.11). The reservoir supplies water to the infiltration disc, and the bubble tower determines the water pressure head, ψ_0 [L], on the infiltration disc ($\approx -0.2 \, \mathrm{m} \leq \psi_0 \leq +0.02 \, \mathrm{m}$). When the infiltrometer is placed on the soil, capillarity "sucks" the water out of the infiltration disc such that water infiltrates the soil under pressure head, ψ_0 (Figure 6.11). A layer of contact sand is frequently placed under the infiltrometer to ensure good hydraulic contact between the infiltration disc and the soil surface. Tension infiltrometer measurements can usually be obtained within a few minutes to several hours, depending on the type of analysis used and number of pressure heads (ψ_0) set on the infiltration disc. Details on tension infiltrometer equipment and procedures can be found in Topp et al.

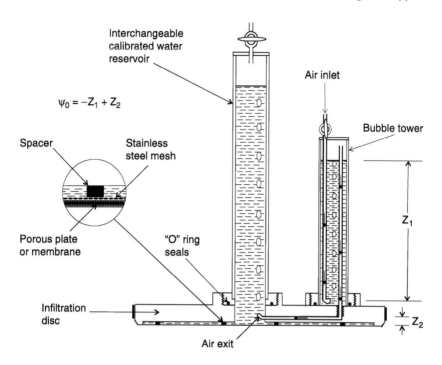

FIGURE 6.11 Schematic of the tension infiltrometer. (Adapted from Perroux and White, 1988.)

(1992), Reynolds (1993b), Reynolds and Zebchuk (1996), Clothier (2000), and Clothier and Scotter (2002).

6.3.3.1 Tension Infiltrometer Flow Theory

All analytical descriptions of flow from a tension infiltrometer assume a homogeneous, isotropic, rigid soil with uniform antecedent water content. Several steady flow and transient flow descriptions are available which involve single, twin or multiple infiltration discs, and single or multiple pressure heads ψ_0) (e.g., Elrick and Reynolds, 1992; White et al., 1992; Clothier, 2000; Smettem and Smith, 2002). We will focus here, however, on three physically based and well-established approaches, which can be applied to a single disc and one or more pressure heads.

Steady constant head infiltration from the tension infiltrometer can be described using the Wooding (1968) "shallow pond" relationship, written in the form:

$$Q(\psi_0) = \pi a^2 K(\psi_0) + \frac{a}{G}\phi(\psi_0) \qquad (6.41$$

where $Q(\psi_0)$ $[L^3T^{-1}]$ is the steady flow rate out of the infiltrometer and into the soil when $\psi = \psi_0$ $[L]$ on the infiltration disc, a $[L]$ is the radius of the infiltration disc, $K(\psi_0)$ [LT^{-1}] is the soil hydraulic conductivity at the infiltration surface, $G = 0.25$ is a shape factor constant, and $\phi(\psi_0)$ is the soil "tension" flux potential defined by

$$\phi(\psi_0) = \int_{\psi_i}^{\psi_0} K(\psi)d\psi; \quad -\infty < \psi_i \le \psi \le \psi_0 \le 0 \tag{6.42}$$

where ψ_i [L] is the antecedent pore water pressure head in the soil. Substituting Eq. (6.3a) into Eq. (6.42) produces

$$\phi(\psi_0) = \frac{K_{fs}}{\alpha}[\exp(\alpha\psi_0) - \exp(\alpha\psi_i)] = \frac{1}{\alpha}[K(\psi_0) - K(\psi_i)] \tag{6.43}$$

where $K(\psi_i)$ [LT^{-1}] is the antecedent soil hydraulic conductivity. When $K(\psi_i)$ is $\le 0.25K(\psi_0)$ to $0.5K(\psi_0)$ (Reynolds and Zebchuk, 1996), Eq. (6.43) can be accurately simplified to

$$\phi(\psi_0) = K(\psi_0)/\alpha \tag{6.44}$$

which in turn allows Eq. (6.41) to be simplified to

$$Q(\psi_0) = \left(\pi a^2 + \frac{a}{\alpha G}\right)K(\psi_0) \tag{6.45}$$

or

$$Q(\psi_0) = \left(\pi a^2 + \frac{a}{\alpha G}\right)K_{fs}\exp(\alpha\psi_0) \tag{6.46}$$

Equations (6.45) and (6.46) indicate that the steady flow rate out of the tension infiltrometer, $Q(\psi_0)$, depends on the radius of the infiltration disc, a, the soil hydraulic conductivity at the infiltration surface, $K(\psi_0)$, and the shape of the exponential $K(\psi)$ relationship, α. The first term on the right of Eq. (6.45) or (6.46) represents the gravitational component of flow out of the infiltrometer, and the second term represents flow due to soil capillarity plus the interaction effects of gravity, capillarity, and source geometry (disc radius).

Transient constant-head infiltration from the tension infiltrometer can be described using a two-term infiltration equation similar in form to that developed by Philip (1957) for one-dimensional infiltration:

$$I(\psi_0) = E_1 t^{1/2} + E_2 t \tag{6.47}$$

where $I(\psi_0)$ [L^{3}L^{-2}] is cumulative infiltration from the infiltration disc at $\psi = \psi_0$, t [T] is time, and E_1 and E_2 are constants related to soil hydraulic properties. Equation (6.47) is best applied by differentiating with respect to $t^{1/2}$ to produce (Vandervaere et al., 2000a),

$$\frac{dI(\psi_0)}{dt^{1/2}} = E_1 + 2E_2 t^{1/2} \tag{6.48}$$

which implies that a plot of $dI(\psi_0)/dt^{1/2}$ versus $t^{1/2}$ should be linear with a y-axis intercept of E_1 and a slope of $2E_2$. Haverkamp et al. (1994) showed that for three-dimensional infiltration at short to medium times (not approaching steady state),

$$E_1 = S(\psi_0) \tag{6.49}$$

$$E_2 = \left(\frac{2-\beta}{3}\right) K(\psi_0) + \frac{\omega S(\psi_0)^2}{a[\theta(\psi_0) - \theta(\psi_i)]} \tag{6.50}$$

where $0 \leq \beta \leq 1$ (Haverkamp et al., 1994), $\omega = 0.75$ (Smettem et al., 1994), $S(\psi_0$ $[LT^{-1/2}]$ is soil sorptivity at $\psi = \psi_0$, $\theta(\psi_0)$ is soil volumetric water content at $\psi = \psi_0$, and $\theta(\psi_i)$ is the antecedent soil volumetric water content at the antecedent pore water pressure head, ψ_i. Inspection of Eqs. (6.49) and (6.50) indicate that flow due to soil capillarity is represented by Eq. (6.49), flow due to gravity is represented by the first term on the right of Eq. (6.50), and the gravity–capillarity–geometry (disc radius) interaction effects are represented by the second term on the right of Eq. (6.50).

Because tension infiltrometers can measure both field-saturated and near-saturated soil hydraulic properties, they have the ability to distinguish the hydraulic properties of soil matrix pores and soil macropores. Soil matrix pores are defined as all pores that are small enough to remain water-filled at a specified negative pore water pressure head, ψ_m [L], whereas macropores (e.g., large cracks, wormholes, root channels, large interaggregate spaces, etc.) are pores that are too large to remain water-filled at ψ_m. The value of ψ_m is not fixed at present, but values of -0.03, -0.06 and -0.1 m have been used (e.g., Watson and Luxmore, 1986; Timlin et al., 1994; Jarvis et al., 2002) (see also Chapter 8). The field-saturated water content, hydraulic conductivity, flux potential, sorptivity and sorptive number of the soil matrix domain are, respectively:

$$\theta_m = \theta(\psi_m); \quad \theta_m \leq \theta_{fs} \tag{6.51}$$

$$K_m = K(\psi_m); \quad K_m \leq K_{fs} \tag{6.52}$$

$$\phi_m = \phi_S(\psi_m); \quad \phi_m < \phi_S \tag{6.53}$$

$$S_m = [\gamma(\theta_m - \theta_i)\phi_m]^{1/2} \tag{6.54}$$

$$\alpha_m^* = K_m/\phi_m \tag{6.55}$$

where the subscript m indicates the soil matrix domain. The ϕ_m parameter is defined by

$$\phi_m = \phi_S(\psi_m) = \int_{\psi_i}^{\psi_m} K(\psi)d\psi; \quad -\infty \leq \psi_i < \psi_m < 0 \tag{6.56}$$

and it is assumed that γ in Eq. (6.54) remains constant at ≈ 1.818 (White and Sully, 1987). For the soil macropore domain (indicated by the subscript, p), the corresponding values are

$$\theta_p = \theta_{fs} - \theta_m; \quad \theta_p \geq 0 \tag{6.57}$$

$$K_p = K_{fs} - K_m; \quad K_p \geq 0 \tag{6.58}$$

$$\phi_p = \phi_S - \phi_m; \quad \phi_p > 0 \tag{6.59}$$

$$S_p = [\gamma(\theta_{fs} - \theta_m)\phi_p]^{1/2} \tag{6.60}$$

$$\alpha_p^* = K_p/\phi_p \tag{6.61}$$

Given that ψ_m is close to zero, soil macropores tend to be large in size (i.e., equivalent diameters on the order of 0.3–1.0 mm) but small in total volume relative to matrix pores. Consequently, θ_p is usually small relative to θ_m, while K_p is often one or more orders of magnitude greater than K_m (Beven and Germann, 1982; Reynolds et al., 1997).

Another feature of the tension infiltrometer is that it can be used to determine a "flow-weighted mean pore size," $R(\psi_0)$ [L], which is given by (Philip, 1987);

$$R(\psi_0) = \frac{\sigma K(\psi_0)}{\rho g \phi_S(\psi_0)} = \frac{\sigma \alpha^*(\psi_0)}{\rho g} \tag{6.62}$$

where σ [MT^{-2}] is the air/pore-water interfacial surface tension, ρ [ML^{-3}] is the pore water density, g [LT^{-2}] is the gravitational acceleration constant, and $K(\psi_0)$, $\phi_S(\psi_0)$ and $\alpha^*(\psi_0)$ are hydraulic conductivity, matric flux potential, and sorptive number, respectively, at pressure head, ψ_0. The $R(\psi_0)$ value has been referred to as an effective "equivalent mean" pore radius that conducts water when infiltration occurs at $\psi = \psi_0$ (White and Sully, 1987). It has also been referred to as an index parameter (rather than a physical pore radius) that represents the mean "water-conductiveness" of the hydraulically active pores (Reynolds et al., 1997). The term water-conductiveness was coined because $R(\psi_0)$ is derived from water flow [via Eq. (6.62)] and must therefore represent in some fashion the combined size, tortuosity, roughness and connectivity of the water-conducting soil pores (Reynolds et al., 1997). As with θ, K, ϕ, S, and α^*, the R value can be determined for soil matrix pores (R_m) and soil macropores (R_p) by substituting the appropriate K, ϕ, or α^* values into Eq. (6.62).

Measurements of θ, K, ϕ, S, α^*, and R for the matrix and macropore domains have so far been used primarily to assess the impacts of cropping and land management on soil water storage and transmission parameters. For example, Table 6.3 compares some of these parameters for a loam soil under long-term no-tillage, long-term conventional moldboard plow tillage, and virgin soil (native forest). Note that θ_p comprises only 3–10% of the total pore space (porosity), but K_p is 2–3 orders of magnitude greater than K_m, and R_p is 1–2 orders of magnitude greater than R_m. The parameters also show that changes in land management can result in substantial changes in the way water is stored and transmitted in soil.

6.3.3.2   Steady Flow — Multiple Head Tension Infiltrometer Analyses

The steady flow, multiple head analysis of Ankeny et al. (1991) assumes that α in Eq. (6.45) is effectively constant between any two adjacent pressure heads set on the infiltration disc, provided that the pressure heads are not too different in magnitude. This allows Eq. (6.45) to be written as

$$Q(\psi_x) = [\pi a^2 + (a/\alpha_{x,\,x+1}G)]K(\psi_x) \qquad (6.63$$

TABLE 6.3
Selected Water Storage and Transmission Parameters for Soil Matrix and Soil Macropore Domains

Parameter[†]	Unit	Virgin soil (native forest)	Long-term, no tillage	Long-term, plow tillage
θ_m	$m^3 m^{-3}$	0.52 a[‡]	0.45 b	0.39 c
θ_p	$m^3 m^{-3}$	0.09 a	0.05 b	0.03 c
K_m	ms^{-1}	3.2×10^{-7} a	1.3×10^{-7} b	3.4×10^{-7} a
K_p	ms^{-1}	3.2×10^{-4} a	4.6×10^{-5} b	2.3×10^{-5} b
R_m	mm	0.077 a	0.047 b	0.050 b
R_p	mm	5.3 a	1.6 b	0.61 c

Parameter definitions given in text. Matrix domain applies for $\psi \leq -0.1\,m$ (subscript m); macropore domain applies for $\psi > -0.1\,m$ (subscript p).

Values within a row followed by the same letter are not significantly different at $p < 0.05$ ($n = 10$).

Source: Adapted from Reynolds et al., 1997.

$$Q(\psi_{x+1}) = [\pi a^2 + (a/\alpha_{x,x+1}G)]K(\psi_{x+1}) \tag{6.64}$$

where $x = 1, 2, 3, \ldots, n$, ψ_x and ψ_{x+1} are pressure heads ($\psi_x \neq \psi_{x+1}$) set in succession on the infiltration disc (i.e., first ψ_x, then ψ_{x+1}), and $\alpha_{x,x+1}$ is a constant that applies for the pressure head range, $|\psi_x - \psi_{x+1}|$. They further showed that

$$\frac{[K(\psi_x) - K(\psi_{x+1})]}{\alpha_{x,x+1}} \approx \frac{[K(\psi_x) + K(\psi_{x+1})]}{2}(\psi_x - \psi_{x+1}) \tag{6.65}$$

which increases in accuracy as the difference between ψ_x and ψ_{x+1} decreases. Equations (6.63), (6.64) and (6.65) thus form a system of three equations that can be solved simultaneously for the three unknowns, $K(\psi_x)$, $K(\psi_{x+1})$ and $\alpha_{x,x+1}$. This approach is applied in piece-wise fashion to successive two-head pairs, (ψ_x, ψ_{x+1}), and near-saturated $K(\psi)$ is calculated using

$$K(\psi_x) = \frac{[K(\psi_x)_{x-1,x} + K(\psi_x)_{x,x+1}]}{2}; \quad x = 2, 3, 4 \ldots, n-1 \tag{6.66}$$

$$K(\psi_1) = K(\psi_1)_{1,2} \tag{6.67}$$

$$K(\psi_n) = K(\psi_n)_{n-1,n} \tag{6.68}$$

where $K(\psi_x)_{x-1,x}$ is obtained from the pressure head pair (ψ_{x-1}, ψ_x), $K(\psi_x)_{x,x+}$ is obtained from the pressure head pair (ψ_x, ψ_{x+1}), and n is the number of pressure heads set on the infiltration disc. The $\alpha_{x,x+1}$ value is obtained directly from Eq. (6.65).

Reynolds and Elrick (1991) also assumed constant α between adjacent pressure heads set on the infiltration disc, but used Eq. (6.46) in the form,

$$\ln Q(\psi_0) = \alpha'_{x,x+1}\psi_0 + \ln\left[\left(\pi a^2 + \frac{a}{\alpha'_{x,x+1}G}\right)K'_{x,x+1}\right] \tag{6.69}$$

which describes a piece-wise linear plot (Figure 6.12) from which $\alpha'_{x,x+1}$ is determined from the piece-wise slope,

$$\alpha'_{x,x+1} = \frac{\ln[Q(\psi_x)/Q(\psi_{x+1})]}{(\psi_x - \psi_{x+1})}; \quad x = 1, 2, 3, \ldots \tag{6.70}$$

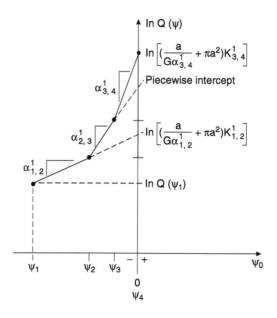

FIGURE 6.12 Piece-wise fitting of the exponential $K(\psi)$ relationship to $\ln Q(\psi_0)$ vs. ψ data. (Adapted from Reynolds and Elrick, 1991.)

and $K'_{x,x+1}$ is determined from the piece-wise intercept,

$$K'_{x,x+1} = \frac{G\alpha'_{x,x+1}Q(\psi_x)}{a(1 + G\alpha'_{x,x+1}\pi a)[Q(\psi_x)/Q(\psi_{x+1})]^P} \tag{6.71}$$

where $P = \psi_x/(\psi_x - \psi_{x,x+1})$, and the pressure heads, ψ_x and ψ_{x+1}, are as defined above. The $\alpha'_{x,x+1}$ and $K'_{x,x+1}$ values are then substituted into

$$K(\psi) = K'_{x,x+1}\exp(\alpha'_{x,x+1}\psi) \tag{6.72}$$

As in Ankeny et al. (1991), Eq. (6.72) is applied in a piece-wise fashion to successive two-head pairs, (ψ_x, ψ_{x+1}), and $K(\psi_x)$ is determined using

$$K(\psi_x) = \frac{[K'_{x-1,x}\exp(\alpha'_{x-1,x}\psi_x) + K'_{x,x+1}\exp(\alpha'_{x,x+1}\psi_x)]}{2} \tag{6.73}$$

$$K(\psi_1) = K'_{1,2}\exp(\alpha'_{1,2}\psi_1) \tag{6.74}$$

$$K(\psi_n) = K'_{n-1,n}\exp(\alpha'_{n-1,n}\psi_n) \tag{6.75}$$

where $x = 2, 3, 4, \ldots, n-1$, and n is the number of pressure heads set on the infiltration disc. Improved estimates of the α' value can then be obtained using

$$\alpha'_{x,x+1} = \frac{\ln K(\psi_x) - \ln K(\psi_{x+1})}{(\psi_x - \psi_{x+1})} \tag{6.76}$$

Reynolds and Elrick (1991) tested the accuracy of Eqs. (6.73)–(6.75) using a numerical simulation model based on Richards' equation and the empirical $K(\psi)$ function of Haverkamp et al. (1977). The $K(\psi)$ function was fitted to data for Yolo clay (Haverkamp et al., 1977), a structureless sand (Haverkamp et al., 1977), and a highly structured soil based on data in Germann and Beven (1981). They found that Eqs. (6.73)–(6.75) were able to predict the actual $K(\psi)$ within $\pm 7\%$, regardless of whether the $K(\psi)$ function was very steep and without an air-entry value (the structured soil), somewhat flat with an indistinct air-entry value (Yolo clay), or steep with a distinct air-entry value (the structureless sand). In addition, the estimates of $K(\psi)$ remained accurate even when $K(\psi_0)$ was within a factor of 2–4 of $K(\psi_i)$, depending on the soil [remember that $K(\psi_i) \ll K(\psi_0)$ is assumed in the analysis]. To maintain $\pm 7\%$ accuracy, the size of $|\psi_x - \psi_{x+1}|$ had to be on the order of 0.01 m when the $K(\psi)$ function was very steep (as in the wet end of the structured soil), but could be as large as 0.10–0.20 m or more when the $K(\psi)$ function was relatively flat (as in the Yolo clay, the structureless sand, or the dry end of the structured soil).

Reynolds and Elrick (1991) also developed a procedure for steady tension infiltration into a ring of radius, a, inserted into the soil to a depth, d. The flow equation is the same form as Eq. (6.41), but the shape factor (G) is now a function of α, d, and a (rather than a constant equal to 0.237). Using numerical simulations, Reynolds and Elrick (1991) developed a plot of G vs. α for realistic ranges of d and a values. They then used the $\alpha'_{x,x+1}$ value from Eq. (6.70) to obtain the appropriate G value from the G vs. α plot before calculating $K'_{x,x+1}$ from Eq. (6.71). Tests based on numerical simulations showed that $K(\psi)$ could be determined within the same $\pm 7\%$ accuracy as achieved when no infiltration ring was used.

A Fortran computer program (TIDAP) is available from the senior author for application of Eqs. (6.69)–(6.76) to steady flow, multiple head tension infiltrometer data. The program applies for tension infiltration with or without an inserted ring. It also has built-in checks to ensure that the input data do not violate the assumptions of the theory (i.e., rigid, homogeneous, isotropic soil with uniform antecedent water content), and it can account for the presence of a contact sand layer (discussed in Sec. 6.3.3.4).

Šimůnek and van Genuchten (1996, 1997) developed an inverse modeling technique (numerical inversion of Richards' equation) to determine near-saturated $K(\psi)$ and $\theta(\psi)$ from multiple-head tension infiltrometer data (see also Chapter 20). They concluded that cumulative infiltration curves [$I(\psi_0)$ vs. t measured at several consecutive pressure heads (ψ_0), plus the initial and final soil water contents immediately below the infiltration disc [$\theta(\psi_0)$ and θ_i respectively], are required to obtain unique estimates of the fitting parameters α_v, n_v, θ_{fs} and K_{fs}) of the van Genuchten (1980) $K(\psi)$ and $\theta(\psi)$ functions. Šimůnek et al. (1998) compared the inverse method to a Wooding (1968)–based analysis for a loamy sand soil where tension infiltration data had been collected using a 0.25 m diameter infiltration disc and pressure heads (ψ_0) of -0.115, -0.09, -0.06, -0.03, -0.01, and -0.001 m. The Wooding analysis was applied for the average pressure heads, -0.1025, -0.075, -0.045, -0.02, and -0.0055 m. The two approaches gave very similar $K(\psi_0)$ values in the pressure head interval, -0.1025 m to -0.02 m. However, a factor of 2 discrepancy occurred between the two $K(\psi_0)$ estimates at $\psi_0 = -0.0055$ m, and between the two extrapolated K_{fs} values. The reasons for the discrepancies were not determined, and it was not clearly established which method produced more accurate wet-end results, although the Wooding–based estimate of K_{fs} was virtually identical to an independent estimate obtained using an empirical regression model and soil texture data. It is possible that experimental artifacts affected both analyses, as the contact sand layer under the infiltration disc was apparently not accounted for (impacts of contact materials discussed in Sec. 6.3.3.4), and hydraulic contact between the infiltration disc and the soil was broken several times to refill the infiltrometer with water (which may have caused perturbations in the soil water flow regime). Public domain software (DISC) for analyzing tension infiltrometer data via the inverse modeling procedure is available from the U.S. Salinity Laboratory, Riverside, CA (www.ussl.ars.usda.gov), and further information on the inverse technique can be found in Hopmans et al. (2002) and Chapter 20.

6.3.3.3 Transient Flow — Single Head Tension Infiltrometer Analysis

At present, the most physically realistic transient analysis seems to be that of Vandervaere et al. (2000a), which makes use of Eqs. (6.48)–(6.50). They discretize Eq. (6.48) to produce

$$\frac{dI(\psi_0)}{dt^{1/2}} \approx \frac{\Delta I(\psi_0)}{\Delta t^{1/2}} = \frac{(I_{j+1} - I_j)}{(t_{j+1}^{1/2} - t_j^{1/2})} = E_1 + 2E_2\left(t_j^{1/2} \times t_{j+1}^{1/2}\right)^{1/2} \tag{6.77}$$

where $j = 1, 2, 3, \ldots, (n-1)$, n is the number of $\Delta I(\psi_0)/\Delta t^{1/2}$ versus $t^{1/2}$ data points, and $t^{1/2}$ is calculated as the geometric mean of $t_j^{1/2}$ and $t_{j+1}^{1/2}$. Then, simple linear regression is used to determine both the magnitudes and standard deviations of E_1 and E_2. An important and elegant feature of Eq. (6.77) is the ease with which the validity of transient tension infiltrometer data can be determined; i.e., a plot of $\Delta I(\psi_0)/\Delta t^{1/2}$ versus $t^{1/2}$ must be linear with both E_1 (intercept) and $2E_2$ (slope) positive (Figure 6.13). If these criteria are not met, the basic assumptions of the method (i.e., rigid, homogeneous, isotropic soil with uniform antecedent water content) are seriously violated and the calculated E_1 and E_2 values are not likely to have physical meaning.

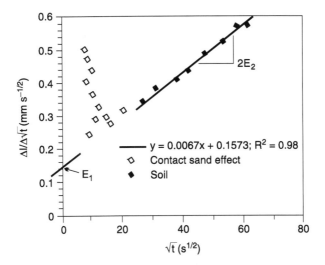

FIGURE 6.13 Application of Eq. (6.77) to transient tension infiltrometer data. (Adapted from Smettem and Smith, 2002.)

Vandervaere et al. (2000a) also showed that Eq. (6.77) provides a convenient and sensitive means for detecting loss of hydraulic contact between the infiltration disc and the soil and for eliminating flow data perturbed by contact sand (see Sec. 6.3.3.4).

Vandervaere et al. (2000b) provide guidelines for determination of $K(\psi_0)$ and $S(\psi_0)$ via Eqs. (6.77), (6.49), and (6.50). They show that $K(\psi_0)$ can be determined reliably from Eq. (6.50) only under the condition

$$\frac{\omega E_1^2}{a[\theta(\psi_0) - \theta(\psi_i)]} < \frac{E_2}{2} \tag{6.78}$$

and that accurate determination of $S(\psi_0)$ from Eq. (6.49) requires the soil sorptivity be close to a so-called "optimal" value given by

$$S(\psi_0)_{opt} = \sqrt{\frac{a[\theta(\psi_0) - \theta(\psi_i)](2 - \beta)K(\psi_0)}{3\omega}} \tag{6.79}$$

Both of these criteria are designed to minimize the interference of source geometry (finite disc radius) in the $K(\psi_0)$ and $S(\psi_0)$ calculations. Equation (6.78) applies when gravity dominates geometry effects during the infiltration process, while Eq. (6.79) applies when capillarity dominates geometry. To help ensure gravity dominated infiltration for accurate $K(\psi_0)$ calculation, one should use a large infiltration disc to increase the magnitude of the denominator on the left of Eq. (6.78) and thereby decrease the geometry effects. One might also choose a relatively wet antecedent soil moisture condition, which would both reduce soil capillarity (sorptivity) and further decrease geometry effects by reducing the numerator on the left of Eq. (6.78). For accurate determination of $S(\psi_0)$, a large infiltration disc is also recommended, plus dry antecedent soil moisture conditions to maximize the sorptivity value. Vandervaere et al. (2000b) further showed that the traditional approach of assuming $S(\psi_0) = dI(\psi_0)/dt^{1/2}$ from early time $I(\psi_0)$ versus $t^{1/2}$ data (e.g., White et al., 1992) often yields inaccurate results. This occurs because gravity-capillarity-geometry interaction effects can be important even at early times [Eqs. (6.48) and (6.50)], and because contact sand effects can be very difficult to detect in $I(\psi_0)$ versus $t^{1/2}$ plots (Vandervaere et al., 2000b). Although the inverse modeling approach (see previous section) applies to both transient and steady tension infiltrometer flow, little testing of transient flow has occurred so far.

A currently unresolved weakness of the transient method is that the β parameter in Eqs. (6.50) and (6.79) is not known *a priori*, and can have a value anywhere between the theoretical limits of 0 and 1 (Haverkamp et al., 1994). At present, $\beta = 0.6$ is usually assumed, but there appears to be no theoretical or practical basis for this value. It is also not clear at this point if the transient method can be used with multiple heads to characterize hydraulic property versus ψ_0 relationships at a particular location. Vandervaere (2002) provides additional information on practical application of the transient flow method.

6.3.3.4 Accounting for Contact Sand

A layer of "contact sand" (usually natural sand, uniform glass beads, or some other fine particulate material) should be placed under the tension infiltrometer to establish and maintain good hydraulic contact between the infiltration disc and the soil surface. This should be done regardless of whether the soil surface has been leveled or left undisturbed (e.g., Perroux and White, 1988; Reynolds, 1993b; Vandervaere, 2002; Smettem and Smith, 2002) and regardless of whether steady state or transient analyses are used (Vandervaere et al., 2002a). The contact sand layer can introduce artifacts, however, which must be accounted for in tension infiltrometer analyses (Reynolds and Zebchuk, 1996).

For both steady-state and transient flow, the saturated hydraulic conductivity of the contact sand, K_{cs} [LT^{-1}], must be greater than the maximum measured $K(\psi_0)$ of the soil. Also, the water-entry pressure head of the contact sand, ψ_w [L] (i.e., the pressure head at which the contact sand spontaneously saturates from a dry state) must be less than the minimum pressure head set on the infiltration disc, ψ_0. If these criteria are not met, the hydraulic conductivity of the contact sand may fall below that of the soil at one or more of the set ψ values, and the sand layer may consequently restrict flow and cause the infiltrometer measurements to be unrepresentative of the soil. Reynolds and Zebchuk (1996) proposed a uniform glass bead material that has K_{cs} and ψ_w values that make it widely applicable as a contact sand.

For steady flow, hydraulic head loss across the contact sand layer can produce an important difference, or "offset," between the pressure head set on the infiltration disc (ψ_0) and the pressure head actually applied to the soil surface (ψ_s). If the contact sand layer is contained within a retaining ring, flow through the sand layer is steady, saturated and rectilinear, and the ψ_s value can be estimated accurately using Darcy's law in the form (Reynolds and Zebchuk, 1996):

$$\psi_s = \psi_0 + \left[1 - \frac{Q(\psi_0)}{\pi a^2 K_{cs}}\right] T_{cs} \tag{6.80}$$

where $Q(\psi_0)$ [L^{3}T^{-1}] is the steady flow rate out of the infiltration disc at $\psi = \psi_0$ K_{cs} [LT^{-1}] is the saturated hydraulic conductivity of the contact sand layer, and T_{cs} [L] is the thickness of the contact sand layer. The value of ψ_s thus depends on ψ_0, $Q(\psi_0)$, K_{cs} and T_{cs}. Using numerical simulations and controlled laboratory experiments, Reynolds and Zebchuk (1996) determined that ψ can range between $\psi_s \approx \psi_0$ and $\psi_s \approx (\psi_0 + T_{cs})$ when $K_{cs} \geq K_{fs}$. Given that $T_{cs} = 10$ mm is often considered a practical minimum for field studies due to the roughness of undisturbed soil surfaces (Thony et al., 1991), the ψ_s value can differ from ψ_0 by as much as 10 mm or more. Tension infiltrometer analyses based on steady flow must consequently use ψ_s in place of ψ_0 to maintain accuracy when contact sand is used. The TIDAP computer program (see Sec. 6.3.3.2) contains a subroutine that uses Eq. (6.80) to account for the effects of contact sand in the steady flow analysis of Reynolds and Elrick (1991) [Eqs. (6.63)–(6.76)]. According to the numerical simulation studies of Reynolds and Zebchuk (1996), this approach can generally determine both ψ_s and the $K(\psi_s)$ of the soil within ±10%.

The contact sand layer can also have a substantial impact on transient infiltration from the tension infiltrometer, although this may be difficult to detect from plots of cumulative infiltration versus time or infiltration flux versus time (Vandervaere et al., 2000a). Plotting infiltration data according to Eq. (6.77), however, reveals a distinct early-time negative slope region followed by a later-time positive slope region (Figure 6.13). The nonlinear negative slope region represents infiltration into the contact sand layer, while the linear positive slope region represents infiltration into the soil (Vandervaere et al., 2000a; Vandervaere, 2002). Thus, only the data in the linear positive slope region are used for the determination of $S(\psi)$ and $K(\psi)$ via Eqs. (6.77), (6.49), and (6.50). Inclusion of the contact sand–affected data, even when the sand layer is only a few millimeters thick, can result in large errors and/or physically meaningless (e.g., negative) parameter values (Vandervaere et al., 2000a). Equation (6.77) consequently provides an easy and convenient means for identifying early-time infiltration into the contact sand and for eliminating these data from the transient analysis. As with the steady state analyses, ψ should be used in place of ψ_0 to account for hydraulic head loss across the contact sand layer, especially if the layer must be thick due to surface roughness. Precise determination of ψ_s may be problematic, however, as Eq. (6.80) suggests that ψ_s may change with time (t) during transient infiltration because $Q(\psi_0)$ changes with time. Although it should be possible to estimate ψ

versus t using Eq. (6.80) and measurements of $Q(\psi_0)$ versus t, the values of $S(\psi_s)$ and $K(\psi_s)$ obtained from Eqs. (6.49) and (6.50) may still be imprecise because ψ_s may not be constant over the measurement period. Further investigation of this issue appears necessary.

6.3.3.5   Strengths and Weaknesses of the Tension Infiltrometer Method

The tension infiltrometer method has many strengths related to the versatility of both apparatus and theory. The apparatus is simple, inexpensive, portable, easily applied in both field and laboratory studies, and requires only small volumes of water. Hence, laboratory/greenhouse studies, detailed investigations of spatial variability, field studies in areas with difficult access, and large-scale surveys over entire watersheds are more feasible with this method than with many other methods (see also Chapter 2). The apparatus also does not require soil disturbance, such as augering wells or inserting rings and probes, and can thus provide highly plausible estimates of the hydraulic properties of fragile aggregates and soil macropores (see also Chapter 8).

Perhaps the main theoretical strength of the method is its ability to determine a number of important water transmission parameters [e.g., $K(\psi_0)$, $S(\psi_0)$, $\alpha^*(\psi_0)$, $R(\psi_0)$] in the near-saturated range, $\approx -0.20\,\mathrm{m} \le \psi_0 \le 0$, where both parameter values and water/solute movement can change dramatically with even small changes in ψ_0. The tension infiltrometer method consequently has the ability to relate "macropore" and "matrix pore" flow parameters to changes in soil condition or soil management. For example, changes in $R(\psi_0)$ have been used to quantify the effects of macrostructure collapse and macropore infilling (White et al., 1992), soil cracking (Thony et al., 1991), tillage practices (Sauer et al., 1990; White et al., 1992; Reynolds et al., 1995), root growth (White et al., 1992), and sediment erosion-deposition (White et al., 1992). Measurements of near-saturated $K(\psi_0)$ and $S(\psi_0)$ have been used to quantify changes in water transmission as a result of different tillage procedures (Sauer et al., 1990; Reynolds et al., 1995), faunal activity (Clothier et al., 1985), soil structural changes during the growing season (Messing and Jarvis, 1993), soil textural changes (Jarvis and Messing, 1995), wheel trafficking (Ankeny et al., 1991), and development of soil hydrophobicity (Clothier et al., 1996). Tension infiltrometers have also been used to characterize near-saturated mobile-immobile soil water contents (Clothier et al., 1992; Angulo-Jaramillo et al., 1997), solute transport characteristics (Jaynes et al., 1995; Clothier et al., 1996; Vogeler et al., 1996), and solute sorption isotherms (Clothier et al., 1995) (see also Chapters 9, 13).

Perhaps the primary weakness of the tension infiltrometer method is the need to use a layer of "contact sand" to establish and maintain good hydraulic contact between the infiltration disc, which is flat and rigid, and the undisturbed soil surface, which is usually rough, undulating, and sloping. The contact sand must meet specific and somewhat restrictive performance criteria, and the presence of contact sand must be accounted for in the data analysis procedures. Specifically, the saturated hydraulic conductivity, K_{cs}, and water-entry pressure head, ψ_w, of the contact sand must be such that the sand never restricts flow into the soil. Recommended values are $K_{cs} \geq 10^{-4} \, \text{ms}^{-1}$ which is greater than the K_{fs} of most unstructured agricultural soils, and $\psi_w \leq -0.2 \, \text{m}$, which is less than the minimum pressure head set on most infiltration discs. The K_{cs} and ψ_w values should also be stable with a narrow standard deviation to minimize contact sand–induced variations. The contact material should be strongly hydrophilic and single grain with a narrow particle size distribution so that it levels easily and readily establishes good hydraulic contact. The material should also be easily obtained, inexpensive, and reusable. Most natural soil materials cannot meet all of the above performance criteria; however, a glass bead material proposed by Reynolds and Zebchuk (1996) appears promising. Provision must also be made to avoid wicking effects due to the infilling of surface-vented macropores by the contact sand and due to lateral slumping of the contact sand beyond the circumference of the infiltration disc. Reynolds and Zebchuk (1996) placed a fine-mesh "guard-cloth" on the soil surface to prevent macropore infilling, and they placed the sand inside a surface "retaining ring" to prevent lateral slumping. Finally, data analysis procedures must account for the fact that the pressure head at the soil surface can differ by varying amounts from that set on the infiltration disc because of hydraulic head loss through the contact sand. The difference can be substantial, and it depends on the flow rate out of the infiltrometer, the hydraulic conductivity of the sand layer, and the thickness of the sand layer.

The various single disc tension infiltrometer analyses also have strengths and weaknesses. Important strengths of the steady flow, multiple head approaches [Eqs. (6.63)–(6.76)] include well-established and tested theory, robustness, provision of measurements at several pressure heads, relatively large (and thereby more representative) sample volumes, and the ability to accurately account for the effects of a contact sand layer. Weaknesses of the steady flow analyses include potentially long equilibration times, the need to independently measure antecedent and final water contents to obtain estimates of sorptivity, and potentially greater susceptibility to error associated with soil heterogeneities (layering) and nonuniform water contents (due to large sample volume). Some recent simulation studies suggest, however, that the steady flow approach may be far less sensitive to near-surface soil layering than previously supposed (Smettem and Smith, 2002). The main strengths of the transient analysis [Eqs. (6.77)–(6.79)] include shorter measurement times (because steady flow is not required), the ability to obtain sorptivity without independent measurement of soil water content, simple but effective procedures for determining valid results and for eliminating flow perturbations caused by contact sand, and straightforward use of linear regression to obtain the sorptivity and hydraulic conductivity values plus their standard deviations. On the other hand, important weaknesses of the transient analysis include imprecise knowledge of both the β parameter and the pressure head at the soil–contact sand interface and an unknown ability to determine hydraulic property versus pressure head relationships at a particular location (through application of multiple heads to a single infiltration disc). The main advantages of the inverse modeling (numerical inversion) technique are the ability to provide estimates of both $K(\psi)$ and $\theta(\psi)$ at near-saturation [if the van Genuchten (1980) functions are assumed] and the ability to analyze single head or multiple head data from either transient flow or steady flow. Disadvantages of the inverse modeling approach include potential nonuniqueness of the results, complicated analysis procedures, and the need to assume a specific $K(\psi)$ function [e.g., the van Genuchten (1980) function], which may not be optimal for the soil in question. It should also be remembered that the $K(\psi)$ and $\theta(\psi)$ relationships provided by the inverse method may not be accurate outside the "near-saturated" pressure head range of the tension infiltrometer (i.e., $\approx -0.20\,\text{m} \le \psi \le 0$).

The main strengths and weaknesses of tension infiltrometer methods are compared to other field methods in Table 6.2.

6.3.4 OTHER METHODS

Other useful methods for determining *in situ* soil hydraulic properties include unconfined irrigation drippers (Shani et al., 1987), rainfall simulators (Amerman, 1979), and the instantaneous profile technique (e.g., Vachaud and Dane, 2002). These methods tend to be used less frequently than those described above, as the data they provide can be more difficult to interpret, or they are more expensive and/or complicated and/or time consuming. Of the three methods, perhaps the instantaneous profile method is the most attractive because it can provide $K(\theta)$ and $\theta(\psi)$ relationships for subsurface soil layers, and it is amenable to analysis using inverse modeling techniques. This method is briefly reviewed below.

6.3.4.1 Instantaneous Profile Method

The instantaneous profile method (also known as the internal drainage method) usually involves installing pairs of probes for *in situ* measurement of θ (e.g., TDR probes) (see Chapter 5) and ψ (tensiometers) at selected depths in the soil profile. Depth increments of about 0.15 m are often recommended, with at least one pair of probes per soil layer or soil horizon. The water content of the soil profile should be low enough that a wide range of θ and ψ can be measured between antecedent and field-saturated conditions, but not so dry that the tensiometer cups lose hydraulic contact with the soil (which usually occurs when the antecedent ψ of the soil falls below about -8 to $-9\,\text{m}$). An area of the soil surface centered around the probes is flood irrigated until the soil profile is either field-saturated (i.e., $\theta = \theta_{fs}$, $\psi = 0$) or θ and ψ are constant at all instrumented depths. The flooded area should be level, free of vegetation, and large enough (e.g., $\geq 12\,\text{m}^2$) that the initial infiltration and subsequent drainage are vertical in the vicinity of the θ and ψ probes. After the soil profile is sufficiently wetted, irrigation is stopped and the flooded area is insulated and covered (e.g., straw/bark mulch laid down and a plastic sheet placed on top) to impose gravity drainage under a surface boundary condition of zero-flux and constant temperature. Drainage causes θ and ψ to decrease with time, and θ versus time and ψ versus time data are collected at all depths from the time irrigation stops until θ and ψ become effectively constant. Monitoring usually needs to be continuous for about the first 3 hours after irrigation is stopped, as initial drainage rates are often quite rapid. Monitoring can be much less frequent as drainage slows, however, and it is recommended that the later-time monitoring intervals follow geometric time increments (e.g., at 3, 6, 12, 24, ... h after irrigation is stopped) to better define the often exponential-like decline of θ versus time.

The amount of water, W [L], stored in the soil profile at time, t [T], is given by

$$W(z, t) = \int_0^z \theta(z, t)dz; \quad 0 \le z \le L \tag{6.81}$$

where z [L] is depth below the soil surface ($z = 0$ at soil surface) and L [L] is the depth to the deepest pair of θ and ψ probes. Given that a no-flow boundary is imposed on the soil surface, the flux density, q [LT^{-1}], for soil profile drainage at any depth and time in the measurement zone is

$$q(z, t)_z = \frac{\partial \int_0^z \theta(z, t)dz}{\partial t} = \frac{\partial W(z, t)}{\partial t}\Big|_z. \tag{6.82}$$

which states that the time rate of decrease in water storage between the soil surface and depth, z, equals the drainage flux density at depth, z. Darcy's law for water flow at any depth and time in the soil profile can be written as

$$q(z, t)_z = K(\theta)_z \frac{\partial h(z, t)}{\partial z}\Big|_z \tag{6.83}$$

where $h(z, t) = \psi(z, t) - z$ is hydraulic head, and $K(\theta)_z$ [LT^{-1}] is the hydraulic conductivity–water content relationship at depth, z. Substituting Eq. (6.82) into (6.83) produces

$$K(\theta)_z = \frac{\partial W(z, t)}{\partial t}\Big|_z \Big/ \frac{\partial h(z, t)}{\partial z}\Big|_z \tag{6.84}$$

which indicates that $K(\theta)_z$ at depth, z, can be determined from the time rate of decrease in water storage between the soil surface and depth, z, divided by the hydraulic head gradient at depth, z. The $\theta(\psi)_z$ and $K(\psi)_z$ relationships can also be determined because both $\theta(z, t)$ and $\psi(z, t)$ are measured at each depth.

The accuracy of Eq. (6.84) is obviously dependent on the accuracy with which $\partial h(z, t)/\partial z$ and $\partial W(z, t)/\partial t$ can be determined. The accuracy of $\partial h(z, t)/\partial z$ is improved by careful calibration of the tensiometers and by minimizing the depth increments between them. The accuracy of $\partial W(z, t)/\partial t$, on the other hand, is improved by curve-fitting time-differentiable expressions to $W(z, t$ versus t data in order to better describe both the very rapid decrease in $W(z, t$ at early time and the very slow decrease in $W(z, t)$ at late time. It has been found, for example, that the empirical relationships

$$W(t) = a \ln t + b \qquad (6.85$$

or

$$W(t) = ct^d \qquad (6.86$$

can often provide good fits to a wide range of soil data, where a, b, c and d are curve-fitting parameters (Vachaud and Dane, 2002). These expressions are then differentiated with respect to time and substituted into Eq. (6.84) to produce

$$K(\theta)_z = [a/t]_z \Big/ \left[\frac{\partial h(z, t)}{\partial z}\Big|_z \right] \qquad (6.87$$

or

$$K(\theta)_z = [cdt^{(d-1)}]_z \Big/ \left[\frac{\partial h(z, t)}{\partial z}\Big|_z \right] \qquad (6.88$$

The determination of $K(\theta)$ can be simplified substantially if the soil profile is homogeneous over the depth range of interest. In this special case, drainage produces $\partial h(z, t)/\partial z \approx -1$ and thus tensiometer measurements of $\psi(z, t)$ are not required. In addition, $\partial W(z, t)/\partial t$ can be represented by

$$\frac{\partial W(z, t)}{\partial t}\Big|_z = z \frac{\partial \bar{\theta}(t)}{\partial t}\Big|_z = zm \frac{\partial \theta(t)}{\partial t}\Big|_z \qquad (6.89$$

where $\bar{\theta}(t)$ is the average water content from the soil surface to depth, z, which can be represented by the empirical relationship (Libardi et al., 1980):

$$\bar{\theta}(t)_z = m\theta(t)_z + n \qquad (6.90)$$

where $\theta(t)_z$ is the measured water content at depth, z, and m and n are curve-fitting parameters (m is generally close to 1). It has also been established that the empirical relationship (Vachaud and Dane, 2002)

$$\theta(t)_z = \tau \times t^\mu \qquad (6.91)$$

applies for drainage in homogeneous soil profiles, where τ and μ are coefficients obtained from a regression analysis of θ versus t data at each depth, z. Substituting Eqs. (6.91), (6.90) and (6.89) into Eq. (6.84) and remembering that $\partial h(z, t)/\partial z \approx -1$ produces

$$K(\theta)_z = zm\tau^{(1/\mu)}\mu\theta_z^{[(\mu-1)/\mu]} \tag{6.92}$$

which provides a simple analytic relationship between $K(\theta)_z$ and θ_z at any depth, z, in a homogeneous soil profile and avoids direct estimation of uncertain space and time derivatives. Further detail on the justification and derivation of Eqs. (6.89)–(6.92) can be found in Libardi et al. (1980) and Vachaud and Dane (2002).

Inverse modeling (numerical inversion of Richards' equation) (see also Chapter 20) can be applied successfully to the instantaneous profile method provided that the soil profile is relatively simple with few soil layers/horizons or preferential flow paths (multilayered and heterogeneous soils often cause the inverse problem to become ill-posed or inaccurate). The input data are θ and ψ measurements at multiple times and depths, and the preferred (most convenient and successful) bottom boundary condition is unit hydraulic head gradient. The inverse modeling approach can be particularly useful for determining the "effective/overall" $K(\theta)$ and $\theta(\psi)$ functions for relatively uniform soil profiles and/or large areas of uniform soil. Details on the application of inverse modeling to the instantaneous profile method can be found in Libardi et al. (1980) and Hopmans et al. (2002).

6.3.4.2   Strengths and Weaknesses of the Instantaneous Profile Method

The main advantage of the instantaneous profile method is that it can yield simultaneous *in situ* estimates of $K(\theta)$, $K(\psi)$ and $\theta(\psi)$ for a number of depths and horizons in the soil profile. No other method is capable of doing this. On the other hand, important constraints of the method include the need for complex and delicate equipment (e.g., TDR probes, tensiometers) installed at various depths below the soil surface; measurement times that can run from days to months; the potential need for large volumes of water to effect profile saturation; inability to replicate measurements easily (because of extensive equipment, time and water requirements); determination of only the wet end ($\approx -8\,\text{m} \leq \psi \leq 0$) of the $K(\theta)$ and $\theta(\psi)$ relationships; and inability to determine $K(\theta)$ within about 0.2–0.3 m of the soil surface [near-surface $\partial h(z, t)/\partial z$ is very low and goes to zero at the soil surface because $q(z, t) = 0$ is imposed at the surface]. Another concern is that the method may not yield realistic representations of $K(\theta)$ and $\theta(\psi)$ in highly structured or low permeability soils. In highly structured soils, flow through preferential flow zones (e.g., macropores, fingers) (Chapter 8) may be missed partially or completely by the relatively small θ and ψ probes. In low-permeability soils, the times required for profile saturation and drainage can be impractically long.

Particular advantages of the simplified approach for homogeneous soil profiles [i.e., Eqs. (6.89) and (6.92)] include avoidance of "finicky" tensiometers, no requirement to evaluate uncertain space and time derivatives, and provision of a straightforward analytical relationship between $K(\theta)$ and θ_z [Eq. (6.92)], which applies at all measurement depths including the top 0.2–0.3 m. Disadvantages of the simplified approach include, loss of useful $\theta(\psi)$ data (because ψ is not measured), reliance on specific empirical expressions which may not fit the data [Eqs. (6.90)–(6.92)], and applicability to only uniform soil profiles (i.e., the approach cannot be used in layered soils).

Despite the difficulties and limitations of the instantaneous profile method, it is often considered a "benchmark" of precision and relevance against which other field methods are evaluated (Vachaud and Dane, 2002). The main strengths and weaknesses of the instantaneous profile method are compared to other field methods in Table 6.2.

6.4 RECOMMENDATIONS FOR FURTHER RESEARCH

With respect to the methods discussed here, a number of areas could benefit from further research and development. Some of the more important of these areas are briefly outlined below.

The transient ring infiltrometer and well permeameter analyses [Eqs. (6.10), (6.18), (6.19), (6.39), (6.40)] assume a uniform, step-function (sharp) wetting front (i.e., Green–Ampt soil). This assumption clearly does not hold in soils with macrostructure (e.g., cracks, wormholes, root channels, large interaggregate spaces, etc.), finger zones, or distinct horizontal layers. The assumption is also problematic for three-dimensional infiltration from well permeameters, where the wetting front is characteristically diffuse rather than sharp. There is consequently a need to determine how sensitive these transient analyses are to violation of the Green–Ampt assumption and perhaps to also develop analyses that do not require the Green–Ampt assumption. There is also a need to develop transient well permeameter analyses that do not require a water-tight seal between a casing and the well wall [Eqs. (6.39) and (6.40)] or prior equilibration to steady flow [Eq. (6.38)]. Establishing and maintaining a water-tight seal between casing and soil can be difficult (sometimes impossible), and the need to obtain steady flow before measurements begin can lead to impractically long measurement times. For the transient tension infiltrometer analyses [Eqs. (6.49), (6.50), and (6.77)], further research is required to more accurately determine the β parameter and the effective pressure head at the soil–contact sand interface.

From a more general perspective, improved methods are needed for *in situ* determination of pore water pressure head (ψ), and new analytic treatments are required that can account for soil heterogeneity and nonuniform initial conditions. The currently used tensiometer method for *in situ* determination of ψ is severely limited by a small measurement range ($\approx -8 \, \text{m} \leq \psi \leq 0$); a relatively slow response time; dubious ability to monitor isolated macropore and finger flow; and probes that are fragile (porous ceramic tips), difficult to install (access holes required), and difficult to maintain (e.g., air accumulation). Analytic determination of soil hydraulic properties [e.g., Eqs. (6.7)–(6.80)] currently requires the restrictive assumption of rigid, homogeneous, isotropic soil with uniform initial water content. New analytic approaches are needed that do not require these idealized soil conditions, which never truly occur in the field. New approaches are also needed for characterizing the spatial and temporal variation of soil hydraulic properties, for scaling soil hydraulic properties to different soil areas or volumes, and for matching the scale of hydraulic property measurements to the scale of the flow process under consideration (see also Chapters 1–3, 17).

6.5 CONCLUDING REMARKS

It has become clear over the past two to three decades that *in situ* measurement of soil hydraulic properties in space and time are essential for dealing with the extreme complexity of water and solute movement in the field. It is therefore essential that accurate and practical field methods for measuring these properties be developed. Unfortunately, the complexities of the field environment, the myriad of objectives and uses for the data, and the various theoretical and practical constraints of each method seem to ensure that no single method or approach will ever be suitable for all soil types or project objectives. It appears instead that field programs will require a "suite" of methods which are complementary in terms of accuracy, practicality, parameters measured, flow geometry, depth range, and versatility. Although by no means all-inclusive, the infiltrometer, permeameter, and instantaneous profile methods discussed here form an important core to such a suite of methods.

REFERENCES

Amerman, C.R. 1979. Rainfall simulation as a research tool in infiltration. P. 85–90. *In* Amerman et al., (ed.), Proc. of the Rainfall Simulator Workshop. USDA-SEA-AR, Sidney, Montana.

Amoozegar, A. and G.W. Wilson. 1999. Methods for measuring hydraulic conductivity and drainable porosity. P. 1149–1205. *In* R.W. Skaggs and J. Van Schilfgaarde (ed.), Agricultural Drainage. Agronomy No. 38, American Society of Agronomy, Inc., Madison, WI.

Angulo-Jaramillo, R., F. Moreno, B.E. Clothier, J.L. Thony, G. Vachaud, E. Fernandez-Boy, and J.A. Cayuela. 1997. Seasonal variation of hydraulic properties of soils measured using a tension disk infiltrometer. Soil Sci. Soc. Am. J. 61:27–32.

Ankeny, M.D., M. Ahmed, T.C. Kaspar, and R. Horton. 1991. Simple field method for determining unsaturated hydraulic conductivity. Soil Sci. Soc. Am. J. 55:467.

Bagarello, V., M. Iovino, and D.E. Elrick. 2003. A simplified falling-head technique for rapid determination of field-saturated hydraulic conductivity. Soil Sci. Soc. Am. J. 68: 66–73.

Beven, K. and P.F. Germann. 1982. Macropores and matric flow in soils. Water Resour. Res. 18:1311.

Bouma, J. 1982. Measuring the hydraulic conductivity of soil horizons with continuous macropores. Soil Sci. Soc. Am. J. 46:438.

Bouma, J. 1985. Soil variability and soil survey. P. 130–149. *In* J. Bouma and D.R. Nielsen (ed.), Proc. Soil Spatial Variability Workshop. PUDOC, Wageningen, The Netherlands.

Bouwer, H. 1966. Rapid field measurement of air-entry value and hydraulic conductivity of soil as significant parameters in flow system analysis. Water Resour. Res. 2:729.

Bouwer, H. 1978. Groundwater Hydrology. McGraw-Hill Book Company, Toronto, Canada, 480 pp.

Bouwer, H. 1986. Intake rate: cylinder infiltrometer. P. 825–844. *In* A. Klute (ed.), Methods of Soil Analysis: Part 1 — Physical and Mineralogical Methods, 2nd ed. Soil Science Society of America, Inc., Madison, WI.

Clothier, B.E. 2002. Infiltration. P. 239–280. *In* K.A. Smith and C.E. Mullins (ed.), Soil and Environmental Analysis: Physical Methods. Marcel Dekker, Inc., New York.

Clothier, B.E. and D.R. Scotter. 2002. Unsaturated water transmission parameters obtained from infiltration. P. 879–898. *In* J.H. Dane and G.C. Topp (ed.), Methods of Soil Analysis: Part 4 — Physical Methods. Soil Science Society of America, Inc., Madison, WI.

Clothier, B.E., D. Scotter, and E. Harper. 1985. Three-dimensional infiltration and trickle irrigation. Trans. Am. Soc. Agric. Eng. 28:497.

Clothier, B.E., M.B. Kirkham, and J.E. McLean. 1992. In situ measurement of the effective transport volume for solute moving through soil. Soil Sci. Soc. Am. J. 56:733.

Clothier, B.E., S.R. Green, and H. Katou. 1995. Multidimensional infiltration: points, furrows, basins, wells, and disks. Soil Sci. Soc. Am. J. 59:286.

Clothier, B.E., G.N. Magesan, L. Heng, and I. Vogeler. 1996. In situ measurement of the solute adsorption isotherm using a disc permeameter. Water Resour. Res. 32:771.

Constantz, J., W.N. Herkelrath, and F. Murphy. 1988. Air encapsulation during infiltration. Soil Sci. Soc. Am. J. 52:10.

Elrick, D.E. and W.D. Reynolds. 2002. Measuring water transmission parameters in vadose zone using ponded infiltration techniques. Agric. Sci. 7:17.

Elrick, D.E., R. Angulo-Jaramillo, D.J. Fallow, W.D. Reynolds, and G.W. Parkin. 2002. Infiltration under constant head and falling head conditions. P. 47–53. *In* P.A.C. Raats, D. Smiles, and A.W. Warrick (ed.), Environmental Mechanics: Water, Mass and Energy Transfer in the Biosphere. Geophysical Monograph 129, American Geophysical Union, Washington, DC.

Elrick, D.E. and W.D Reynolds. 1986. An analysis of the percolation test based on three-dimensional, saturated-unsaturated flow from a cylindrical test hole. Soil Sci. 142:308.

Elrick, D.E. and W.D. Reynolds. 1992. Infiltration from constant-head well permeameters and infiltrometers. P. 1–24. *In* G.C. Topp, W.D. Reynolds and R.E. Green (ed.), Advances in Measurement of Soil Physical Properties: Bringing Theory into Practice, SSSA Spec. Pub. No. 30, Soil Science Society of America, Inc., Madison, WI.

Elrick, D.E., W.D. Reynolds, and K.A. Tan. 1989. Hydraulic conductivity measurements in the unsaturated zone using improved well analyses. Ground Water Monit. Rev. 9:184.

Gardner, W.R. 1958. Some steady-state solutions of the unsaturated moisture flow equation with application to evaporation from a water table. Soil Sci. 85:228.

Germann, P. and K. Beven. 1981. Water flow in soil macropores: I. An experimental approach. J. Soil Sci. 32:1.

Green, W.H. and G.A. Ampt. 1911. Studies in soil physics. I. The flow of air and water through soils. J. Agric. Sci. 4:1.

Guyonnet, D., N. Amraoui, and R. Kara. 2000. Analysis of transient data from infiltrometer tests in fine-grained soils. Ground Water 38:396.

Haverkamp, R., M. Vauclin, J. Touma, P.J. Wierenga, and G. Vachaud. 1977. A comparison of numerical simulation models for one-dimensional infiltration. Soil Sci. Soc. Am. J. 41:285.

Haverkamp, R., P.J. Ross, K.R.P. Smettem, and J.-Y. Parlange. 1994. Three-dimensional analysis of infiltration from the disc infiltrometer: 2. Physically based infiltration equation. Water Resour. Res. 30:2931.

Hopmans, J.W., J. Šimůnek, N. Romano, and W. Durner. 2002. Inverse methods. P. 963–1008. *In* J.H. Dane and G.C. Topp (ed.), Methods of Soil Analysis: Part 4 — Physical Methods. Soil Science Society of America, Inc., Madison, WI.

Jarvis, N.J. and I. Messing. 1995. Near-saturated hydraulic conductivity in soils of contrasting texture measured by tension infiltrometers. Soil Sci. Soc. Am. J. 59:27.

Jarvis, N.J., L. Zavattaro, K. Rajkai, W.D. Reynolds, P.-A. Olsen, M. McGechan, M. Mecke, B. Mohanty, P.B. Leeds-Harrison, and D. Jacques. 2002. Indirect estimation of near-saturated hydraulic conductivity from readily available soil information. Geoderma 108:1.

Jaynes, D.B., S.D. Logsdon, and R. Horton. 1995. Field method for measuring mobile/immobile water content and solute transfer rate coefficient. Soil Sci. Soc. Am. J. 59:352.

Kirkham, D. and W.L. Powers. 1972. Advanced Soil Physics. Wiley-Interscience, New York.

Lauren, J.G., R.J. Wagenet, J. Bouma, and J.H.M. Wosten. 1988. Variability of saturated hydraulic conductivity in a glossaquic hapludalf with macropores. Soil Sci. 145:20.

Lee, D.M., W.D. Reynolds, D.E. Elrick, and B.E. Clothier. 1985. A comparison of three field methods for measuring saturated hydraulic conductivity. Can. J. Soil Sci. 65:563.

Leeds-Harrison, P.B. and E.G. Youngs. 1997. Estimating the hydraulic conductivity of soil aggregates conditioned by different tillage treatments from sorption measurements. Soil Tillage Res. 41:141.

Libardi, P.L., K. Reichardt, D.R. Nielsen, and J.W. Biggar. 1980. Simple field methods for estimating soil hydraulic conductivity. Soil Sci. Soc. Am. J. 44:3.

Lukens, R.P. (ed.),. 1981. Annual Book of ASTM Standards, Part 19: Soil and Rock. American Standards of Materials and Testing. Washington, DC, pp. 509–514.

Mein, R.G. and D.A. Farrell. 1974. Determination of wetting front suction in the Green–Ampt equation. Proc. Soil Sci. Soc. Am. 38:872.

Messing, I. and N.J. Jarvis. 1993. Temporal variation in the hydraulic conductivity of a tilled clay soil as measured by tension infiltrometers. J. Soil Sci. 44:11.

Muñoz-Carpena, R., C.M. Regalado, J. Alvarez-Benedi, and F. Bartoli, F. 2002. Field evaluation of the new Philip-Dunne Permeameter for measuring saturated hydraulic conductivity. Soil Sci. 167:9.

Perroux, K.M. and I. White. 1988. Designs for disc permeameters. Soil Sci. Soc. Am. J. 52:1205.

Philip, J.R. 1957. The theory of infiltration. 4. Sorptivity and algebraic infiltration equations. Soil Sci. 84:257.

Philip, J.R. 1968. Steady infiltration from buried point sources and spherical cavities. Water Resour. Res. 4:1039.

Philip, J.R. 1985. Approximate analysis of the borehole permeameter in unsaturated soil. Water Resour. Res. 21:1025.

Philip, J.R. 1987. The quasilinear analysis, scattering analog, and other aspects of infiltration and seepage. P. 1–27. In Y.-S. Fok (ed.), Proc. Infiltration Development and Application. Water Resources Research Center, Honolulu, Hawaii.

Philip, J.R. 1992. Falling head ponded infiltration. Water Resour. Res. 28:2147.

Philip, J.R. 1993. Approximate analysis of falling-head lined borehole permeameter. Water Resour. Res. 29:3763.

Raats, P.A.C. 1976. Analytical solutions of a simplified flow equation. Trans. ASAE 19:683.

Reynolds, W.D. 1993a. Saturated hydraulic conductivity: field measurement. P. 599–613. In M.R. Carter (ed.), Soil Sampling and Methods of Analysis. Canadian Society of Soil Science, Lewis Publishers, Boca Raton, FL.

Reynolds, W.D. 1993b. Unsaturated hydraulic conductivity: field measurement. P. 633–644. In M.R. Carter (ed.), Soil Sampling and Methods of Analysis. Canadian Society of Soil Science, Lewis Publishers, Boca Raton, FL.

Reynolds, W.D., D.E. Elrick, and G.C. Topp. 1983. A re-examination of the constant head well permeameter method for measuring saturated hydraulic conductivity above the water table. Soil Sci. 136:250.

Reynolds, W.D., D.E. Elrick, and B.E. Clothier. 1985. The constant head well permeameter: effect of unsaturated flow. Soil Sci. 139:172.

Reynolds, W.D. and D.E. Elrick. 1986. A method for simultaneous in situ measurement in the vadose zone of field-saturated hydraulic conductivity, sorptivity and the conductivity-pressure head relationship. Ground Water Monit. Rev. 6:84.

Reynolds, W.D. and D.E. Elrick. 1987. A laboratory and numerical assessment of the Guelph permeameter method. Soil Sci. 144:282.

Reynolds, W.D. and D.E. Elrick. 1990. Ponded infiltration from a single ring: I. Analysis of steady flow. Soil Sci. Soc. Am. J. 54:1233.

Reynolds, W.D. and D.E. Elrick. 1991. Determination of hydraulic conductivity using a tension infiltrometer. Soil Sci. Soc. Am. J. 55:633.

Reynolds, W.D., E.G. Gregorich, and W.E. Curnoe. 1995. Characterization of water transmission properties in tilled and untilled soils using tension infiltrometers. Soil Tillage Res. 33:117.

Reynolds, W.D., B.T. Bowman, and A.D. Tomlin. 1997. Comparison of selected water and air properties in soil under forest, no-tillage, and conventional tillage. P. 235–248. In J. Caron, D.A. Angers, and G.C. Topp (ed.), Proc. 3rd Eastern Canada Soil Structure Workshop. Université Laval, Sainte-Foy, Quebec, Canada.

Reynolds, W.D., S.R. Vieira, and G.C. Topp. 1992. An assessment of the single-head analysis for the constant head well permeameter. Can. J. Soil Sci. 72:489.

Reynolds, W.D., D.E. Elrick, and E.G. Youngs. 2002. Ring or cylinder infiltrometers (vadose zone). P. 818–843. *In* J.H. Dane and G.C. Topp (ed.), Methods of Soil Analysis: Part 4 — Physical Methods. Soil Science Society of America, Inc., Madison, WI.

Reynolds, W.D. and W.D. Zebchuk. 1996. Use of contact material in tension infiltrometer measurements. Soil Technol. 9:141.

Richards, L.A. 1931. Capillary conduction of liquids through porous mediums. Physics 1: 318.

Richards, N.E. 1987. A comparison of three methods for measuring hydraulic properties of field soils. M.Sc. thesis, University of Guelph, Guelph, Canada.

Sauer, T.J., B.E. Clothier, and T.C. Daniel. 1990. Surface measurements of the hydraulic properties of a tilled and untilled soil. Soil Tillage Res. 15:359.

Scotter, D.R., B.E. Clothier, and E.R. Harper. 1982. Measuring saturated hydraulic conductivity using twin rings. Aust. J. Soil Res. 20:295.

Shani, U., R.J. Hanks, E. Bresler, and C.A.S. Oliveira. 1987. Field method for estimating hydraulic conductivity and matric potential–water content relations. Soil Sci. Soc. Am. J. 51:298.

Simůnek, J. and M.Th. van Genutchen. 1996. Estimating unsaturated soil hydraulic properties from tension disc infiltrometer data by numerical inversion. Water Resour. Res. 32:2683.

Simůnek, J. and M. Th. van Genutchen. 1997. Estimating unsaturated soil hydraulic properties from multiple tension disc infiltrometer data. Soil Sci. 162:383.

Simůnek, J., R. Angulo-Jaramillo, M.G. Schaap, J.-P. Vandervaere, and M.Th. van Genuchten. 1998. Using an inverse method to estimate the hydraulic properties of crusted soils from tension-disc infiltrometer data. Geoderma 86:61.

Smettem, K.R.J. and R.E. Smith. 2002. Field measurement of infiltration parameters. P. 135–157. *In* R.E. Smith (ed.), Infiltration Theory for Hydrologic Applications. Water Resources Monograph 15, American Geophysical Union, Washington, DC.

Smettem, K.R.J., J.-Y. Parlange, P.J. Ross, and R. Haverkamp. 1994. Three-dimensional analysis of infiltration from the disc infiltrometer: 1. a capillary-based theory. Water Resour. Res. 30:2925.

Stephens, D.B., K. Lambert, and D. Watson. 1987. Regression models for hydraulic conductivity and field test of the borehole permeameter. Water Resour. Res. 23:2207.

Swartzendruber, D. and T.C. Olsen. 1961. Model study of the double ring infiltrometer as affected by depth of wetting and particle size. Soil Sci. 92:219.

Thony, J.-L., G. Vachaud, B.E. Clothier, and R. Angulo-Jaramillo. 1991. Field measurement of the hydraulic properties of soil. Soil Technol. 4:111.

Timlin, D.J., L.R. Ahuja, and M.D. Ankeny. 1994. Comparison of three field methods to characterize apparent macropore conductivity. Soil Sci. Soc. Am. J. 58:278.

Topp, G.C. and M.R. Binns. 1976. Field measurements of hydraulic conductivity with a modified air-entry permeameter. Can. J. Soil Sci. 56:139.

Topp, G.C., W.D. Reynolds, and R.E. Green (ed.),. 1992. Advances in Measurement of Soil Physical Properties: Bringing Theory into Practice. SSSA Special Pub. 30, Soil Science Society of America, Inc., Madison, WI.

Vachaud, G. and J.H. Dane. 2002. Instantaneous profile. P. 937–945. *In* J.H. Dane and G.C. Topp (ed.), Methods of Soil Analysis: Part 4 — Physical Methods. Soil Science Society of America, Inc., Madison, WI.

Vandervaere, J.-P. 2002. Early-time observations. P. 889–894. *In* J.H. Dane and G.C. Topp (ed.), Methods of Soil Analysis: Part 4 — Physical Methods. Soil Science Society of America, Inc., Madison, WI.

Vandervaere, J.-P., M. Vauclin, and D.E. Elrick. 2000a. Transient flow from tension infiltrometers: I. The two-parameter equation. Soil Sci. Soc. Am. J. 64:1263.

Vandervaere, J.-P., M. Vauclin, and D.E. Elrick. 2000b. Transient flow from tension infiltrometers: II. Four methods to determine sorptivity and conductivity. Soil Sci. Soc. Am. J. 64:1272.

van Genuchten, M.Th. 1980. A closed-form equation for predicting the hydraulic conductivity of unsaturated soils. Soil Sci. Soc. Am. J. 44:892.

Vogeler, I., B.E. Clothier, S.R. Green, D.R. Scotter, and R.W. Tillman. 1996. Characterizing water and solute movement by time domain reflectometry and disk permeametry. Soil Sci. Soc. Am. J. 60:5.

Warrick, A.W. and D.R. Nielsen. 1980. Spatial variability of soil physical properties in the field. P. 319–344. *In* D. Hillel (ed.), Applications of Soil Physics. Academic Press, Toronto.

Watson, K.W. and R.J. Luxmore. 1986. Estimating macroporosity in a forest watershed by use of a tension infiltrometer. Soil Sci. Soc. Am. J. 50:578.

White, I. and M.J. Sully. 1987. Macroscopic and microscopic capillary length and time scales from field infiltration. Water Resour. Res. 23:1514.

White, I., M.J. Sully, and K.M. Perroux. 1992. Measurement of surface-soil hydraulic properties: disk permeameters, tension infiltrometers, and other techniques. P. 69–103. *In* G.C. Topp, W.D. Reynolds, and R.E. Green (ed.), Advances in Measurement of Soil Physical Properties: Bringing Theory into Practice. SSSA Spec. Pub. No. 30, Soil Science Society of America, Inc., Madison, WI.

Wooding, R. 1968. Steady infiltration from a shallow circular pond. Water Resour. Res. 4:1259.

Wu, L., L. Pan, M.L. Robertson, and P.J. Shouse. 1997. Numerical evaluation of ring-infiltrometers under various soil conditions. Soil Sci. 162:771.

Youngs, E.G., 1987. Estimating hydraulic conductivity values from ring infiltrometer measurements. J. Soil Sci. 38:623.

Youngs, E.G., D.E. Elrick, and W.D. Reynolds. 1993. Comparison of steady flows from infiltration rings in "Green and Ampt" soils and "Gardner" soils. Water Resour. Res. 29:1647.

Youngs, E.G., G. Spoor, and G.R. Goodall. 1996. Infiltration from surface ponds into soils overlying a very permeable substratum. J. Hydrol. 186:327.

Zangar, C.N. 1953. Theory and problems of water percolation. Eng. Monogr. No. 8, U.S. Department of the Interior, Bureau of Reclamation, Denver.

Zhang, Z.F., P.H. Groenevelt, and G.W. Parkin. 1998. The well shape-factor for the measurement of soil hydraulic properties using the Guelph permeameter. Soil Tillage Res. 49:219.

7 Unraveling Microscale Flow and Pore Geometry: NMRI and X-Ray Tomography

Markus Deurer
University of Hannover, Hannover, Germany

Brent E. Clothier
HortResearch Institute, Palmerston North, New Zealand

CONTENTS

1-5667-0657-2/05/$0.00+$1.50
© 2005 by CRC Press

7.1 INTRODUCTION

Soil, a topologically complex three-phase porous medium, is the thin and fertile skin of planet Earth. As the interface between the atmosphere and the subterranean realm, soil is subject to massive inputs of mass and energy, which are either stored, reflected, reemitted, or transmitted. To describe the fluxes of mass and energy in soil, a theory has been developed using equations based on a representative elementary volume of a scale that is much larger than that of the individual pores. At this so-called Darcy scale, the soil has measurable hydrodynamic properties that permit description of water flow and solute transport using partial differential transport equations.

However, local disequilibria at the pore scale may invalidate any description of the process at the Darcy scale. It has been difficult to observe pore-scale processes, yet macroscale observations confirm the effects of disequilibria. Textural layers in soil, connected macropores, intra-aggregate microporosity, surface hydrophobicity, and distributed uptake across plant-root systems are all common factors that can lead to local, pore-scale disequilibrium processes. Through measurements in the field these phenomena are observed to confound the predictions made using uniform descriptions based at the Darcy scale.

New technology such as nuclear magnetic resonance imaging (NMRI) and x-ray computed tomography (CT) is providing us with new vision at the microscopic scale of pore geometry and topology and of water and solute transport. With x-ray tomography it is possible to reconstruct an entire pore network ranging from micropores to macropores. The dynamics of water flow in such a system can be analyzed with NMRI that can directly measure the distribution of pore water velocities and volumetric water contents. Both methods are capable of yielding three-dimensional information. NMRI and x-ray tomography are noninvasive and enable the repetitive measurement of a sample, for example, to capture the time evolution of processes or to investigate the influence of varying initial and boundary conditions. This new understanding will allow us to better describe the disequilibrium phenomena that we will need to incorporate into new theories to better predict the fate of the mass and energy that are incident upon the soil of the Earth's skin.

7.2   NUCLEAR MAGNETIC RESONANCE IMAGING

For detailed introductory accounts on the theory of NMRI we also refer the reader to Callaghan (1993), Gadian (1995), Farrar and Becker (1971), Farrar (1987), Carrington and McLachlan (1967), Freeman (1988), and Fukushima and Roeder (1981), and recommend the reviews on imaging by Price (1998) and on the quantification of flow by Pope and Yao (1993).

7.2.1   MEASUREMENT PRINCIPLE: THE BEHAVIOR OF SPINS IN MAGNETIC FIELDS

Many atomic nuclei have unpaired protons, electrons, neutrons or a combination of those and behave as microscopic magnetic dipoles. These charged nuclei simultaneously precess with intrinsic spins that are proportional to the magnitude of the magnetic dipoles and produce microscopic magnetic fields (= nuclear magnetism). The proportionality constant between the microscopic magnetic (dipole) moment and the spin (angular) momentum is termed the magnetogyric ratio, γ [rad $T^1 s^{-1}$]. In a thermodynamic equilibrium, microscopic magnetic moments are randomly orientated and create no bulk magnetization. NMRI can focus the microscopic magnetic moments by applying external magnetic fields to yield a measurable bulk magnetization. This macroscopic magnetization is proportional to the density of microscopic magnetic moments per unit volume, which in the context of NMRI is called spin density.

In a first step the nuclei are placed in a strong static magnetic field B_0 [T] that is directed along the z-axis (see Figure 7.1A). As a consequence of quantum physics, the magnetic moments of the nuclei orientate themselves in $(2I + 1)$ directions, where I is the spin quantum number (see Table 7.1). For example, the magnetic moments of ^1H, ^{31}P, ^{13}C, and ^{19}F with a spin-quantum number of ½ either align or antialign, with the direction of B_0 representing a low- or a high-energy state, respectively. Following the Boltzmann probability distribution, the probability slightly favors the low-energy state and leads to a very small additional net bulk magnetization in the direction of B_0. Simultaneously the magnetic moments precess in a clockwise direction about the static magnetic field like on the surface of cones (see Figure 7.1A). The constant angular frequency, ω_0 [rad s^{-1}], of this motion is called the Larmor frequency,

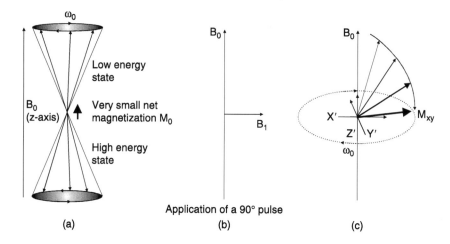

FIGURE 7.1 The behavior of intrinsic microscopic magnetic moments in magnetic fields. (A) Precession of microscopic magnetic moments about a static magnetic field B_0 Random incoherent phases result in a very small and not measurable net magnetization M_0. (B) With the application of a B_1 ($=90°$) pulse. (C) the microscopic magnetic moments tip into the x-y plane. Their now coherent phases rotating at the Larmor frequency create a measurable transverse magnetization M_{xy}.

and is atom-specific and proportional to the strength of B_0 (Xia, 1998):

$$\omega_0 = -\gamma B_0 \tag{7.1}$$

where γ is the atom-specific magnetogyric ratio (see Table 7.1). For example, using a magnet with strength B_0 of 1.5 T, the Larmor frequency of ^1H would be 63.9 Mhz.

In the second step to detect the very small net bulk magnetization within the sample, it has to be separated from the applied B_0 field. Therefore, the net bulk magnetization is tilted into the x-y plane by applying another magnetic field B_1 in the form of an rf ($=$radiofrequency) pulse. The B_1 field is perpendicular to B_0 (see Figure 7.1B). With a receiver coil (see Figure 7.7B, below) the bulk magnetization M_{xy} is then detected in the x-y plane (see Figure 7.1C). The frequency of the radiofrequency pulse creating the B_1 field

TABLE 7.1

Spin Quantum Numbers and Magnetogyric Ratios of Isotopes of Different Elements Used for NMR Measurements

	^1H	^2H	^{31}P	^{23}Na	^{14}N	^{13}C	^{19}F
Spin quantum number, I [–]	1/2	1	1/2	3/2	1	1/2	1/2
Magnetogyric ratio, γ [Mhz T^1]	42.58	6.54	17.25	11.27	3.08	10.71	40.08

exactly matches the Larmor frequency. The consequences for the microscopic magnetic moments can be more easily understood from a nuclear point of view. The B_1 field exerts a torque to the microscopic magnetic moments that are aligned with B_0. This torque induces the energy transition (= excitation) of the microscopic magnetic moments from a state of low to a state of high energy. During the process, the nuclei absorb the energy amount ΔE [J], and because of this they resonate. The crucial difference necessary to induce the transition from a low- to a high-energy state is given by (Carrington and McLachlan, 1967):

$$\Delta E = \hbar \omega_0 \qquad (7.2)$$

where $\hbar$ [J s] is Planck's constant. This is the exact energy amount that has to be supplied by B_1. The B_1 field is applied for a specific duration t_p [s]. Macroscopically, the bulk magnetization can assume any tip-angle θ between the low- and high-energy states and that is determined by the pulse duration t_p. For example, the time t_p to achieve a tip-angle of 90° can be calculated as (Gadian, 1995):

$$t_p = \frac{\pi/2}{\gamma B_1} \qquad (7.3)$$

This particular pulse, is termed a 90° pulse, and the respective signal can be detected in the x-y plane.

In the following, a cartesian coordinate system rotating with the Larmor frequency ω_0 (symbolized with x', y', z' — rotating frame of reference) (see Figure 7.1C) will be used instead of the usual static one (symbolized with x, y, z — laboratory frame of reference). In the absence of the B_1 field, the nuclear magnetic moments precess with ω_0 randomly about B_0. At any instant the bulk magnetization within the x-y plane, M_x and M_y, is zero. The 90° pulse tilts the net nuclear magnetization M_0 from the z'-axis into the x'-y' plane. Note that M_0 is directly proportional to the density of the spins in the sample denoted by ρ. After t_p, the B_1 field is switched off and only B_0 continues to act on the nuclear spins. But immediately after the application of B_1 the precession of the spins about B_0 is not random. Instead, the net component of magnetization that has been generated in the x'-y' plane (= transverse magnetization) rotates coherently about B_0 with frequency ω_0 (see Figure 7.1C). The rotation induces an electromotive force (e.m.f.) that oscillates with ω_0 and is detected with a receiver coil (see Figure 7.7B), amplified, and processed. It constitutes the NMR signal. Due to the resonance condition [see Eq. (7.2)], only specific atoms are selected and contribute to the measured signal. The selection is determined by the magnetogyric ratio of the atoms (see Table 7.1).

After the B_1 field is switched off, the bulk spin-system of the sample returns back to the equilibrium (= relaxation). The corresponding signal decay is termed the free induction decay (FID)(see Figure 7.2, below). Two different relaxation processes occur with characteristic times.

In a first fast process, energy is transferred to neighboring spins (= spin-spin relaxation) and the coherence of spins in their precession about B decays. The spins diphase and consequently the transverse magnetization vanishes with a time T_2 [s]. Assuming we applied a 90° pulse at time $t = 0$, the decaying transverse magnetization M_{xy} as a function of time in the laboratory frame of reference is given by (Gadian, 1995):

$$M_x(t) = M_0 \cos(\omega_0 t) \exp(-t/T_2)$$ (7.4

$$M_y(t) = M_0 \sin(\omega_0 t) \exp(-t/T_2)$$

Using the technical method of quadrature detection (Callaghan, 1993), $M_x(t$ and $M_y(t)$ can be simultaneously measured. This can be mathematically expressed with complex number notation where $M_x(t)$ denotes the real part = absorption spectrum) and $M_y(t)$ the imaginary part (= dispersion spectrum) of the magnetization (see Figure 7.4, below):

$$M(t) = M_0 \exp(i\omega_0 t) \exp(-t/T_2)$$ (7.5

where $i = \sqrt{-1}$.

A second relaxation process, the spin-lattice relaxation, contributes to the return of the spin system back to its energy distribution at the time before the perturbation by the B_1 field with a characteristic time T_1 [s]. The surplus energy is emitted in the form of heat to the surroundings, and the magnetization along the longitudinal z-axis is reestablished:

$$M_z(t) = M_0(1 - \exp(-t/T_1))$$ (7.6

By fitting Eq. (7.5) to the data of the complex signal, the free induction decay, it is now possible to quantify the behavior of the spin ensemble of the sample. The initial magnetization M_0, the waveforms of the transverse magnetization M_x and M_y, and the characteristic relaxation times T_1 and T_2 represent different sources of information. They can be used, for example, to analyze the density of spins (Callaghan, 1993) or distinguish between different materials via their characteristic relaxation times.

To probe the flow velocities of the spins (Pope and Yao, 1993; Callaghan, 1993; Price, 1998) the phase information of the signal waveform can be exploited. After the application of a 90° pulse, the magnetization M_{xy} rotates about the z-axis with $\omega_0 = 2\pi\nu_0$, where ν_0 [s^{-1}] is the frequency. It rotates through 2π radians (= 360°) in a time $1/\nu_0$, and therefore the phase angle φ of the magnetization is:

$$\varphi = 2\pi\nu_0 t = \omega_0 t$$ (7.7

We will show in the next section how Eq. (7.7) is used for the measurement of pore water velocities.

7.2.2 FOURIER IMAGING

Methods of Fourier transformation (Jennison, 1961; Brigham, 1974; Kak and Slaney, 1988) are commonly applied to analyze NMRI signals. The detected oscillating bulk magnetization is Fourier-transformed and further processed to yield an image of the magnetization property of interest. The spatial distribution of spin densities or the characteristic relaxation times (T_1 and T_2) are examples.

The NMRI signal is created by applying rf pulses to the sample in an ordered sequence and for specific times. The pulse sequence determines what kind of information is stored in the NMRI signal. The static B_0 field is switched on for the entire experiment irrespective of the chosen pulse sequence.

7.2.2.1 Pulse Sequence Design

The basic step of every sequence is to focus the microscopic magnetic moments to create a bulk magnetization that is detectable within a strong static B_0 field. One possibility is to apply a single 90° rf pulse and then acquire the resulting free induction decay (FID). To increase the signal-to-noise ratio, this step is usually repeated N times. The resulting N FIDs are summed, and the signal-to-noise ratio is improved on the order $\sqrt{N}$. This simple series of pulses is called a "pulse and acquire" sequence. The time between the elementary pulse units of the experiment (see Figure 7.2) is the time of repetition (TR). It should be long enough to guarantee that the transversal and longitudinal relaxation processes are finished. The Fourier transform of the FID yields a bell-shaped curve. The area under the curve is directly proportional to the spin density within the imaged sample section. In the case of ^{1}H-spins, this is proportional to the volumetric water content.

The one-dimensional bulk magnetization is spatially resolved (= imaging) with additional specific rf pulses, the magnetic field gradients. A magnetic field

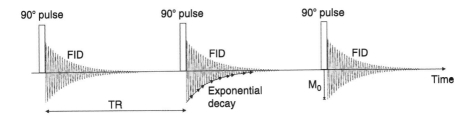

FIGURE 7.2 Pulse and aquire sequence. 90° pulses are applied within periods of TR. The resulting signals (= FIDs) decay exponentially mainly due to spin-spin relaxation. The signals shown here for illustration purposes contain only one frequency; in reality they contain multiple frequencies.

gradient is a systematic variation (increase or decrease) in the magnetic field strength with distance that is added to the field. The magnetic field gradients G_x, G_y, and G_z [T m^{-1}] spatially encode the spins in three dimensions (x, y, z).

G_z, the slice gradient, encodes for the position and thickness of the imaged sample cross section. It is applied in the form of an rf pulse that is switched on and off simultaneously with the 90°-pulse (see Figure 7.3A). The strength of G increases linearly with z and, thereby, modifies the static magnetic field B_0. The resulting effective magnetic field $B_{eff}(z)$ is now a function of z (see Figure 7.3B):

$$B_{eff}(z) = B_0 + G_z z \qquad (7.8$$

As a consequence, the Larmor (= resonance) frequency is modified and now also a function of z (see Figure 7.3B):

$$\omega_{eff}(z) = \gamma(B_0 + G_z z) \qquad (7.9$$

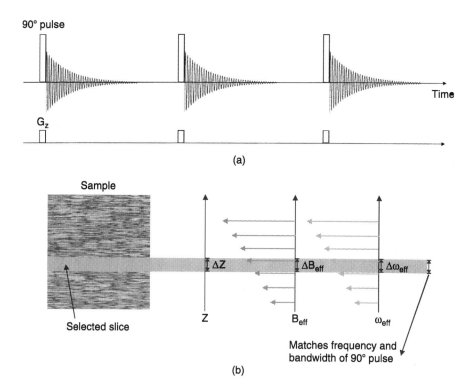

FIGURE 7.3 The application of the magnetic field gradient G_z (= slice gradient) in the framework of a pulse and aquire sequence. (A) Basic sequence. (B) The principle of the slice selection.

Therefore, the B1 field with its preselected frequency will only excite spins within a slice of the sample where it matches ω_{eff}. The thickness of the slice Δz is determined by the strength of G_z and the bandwidth of the B_1 pulse, $\Delta\omega$ (Gadian, 1995) (see Figure 7.3B):

$$\Delta z = \frac{\Delta\omega_{eff}}{\gamma G_z} \tag{7.10}$$

The spatial encoding of the plane, the x and y coordinate, with Gx and Gy can be easier understood with the concept of k-space (Mansfield and Grannell, 1975). This concept shows the connection between real and Fourier space for a three-dimensional NMR measurement. The superimposition of three-dimensional magnetic field gradients, denoted by the vector $\mathbf{G}$, onto the signal transforms the waveform of the transverse magnetization $M(t)$ [see Eq. (7.5)] into a gradient and time-dependent signal, $S(\mathbf{G}; t)$. The small change of the signal, $dS(\mathbf{G}; t)$, within a small volume element, dV, that is induced by the transverse magnetization of the spins having a specific density ρ within dV is:

$$\frac{dS(\mathbf{G}; t)}{dV} = \rho(\mathbf{s})\exp(i\omega t) = \rho(s)\exp[i(\gamma B_0 + \gamma \mathbf{G}\mathbf{s})t] \tag{7.11}$$

where $\mathbf{s}$ denotes a three-dimensional space vector. For convenience, the received signal is commonly mixed with a reference oscillation, which practically cancels out the term γB_0 in Eq. (7.11). In Eq. (7.11) the signal decay due to spin-spin relaxation is neglected. This is justified because the dephasing of

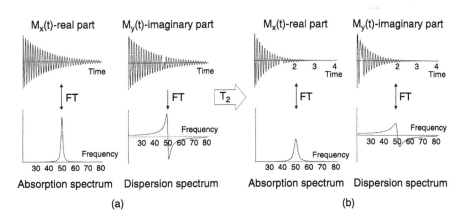

FIGURE 7.4 The free induction decays (FID) of the x and y components of the signal analyzed with quadrature detection in the time domain and after Fourier transformation (FT) the corresponding spectra in the frequency domain. (A) Sample with long transverse relaxation time (T_2). (B) Sample with 50% shorter transverse relaxation time (T_2) than in A.

the magnetization due to the spread in $\gamma \mathbf{G}s$ is much more rapid than due to transverse relaxation (Callaghan, 1993). Introducing the vector $\mathbf{k}$ (Mansfield and Grannell, 1975):

$$\mathbf{k} = \frac{1}{2\pi}\gamma \mathbf{G}t \qquad (7.12$$

and substituing it into Eq. (7.11), the signal waveform and the spin density can be formulated as a mutually conjugate Fourier pair:

$$S(\mathbf{k}) = \int \rho(\mathbf{s})\exp(i2\pi\mathbf{k}s)\,d\mathbf{s} \qquad (7.13$$

$$\rho(\mathbf{s}) = \int S(\mathbf{k})\exp(-i2\pi\mathbf{k}s)\,d\mathbf{k}$$

Equation (7.13) is the fundamental relationship of NMR imaging (Callaghan, 1993). The spin density $\rho(\mathbf{s})$ is the integral of the Fourier transformed signal (see Figure 7.4). The (spectral) width of this bell-shaped curve in the form of its FWHM (= Full Width at Half Maximum) is inversely proportional to T_2 $FWHM = 1/(\pi T_2)$(compare Figure 7.4A with Figure 7.4B). The features of the image itself are mainly a function of the x-y plane information, while the z-axis gives an average of the spins over the thickness of the slice.

The components k_x and k_y of the $\mathbf{k}$ vector correspond to the x and y components of $\mathbf{s}$. As can be seen from Eq. (7.12), two-dimensional k-space can be traversed (= sampled) in a series of pulse sequences by either incrementing the gradients, G_x and G_y, or incrementing the times, t_x and t_y, during which the gradients are switched on. In Fourier imaging k-space is sampled on a regular k_x-k_y grid (Price,1998) (see Figure 7.5B). After the slice gradient (G_z) is switched off, the gradient G_y (phase gradient) is switched on for a fixed time t_y (see Figure 7.5B). To traverse along k_y the phase gradient G_y is incremented (see Figure 7.5) according to (Price, 1998):

$$\Delta k_y = \frac{1}{2\pi}\gamma(\Delta G_y)t_y \qquad (7.14$$

where the direction of the gradient G_y also gives the orientation of the k_y coordinate.

When G_y is switched off, the signal is aquired and simultaneously G_x (read-out gradient) is switched on (see Figure 7.5A). G_x has a fixed strength, and to traverse along k_x the time t_x is incremented (Price, 1998) (see Figure 7.5):

$$\Delta k_x = \frac{1}{2\pi}\gamma G_x(\Delta t_x) \qquad (7.15$$

The number of increments, Δk_x and Δk_y, and thus sequence blocks within a NMRI measurement are proportional to the desired spatial resolution of

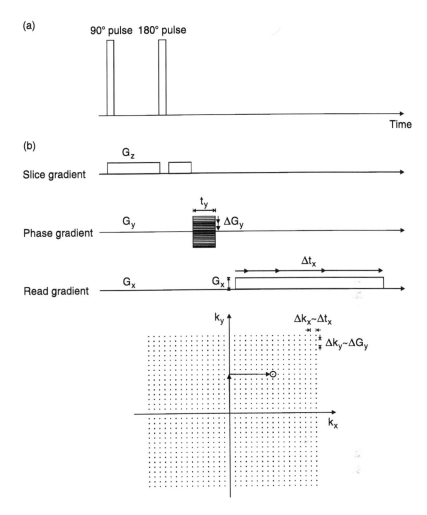

FIGURE 7.5 The basic pulse sequence and concept of two-dimensional Fourier imaging. (A) Scheme of the different building blocks of the sequence. (B) The principle of sampling k-space on a regular grid. In this example the resolution is 32×32 pixel.

the image (e.g., 64×64 pixels or 256×256 pixels). Technically, the signal is incrementally mapped in k-space and the respective information is stored. After completion of the experiment, the entire k-space data set can be directly transformed into its real-space conjugate [see Eq. (7.13)] yielding the image. Conceptually, the FIDs of small values of $|\mathbf{k}|$ provide the shapes and outlines (low-frequency information) and the FIDs of high values of $|\mathbf{k}|$ provide the fine details (high-frequency information) of the image.

Another commonly used NMRI building block is the spin-echo pulse sequence (Hahn, 1950). First, a $90°$ pulse rotates the magnetization into the $x'y'$-plane (see Figure 7.6B,a). After the $90°$ pulse is switched off, the coherence of microscopic magnetic moments decays and the transverse magnetization dephases (see Figure 7.6B,b). Looking down from the z'-axis, the nuclei precessing faster than the average appear to move clockwise and the slower ones anti-clockwise (see Figure 7.6B,b). Some time later (at $TE/2$) a 180° pulse is applied rotating the transverse magnetization 180° about the x'-axis (see Figure 7.6A). Looking down from the z'-axis, the nuclei precessing faster than the average still appear to move clockwise and the slower ones still anticlockwise. So, by this movement, they now rephase (see Figure 7.6B,c). When the rephasing is finished and the signal is at a maximum, the signal FID is acquired. The time of signal acquisition is called the echo time TE (see Figure 7.6A and B,d). One application is to repeat the sequence incrementing the echo times TE. This makes it possible to analyze the signal magnitude as a function of time ($=TEs$) and to derive, for example, the spatial distribution of the characteristic T_2 times within the slice of the sample.

However, an unbiased estimate of T_2 requires that the magnitude of the echo-signal at the shortest (and first) TE facilitates an unbiased extrapolation to the signal immediately after the 90° pulse. To do so the first TE should be as short as technically feasible. Biased estimates of T_2 can be a result of the spin-spin relaxation properties of the sample (extremely short T_2), of the technical limitations of the NMR spectrometer (relatively long first TE), or both. As a rule of thumb, the irreversible signal loss with time TE can be estimated to be

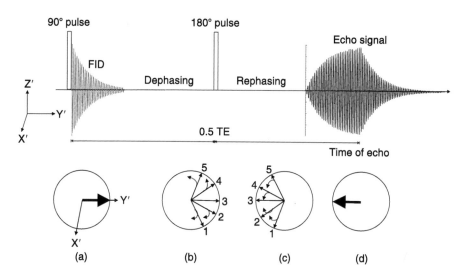

FIGURE 7.6 Hahn's (1950) spin-echo sequence. (A) The basic sequence and the resulting signals. (B) Behavior of microscopic magnetic moments in the x'-y' plane of the rotating frame of reference during the sequence (a) maximum of the transverse magnetization immediately after the 90° pulse is switched off; (b) dephasing of individual microscopic magnetic moments (labeled 1–5): (c) rephasing of individual microscopic moments after the application of a 180° pulse: (d) The rephasing is finished after a time TE. The signal shows again a maximum transverse magnetization.

proportional to e^{-TE/T_2}. The rapid signal loss regularly poses a problem when water-spins in unsaturated soils are imaged. T_2 times as short as 1 ms (Hall et al., 1997) are encountered, especially when the soils contain para- and/or ferromagnetics that enhance the signal loss. If, for example, TE cannot be shortened to less than 2 ms and T_2 is 1 ms, then roughly only 14% of the signal immediately after the 90° pulse is left at the time of measurement TE. As a consequence the derived T_2 is highly biased. Therefore, images measured with a spin-echo sequence at different times TE are said to be T_2 weighted, meaning, that the local intensity of the signal is proportional not only to the spin density but also to its "visibility" due to rapid spin-spin relaxation with characteristic time T_2.

To monitor longitudinal relaxation processes, the inversion recovery sequence can be used. A 180° pulse inverts the longitudinal magnetization rotating it from the $+z'$-axis onto the $-z'$-axis. After the 180° pulse is switched off, the longitudinal magnetization decays due to spin-spin relaxation processes. The longitudinal magnetization that remains during this process can be measured as an equivalent transverse magnetization by applying 90° pulses at a specific time TI ($=$ time for inversion). Repeating the experiment at successively smaller TI, the cross-over time t_0 at the point of zero longitudinal magnetization ($=$ only transverse magnetization exists) can be identified. From the relation $t_0 = 0.6931\ T_1$ (Callaghan, 1993) the characteristic time T_1 can be derived. The T_1 times are, for example, used to derive the pore size distribution of a water-saturated sample (Hinendi et al., 1997). The rates of both spin-spin but especially of spin-lattice relaxation of ^{1}H are enhanced at solid-water interfaces. Therefore, the T_1 times of ^{1}H spins in small-diameter pores will be on average shorter than within larger-diameter pores.

Several pulse sequences can be used to detect the translational velocities (propagation) of spins (Pope and Yao, 1993; Price, 1998). One is the method of phase encoding. At a specific time during the sequence, an additional velocity-encoding gradient, g_z [T m^{-1}], is applied along the main waterflow direction z If a positive velocity encoding magnetic field gradient g_z is applied for a time δ [s] along z, then, the spins aquire a z-dependent phase shift. According to Eq. (7.7), this phase shift $\varphi_1(z)$ can be calculated as:

$$\varphi_1(z) = \gamma g_z z \delta \tag{7.16}$$

After a time interval Δ [s], the gradient g_z is reapplied in the opposite direction (reversed). Meanwhile, the spins have moved along the z-axis to $z = v_l\Delta$, where v_l [m s^{-1}] is the effective longitudinal (z) velocity of the spins. They now rephase by a phase shift $\varphi_2(z)$ that is only dependent on their present position along z. The resultant phase shift is directly proportional to v_l and independent of the intrinsic position of the travelling spin (Callaghan, 1993), so:

$$\varphi_1(z) - \varphi_2(z) = \gamma g_z v_l \delta \Delta \tag{7.17}$$

A spin that did not move along z shows a zero resultant phase shift. For observing flow processes, this sequence offers a Lagrangian framework (Seymour and Callaghan, 1997). Methodologically, the contribution of transverse flow is not separated, as NMRI only maps the effective relative propagation of the spins along the axis of main flow. The effective longitudinal velocity can be analyzed as a function of Δ with spins sampling a different length of flow paths. With this it is possible to indirectly investigate the microscopic tortuosity and connectivity of pores.

7.2.2.2   Key Hardware Components

The NMRI system consists of many individual components. Their technical specifications strongly depend on the field of application. In the following an overview of the central parts of the imaging system will be given. For further details we refer the reader to Chen and Hoult (1989) or Callaghan (1993).

7.2.2.2.1   NMR Magnet

A crucial requirement for NMR imaging is a highly uniform and strong static magnetic B_0 field (see Figure 7.7A). Permanent magnets are often used in low field systems for small sample sizes. They are susceptible to temperature-induced field fluctuations, which can be corrected with additional small

FIGURE 7.7 Key hardware components of the NMR sytem. (A) Superconducting magnet. (B) NMR probe.

electromagnets. With rare-earth magnetic materials such as neodymium-iron-boron, magnetic fields as strong as 1.5 T can be generated.

Iron-core electromagnets can attain field strengths of up to about 2 T. The power supply requirements are substantial, as is the associated cooling system, especially for larger samples.

Fields larger than 2 T can be produced either with air-cored electromagnets or a superconducting system. Air-cored electromagnets are very expensive, as large amounts of electric power and cooling water are needed. Most commercial magnets used in NMRI are nowadays superconducting systems and use solenoids made from a superconducting wire (e.g., niobium-titanium). At liquid helium temperatures, these wires can conduct very large currents without dissipating power. By applying a high current, uniform magnetic fields as high as 21 T can be produced. Stronger fields cannot be achieved due to the critical field limitation on current superconductors. The superconducting magnet must be kept cold at all times and, therefore, requires a regular and expensive supply of the cryogens liquid helium and nitrogen.

All NMR magnet systems need the addition of shim coils, which are used to remove any irregularities from the B_0 field over the sample region.

7.2.2.2.2 NMR Probe

The NMR probe (see Figure 7.7B) is the central part of the NMR imaging system. It contains the sample and consists of the coil that can simultaneously transmit the B_1 field to the sample and subsequently detect the FID. A duplexor connects this radiofrequency coil to either the transmitting (for the B field) or the receiving (detection of the FID) part of the spectrometer. Many designs for rf coils exist that are especially tailored to the field of application and the technical specifications of the NMRI system (Callaghan, 1993). They are, for example, optimized for the sample geometry and the type of applied B_0 field. The closest possible contact between coil and sample provides the best results, as the sensitivity of the NMR probe is inversely proportional to the radius of the coil.

7.2.2.2.3 Magnetic Field Gradient Coils

The NMR probe is encased in a set of gradient coils, which supply the orthogonal magnetic field gradients G_x, G_y, and G_z to encode the spatial dimensions of the measurements. The accuracy of the spatial encoding depends on the linearity of the gradients. Any nonlinearity leads to a positional distortion of the image. The spatial resolution of the image is directly proportional to the strength of the gradients. Especially for velocity imaging, very large gradients in the order of several $T\,m^{-1}$ are necessary. Different designs for gradients exist that depend, among others, on the type of applied B_0 field (Callaghan, 1993; Eccles et al., 1994).

7.2.2.2.4 NMR Imaging Spectrometer

The NMR spectrometer is the heart of the NMRI system. The spectrometer coordinates the acquisition and processing of the measured FIDs and controls the transmission of the rf pulses and the switching of the magnetic field gradients. Separate gradient and rf waveform generators create the shapes of the $B_1(t)$ and $\mathbf{G}(t)$ fields to initiate the measurement and encode for the spatial dimensions. The measured FID is preamplified and then processed with a quadrature receiver. Basically, the receiver manipulates the preamplified FIDs in three steps. The signal is further amplified, and then its frequencies are filtered down from the MHz to the KHz range. In the last step the receiver downconverts the FID by shifting the Larmor frequency to zero frequency. This greatly reduces the data storage requirements and increases the processing efficiency when digitizing the received signals. A computer with specific software controls the overall architecture of rf and gradient pulses of the NMRI sequence and facilitates the subsequent image analysis.

7.2.3 APPLICATIONS OF NMRI TO SOIL-PLANT-WATER PROCESSES

NMRI is mostly applied to measure the different components of water flow such as volumetric water contents and pore water velocities.

So far, the majority of NMRI studies concerning pore structure, hydraulic properties, and water and solute transport have been done in the fields of petroleum engineering and fluid dynamics. But although many investigations used only sandstones and artificial media such as glass beads for their analyses, fundamental processes and principles of water flow could be derived that apply for porous media in general. Due to its noninvasiveness, NMRI has been successfully applied to many soil-plant-water–related questions.

The measurement of the spatial distribution of water velocities within slices through a cylinder filled with glass beads and subject to steady state flow is a typical NMRI application (see Figure 7.8). The analysis of the scale dependency of the variability of water velocities is an example of how to interpret such images. The water velocity of every pixel was averaged over the third dimension, the slice dimension. With increasing slice thickness, the effectively two-dimensional distribution of velocities changes (see Figure 7.8). At the scale of the beads (see Figure 7.8A) the variability of velocities is highest and decreases as a function of the slice thickness (see Figure 7.8B,C). This was attributed to the formation of a network of flow paths (Deurer et al., 2004). In the same context it could be shown that the transverse correlation length of longitudinal pore water velocities is constant with slice thickness. This experimentally confirmed that transverse dispersivity is scale invariant (Deurer et al., 2004).

Both water velocities and volumetric water contents can be directly quantified with two subsequent measurement sequences. By multiplying water velocities and contents, the water flux can be directly derived and then analyzed (e.g., Deurer et al., 2004). NMRI renders the traditional detour of a chemical tracer for the study of water flow unnecessary.

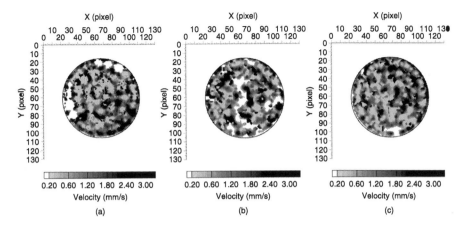

FIGURE 7.8 Scale dependency of two-dimensional distribution of voxel velocities in randomly packed and water-saturated glass beads (2 mm diameter) with a porosity of 0.45. One voxel has the dimensions 0.125 mm (x-pixel) × 0.125 mm (y-pixel) × slice thickness. The slices are taken across the center of a cylindrical column (46 mm × 14.1 mm) during horizontal flow. The mean flow velocity along the main direction of flow was 1.2 mm/s. (A) 2 mm thick slice. Note that at this scale the structure of the glass beads as flow obstacles can be clearly seen. (B) 8 mm thick slice. (C) 15 mm thick slice.

The distributions of either water contents, water velocities, or both have been analyzed in many NMR and NMRI studies that focused on water and solute transport (van As and van Dusschoten, 1997). Such studies helped to better understand and describe the dispersion process that is partly caused by the variability of water velocities (e.g., Seymour and Callaghan, 1997). Mostly porous rocks (e.g., Mansfield and Issa 1994, 1996) and bead packs (Kutskovsky et al., 1996; Seymour and Callaghan, 1997) were used as porous media. Indirect methods such as paramagnetic tracers can help to characterize water flow through porous (Herrmann et al., 2002) or aggregated media (Greiner et al., 1997), but also to observe two-phase flow or saltwater intrusion (Chen et al., 2002a). For that purpose NMRI sequences can be designed where the local concentration of a paramagnetic tracer, such as $CuSO_4$, is directly related to the spatially varying signal strength (e.g., Greiner et al., 1997).

Another application is the direct verification of the capability of fluid-dynamics models to describe water transport in bead packs (Seymour and Callaghan, 1997; Maier et al., 1998). For example, the distribution of water velocities within a column of glass beads during steady-state flow was numerically predicted with the lattice Boltzmann method. The results were validated with NMR measurements (Maier et al., 1998).

NMRI can help to develop a better understanding of the heterogeneity of water and solute transport. Properties relevant for the heterogeneity of transport such as the distribution of water contents (Amin et al., 1993), surface hydrophobicity (Allen et al., 1997), or the transport process itself, such as infiltration into a heterogeneous soil (Amin et al., 1997), can be directly studied. Preferential flow is a key process of heterogeneous water and solute flows and has been investigated with NMRI (Pearl et al., 1993; Cislerova et al., 1999; Baumann et al., 2002). In a box filled with analytical sand, Pearl et al. (1993) captured and analyzed the time evolution of the fingering of a high-density paramagnetic aqueous solution ($NiCl_2$) into one of low density (distilled water).

So far, the quantitative description of soil structure and hydraulic properties with NMRI has often been one-dimensional, as without the extra imaging gradients the pulse sequence can be shortened and the intrinsic spatial resolution enhanced. Pore systems can be analyzed with NMR by their correlation with relaxation processes. In the vicinity of a solid surface, such as a pore, the longitudinal and the transverse relaxation times are shorter in comparison to the bulk water (Brownstein and Tarr, 1979). In an individual water-filled pore, the effective relaxation time results from both the short relaxation time of the pore surface–influenced water and the longer relaxation time of the bulk water. A multiexponential decay of magnetization with multiple characteristic relaxation times reflects a multimodal pore size distribution (Hinendi et al., 1993). However, the reconstruction of a pore size distribution from relaxation times additionally depends on a pore shape factor (D'Orazio et al., 1989; Hinendi et al., 1993) that cannot be directly inferred from NMR measurements. Porosities and pore size distributions were characterized by means of NMR in sandstones (Lipsicas et al., 1986; Schmidt et al., 1986), porous silica beads (Hinendi et al., 1993), and soil and aquifer material (Hinendi et al., 1993). The NMR-inferred pore size distributions often agreed well with the results of alternative traditional measurement methods such as mercury porosimetry (Schmidt et al., 1986; Hinendi et al., 1993). Unlike other techniques, such as mercury porosimetry, NMR does not depend on the interconnectivity of pore spaces. All pores that are filled with water are measured. However, this also means that only pores that contain water can be detected and dry porous media cannot be characterized.

The permeability of sandstone to water (deuterium) was successfully predicted as a function of the porosity and the measured bulk T_1 (e.g., Thompson et al., 1989). The permeability and the pore size distribution as a function of pressure have been analyzed by means of NMR (Chen et al., 2002b).

Water and solute uptake and transformation processes of, for example, carbon, nitrogen, or pesticides take place in the rhizosphere. Fundamental research of the interactions between roots and the soil environment critically depends on undisturbed measurements with a high spatial resolution. NMRI can image roots in soil and artificial porous media (Rogers and Bottomeley, 1987). It has also been used to analyze soil water–depletion patterns of seedlings (MacFall et al., 1991). The understanding of root water and solute uptake is crucial for the development of new, more efficient irrigation methods or phytoremediation techniques. The general feasibility of NMRI in that respect was reviewed by MacFall and Johnson (1994).

7.2.4 STRENGTHS AND WEAKNESSES OF NMR IMAGING

7.2.4.1 Strengths

NMRI is the only well-established method that can directly measure microscale flow processes. Pore water velocities and diffusion can be imaged with a relatively high spatial resolution of up to about 35 μm. It is possible to analyze the spatial distribution of pore water velocities in a Lagrangian framework, and by using a paramagnetic tracer it is feasible to employ an Eulerian framework. Both the spectrum of pore water velocities and the spatial pattern of volumetric water contents can be measured to monitor the individual water flow components.

NMRI is a noninvasive method. This enables the study of the long-term evolution of the water dynamics in the soil-plant-atmosphere continuum as NMRI has no adverse effects on biological tissue. In general, repetitive measurements of these water flow processes allow observation over time.

The versatility of the scale and the dimensionality of the image are other principal advantages of NMRI. Any section of the sample that should be imaged can be easily selected, including its thickness and spatial resolution. Depending on the measurement sequence and the apparatus used, NMRI is capable of yielding three-dimensional information.

7.2.4.2 Weaknesses

Sensitivity to para- and ferromagnetics is inherent in the measurement principle of NMRI. So, from the perspective of a pedologist or a hydrologist, this is a major disadvantage. Para- and ferromagnetics are ubiquitous in natural sediments, and they can cause rapid local-signal decay (Hall et al., 1997). As a consequence, the received signal represents only a small part of the originally exited water molecules, and the observations are accordingly biased. The severity of the effect increases with the resonance frequency, which itself results from a strong static magnetic field and large magnetic gradients. Therefore, it is always a problem when a high spatial resolution is needed. Extensive calibration measurements (Hall et al., 1997; Deurer et al., 2002) and the development of new pulse sequences and measurement techniques (Kinchesh et al., 2002) seem to be promising strategies to address this problem.

For the study of unsaturated or vertically oriented flow within a sample, a vertical bore installation with the static magnetic field is necessary. Unfortunately, this requirement renders unsuitable most of the NMRI machines routinely used in hospitals.

With NMRI, the pores and their porous structure can be imaged, but only as long as water or at least water films make them visible. Dry porous media cannot be characterized.

The spatial resolution of NMRI is negatively correlated with the sample size. It is, at best, about 35 μm, limiting the suitability of NMRI to the imaging of pore geometries and porous topologies at the microscale.

7.3 X-RAY COMPUTED TOMOGRAPHY

X-ray tomography is the cross-sectional imaging of an object from either transmission or reflection data collected by illuminating the object from many different directions (Kak and Slaney, 1988). In the case of x-ray tomography, the radiation consists of x-rays.

7.3.1 MEASUREMENT PRINCIPLE: ATTENUATION OF X-RAY PHOTON ENERGY

X-ray radiation is a high-energy electromagnetic radiation with a wavelength of 0.3 pm–3.0 nm. X-rays are nondiffracting and propagate along straight lines from the radiation source through the object to the detector (see Figure 7.9). On their way through an object along the beam path, the x-ray photons are (partly) attenuated. The degree of attenuation depends on both the physicochemical properties of the object, such as its density and atomic constituents, and the frequency (energy) of the x-ray radiation. The variation in

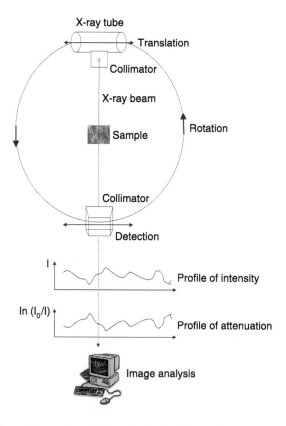

FIGURE 7.9 Schematic overview of the principal and most important components of an x-ray measurement.

attenuation constitutes the contrast that is used to reconstruct the spatial distribution (image) of physicochemical properties of an object.

For the range of photon energies commonly used for imaging purposes, 20–450 keV, two mechanisms contribute to the attenuation of x-ray energy. First, the x-ray photon can transfer its energy to a tightly bond inner electron of an atom of the imaged object. Having absorbed the energy (= photoelectric absorption), this electron is set free from its electron shell and leaves its atomic environment. Second, an x-ray photon can interact with a free electron of the imaged object or one that is only loosely bound in an outer atomic shell. Consequently, the x-ray photon is deflected from its path and loses energy (= Compton scattering). The contributions to attenuation by the two mechanisms are element-specific and a function of the applied photon energy. Hubbell (1969) gives tables with the respective values.

7.3.2 MEASUREMENT COMPONENTS

Many different technical configurations exist that are optimized for medical, industrial, or scientific applications. Details of technical specifications are described and discussed elsewhere (e.g., see Haddad et al., 1994; Kalender, 2000).

By irradiating a specific material (e.g., wolfram) with electrons, x-ray photons are generated by two mechanisms. First, the irradiation of wolfram atoms with electrons causes electron holes within the inner electron shells of those atoms. The holes are refilled with electrons of outer electron shells, and the resulting energy excess is emitted in the form of x-ray photons. This radiation leads to a narrow and characteristic (energy) line in the emitted x-ray spectrum. Second, the incident electrons can be deflected and decelerated in the coulomb fields of the wolfram atoms. The resulting energy excess is emitted in form of an x-ray radiation that is termed the bremsstrahlung. This is a continuous x-ray spectrum where the maximum x-ray energy is smaller than or equal to the kinetic energy of the incident electrons. The x-ray spectrum shows, therefore, the contribution by excitation as distinct peaks superimposed on the continuous bremsstrahlung spectrum.

The x-ray radiation is resolved into individual beams. The focal area of individual beams ranges from around 1–2 µm (Haddad et al., 1994; Spanne et al., 1994) for scientific scanners to about 0.5–2.0 mm for medical scanners. Generally, the smaller the focal area, the smaller is the potential penetration depth of x-rays. The set of beams is usually arranged either in the form of a two-dimensional fan or a three-dimensional cone. One scan results in a two- or three-dimensional projection of the entire object. The radiation source rotates in small angular increments around (360°) the object and usually 800–1500 projections are taken to reconstruct an image.

Collimators, screens, and filters filter the x-ray spectrum, focus the x-ray beams, define the thickness of the scanned section, shield the detectors against scattering radiation, and serve as protection against radiation.

Detection comprises several steps. After the passage of the imaged object, the remaining x-ray photon energy is transformed into an electric signal, amplified, and finally transformed from an analog into a digital signal. The first step, the crucial one, is usually achieved with either a scintillation detector or an ionization chamber.

A scintillation detector consists of crystals such as cesium-iodide, cadmium-wolframate, or ceramic materials (e.g., gadolinium oxysulfide). Subject to x-ray radiation, they emit visible light that is detected with a photodiode and transformed into an electric signal.

An ionization chamber is mostly filled with xenon at high pressure. Incident photons produce free electrons and positive ions when traveling through the gas. An electric field is applied across the chamber. As a result, the electrons drift towards the cathode, generating an electric signal.

7.3.3   ANALYSIS OF MEASURED ATTENUATION

7.3.3.1   Interpretation of Attenuation Coefficients

A measurement consists of many line integrals of the attenuated x-ray energy taken from different beam angles. The attenuation coefficient either can be directly calculated from the projection data (= profiles of attenuation) or has to be derived via image analysis (see below). Theoretically, three cases have to be considered.

7.3.3.1.1   Homogeneous Object and Monochromatic X-Rays

A specific flux density of photons N [$m^2 s^{-1}$] with the same energy = monochromatic) propagate along a beam path, s [m], through a physico-chemically homogeneous object of thickness d [m] (Figure 7.10A). Due to photoelectric absorption, τ [m^{-1}], and Compton scatterning, σ [m^{-1}], the flux density is reduced to $N + \Delta N$ when the photons emerge on the other side. This is readily formulated as (Kak and Slaney, 1988):

$$\frac{\Delta N}{N}\frac{1}{d} = -\tau - \sigma = -\mu \qquad (7.18$$

where μ [m^{-1}] is the linear attenuation coefficient of the material that is the product of the mass attenuation coefficient μ' [$m^2 kg^{-1}$] and the object's density ρ [$kg\ m^{-3}$]. The *photoelectric absorption* strongly depends on the atomic number Z of the scanned object:

$$\frac{\tau}{\rho} \approx \frac{Z^4}{(hf)^3} \qquad (7.19$$

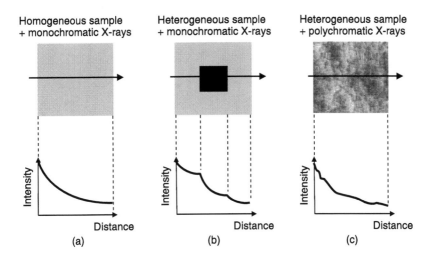

FIGURE 7.10 Interpretation of attenuation coefficients. (A) Homogeneous object and monochromatic x-rays. (B) Heterogeneous object and monochromatic x-rays. (C) Heterogeneous object and polychromatic x-rays.

where the product hf denotes the photon energy [eV]. The *Compton scattering* is inversely proportional to the applied photon energy:

$$\frac{\sigma}{\rho} \approx \frac{1}{hf} \tag{7.20}$$

Therefore, the linear attenuation coefficient depends on the physicochemical properties of the object (Z, ρ) as well as on the applied photon energy.

As the material is homogeneous and the x-rays are monochromatic, μ has the same value throughout the imaged object. Integrating both sides of Eq. (7.18) gives for the limit $\Delta s \to 0$:

$$\int_{N_0}^{N} \frac{dN}{N} = -\mu \int_0^d ds \tag{7.21}$$

where N_0 is the photon flux density entering the object.

From Eq. (7.21) the photon flux density can be derived as a function of the position along the beam path:

$$N(s) = N_0 e^{-\mu s} \tag{7.22}$$

Knowing the measured $N(s) = N_{exit}$ and the thickness d of the object, the attenuation coefficient of the imaged object can be derived as:

$$\mu = \frac{1}{d} \ln \frac{N_0}{N_{exit}} \tag{7.23}$$

The attenuation of the photon energy by the object is just μd.

7.3.3.1.1   Heterogeneous Object and Monochromatic X-Rays

Most natural objects are not homogeneous, but consist of a mixture of physicochemically different materials. Propagating through a heterogeneous object (Figure 7.10B), the attenuation of x-ray photons is now an integral over the different materials encountered along the beam path s:

$$\int_{ray} \mu(x; y; z)\, ds = \ln \frac{N_{in}}{N_{exit}} \tag{7.24}$$

To infer the local attenuation coefficient of a specific heterogeneous material, $\mu(x; y; z)$, many two-dimensional measurements (projection data) and image analysis (see below) are needed.

The value $\mu(x; y; z)$ represents the average of a voxel (volume element). A voxel of a soil can, for example, contain a mixture of the solid, water, and air phase. Therefore, the measured value of $\mu(x; y; z)$ is an unknown combination of the three phases. With dual-energy (monochromatic) imaging, the fractions can be separated (Phogat et al., 1991; DiCarlo et al., 1997). The two different monochromatic photon energies can be generated, for example, with a double gamma ray source such as ^{137}Cs and ^{169}Yb (Phogat et al., 1991) or by filtering a synchrotron x-ray radiation (DiCarlo et al., 1997).

The ranges of attenuation coefficients for different materials of a scanned object like roots or soil matrix within a soil core are often not very distinct and overlap (Heeraman et al., 1997). The contrast generally increases at lower photon energy (Sivers and Silver, 1990), but the unfavorable effects of beam hardening (see below) simultaneously increase (Hopmans et al., 1994). Therefore, a specific material component is mostly determined by identifying and separating distinct domains of values in the frequency distribution of the measured attenuation coefficients (Heeraman et al., 1997; Pierret et al., 1999).

7.3.1.1.3.   Heterogeneous Object and Polychromatic X-Rays

Commercial x-ray sources that are used for medical and industrial purposes do not produce monochromatic photon energies but emit a polychromatic x-ray spectrum (Figure 7.10C) (examples of spectra, are given in Kak and Slaney, 1988, and Clausnitzer and Hopmans, 2000). Within the energy range commonly used for imaging, the attenuation coefficients μ of most imaged materials decrease with increasing photon energy. The low-energy ("soft") photons of a polychromatic x-ray are preferentially absorbed. The remaining beam becomes proportionately richer in high-energy ("hard") photons. This process is termed beam hardening. If a polychromatic x-ray propagates through a homogeneous material, the mean energy of the spectrum increases and the attenuation coefficient decreases. A precise identification of individual material components simply by their measured attenuation coefficients is not possible, and Eq. (7.24) is not valid in this context. The detected attenuated photon flux density N_{exit} is now additionally a function of the energy spectrum of the beam:

$$N_{exit} = \int_0^{E_{max}} N_0(E) \exp\left[-\int_0^d \mu(x; y; z; E)\, ds \right] dE \qquad (7.25)$$

The incident photon flux density, N_0, is now given by $\int_0^{E_{max}} N_0(E)\, dE$.

To separate the effect of the changing physicochemical properties from changes in the photon energy for a specific value of $\mu(x; y; z; E)$, additional steps are necessary. Three major strategies exist (Kak and Slaney, 1988).

Preprocessing of projection data. The effect of beam hardening can be quantified and eliminated with additional calibration measurements. The imaging and quantitative analysis of a solute transport experiment is one instance where such a correction is necessary. Beam hardening would otherwise lead to a systematic bias of the reconstructed solution concentrations. Clausnitzer and Hopmans (2000) give a detailed description of a respective calibration procedure.

Postprocessing of the reconstructed image with image analysis methods [see, e.g., Kijewski and Bjarngard (1978) for the method and examples].

Dual energy imaging [see Alvarez and Macovski (1976); Duerinckx and Macovski (1978); Kalender et al. (1987)] can also be applied. The sample is imaged using two different energy spectra. The resulting two sets of projection data can then be used to separate the contrast caused by changing physicochemical properties from the one caused by a change in photon energy.

Alternatively, using radioactively decaying sources, monochromatic x-ray radiation can be produced and the problem of beam hardening can be prevented. However, the photon flux is usually low and long acquisition times are required (Brown et al., 1993).

The final output of an x-ray tomography image is the spatial distribution of the attenuation coefficient, $\mu(x; y; z; E)$. To compare the results across different technical settings or applied x-ray photon energy spectra, a conversion of attenuation coefficients into Hounsfield units, [HU], is customary:

$$H = \frac{\mu(x; y; z; E) - \mu_{water}(x; y; z; E)}{\mu_{water}(x; y; z; E)} 1000 \qquad (7.26)$$

where $\mu_{water}(x; y; z; E)$ has to be measured with identical technical settings as $\mu(x; y; z; E)$.

The values of H usually range from -1024 to $+3071$, where $H=0$ corresponds to water and $H=-1000$ to air.

7.3.3.2 Image Reconstruction

In radiography the radiograph shows an image of the attenuation of x-rays at one specific beam angle. This is one projection of the object, and the individual points of the projection are line integrals of the attenuation coefficients along the beam path. In x-ray computed tomography, the actual spatial distribution of attenuation coefficients is reconstructed from many projections with image analysis.

The algorithm commonly applied is the filtered backprojection. For an explanation of the algorithm, parallel-beam projection data will be used and the coordinate of the slice position z will be dropped. The extension to the reconstruction of fan- and cone-beam projection data is elaborated elsewhere (e.g., Kak and Slaney, 1988).

A projection $P_\theta(t)$ is generated by scanning the object that is represented by its spatially varying attenuation coefficient $f(x; y)$ at a specific beam angle θ. The translational position t describes the position of an individual beam relative to the center of the object. A projection consists of a number of points, each of which is a ray (attenuation) line integral of the object. The projection $P_\theta(t)$ is also called the Radon transform (Radon, 1917) of the object and combines these line integrals:

$$P_\theta(t) = \int_{(\theta; t)} f(x; y) \, ds = \int_{-\infty}^{\infty} \int_{-\infty}^{\infty} f(x; y) \delta(x \cos \theta + y \sin \theta - t) \, dx dy \quad (7.27$$

where δ is Dirac's delta function. The projection is Fourier-transformed to give $S_\theta(w)$:

$$S_\theta(w) = \int_{-\infty}^{\infty} P_\theta(t) e^{-2j\pi wt} \, dt \quad (7.28$$

where j is $\sqrt{-1}$ and w the spatial frequency. Every such one-dimensional Fourier-transformed projection $S_\theta(w)$ gives a line of the two-dimensional Fourier transform of the object, $F(w; \theta)$. This constitutes the *Fourier slice theorem*. Therefore, with an infinite number of $S_\theta(w)$ the entire Fourier transform of the object is known, and the object $f(x; y)$ may directly reconstructed by inverse Fourier transform:

$$f(x; y) = \int_0^{2\pi} \int_0^{\infty} F(w; \theta) e^{j2\pi w(x \cos \theta + y \sin \theta)} w \, dw d\theta \quad (7.29$$

Because the results of the x-ray projections are independent of the beam direction, Eq. (7.29) can be simplified to:

$$f(x; y) = \int_0^\pi \left[\int_{-\infty}^\infty F(w; \theta)|w|e^{j2\pi wt}dw \right] d\theta \qquad (7.30)$$

However, with x-ray tomography only a finite number of projections are measured, and $F(w; \theta)$ can only be estimated using the limited number of $S_\theta(w)$. Each $S_\theta(w)$ is weighted (= interpolation in Fourier space) with a *filtering* function that is, in the simplest case, the ramp function $|w|$ as in Eq. (7.30). Often $|w|$ is extended with additional filtering functions that can reduce high-frequency noise (e.g., Shepp and Logan, 1974). Then,

$$f(x; y) = \int_0^\pi \left[\int_{-\infty}^\infty S_\theta(w)|w|e^{j2\pi wt}dw \right] d\theta \qquad (7.31)$$

For computational efficiency, $S_\theta(w)$ is directly *backprojected* (= inverse Fourier transform) after filtering. The (filtered) backprojections are added in real space to reconstruct $f(x; y)$:

$$f(x; y) = \int_0^\pi Q_\theta(t) \, d\theta = \int_0^\pi Q_\theta(x \cos \theta + y \sin \theta) \, d\theta \qquad (7.32)$$

where $Q_\theta(t)$ is the filtered backprojection of $S_\theta(w)$:

$$Q_\theta(t) = \int_0^\infty S_\theta(w)|w|e^{j2\pi wt}dw \qquad (7.33)$$

The reconstructed $f(x; y)$ is the spatial distribution of the attenuation coefficients that can be used for further interpretation.

7.3.4 APPLICATIONS OF X-RAY TOMOGRAPHY TO SOIL-PLANT-WATER PROCESSES

X-ray CT is mostly applied to reconstruct the pore geometry and topology of samples.

The transport of water and solutes, erosion, microbial dynamics, and many other processes depend on the soil structure, which consists of the soil matrix and the pore space. The reconstruction of the pore system is a typical CT application. CT of a dry soil sample yields a distribution of attenuation coefficients (see Figure 7.11). In the next step the varying values are attributed

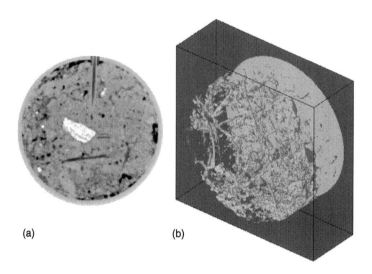

(a)   (b)

FIGURE 7.11 Distribution of attenuation coefficients and reconstructed macropore network of an undisturbed soil sample (sample diameter 0.16 m and height 0.10 m). The sample was taken from the Ap horizon of a calcareous soil in the Paris basin in France. (A) Distribution of attenuation coefficients (view onto the top of the sample). The dark, small vertically oriented elipsoid at the top of the picture is an installed mini-tensiometer. The black spots close to the insertion point of the tensiometer at the sample wall indicate local disturbances (artificially created air gaps). The white area in the center of the picture represents limestone. (B) The reconstructed macropore network of the entire sample. The light disk in the back of the cuboid marks the bottom of the sample. (Both images courtesy of H. J. Vogel.)

to the soils matrix or to the pore space. Then, the pore space can be separated into specific pore domains such as the one of macropores (see Figure 7.11).

Three-dimensional macropore networks (see Figure 7.11) can be analyzed and their topology, like their connectivity, tortuosity, average length, and density, derived (Perret et al., 1999). The latter seems to be a breakthrough to develop deterministic numerical models for macropore flow.

Our ability to predict water and solute transport is still somewhat restricted to cases where preferential flow phenomena do not occur (Flury, 1996). A better representation of the heterogeneous pattern of hydraulic properties and especially of the distribution of macropores seems to be a promising solution strategy to this problem. Kasteel et al. (1999) successfully simulated water and solute transport at the column scale based on the distribution of CT-measured bulk densities as a proxy for the heterogeneity of hydraulic properties. Rasiah and Aylmore (1998b) derived the distribution of microscale hydraulic conductivities partly based on CT measurements.

The observation of solute transport is, in principle, possible by the CT measurement of solution concentrations (Anderson et al., 1992; Clausnitzer and Hopmans, 2000). However, this remains an indirect approach.

For example, a main cause of solute dispersion, the variability of pore water velocities, cannot be directly inferred. Still, it enables a more rigorous test of transport models as with traditional column-scale solute breakthrough experiments. Also, transport can be directly linked to the soil structure, for example, its aggregate size classes (Anderson et al., 1992).

With CT, the spatial distribution of different pore structure–related parameters, such as bulk density (e.g., Petrovic et al., 1982; Anderson et al., 1988), porosity (e.g., Phogat and Aylmore, 1989; Warner et al., 1989), and water content (e.g., Crestana et al., 1985; Anderson et al., 1988; Hopmans et al., 1992), can be derived. With dual-energy CT, several characteristics can be simultaneously analyzed (Rogasik et al., 1999).

CT is a noninvasive method and can be used to study the dynamics of soil structure. Porosity changes during wetting and drying cycles (Rasiah and Aylmore, 1998a) and due to different tillage practices (Gantzer and Anderson, 2002) are examples. The implementation of time-dependent hydraulic properties in water flow and solute transport models, the evaluation of agronomic measures, and the optimization of irrigation efficiency could be typical applications, especially for clay-rich soils. The increase in bulk density at the soil surface was measured with CT and attributed to the impact of rainfall (Fohrer et al. 1999). The methods of fractal geometry are often used in the context of the dynamics of soil structure to interpret CT images. The fractal dimension is an example (Peyton et al., 1994; Zeng et al., 1996; Rasiah and Aylmore, 1998a; Gantzer and Anderson, 2002).

CT is also a viable means to evaluate the microscale assumptions of bulk-scale measurement protocols, such as one-step outflow experiments (Hopmans et al., 1992) or to monitor how undisturbed the installation of specific measurement equipment was (e.g., the installation of a mini-tensiometer) (see Figure 7.11).

For irrigation or bioremediation purposes, the soil-plant-water interactions in the rhizosphere are crucial. But many aspects of it remain unclear. The three-dimensional distribution of the roots of seedlings (Heeraman et al., 1997) and trees (Pierret et al., 1999) in soil has been reconstructed using CT, and the water movement around roots also been considered (Aylmore, 1993).

7.3.5 Strengths and Weaknesses of X-Ray Tomography

7.3.5.1 Strengths

With x-ray tomography, the pore structure–related parameters such as bulk density and volumetric water content can be imaged with a very high spatial resolution of up to 2 μm. Pore geometry and topology can be observed over several orders of magnitude, ranging from micropores to macropores. The root system geometry, including root hairs and the architecture of soil-plant root systems, can be tomographically investigated. Unlike NMRI, with x-rays it is feasible to analyze soil samples or aggregates containing para- or ferromagnetics.

X-ray tomography is also noninvasive, and it allows for repetitive measurements. It is possible to study the time evolution of pore structures subject to specific initial and boundary conditions or to observe the dynamics of soil-plant root systems.

X-ray tomography is also capable of providing three-dimensional information.

7.3.5.2   Weaknesses

Most x-ray tomographs use polychromatic x-ray sources, which lead to the phenomenon of "beam hardening." A dependency of the analyzed property upon the energy spectrum of incident x-rays is the consequence. This can only be avoided by applying a sophisticated calibration procedure.

Soil pores are often imaged to better understand the pathways for water and solute transport. Pore diameters range over orders of magnitude, and different pore domains, such as micropores or macropores, might have quite different representative elementary volumes (Vogel et al., 2002). Therefore, the reconstruction of the pore system that is representative for water and solute flow requires imaging large soil samples with high spatial resolution. But for a given x-ray tomography the spatial resolution is negatively correlated with the sample size. Therefore, only specific pore domains can be separately imaged with sufficient spatial resolution within their representative volume. Also, the sample geometry must be isometric. Yet there are strongly anisotropic pore topologies in soil such as those of macropore networks, which are caused by earthworms.

The dynamics of water flow, such as the distribution of pore water velocities, can only be indirectly measured with x-rays, and this limits observations of the time evolution of volumetric water content or tracer concentrations.

7.4.   USE OF NMRI AND X-RAY TOMOGRAPHY FOR PRACTICAL ENGINEERING PURPOSES

The transfer of NMRI and x-ray tomography techniques from the area of fundamental research to the solution of practical engineering problems is only beginning. Three factors hamper their frequent use by practitioners: (1) both methods have high investment costs, and for NMRI the maintenance costs are also substantial; (2) access to hospitals or research centers where the techniques could be publicly used is restricted; (3) practical soil-plant-water–related problems mostly arise at the macroscale of entire agricultural fields or landscapes. But research currently offers few recommendation as to how to upscale and apply the microscopic results of NMRI and x-ray tomography.

At the moment, microscopic imaging techniques already seem to be valuable tools for the risk assessment of macropore flow and soil degradation. For example, the measured pore topology can be used to derive effective parameters for preferential flow models such as MACRO (Jarvis et al., 1991) or to quantify the change in soil structure due to compaction.

For most practical applications, NMRI and x-ray tomography methods are not yet cost-effective, but with the rapid advance of technology and research, they might be in the near future.

7.5 PROSPECTS AND FUTURE RESEARCH IMPERATIVES

It would be fruitful to combine NMRI and x-ray tomography. The strengths of x-ray tomography are to reconstruct the pore geometry and topology, whereas those of NMRI are the ability to measure the components of flow through porous structures. The two are complementary. A combined capability could, for example, enable a rigorous test of existing models of heterogeneous water flow and solute transport at the pore scale and thereby help to develop better transport models in complex porous continua.

In the future, combined NMRI and x-ray tomography could help to tackle issues both at the microscale of the pores as well as at the macroscale of soil. These are discussed briefly in this conclusion.

7.5.1 MICROSCALE

The temporal dynamics of pore structures needs to be better understood, as the processes of swelling and shrinking, the growth of plant roots, the activity of soil organisms, and the transport of soil particles can change the biophysical functioning of porous networks. Whenever the transport of reactive contaminants, such as pesticides, is considered, not only are the topology of pores and the distribution of water fluxes important, but also the interaction of these substances with the pore surfaces during flow. Here, the methods of NMRI and x-ray tomography would complement existing methods that assess the physicochemical exchange properties of pore surfaces. The interaction of plant roots, especially of the physiologically active root hairs, with the surrounding soil is a research topic worthy of study. Plant nutrition, irrigation management, and phytoremediation would benefit from such new understanding that might come from the use of NMRI and x-ray imaging at the microscale.

7.5.2 MACROSCALE

A soil-scientific challenge is to upscale and integrate our microscale understanding to spatial scales that are relevant for the description of hydraulic properties at the field or catchment scale (Vogel and Roth, 2003). Here, NMRI and x-ray tomography might enhance large-scale geophysical methods. They have, for example, the potential to be able to observe the effective hydraulic parameters of larger scale units, such as soil horizons or macropore networks, and to establish the hydraulic correlation between such units.

REFERENCES

Allen, S.G., P.C.L. Stephenson, and J.H. Strange. 1997. Morphology of porous media studied by nuclear magnetic resonance. J. Chem. Phys. 106:7802–7809.

Alvarez, R.E., and A. Macovski. 1976. Energy-selective reconstruction in x-ray computerized tomography. Phys. Med. Biol. 21:733–744.

Amin, M.H.G., R.J. Chorley, K.S. Richards, B.W. Bache, L.D. Hall, and T.A. Carpenter. 1993. Spatial and temporal mapping of water in soil by magnetic resonance imaging. Hydrol. Proc. 7:279–286.

Amin, M.H.G., R.J. Chorley, K.S. Richards, L.D. Hall, T.A. Carpenter, M. Cislerová and T. Vogel. 1997. Study of infiltration into a heterogeneous soil using magnetic resonance imaging. Hydrol. Proc. 11:471–483.

Anderson, S.H., C.J. Gantzer, J.M. Boone, and R.J. Tully. 1988. Rapid nondestructive bulk density and soil-water content determination by computed tomography. Soil Sci. Soc. Am. J. 52:35–40.

Anderson, S.H., R.L. Peyton, J.W. Wigger, and C.J. Gantzer. 1992. Influence of aggregate size on solute transport as measured using computed tomography Geoderma 53:387–398.

Aylmore, L.A.G. 1993. Use of computer-assisted tomography in studying water movement around plant roots. Adv. Agron. 49:1–54.

Baumann, T., R. Petsch, G. Fesl, and R. Niessner. 2002. Flow and diffusion measurements in natural porous media using magnetic resonance imaging. J. Environ. Qual. 31:470–476.

Brigham, E.O. 1974. The Fast Fourier Transform. Prentice-Hall, Englewood Cliffs, NJ.

Brown, G.O., M.L. Stone, and J.E. Gazin. 1993. Accuracy of gamma ray computerized tomography in porous media. Water Resour. Res. 29:479–486.

Brownstein, K.R., and C.E. Tarr. 1979. Importance of classical diffusion in NMR studies of water in biological cells. Phys. Rev. A 19:2446–2453.

Callaghan, P.T. 1993. Principles of Nuclear Magnetic Resonance Microscopy. 2nd ed. Clarendon Press, Oxford.

Carrington, A., and A.D. McLachlan. 1967. Introduction to Magnetic Resonance. Chapman and Hall, London.

Chen, C.N., and D.I. Hoult. 1989. Biomedical Magnetic Resonance Technology. Adam Hilger, Bristol and New York.

Chen, Q., W. Kinzelbach, and S. Oswald. 2002a. Nuclear magnetic resonance imaging for studies of flow and transport in porous media. J. Environ. Qual. 31:477–486.

Chen, Q., W. Kinzelbach, C. Ye, and Y. Yue. 2002b. Variations of permeability and pore size distribution of porous media with pressure. J. Environ. Qual 31:500–505.

Cislerová, M., J. Votrubuvá, T. Vogel, M.H.G. Amin, and L.D. Hall. 1999. Magnetic resonance imaging and preferential flow in soils. p. 397–411. In Proc. of the Int. Workshop Characterization and Measurement of the Hydraulic Properties of Unsaturated Porous Media. Part 1. Riverside, CA, Oct. 22–24, 1999. USDA, Riverside, CA.

Clausnitzer, V., and J.W. Hopmans. 2000. Pore-scale measurements of solute breakthrough using microfocus x-ray computed tomography. Water Resour. Res. 36:2067–2079.

Crestana, S., S. Mascarenhas, and R.S. Pozzi-Mucelli. 1985. Static and dynamic three dimensional studies of water in soil using computed tomographic scanning. Soil Sci. 140:326–332.

Deurer, M., I. Vogeler, B.E. Clothier, and D.R. Scotter. 2004. Magnetic resonance imaging of hydrodynamic dispersion in a saturated porous medium. Transp. Porous Media 54(2):145–166.

Deurer, M., I. Vogeler, A. Khrapitchev, and D. Scotter. 2002. Imaging of water flow in porous media by magnetic resonance imaging microscopy. J. Environ. Qual 31:487–493.

DiCarlo, D.A., T.W. Bauters, T.S. Steenhuis, J.-Y. Parlange, and B.R. Bierck. 1997. High-speed measurements of three-phase flow using synchrotron x-rays. Water Resour. Res. 33:569–576.

D'Orazio, F., J.C. Tarczon, W.P. Halperin, K. Eguchi, and T. Mizusaki. 1989. Application of nuclear magnetic resonance pore structure analysis to porous silica glass. J. Appl. Phys. 65:742–750.

Duerinckx, A.J., and A. Macovski. 1978. Polychromatic streak artifacts in computed tomography images. J. Comput. Assist. Tomog. 2:481–487.

Eccles, C.D., S. Crozier, W. Roffman, D.M. Doddrell, P. Back, and P.T. Callaghan. 1994. Practical aspects of shielded gradient design for localised *in vivo* NMR spectroscopy and small scale imaging. Magn. Res. Imaging 12:621–630.

Farrar, T.C. 1987. An Introduction to Pulse NMR Spectroscopy. Farragut Press, Chicago.

Farrar, T.C., and E.D. Becker. 1971. Pulse and Fourier Transform NMR. Introduction to Theory and Methods. Academic Press, New York.

Flury, M. 1996. Experimental evidence of transport of pesticides through field soils — a review. J. Environ. Qual. 25:25–45.

Fohrer, N., J. Berkenhagen, and A.P.J. de Roo. 1999. Changing soil and surface conditions during rainfall/single rainstorm/subsequent rainstorms. Catena 37:355–375.

Freeman, R. 1988. A Handbook of Nuclear Magnetic Resonance. Longman Scientific & Technical, Essex.

Fukushima, E., and S.B.W. Roeder. 1981. Experimental Pulse NMR. Addison-Wesley, Reading, MA.

Gadian, D.G. 1995. NMR and Its Applications to Living Systems. Oxford University Press, Oxford.

Gantzer, C.J., and S.H. Anderson. 2002. Computed tomographic measurement of macroporosity in chisel-disk and no-tillage seedbeds. Soil and Tillage Res 64:101–111.

Greiner, A., W. Schreiber, G. Brix, and W. Kinzelbach. 1997. Magnetic resonance imaging of paramagnetic tracers in porous media: quantification of flow and transport parameters. Water Resour. Res. 33:1461–1473.

Haddad, W.S., I. McNulty, J.E. Trebes, E.H. Anderson, R.A. Levesque, and L. Yang. 1994. Ultrahigh-resolution X-ray tomography. Science 266:1213–1215.

Hahn, E. L. 1950. Spin echoes. Phys. Rev. 80:580–594.

Hall, D.L., M.H.G. Amin, E. Dougherty, M. Sanda, J. Votrubová, K.S. Richards, R.J. Chorley, and M. Cislerová. 1997. MR properties of water in saturated soils and resulting loss of MRI signal in water content detection at 2 tesla. Geoderma 80:431–448.

Heeraman, D.A., J.W. Hopmans, and V. Clausnitzer. 1997. Three dimensional imaging of plant roots *in situ* with X-ray computed tomography. Plant Soil 189:167–179.

Herrmann, K.-H., A. Pohlmeier, S. Wiese, N.J. Shah, O. Nitzsche, and H. Vereecken. 2002. Three-dimensional nickel ion transport through porous media using magnetic resonance imaging. J. Environ. Qual. 31:506–514.

Hinendi, Z.R., A.C. Chang, M.A. Anderson, and D.B. Borchardt. 1997. Quantification of microporosity by nuclear magnetic resonance relaxation of water imbibed in porous media. Water Resour. Res. 33:2697–2704.

Hinendi, Z.R., Z.J. Kabala, T.H. Skaggs, D.B. Borchardt, R.W.K. Lee, and A.C. Chang. 1993. Probing soil and aquifer material porosity with nuclear magnetic resonance. Water Resour. Res. 29:3861–3866.

Hopmans, J.W., M. Cislerová, and T. Vogel. 1994. X-ray tomography of soil properties. p. 17–28. *In* S.E. Anderson and J.W. Hopmans (eds.) Tomography of Soil-Water Root Processes. ASA Spec. Publ. 50. ASA, CSSA, and SSSA, Madison, WI.

Hopmans, J.W., T. Vogel, and P.D. Koblik. 1992. X-ray tomography of soil water distribution in one-step outflow experiment. Soil Sci. Soc. Am. J. 56:355–362.

Hubbell, J.H. 1969. Photon cross sections, attenuation coefficients, and energy absorption coefficients from 10 keV to 100 GeV. Natl. Stand. Ref. Data Ser. 29. Natl. Inst. of Stand. and Technol., Gaithersburg, MD.

Jarvis, N.J., P.-E. Jansson, P.E. Dik, and I. Messing. 1991. Modeling water and solute transport in macroporous soil. 1. Model description and sensitivity analysis. J. Soil Sci. 42:59–70.

Jennison, R.C. 1961. Fourier Transforms and Convolutions. Pergamon Press, New York.

Kak, A.C., and M. Slaney. 1988. Principles of Computerized Tomographic Imaging. EE Press, New York.

Kalender, W.A. 2000. Computertomographie: Grundlagen, Gerätetechnologie, Bildqualität, Anwendungen (in German). Pulicis MCD Verlag, München.

Kalender, W.A., W. Bautz, D. Felsenberg, C. Süß, and E. Klotz. 1987. Materialselektive Bildgebung und Dichtemessung mit der Zwei-Spektren Methode. I. Grundlagen und Methodik (in German). Digitale Bilddiagn. 7:66–72.

Kasteel, R., H.-J. Vogel, and K. Roth. 1999. From local hydraulic properties to effective transport in soil. Eur. J. of Soil Sci. 51:81–91.

Kijewski, D.K., and B.E. Bjarngard. 1978. Correction for beam hardening in computed tomography. Med. Phys. 5:209–214.

Kinchesh, P., A.A. Samoilenko, A.R. Preston, and E.W. Randall. 2002. Stray field nuclear magnetic resonance of soil water: development of a new large probe and preliminary results. J. Environ. Qual. 31:494–499.

Kutsovsky, Y.E., V. Alvarado, H.T. Davis, L.E. Scriven, and B.E. Hammer. 1996. Dispersion of paramagnetic tracers in bead packs by T1 mapping: experiments and simulations. Magn. Reson. Imag. 14:833–839.

Lipsicas, M., J.R. Banavar, and J. Willemsen. 1986. Surface relaxation and pore sizes in rocks-a nuclear magnetic resonance analysis. Appl. Phys. Lett. 48:1544–1546.

MacFall, J.S., and G.H. Johnson. 1994. Use of magnetic resonance imaging in study of plants and soil. p. 99–112. *In* S.E. Anderson and J.W. Hopmans (eds.), Tomography of Soil-Water-Root Processes. SSSA Spec. Publ. 36. SSSA, Madison, WI.

MacFall, J.S., G.H. Johnson, and P.J. Kramer. 1991. Comparative water uptake by roots of different ages in seedlings of loblolly pine (*Pinus taeda* L.). New Phytol 119:551–560.

Maier, R.S., D.M. Kroll, Y.E. Kutskovsky, H.T. Davis, and R.S. Bernard. 1998. Simulation of flow through bead packs using the lattice Boltzmann method. Phys. Fluids 10:60–74.

Mansfield, P., and P.K. Grannell. 1975. 'Diffraction' and microscopy in solids and liquids by NMR. Phys. Rev. B. 12:3618–3634.

Mansfield, P., and B. Issa. 1994. Studies of fluid transport in porous rocks by echo-planar MRI. Magn. Reson. Imag. 12:275–278.

Mansfield, P., and B. Issa. 1996. Fluid transport in porous rocks. I. EPI studies and a stochastic model of flow. J. Magn. Reson. A 122:137–148.

Pearl, Z., M. Magaritz, and P. Bendel. 1993. Nuclear magnetic resonance imaging of miscible fingering in porous media. Transp. Porous Media 12:107–123.

Perret, J., S.O. Prasher, A. Kantzas, and C. Langford. 1999. Three-dimensional quantification of macropore networks in undisturbed soil cores. Soil Sci. Soc. Am. J. 63:1530–1543.

Petrovic, A.M., J.E. Siebert, and P.E. Riecke. 1982. Soil bulk density analysis in three dimensions by computer tomographic scanning. Soil Sci. Soc. Am. J. 46:445–450.

Peyton, R.L., C.J. Gantzer, S.H. Anderson, B.A. Haeffner, and P. Pfeiffer. 1994. Fractal dimension to describe soil macropore structure using x-ray computed tomography. Water Resour. Res. 3:691–700.

Phogat, V.K., and L.A.G. Aylmore. 1989. Evaluation of soil structure by using computer assisted tomography. Aust. J. Soil Res. 27:313–323.

Phogat, V.K., L.A.G. Aylmore, and R.D. Schuller. 1991. Simultaneous measurement of the spatial distribution of soil water content and bulk density. Soil Sci. Soc. Am. J. 55:908–915.

Pierret, A., Y. Capowiez, C.J. Moran, and A. Kretschmar. 1999. X-ray computed tomography to quantify tree rooting spatial distributions. Geoderma 90:307–326.

Pope, J.M., and S. Yao. 1993. Quantitative NMR imaging of flow. Concepts Magnetic Resonance 5:281–302.

Price, W.S. 1998. NMR imaging. Annual reports on NMR spectroscopy 35:139–216.

Radon, J.H. 1917. Über die Bestimmung von Funktionen durch ihre Integralwerte längs gewisser Mannigfaltigkeiten (in German). Ber. Sächsi. Akad. Wiss. 69:262.

Rasiah, V., and L.A.G. Aylmore. 1998a. Characterizing the changes in soil porosity by computed tomography and fractal dimension. Soil Sci. 163:203–211.

Rasiah, V., and L.A.G. Aylmore. 1998b. Estimating microscale spatial distribution of conductivity and pore continuity using computed tomography. Soil Sci. Soc. Am. J. 62:1197–1202.

Rogasik, H., J.W. Crawford, O. Wendroth, I.M. Young, M. Joschko, and K. Ritz. 1999. Discrimination of soil phases by dual energy x-ray tomography. Soil Sci. Soc. Am. J. 63:741–751.

Rogers, H.H., and P.M. Bottomely. 1987. In situ nuclear magnetic resonance imaging of roots: influence of soil type, ferromagnetic particles content and soil water. Agron. J. 79:957–965.

Schmidt, E.J., K.K. Velasco, and A.M. Nur. 1986. Quantifying solid fluid interfacial phenomena in porous rocks with proton nuclear magnetic resonance. J. Appl. Phys. 59:2788–2797.

Seymour, J.D., and P.T. Callaghan. 1997. Generalized approach to NMR analysis of flow and dispersion in porous media. AICHE J. 43:2096–2111.

Shepp, L.A., and B.F. Logan. 1974. The Fourier reconstruction of a head section. EE Trans. Nuclear Sci. 21:21–43.

Sivers, E.A., and M.D. Silver. 1990. Performance of x-ray computed tomographic imaging systems. Mater. Eval. 48:706–713.

Spanne, P., K.W. Jones, L. Prunty, and S.H. Anderson. 1994. Potential applications of synchrotron computed microtomography to soil science. p. 43–57. *In* S.E. Anderson and J.W. Hopmans (eds.), Tomography of Soil-Water-Root Processes. SSSA Spec. Publ. 36. SSSA, Madison, WI.

Thompson, A.H., S.W. Sinton, S.L. Huff, A.J. Katz, R.A. Raschke, and G.A. Gist. 1989. Deuterium magnetic resonance and permeability in porous media. J. Appl. Phys. 65:3259–3263.

Van As, H., and D. van Dusschoten. 1997. NMR methods for imaging of transport processes in micro-porous systems. Geoderma 80:389–403.

Vogel, H.-J., I. Cousin, and K. Roth. 2002. Quantification of pore structure and gas diffusion as a function of scale. Eur. J. Soil Sci. 53:465–473.

Vogel, H.-J., and K. Roth. 2003. Moving through scales of flow and transport in soil. J. Hydrol. 272:95–106.

Warner, G.S., J.L. Nieber, I.D. Moore, and R.A. Geise. 1989. Characterizing macropores in soil by computed tomography. Soil Sci. Soc. Am. J. 53:653–660.

Xia, Y. 1998. Introduction to magnetic resonance. p. 713–739. *In* P. Blümler, B. Blümich, R.F. Botto, and E. Fukushima (eds.), Spatially Resolved Magnetic Resonance. Methods, Materials, Medicine, Biology, Rheology, Geology, Ecology, Hardware. Wiley-VCH, Weinheim.

Zeng, Y., C.J. Gantzer, R.L. Peyton, and S.H. Anderson. 1996. Fractal dimension and lacunarity of bulk density determined with x-ray computed tomography. Soil Sci. Soc. Am. J. 60:1718–1724.

8 Preferential Flow: Identification and Quantification

Adel Shirmohammadi and H. Montas
University of Maryland College Park, College Park,
Maryland, U.S.A.

Lars Bergström
Swedish University of Agricultural Sciences, Uppsala, Sweden

Ali Sadeghi
USDA-ARS, Beltsville, Maryland, U.S.A.

David Bosch
USDA-ARS, Tifton, Georgia, U.S.A.

CONTENTS

1-5667-0657-2/05/$0.00 + $1.50

8.1   INTRODUCTION

Concern over chemical loadings to unconfined aquifers and into the surface water sources through drain tiles and subsurface groundwater flow has directed researchers to focus on the pathways that speed up the pollutant arrival to such sources. Preferential flow of water and chemical transport though porous media has attracted the attention of scientists and engineers working in the environmental field. Research has indicated that structured soils promote bypass flow (a form of preferential flow induced by macropores formed due to shrinking and swelling of soils) that results in fast movement of solutes, whereas "piston flow" is mostly responsible for the flow of water and solutes in nonstructured homogeneous (e.g., homogeneous sand-textured) soils (Skopp, 1981; Schumacher, 1864; Bergström and Shirmohammadi, 1999). Nieber (2001) stated that preferential flow includes macropore flow, gravity-driven unstable flow, heterogeneity-driven flow, oscillatory flow, and depression-focused recharge. Flow systems such as fingering caused by a sequence of texturally different layers (Hill and Parlange, 1972), hydrophobicity (Ritsema et al., 1983), and funnel flow due to texturally different lenses (Kung, 1990) are examples of preferential flow mechanisms. Such multiple transport behavior has created a multitude of difficulties in modeling solute leaching in vadose zone.

Most transport models assume that soils are homogeneous porous media and that water moves downward as a well-defined coherent wetting front. Models that are based strictly on Darcian relationships have been used as transport models in porous media or as part of the large-scale models that evaluate the impacts of agricultural best management practices (BMPs) for decades. These models have been developed either for environmental screening of organic chemicals through soil profile (Lindstrom et al., 1968; Davidson and McDougal, 1973; Enfield et al., 1982; Nofziger and Hornsby, 1984; Wagenet and Hutson, 1986) or for BMP evaluation (Carsel et al., 1985; Leonard et al., 1987; Shirmohammadi et al., 1989). These or other models based on Darcian theory are under question relative to the importance and role of large continuous openings (macropores) on water and solute movement through the soil profile (Beven and Germann, 1982).

The concept of bypass flow through macropores and preferred routes and the fact that they will permit rapid movement of water and chemicals dates back to studies by Schumacher (1864), who stated that "the permeability of a soil during infiltration is mainly controlled by big pores, in which the water is not held under the influence of capillary forces." Lawes et al. (1882) later observed that drainage water in a clay soil consisted of "muddy" surface water (that apparently bypassed the soil matrix at early times followed by clearer water (filtered by the soil matrix) at later times. Studies by Shirmohammadi and Skaggs (1985) showed that a 2-year-old fescue root system caused a fivefold increase in saturated hydraulic conductivity of a fine sandy soil. They attributed this increase to the biological loosening effect of the fescue roots. They also showed that using original saturated hydraulic conductivity of 9 cm/h instead of increased conductivity values during the growing season in the Green and Ampt infiltration model resulted in significant underprediction of infiltration. Obviously, in its classical form, the Green and Ampt (1911) infiltration model is a Darcian-based model and considers piston flow and neglects the macropores or channels created by biological activities in the root zone.

Quantification of water and chemical transport through preferential pathways has been offering a great challenge to researchers in the last two decades. Using Darcian-based flow in solute transport models, generally derived for homogeneous soils, has produced difficulty in their application to structured soils and soils affected by biological activities (van Genuchten and Wierenga, 1976; Beven and Germann, 1982; Gish and Jury, 1983). In addition, the heterogeneous nature of soil pores with respect to their occurrence, size, shape, and spatial distribution poses another challenge in devising models capable of handling such variability in transport media.

This chapter provides background on nature and occurrence of macropores and presents some of the most recent experimental and theoretical approaches suggested to handle preferential flow in assessing water and solute transport through porous media.

8.2 BACKGROUND ON PREFERENTIAL FLOW PROCESSES AND IDENTIFICATION

Beven and Germann (1982) categorized macropores into four groups, based on their formation and geometrical shape, as:

1. Pores formed by soil fauna that are tubular in shape and may range in size from 1 mm to > 50 mm in diameter. These pores are formed by burrowing animals such as ants, earthworms, moles, gophers, and wombats.
2. Pores formed by plant roots that may be formed by alive or decayed roots and are tubular in shape. Their size and extension depends on the type of plant. Living roots can both reduce the volume of voids available for flow (by growing in existing voids) and, conversely, develop new macropores along a growing soil-root interface (Gish and Jury, 1983).
3. Cracks and fissures that are formed by shrinkage in clay-abundant soils or by chemical weathering of the bedrock material. They may also be formed by a freeze/thaw cycle and certain tillage practices, such as subsoiling.
4. Natural soil pipes that may form due to the erosive action of subsurface flows in highly permeable and relatively noncohesive materials that are subject to high hydraulic gradients.

Technologies such as ground-penetrating radar (GPR) have been used to characterize the occurrence and extent of preferential flow pathways due to texturally differing lenses under field conditions (Kung and Lu, 1993). On a column scale, Capowiez et al. (2003) used x-ray computed tomography (CT) to analyze the three-dimensional macropore networks of earthworm barrows, which could be very useful in modeling preferential flow systems. In addition, nuclear magnetic resonance imaging (NMRI) has been used to measure the different parameters of the flow and transport system, including pore water velocities, volumetric water contents, and dispersion rates (Van As and Van Dusschoten, 1997). However, methods such as CT and NMRI can be cost prohibitive on an applied level. For further details on CT and NMRI, the reader is referred to Chapter 7.

Several studies have examined macropore characteristics in near-surface soils located on sites where both conservation and conventional tillage practices are followed. Edwards et al. (1988) studied the size, frequency, vertical and lateral distribution, and continuity of macropores and emphasized their importance in the assessment of the hydrologic characterization of the soil matrix. Their study and others indicate increased worm activity near the surface (to a depth of about 0.5 m) in no-till fields and sites where machinery and cattle were prohibited (Smettem, 1984; Edwards et al., 1988). Soil compaction by heavy equipment and cattle increases the density of the upper soil layer and destroys macropores, leading to potentially lower infiltration (Beven and Germann, 1982). Edwards et al. (1988) also reported that the number of pores varies inversely with pore diameter and that more pores exist at a lower depth than near the surface. The study found that nearly all of the wormholes were continuous, vertical, and did not branch or interconnect with other macropores. Many of the wormholes could be traced from the surface all the way to the bedrock, approximately 1 m below the surface.

Several researchers have attempted to define macropores in terms of specific pore sizes or equivalent moisture tension. These definitions are important both in experimental and theoretical quantification of the preferential flow. Beven and Germann (1982) and Shirmohammadi et al. (1991) defined the macropores and macroporosity as identified by different researchers (Table 8.1). For example, infiltration measurement techniques such as the tension infiltrometer method use capillary potential of -5 cm as a boundary potential value to separate macropore flow ($h > -5$ cm) and micropore flow ($h < -5$ cm) as reported by Wei (1999). It should be noted that setting such a boundary is arbitrary at best, as indicted in Table 8.1. Both texture and structure of soil may affect this boundary potential value, and more than two classes of soil pore sizes (macro-, meso-, and micropores) have also been suggested.

Irrespective of the specific size threshold used to define them, it is clear that macropores have a significant effect on rapid localized flow and transport in soil. A study by Watson and Luxmoore (1986) showed that $< 1\%$ of the soil volume was used in conducting 96% of the water flow. Bergström and Shirmohammadi (1999) showed that, depending on the profile depth, the structured clay soil had 25–61% less cross-sectional area contributing to the flow process under ponded flow conditions than the sandy soil with which comparison was made (Figures 8.1 and 8.2). Their results indicated that models

TABLE 8.1
Some Quantitative Definitions of Macropores and Macroporosity

Ref.	Capillary potential (kPa)	Equivalent diameter (μM)
Nelson and Baver (1940)	> −3.0	
Marshall (1959)	> −10.0	> 30
Brewer (1964)		
Coarse macropores		5000
Medium macropores		2000–5000
Fine macropores		1000–5000
Very fine macropores		75–1000
McDonald (1967)	> −6.0	
Webster (1979)		
Quoted in Mosley (1979)	> −5.0	
Ranken (1974)	> −1.0	
Bullock and Thomasson (1979)	> −5.0	> 60
Reeves (1980)		
Enlarged macrofissures		2000–10,000
Macrofissures		200–2000
Luxmoore (1981)	> −0.3	> 1000
Beven and Germann (1982)	> −0.1	> 3000

describing water and solute transport through structured soils must consider preferential flow–induced bimodal transport instead of using classical concepts of "piston flow" and "convective-dispersive" transport.

8.3 QUANTIFICATION OF PREFERENTIAL FLOW

8.3.1 EXPERIMENTAL

Both experimental and theoretical approaches have been used to quantify the preferential movement of water and solutes through porous media. Experimental approaches have used dye-staining techniques, tracer studies, and tension infiltrometer measurements to evaluate distribution and magnitude of macropore flow (Bouma, 1981; Watson and Luxmoore, 1986; Sollins and Radulovich; 1988; Quisenberry et al., 1994; Trapp et al., 1995). Many of these studies also refer to spatial distribution of macropores. For example, Trapp et al. (1995) used a cell lysimeter [modified form of a wick sampler used by Poletika and Jury (1994)] to evaluate the spatial and temporal distribution of water and solute movement in an undisturbed sandy soil in northern Germany. Their study indicated the occurrence of spatial variability of macropore flow even in small 0.64 m² area samples. They also showed that about 50% of the outflow was limited to 20% of the lysimeter cross-sectional area. The outflow from each of the 25 cells at the bottom of the sampler varied from < 2 to 500%

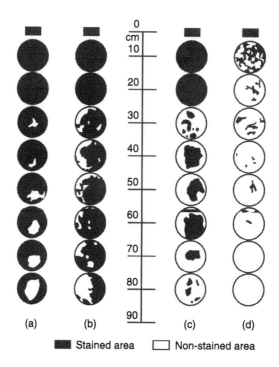

Stained area   ☐ **Non-stained area**

FIGURE 8.1 Cross-sectional representation of areal extent and location of flow pathways in Melby sand Monoliths [(a): ponded flow conditions; (b): transient flow conditions] and in Lanna clay monoliths [(c): ponded flow conditions; (d): transient flow conditions].

of the mean discharge (48 mm), indicating strong spatial variability. Wilson and Luxmoore (1988) summarized estimates of hydrologically active macroporosity (pore sizes > 1000 µm) and mesoporosity (pore sizes 10–1000 µm) from measurements with a tension infiltronmeter made on two forested sites. They concluded that regardless of the tension under which infiltration tests were conducted, infiltration is a stochastic process with respect to spatial variability. They also concluded that although macroporosities play a major role in transport processes, mesopores may still conduct most rainfall during a storm event, without macropores filling with water and contributing. However, their study and other previously referenced work failed to relate the areal extent of macropore flow.

Bergström and Shirmohammadi (1999) used Acid Red–stained water with a concentration of 3.8 g/L to evaluate the areal extent of the macropores in both a sandy soil and a structured clay soil. They used monolith lysimeters to conduct this study. Their results showed that under the ponded conditions the structured clay soil displayed 25–61% less cross-sectional area contributing to the flow process than the sandy soil (Figures 8.1 and 8.2). Their study also showed that despite lower infiltration rates in the clay soil than in the

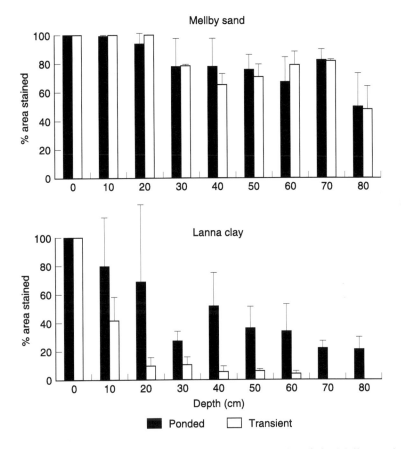

FIGURE 8.2 Average percent stained area at different depths of the Melby sand and Lanna clay monoliths for ponded and transient flow conditions. Bars represent standard deviations ($n = 2$–3).

sand, arrival of the dye tracer at the percolation boundary (at 1 m depth) was very fast in the clay soil.

Wei (1999) used a tension infiltrometer to quantify the significance of macropore flow in a fine sandy soil in the coastal plain of Maryland. Her measurements were conducted in both conventionally tilled and untilled fields. She measured the infiltration at both 0 and -5 cm tensions. At -5 cm, the effect of macropores was excluded, while at 0 tension all the pores, including macropores, were contributing to the flow. Her results indicated that macropore flow was significant, especially in the untilled field.

8.3.2 THEORETICAL

Mechanistic versus empirical, deterministic versus stochastic, single domain versus multidomain, and laboratory versus field — each pair of nominally exclusive categories may coalesce in limiting cases. As such, theoretical methods used to model preferential flow may be categorized into mechanistic, mechanistic-deterministic, mechanistic-stochastic, empirical-deterministic, and empirical-stochastic (Clark, 1973; Nieber, 2001). Methods using mechanistic approaches have often failed to consider their applicability to field conditions and have just relied on describing the process based on the principles of fluid dynamics. Such methods have also failed to consider the parameter uncertainty in their application to spatially heterogeneous porous media (Nieber, 2001; Vanderborght et al., 1997; McCoy et al., 1994).

8.3.2.1 Mechanistic, Single-Domain, Derived Stochastically (Averaging) with Deterministic Result

In an overview paper, Nieber (2001) proposed the use of fundamental fluid mechanics, combining the principle of conservation of mass and conservation of momentum (Navier-Stokes equation). These equations may be written for flow in individual pores of the porous media and then integrated over a representative elementary volume (REV) (see Chapter 2) of porous media to yield volume-averaged results. In his review, the volume-averaged equations for the conservation of mass and Navier-Stokes are as follows:

$$\frac{\partial \theta}{\partial t} = -\left(\frac{\partial q_x}{\partial x} + \frac{\partial q_y}{\partial y} + \frac{\partial q_z}{\partial z}\right) \tag{8.1}$$

$$\rho\left(\frac{\partial q_z}{\partial t} + q_x\frac{\partial q_z}{\partial x} + q_y\frac{\partial q_z}{\partial y} + q_z\frac{\partial q_z}{\partial z}\right) = -\frac{\partial P}{\partial x} + \mu_{eff}\left(\frac{\partial^2 q_z}{\partial x^2} + \frac{\partial^2 q_z}{\partial y^2}\right) - \frac{\mu_f}{k}q_z \tag{8.2}$$

where

θ	= volumetric water content $[L^3\ L^{-3}]$
x, y, and z	= spatial coordinates $[L]$
t	= time $[T]$
q_x, q_y, and q_z	= flux of water in x, y, and z coordinates, respectively $[L\ T^{-1}]$
ρ	= density of water $[M\ L^{-3}]$
P	= water pressure $[N\ L^{-2}]$
μ_{eff}	= effective dynamic viscosity of water $[M\ T^{-1}\ L^{-1}]$
μ_f	= dynamic viscosity of water $[M\ T^{-1}\ L^{-1}]$
g	= acceleration due to gravity $[L\ T^{-2}]$
k	= intrinsic permeability of porous media $[L^2]$

Equation (8.2) is the Navier-Stokes equation in vertical (z) direction volume averaged over a REV with x, y, and z coordinates. The terms on the left-hand side of this equation are the inertia terms, and the terms on the right are the force terms composed of pressure, viscosity, and gravity. In deriving Darcy's (1856) law from the Navier-Stokes equation, it is assumed that the inertia terms to the left of Eq. (8.2) are zero, and in effect pressure force and gravitational force balance the viscous force imposed between the fluid and the solid (Bear, 1972). Such assumptions lead to the classical simplified Darcy's equation and, if substituted into the conservation of mass equation [i.e., Eq. (8.1)], results in the Richards (1931) equation. While the resulting classical form of Darcy's equation is applicable to homogeneous media displaying piston flow, its application to soils displaying preferential flow fails. Additional research is needed to apply this strategy to the development of more general flow equations applicable to macroporous systems.

It should be noted that Nieber (2001) points out several cases where Richards equation has been successful in modeling preferential flow (e.g., Gerke and van Genuchten, 1993; Nieber, 1996; Ju and Kung, 1997). However, he relates these successes to the adjustment of model parameters representing the medium instead of proper mathematical representation. He further points out that using the full Navier-Stokes equations may improve the general representation of the preferential flow. His suggestion may be true if we were able to quantify the necessary input parameters in x, y, and z flux directions so that they can represent spatial and temporal variability. Such detailed parameterization being almost impossible for field conditions due to the required cost and time, quantification of preferential flow with even full-form Navier-Stokes equations seems a distant possibility.

8.3.2.2 Empirical Single-Domain, Deterministic

Germann and DiPietro (1999) represented the flux law by a power law equation rather than Darcy's law, substituting the power law equation for flux into Eq. (8.1) and obtaining the kinematic wave equation. The power law function for flux is presented as

$$q = bw^a \qquad (8.3)$$

where a is a dimensionless exponent and conductance (b) is presented as:

$$b = \frac{g}{2a\eta} \left[\frac{A}{\ell}\right]^2 \qquad (8.4)$$

where g is the acceleration due to gravity, η ($\approx 10^6\, m^2\, s^{-1}$) is kinematic viscosity of water, and ℓ is the constant length of the mobile water per cross-sectional area, A. The parameter w is defined as:

$$w = \frac{\ell F}{A} \qquad (8.5)$$

where F is the average thickness of the mobile water film. It is obvious that defining flux by this power function introduces empiricism into the algorithms, thus requiring a great deal of parameterizations and assumptions. The kinematic wave form of the equation has successfully been used in modeling preferential flow in soils (Germann, 1985, 2001). However, Germann (2001) concluded that even the kinematic wave equation has limited predictive capability because of our lack of knowledge on interaction between diffusive and dissipative flow at practical macropore scale due to limitations of *in situ* parameterization. He goes ahead and raises hope for empirical representation of preferential flow, which may itself be a cost-prohibitive approach.

8.3.2.3 Mechanistic, Bidomain and Multidomain, Deterministic

A major effort has been placed into the application of the convective-dispersive equation to quantify the solute transport through soils displaying preferential flow. For chemically conservative, nonreactive chemicals, Van Genuchten and Wierenga (1976) improved the solute transport simulations by separating the flow regime into two components of mobile (that portion of flow and chemical moving through the pores of soil aggregates) and immobile (that portion of flow and chemical residing in the soil matrix). Gish and Jury (1983) extended the mobile-immobile concept introduced by Van Genuchten and Wierenga (1976) to represent the flow process for a nonconservative, chemically reactive compound in soils containing root channels. Finally, Gish et al. (1991) expressed the two-domain solute transport equation as follows:

$$\left(\theta_m + f_m \rho_b K_d\right)\frac{\partial C_m}{\partial t} + \left[\theta_{im} + (l - f_m)\rho_d K_d\right]\frac{\partial C_{im}}{\partial t}$$
$$= \theta_m D_m \frac{\partial^2 C_m}{\partial x^2} - \theta_m V_m \frac{\partial C_m}{\partial x} - \mu \theta_v C_m \qquad (8.6$$

where

θ_m = mobile volumetric water content [$L^3\ L^{-3}$]
θ_{im} = immobile volumetric water content [$L^3\ L^{-3}$]
C_m = solute concentration in the mobile region [$M\ L^{-3}$]
C_{im} = solute concentration in the immobile region [$M\ L^{-3}$]
f_m = fraction of the soil matrix in the macropore region
V_m = average water velocity in the mobile region [$L\ T^{-1}$]
K_d = linear adsorption coefficient

Steenhuis et al. (1990) divided the porous media in different pore groups in such a way that velocity of water in pore group, say P, is a factor, f, times greater than in pore group $P - 1$ as:

$$f = V_P/V_{p-1}; \quad P = 1,\ldots,N \qquad (8.7$$

where f is an integer with a value of two or greater and V_P is the velocity for pore group P. This procedure uses piecewise linear function to compute hydraulic conductivity. Nijssen et al. (1991) used piecewise linear function up to a moisture content θ_f with hydraulic conductivity, K_f. However, they defined the velocity for the macropore group as:

$$V_N = \frac{K_s - K_f}{\theta_s - \theta_f} \tag{8.8}$$

where K_s and θ_s are hydraulic conductivity and moisture content at saturation, respectively, and K_f and θ_f are hydraulic conductivity and moisture content at about field capacity level termed as the highest moisture content, respectively.

For solute transport, Steenhuis et al. (2001) divided the porous media into a distribution zone and a conveyance zone. Then, they applied the convective-dispersive equation to the conveyance zone with the velocity of solute defined by an adjusted flux (q) considering the mobile region fraction coefficient.

Flow in structured soils at the macroscopic scale is dominated by either the capillary equilibrium or momentum (Germann, 2001). The capillary equilibrium concept has been used in the MACRO model by Jarvis et al. (1991), in which the water content above a threshold value of a highly saturated soil (e.g., water content corresponding to the matrix potential greater than $-5\,\mathrm{cm}$) may drain freely and with high speed. In the momentum-dominated flow regime, water is lost to the smallest pores and these losses depend directly on the pressure gradient between macropore surfaces ($\Psi \approx \mathrm{o}$) and the soil matrix and inversely on the momentum of preferential flow. This phenomenon causes water to partially overshoot capillarity.

Montas and Shirmohammadi (2001) used a first-order formulation of nonequilibrium potential and conductivity decay to describe the redistribution of water after a preferential infiltration event. They used mean, variance, standard deviation, skewness, and lag auto-covariance to describe the unsaturated hydraulic conductivity over a REV. Such an approach seemed to replicate the trends of detailed simulations in moisture redistribution in a stochastic soil well, but underestimated the initial decay rate during nonequilibrium cases. They concluded that this approach may overestimate the preferential flow. Montas et al. (2000) demonstrated that deterministic bicontinuum and volume-averaged stochastic approaches to solve transport are equivalent to one another. However, the formulation of equivalence in their study holds strictly only for steady, saturated flow conditions in a media where heterogeneities are in a plane orthogonal to the flow direction.

Shirmohammadi et al. (1991) modeled infiltration by considering the observed infiltration (i_o) rate to be the sum of the infiltration rate through soil matrix (i_m) and the rate through the macropores (i_p). Then, they used the modified Darcian flux equation and defined the matrix infiltration (i_m). For the macropore component, they used the Darcian flux as well, but let the macropore-saturated hydraulic conductivity be determined by Hagen-Poiseville's definition as:

$$K_{sP} = \frac{\beta R^2 \rho g}{8\mu}$$ (8.9

where β is a coefficient that takes the effect of variations in R (radius of individual pore) and channel tortousity into account, ρ is the density of liquid, g is acceleration due to gravity, and μ is the dynamic viscosity of the liquid. However, definition of R (e.g., size and distribution) may require the use of a stochastic function of some sort of field measurements. One may also use the pore diameters and classification given by Beven and Germann (1982).

8.3.2.4 Mechanistic, Single-Domain, Stochastic

Stochastic approaches have also been used to quantify water and solute transport through soils displaying preferential flow (Jury, 1982). In his study, Jury used the transfer function method (TFM) to predict mean values of solute concentration as a function of depth and time. He used a Monte Carlo simulation approach to develop a probability density function representing water input at the surface of the profile. Coyne (1999) and Shirmohammadi et al. (2001) used a modified form of the Jury (1982) method and described atrazine concentration on a field-scale at different depths (up to 180 cm) and at different times of the year. They showed that using the modified TFM provided more reasonable results in predicting atrazine transport through a sandy loam soil than the GLEAMS model (a deterministic model without consideration for preferential flow) and the MACRO model (a deterministic model with consideration of preferential flow). They concluded that amongst the two deterministic models, MACRO performed more reasonably than GLEAMS. They also concluded that a stochastic approach is much stronger in capturing the field heterogeneity than both of the deterministic models.

8.3.2.5 A New Three-Domain Infiltration Concept for
Structured Soils

Figure 8.3 shows a new concept whereby the flow through the vadose zone is considered to take place in three domains: i.e., micropores (immobile zone), mesopores (slow flow region) and macropores (fast flow region), Shirmoham-madi and Montas (2003). This concept was first presented by Shrimohammadi et al. (1991) for flow in two domains. Dividing the vadose zone in this way is similar to the concept used in the pesticide leaching model (PLM) by Hall (1994). However, PLM considers that the infiltrated component of rainfall goes into either mesopores or macropores and allows the water to move into the immobile zone through diffusion.

This new concept assumes that rainfall infiltrates into the portion of soil matrix marked as the immobile region until the moisture content reaches field

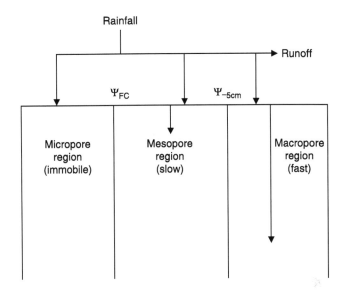

FIGURE 8.3 Concept of three-domain flow.

capacity at a pressure potential equal to Ψ_{FC}. For pressure potentials greater than Ψ_{FC}, mesopores take over and dominate the flow until the pressure potential in the soil reaches about $-5\,\text{cm}$ ($\Psi = \Psi_b = -5\,\text{cm}$). Finally, macropore flow dominates when the pressure potential becomes greater than $-5\,\text{cm}$. The preferential infiltration process is thus under the control of surface conditions.

Mathematically and mechanistically, the three-domain concept may be presented as follows:

For rainfall rates $I <$ matrix infiltration rate (i_m), all the rain will infiltrate into the soil profile and the infiltration rate will be equal to the rainfall rate. It is hypothesized that all the rain in this stage will infiltrate through micropores and will occupy the immobile region. Of course, this is only true if the pressure potential in the profile is less than Ψ_{FC}.

For $R > i_m$ and $\Psi < \Psi_{FC}$, infiltration will be distributed into both micropores (immobile region) and will also be conducted through mesopores (slow-flow region). Micropore flow takes place for as long as $\Psi < \Psi_{FC}$, and then mesopores will dominate when $\Psi > \Psi_{FC}$. Under such conditions, the total infiltration rate at any time will be equal to:

$$I(t) = i_m + i_{mes} \tag{8.10}$$

$$i_t = \left[K_m + \frac{K_m(\theta_{FC} - \theta_i)S_a}{(\theta_{FC} - \theta_i)Y_m} \right] + \left[K_{mes} + \frac{K_{mes}(\theta_b - \theta_{FC})S_a}{(\theta_b - \theta_{FC})Y_{mes}} \right] \tag{8.11}$$

If $R > i_m + i_{mes}$, then complete ponding will occur and infiltration is mostly dominated by the macropores (fast-flow region). The total infiltration rate for this condition may be defined as:

$$i_t = i_m + i_{mes} + i_{mac} \qquad (8.12$$

or

$$i_t = \left[K_m + \frac{K_m(\theta_{FC} - \theta_i)}{(\theta_{FC} - \theta_i)Y_m} \right] + \left[K_{mes} + \frac{K_{mes}(\theta_b - \theta_{FC})S_a}{(\theta_b - \theta_{FC})Y_{mes}} \right]$$
$$+ \left[K_{mac} + \frac{K_{mac}(\theta_s - \theta_b)S_a}{(\theta_s - \theta_b)Y_{mac}} \right] \qquad (8.13$$

where

$i_t, i_m, i_{mes}, i_{mac}$	= total, matrix, mesopore, and macropore infiltration rate
K_m	= saturated hydraulic conductivity of the immobile region and is assumed to be equal to K_{FC} (hydraulic conductivity at field capacity because complete saturation does not occur)
K_{mes}	= hydraulic conductivity of the mesopore region, which may be considered as the average hydraulic conductivity between Ψ_{FC} (pressure potential at field capacity) and Ψ_b (pressure potential at $-5\,\text{cm}$), the potential separating mesopore flow from macropore flow
K_{mac}	= hydraulic conductivity between $\Psi = \Psi_b$ and $\Psi = 0$
$\theta_i, \theta_{FC}, \theta_b,$ and θ_s	= volumetric water contents at the initial condition, field capacity, $\Psi = -5\,\text{cm}$, and $\Psi = 0$, respectively
$Y_m, Y_{mes},$ and Y_{mac}	= depth to the wetting front in micropores (immobile region), mesopores (slow-flow region), and macropores (fast-flow region), respectively
S_a	= average suction ahead of the wetting front

The method suggested here is in fact a modification of the approach proposed by Shirmohammadi et al. (1991), which will require proper parameterization for all three flow regions. One approach would be to define the micro-, meso-, and macropore zones by discrete distribution of each instead of a probability distribution function.

Table 8.2 presents the summary of the theoretical methods used to quantify preferential flow of water and solutes. The information presented in Table 8.2 may help the reader to select any given model based on the domain of use and availability of input parameters. It also provides proper references

TABLE 8.2
Prediction Methods Quantifying Preferential Flow

Method	Type	Sample input parameters	Ref.
Navier-Stokes equation	Mechanistic, single-domain, deterministic	Inertia terms and force terms (pressure, viscosity, and gravity)	Nieber (2001)
Power law function	Empirical, single-domain, deterministic	Exponent a and coefficient b based on the Kinenatic viscosity (η) and average thickness of mobile water (ℓ)	Germann and DiPietro (1999)
Extended convective-dispersive equation	Mechanistic, bi-domain (mobile-immobile) and multi-domain (different pore groups), deterministic	Fraction of soil matrix in the mobile region (f_m), average water velocity in the mobile region (V_m), density of liquid (ρ) and linear adsorption coefficient (K_d)	Van Genuchten and Wierenga (1976), Gish et al. (1991), Steenhuis et al. (1990), Montas et al. (2000)
Transfer function model (TFM)	Empirical, single-domain, stochastic	Probability density functions (pdfs) describing flow and chemical concentration means and variance in a control volume	Jury (1982), Coyne (1999), Shirmohammadi et al. (2001)
Modified Green and Ampt model	mechanistic, two- and three-domain, deterministic	Saturated hydraulic conductive of micropores (K_m), mesopores (K_{mes}) and macropores (K_{mac}); water content (initial θ_i; at field capacity, θ_{Fc}, at $\Psi = -5\,cm$; and at $\Psi = 0$), average suction ahead of wetting front (S_a)	Shirmohammadi et al. (1991), Shirmohammadi et al. (2003)

for each method, which may help the reader to obtain further details on each method.

8.4 SUMMARY AND CONCLUSIONS

Speedy arrival of water and chemicals through preferential pathways has attracted the attention of researchers in its quantification through both experimental and theoretical approaches. This chapter provides a review of some of the experimental and mathematical approaches used to quantify flow of water and chemicals in soils displaying preferential transport. An overview is given of the experimental techniques such as staining, tension infiltrometer, and lysimeter for identification and quantification of preferential flow and solute transport, as well as state-of-the-art mechanistic methods considering hydrodynamic principles. Mechanistic-empirical approaches such as the method by Germann (2001) and stochastic approaches such as the method by Jury (1982) are also discussed. Finally, the review provides a glance at the manipulated Darcian approaches (e.g., Shirmohammadi et al., 1991) and provides a new three-domain concept for quantifying the infiltration rate for soils displaying preferential flow. It is concluded that our handling of the preferential flow either fails the proper mathematical representation or fails the proper parameterization for proper representation of the system. One may use limited data methods, such as artificial neural networks (ANNs), to provide a proper parameter distribution for use in either mechanistic or stochastic models. However, using a black-box approach such as the ANN method may limit our understanding of the dynamic processes. It may be appropriate to quote the old Indian saying: "The one who rides a tiger feels strong but cannot get down; you wonder whether he is the captor or the captive." Similarly, our state of knowledge and handling of preferential flow is at the state of "wondering" at best; thus further attention is needed to develop appropriate mathematical algorithms and easy and cost-effective parameter-quantification techniques. Parameter quantification should be performed on a field scale for consideration of the type of preferential flow mechanism, its spatial and temporal extent, and variability. Also, a potential for using new technologies such as nanotechnology exists for proper and detailed identification and quantification of the preferential flow mechanism. Such technologies may lend a hand in parameter evaluation, thus improving the predictive capability of our models. It could also help to remediate the contaminated soils affected by preferential flow.

REFERENCES

Bear, J. 1972. Dynamics of fluids in porous media. American Elsevier, New York, NY, 764p.

Bergström, L. F. and A. Shirmohammadi. 1999. Areal extent of preferential flow with profile depth in sand and clay monoliths. J. Soil Contam., 8(6):637–651.

Beven, K. and P. Germann. 1982. Macropores and water flow in soils. Water Resour. Res., 18(5):1311–1325.

Brewer, R. 1064. Fabric and Mineral Analysis of Soils. John Wiley, New York.

Bouma, J. 1981. Soil morphology and preferential flow along macropores. Agric. Water Manage., 3:235–250.

Bullock, P., and A.J. Thomasson. 1979. Rothamsted studies of soil structure, 2, Measurement and characterization of macroporosity by image analysis and comparison with data from water retention measurements. J. Soil Sci., 30(3):391–414.

Capowiez, Y., A. Pierret, and C.J. Moran. 2003. Characterization of the three-dimensional structure of earthworm barrow systems using image analysis and mathematical morphology. Biol. Fert. Soils, 38:301–310.

Carsel, R.F., C.N. Smith, L.A. Mulkey, M.N. Larbor, and L.B. Baskin. 1985. The pesticide root zone model (PRZM): A procedure for pesticide leaching threats to groundwater. Ecological Modeling, 30:49–69.

Clark, R.T., 1973. A review of some mathematical models used in hydrology, with observations on their calibration and use. J. Hydrol., 19:1–20.

Coyne, K.J. 1999. Prediction of pesticide transport through the vadose zone using stochastic modeling. Unpublished M.S. thesis, University of Maryland, College Park, MD.

Darcy, H. (1856). Les Fontaines Publiques de la Ville de Dijion. Dalmont, Paris.

Davidson, J.M. and J.R. McDougal. 1973. Experimental and predicted movement of three herbicides in a water saturated soil. J. Environ. Qual., 2:428–433.

Edwards, W.M., L.D. Norton, and C.E. Redmond. 1988. Characterizing macropores that affect infiltration into notilled soil. Soil Sci. Soc. Am. J., 52:483–487.

Enfield, C.G., R.F. Carsel, S.Z. Cohen, T. Phan, and D.M. Walters. 1982. Approximating pollutant transport to ground water. Ground Water 20:711–722.

Gerke, H. and T.H.M. van Genuchten. 1993. A dual-porosity model for simulating the preferential movement of water and solutes in structured porous media. Water Resour. Res., 29:305–319.

Germann, P.F. 1985. Kinematic wave approach to infiltration and drainage into and from soil macropores. Trans. ASAE, 28:745–749.

Germann, P.F. 2001. Preferential flow in field soils. Proceedings of the International Symposium on Preferential Flow, Honolulu, Hawaii. American Society of Agric. Eng. (ASAE). pp. 1–9.

Germann, P.F. and L. DiPietro. 1999. Scales and dimensions of momentum dissipation during preferential flow in soils. Water Resour. Res., 35:1443–1454.

Gish, T.J., and W.A. Jury. 1983. Effect of plant roots and root channels on solute transport. Trans. ASAE, 26(2):440–444, 451.

Green, W.H. and G. Ampt. 1911. Studies of soil physics, Part I — The flow of air and water through soils. J. Agric. Sci. 4: 1–24.

Hall, D.G.M. 1994. Simulation of dichlorprop leaching in three texturally distinct soils using the pesticide leaching model. J. Environ. Sci. Health, A29(6): 1211–1230.

Hill, D.E. and J.-Y. Parlange. 1972. Wetting front instability in layered soils. Soil Sci. Soc. Am. Proc., 36:697–702.

Jarvis, N.J., P.-E. Jansson, P.E. Dik, and I. Messing. 1991. Modeling water and solute transport in macroporous soil. 1. Model description and sensitivity analysis. J. Soil Sci., 42:59–70.

Ju, S.-H. and K.-J.S. Kung. 1997. Steady-state funnel flow: Its characteristics and impact on modeling. Soil Sci. Soc. Am. J., 61:416–427.

Jury, W.A. 1982. Simulation of solute transport using a transfer function model. Water Resour. Res., 18:363–368.

Kung, K.J.S. 1990. Preferential flow is a sandy vadose zone. I. Field Observation. Geoderma, 46:51–58.

Kung, K.J.S. and Z.B. Lu. 1993. Using ground penetrating radar to detect layers of discontinuous dielectric constant. Soil Sci. Soc. Am. J., 57:335–340.

Lawes, J.B., J.H. Gilbert, and R. Warington. 1882. On the amount and composition of the rain and drainage water collected at Rothamsted. Williams, Clowes and Sons Lt., London.

Leonard, R.A., W.G. Knisel, and D.A. Still. 1987. GLEAMS: Groundwater loading effects of agricultural management systems. TRANSACTIONS of ASAE, 30(5):1403–1418.

Lindstrom, F.T., L. Boersma, and H. Gardiner. 1968. 2,4-D diffusion in saturated soils. A mathematical theory. Soil Sci., 105:107–113.

Luxmoore, R.J. 1981. Micro-, meso- and macraporosity of soil. Soil Sci. Soc. Am. J., 45, 671.

Marshall, T.J. 1959. Relations between water and soil. Tech. Comm. 50, Commonwealth Bur. Soils, Haprenden, U.K.

McCoy, H.J., C.W. Boast, R.C. Stehouwer and E.J. Kladivko. 1994. Macropore hydraulics: taking a sledgehammer to classical theory. In: R. Lal and B.A. Stewart (eds.), Soil Processes and Water Quality. Advances in Soil Science. Lewis Pub., CRC Press, Boca Raton, FL, pp. 303–348.

McDonald, P.M. 1967. Disposition of soil moisture held in temporary storage in large pores. Soil Sci., 103(2):139–143.

Montas, H.J. and A. Shirmohammadi. 2001. First-order modeling of conductivity decay during lateral redistribution in a stochastic soil. Proceedings of the International Symposium on Preferential Flow, Honolulu, Hawaii, January 3–5, 2001. Published by American Society of Agric. Eng. (ASAE), St. Joseph, MI, pp. 133–136.

Montas, H.J., A. Shirmohammadi, K. Haghighi, and B. Engel. 2000. Equivalence of bicontinuum and second-order transport in heterogeneous soils and aquifers. Water Resour. Res., 36(2):3427–3446.

Nelson, W.R. and L.D. Baver. 1940. Movement of water through soils in relation to the nature of the pores. Soil Sci. Soc. Am. Proceedings, 5:69–76.

Nieber, J.L. 1996. Modeling finger development and persistence in initially dry porous media. Geoderma, 70:209–229.

Nieber, J.L. 2001. The relation of preferential flow to water quality, and its theoretical and experimental quantification. Proceedings of the International Symposium on Preferential Flow, Honolulu, Hawaii, January 3–5, 2001. Published by American Society of Agric. Eng. (ASAE), St. Joseph, MI, pp. 1–9.

Nijssen, B.M., T.S. Steenhuis, G.J. Kluitenberg, F. Stagnitti, and J.-Y. Parlange. 1991. Moving water and solutes through the soil: testing of a preferential flow model. Proceedings of the National Preferential Flow Symposium (eds: Gish, T.J. and A. Shirmohammadi), Chicago, IL. December 16–17, 1991. Published by American Society of Agric. Eng. (ASAE), St. Joseph, MI, pp. 223–232.

Nofziger, D.L., and A.G. Hornsby. 1984. Chemical movement in soil. User's Guide, University of Florida, Gainesville, Florida.

Poletika, N.N. and Jury, W.A. 1994. Effect of soil surface management on water flow distribution and solute dispersion. Soil Sci. Soc. Am. J., 58, 999–1006.

Quinseberry, V.L., R.E. Phillips and J.M. Zelenik. 1994. Spatial distribution of water and chloride macropore flow in a well-structured soil. Soil Sci. Soc. Am. J., 58, 1294–1300.

Ranken, D.W. 1974. Hydrologic properties of soil and subsoil on a steep forested slope, Masters thesis, Oreg. State Univ., Corvallis, OR.

Reeves, M.J. 1980. Recharge of the English chalk, a possible mechanism. Eng. Geol., 14(4):231–240.

Richards, L.A. 1931. Capillary conduction of liquid in porous medium. Physics, 1:318–333.

Ritsema, C.J., L.W. Dekker, J.M.H. Hendrickx, and W. Hamminga. 1993. Preferential flow mechanism in a water repellent sandy soil. Water Resour. Res., 29:2183–2193.

Schumacher, W. 1864. Die Physik des Bodens. Wiegandt and Hempet, Berlin, 1864.

Shirmohammadi, A. and H.J. Montas. 2003. Mathematical representation of preferential flow: a review. ASAE Paper No. 033106, American Society of Agric. Eng. (ASAE), St. Joseph, MI.

Shirmohammadi, A., T.J. Gish, A. Sadeghi, and D.A. Lehman. 1991. Theoretical representation of flow through soils considering macropore effect. Proceedings of the National Preferential Flow Symposium (eds: Gish, T.J. and A. Shirmohammadi), Chicago, IL, December 16–17, 1991. Published by American Society of Agric. Eng. (ASAE), St. Joseph, MI, pp. 233–243.

Shirmohammadi, A., H. Montas, L. Bergström, K. Coyne, S. Wei, and T.J. Gish. 2001. Deterministic and stochastic prediction of atrazine transport in soils displaying macropore flow. Proceedings of the International Symposium on Preferential Flow, Honolulu, Hawaii, January 3–5, 2001. Published by American Society of Agric. Eng. (ASAE), St. Joseph, MI, pp. 133–136.

Shirmohammadi, A., T.J. Gish, D.E. Lehman, and W.L. Magette. 1989. GLEAMS and the vadose zone modeling of pesticide transport. ASAE Paper No. 89-2071, American Society of Agricultural Engineers, St. Joseph, MI 49085-9659.

Shirmohammadi, A., and R.W. Skaggs. 1985. Predicting infiltration for shallow water table soils with different surface covers. TRANSACTIONS of ASAE, 28(6):1829–1837.

Skopp, J. 1981. Comment of "micro-, meso- and macroporsity" of soil. Soil. Sci. Soc. Am. J. 45:1246.

Smettem, K.R.J. 1984. Soil water residence time and solute uptake, 3. Mass transfer in undisturbed soil cores. J. Hydrol., 67:235–248.

Sollins, P. and R. Radulovich. 1988. Effects of soil physical structure on solute transport in a weathered tropical soil. Soil Sci. Soc. Am. J., 52(4):1168–1173.

Steenhuis, T.S., J.-Y. Parlange, and M.S. Andreini. 1990. A numerical model for preferential solute movement in structured soils. Geoderma, 46:193–208.

Steenhuis, T.S., Y.-J. Kim, J.-Y. Parlange, M.S. Akhtar, B.K. Richards, K.-J.S. Kung, T.J. Gish, L.W. Dekker, C.J. Ritsema, and S.O. Aburime. 2001. An equation for describing solute transport in field soils with preferential flow paths. In: Proceedings of the International Symposium on Preferential Flow, Honolulu, Hawaii, January 3–5, 2001. American Society of Agric. Eng. (ASAE), St. Joseph, MI, pp. 137–140.

Trapp, G., Meyer-Windel, S., and Lennartz, B. 1995. Cell lysimeter for studying solute movement as influenced by soil heterogeneity. In: Pesticide Movement to Water, pp. 123–134 (Walker, A., Allen, R., Bailey, S. W., Blair, A.M., Brown, C.D., Gunther, P., Leake, C.R., and Nicholls, P.H., eds.). BCPC Monograph No. 62, Farham, UK.

Van As, H. and Van Dusschoten, D. 1997. NMR methods for imaging of transport processes in micro-porous systems. Geoderma, 80, 389.

Van Genuchten, M.T., and P.J. Wierenga. 1976. Mass transfer studies in sorbing porous media I. Analytical solutions. Soil Sci. Soc. Am. J., 40(4):473–480.

Vanderborght, J., D. Mallants, M. Vanclooster, and J. Feyen. 1997. Parameter uncertainty in the mobile-immobile solute transport model. J. Hydrol., 190: 75–101.

Wagenet, R.J. and J.L. Hutson. 1986. Predicting the fate of nonvolatile pesticides in the unsaturated zone. J. Environ. Qual., 15:315–322.

Watson, K.W. and R.J. Luxmoore. 1986. Estimating macroporsity in a forest watershed by use of a tension infiltrometer. Soil Sci. Soc. Am. J., 50:578–582.

Webster, J. 1974. The hydrologic properties of the forest floor under beech/podocarp/ hardwood forest, North Westland. Master's thesis, 77 pp., Univ. Canterbury, New Zealand.

Wei, S. 1999. Pesticide transport and significance of macropore flow in the coastal plain soils of Maryland. Unpublished Ph.D. dissertation, University of Maryland, College Park.

Wilson, G.V. and Luxmoore, R.J. 1988. Infiltration, macroporosity, and mesoporosity distributions on two forested watersheds. Soil Sci. Soc. Am. J., 52:329–335.

9 Field Methods for Monitoring Solute Transport

Markus Tuller and Mohammed R. Islam
University of Idaho, Moscow, Idaho, U.S.A.

CONTENTS

9.1   INTRODUCTION

Rapidly progressing problems with contamination of natural resources and associated public health concerns emphasize the growing demand for advanced capabilities to quantify fate and transport of contaminants in the vadose zone. Automated measurement technologies with high spatial resolution are required for continuous monitoring of subsurface chemical transport below potential point and nonpoint contaminant sources such as industrial and municipal waste disposal sites, accidental chemical spills, or agrochemicals applied over extended areas. Basic understanding of complex interactions between physical, chemical, and biological processes and the ability to early detect and assess the extension of migrating contaminant plumes is crucial for initiation of advanced remediation measures. Despite significant new developments of advanced measurement technologies and progress in characterizing solute transport processes over the last few decades, prediction of field-scale phenomena remains a challenge due to the inherent heterogeneity of the vadose zone.

In this chapter we (1) present commonly applied direct methods for *in situ* solute extraction, (2) provide a comprehensive overview of common and advanced indirect methods and sensors for monitoring solute transport under field conditions, and (3) discuss their application based on two case studies. A comparison of various methods and an outlook regarding future research opportunities are provided throughout and at the end of the chapter.

9.2   DIRECT EXTRACTION OF SOIL SOLUTION

The importance of collecting soil solution for environmental studies was recognized long ago by Joffe (1932), who described the soil solution as the "blood circulating in the soil body." Soil scientists, hydrologists, geochemists, ecologists, engineers, and health safety specialists have a major interest in the chemical composition of the soil solution, as it provides crucial information regarding distribution of plant nutrients and hazardous chemicals in the soil profile. Water quality monitoring below waste disposal sites, for example, is important for the detection of contaminant plumes migrating from leaking liners towards the groundwater table and allows early initiation of remedial measures to prevent extended pollution of aquifers. In order to determine the chemical composition of the soil solution, a wide variety of extraction techniques and devices have been developed in recent decades. In the following sections we will focus on the equipment and techniques used to extract solution. For chemical analyses, interested readers are referred to Chapter 14.

9.2.1   Field Methods for *In Situ* Extraction of Soil Solution

9.2.1.1   Suction Cups

Briggs and McCall (1904) were among the first to introduce a soil-water extraction method through porous ceramic cups. Numerous modifications to the initial design of the suction cup were developed since its invention almost one century ago. Among those modifications was the introduction of automated soil solution samplers by Cole (1968). Chow (1977) developed a vacuum sampler that automatically shuts down after collecting a specific volume of soil solution. Further improved samplers were introduced by Parizek and Lane (1970), Wood (1973), and Stone and Robl (1996), who designed a heavy duty device to withstand soil compaction due to farm equipment.

The most commonly applied devices for collection of solution from unsaturated soils are vacuum soil water samplers (Rhoades and Oster, 1986), such as suction cups or suction lysimeters. These instruments operate under the same principle where a porous material (cup or plate) is brought in hydraulic contact with the surrounding soil, and evacuation of the sampler to a pressure slightly below the soils matric potential induces a pressure gradient and flow of solution into the sampler and collection containers. It is important to manually or automatically adjust the applied vacuum based on tensiometer measurements to prevent high gradients and the development of preferential flow paths towards the cup. Therefore, soil water samplers are commonly installed in combination with tensiometers, or sampler and tensiometer are combined in one instrument, as discussed shortly. The potential field that develops around a suction cup was measured with tensiometers by Krone et al. (1951).

The time requirement for collection of soil solution depends on the volume necessary for chemical analysis, the hydraulic conductivity and water content (matric potential) of the soil, and the applied gradient (Rhoades and Oster, 1986). A sandy soil close to field capacity will provide sufficient sample volume within a few hours. Note that automated sampling stations (Cepuder and Tuller, 1996) may be conceptualized for continuous sampling within the limitations discussed below. A typical soil water collection system contains three main functional units: the suction cups or plates, sampling bottles, and a vacuum container connected to a vacuum pump (Figure 9.1).

The applicable range of soil water samplers is limited to suction values (vacuum) of less than 100 kPa, i.e., 1 bar or 10 m head of water. The application range mainly depends on the bubbling pressure (air entry value) of the porous material. The bubbling pressure is defined as the pressure required for evacuating the largest pore of the cup or plate. If the applied vacuum is less (more negative) than the bubbling pressure, air enters the sampler and the hydraulic contact with the surrounding soil ceases. The largest possible pore radius (r) for the material to remain saturated to a certain pressure (vacuum) is simply obtained by rearranging the capillary rise equation:

$$r = \frac{2\sigma \cos \gamma}{\rho_w gh} \tag{9.1}$$

where σ is the liquids surface tension (N m^{-1}), γ is the contact angle, ρ_w is the liquid density (kg m^{-3}), g is the acceleration of gravity (m s^{-2}), and h is the bubbling pressure (m).

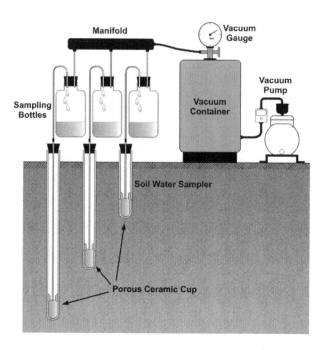

FIGURE 9.1 Sketch showing a common setup with soil water samplers (suction cups).

The primary differences between various suction soil water samplers are shape and size of the devices and the chemical and physical properties of the porous materials used to establish hydraulic contact with the surrounding soil. A vast number of porous materials, such as ceramic (widely used), polytetra-fluoroethylene (PTFE), polyethylene (PE), stainless steel, nylon, PVC, PP, PVDF, Teflon, or glass, may be used for the suction cups or plates. There are no general guidelines for the choice of material; therefore, users should be aware of certain limitations associated with porous materials previously used.

The results of different studies assessing the applicability of ceramic samplers are somewhat conflicting. This is probably related to differences in chemical composition and physical properties of the ceramic material used for construction of the sampling devices and the difference in pH and chemical composition of the ambient soil solution.

Wenzel et al. (1997) reported adsorption of significant amounts of Cd, Co, Mn, Ni, or Zn on ceramic materials. Less soluble elements, such as Cu, Cr, As, and Pb, were almost completely removed from the solution at pH values ranging from 4 to 6. In contrast, nylon membranes did not bias trace element concentrations within this pH range. Beier et al. (1992) did not find any significant differences for Na^+, K^+, Ca^{2+}, Al^{3+}, and NH_4^+ in low concentrated solutions that were collected under controlled field conditions with ceramic and PTFE cups. PTFE is assumed to be chemically inert, which means that the material should not cause changes of the sampled solution. Other studies (Bottcher et al., 1984; Grover and Lamborn, 1970; Hansen and Harris, 1975;

Zimmermann et al., 1978) indicate significant alteration of the solution due to adsorption of NH_4^+, PO_4^{3-}, and K^+ on ceramic materials. Silica flour, often applied to establish hydraulic contact between sampler and surrounding soil (Linden, 1977), adsorbs trace metals as reported by James and Healy (1971). Ceramic or aluminum oxide cups are dissolved at pH values lower than 4 and subsequently release Al and other substances into the solution sample (Beier et al., 1989; Neary and Tomassini, 1985; Raulund-Rasmussen, 1989).

New suction cups often release contaminants left from the production process. Therefore, they should be thoroughly cleaned with diluted acid (i.e., HCL) as recommended by Grossmann et al. (1987) and Litaor (1988). Cleaning is crucial particularly for trace element investigations as reported in a number of studies (Creasey and Dreiss, 1988; Neary and Tomassini, 1985; Wolff, 1967; Wood, 1973). Soil microorganisms colonizing sampling devices may clog pores with biofilms that can adsorb both biodegradable and nonbiodegradable chemicals. Sampler pretreatment with a biocide (sodium hypochlorite) or a bacteriostat (copper salt) helps in preventing microbial activity and leads to more accurate results (Lewis et al., 1992).

Before installation, suction devices should be thoroughly tested under controlled laboratory conditions. To facilitate undisturbed operation of sampling stations containing arrays of samplers, it is recommended to only use suction cups with uniform hydraulic properties. It is also important to saturate the porous cups for 24–48 hours prior to installation. Before a suction cup can be put in place, a hole with a diameter slightly larger than the sampler diameter is cored to the intended sampling depth (Stone and Robl, 1996). Soil material collected close to the bottom of the hole is sieved and mixed to a slurry that is poured back to refill the first 10–20 cm. Now the sampler is gently pushed into the slurry that establishes tight hydraulic contact between the saturated porous cup and the surrounding soil (Alberts et al., 1977; Barbee and Brown, 1986; Starr, 1985; Wood, 1973). For exact sampler placement, it is advantageous to mark the sampler and auger beforehand. As mentioned above, in some cases silica flour is used instead of the ambient soil to improve the contact (Morrison and Szecsody, 1987; Parizek and Lane, 1970; Smith and Carsel, 1986). Especially in expansive soils, caution should be taken to prevent water from seeping through gaps between the sampler and auger hole. In such cases it is recommended to pour and compact a bentonite collar around the top portion of the sampler. Note that suction cups might be buried or installed horizontally from a trench (Cepuder and Tuller, 1996) that is refilled. After setting up the sampling station and evacuating the samplers, it is recommended to allow a stabilization phase and discard the first few samples.

Macropores or highly structured coarse soils may cause significant problems for the application of suction cups. Preferential flow might bypass the samplers and prevent detection of contaminant fronts especially after high-intensity precipitation events (Grossmann and Udluft, 1991). Under such conditions it is advantageous to use suction lysimeters, which cover a larger soil volume. Several investigators (Barbee and Brown, 1986; Haines et al., 1982; Shaffer et al., 1979; Shuford et al., 1977) report considerable differences between samples from suction cups and lysimeters under such highly hetero-geneous conditions.

9.2.1.2 Combined Solution Sampling — Tensiometer Probes

Characterization of solute transport in soils requires measurement of spatial and temporal changes of the soil solute concentration and soil water status (matric potential), which is commonly achieved with suction samplers, such as introduced in the previous section, and tensiometers, respectively. Due to a similar basic design of tensiometers and suction samplers, it is convenient to combine these devices into one individual probe.

A commonly applied device for monitoring soil water status is the tensio-meter, which was introduced as early as 1908 by Burton E. Livingston with advanced implementation of similar concepts for "measuring the capillary lift of soils" by Lynde and Dupre in 1913 (Or, 2001). A typical tensiometer consists of a porous cup (usually made of ceramic with very fine pores) connected to a vacuum gauge (mechanical or electronic transducer) through a rigid water-filled tube (Figure 9.2). The porous cup is placed in intimate contact with the bulk soil at the depth of measurement. When the matric potential of the soil is lower (more negative) than the equivalent pressure inside the tensiometer cup, water moves from the tensiometer along a potential energy gradient to

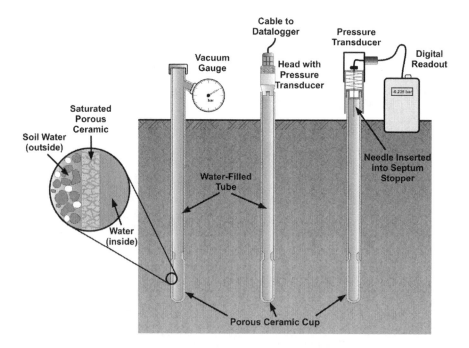

FIGURE 9.2 Sketch illustrating commonly applied tensiometers for matric potential measurement with vacuum gauges and electronic pressure transducers.

the soil through the saturated porous cup, thereby creating suction sensed by the gauge. Water flow into the soil continues until equilibrium is reached and the suction inside the tensiometer equals the soil matric potential (i.e., when the driving force dissipates). When the soil is wetted, flow may occur in the reverse direction; that is, soil water enters the tensiometer until a new equilibrium is attained. The field installation of tensiometers and combined probes closely follows the procedure discussed for suction cups (see above).

Moutonnet et al. (1989) introduced a modified tensiometer termed Tensionic that allows measurement of matric potential and extraction of soil solution (Figure 9.3a). The ceramic cup is sealed immediately after entering the PVC shaft portion of the tensiometer, with two tubes guided to the soil surface for priming and sample extraction and a third tube leading to a sealed tensiometer compartment in the upper portion of the shaft (Moutonnet et al., 1993). Following installation the tensiometer is primed with deaired and deionized water. After equilibration of the pressure inside the tensiometer with the matric potential of the surrounding soil the Tensionic operates in tensiometer mode, and the matric potential is manually or automatically recorded using vacuum gauges or pressure transducers in combination with dataloggers, as depicted in Figure 9.2. At the same time, ions present in the soil solution diffuse through the porous cup, and after some time

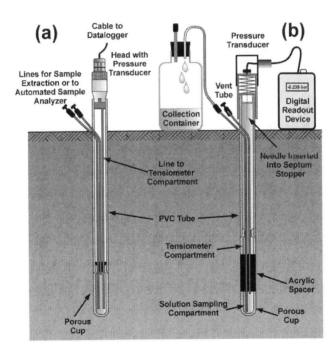

FIGURE 9.3 Sketch showing (a) the Tensionic probe (Moutonnet et al., 1993), and (b) a probe with separated sampling and tensiometer compartments (Essert and Hopmans, 1998).

(Moutonnet et al. 1993 report 8–10 days for NO_3^-) the water inside the tensiometer is in chemical equilibrium with the surrounding pore water. After chemical equilibrium is attained the sample is extracted, and the tensiometer flushed and refilled with deaired and deionized water. Moutonnet et al. (1993) and Moutonnet and Fardeau (1997) used automated Tensionic probes to determine concentrations of NO_3^-–N, NO_2^-–N, and NH_4^+–N under maize. Similar devices were applied by Morrison et al. (1983) and Rehm et al. (1986) to characterize contaminant migration under waste disposal sites. Major drawbacks of the Tensionic are the lack of control over the sample volume that is largely dependent on soil water content (Saragoni et al., 1990) and uncertainty regarding the required time for chemical equilibration.

Essert and Hopmans (1998) took a different design approach when combining tensiometer and soil water sampler. They separated the porous cup with an acrylic barrier into two compartments (Figure 9.3b). The bottom compartment is used for sample extraction, while the top compartment operates as a tensiometer. Interactions between the two compartments and operation modes may occur due to (1) reduction of the matric potential caused by solute extraction and (2) change of solute concentration due to diffusion of solutes between soil and tensiometer compartment. To overcome these biases Essert and Hopmans (1998) recommend separating the compartments with a 10 cm spacer. Though this distance is somewhat arbitrary, it is supported by theoretical considerations (Warrick and Amoozegar-Fard, 1977).

9.2.1.3 Suction Lysimeters

Lysimeter studies date back to the late eighteenth century when scientists investigated the fate of precipitation in soils (Joffe, 1932). The term lysimeter, originating from the Greek words *lysi* (loosening) and *meter* (measuring), is somewhat misused today. Lysimeters were originally developed to study the complex soil-plant-atmospheric relationships, with solute transport being only one component. In its original definition a lysimeter is a large soil block surrounded by a casing, with its lower boundary separated from the parent material (Bergström, 1990) and commonly mounted on a large balance for monitoring evaporation or evapotranspiration as a function of atmospheric conditions and soil water status. Numerous different lysimeter designs and sizes with varying boundary conditions and application areas are reported in the literature. In this section we will briefly describe lysimeters designed for collection of soil solution and refer interested readers to additional publications when appropriate.

We distinguish between two basic types of lysimeters that either contain a soil monolith (undisturbed) or are filled with (disturbed) soil. They may reach from the soil surface to depths of 2–3 m or may be buried (Cepuder and Tuller, 1996) or installed from the sidewall of a trench. They may be isolated from the surrounding soil via impermeable sidewalls or in hydraulic contact with the parent material. Casings may be round, square, or rectangular, and made of concrete, steel, fiberglass, or PVC (ASTM, 1998; Best and Weber,

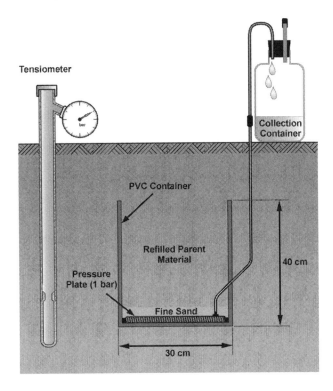

FIGURE 9.4 Sketch showing a refilled suction lysimeter installed in combination with a tensiometer for monitoring soil matric potential. A similar design was used by Cepuder and Tuller (1996) in nitrate leaching studies.

1974; Furth, 1985; Weber, 1995). Collection of soil solution is commonly achieved by means of gravity drainage or through a porous material via suction (Figure 9.4), similar to the previously discussed suction samplers (note that suction samplers are sometimes referred to as lysimeters).

Suction lysimeters (Figure 9.4) as well as gravity drainage lysimeters have been applied in numerous studies, ranging from monitoring transport of agrochemicals and water movement (Bergström, 1990; Cepuder and Tuller, 1996; Dolan et al., 1993; Jemison and Fox, 1994; Joffe, 1932; Karnok and Kucharski, 1982; Kilmer et al., 1944; Kohnke et al., 1940; McMahon and Thomas, 1974; Moyer et al., 1996; Tyler and Thomas, 1977; Winton and Weber, 1996) to colloid-facilitated transport of organic compounds and heavy metals (Thompson and Scharf, 1994), and fate and cycling of N^{15} (Reeder, 1986).

9.2.1.4 Passive Capillary Samplers

Passive capillary samplers (PCAPS) utilize tension exerted by a hanging wick to passively extract solution from the soil above a sampling pan (Figure 9.5). PCAPS first introduced by Brown et al. (1986) show distinct advantages, such

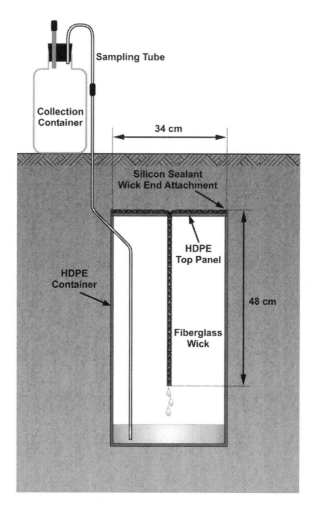

FIGURE 9.5 Sketch showing the setup of a passive capillary sampler. (From Louie et al., 2000.)

as potential for measuring water flux density, when compared to suction cups or lysimeters (Selker, 2002). Recent progress of the PCAPS technique includes the development of advanced wick selection design equations (Boll et al., 1992; Knutson and Selker, 1994; Rimmer et al., 1995) and wick pretreatment methods (Knutson et al., 1993).

PCAPS are designed for long-term operation using environmentally stable, nonadsorbing materials such as stainless steel, fiberglass, and HDPE (Topp and Smith, 1992). The major component is a fiberglass or HDPE container that supports a stainless steel or HDPE top panel that is divided into multiple compartments with a circular opening for the wick in the center of each section (Louie et al., 2000). The wick is cut to the desired length, and one end is separated into individual strands and cleaned by kiln combustion according to Knutson et al. (1993). The wick is guided through the center hole, and the filaments of the open end are spread out radially on the top of the panel and the ends are glued into place with silicone (Figure 9.5)

The wicks utilized for PCAPS are custom products for furnace isolation. They are available in a wide range of dimensions, weaves, and densities that can be optimized for a wide range of soil textures (Selker, 2002). The applicable wick materials show exponential relationships between hydraulic conductivity and pressure that were tabulated by Knutson and Selker (1994). From these tabulated values and the determined unsaturated hydraulic conductivity of the parent soil, design criteria for PCAPS (e.g., length, type, and number of wicks for sampling a given area) can be computed. Some caution regarding the wick material is required (Selker, 2002). Binding agents (e.g., starch) applied during the manufacturing process may reduce the wettability of wicks and lead to problems with inducing tension to the soil water. Kiln combustion at 450°C was found to be the most effective procedure for coating removal (Knutson et al., 1993). This, however, might induce contamination of the wick surface with ash, which is undesirable especially if the collected solution is intended to be analyzed for trace elements. As for all other samplers, thorough laboratory testing and cleaning of the wick with acid and deionized water is imperative before field installation.

PCAPS are commonly installed from trenches. A tunnel only slightly larger than the sampling device is excavated perpendicular to the trench at the desired sampling depth. After filling the top panel with slightly compacted native soil, the PCAP is carefully pushed into the tunnel to its desired location and elevated with wedges to achieve tight hydraulic contact between the soil layer on top of the sampler and the tunnel ceiling (Louie et al., 2000). A bentonite layer is applied to hydraulically isolate the sampler from the trench. After guiding the tubing for sample extraction to the soil surface the trench is refilled and compacted. The soil solution collected on the bottom of the sampler is extracted with a manual or battery-operated vacuum pump at predetermined time intervals or dependent on monitored matric potential or soil water content.

Passive capillary samplers were successfully tested in laboratory experiments (Knutson and Selker, 1996; Rimmer et al., 1995) and applied in a number of field trials (Brandi-Dohrn et al., 1996; Louie et al., 2000; Holder et al., 1991; Boll et al., 1991).

9.2.1.5 Capillary Absorbers

When two porous materials with differing water potential energy (e.g., filter paper and soil) are brought in close hydraulic contact, water will flow from the medium with higher potential energy to the medium with lower potential energy in pursuit of equilibrium state. The driving force is the gradient due to differences in matric and osmotic potentials between soil and filter paper. This physical phenomenon is utilized in capillary absorbers, where a porous

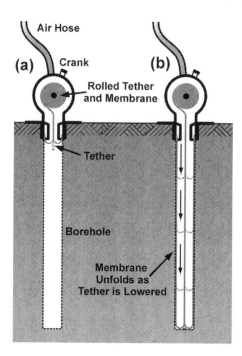

FIGURE 9.6 Deployment sequence of capillary absorbers: (a) canister placed on surface casing, and (b) membrane lowered into the borehole and pushed against the borehole wall using pressurized air. (From Koglin et al., 1995.)

membrane (absorber) is brought in close hydraulic contact with the wall of a borehole (Figure 9.6). Due to the difference in water potential, soil solution is wicked into the absorber. After allowing equilibration with the surrounding soil, the membrane is retrieved and the solution extracted for chemical analyses (Keller and Hendrickx, 2002).

The membrane-soil system effective permeability controls the flow rate of the liquid into the absorbing material. This leads to low transfer rates and extended equilibration time under dry conditions (Keller and Travis, 1993). In moist soils with high hydraulic conductivity, on the other hand, the absorber may quickly wick a sample. To estimate equilibration time, it is useful to attach a wire pair to the absorber and monitor changes in electrical resistance as the absorber wicks the soil solution. From the slope of the resistance-time relationship, one can approximately estimate when equilibrium is reached (Keller and Hendrickx, 2002).

A number of issues require consideration for successful operation of capillary absorbers. It is crucial to establish tight contact between soil and absorber to prevent evaporation losses during equilibration, which could cause an increase in solute concentration. Furthermore, it is important to prevent evaporation losses and cross-contamination during absorber retrieval.

An advanced installation method that eliminates potential problems with absorber placement, borehole isolation, and absorber retrieval employs an impermeable balloon-shaped liner with the absorber material glued to the liner surface (Keller and Hendrickx, 2002; Koglin et al., 1995). (Note that absorbers may be organized as circular or annular patches or cover the entire liner surface.) The inverted liner and attached tether (i.e., turned outside in) are wound onto a reel (Figure 9.6a). When the liner unrolls from the reel as it is lowered into the borehole with the tether, it is everted so that the absorber faces the borehole wall (Figure 9.6b). Pressurized air or in certain cases water is used to establish tight contact between absorber and wall and to isolate individual absorbers if patches are used. When retrieving the absorber, the liner is inverted again. This avoids cross-contamination and exposure of personnel to hazardous chemicals.

9.2.2   SOLUTION EXTRACTION FROM SOIL SAMPLES

For completeness this section contains a brief discussion regarding laboratory solute extraction from soil samples collected in the field by means of coring or simple excavation. Note that a distinction is made between gravimetric solute content (mass of solute related to mass of oven-dry soil in kg kg^{-1}) obtained from disturbed samples, and volumetric solute content (mass of solute related to volume of soil in kg m^{-3}) determined from undisturbed samples with known volume (e.g., soil cores). A vast variety of single and sequential extraction techniques, including column displacement and centrifugation, or a combination of both are described in literature (Adams et al., 1980; Fuentes et al., 2004; Martens, 2002; Pueyo et al., 2003; Villar-Mir et al., 2002). The method selection is mainly based on the chemical species under investigation and the sample volume required for a particular analysis technique.

Adams et al. (1980) compare the ionic composition of solutions extracted from loam and clay soils by means of column displacement with a $CaSO_4$–KCNS solution, centrifugation of moist soil with carbon tetrachloride (CCl_4) added, and simple centrifugation of moist soil. They conclude that the composition of the extracted solution was not affected by the employed method. Alberts et al. (1977) and Villar-Mir et al. (2002) describe extraction techniques suitable for NO_3^-–N determination. Fuentes et al. (2004) and Pueyo et al. (2003) present single and sequential extraction techniques to determine heavy metals in sewage sludge and contaminated soils.

9.3   INDIRECT FIELD METHODS FOR DETERMINING SOLUTE CONCENTRATION

9.3.1   TIME DOMAIN REFLECTOMETRY

Time domain reflectometry (TDR), introduced to soil science by Topp et al. (1980), is a relatively new technique that allows inference of volumetric water content (θ_v) and bulk soil electrical conductivity (EC_b) from measurements of the soil bulk dielectric constant (ε_b).

The basic principle of TDR is based on measuring the propagation velocity
v) of an electromagnetic pulse generated by a TDR cable tester along a
transmission line (wave guide) that is embedded in the soil. The propagation
velocity is a function of the soil bulk dielectric constant given as:

$$\varepsilon_b = \left(\frac{c}{v}\right)^2 = \left(\frac{ct}{2L}\right)^2 \qquad (9.2$$

where c is the velocity of an electromagnetic wave in vacuum ($3 \times 10^8\,\mathrm{m/s}$)
and t is the travel time for the pulse to traverse the length (L) of the embedded waveguide (forth and back $= 2L$). Equation 9.2 simply states that the
bulk dielectric constant of a medium is the propagation velocity in vacuum
relative to velocity in the porous medium squared. The soil bulk dielectric
constant (ε_b) is governed by the dielectric of liquid water $\varepsilon_w = 81$, as the
dielectric constants of other soil constituents are much smaller (e.g., soil
minerals $\varepsilon_s = 3$–5, frozen water (ice) $\varepsilon_i = 4$, and air $\varepsilon_a = 1$). This large disparity
between the dielectric constants makes the method relatively insensitive to
soil composition and texture and thus a good method for liquid soil water
measurement.

Two basic approaches are currently applied to establish the relationships
between ε_b and volumetric soil water content (θ_v); an empirical approach, where
mathematical expressions are simply fitted to observed data (Topp et al.,1980),
or a physically based approach, the so-called mixing model that uses the
dielectric constants and volume fractions of each of the soil components to
derive the ε_b-θ_v relationship (Birchak et al., 1974; Dobson et al., 1985; Roth
et al., 1990). Readers interested in the application of TDR for measuring
volumetric soil water content are referred to Chapter 5 and Jones et al. (2002).
An overview of available TDR systems is depicted in Figure 9.7.

The measurement of the apparent soil bulk electrical conductivity (EC_a) is
based on attenuation of the TDR signal as a function of EC_a. As a transverse

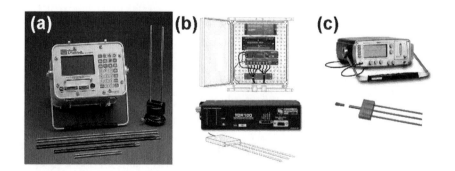

FIGURE 9.7 Overview of commonly applied TDR systems: (a) trace system
(Soilmoisture Equipment Corp.); (b) TDR100 (Campbell Scientific Inc.); and
(c) Tektronix 1502C general purpose cable tester.

electromagnetic wave (TEM) propagates along the buried wave guide, the signal energy is attenuated in proportion to the electrical conductivity along the travel path. This proportional reduction in signal voltage may serve as a basis for measuring EC_a as was first recognized by Dalton et al. (1984), who proposed a "lumped circuit load" transmission line analogy for EC_a measurement by TDR. The soil – wave guide system is assumed to comprise a lumped circuit having a load impedance Z_L at the end of a low loss transmission line (impedance of the soil – wave guide) and a known cable impedance Z_C (for typical TDR applications this is a coaxial cable with 50Ω). A reflection coefficient (ϕ) may be defined in terms of the cable and load (i.e., soil-probe) impedances as:

$$\phi = (Z_L - Z_C)/(Z_L + Z_C) \qquad (9.3)$$

The reflection coefficient ϕ can be determined from analyses of the attenuated signal (Figure 9.8) as:

$$\phi = \frac{V_f - V_0}{V_0} \qquad (9.4)$$

With known reflection coefficient (ϕ) and cable impedance (Z_C) the load impedance (Z_L) is calculated by rearranging Eq. (9.3). Following the "probe constant" approach proposed by Dalton et al. (1990) and Nadler et al. (1991), an experimentally obtained constant (K) may be used to relate EC_a to load impedance:

$$EC_a = \frac{K}{Z_L} \qquad (9.5)$$

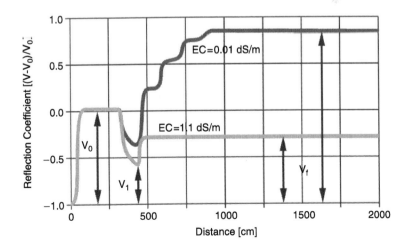

FIGURE 9.8 Analyses of attenuated TDR signals for determination of bulk soil electrical conductivity (EC_a).

The probe constant (K) can be determined based on measurement of the load impedance for one or preferably more reference solutions with known EC according to:

$$K = \frac{EC_{ref}Z_L}{f_T}$$
(9.6

where EC_{ref} is electrical conductivity of the reference solution and f_T is a temperature correction coefficient that relates the measured reference solution to a desired standard temperature. Heimovaara et al. (1995) found that the relationship $f_T = 1/(1 + 0.019[T-25])$ was appropriate for a variety of saline solutions using 25°C as the standard temperature.

Another common method to derive the bulk electrical conductivity from signal analyses is the "thin-section" approach, originally proposed by Giese and Tiemann (1975). The Giese and Tiemann equation is given as:

$$EC_a = \frac{\varepsilon_0 c}{L}\frac{Z_0}{Z_C}\left(\frac{2V_0}{V_f} - 1\right) \quad [S/m]$$
(9.7

where ε_0 is the permittivity of free space (8.9×10^{-12} F/m), c is the speed of light in vacuum (3×10^8 m/s), L is probe length (m), Z_0 is characteristic probe impedance (Ω), Z_C is the TDR cable tester output impedance (usually 50Ω), V is the incident pulse voltage, and V_f is the return pulse voltage after multiple reflections have died out (Figure 9.8). The characteristic probe impedance is determined by immersion of the probe in deionized water or another liquid with known dielectric constant or may be calculated based on wave guide geometry. For a detailed discussion of various available methods for derivation of bulk soil EC_a, interested readers are referred to Chapter 10.

The electrical conductivity of the bulk soil (EC_a) may be regarded as resulting from two parallel conductors, a bulk liquid phase conductivity (EC_b) associated with the free ions within the liquid filled pores, and a bulk surface conductivity (EC_s) related with the exchangeable ions at the solid liquid interface (Rhoades et al., 1976):

$$EC_a = EC_b + EC_s$$
(9.8

Assuming a linear dependence of EC_b on the electrical conductivity of soil water (EC_w), and that only the portion of the total cross-sectional area that is occupied by the liquid phase conducts current, EC_a can be expressed as:

$$EC_a = T\theta_v EC_w + EC_s$$
(9.9

where θ_v is the volumetric soil water content and T is a transmission coefficient that accounts for the tortuosity of the current path and any decreases in ion mobility near the solid-liquid and liquid-gas interfaces, and thus changes in response to soil water content θ_v and alteration of the particle or pore arrangements. The transmission coefficient (T) and surface conductivity (EC_s) need to be determined for each soil from calibration experiments as outlined in Rhoades et al. (1976). Once known, the solution electrical conductivity (EC_w) can be determined from TDR-measured θ_v and EC_a by solving Eq. (9.9). A comparison between EC_w estimated with Eq. (9.8) from TDR measured EC_b and direct measurements of soil solution EC_w with a conductivity meter is depicted in Figure 9.9 (Mmolawa and Or, 2000).

Other commonly applied empirical calibration relationships that often require laborious experimental determination of soil-specific parameters were proposed by Nadler et al. (1984), Vogeler et al. (1997), and Hart and Lowery (1998), among others. An alternative approach is the application of physical parameters, such as soil hydraulic properties to relate EC_a and EC_w (Mualem and Friedman, 1991; Heimovaara et al., 1995). For a detailed discussion of commonly applied calibration procedures for relating soil bulk EC_a and solution EC_w readers are referred to Chapter 10.

Recent applications of TDR include estimation of steady-state and transient transport of conservative and reactive solutes through homogeneous and layered soil columns (Vanclooster et al., 1993; Wraith et al., 1993; Mallants et al., 1994; Ward et al., 1994; Risler et al., 1996; Vogeler et al., 1996; Hart and Lowery, 1998; Vogeler et al., 2000), and estimation of transport parameters under field conditions (Kachanoski et al., 1992; Caron et al., 1999; Das et al., 1999; Mmolawa and Or, 2000, 2003; Campbell et al., 2002).

Summarizing, TDR is a promising method to characterize and monitor solute transport under field conditions. Advantages of automated TDR over

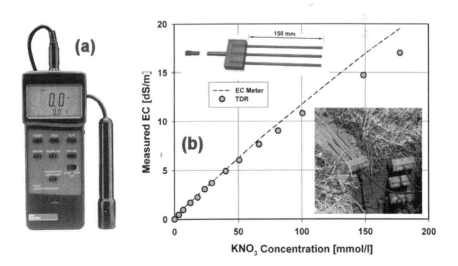

FIGURE 9.9 (a) Handheld electrical conductivity (EC) meter. (b) TDR probes and solution EC vs. concentration measured with TDR and EC meter. (From Mmolawa and Or, 2000.)

time-consuming soil solution analyses include the nondestructive nature, the ability to provide highly detailed breakthrough curves (BTCs), and the capability for continuous and unattended operation (Jones et al., 2002). However, TDR measures the effects of total ionic solute concentration and cannot distinguish between soil solution constituents.

9.3.2 ELECTRICAL RESISTIVITY METHODS

The basic principle of electric resistivity measurements is best explained by means of the relationship between electric resistance (R) and resistivity (ρ) of a wire that is given as:

$$R = \rho \frac{L}{A}$$

(9.10

where R is the electric resistance of the wire (Ω), L is the wire length (m), and A is the cross-sectional area of the wire (m^2). Substituting Ohm's law that relates the electromotive force V (volts) to current flow I (amperes) ($V = I \cdot R$) and rearranging Eq. (9.10) yields an explicit expression for resistivity:

$$\rho = \frac{A}{L} \cdot \frac{V}{I} [\Omega m]$$

(9.11

Equation (9.11) illustrates that resistivity is a function of the ratio of voltage drop to current and the dimensions of the conductor (wire). This principle can be applied to measure resistivity of soil that acts as a conductor between two or more electrodes. The resistivity of natural porous media is highly dependent on water content, solute concentration, texture, and structure. In general, increasing water content leads to a decrease in resistivity as water fills up the pore space and displaces air. Course-textured soils usually have higher resistivity than fine-textured soils at the same water content due to smaller grain to grain contact area.

The first soil electrical resistivity measurements date back to geophysical prospecting in the 1920s, when two current electrodes (outer pair) and two potential electrodes (inner pair) were brought in contact with the soil and lined to form an array. The inner pair is used to measure the potential while constant current is passed through the outer pair (Figure 9.10). Two basic array configurations are commonly employed: the Wenner array (WA) with equally spaced electrodes and the Schlumberger array (SA) where the distance between the two potential electrodes is small in comparison with the distance of the current electrodes (Figure 9.10). The measured resistivity is a function of electrode spacing (a, b) (Figure 9.10) given as:

$$\text{WA:} \quad \rho = 2\pi a R$$

(9.12

$$\text{SA:} \quad \rho = \pi(a^2 + ab)R$$

(9.13

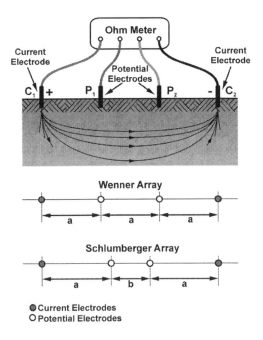

FIGURE 9.10 Sketch showing the measurement principle and basic electrode configurations for surface electrical resistivity measurements.

The bulk soil electrical conductivity EC_a is the inverse of the resistivity as illustrated for the Wenner array:

$$EC_a = \frac{1}{2\pi a R} \qquad (9.14)$$

The depth of current penetration for the WA probe configuration is approximately equal to the inner electrode spacing (Rhoades and Ingvalson, 1971; Rhoades and van Schilfgaarde, 1976; Rhoades, 1978). Nadler (1980) theoretically and experimentally determined the soil volume influenced by the applied electrical field and found that the depth sensed with a WA is deeper than the inner electrode spacing for most field situations, especially under highly variable conductivity conditions with depth, caused by varying water content and solute concentrations.

As with all other methods that measure the soil bulk electrical conductivity (e.g., TDR, electromagnetic induction), a calibration relationship is required to determine solution EC_w (see Sec. 9.3.1 and Chapter 10).

Frame-mounted electrical resistivity systems equipped with GPS and dataloggers are commercially available from a number of companies (e.g., GEOSCAN Research, VERIS Technologies, GEONICS Limited) but can be easily assembled from scratch (Rhoades and Halvorson, 1977). The basic components consist of a battery-powered constant current source resistance meter with a range from 0.1 to 1000 Ω, four metal electrodes, and connecting wire.

FIGURE 9.11 Field setup of a Wenner array configuration. (From Rhoades and Oster, 1986.)

The electrodes are pushed into the soil to a depth of approximately 2.5 cm, lined, and spaced based on the applied configuration scheme (Figure 9.11). The separation distance is determined based on the desired depth and volume of influence (Rhoades, 1978; Nadler, 1980). For large-scale monitoring with fixed electrode spacing, it is advantageous to mount the electrodes, the current source resistance meter, and datalogger on a nonconducting frame. This allows rapid repositioning and substantial time savings when a large number of measurements are required.

By successively increasing the inner electrode separation distance around a point of interest, EC_a can be determined for discrete depth intervals (Halvorson and Rhoades, 1974). Assuming that the penetration depth is equal to the separation distance, the volume measured may be considered as a uniform lateral layer to which deeper layers are added successively as the separation distance increases. These layers may be treated as parallel resistors, and the bulk electrical conductivity (EC_x) for each layer calculated as (Barnes, 1952):

$$EC_x = EC_{(a_i - a_{i-1})} = \frac{EC_{a_i} \cdot a_i - EC_{a_{i-1}} \cdot a_{i-1}}{a_i - a_{i-1}} \tag{9.15}$$

where a_i is the sampling depth and a_{i-1} is the prior sampling depth.

An alternative to the surface-based electrical resistivity method for determination of depth-dependent EC_a was introduced by Rhoades and van Schilfgaarde (1976). They developed a single probe with four equally spaced electrodes mounted as annular rings. The probe that is pushed into the soil to the desired depth provides higher measurement resolution. A drawback, however, is the small sampling volume that requires a large number of measurements to obtain representative values of EC_a. Therefore, application of this probe is advantageous when a precise measurement of salinity within a small localized region is required. To determine a soil salinity index for extended areas, surface positioned probes are better suited.

Observations of soil electrical conductivity to greater depths can be achieved with borehole electrical resistivity tomography (ERT), where arrays of current and potential electrodes (pole-dipole) mounted on cables are lowered into boreholes. Low-frequency electrical current is injected into the subsurface, and the resulting potential distribution is measured for a vast number of different current and potential electrode orientations. Robust regularized nonlinear inverse methods (Binley et al., 2002) allow the reconstruction of 2-D and 3-D electrical resistivity distributions within the soil volume between two or more boreholes (cross-borehole ERT). In special cases the current and potential electrodes are placed in a single borehole (in-line ERT). While ERT was successfully applied to qualitatively study flow and transport in porous (Daily et al., 1992, 1995) and fractured porous (Slater et al., 1997) media, quantitative assessment of transport characteristics in soils and rocks from ERT data is poorly documented (Binley et al., 1996; Slater et al., 2000).

9.3.3 ELECTROMAGNETIC INDUCTION

Another widely applied method to measure soil apparent electrical conductivity is based on electromagnetic induction. In contrast to the electrical resistivity methods discussed in the previous section, current is applied to the soil through electromagnetic induction, and thus no direct contact with the soil surface is required (Corwin and Lesch, 2003). Common instruments consist of a transmitter coil that, when energized with alternating current (AC) at audio frequency, produces an electromagnetic field (Figure 9.12). The time-varying electromagnetic field emitted from the transmitter coil induces weak circular eddy current loops in the conducting soil, which in turn generate a secondary electromagnetic field that differs in amplitude and phase from the primary field (McNeill, 1980). The magnitude of amplitude and phase differences between primary and secondary field depends on soil properties such as texture, structure, water content, and solute concentration as well as spacing between transmitter and receiver coil, distance between coils and soil surface, and coil orientation (parallel or perpendicular to the soil surface). The effect of magnetic permeability of the soil seems to be negligible according to De Jong et al. (1979). The primary and secondary fields are sensed as apparent conductivity at the receiver coil in Siemens per meter according to McNeill (1980):

$$m = \frac{4}{\omega \mu_0 s^2} \left(\frac{H_s}{H_p} \right) \tag{9.16}$$

where ω is the angular operating frequency (radians per second) of the instrument, μ_0 is the permeability of free space (1.2566×10^{-6} H m^{-1}; H = Henries), s is the coil spacing (m), and H_s and H_p are the sensed intensities of the primary and secondary fields at the receiver coil (A m^{-1}). Note that the linear relationship in Eq. (9.16) is only valid under the assumptions that the distance between the coils and the soil surface is zero, the soil is homogeneous and has uniform EC_a, and the induction number N_B is much smaller than 1 ($N_B \ll 1$). The induction number is defined as:

$$N_B = \frac{s}{\delta} \tag{9.17}$$

where s is the coil spacing (m) and δ is the skin depth, defined as the depth where the primary magnetic field has been attenuated to $1/e$ (i.e., 37%) of its original strength (e is the base of the natural logarithm) (Hendrickx et al., 2002).

Assuming that $N_B \ll 1$, depth-dependent bulk soil electrical conductivity $EC_a(z)$ can be calculated for horizontal and vertical coil orientation by solving the following Fredholm integral equations of first kind (McNeill, 1980; Borchers et al., 1997; Hendrickx et al., 2002):

$$m^H(h) = \int_0^\infty \phi^H(z+h) EC_a(z) dz \tag{9.18}$$

with the sensitivity function $\phi^H(z)$ given as:

$$\phi^H(z) = 2 - \frac{4z}{(4z^2 + 1)^{1/2}} \tag{9.19}$$

and

$$m^V(h) = \int_0^\infty \phi^V(z+h) EC_a(z) dz \tag{9.20}$$

with the sensitivity function $\phi^V(z)$ given as:

$$\phi^V(z) = \frac{4z}{(4z^2 + 1)^{3/2}} \tag{9.21}$$

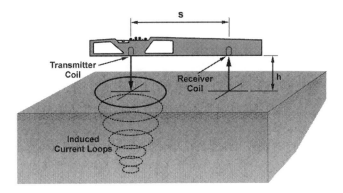

FIGURE 9.12 Sketch illustrating the basic principle of electromagnetic induction measurements.

where superscripts H and V indicate horizontal and vertical coil orientation, respectively, h is the distance between the coils and the soil surface (Figure 9.12), and $m(h)$ is the instrument reading $(\text{S}\,\text{m}^{-1})$. The sensitivity functions represent the relative contribution of the electrical conductivity at depth z to the instrument reading $m(h)$. In practice it might be difficult to solve the inverse problem given in Eqs. (9.18–9.21), because the continuous functions $m(h)$ are not known (only finite sets of measurements at different heights h are available), and small variations in $m(h)$ might lead to large changes in $EC_a(z)$. Borchers et al. (1997) applied a second-order Tikhonov regularization method to solve the inverse problem for EC_a profiles in layered soils.

For cases where the assumption $N_B \ll 1$ is not applicable, Hendrickx et al. (2002) present a nonlinear model based on solutions of Maxwell's equations in the frequency domain to relate electromagnetic induction measurements to depth-dependent EC_a.

An important issue is whether the physical models discussed above that describe the electromagnetic response for homogeneous media are applicable for heterogeneous field soils. As stated in Hendrickx et al. (2002), at this point it is not entirely clear whether the attempts to use electromagnetic measurements (EM) for determination of vertical EC_a distributions are only thwarted by the problem of nonuniqueness inherent to inverse procedures or also by the lack of understanding of physical relationships between the vertical distribution of soil EC and the response of electromagnetic induction EM ground conductivity meters under heterogeneous field conditions.

Electromagnetic induction meters are commercially available from a vast number of geophysical instrumentation companies. A comprehensive literature review indicates that the most commonly applied systems in vadose zone hydrology and soil science are the GEONICS Limited EM-31 and EM-38 (Figure 9.13) ground conductivity meters. The EM-31 has a coil spacing of 3.66 m, which results in a penetration depth of approximately 3 m when the coils are oriented parallel to the soil surface (horizontal), and 6 m when the coils are perpendicular to the surface (vertical orientation). The coil spacing of

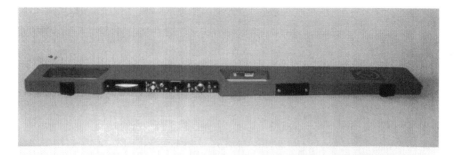

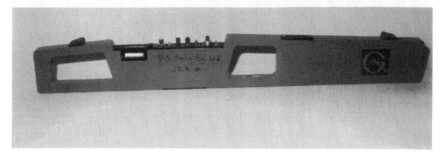

FIGURE 9.13 Handheld Geonics EM-38 ground conductivity meter with horizontal coil orientation (top) and vertical coil orientation (bottom). (From Corwin and Lesch, 2003.)

the EM-38 is exactly 1 m, which leads to penetration depths of 0.75 and 1.0 m, respectively, when operated in horizontal and vertical modes (Figure 9.13). Note that the horizontal mode is obtained by simply rotating the instrument 90°. Both instruments are lightweight and can be easily operated by a single person. For large-scale salinity surveys it might be more convenient to mount the instrument on a sledge that is pulled by an all-terrain vehicle, as shown in Corwin and Lesch (2003).

Recent applications of the electromagnetic induction method in soil and environmental science are reported in Hendrickx et al. (1992, 2002), Triantafilis et al. (2000), Lesch and Corwin (2003), Corwin and Lesch (2003), and Sudduth et al. (2003).

9.3.4 PÖROUS MATRIX SENSORS

Porous matrix or salinity sensors are comprised of a porous ceramic matrix with embedded platinum electrodes that allow direct measurement of solution electrical conductivity (EC_w). When the porous ceramic, which needs to be saturated with deionized water for several days prior to installation, is brought in close contact with the surrounding soil, the pressure of the liquid within the ceramic equilibrates with the soils matric potential. Simultaneously ions diffuse from the soil solution to the water within the ceramic until the electrolyte concentration is in equilibrium. Note that the choice of ceramic material is based on the requirement that it remains completely saturated for the matric potential range under consideration (see the bubbling pressure concept in Sec. 9.2.1.1). The electrical conductivity of the solution is then measured based on Ohm's law:

$$V = I \cdot R \tag{9.22}$$

where V is the electromotive force (volts), I is the current flow (amperes), and R the resistance (ohms). For constant voltage, the current flowing through a conductor is inversely proportional to the resistance or directly proportional to the electrical conductance. The electrical conductivity of the solution within the porous matrix (EC_w) is thus determined from the applied constant voltage, known electrode geometry, and measurement of the electric current.

Porous matrix sensors were first introduced by Kemper (1959), who embedded two parallel platinum wire electrodes in a cylindrical porous ceramic block. Though the cylindrical ceramic element ensured constant geometry of the immediate conductance paths between the electrodes, a considerable field developed outside the sensor leading to unwanted variations of measurements. Richards (1966) developed a sensor consisting of a relatively small, square-shaped (6×6 mm) ceramic element with platinum wire-mesh electrodes of dissimilar area embedded on the opposing surfaces of the 1 mm thick plate. Arrangement and shielding of the electrodes prevented the development of appreciable external fields. The sensor is spring-loaded to ensure good contact between ceramic and surrounding soil and equipped with a thermistor for temperature compensation (note that a temperature correction needs to be applied to field measurements because the sensor is commonly calibrated at room temperature). This Richards-type design is probably the most widely used. A slightly modified version depicted in Figure 9.14 is commercially available from Soilmoisture Equipment Corporation. Other salinity sensors that are based on the same principle but differ in geometry and employed

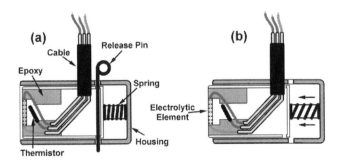

FIGURE 9.14 Schematic sketch depicting a porous matrix sensor (Soilmoisture Equipment Corp.) prior to installation (a) and after installation with removed release pin (b).

porous materials have been designed by Enfield and Evans (1969) and Reicosky et al. (1970).

For installation it is recommended to mount the sensor at the end of a tube, which is used to guide the cable and the string attached to the release pin to the surface. After coring a hole to the desired placement depth, the first few centimeters are refilled with slurry, and the tube holding the sensor is carefully lowered. The sensor is pushed into place, and the release pin removed by pulling the attached string. The loaded spring pushes the sensor (electrolytic element) out of the casing towards the soil, thereby establishing good hydraulic contact. Then the tube is removed and the hole refilled with parent material. As mentioned above, the sensor needs to be saturated for several days prior to installation.

Each individual sensor requires calibration due to variations in retention characteristics of the used porous materials, even if they originate from the same batch. The calibration should be conducted with the measurement bridge used for the field readings to avoid errors emanating from differences in resistance of the bridge electrical circuitry between two instruments. Calibration is required to establish relationships between measured conductance and solution electrical conductivity (EC_w) and thermistor resistance and temperature.

The calibration of the electrolytic element is achieved by immersing the sensor in solutions of known electrical conductivity (e.g., mixtures of NaCl and $CaCl_2$). Note that it may require several hours for the sensor to equilibrate for each calibration step. The resulting calibration curve is linear for conductivity values ranging from approximately 1 to 40 dS m^{-1}, therefore 3–4 calibration points are adequate for this region. If soils with very low EC_w are expected, additional calibration points need to be added to establish the curve-linear portion below 1 dS m^{-1}.

Calibration of the thermistor is simply achieved by immersing the sensor in a constant temperature bath and taking measurements over a range of temperatures that might be expected under field conditions. A typical thermistor calibration equation is given as (Corwin, 2002):

$$T = a\left(\frac{R}{R_{25}}\right)^2 - b\left(\frac{R}{R_{25}}\right) + c \qquad (9.23$$

where T is the temperature (°C), R is the thermistor resistance (Ω), R_{25} is the thermistor resistance at 25°C, and a, b, and c are regression coefficients obtained from fitting a polynomial function to measured data points.

It is important to note that the calibration relationships for electrolytic elements do not remain stable with time. A study conducted by Wood (1978) reveals significant deviations between initial calibration and calibration relationships that were established after sensor removal following 3 and 5 years of field deployment. The observed shifts vary in direction and magnitude. Wood (1978) also documented an increase in thermistor resistance with time leading to measurement errors ranging from 1 to 3°C. Wood (1978) concluded that the observed calibration changes are tolerable for most practical field situations. For precise salinity measurements, however, he recommended recalibration after 3 years.

Another important issue is the response time of salinity sensors, which is important under rapidly changing conditions (e.g., irrigation). The response time is mainly determined by the diffusion rate of the solutes, i.e., the time required for chemical equilibration. Wood (1978) reports response times due to changes in salinity of 2–5 days for a matric potential range from -0.5 to -1.5 bars. This renders salinity sensors impractical for monitoring artificial salinity management practices (e.g., salt leaching). Wesseling and Oster (1973) presented a mathematical description of sensor response and experimentally verified that a single response factor can be used to adjust sensor readings even under rapidly changing conditions.

9.3.5 FIBER OPTIC SENSORS

The application of fiber optic chemical sensors for monitoring solute transport is a promising evolving technique alternative to other direct and indirect measurement methods. The technique is based on directing a constant light beam through optical fibers (input leg) to a target location within the soil matrix, where it is partially adsorbed and partially reflected back into the probe. The reflected light is guided through a separate fiber bundle (output leg) from the probe to a photo detector that quantifies its intensity and converts the optical to an electrical signal that is recorded with a computer or datalogger (Figure 9.15). Narrow and broad band filters are employed to condition the outgoing and reflected light beams, respectively (Ghodrati, 1999). Given that the intensity of the ingoing light remains constant with time, the intensity of

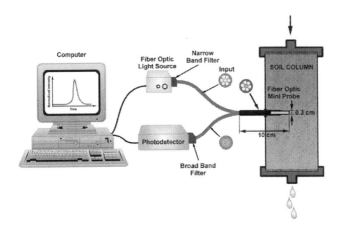

FIGURE 9.15 Schematic illustration of a fiber optic miniprobe measurement system. (From Ghodrati, 1999.)

the reflected beam will be constant if the system under investigation is in equilibrium (Krohn, 1988). Perturbations of the equilibrium state will cause a change in output light intensity that can be analyzed to quantify the perturbation causing process based on calibration relationships that need to be established for each probe and the process of interest (e.g., change in solute concentration or water content).

Though fiber optic sensors were previously applied to measure soil water content (Alessi and Prunty, 1986; Garrido et al., 1999), the most promising application for soil and environmental science is the characterization of solute transport phenomena in soils via fluorescent tracers (Ghodrati, 1999).

Assuming a rigid and stable soil matrix (no particle rearrangement or swelling), Garrido et al. (2000) provide calibration procedures for laboratory and field experiments that relate tracer concentration to output light intensity. The conventional calibration procedure for laboratory experiments consists of stepwise leaching of a soil column with several pore volumes of tracer with known concentration. After the photodetector indicates stable output intensity for a certain tracer concentration, the column is flushed with $CaCl_2$ and the procedure repeated for the next tracer concentration until a few points are established on the calibration curve. A second-order polynomial function may be fitted to the measurements to establish a continuous calibration relationship Ghodrati (1999).

Due to the presumably large amount of tracer that would be required, this procedure is not practical for field application. Therefore, Garrido et al. (2000) developed a point calibration device that allows site-specific calibration of fiber optic sensors. The device consists of a stainless steel tube that either forms a jacket around or is attached to the outside of the miniprobe and allows injection of a small amount of tracer directly into the soil in front of the fiber optics. For tracer injection the tube is connected to a peristaltic pump or a syringe. The calibration curve is constructed in the same manner as for the conventional method after the concentration–light intensity relationship is established for a few points. A comparison of both methods in the laboratory shows good agreement (Garrido et al., 2000).

Though studies by Kulp et al. (1988), Nielsen et al. (1991), Campbell et al. (1999), Ghodrati (1999), and Ghodrati et al. (2000) and a comprehensive review of fiber optic sensors and applications for environmental monitoring by Rogers and Poziomek (1996) reveal great potential for this technique, extensive testing and calibration under field conditions and further sensor development is required to make it feasible for continuous real-time monitoring of solute transport phenomena.

9.4   COMPARISON OF DIRECT AND INDIRECT METHODS

There are no general guidelines for choosing the right method or sensor for a given objective. Factors that need to be considered when planning a field study are available funds, expected spatial and temporal variability of the properties under investigation, location, accessibility, and areal extent of the planned experiment, climatic conditions (e.g., humid or arid), required spatial and temporal measurement resolution, and site-specific management operations (e.g., tillage), only to name a few. Another important aspect to consider is the difference in influence volume associated with each individual method. Since modeling and monitoring solute transport requires simultaneous measurement of water content or soil water potential and concentration of the soil solution, the influence volume needs to be considered when pairing different sensors. This can be avoided with technologies such as TDR capable of measuring solute concentration and water content with the same sensor. Table 9.1 gives a brief overview of the documented shortcomings and advantages of the various methods discussed in the previous sections.

9.5 CASE STUDIES AND RECOMMENDATIONS FOR FUTURE RESEARCH

9.5.1 DETAILED CHARACTERIZATION OF SOLUTE TRANSPORT IN A HETEROGENEOUS FIELD SOIL WITH FIBER OPTIC MINI PROBES AND TIME DOMAIN REFLECTOMETRY

Garrido et al. (2001) tested the applicability of time domain reflectometry (TDR) and fiber optic miniprobes (FOMPs) for characterizing solute transport in heterogeneous clay loam soil. They equipped a small (82 × 82 cm) field plot with arrays of TDR waveguides and FOMPs. Eight 20 cm TDR waveguides were installed horizontally in 10 and 20 cm depths, and eight 5 cm waveguides were installed vertically as outlined in Figure 9.16. In addition, 20 FOMPs were installed vertically with 10 probes at each depth. After environmentally isolating the plot with a tent, Garrido and coworkers conducted a steady-state miscible displacement experiment and monitored solute break trough curves (BTCs).

The plot was drip-irrigated at a rate of 1.5 cm/h for more than one week to establish steady-state flow conditions. Once steady state flow was indicated by means of TDR-measured uniform soil water content, a 2 cm pulse of fluorescent pyranine tracer at a concentration of 4 g/l was applied and irrigation with water resumed for 5 days. The TDR and FOMP systems recorded measurements at 10- and 5- minute intervals, respectively. The retrieved data were then converted to solute concentration using predetermined calibration functions and plotted as a function of displaced pore volumes to obtain BTCs.

The "probe constant" approach (Dalton et al., 1990; Nadler et al., 1991) outlined in Sec. 9.3.1 was used to relate soil bulk EC_a to the load impedance of the soil-TDR waveguide system. Linear relationships (Rhoades et al., 1976) with linearity coefficients determined from calibration

TABLE 9.1
Comparison of Methods for Estimation of Solute Concentration

Methods	Advantages	Limitations
Suction cups	Relatively inexpensive	Limited application range from saturation to approximately −1 bar
	Easy to install and operate	Relatively small volume of influence requires many suction cups to spatially resolve solute concentration, therefore not suitable for highly structured soils
	No calibration required	
	Extraction of soil solution and chemical analyses allows quantitative specification of constituents	Potential interactions between porous material and soil solution may lead to erroneous results
		It is difficult to keep the applied suction slightly below the soil's matric potential, which may lead to creation of preferential flow paths towards the cup
		Limited capability of continuous monitoring may leave major events undetected (e.g., after heavy rain)
Combined solution sampling tensiometer probes	See suction cups above	See suction cups above
	Allows extraction of solute samples and continuous measurement of matric potential	Potential disadvantageous interactions between tensiometer and sampling operation mode
		Long equilibration times
Suction lysimeters	Better spatial resolution due to larger influence volume when compared to suction cups	Installation usually causes extended disturbance of the parent soil, unless monoliths are used
	No calibration required	Limited application range from saturation to approximately −1 bar
	Extraction of soil solution and chemical analyses allows quantitative specification of constituents	Potential interactions between porous material and soil solution may lead to erroneous results
		It is difficult to keep the applied suction slightly below the soil's matric potential, which may lead to creation of preferential

Method	Advantages	Disadvantages
Passive capillary samplers	Passive sampling mode prevents development of preferential flow paths Provides information regarding solute flux and concentration Suitable for long-term studies Minimal maintenance required Method allows quantitative specification of constituents	Limited capability of continuous monitoring may leave major events undetected (e.g., after heavy rain); See suction cups above Quite expensive Laborious and complicated installation Large disturbance of parent soil Potential interactions between wick material and soil solution Limited automation capability
Capillary absorbers	Passive sampling mode prevents development of preferential flow paths Method allows quantitative specification of constituents	Quite expensive Laborious and complicated installation Potential interactions between absorber material and soil solution No automation capability Potential for sample cross-contamination
Solute extraction from soil samples	Inexpensive Allows quantitative specification of extract constituents	Destructive method that causes substantial disturbance of the parent soil Method does not allow repeated measurements at the same location
TDR	Excellent spatial and temporal resolution within the volume of influence Allows simultaneous measurement of volumetric water content and bulk soil EC Excellent automation and multiplexing capability	Quite expensive Relatively small volume of influence makes it impractical for large scale observations Calibration required Cannot specify solution constituents
Electrical resistivity	Relatively cheap and easy to install	Calibration required

TABLE 9.1
Continued

Methods	Advantages	Limitations
	Large volume of influence and sampling depth, especially with electrical resistivity tomography (ERT) Good for large-scale mapping Good automation capability	Cannot specify solution constituents Provides qualitative estimates of solute migration
Electromagnetic induction	Inexpensive and easy to operate Large volume of influence and sampling depth Good for large-scale mapping Good automation capability	Calibration required Cannot specify solution constituents Provides qualitative estimates of solute migration
Salinity sensors	Relatively cheap and easy to operate Automation capability	Extremely small volume of influence makes the sensor impractical for large-scale observations Slow response time makes the sensors impractical under rapidly changing salinity condition Calibration required Calibration relationships do not remain stable with time
Fiber optic sensors	Excellent spatial and temporal resolution within the volume of influence Allows simultaneous measurement of water content and solution EC Excellent automation and multiplexing capability Small sensor size makes technology especially useful for monitoring small-scale processes	Required photodedector and light source is expensive Extremely small volume of influence makes it impractical for large-scale observations To date there are only a few feasibility studies for field applications available Technique is not fully developed for field applications

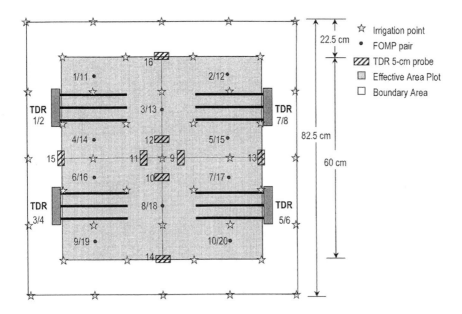

FIGURE 9.16 Sketch illustrating design and instrumentation of the field plot. (From Garrido et al., 2001.)

experiments were employed to relate EC_a to solution EC_w and pyranine tracer concentration.

The FOMPs were calibrated *in situ* with the point calibration device (Garrido et al., 2000) discussed in Sec. 9.3.5. The yielded calibration points relating output light intensity to pyranine concentration were then approximated with a second-order polynomial function to obtain a continuous curve.

The measured BTCs and calculated tracer mass recoveries are quite different for TDR and FOMPs, as illustrated in Figure 9.17 for the probes located 20 cm deep at the right bottom quarter of the field plot (see Figure 9.16). This is most likely related to the heterogeneous soil conditions and differing sampling volumes of TDR and fiber optic techniques. While the calculated tracer mass recovery of TDR waveguide 6 was 44%, the mass recovery of the surrounding FOMPs ranges from 15% (FOMP 20) to 66% (FOMP 18). The rather distinct difference in mass recovery between individual FOMPs indicates preferential transport of the pyranine tracer. This case study highlights the importance of influence volumes associated with different measurement techniques and the difficulties and challenges to spatially resolve solute transport phenomena in heterogeneous soils. Despite excellent automation capabilities and measurement resolution (within the respective volume of influence) of these advanced measurement techniques, it is extremely difficult to monitor and characterize solute transport, even within this relatively small field plot.

For further details interested readers are referred to Garrido et al. (2001).

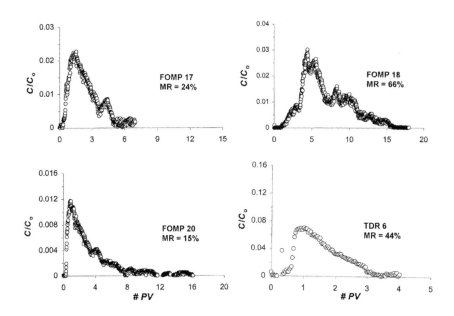

FIGURE 9.17 Breakthrough curves and mass recovery measured with TDR and FOMPs installed in close proximity at a depth of 20 cm. (From Garrido et al., 2001.)

9.5.2   MONITORING SNOWMELT-INDUCED UNSATURATED FLOW AND TRANSPORT USING ELECTRICAL RESISTIVITY TOMOGRAPHY AND SUCTION SAMPLERS

This study was carried out by French et al. (2002) at Oslo Gardermoen Airport in Norway to determine the potential risks of aquifer pollution during snowmelt and after heavy precipitation due to airport operations (e.g., deicing chemicals). Though the airport is equipped with 60 suction cup samplers that are installed along the runways to monitor solute concentrations in the vadose zone, there is high risk of missing significant preferential transport events caused by nonuniform distribution of ground frost and basal ice. To test the applicability of cross-borehole electrical resistivity tomography (ERT) (see Sec. 9.3.2) for enhanced monitoring of chemical migration originating from the airport, French and coworkers (2002) set up a heavily instrumented field site containing arrays of horizontally installed Teflon suction cup samplers and two boreholes (SEL 2 and SEL 3) housing the ERT electrode arrays (Figure 9.18). A sodium bromide tracer ($1.86 \, \mathrm{g} \, \mathrm{l}^{-1}$ NaBr with an electrical conductivity of $2.15 \, \mathrm{mS} \, \mathrm{cm}^{-1}$) was applied in five frozen columns placed below the snow cover prior to snow melt. Monitoring started immediately at the onset of snow melt, with soil water samples extracted and analyzed on a bi-daily basis. Electrical resistivity between the boreholes was measured once a week. One ERT dataset was obtained prior to snow melt at the date of tracer application and used as a reference state to produce consecutive images

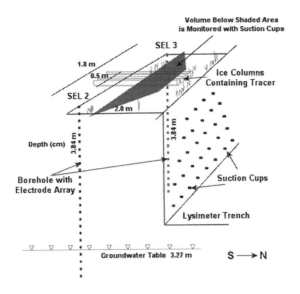

FIGURE 9.18 Instrumentation and setup of the field experiment. (French et al., 2002.)

showing changes in bulk soil electrical conductivity. The "Occams" approach (LaBrecque et al., 1996) was used for ERT data inversion, i.e., inversion of ERT measurements yield resistivity distributions that should be consistent with measured resistance distributions (see Sec. 9.3.2).

Figure 9.19 shows a series of images of Br^- concentrations that were obtained by employing an inverse distance weighing procedure to the suction cup measurements at the 28 locations within the observed region. In early April there are no noticeable changes in bromide concentration throughout the profile. On April 15 an increase in concentration becomes apparent at a depth of approximately 1.9 m, with no visible increases above. This indicates preferential flow bypassing the top soil layers. At the end of April concentrations increase in the top layers and the sodium bromide plume consecutively spreads and migrates downward. Figure 9.20 shows the same time sequence obtained from ERT observations. Note that changes in bulk soil electrical conductivity obtained with ERT are due to varying soil moisture and solute concentration. The ERT image reconstructed from measurements on April 4 already shows a distinctive relative change of soil bulk electrical conductivity in the top layer in close proximity to the tracer columns. The image from April 8 shows a relative conductivity increase at the left side, indicating nonuniform tracer infiltration at the surface, probably due to ice lenses or nonuniform thawing. It is interesting to note that a similar conductivity pattern as observed with the suction cups on April 15 shows up in the ERT image one week earlier on the right side at ~1.5–2.0 m depth. After thawing it is apparent that the tracer spreads more uniformly. The most pronounced change in electrical conductivity takes place on May 1 within the

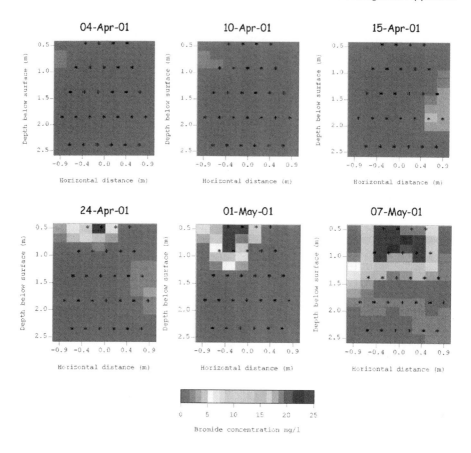

FIGURE 9.19 Time sequence showing the bromide concentration distributions obtained from chemical analyses of suction cup extracts. (French et al., 2002.)

soil layer from 0.5 to 1.5 m. It appears that the tracer is leaching to deeper regions due to an increased supply of melt water.

The results of this study indicate that both methods were capable of capturing preferential tracer transport reasonably well, though ERT measurements provided a more detailed picture of preferential flow patterns throughout the profile despite the relatively narrow array and large number of suction cups. It is clear that under practical conditions it is not feasible to equip extended areas with hundreds of suction cup samplers, therefore cross-borehole ERT seems to be a viable option to monitor migration of contaminant plumes in highly structured soils and rock. However, we need to keep in mind that ERT only provides a qualitative estimate of contaminant migration. To determine chemical composition and concentration of the contaminant, we still rely on direct methods to extract soil solution. Chapter 18 also shows the application of ER for characterizing soil spatial variability.

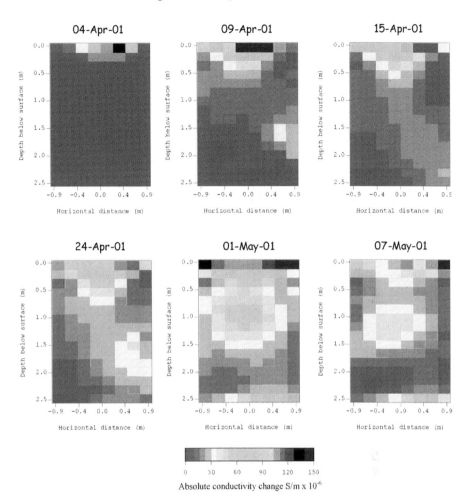

FIGURE 9.20 Time sequence showing changes in absolute conductivity obtained with electrical resistivity tomography. (French et al., 2002).

9.5.3 RECOMMENDATIONS FOR FUTURE RESEARCH

Existing and advancing problems with contaminated natural resources clearly indicate that our current understanding of solute transport phenomena and available techniques to monitor contaminant migration in the vadose zone are inadequate. Our ability to measure and characterize the spatial distribution of chemicals and preferential migration pathways is restricted due to the application of *in situ* point-measurement devices with limited influence volume and geophysical techniques that only indirectly or qualitatively capture contaminant migration. To date there is no reliable technology that ensures detection of solute migration in discrete, narrow fingers and localized zones. In addition, field investigations often reveal that the extent of chemical migration is poorly predicted with existing mathematical models. This perturbational situation may be attributed to incomplete understanding or unrecognized physical, chemical, and biological processes, insufficient or inaccurate characterization of site properties, and inadequate or inaccurate numerical modeling capabilities. This opens a vast area for potential future research. From a monitoring point of view, further development of existing methods including data analysis techniques and introduction of new technologies are imperative to improve spatial resolution and allow quantitative assessments of solute concentrations within the vadose zone. Electrical methods such as TDR, ERT, and magnetic induction seem to be the most promising for large-scale and real-time monitoring.

ACKNOWLEDGMENTS

The authors gratefully acknowledge the support of the Idaho Agricultural Experimental Station (IAES). Special thanks go to Dani Or (UConn) and Jon Wraith (MSU) for their support and insightful advice regarding the application of time domain reflectometry, and to Helen French (Norwegian Center for Soil and Environmental Research) for providing high resolution graphics for section 9.5.2. Approved as IAES publication 4706.

NOTATION

c	velocity of an electromagnetic wave in vacuum [3×10^{-8} m s^{-1}]
f_T	temperature correction coefficient used to determine K
g	acceleration of gravity [m s^{-2}]
r	pore radius [m]
s	coil spacing [m]
	travel time of an electromagnetic wave [s]
V	propagation velocity of an electromagnetic wave along a transmission line [m s^{-1}]
A	cross-sectional area [m^2]
EC_a	soil bulk electrical conductivity [dS m^{-1}]
EC_b	bulk liquid phase electrical conductivity [dS m^{-1}]
EC_s	bulk surface electrical conductivity [dS m^{-1}]
EC_{ref}	electrical conductivity of a reference solution used to determine K [dS m^{-1}]
EC_w	electrical conductivity of the soil solution [dS m^{-1}]
FOMP	fiber optic miniprobe
I	current [amperes]
K	TDR probe constant
L	length [m]
N_B	induction number
PCAPS	passive capillary sampler
R	electric resistance [Ω]
T	transmission coefficient
TDR	time domain reflectometry

V	electromotive force [volts]
V_0, V_1, V_f	return pulse voltages at different locations of the TDR trace used to calculate EC_a
Z_0	characteristic probe impedance [Ω]
Z_C	cable impedance [Ω]
Z_L	load impedance of the soil-waveguide system [Ω]
γ	contact angle [radians]
δ	skin depth
ε_0	permittivity of free space [89×10^{-12} F m^{-1}]
θ_v	volumetric soil water content [m^3 m^{-3}]
μ_0	permeability of free space [1.2566×10^{-6} H m^{-1}]
ρ	electric resistivity [Ωm]
ρ_w	liquid density [kg m^{-3}]
σ	liquid surface tension [N m^{-1}]
ϕ	reflection coefficient
ω	angular radiation frequency [radians s^{-1}]

REFERENCES

Adams, F., C. Burmester, N.V. Hue, and F.L. Long, 1980. A comparison of column-displacement and centrifuge methods for obtaining soil solutions. Soil Sci Soc Am. J, 44:733–735.

Alberts, E.E., R.E. Burwell, and G.E. Schuman, 1977. Soil nitrate-nitrogen determined by coring and solution extraction techniques. Soil Sci Soc Am J, 41:90–92.

Alessi, R.S., and G.A. Prunty, 1986. Soil water determination using fiber optics. Soil Sci Soc Am J, 50:860–863.

ASTM, 1998. Standard Guide for Conducting a Terrestrial Soil-Core Microcosm Test. E1197-87, American Society for Testing and Materials, Philadelphia.

Barbee, G.C. and K.W. Brown, 1986. Comparison between suction and free-drainage soil solution samplers. Soil Sci, 141(2):149–154.

Barnes, H.E., 1952. Soil investigation employing a new method of layer-value determination for earth resistivity interpretation. Highway Res Board Bull, 65:26–36.

Beier, C., K. Hansen, P. Gundersen, and B.R. Andersen, 1992. Long-term field comparison of ceramic and poly (tetrafluoroethene) porous cup soil-water samplers. Environ Sci Technol, 26(10):2005–2011.

Beier, C., M. Butts, M. von Freiesleben, N.E. Jensen, and L. Rasmussen, 1989. Monitoring of soil water and ion fluxes in forests. In Methods for Integrated Monitoring in the Nordic Countries 11, pp. 63–138. The Nordic Council of Ministers, Copenhagen.

Bergström, L.F., 1990. Use of lysimeters to estimate leaching of pesticides in agricultural soils. Environ Pollut, 67:325–347.

Best, J.A., and J.B. Weber, 1974. Disappearance of s-triazines as affected by soil pH using a balance-sheet approach. Weed Sci, 22:364–373.

Binley, A., G. Cassiani, R. Middleton, and P. Winship, 2002. Vadose zone flow model parameterization using cross-borehole radar and resistivity imaging. J Hydro, 267(3–4):147–159.

Binley, A., S. Henry-Poulter, and B. Shaw, 1996. Examination of solute transport in an undisturbed soil column using electrical resistance tomography. Water Resour Res, 32(4):763–769.

Birchak, J.R., C.G. Gardner, J.E. Hipp, and J.M. Victor. 1974. High dielectric constant microwave probes for sensing soil moisture. Proc Inst Elec Electron Eng, January 1974, 62:93–98.

Boll, J., J.S. Selker, B.M. Nijssen, T.S. Steenhuis, J. Van Winkle, and E. Jolles, 1991. Water quality sampling under preferential flow conditions. pp. 290–298. In R.G. Allen et al. (ed) Lysimeters for evapotranspiration and environmental measurement. Proc. ASCE Int. Symp. Lysimetry, Honolulu, HI, 23–25 July 1991. ASCE, New York.

Boll, J., T.S. Steenhuis, and J.S. Selker, 1992. Fiberglass wicks for sampling of water and solutes in the vadose zone. Soil Sci Soc Am J, 56:701–707.

Borchers, B., T. Uram, and J.M.H Hendrickx, 1997. Tikhonov regularization of electrical conductivity depth profiles in field soils. Soil Sci Soc Am J, 61:1004–1009.

Bottcher, A.B., L.W. Miller, and K.L. Campbell, 1984. Phosphorus adsorption in various soil-water extraction cup materials: effect of acid wash. Soil Sci, 137(4):239–244.

Brandi-Dohrn, F.M., R.P. Dick, M. Hess, and J.S. Selker, 1996. Field evaluation of passive capillary samplers. Soil Sci Soc Am J, 60:1705–1713.

Briggs, L.J., and A.G. McCall, 1904. An artificial root for inducing capillary movement of the soil moisture. Science, 20:566–569.

Brown, K.W., J.C. Thomas, and M.W. Holder, 1986. Development of a capillary unsaturated zone water sampler. Cooperative Agreement CR812316–01–0, USEPA Environmental Monitoring Systems Laboratory, Las Vegas, NV.

Campbell, C.G., M. Ghodrati, and F. Garrido, 1999. Comparison of time domain reflectometry, fiber optic mini-probes, and solution samplers for real time measurement of solute transport in soil. Soil Sci, 164(3):156–170.

Campbell, C.G., M. Ghodrati, and F. Garrido, 2002. Using time domain reflectometry to characterize shallow solute transport in an oak woodland hillslope in northern California, USA. Hydro Proc, 16(15):2921–2940.

Caron, J., S.B. Jemia, J. Gallichand, and L. Trépanier, 1999. Field bromide transport under transient-state: monitoring with time domain reflectometry and porous cup. Soil Sci Soc Am J, 63:1544–1553.

Cepuder, P., and M. Tuller, 1996. Simple field-testing sites to determine the extent of nitrogen leaching from agricultural areas. In N. Ahmad, Nitrogen Economy in Tropical Soils, Kluwer Academic Publishers, Dordrecht, pp. 355–361.

Chow, T.L., 1977. A porous cup soil-water sampler with volume control. Soil Sci, 124(3):173–176.

Cole, D.W., 1968. A system for measuring conductivity, acidity, and rate of water flow in a forest soil, Water Resour Res, 4(5):1127–1136.

Corwin, D.L., 2002. Porous matrix sensors. p.1269–1273. In: W.A. Dick et al. (ed.), Methods of Soil Analysis Part 4 — Physical Methods. Soil Science Society of America Inc., Madison, WI.

Corwin, D.L., and S.M. Lesch, 2003. Application of soil electrical conductivity to precision agriculture: theory, principles, and guidelines. Agron J, 95:455–471.

Creasey, C.L. and S.J. Dreiss, 1988. Porous cup samplers: cleaning procedures and potential sample bias from trace element contamination. Soil Sci, 145(2): 93–101.

Daily, W.D., A.L. Ramirez, D.J LaBrecque, and W. Barber, 1995. Electrical resistance tomography experiments at the Oregon Graduate Institute. J Appl Geophys, 33:227–237.

Daily, W.D., A.L. Ramirez, D.J. LaBrecque, and J. Nitao, 1992. Electrical resistivity tomography of vadose water movement. Electrical resistivity tomography of vadose water movement. Water Resour Res, 28(5):1429–1442.

Dalton, F.N., J.A. Poss, T.J. Heimovaara, R.S. Austin, W.J. Alves, and W.B. Russell, 1990. Principles, techniques and design considerations for measuring soil water content and soil salinity using time-domain reflectometry. U.S. Salinity Lab. Res. Rep. 122. U.S. Salinity Lab., Riverside, CA.

Dalton, F.N., W.N. Herkelrath, D.S. Rawlins, and J.D. Rhoades, 1984. Time-domain reflectometry: simultaneous measurement of soil water content and electrical conductivity with a single probe. Science, 224:989–990.

Das, B.S., J.M. Wraith, and W.P. Inskeep, 1999. Nitrate concentrations in the root zone estimated using time domain reflectometry. Soil Sci Soc Am J, 63:1561–1570.

De Jong, E., A.K. Ballantyne, D.R. Cameron, and D.W.L. Read, 1979. Measurement of apparent electrical conductivity of soils by an electromagnetic induction probe. Soil Sci Soc Am J, 43:810–812.

Dobson, M.C., F.T. Ulaby, M.T. Hallikainen, and M.A. El-Rayes, 1985. Microwave dielectric behavior of wet soil: II. Dielectric mixing models. Inst Elec Electron Eng Trans Geosci Remote Sensing, 23:35–46.

Dolan, P.W., B. Lowery, K.J. Fermanich, N.C. Wollenhaupt, and K. NcSweeney, 1993. Nitrogen placement and leaching in a ridge-tillage system. pp. 176–183. In Conference Proceedings on Agricultural Research to Protect Water Quality, 21–24 February 1993, Minneapolis, MN. Soil Water Conserv. Soc., Ankeny, IA.

Enfield, C.G., and D.D. Evans, 1969. Conductivity instrumentation for in situ measurement of soil salinity. Soil Sci Soc Am Proc., 33:787–789.

Essert, S., and J.W. Hopmans, 1998. Combined tensiometer-solution sampling probe. Soil Tillage Res, 45:299–309.

French, H.K., C. Hardbattle, A. Binley, P. Winship, and L. Jakobsen, 2002. Monitoring snowmelt induced unsaturated flow and transport using electrical resistivity tomography. J Hydro, 267:273–284.

Fuentes, A., M. Lloréns, J. Sáez, A. Soler, M.I. Aguilar, J.F. Ortuño, and V.F. Meseguer, 2004. Simple and sequential extractions of heavy metals from different sewage sludges. Chemosphere 54:1039–1047.

Furth, R., 1985. Application of ^{14}C-labeled herbicides in lysimeter studies. Weed Sci, 33 (suppl 2):11–17.

Garrido, F., M. Ghodrati, and C.G. Campbell, 2000. Method for in situ field calibration of fiber optic miniprobes. Soil Sci Soc Am J, 64:836–842.

Garrido, F., M. Ghodrati, and M. Chendorain, 1999. Small-scale measurement of soil water content using a fiber optic sensor. Soil Sci Soc Am J, 63:1505–1512.

Garrido, F., M. Ghodrati, C.G. Campbell, and M. Chendorain, 2001. Detailed characterization of solute transport in a heterogeneous field soil. J Environ Qual, 30:573–583.

Ghodrati, M., 1999. Point measurement of solute transport processes in soil using fiber optic sensors. Soil Sci Soc Am J, 63: 471–479.

Ghodrati, M., F. Garrido, C.G. Campbell, and M. Chendorain, 2000. A multiplexed fiber optic miniprobe system for measuring solute transport in soil. J Environ Qual, 29:540–550.

Giese, K., and R. Tiemann, 1975. Determination of the complex permittivity from thin-sample time domain reflectometry: improved analysis of the step response waveform. Adv Mol Relax Proc, 7:45–59.

Grossmann, J., and P. Udluft, 1991. The extraction of soil water by the suction cup-method: a review. J Soil Sci, 42:83–93.

Grossmann, J., K.-E. Quentin, and P. Udluft, 1987. Sickerwassergewinnung mittles Saugkerzen-eine Literaturstudie. Z Pflanzenernähr Bodenkd, 150:258–261.

Grover, B.L., and R.E. Lamborn, 1970. Preparation of porous ceramic cups to be used for extraction of soil water having low solute concentrations. Soil Sci Soc Am Proc, 34:706–708.

Haines, B.L., J.B. Waide, and R.L. Todd, 1982. Soil solution nutrient concentrations sampled with tension and zero-tension lysimeters: report of discrepancies. Soil Sci Soc Am J, 46:658–661.

Halvorson, A.D., and J.D. Rhoades, 1974. Assessing soil salinity and identifying potential saline-seep areas with field soil resistance measurements. Soil Sci Soc Am Proc, 38:576–581.

Hansen, E.A., and A.R. Harris, 1975. Validity of soil-water samples collected with porous ceramic cups. Soil Sci Soc Am Proc, 39: 528–536.

Hart, G.L., and B. Lowery, 1998. Measuring instantaneous solute flux and loading with time domain reflectometry. Soil Sci Soc Am J, 62:23–35.

Heimovaara, T.J., A.G. Focke, W. Bouten, and J.M. Verstraten, 1995. Assessing temporal variations in soil water composition with time domain reflectometry. Soil Sci Soc Am J, 59:689–698.

Hendrickx, J.M.H., B. Baerends, Z.I. Raza, M. Sadig, and M. Akram Chaudhry, 1992. Soil salinity assessment by electromagnetic induction of irrigated land. Soil Sci Soc Am J, 56:1933–1941.

Hendrickx, J.M.H., B. Borchers, D.L. Corwin, S.M. Lesch, A.C. Hilgendorf, and J. Schlue, 2002. Inversion of soil conductivity profiles from electromagnetic induction measurements: theory and experimental verification. Soil Sci Soc Am J, 66:673–685.

Holder, M., K.W. Brown, J.C. Thomas, D. Zabcik, and H.E. Murray, 1991. Capillary-wick unsaturated zone soil pore water sampler. Soil Sci Soc Am J, 55(5):1195–1202.

James, R.O., and T.W. Healy, 1971. Adsorption of hydrolyzable metal ions at the oxide-water interface: Co(II) adsorption on SiO_2 and TiO_2 as model systems. J Coll Interface Sci, 40:42–52.

Jemison, J.M., and R.H. Fox, 1994. Nitrate leaching from nitrogen fertilized and manured corn measured with zero-tension pan lysimeters. J Environ Qual, 23:337–343.

Joffe, J.S., 1932. Lysimeter studies. I. Moisture percolation through soil profiles. Soil Sci, 34:123–143.

Jones, S.B., J.M. Wraith, and D. Or, 2002, Time domain reflectometry measurement principles and applications, Hydrol Proc, 16:141–153.

Kachanoski, R.G., E. Pringle, and A. Ward, 1992. Field measurement of solute travel times using time domain reflectometry. Soil Sci Soc Am J, 56:47–52.

Karnok, K.J., and R.T. Kucharski, 1982. Design and construction of rhizotron-lysimeter facility at the Ohio State University. Agron J, 74:152–156.

Keller, C., and B. Travis, 1993. Evaluation of the potential of fluid absorber mapping of contaminants in ground water. p.421. In Proceedings of the Seventh National

Outdoor Action Conference and Exposition. National Groundwater Association, Dublin, OH.

Keller, C., and J.M.H. Hendrickx, 2002. Capillary absorbers. pp. 1308–1311. In: W.A. Dick et al. (ed.), Methods of Soil Analysis Part 4 — Physical Methods. Soil Science Society of America Inc., Madison, WI.

Kemper, W.D., 1959. Estimation of osmotic stress in soil water from the electrical resistance of finely porous ceramic units. Soil Sci, 87:345–349.

Kilmer, V.J., O.E. Hays, and R.J. Muckenhirn, 1944. Plant nutrients and water losses from Fayette silt loam as measured by monolith lysimeters. J Am Soc Agron, 36:249–263.

Knutson, J.H. and J.S. Selker, 1996. Fiberglass wick sampler effects on measurement of solute transport in the vadose zone. Soil Sci Soc Am J, 60:420–424.

Knutson, J.H., and J.S. Selker, 1994. Unsaturated hydraulic conductivities of fiberglass wicks and designing capillary pore-water samplers. Soil Sci Soc Am J, 58:721–729.

Knutson, J.H., S.B. Lee, W.Q. Zhang, and J.S. Selker, 1993. Fiberglass wick preparation for use in passive capillary wick soil pore-water samplers. Soil Sci Soc Am J, 57:1474–1476.

Koglin, E.N., E.J. Poziomek, and M.L. Kram, 1995. Emerging technologies for detecting and measuring contaminants in the vadose zone. p. 657–700. In L.G. Wilson et al. (ed.), Handbook of Vadose Zone Characterizing and Monitoring. Lewis Publishers, Boca Raton, FL.

Kohnke, H., F.R. Dreibelbis, and J.M. Davidson, 1940. A survey and discussion of lysimeters and a bibliography on their construction and performance. Misc. Publ. no. 374, Department of Agriculture, Washington, DC.

Krohn, D.A., 1988. Chemical analysis. p.193–209. In Fiber Optic Sensors: Fundamentals and Applications. Instrument Society of America, Research Triangle Park, NC.

Krone, R.B., H.F. Ludwig, and J.F. Thomas, 1951. Porous tube device for sampling soil solutions during water spreading operations. Soil Sci, 73:211–219.

Kulp, T.J., D. Bishop, and S.M. Angel, 1988. Column-profile measurements using fiber optic spectroscopy. Soil Sci Soc Am J, 52:624–627.

Lesch, S.M., and D.L. Corwin, 2003. Using the dual-pathway conductance model to determine how different soil properties influence conductivity survey data. Agron J, 95:365–379.

Lewis, D.L., A.P. Simons, W.B. Moore, and D.K. Gattie, 1992. Treating soil solution samplers to prevent microbial removal of analytes. Appl Environ Microbiol, 58(1):1–5.

Linden, D.R., 1977. Design, installation and use of porous ceramic samplers for monitoring soil-water quality. U.S. Dep. Agric. Tech. Bull., 1562.

Litaor, M.I., 1988. Review of soil solution samplers. Water Resour Res, 24(5): 727–733.

Livingston, B.E. 1908. A method for controlling plant moisture. Plant World, 11:39–40

Louie, M.J., P.M. Shelby, J.S. Smesrud, L.O. Gatchell, and J.S. Selker, 2000. Field evaluation of passive capillary samplers for estimating groundwater recharge. Water Resour Res, 36(9):2407–2416.

Lynde, C.J., and H.A. Dupre, 1913. On a new method of measuring the capillary lift of soils. J Am Soc Agron, 5:107–116.

Mallants, D., M. Vanclooster, M. Meddahi, and J. Feyen, 1994. Estimating solute transport in undisturbed soil columns using time domain reflectometry. J Contam Hydrol, 17:91–109.

Martens, D.A., 2002. Identification of phenolic acid composition of alkali-extracted plants and soils. Soil Sci Soc Am J, 66:1240–1248.

McMahon, M.A., and G.W. Thomas, 1974. Chloride and tritiated water flow in disturbed and undisturbed soil cores. Soil Sci Soc Am Proc, 38: 727–732.

McNeill, J.D., 1980. Electromagnetic terrain conductivity measurement at low induction numbers. Technical Note TN-6, Geonics Limited, Mississauga, Ontario, Canada.

Mmolawa, K., and D. Or, 2000. Root zone solute dynamics under drip irrigation: a review. Plant Soil, 222:163–190.

Mmolawa, K., and D. Or, 2003. Experimental and numerical evaluation of analytical volume balance model for soil water dynamics under drip irrigation. Soil Sci Soc Am J, 67:1657–1671.

Morrison, R.D., and J.E. Szecsody, 1987. A tensiometer and pore water sampler for vadose zone monitoring. Soil Sci, 144(5):367–372.

Morrison, R.D., K. Lepic, and J. Baker, 1983. Vadose zone monitoring at a hazardous waste landfill. Proceedings of the Characterization and Monitoring of the Vadose Zone. National Well Association, Las Vegas, NV, pp.517–528.

Moutonnet, P., and J.C. Fardeau, 1997. Inorganic nitrogen in soil solution collected with Tensionic samplers. Soil Sci Soc Am J, 61:822–825.

Moutonnet, P., G. Guiraud, and C. Marol, 1989. Le tensiomètre et la teneur en nitrates de la solution de sol. Milieux Poreux Transferts Hydriques, 26:11–29.

Moutonnet, P., J.F. Pagenel, and J.C. Fardeau, 1993. Simultaneous field measurement of nitrate-nitrogen and matric pressure head. Soil Sci Soc Am J, 57:1458–1462.

Moyer, J.W., L.S. Saporito, and R.R. Janke, 1996. Design, construction, and installation of an intact soil core lysimeter. Agron J, 88:253–256.

Mualem, Y., and S.P. Friedmann, 1991. Theoretical prediction of electrical conductivity in saturated and unsaturated soil. Water Resour Res, 27:2771–2777.

Nadler, A., 1980. Determining the volume of sampled soil when using the four-electrode technique. Soil Sci. Soc. Am. J., 44:1186–1190.

Nadler, A., H. Frenkel, and A. Mantell, 1984. Applicability of the four-probe technique under extremely variable water contents and salinity distributions. Soil Sci Soc Am J, 48:1258–1261.

Nadler, A., S. Dasberg, and I. Lapid, 1991. Time domain reflectometry measurements of water content and electrical conductivity of layered soil columns. Soil Sci Soc Am J, 55:938–943.

Neary, A.J., and F. Tomassini, 1985. Preparation of alundum/ceramic plate tension lysimeters for soil water collection. Can J Soil Sci, 65:169–177.

Nielsen, J.M., G.F. Pinder, T.J. Kulp, and S.M. Angel, 1991. Investigation of dispersion in porous media using fiber-optic technology. Water Resour Res, 27(10):2743–2749.

Or, D., 2001. Who invented the tensiometer? Soil Sci Soc Am J, 65:1–3.

Parizek, R.R., and R.E. Lane, 1970. Soil-water sampling using pan and deep pressure-vacuum lysimeters. J Hydrol 11(1):1–21.

Pueyo, M., J. Sastre, E. Hernández, M. Vidal, J.F. López-Sánchez, and G. Rauret, 2003. Prediction of trace element mobility in contaminated soils by sequential extraction. J Environ Qual, 32:2054–2066.

Raulund-Rasmussen, K., 1989. Aluminum contamination and other changes of acid soil solution isolated by means of porcelain suction cups. J Soil Sci, 40:95–101.

Reeder, J.D., 1986. A nonweighing lysimeter design for field studies using nitrogen-15. Soil Sci Soc Am J, 50:1224–1227.

Rehm, B., B. Christel, T. Stolzenberg, D. Nichols, B. Lowery, and B. Andraski, 1986. Field evaluation of instruments of unsaturated hydraulic properties of fly ash. Electric Power Research Institute, Palo Alto, CA.

Reicosky, D.C., R.J. Millington, and D.B. Peters, 1970. A salt sensor for use in saturated and unsaturated soils. Soil Sci Soc Am Proc, 34:214–217.

Rhoades, J.D., 1978. Monitoring soil salinity: a review of methods. pp.150–165. In L.G. Everett and K.D. Schmidt (ed.), Establishment of Water Quality Monitoring Programs. American Water Resources Association, St. Paul, MN.

Rhoades, J.D., and A.D. Halvorson, 1977. Electrical conductivity methods for detecting and delineating saline seeps and measuring salinity in northern Great Plains soils. USDA-ARS-42. U.S. Government Printing Office, Washington, DC.

Rhoades, J.D., and J. van Schilfgaarde, 1976. An electrical conductivity probe for determining soil salinity. Soil Sci Soc Am J, 40:647–651.

Rhoades, J.D., and J.D. Oster, 1986. Solute Content. In: Methods of Soil Analyses Part 1 — Physical and Mineralogical Methods 2nd ed. A. Klute (ed.), Soil Science Society of America, Inc., Madison, WI.

Rhoades, J.D., and R.D. Ingvalson, 1971. Determining salinity in field soils with soil resistance measurements. Soil Sci Soc Am Proc, 35:54–60.

Rhoades, J.D., P.A.C Raats, and R.J. Prather, 1976. Effects of liquid-phase electrical conductivity, water content, and surface conductivity on bulk soil electrical conductivity. Soil Sci Soc Am J, 40:651–655.

Richards, L.A., 1966. A soil salinity sensor of improved design. Soil Sci Soc Am Proc, 30:333–337.

Rimmer, A., Steenhuis, T.S. and Selker, J.S., 1995. Wick samplers: an evaluation of solute travel times. Soil Sci Soc Am J, 59:235–243.

Rimmer, A., T.S. Steenhuis, and J.S. Selker, 1995. One-dimensional model to evaluate the performance of wick samplers in soils. Soil Sci Soc Am J, 59:88–92.

Risler, P.D., J.M. Wraith, and H.M. Gaber, 1996. Solute transport under transient flow conditions estimated using time domain reflectometry. Soil Sci Soc Am J, 60:1297–1305.

Rogers, K.R., and E.J. Poziomek, 1996. Fiber optic sensors for environmental monitoring. Chemosphere, 33(6):1151–1174.

Roth, K., R. Schulin, H. Flühler, and W. Attinger. 1990. Calibration of time domain reflectometry for water content measurement using composite dielectric approach. Water Resour Res, 26(10):2267–2273.

Saragoni, H., R. Poss, and R. Oliver, 1990. Dynamique et lixiviation des élé ments minéraux dans les terres de barre du sud du Togo. Agron Trop (Paris), 45:259–273.

Selker, J.S., 2002. Passive Capillary Samplers. p.1266–1269. In: W.A. Dick et al. (ed.), Methods of Soil Analysis Part 4 — Physical Methods. Soil Science Society of America Inc., Madison, WI.

Shaffer, K.A., D.D Fritton, and D.E. Baker, 1979. Drainage water sampling in a wet, dual-pore soil system. J Environ Qual, 8(2):241–246.

Shuford, J.W., D.D. Fritton, and D.E. Baker, 1977. Nitrate–nitrogen and chloride movement through undisturbed field soil. J Environ Qual, 6:255–259

Slater, L., A. Binley, and D. Brown, 1997. Electrical imaging of fractures using ground-water salinity change. Ground Water, 35:436–442.

Slater, L., A. Binley, W. Daily, and R. Johnson, 2000. Cross-hole electrical imaging of a controlled saline tracer injection. J Appl Geophys, 44:85–102.

Smith, C.N., and R.F. Carsel, 1986. A stainless-steel soil solution sampler for monitoring pesticides in the vadose zone. Soil Sci Soc Am J, 50:263–265.

Starr, M.R., 1985. Variation in the quality of the tension lysimeter soil water samples from Finnish forest soil. Soil Sci, 140(6):453–461.

Stone, D.M., and J.L. Robl, 1996. Construction and performance of rugged ceramic cup soil water samplers. Soil Sci Soc Am J, 60:417–420.

Sudduth, K.A., N.R. Kitchen, G.A. Bollero, D.G. Bullock, and W.J. Wiebold, 2003. Comparison of electromagnetic induction and direct sensing soil electrical conductivity. Agron Jl, 95:472–482.

Thompson, M.L., and R.L., Scharf, 1994. An improved zero-tension lysimeter to monitor colloid transport in soils. J Environ Qual, 23:378–383.

Topp, E., and W. Smith, 1992. Sorption of herbicides atrazine and metolachlor to selected plastics and silicone rubber. J Environ Qual, 8(2):316–317.

Topp, G.C., J.L. Davis, and A.P. Annan, 1980. Electromagnetic determination of soil water content: measurements in coaxial transmission lines. Water Resour Res, 16:574–582.

Triantafilis, J., G.M. Laslett, and A.B. McBratney, 2000. Calibrating an electromagnetic induction instrument to measure salinity under irrigated cotton. Soil Sci Am J, 64:1009–1017.

Tyler, D.D., and G.W. Thomas, 1977. Lysimeter measurements of nitrate and chloride losses from soil under conventional and no-tillage corn. J Environ Qual, 6:63–66.

Vanclooster, M., D. Mallants, J. Diels, and J. Feyen, 1993. Determining local–scale solute transport parameters using time domain reflectometry (TDR). J Hydr, 148:93–107.

Villar-Mir, J.M., P. Villar-Mir, C.O. Stockle, F. Ferrer, and M. Aran, 2002. On-farm monitoring of soil nitrate-nitrogen in irrigated cornfields in the Ebro valley (northeast Spain). Agron J 94:373–380.

Vogeler, I., B.E Clothier, S.R. Green, D.R. Scotter, and R.W. Tillman, 1996. Characterizing water and solute movement by TDR and disk permeametry. Soil Sci Soc Am J, 60:5–12.

Vogeler, I., B.E. Clothier, and S.R. Green, 1997. TDR estimation of electrolyte in the soil solution. Aust Soil Res, 35:515–526.

Vogeler, I., C. Duwig, B.E. Clothier, and S.R. Green, 2000. A simple approach to determine reactive solute transport using time domain reflectometry. Soil Sci Soc Am J, 64:12–18.

Ward, A.L., R.G. Kachanoski, and D.E. Elrick, 1994. Laboratory measurements of solute transport using time domain reflectometry. Soil Sci Soc Am J, 58:1031–1039.

Warrick, A.W., and A. Amoozegar-Fard, 1977. Soil water regimes near porous cup water samplers. Water Resour Res, 13(1):203–207.

Weber, J.B., 1995. Physicochemical and mobility studies with pesticides. pp. 99–115. In: M.L. Leng, E.M.K. Leovey and P.L. Zubkoff (ed.), Agrochemical Environmental Fate Studies: State of the Art. CRC Press, Boca Raton, FL.

Wenzel, W.W., R.S. Sletten, A. Brandstetter, G. Wieshammer, and R. Stingeder, 1997. Adsorption of trace metals by tension lysimeters: nylon membrane vs. porous ceramic cup. J Environ Qual, 26:1430–1434.

Wesseling, J., and J.D. Oster, 1973. Response of salinity sensors to rapidly changing salinity. Soil Sci Soc Am Proc, 37:553–557.

Winton, K., and J.B. Weber, 1996. A review of field lysimeter studies to describe the environmental fate of pesticides. Weed Technol, 10:202–209.

Wolff, R.G., 1967. Weathering Woodstock granite, near Baltimore, Maryland. Am J Sci, 265:106–117.

Wood, J.D., 1978. Calibration stability and response time for salinity sensors. Soil Sci Soc Am J, 42:248–250.

Wood, W.W., 1973. A technique using porous cups for water sampling at any depth in the unsaturated zone. Water Resour Res, 9(2):486–488.

Wraith, J.M., S.D. Comfort, B.L. Woodbury, and W.P. Inskeep, 1993. A simplified analysis approach for monitoring solute transport using time-domain reflectometry. Soil Sci Soc Am J, 57:637–642.

Zimmermann, C.F., M.T. Price, and J.R. Montgomery, 1978. A comparison of ceramic and Teflon *in situ* samplers for nutrient pore water determinations. Estuarine Coastal Marine Sci, 7:93–97.

10 Time Domain Reflectometry as an Alternative in Solute Transport Studies

Iris Vogeler, Steve Green, and Brent E. Clothier
Environment and Risk Management Group, HortResearch
Institute, Palmerston North, New Zealand

CONTENTS

1-5667-0657-2/05/$0.00 + $1.50
© 2005 by CRC Press

10.1 INTRODUCTION

Knowledge of the mechanisms of water and solute transport in soils and other porous media is fundamentally important in many diverse fields, from agriculture, through hydrology and petroleum engineering, to environmental science. Better observation and measurement technology is a key component for achieving better knowledge. Techniques are needed that are reliable, robust, automated, and can be used across a wide range of spatial and temporal scales. Such techniques not only will improve our ability to manage soil and water resources, they are also a prerequisite for achieving advances in our fundamental understanding of water and solute transport through soils in both the laboratory and the field.

A common technique for monitoring solute movement is to perform column leaching experiments and measuring the concentration of solute quitting the base of the column, which can be presented as a breakthrough curve (BTC). The BTC can then be used to assess solute-transport parameters (van Genuchten, 1981). Column leaching experiments are, however, time consuming, and they do not provide information about how the transport parameters may vary with depth. As Jury and Roth (1990) pointed out, transport models based on quite different hypotheses, e.g., the convective-dispersive model or the stochastic convective scheme, can both achieve good agreement with solute outflow concentration data obtained at a single depth. Yet, transport models should also be tested using observations at different distances from the surface, especially in structured or layered soils (Dyson and White, 1987; Jury and Roth, 1990).

Over the last two decades, the technique of time domain reflectometry (TDR) has been used to monitor not only the soil's water content, but also the bulk soil electrical conductivity (σ). The latter can be related to the electrolyte concentration in the soil solution. TDR for measuring soil water content, θ, has now become a well-accepted method (Topp and Davis, 1985; Zegelin et al. 1989; Dalton, 1992). TDR has also proved successful for measuring the electrolyte concentration in the soil solution, provided that appropriate calibrations are made (Topp and Davis, 1985; Dasberg and Dalton, 1985; Heimovaara, 1993). With this dual capacity, TDR offers great scope for monitoring solute transport under both steady-state and transient-flow conditions. Increasingly, researchers are now using this method because it is the easiest and most reliable way to measure both θ and σ continuously and *in situ*. The probes themselves are simple and can vary in size and length and be developed to cover a wide range of experimental conditions. In addition, accurate TDR data can also be used to test solute transport models and/or characterize their transport parameters.

In this chapter we provide a practical introduction to the principles of TDR as it is used to measure both the volumetric water content and the resident solute concentration. We describe different calibration approaches that have been used to relate the TDR-measured impedance (Z_L) to the soil's electrical conductivity (σ), and to the resident concentration in the soil solution (C_r). We review the laboratory studies of solute transport that have been made under a range of conditions, including steady-state and transient water flow, and using both inert and reactive solutes. Transport models and parameterization using TDR data are described briefly. Practical guidance for setting up TDR transport studies is also given.

10.2   THE TDR SYSTEM FOR MONITORING WATER AND SOLUTE TRANSPORT

10.2.1   THE MEASUREMENT SYSTEM

The main components of a TDR system are a cable tester in some form (e.g., 1502C Tektronix, Beaverton, OR), a coaxial connection cable, and the TDR probe itself. In some commercial TDR units the cable testing role is incorporated in a tailor-made system (e.g., trace, soil moisture equipment). We will now only consider the cable testing case, as this allows a discussion of the fundamental principles. For automated measurements using a cable tester, this device can be controlled via a computer. A multiplexer can be used to sequentially select an array of probes that are embedded in the soil. A typical setup with three-wire probes is shown in Figure 10.1. The computer is used to control the cable tester, and the multiplexer for automatic collection and analysis of the TDR trace for water content (θ) and impedance (Z_L). Special software, such as the one of Baker and Allmaras (1990), can then be used for the waveform analysis of θ and Z_L. The bulk soil electrical conductivity σ) is derived from a measure of Z_L. The electrical conductivity of the soil solution (σ_w) and the solute resident concentration (C_r) are determined via an empirical calibration. The impedance Z_L can also be directly read from the TDR unit.

10.2.2   TDR OPERATION

The TDR cable tester, or its commercial equivalent, sends out an electromagnetic pulse, which travels down the coaxial cable and enters the TDR probe, which is embedded in the soil. At the end of the probe, the pulse is reflected back. The TDR device measures the propagation and reflection of the pulse, as well as its attenuation.

The entry and return travel time of the pulse depends on the dielectric constant (ε) of the soil, which comprises a volume fraction of soil solids, water, and air. Representative values of ε are 80.36 for liquid water, 3–5 for soil minerals, 6–8 for organic matter, and 1 for air at a frequency of 1 GHz and a temperature of 20°C (Weast, 1965). Thus, ε of a soil is dominated by the volumetric content of water because of its high ε value. The dielectric constant of the soil can be related, by calibration, to the volumetric water content of the soil. The impedance, Z_L, is a measure of the total resistance of a conductor to AC current and is measured in ohms [Ω].

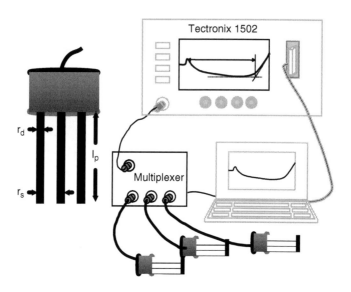

FIGURE 10.1 Schematic diagram of a TDR measurement system.

Any change in the impedance along the probe causes a partial reflection of the incident wave, and this causes a change in voltage between the conductor and the shield of the coaxial cable. The TDR cable tester measures the change in voltage. This is displayed on the oscilloscope screen. The reflection coefficient (r) is the ratio of the voltage reflected back to the receiver $V_f - V_0$), divided by the output voltage applied from the TDR unit (V_0). The entire curve can be simply called the trace. The dielectric constant and the impedance or bulk soil electrical conductivity can be calculated from this TDR trace. Such measurements provide the basis for calculating the volumetric water content in the zone of influence around the probes and the corresponding electrical conductivity of the soil solution sensed by the probes.

An idealized TDR trace is illustrated in Figure 10.2. The Tektronix cable tester sends out an electromagnetic pulse that has an amplitude of about 0.225 V. The output pulse then enters the coaxial cable at point A, and travels down the cable to the TDR probe at point B, which is the start of the probe. At this point some of the voltage pulse is reflected back down the coaxial cable. The remainder of the pulse is transmitted down the probe. Any decrease in impedance beyond point B results in a counterphase reflection and produces a drop in amplitude compared to the output voltage pulse. At point C, which corresponds to the end of the probe, the remainder of the voltage pulse is reflected, in phase, so that the total amplitude becomes twice the transmitted voltage pulse in the absence of any further signal loss. On the way back to point B some of the reflected signal is again transmitted through to point A, and the remainder is again reflected back towards point C.

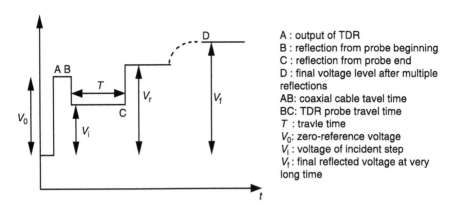

A : output of TDR
B : reflection from probe beginning
C : reflection from probe end
D : final voltage level after multiple reflections
AB: coaxial cable tavel time
BC: TDR probe travel time
T : travle time
V_0: zero-reference voltage
V_i : voltage of incident step
V_f: final reflected voltage at very long time

FIGURE 10.2 An idealized TDR voltage trace vs. time.

This continues until the multiple reflections finally die away. The final voltage level D is reached.

10.2.3 EXPERIMENTAL SETUP FOR LABORATORY EXPERIMENTS

For studying solute transport in the laboratory, TDR can be either combined with a disk permeameter, a rainfall simulator, or ponded infiltration. These can be used to study solute transport under either saturated or unsaturated conditions, as might prevail in the field. While disk permeameters are generally easier to use, rainfall simulators can be used to study solute transport in intact columns growing plants. Figure 10.3 shows a setup used for a laboratory study (Vogeler et al., 1997) with a disk permeameter on top of the soil column. The disk permeameter can be set to pressure potential heads between saturation and an h_0 of $-150\,mm$, allowing transport studies under unsaturated conditions (Figure 10.3). An automated leachate collector under vacuum allows the same pressure potential to be applied at the bottom of the column and permits regular sampling of the leachate. TDR probes in this setup were installed horizontally into the soil column at depths of 50, 100, and 150 mm. If undisturbed soil columns are used, it is recommended to predrill undersized holes to avoid soil disturbance and ensure good contact. Vogeler et al. (1997) used three-rod probes of length 100 mm, diameter 2 mm, and spacings between the center and outer rods of 12.5 mm. The probes were connected, via a $50\,\Omega$ coaxial cable and a multiplexer, to the cable tester. TDR measurements of θ and σ were made regularly, with time increment depending on the flow rate. If the temperature in the laboratory is not constant during the experiment, temperature measurements need to be made. The TDR-measured impedance can then be standardized to 25°C.

Solute transport studies with TDR can also be carried out without collecting the leachate. However combining the two techniques of TDR and

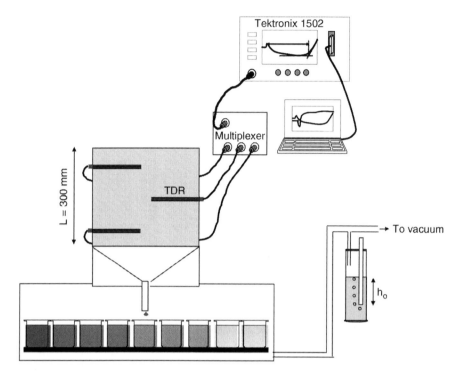

FIGURE 10.3 Experimental setup for laboratory studies with disc permeameter, soil column, leachate collector, TDR cable tester, multiplexer, and computer.

BTCs offers a way of employing the synergy that comes from using both new technologies for measurement and modeling for prediction and parameterization.

10.2.4 PROBE DESIGN AND PLACEMENT

The first application of TDR to measure soil water content used coaxial transmission lines, consisting of a steel cylinder filled with disturbed soil (Topp et al., 1980). In later studies Topp et al. (1982) used parallel probes (two-wire probes) with an impedance-matching pulse transformer (balun). The balun was needed to go from the coaxial cable to the parallel probe. Zegelin et al. (1989) were the first to introduce a multiwire transmission line that overcomes the need for a balun. The characteristic impedance of multiwire probes approach that of a coaxial probe and they produce a much 'cleaner' signal than a two-wire probe. For the determination of electrical conductivity, it is essential that the impedance and the electrical geometry of the probe be as closely matched as possible to the coaxial cable that connects the probe to the TDR device. Any impedance mismatches between the probe and the cable cause a partial reflection and some loss of pulse.

The minimum number of wires needed to simulate a coaxial transmission line is three. Whereas probes with more than three wires will more closely approximate a coaxial transmission line, they will also result in a greater soil disturbance. Three-wire probes are a good compromise, and they are now the most widely used. The center conductor of the coaxial cable is connected to the central wire of the three-wire probe. The two outer wires of the probe are connected to the coaxial shield. Recently, Vogeler et al. (2002) used a dielectric lysimeter to study solute transport through soil columns. The dielectric lysimeter consists of a copper cylinder 200 mm in length and 150 mm in diameter, with a central rod going vertically through the axis of the cylinder. The conductor of the coaxial cable from the TDR is connected to the central rod, and the shield is connected to the outer cylinder. This design is similar to the transmission line used in the study of Topp et al. (1980).

For practical purposes, the length of the probe (l_p) (see Figure 10.1), is limited by the strength of the TDR signal (the voltage step rise time), the resolution of the TDR instrument, and the attenuation of the TDR signal by dielectric losses as the pulse propagates through the soil. The upper limit for l_p is imposed by signal loss in the soil, which can be severe in saline or heavy clay soils. In the field, probes with up to 2 m length have been used. In the laboratory the maximum length is limited by the column size. The lower limit for l_p is imposed by the time resolution of the TDR device. This is currently in the order of 0.1 ns. This limits the minimum probe length to about 50 mm, although Ren et al. (1999) suggested that probes as small as about 20 mm could be used. Short probes of 50 mm should not be used with cables longer than about 3 m, as long cables increase the rise time of the voltage pulse, which results in reflections with smaller amplitudes and smaller slopes.

Very small rod-to-rod spacings, r_s (see Figure 10.1), will result in a reduced measurement zone, which reduces the representative sample volume. On the other hand, wider rod separations will affect the rising time of the propagated signal. Zeglin et al. (1992) suggested that the upper limit of r_s for the signal propagation should be about 0.1 m.

The size of the rod diameter, r_d (see Figure 10.1), is also limited because of problems with probe insertion and soil disturbance. The lower limitation to $_d$ arises because of measurement volume sensitivity (Knight, 1992). The most important factor in designing TDR probes is the ratio of r_s to r_d. Knight (1992) recommended that the ratio r_s/r_d should not be greater than 10, and $_d$ should be as large as possible as long as there is no significant compaction and local disturbance by insertion.

Of great importance to TDR measurements is the volume of the soil sensed by the probe and the sensitivity of the measurement to the spatial distribution of ε and σ in the soil. This sensitivity was illustrated by Zegelin et al. (1989) using the approximate form of the electric field distribution around TDR probes. The main part of the field is concentrated close to the transmission line so that TDR measurements are most sensitive to the region close to the wires. Thus, TDR weights σ nonlinearly in the plane perpendicular to the long axis of the probe, with local maximums at the surface of the probe. This can result in TDR-induced diffusion in the case of steep concentration gradients within the profile, as was demonstrated by Nissen et al. (2001). This feature of TDR needs to be taken into account when comparing TDR measurements of water contents or electrical conductivity with those obtained from other methods. The TDR zone of influence can be approximated with a width 1.4 times the interrod distance (Knight, 1992) and a length equal to the TDR probe length. Cracks, root channels, or other kinds of heterogeneity in the soil can therefore result in errors of observation relative to the bulk soil. Because of the heightened sensitivity close to the wires, special care needs to be taken with insertion of the probe. Also, for shrink-swell soils, it may not be possible to rely on measurements from probes that remain in the soil permanently.

For practical purposes TDR probes can be installed either vertically or horizontally into the soil. In either case the measured values of ε and σ are representative of the average value over the total length of the probe (Nadler et al., 2002). Thus, it is assumed that vertically installed probes measure the average θ and σ over their entire length, integrating any changes with depth (Kachanoski et al., 1994). Horizontally installed probes, on the other hand are assumed to integrate horizontally, and they can provide a precise depth-wise measurement of θ and σ profiles. However, in the field their insertion requires that a pit be dug, and this can lead to significant disturbance of the soil profile. With vertically installed TDR probes, analysis of the TDR trace is often problematic in multilayered heterogeneous soils because of multiple reflections. To obtain representative measurements of resident concentrations for each measurement depth in the soil profile, horizontal TDR probes are thus preferable in well-structured heterogeneous soils. Nonetheless, a great advantage of TDR is that probes can be installed vertically, horizontally, or even at an angle (Lee et al., 2002) very close to the soil surface with no loss in accuracy. This was shown by Baker and Lascano (1989), who measured water contents with probes placed just 20 mm below the soil surface.

10.2.5 TDR DATA ANALYSIS

10.2.5.1 Soil Moisture Content

The measurement of the water content by TDR is based on the determination of the dielectric constant (ε). The latter is determined using:

$$\varepsilon = \left(\frac{l_a}{l_p v_p}\right)^2 \tag{10.1}$$

where l_a is the apparent length obtained from the TDR trace and v_p the relative velocity of propagation to the speed of light, which accounts for the specific dielectric of the coaxial cable.

The volumetric water content, θ, of the soil can then be determined by calibration based on a simple measurement of ε. Different approaches have been proposed to relate ε and θ. Some employ mixing laws (Roth et al., 1990; Dirksen and Dasberg, 1993), while others are empirical (Topp et al., 1980). The empirical relationship found by Topp et al. (1980) was initially claimed to be universal and independent of soil texture, soil density, and salinity influences. For this reason it is often referred to as the "universal Topp relation" (Zegelin et al., 1992) and is given by:

$$\theta = -5.3 \times 10^{-2} + 2.92 \times 10^{-2}\varepsilon - 5.5 \times 10^{-4}\varepsilon^2 + 4.3 \times 10^{-6}\varepsilon^3 \qquad (10.2)$$

This empirical relationship has been substantiated by other investigators (Dalton and van Genuchten, 1986; Zeglin et al., 1989). Roth et al. (1992) found that Eq. (10.2) can be used for mineral soils with an absolute error in θ of about $0.015\,m^3\,m^{-3}$. If higher accuracy is required, individual calibrations were considered necessary. The influence of other soil factors on the θ-ε relation have also been studied. These include bulk density (Ledieu et al., 1986), soil temperature (Roth et al., 1990), clay content, and organic matter (Jacobsen and Schjønning, 1993). For organic soils the "universal" relationship has often been found inappropriate (Vogeler et al., 1996). This discrepancy has been attributed to the higher amount of bound water associated with organic matter. Bound water has a lower dielectric constant then free water, due to the restricted rotational freedom of the liquid water molecules. The relationship between ε and θ developed by Topp et al. (1980) has now been widely adopted as a calibration standard for soils with low organic matter content.

The use of mixing models is an alternative to Eq. (10.1). Here ε is obtained from the individual ε values as associated with three soil components: the soil, the enclosed air, and the remaining volume fraction of water. The ε of a soil is dominated by the volumetric content of water because of its high ε value. Major constraints for the use of mixing models include the difficulty in determining the required constants and the effect of bound water.

10.2.5.2 Solute Concentration

Dalton et al. (1984) first proposed the use of TDR for measuring the soil's bulk electrical conductivity (σ). They demonstrated that the attenuation of the voltage pulse as it travels along the TDR probe could be used to deduce σ. This σ could then be used to infer the solute resident concentration (C_r). Several different approaches have subsequently been suggested based on the voltages at different points along the TDR trace (Figure 10.2). So far it remains unclear as to which of the alternative expressions is the most appropriate for the calculation of σ.

Dalton et al., (1984) used the ratio of the signal amplitude at the start of the probe, V_i, and the voltage upon its first reflection from the probe ($V_r - V_i$), to find σ (Figure 10.2):

$$\sigma_D = \left[\frac{\varepsilon^{1/2}}{120\pi l_P} \ln\left(\frac{V_r - V_i}{V_i}\right) \right] \qquad (10.3$$

Later work by Topp et al. (1988) demonstrated that multiple reflections of the signal could occur. These multiple reflections were ignored in the approach of Dalton et al. (1984). By using the attenuation voltage at long times (V_f), Yanuka et al. (1988) considered the effect of multiple reflections. They proposed that

$$\sigma_Y = \left[\frac{\varepsilon^{1/2}}{120\pi l_p} \ln\left(\frac{V_i V_f - V_0(V_i + V_f)}{V_0(V_i - V_f)}\right) \right] \qquad (10.4$$

where V_0 is the TDR output voltage.

Zegelin et al. (1989) adapted the method of thin-sample conductivity analysis, originally proposed by Giese and Tiemann (1975), to show good agreement between values of σ measured with an AC conductivity bridge and those measured by TDR. They used the formulation

$$\sigma_Z = \frac{\varepsilon^{1/2}}{120\pi l_p} \left[\frac{V_i}{V_f} \left(\frac{(2V_0 - V_f)}{2V_0 - V_i}\right) \right] \qquad (10.5$$

It is noted that all of these formulations require simultaneous measurement of the dielectric constant to determine a value for σ. Furthermore, as noted by Mojid et al. (1997), electrical conductivities estimated by these different equations can deviate considerably from each other.

Nadler et al. (1991) proposed yet another method of determining σ based on a measurement of the impedance of the TDR probe (Z_L [Ω]). A value for Z_L can either be read directly from the screen or determined using the refelction coefficient r, which is the ratio between the reflected amplitude at long times ($V_f - V_0$) to the incident signal amplitude (V_0). Thus here,

$$r = \frac{Z_L - Z_0}{Z_L + Z_0} \qquad (10.6$$

where Z_0 is the characteristic impedance of the cable (in our case 50 Ω).

Nadler et al. (1991) then converted Z_L to σ using an equation identical to that of Rhoades and van Schilfgaarde (1976):

$$\sigma_{25} = K_G f Z_L^{-1} \qquad (10.7$$

where σ_{25} is the bulk soil electrical conductivity at 25°C, K_G is the probe geometry constant [m^{-1}], and f is a correction coefficient for temperatures other than 25°C (U.S. Salinity Laboratory Staff, 1954, Table 15). The value of K_G is influenced by the length, spacing, and diameter of the TDR probes. This constant can, however, be determined easily by immersing the TDR probe into solutions of various known electrical conductivities and measuring the respective impedances (Nadler et al., 1991):

$$K_G = \sigma_{ref} Z_L f^{-1} \tag{10.8}$$

where σ_{ref} is the electrical conductivity of a reference solution at 25°C. Reference solutions ranging from 0 to 0.5 M KCl have typically been used (Zeglin et al., 1989; Vogeler et al., 1996).

A value for Z_L can also be calculated using the respective voltages (Wraith et al., 1993):

$$Z_L = Z_0 \frac{(V_i - V_0) + (V_f - V_i)}{(V_i - V_0) - (V_f - V_i)} = Z_0 \frac{V_f - V_0}{2V_i - V_f - V_0} \tag{10.9}$$

The bulk soil electrical conductivity is then calculated using Eq. (10.6).

Nadler et al. (1991) compared various methods for obtaining σ and concluded that their method [Eqs. (10.6–10.8)] and the one of Dalton et al. (1984) [Eq. (10.3)] were the most suitable for calculating σ from TDR measurements. The methods of Nadler et al. (1991) and Wraith et al. (1993) provide a practical means for the determination of σ, and they require parameters that are easily measured.

The TDR measured electrical conductivity (σ) is a complex function of the soil water content (θ), the pore-water electrical conductivity (σ_w) due to ionic species in solution, the surface conductance (σ_s) due to the presence of surface charges on the soil minerals, and the tortuosity of the electrical flow paths. Different approaches have been used to obtain σ_w from measured values of σ. (discussed in Sec. 10.2.6.) Once σ_w is determined, the solute resident concentration (C_r) can be calculated from known relationships between σ_w and C_r, providing the chemical composition of the soil solution is also known. Alternatively, as discussed for the indirect calibration method, a relative σ can be used to obtain transport parameters.

The relation between Z_L, σ, σ_w, and C_r is generally assumed to be linear, at least for a σ_w up to 0.8–2 S m^{-1}, for a fixed water content (Dalton and van Genuchten, 1986; Vanclooster et al., 1993). With higher conductivity, the relation becomes nonlinear. The degree of nonlinearity is influenced by the characteristics of the TDR probe, with three-wire sensors being superior to two-wire sensors (Zeglin et al., 1989). At low σ_w concentrations, under 0.1–0.5 S/m, a nonlinearity in the σ-σ_w relation has been observed in some soils (Shainberg et al., 1980; Nadler and Frenkel, 1980; Rhoades et al., 1989). This nonlinearity has been attributed to the effect of clay content of the soil, by the Na^{2+} saturation, the water content, and by probe size (Nadler, 1997). The curvilinear relationship at low concentrations can result in an underestimation of solute in the soil, which in transport studies would result in a less dispersed BTC. This can be avoided by using water with a baseline concentration of σ_w greater then the minimum value for linearity. Thus TDR studies of solute transport should be made using solutions with a range of σ_w from about 0.1 to $2\,S\,m^{-1}$. Note that this is an approximate guide that depends on the soil.

10.2.6 CALIBRATION

TDR calibration has been the topic of many studies. It is already well known that apart from θ, σ_w, and σ_s, the bulk density and soil temperature can also affect the TDR-measured σ. However, the overall effect of these soil factors on σ is not yet fully understood. Various approaches have been adopted to relate σ and C_r (Ward et al., 1994; Mallants et al., 1996). These include both the direct and indirect calibration approaches.

10.2.6.1 Direct Calibration Approach

In the direct calibration approach, the TDR-measured impedance (Z_L), or bulk electrical conductivity (σ), is related to the electrical conductivity of the soil solution (σ_w) and the soil water content (θ). Values of σ_w are converted into solution concentrations assuming a linear relationship exists between σ_w and C_r. This calibration procedure is generally made in a separate experiment. Once a σ-σ_w-θ relationship has been found, it should be possible to study solute transport under transient conditions. Various approaches exist. The first conceptual model was presented by Rhoades et al. (1976) using a four-electrode resistance technique. Their approach considered a constant surface conductivity, a linear relationship between σ and σ_w, plus a tortuosity factor τ) that depended on θ. They suggested that the electrical conductivity of a soil could be approximated as:

$$\sigma = \sigma_s + \sigma_w \theta \tau(\theta) \qquad (10.10$$

using a simple empirical relationship between τ and θ given by

$$\tau = a_1 \theta + b_1 \qquad (10.11$$

Here a_1 and b_1 are soil-dependent constants. Rhoades et al. (1976) presented values for these constants for a few selected soils. For other soils they must be found by calibration. Risler et al. (1996) calculated values for σ_s and τ and found different values for these in different calibration runs. On other occasions negative values for σ_s have been found (Persson and Berndtsson, 1998), and these of course have no physical meaning.

Vogeler et al. (1997) used an alternative approach based on an empirical approach to determine a σ-σ_w-θ relationship. Assuming the surface conductivity is negligible, the pore water electrical conductivity can be obtained from simultaneous measurements of θ and σ using:

$$\sigma_w = \frac{\sigma - (a_2\theta - b_2)}{(c_2\theta - d_2)} \tag{10.12}$$

where a_2, b_2, c_2, and d_2 are assumed to be constant.

Nadler et al. (1984) used the Waxman and Smits (1968) equation, which can be written as:

$$\sigma_w = (\sigma - \delta I_n)F \tag{10.13}$$

where F is a soil-structure–dependent formation factor, δ is the empirical ratio between equivalent conductance of clay counterions to the maximum value of this equivalent conductance, and I_n is the intercept of the linear part of the σ-σ_w curve. The formation factor relates σ to soil texture through the pore size distribution and θ. It is a measure of the tortuosity of the soil with a nonconductive solid phase. Nadler et al. (1984) established a relationship between F and θ for a range of different soils.

Other σ-σ_w-θ relationships have also been developed (Heimovaara et al., 1995; Persson, 1997; Hart and Lowery, 1998). Each approach for finding σ_w from TDR-measured σ and θ requires a determination of soil-specific constants. This remains difficult. Furthermore, as calibrations are generally performed on homogeneously repacked soil columns, these constants might not be appropriate in structured or layered soils.

An alternative approach is the use of physical parameters, such as soil hydraulic properties in the form of tortuosity factors (Mualem and Friedman, 1991; Heimovaara et al., 1995). The advantage of this approach is that the hydraulic parameters are often already measured. However, the above authors and other investigators (Amente et al., 2000) concluded that the approach often fails to accurately estimate soil solution electrical conductivities.

Once the pore water electrical conductivity has been calculated using one of the approaches given above, the solution concentration can then be calculated for the case of a single solute species. For example, if the soil solution contains mainly calcium and chloride, then C_r can be calculated from (Weast, 1965)

$$C_r = -1.2 \times 10^{-3} + 9.4 \times 10^{-2}\sigma_w \tag{10.14}$$

Different linear relationships are needed for other solute species.

We now briefly describe the practical steps involved in determining the σ-σ_w-θ relation based on Eqs. (10.6–10.8) and (10.11). First, the probe

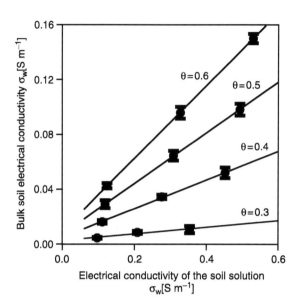

FIGURE 10.4 TDR-measured bulk soil electrical conductivity vs. electrical conductivity of the soil solution measured with a conductivity meter for different water contents of Ramiha silt loam. Also shown are the standard deviations and the fitted relationship. (From Vogeler et al., *Soil Sci. Soc. Am. J.*, 60, 5–12, 1996. With permission.)

geometry constant (K_G) is determined by immersing the TDR probe into various solutions of known concentrations, e.g., KCl solutions ranging from 0.005 to 0.05 M. Then, a calibration for the bulk soil electrical conductivity is performed, generally on homogeneously repacked soil columns, in which both the soil structure and the pore size distribution are disturbed. Soil samples are equilibrated to various θ using different solution concentrations. The pore water electrical conductivity is either set equal to the σ_w of the added solution or measured independently in supernatants obtained by centrifuging soil samples (Vogeler et al., 1996). This latter method is important when exchange reactions are expected between the soil and the added solution as occurs in clay soils. Unfortunately, the centrifuging technique can only be used on reasonably wet and coarse soils.

Figure 10.4 shows a comparison between TDR-measured σ and σ_w as measured with a conductivity meter on supernatants, for the Ramiha silt loam (Vogeler et al., 1996). The soil was wet to average θ of 0.3, 0.4, 0.5, and 0.6 using aqueous KCl solutions of between 0.005 and 0.05 M. The results show that a linear relationship exists between σ and σ_w in the range of about 0.1–0.5 S m^{-1}. The slope and the intercept of the lines depend on θ. From these data values for the constants, a, b, c, and d in Eq. (10.12) were obtained. This direct calibration approach was used to calculate the resident concentration during the invasion of a 0.025 M MgCl$_2$ solution into undisturbed soil columns of Ramiha soil during steady-state water flow. The resulting

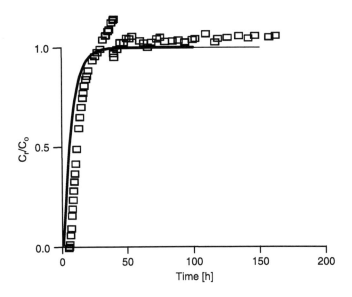

FIGURE 10.5 Normalized resident concentrations calculated from TDR measurements for Ramiha soil at a depth of 50 mm. Also shown are predictions from the CDE with model parameters obtained from the effluent. (From Vogeler et al., *Austr. J. Soil Res.*, 35, 515–526, 1997. With permission.)

concentrations (Figure 10.5) were normalized using the input concentration C_0 and they agree well with the model calculation based on parameters obtained from the effluent concentration. Note that for the $MgCl_2$ solution we used a slightly different equation from the one given above [Eq. (10.14)]. For details on the use of transport models, the reader is referred to Sec. 10.2.7 and Chapters 3 and 11.

However, while the direct calibration approach was appropriate in the strongly aggregated layer of the surface Ramiha silt loam, it failed in the lower profile of the Ramiha soil and in the weakly structured Manawatu fine sandy loam (Vogeler et al., 1997). It seems that the TDR measurement of σ is highly sensitive to the local soil structure, texture, and perhaps bulk density surrounding the probes, and especially to the probe-soil contact. Similar findings with depth-dependent relationships were made by Vanclooster et al. (1995), Mallants et al. (1996), and Persson and Berndtsson (1998). This was attributed to changes in soil structure and physicochemical properties, including the clay content. Persson and Berndtsson (1998) also found a significant difference between the σ-σ_w-θ relationship determined using disturbed or undisturbed soil.

Alternatively, the σ-σ_w-θ relationship can be determined by leaching sufficient pore volumes of an inert solution with various concentrations and then assuming that all the water has been replaced by the inflowing solution. TDR measurements of σ and θ made during several wetting and drying cycles are then related to the concentration of the inflowing solution σ_w

The difficulty using this method is to obtain measurements with uniform water content along the TDR probe over a wide range of θ. Either pressure plates or layers of sand and gravel underneath the soil can be used to obtain low θ values (Persson and Berndtsson, 1998). Recently, Risler et al. (1996) used a vaccum chamber and obtained a θ range of 0.15–0.25 for different soils. The major advantage of this method is that the calibration can be done with undisturbed soil columns (Vanclooster et al., 1995). As demonstrated by Risler et al. (1996), a calibration can be done on the same soil column as the subsequent transport experiment.

The direct calibration approach provides solute concentrations in the soil solution under transient water flow conditions, which often prevail in the field. The method is independent of the type of probe and solute application method. However, the calibration is time-consuming and not unique for a structured soil profile. Many transport issues can be studied under steady-state water flow, and here an indirect calibration approach is useful.

10.2.6.2 Indirect Calibration Approach

If the indirect calibration method is used for solute transport studies, the TDR-measured σ values need not be converted directly into σ_w or C_r values. Instead, the TDR-measured change in impedance, or σ, is related either to the mass (M_0) of the applied tracer (following a pulse application of solute) or to the input concentration (for a step change in solute application). This is possible under steady-state water flow because a linear relationship exists between σ-σ_w-C_r for θ ranging from dry to saturation, and salinity levels from 0.1 to $2\,S\,m^{-1}$. Thus, the relationship between σ and M_0 is also linear. Various indirect approaches exist, depending on the installation of the TDR probes and the type of solute application.

10.2.6.2.1 Pulse Application

The first indirect approach was introduced by Kachanoski et al. (1992) using vertically installed TDR probes and a pulse application of solute infiltrated. The difference in impedance before and after application of solute was equated to the total mass of solute. As the relationship between the impedance and σ is linear, then

$$M(t) = b_3(\sigma(t) - \sigma_i) \tag{10.15}$$

where $M(t)$ is the total mass of solute as seen by the TDR probe at any time t b_3 is a calibration constant, and σ_i is the bulk soil electrical conductivity measured before solute application.

A pulse application of solute results initially in a rise in σ, and at greater times, as the solute pulse moves past the end of the TDR probe, the σ will drop and possibly return back to its starting value, σ_i. Because the total specific mass of added solute is known and because σ and σ_i are both measured by TDR, the value of b_3 can easily be determined. However, if the relative solute mass (M_r) is used to estimate transport parameters, the calibration constant is not needed (Kachanoski et al., 1992):

$$M_r(t) = \frac{\sigma(t) - \sigma_i}{\sigma_f - \sigma_i} \tag{10.16}$$

where σ_f is the bulk soil electrical conductivity after the pulse application. This approach assumes that all of the solute applied at the surface is immediately detected by the probe. Furthermore, bypass or preferential flow of solute is ignored. This might not always be valid.

When horizontally installed TDR probes are used with a pulse solute input under steady-state water flow, a determination of calibration coefficients is necessary and mass recovery for all TDR probes is required. The total mass can be calculated from the TDR-measured σ using (Vogeler et al., 2000)

$$M = q \int_0^\infty C_r(t)dt = a_4 q \int_0^\infty [\sigma(t) - \sigma]dt \tag{10.17}$$

where q is the steady-state water flux, C_r is the resident concentration as measured by TDR at depth z, and a_4 is an empirical constant that provides the integrally correct interpretation of the conductivity measurements. Various other procedures for the pulse input and horizontally installed TDR probes were discussed by Ward et al. (1994), and Mallants et al. (1996). Problems have been found when the applied mass is not recovered in the TDR measurements. Thus it is hard to obtain a unique calibration coefficient.

10.2.6.2.2 Continuous Solute Application

Another indirect calibration approach, this time using horizontally installed TDR probes, involves the continuous solute application. Solute is applied until a final, constant σ_f is reached. This asymptote is then simply equated to the input concentration C_0. The use of relative concentrations, C/C_0 also means that calibration coefficients are not needed. Under steady water flow conditions, C/C_0 can be calculated from the TDR-measured σ as

$$\frac{C(t)}{C_0} = \frac{\sigma(t) - \sigma_i}{\sigma_f - \sigma_i} \tag{10.18}$$

where σ_f is the final soil bulk electrical conductivity when the antecedent soil solution is completely replaced by the invading solution. Both σ_i and σ_f can be depth dependent.

A disadvantage of this method is that in soils with immobile water, the equilibration time between the two water domains might be long, so that this method requires a long application time. Vogeler et al. (1996) have shown that this method can also be used with vertically installed TDR probes. Again, long equilibration times might be needed. Furthermore, the use of a numerical solution of the convection-dispersion equation (CDE) is required (see Chapters 3 and 20).

The indirect calibration approach with a step change in application and the use of relative concentration is the most common and easiest method. TDR probes are generally installed horizontally, which means that a larger area is measured perpendicular to the direction of flow. The installation is no problem in laboratory experiments with columns. However, in structured heterogeneous soils, a long application time might be required. Furthermore, the indirect method is restricted to steady-state conditions. In the field, water and solute transport are often transient. In this case direct calibration is required.

10.2.7 TRANSPORT MODELS LINKED TO TDR MEASUREMENTS

The most commonly used macroscopic models for solute transport studies are the convection-dispersion equation and the convective lognormal transfer function (CLT) (Jury and Roth, 1990). In the CDE [also referred as the advection-dispersion equation, (ADE)] (see Chapter 3), solute movement is due to the combined effect of convective mass flow of water and molecular diffusion. Thus the CDE is process oriented. The CLT approach is consistent with solute movement being solely due to convection in separated stream tubes, with different flow velocities in each stream tube. The velocity distribution can be described by a probability density function. Thus the CLT is a nonmechanistic stochastic approach. A detailed discussion of the differences of the two models, especially the depth-dependence of transport properties, has been given by Jury and Roth (1990).

The general form for the CDE is given by:

$$\frac{\partial(\theta C_r)}{\partial t} = \frac{\partial}{\partial z}\left(\theta D_s \frac{\partial C_r}{\partial z}\right) - \frac{\partial}{\partial z}(q C_r) \tag{10.19}$$

where D_s is the dispersion coefficient [$m^2 s^{-1}$] and q the Darcy water flux density [$m s^{-1}$]. For steady water flow and a constant water content, the CDE simplifies to

$$\frac{\partial C_r}{\partial t} = D_s \frac{\partial^2 C_r}{\partial z^2} - v \frac{\partial C_r}{\partial z} \tag{10.20}$$

where v is the pore water velocity [$m s^{-1}$], given by q/θ. Often D_s is assumed to be linearly related to the pore water velocity, $D_s = \lambda \, v$, where λ is the dispersivity [m].

The CDE can either be solved analytically for steady-state water flow, or it can be solved by numerical means for variable water flow. Analytical solutions for various boundary and initial conditions are given by van Genuchten and Alves (1982). Here we only give the two formulations most commonly used solutions for their use in solute transport studies involving TDR.

The solution for the CDE, with TDR probes installed vertically, for the relative mass from the surface to depth $z = L$, for a pulse input and steady-state water flow [Eq. (10.15)] is given by (Elrick et al., 1992):

$$M_r(t) = 1 - \left[\frac{1}{2}\text{erfc}\left(\frac{l - vt}{2\sqrt{\lambda vt}}\right) - \frac{1}{2}\exp\left(\frac{l}{\lambda}\right)\text{erfc}\left(\frac{l + vt}{2\sqrt{\lambda vt}}\right)\right] \qquad (10.21)$$

The solution for the CDE, with TDR probes installed horizontally, in terms of the relative resident concentration at depth z, for a step change in solute application [Eq. (10.17)] is given by:

$$\frac{C_r}{C_0} = \frac{1}{2}\text{erfc}\left[\frac{z - vt}{2\sqrt{\lambda vt}}\right] + \left(\frac{vt}{\pi\lambda}\right)\exp\left[-\frac{(z - vt)^2}{4\lambda vt}\right] \qquad (10.22)$$
$$- \frac{1}{2}\left(1 + \frac{z + vt}{\lambda}\right)\exp\left(\frac{z}{\lambda}\right)\text{erfc}\left[\frac{z + vt}{2\sqrt{\lambda vt}}\right]$$

Both M_r and the ratio C_r/C_0 can be measured directly using Eqs. (10.16) and (10.18), respectively. Thus, TDR offers the possibility of estimating the various solute transport parameters. The analysis of BTCs is developed in Chapter 11.

Alternatively, the travel-time probability density function (f_t) of the CLT, assuming that a lognormal probability transfer function describes the solute travel times, is given by (Jury et al., 1991):

$$f_t(t,z) = \frac{1}{\sqrt{(2\pi)}st}\exp\left(-\frac{[\ln(tl/z) - \mu]^2}{2s^2}\right) \qquad (10.23)$$

where μ and s^2 are the mean and variance of the distribution of the logarithm of travel time (log t) measured at a reference depth l.

Vanderborght et al. (1996) compared three different approaches to determine the parameters of the CLT using a time series of resident concentrations measured by TDR. They found the following approach, based on a time-integral normalized resident concentration, C^r, most appropriate for horizontally installed TDR probes with a pulse solute input:

$$C^r(t) = \frac{1}{\sqrt{(2\pi)}st}\exp\left(-\frac{(\ln(tl/z) - \mu - s^2)^2}{2s^2}\right) \qquad (10.24)$$

The model parameters can be found by fitting the above equations [Eqs. (10.21), (10.22), or (10.24)] to the measured time series of resident concentrations or bulk soil electrical conductivities using a least-squares optimization. Alternatively, BTCs obtained from TDR and effluent data can be compared independently from a particular transport model, using time moment analysis (Kachanoski et al., 1992; Mallants et al., 1994). For transient water flow conditions the general form of the CDE [Eqs. (10.21), (10.22), (10.24), or (10.25)] can be coupled with a Richards equation for water flow and then solved using numerical methods, or it can be resolved by the use of a solute penetration depth coordinate (Vanderborght et al., 2000).

10.2.8 STRENGTH AND WEAKNESS OF TDR FOR SOLUTE TRANSPORT STUDIES

Time domain reflectometry is a measurement technique with a great potential for water and solute transport studies in the laboratory. The main advantages of TDR over other methods, such as collecting the effluent or suction cup samples, are:

1. The capability to obtain detailed BTCs under unattended operation and without time-consuming soil solution analyis.
2. The ability to measure both the soil water content and the electrolyte concentration in a nondestructive way with minimal disturbance of the flow pattern. With this dual capacity, TDR offers great scope for monitoring solute transport under both steady-state and transient-flow conditions.
3. TDR measurements are simple to obtain and can be taken continuously and automatically, in a non-destructive way. Software can be used to analyze the waveforms coming from the automated TDR measurements for water content and electrical conductivity. This offers the possibility to study the spatial and temporal dynamics of water transport and solute movement.
4. The ease of replication, via multiplexing a large number of probes to the cable tester or TDR device, means that large sets of data can routinely be taken to assess variations of water contents and solute concentrations at multiple locations, e.g., depths. This spatial and temporal resolution of TDR data should enhance the understanding of physics of flow and transport in unsaturated, heterogeneous soils, not only in the laboratory but also in the field. The use of TDR for solute transport studies in the field is discussed in the previous chapter.
5. Horizontal TDR probes can be installed at different depth in the soil profile and very close to the soil surface. Thus, they can be used to evaluate different solute-transport models and/or characterize their transport parameters.

6. The TDR probes themselves are simple and can vary in size and length and be developed to cover a wide range of experimental conditions.
7. TDR can be used for studying the transport of both inert and reactive solutes, provided that the adsorption is due to an increase in the adsorption capacity.

The weaknesses of the TDR system are:

1. The relatively small zone of influence of the TDR probes and the sensitivity to the region immediately adjacent to the probe wires. This makes the measurement sensitive to small-scale variations, e.g., macropores in structured soils, which in turn can lead to quite variable BTCs. Thus, a considerable number of replications are needed in structured soils with nonuniform water flow.
2. TDR is unable to discriminate between different ionic species, for it measures the total ionic solute concentration. It cannot therefore characterize the soil solution constituents.
3. For studying solute transport under transient flow, a σ-σ_w-θ relation has to be obtained prior to the experiment. This relation is not easily obtained for low water contents, unless repacked soil columns are used for the calibration. However, the relationship seems also highly sensitivity to the local soil structure, texture, and perhaps bulk density surrounding the probes.

10.3 APPLICATION OF TDR FOR SOLUTE TRANSPORT STUDIES

The TDR technique has become popular for studying solute transport in the laboratory. Since its introduction by Kachanoski et al. (1992), various studies with different conditions have been made. These include steady-state and transient water flow, inert and reactive solutes, and examination of the depth dependency of transport properties. Table 10.1 summarizes the studies reported in this article.

10.3.1 STEADY-STATE WATER FLOW AND INERT SOLUTES

Kachanoski et al. (1992) used vertically installed TDR probes to monitor the movement of Cl$^-$ ions under steady-state flow conditions in repacked soil columns in the laboratory. The parameters they obtained from TDR measurements were compared with those derived from effluent BTCs. The agreement between the two different methods was good. In their approach they used TDR measurements to determine a travel time probability density function. Vanclooster et al. (1993) successfully adapted the method of Kachanoski et al. (1992) for use with horizontally installed TDR probes on both disturbed and undisturbed soil columns of a sandy material.

TABLE 10.1
Summary of TDR Laboratory Studies on Solute Transport

Ref.	Medium	Flow condition	Calibration	Solute	Installation
Comegna et al., 1999	Intact	Steady state	Indirect	Cl	Vertical
Hart and Lowery, 1998	Repacked	Steady state and transient	Direct	Br	Horizontal
Kachanoski et al., 1992	Repacked	Steady state	Indirect	Cl	Vertical
Lee et al., 2000	Intact	Steady state	Indirect	Cl + fluoro-benzoate	Diagonal
Magesan et al., 2003	Intact	Steady state	Indirect	Br and Cl pulse	Horizontal
Mallants et al., 1994	Intact	Steady state	Indirect	Cl pulse	Horizontal
Persson and Berndtsson, 1998	Intact	Steady state and transient	Indirect	Br	Horizontal
Risler et al., 1996	Repacked	Transient	Direct	Br pulse	
Vanclooster et al., 1993	Repacked +intact	Steady state	Indirect	Cl pulse	Horizontal
Vanclooster et al., 1995	Intact	Steady state	Direct	Cl pulse	Horizontal
Vanderborght et al., 1996	Intact	Steady state	Direct + indirect	Cl pulse	Horizontal
Vanderborght et al., 2000	Intact	Steady state +transient	Indirect	Cl	Horizontal
Vogeler et al., 1996	Repacked	Transient	Direct	Cl pulse	Horizontal + vertical
Vogeler et al., 1997	Intact	Steady state	Direct + indirect	Cl	Horizontal
Vogeler et al., 2000	Repacked	Steady state +transient	Indirect	Br and N pulse	Horizontal
Wraith et al., 1993	Intact	Steady state	Indirect	Br pulse	Horizontal

Mallants et al. (1994) similarly used horizontal TDR probes to study transport in undisturbed soil columns and found quite variable BTCs. They attributed this discrepancy to the relatively small sampling volume of the TDR probes and heterogeneity of the soil. They concluded for structured soils that several TDR probes are needed to obtain reasonable estimates of solute transport behavior at the column scale.

In a leaching experiment using steady flow through undisturbed soil column of Manawatu fine sandy loam, Vogeler et al. (1997) used horizontal TDR probes at two depths to determine the transport properties. Regular TDR measurements of ε and Z_L made during leaching with $MgCl_2$ were used to infer σ, using Eqs. (10.7–10.9) and (10.12). The values of σ were normalized using Eq. (10.18) and fitted to an analytical solution of the CDE for the resident concentration [Eq. (10.22)]. The dispersivity (λ) and water content (θ) were used as fitting parameters. Note that $v = q/\theta$ The values for the dispersivity obtained from the TDR measurements were 43 and 38 mm, which agree well with the value of 38 mm obtained from the effluent. Mobile water contents of 0.80 were obtained from the TDR data, slightly higher than the value inferred from the effluent data of 0.91.

Lee et al. (2000) extended this TDR approach to estimate solute transport parameters for the mobile-immobile (MIM) version of the CDE. The estimates of the immobile water content and the mass exchange coefficient obtained from the TDR were very similar to those obtained from inverse modeling of the effluent BTC. However to estimate, in addition, the dispersion coefficient, they got better results if the other two fitting parameters (the immobile water content and the mass exchange coefficient) were fixed (Lee et al., 2002).

Horizontal TDR probes can also be used to look at changes in transport parameters with depth (Magesan et al., 2003). Figure 10.6 shows results from a leaching experiment on a undisturbed column of Horotiu soil under steady-state water flow. The Horotiu soil is a silt loam with a relative large allophane content of about 10–12%. TDR probes were installed at depths of 50, 100, and 150 mm. A pulse of Cl was applied to the surface, and regular measurements of σ and θ were made at 5-minute intervals. The normalized values of σ were fitted to an equation similar to Eq. (10.21), but including a retardation, R, factor to account for anion adsorption. The dispersivity values decreased with depth, suggesting that the flow became more uniform with depth. However, Vanclooster et al. (1995) found in their topsoil that λ values increased with depth, as expected for stochastic-convective transport. Deeper in the column they found that λ remained constant, suggesting convective-dispersive transport.

Similar observations by Vanderborght et al. (2000) showed depth-invariant λ in both their sandy loam and loam soil at low flow rates. However, at high flow rates, λ increased with depth in the loam soil. They suggested that lateral mixing between regions of different mobilities is incomplete at high

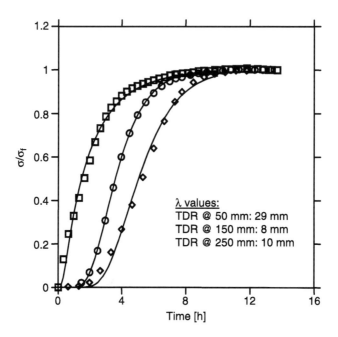

FIGURE 10.6 Normalized bulk soil electrical conductivity as measured by TDR for three different depths in a column of Horotiu soil. Also shown are the fitted curves using the CDE. (From Magesan et al., *J. Environ. Qual.*, 6, 2325–2332, 2003. With permission.)

flow rates, thereby resulting in stochastic-convective flow. Hart and Lowery (1998) found lower values of λ than predicted by the CDE. They suggested that this may be due to the failure of the TDR to measure low concentrations. This problem can be overcome by using a background conductivity larger than the minimum value of σ required for a linear σ-C relation.

Recently Vogeler et al. used a newly developed dielectric lysimeter system, described in Sec. 2.4, to study solute transport on undisturbed soil columns in the laboratory. The dielectric lysimeter measures an average value of θ and σ over the entire volume of the cylinder. Figure 10.7 shows the relative electrical conductivity of the bulk soil as measured by the dielectric lysimeter, which is integrated over the cylinder length of 150 mm. A simulation using the CDE and model parameters obtained from the effluent agrees well with the TDR measurements. The small deviations are probably due to the change in solute transport properties with depth, which cannot be considered, in this case. The dielectric lysimeter appears a useful tool for studying solute transport in the laboratory. In addition, it can also be installed easily in the field either on bare soil or even incorporating a plant. Then the movement of fertilizers through the entire root zone of a plant could also be studied.

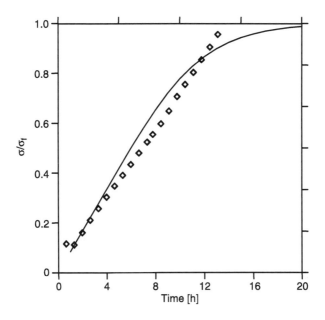

FIGURE 10.7 Normalized bulk soil electrical conductivity as measured by the dielectric TDR lysimeter. Also shown is the fitted curve using the CDE.

10.3.2 Transient Flow and Inert Solutes

The success of TDR techniques for studying solute transport under transient conditions depends on the accuracy of the calibration. Several attempts have been made to use TDR for studying solute transport under transient conditions using the direct calibration approach. Risler et al. (1996) used TDR to study transient transport of a Br pulse through four different soils under transient water flow. The BTCs obtained from the TDR measurements were similar to those obtained from the effluent. However, for two of the soils studied, peak σ_w values were substantially lower for TDR than for the effluent. This is in contrast to our findings in the deeper layer of the Ramiha soil, where we found higher values from TDR compared to the effluent (Vogeler et al., 1997). Both studies emphasize that the TDR-measured σ is highly influenced by soil structure. Hart and Lowery (1998) found similar discrepancies in their study of Br movement through repacked sandy soil columns under steady-state and transient-flow conditions. They used a simple direct calibration to relate σ to C. The recovery of Br in the transient study ranged from 65% to 121%. Thus, even in a repacked sandy soil under controlled conditions, TDR estimates of concentrations are not that accurate.

In some cases, when water infiltrates uniformly into a dry soil, such as a Green-Ampt soil (a soil possessing a Dirac-δ diffusivity function), and solute is leached following a solute step change under steady water flow, an indirect calibration approach can also be used. This is because in already wet soil, the wet front moves ahead of the invading solute front. Thus, an initial rise in TDR-measured σ will first occur due to an increase in θ, and this will be followed by another increase in σ due to the solute front passing the TDR probe. This approach has been demonstrated by Vogeler et al. (1996, 2000) However, for different reasons, namely cation exchange and nonuniform wetting, the analysis was not straightforward to interpret.

Vogeler et al. (2002) also used two different σ_a-σ_w-θ relationships, according to either Eqs. (10.12) and (10.13), as found for the Manawatu soil in the laboratory, to monitor the transport of conservative tracers through field soil under transient water flow. A mixed pulse of chloride and bromide was applied to an orchard soil under a kiwifruit vine. The movement of this solute pulse was monitored by TDR probes, installed vertically into the soil at depths ranging from the surface to a depth of 1 m. A total of 63 TDR probes of 150 mm length were used. With both calibration procedures a well-defined peak of the solute pulse could be seen. However, the TDR-inferred concentrations deviated by up to 50% from those measured directly on subsamples. These differences might again be due to the structure difference in the field soil with roots, compared to the uniform repacked soil columns that were used to obtain the θ-σ_a-σ_w relationships. Thus, care must be taken when TDR calibrations obtained in the laboratory are taken into the field.

Recently, Vanderborght et al. (2000) introduced the use of a solute-penetration depth to study transient solute transport through undisturbed columns of two contrasting soils. Their approach is based on the assumption that transient water flow is primarily a function of the cumulative amount of water flowing past a given depth and not significantly affected by intermittent drying and wetting cycles or the rate at which the water is applied. Vanderborght et al. (2000) then compared TDR results to data from the effluent BTC. Whereas the TDR technique accurately described transport in the sandy loam, it failed in the loam soil. This failure was attributed to preferential flow, which was not detected by the TDR probes. Persson and Berndtsson (1998) followed the same approach and used the TDR technique to describe transient solute transport through soil columns, provided that the σ-σ_w-θ relationship found on undisturbed soil columns was used. Different relationships were found for the different layers in the soil profile.

10.3.3 REACTIVE SOLUTES

TDR has the potential to determine the retardation factors associated with the transport of reactive solutes through soil. However, in contrast to studies where the concentration of anions is measured in the leachate or via suction cups, TDR can only detect anion adsorption that is caused by a change in the anion exchange capacity of the soil with increasing external solution concentration. This is because anion exchange would have only a small effect on σ_w, and thus the TDR-measured σ. The degree of anion

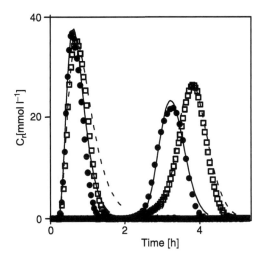

FIGURE 10.8 Bromide (●) and nitrate (□) concentrations obtained from TDR measurements of the bulk soil electrical conductivity. Also shown are the fitted curves using the CDE. (From Vogeler et al., *Soil Sci. Soc. Am.*, 64, 12–18, 2000. With permission.)

adsorption depends on the exchange capacity of the soil. This is generally determined from batch experiments. The limitations of batch experiments are well known.

Figure 10.8 shows TDR measurements from a leaching experiment performed in the laboratory on repacked columns of a ferrasol from Maré in New Caledonia (Vogeler et al., 2000). A pulse of Cl and N was applied to the soil column via a rainfall simulator. The normalized TDR measurements from two different depths were fitted to a solution of the CDE for resident concentrations, similar to Eq. (10.22), but including a retardation factor (R). The fitted parameters were the dispersivity (λ) and R. These were similar to those obtained from the effluent (Table 10.2). Whereas the effluent gives us only an average retardation value over the entire column length, TDR provides us with values for different depths. Thus, changing properties within the soil profile can be determined using TDR. As the soil is known not to possess any immobile water, TDR also gives an appropriate value for the retardation factor. In soils with immobile water, an independent measure of the resident concentration needs to be taken. Otherwise, the TDR estimate of the retardation factor could be too low. Good results for measuring anion retardation in soils by TDR were also found by Wraith et al. (1993) and Comegna et al. (1999).

Cation exchange can also occur during invasion of a solute and may lead to mass balance errors in TDR. As demonstrated by Vogeler et al. (1996) cation exchange between the invading solution and the exchange sites can also affect the TDR-measured σ, as the accompanying cation can alter the conductivity of the soil solution. The use of solutions with potassium as a

TABLE 10.2
Model Parameters Obtained by Fitting Effluent and TDR Measurements of Normalized σ to CDE[a]

	Effluent		TDR 1 @ 50 mm		TDR2 @ 100 mm	
	λ [mm]	R	λ [mm]	R	λ [mm]	R
Br	3	1.3	2	1.3	1	1.4
N	4	1.2	3	1.2	1	1.3

The leaching experiment was performed on a column packed with a ferrasol from Maré.

Source: Vogeler et al., 2000.

cation should be avoided in soils where the cation-exchange sites are calcium dominated. The σ_w of a 0.015 M $CaCl_2$ solution is only 83% of a 0.03 M KCl solution (Vogeler et al., 1996).

10.4. RECOMMENDATIONS AND FUTURE RESEARCH

We have reviewed the use of TDR for monitoring solute transport in soils. TDR is based on the measurement of the soil's dielectric constant (ε) and bulk soil electrical conductivity (σ). These factors can be related, by calibration, to the volumetric water content (θ) and resident soil solution concentration (C_r). Whereas the measurement of θ is generally straightforward, the calibration for the solute concentration still remains a challenge. Solute transport studies with steady water flow can be analyzed in terms of relative values of σ. The use of TDR under these solute-flow conditions seems to be well established. However, for transient studies, a relation between σ and C_r needs to be established. This is soil specific and influenced by soil structure, texture, and bulk density. Despite numerous applications of the technique, this relationship is not yet fully understood. Further work is still required to solve this calibration problem.

It has been shown that TDR can be used equally well to study nonreactive and reactive solute transport, provided that the soil exhibits an increase in anion adsorption capacity with increasing soil solution concentration. The use of horizontal TDR probes enables precise depthwise resolution. This enables us to observe how transport properties change with soil depth.

The strength of the TDR system is in its ability to monitor *in situ* the changes in water content and electrical conductivity both rapidly and nondestructively. This is, however, limited to soils that are non-swelling, as shrink-swell can alter the soil-probe contact. TDR can be automated and multiplexed, enabling continuous measurement of solute transport properties in time and space. This spatial and temporal resolution of TDR data should enhance the understanding of physics of flow and transport in unsaturated, heterogeneous soils, not only at the laboratory but also in the field.

One weakness of the TDR system is the relatively small zone of influence of the TDR probes and the sensitivity to the region immediately adjacent to the probe wires. Thus, a considerable number of replications are needed in structured and swelling soils with nonuniform water flow. Furthermore, TDR is unable to discriminate between different ionic species, for it measures the total ionic solute concentration. It cannot therefore characterize the soil solution constituents.

Finally, we believe that TDR should not be used to replace existing monitoring techniques, such as effluent collection. Rather, TDR should be used as an additional tool, e.g., for providing information on the depth dependence of transport parameters.

We conclude that TDR is a reliable and sound practical technique for studying solute transport in both laboratory and field soils. Hopefully in the future the σ-C_r relationship will be better understood, so that studies can also be done, with confidence, under transient conditions even in the field.

NOTATION

ROMAN

a_x, b_x, c_x, d_x	empirical constants in various equations [–]
f	temperature correction coefficient [–]
f_t	travel time probability density function
a	apparent length [m]
p	length of TDR probe [m]
q	water flux density [m s^{-1}]
	reflection coefficient [–]
d	TDR rod diameter [m]
s	TDR rod spacing [–]
	variance of the lognormal distribution [s]
	time [s]
v	pore water velocity [m s^{-1}]
v_p	relative velocity of propagation [–]
z	depth [m]
C	concentration in soil solution [kg or mol m^{-3}]
$C_0(t)$	time-dependent input solution concentration [kg or mol m^{-3}]
C_r	resident soil solution concentration [kg or mol m^{-3}]
C^r	time-intergral normalised resident concentration [–]
D_s	dispersion coefficient [m^2 s^{-1}]
F	formation factor [–]
I_n	variable in Eq. (10.12) [–]
K_G	geometric constant of TDR probe [m^{-1}]

M_0	pulse of solute applied to soil surface [kg or $mol\,m^{-2}$]
M_r	relative solute mass [–]
R	retardation factor -
V_0	zero reference voltage [V]
V_f	final reflected voltage at very long time [V]
V_i	voltage of incident step [V]
V_r	voltage after reflection from probe end [V]
Z_0	characteristic impedance of TDR cable [Ω]
Z_L	impedance of TDR probe [Ω]

GREEK

ε	dielectric constant -
λ	dispersivity [m]
μ	mean of the lognormal distribution [s]
ρ	soil bulk density [$kg\,m^{-3}$]
σ	bulk soil electrical conductivity [$S\,m^{-1}$]
σ_f	final bulk soil electrical conductivity [$S\,m^{-1}$]
σ_i	initial bulk soil electrical conductivity [$S\,m^{-1}$]
σ_{ref}	electrical conductivity of reference solution [$S\,m^{-1}$]
σ_s	surface conductance [$S\,m^{-1}$]
σ_w	pore water electrical conductivity [$S\,m^{-1}$]
τ	tortuosity factor -
θ	volumetric water content [$m^3\,m^{-3}$]

ABBREVIATIONS

BTC	breakthrough curve
CDE	convection dispersion equation
CEC	cation exchange capacity
CLT	convective lognormal transfer function
TDR	time domain reflectometry

REFERENCES

Amente, G., Baker, J.M., and Reece, C.F., 2000. Estimation of soil solution electrical conductivity from bulk soil electrical conductivity in sandy soils. *Soil Sci. Soc. Am. J.*, 64, 1931–1939.

Baker, J.M., and Allmaras, R.R., 1990. System for automating and multiplexing soil moisture measurements by time domain reflectometry. *Soil Sci. Soc. Am. J.* 378–384.

Baker, J.M., and Lascano, R.J., 1989. The spatial sensitivity of time domain reflectometry. *Soil Sci.*, 147, 378–384.

Comegna, V., Coppola, A., and Sommella, A., 1999. Nonreactive solute transport in variously structured soil materials as determined by time domain reflectometry. *Geoderma*, 92, 167–184.

Dalton, F.N., 1992. Development of time domain reflectometry for measuring soil water content and bulk soil electrical conductivity. In *Advances in Measurement of Soil Physical Properties: Bringing Theory into Practice*, SSSA Special Publication No. 30.

Dalton, F.N., and van Genuchten, M.Th., 1986. The time-domain reflectometry method for measuring soil water content and salinity. *Geoderma*, 38, 237–250.

Dalton, F.N., Herkelrath, W.N., Rawlins, D.S., and Rhoades, J.D., 1984. Time-domain reflectometry: simultaneous measurement of soil water content and electric conductivity with a single probe. *Science*, 224, 989–990.

Dasberg, S., and Dalton, F.N., 1985. Time domain reflectometry field measurements of soil water content and electrical conductivity. *Soil Sci. Soc. Am. J.*, 49, 293–297.

Dirksen, C., and Dasberg, S., 1993. Improved calibration of time domain reflectometry soil water content measurements. *Soil Sci. Soc. Am. J.*, 57, 660–667.

Dyson, J.S., and White, R.E., 1987. A comparison of the convection-dispersion equation and transfer function model for predicting chloride leaching through undisturbed, structured clay soil. *J. Soil Sci.*, 38, 157–172.

Elrick, D.E., Kachanoski, R.G., Pringle, E.A., and Ward, A.L., 1992. Parameter estimates of field solute transport models based on time domain reflectometry measurements. *Soil Sci. Soc. Am. J.*, 56, 1663–1666.

Giese, K., and Tiemann, R., 1975. Determination of the complex permittivity from thin-sample time domain reflectometry: improved analysis of the step response waveform. *Adv. Mol. Relax. Proc.*, 7, 45–49.

Hart, G.L., and Lowery, B., 1998. Measuring instantaneous solute flux and loading with time domain reflectometry. *Soil Sci. Soc. Am. J.*, 62, 23–35.

Heimovaara, T.J., 1993. Design of triple-wire time domain reflectometry probes in practice and theory. *Soil Sci. Soc. Am. J.*, 57, 1410–1417.

Heimovaara, T.J., Focke, A.G., Boutne, W., and Verstaten, J.M., 1995. Assessing temporal variations in soil water composition with time domain reflectomtery. *Soil Sci. Soc. Am. J.*, 59, 689–698.

Jacobsen, O.H., and Schjønning, P., 1993. A laboratory calibration of time domain reflectometry for soil water measurements including effects of bulk density and texture. *J. Hydr.*, 151, 147–157.

Jury, W.A., and Roth, K., 1990. *Transfer Functions and Solute Movement through Soil Theory and Applications*. Birkhäuser Verlag, Basel.

Jury, W.A., and Sposito, G., 1985. Field calibration and validation of solute transport models for the unsaturated zone. *Soil Sci. Soc. Am. J.*, 49, 1331–1341.

Jury, W.A., Gardner, W.R., and Gardner, W.H., 1991. *Soil Physics*, 5th ed. John Wiley and Sons, Inc, New York, pp. 218–267.

Kachanoski, R.G., Pringle, E., and Ward, A., 1992. Field measurement of solute travel times using time domain reflectometry. *Soil Sci. Soc. Am. J.*, 56, 47–52.

Kachanoski, R.G., Ward, A.L., and van Wesenberg, I.J., 1994. Measurement of transport properties at the field scale. *Transactions 15th World Congress of Soil Science*, Vol. 2a, Commission 1, Acapulco, Mexico.

Knight, J.H., 1992. Sensitivity of time domain reflectometry measurements to lateral variations in soil water content. *Wat. Resour. Res.*, 28, 2345–2352.

Ledieu, J., de Ridder, P., de Clerck, P., and Dautrebande, S., 1986. A method of measuring soil moisture by time domain reflectometry. *J. Hydr.*, 88, 319–328.

Lee, J., Horton, R., and Jaynes, D.B., 2000. A time domain reflectometry method to measure immobile water content and mass exchange coefficient. *Soil Sci. Soc. Am. J.*, 64, 1911–1917.

Lee, J., Horton, R., and Jaynes, D.B., 2002. The feasibility of time domain reflectometry to describe solute transport through undisturbed soil cores. *Soil Sci. Soc. Am. J.* 66, 53–57.

Magesan, G.N., Vogeler, I., Clothier, B.E., Green, S., and Lee, R., 2003. Solute movement through an allophanic soil. *Eur. J. Soil Sci.*, 6, 2325–2332.

Mallants, D., Vanclooster, M., Meddahi, M., and Feyen, J., 1994. Estimating solute transport in undisturbed soil columns using time domain reflectometry, *J. Cont. Hydr.*, 17, 91–109.

Mallants, D., Vanclooster, M., Toride, N., Vanderborght, J., van Genuchten. M.Th., and Feyen, J., 1996. Comparison of three methods to calibrate TDR for monitoring solute movement in undisturbed soil. *Soil Sci. Soc. Am. J.*, 60, 747–754.

Mojid, M.A., Wyseure, G.C.L., and Rose, D.A., 1997. Extension of the measurement range of electrical conductivity by time domain reflectometry. *Hydr. Earth Sys. Sci.*, 1, 175–183.

Mualem, Y., and Friedmann, S.P., 1991. Theoretical prediction of electrical conductivity in saturated and unsaturated soil. *Wat. Resour. Res.*, 27, 2771–2777.

Nadler, A., 1997. Discrepancies between soil solute concentration estimates obtained by TDR and aqueous extracts. *Aust. J. Soil Res.*, 35, 527–537.

Nadler, A., and Frenkel, H., 1980. Determination of soil solution electrical conductivity from bulk soil electrical conductivity measurements by the four-electrode method. *Soil Sci. Soc. Am. J.*, 44, 1216–1221.

Nadler, A. Frenkel, H., and Mantell, A.,1984. Applicability of the four-probe technique under extremely variable water contents and salinity distributions. *Soil Sci. Soc. Am. J.*, 48, 1258–1261.

Nadler, A., Dasberg, S., and Lapid, I., 1991. Time domain reflectometry measurements of water content and electrical conductivity of layered soil columns. *Soil Sci. Soc. Am. J.*, 55, 938–943.

Nadler, A., Green, S.R., Vogeler, I., and Clothier, B.E., 2002. Horizontal and vertical TDR measurements of soil water content and electrical conductivity. *Soil Sci. Soc. Am. J.*, 66, 735–743.

Nissen, H.H., Moldrup, P., Olesen, T., and Jensen, O.K., 2001. Time domain reflectometry sensitivity to lateral variations in bulk soil electrical conductivity. *Soil Sci. Soc. Am. J.*, 65, 1351–1360.

Noborio, K., McInnes, K.J., and Heilman, J.L., 1994. Field measurements of soil electrical conductivity and water content by time domain reflectometry. *Comp. Electron Agric.*, 11, 131–142.

Persson, M., 1997. Soil solution electrical conductivity measurements under transient conditions using time domain reflectometry. *Soil Sci. Soc. Am. J.* 61, 997–1003

Persson, M., and Berndtsson, R., 1998. Estimating transport parameters in an undisturbed soil column using time domain reflectometry and transfer function theory. *J. Hydr.*, 205, 232–247.

Ren, T., Noborio, K., and Horton, R., 1999. Measuring soil water content, electrical conductivity, and thermal properties with a thermo-time domain reflectometry probe. *Soil Sci. Soc. Am. J.*, 63, 450–457.

Rhoades, J.D., Manteghi, N.A., Shouse, P.J., and Alves, W.J., 1989. Soil electrical conductivity and soil salinity: new formulations and calibrations. *Soil Sci. Soc. Am. J.*, 53, 433–439.

Rhoades, J.D., Raats, P.A.C., and Prather, R.J., 1976. Effects of liquid-phase electrical conductivity, water content and surface conductivity on bulk soil electrical conductivity. *Soil Sci. Soc. Am. J.*, 40, 651–655.

Rhoades, J.D., and van Schilfgaarde, J., 1976. An electrical conductivity probe for determining soil salinity. *Soil Sci. Soc. Am. J.*, 40, 647–651.

Risler, P.D., Wraith, J.M., and Gaber, H.M., 1996. Solute transport under transient flow conditions estimated using time domain reflectometry. *Soil Sci. Soc. Am. J.*, 60, 1297–1305.

Roth, C.H., Malicki, M.A., and Plagge, R., 1992. Empirical evaluation of the relationship between soil dielectric constant and volumetric water content as the basis for calibrating soil moisture measurements by TDR. *J. Soil Sci.* 43, 1–13

Roth, K., Schulin, R., Flühler, H. and Attinger, W., 1990. Calibration of time domain reflectometry for water content measurement using composite dielectricity approach. *Wat. Resour. Res.*, 26, 2267–2273.

Shainberg, I., Rhoades, J.D., and Prather, R.J. 1980. Effect of exchangeable sodium percentage, cation exchange capacity, and soil solution concentration on soil electrical conductivity. *Soil Sci. Soc. Am. J.*, 44, 469–473.

Topp, G.C., and Davis, J.L., 1985. Measurement of soil water content using time-domain reflectometry (TDR): A field evaluation. *Soil Sci. Soc. Am. J.*, 49, 19–24.

Topp, G.C., Davis, J.L., and Annan, A.P., 1980. Electromagnetic determination of soil water content: measurement in coaxial transmission lines. *Wat. Resour. Res.* 16, 574–582.

Topp, G.C. and Davis, J.L. and Annan, A.P., 1982. Electromagnetic determination of soil water content using TDR: I. Applications to wetting fronts and steep gradients. *Soil Sci. Soc. Am. J.*, 46, 672–678.

Topp, G.C., Yanuka, M., Zebchuk, W.D., and Zeglin, S., 1988. Determination of electrical conductivity using time domain reflectometry: soil and water experiments in coaxial lines. *Wat. Resour. Res.*, 24, 945–952.

United States Salinity Laboratory Staff. 1954. *Diagnosis and Improvement of Saline and Alkali Soils*. U.S. Dep. Agri. Handbook 60.

Vanclooster, M., Mallants, D., Diels, J., and Feyen, J., 1993. Determining local-scale solute transport parameters using time domain reflectometry (TDR). *J. Hydr.* 148, 93–107.

Vanclooster, M., Mallants, D., Vanderborght, J., Diels, J., van Orshoven, J., and Feyen, J., 1995. Monitoring solute transport in a multi-layered sandy lysimeter using time domain reflectometry. *Soil Sci. Soc. Am. J.*, 59, 337–344.

Vanderborght, J., Vanclooster, M., Mallants, D., Diels, J., and Feyen, J., 1996. Determining convective lognormal solute transport parameters from resident concentration data. *Soil Sci. Soc. Am. J.*, 60, 1306–1317.

Vanderborght, J., Timmerman, A., and Feyen, J., 2000. Solute transport for steady-state and transient flow in soils with and without macropores. *Soil Sci. Soc. Am. J.*, 64, 1305–1317.

van Genuchten, M.Th., 1981. Analytical solutions for the movement of solutes. *J. Hydr.* (Amsterdam), 49, 213–233.

van Genuchten, M.Th., and Alves, W.J., 1982. Analyitical solutions to the one-dimensional convective-dispersive solute transport equation. United States Department of Agriculture. *Agricultural Research Service*, Technical Bulletin No. 1661.

Vogeler, I., Clothier, B.E., Green, S.R., Scotter, D.R., and Tillman, R.W., 1996. Characterizing water and solute movement by TDR and disk permeameter. *Soil Sci. Soc. Am. J.*, 60, 5–12.

Vogeler, I., Clothier, B.E., and Green, S.R., 1997. TDR estimation of electrolyte in the soil solution. *Aust. J. Soil Res.*, 35, 515–526.

Vogeler, I., Duwig, C., Clothier, B.E., and Green, S.R., 2000. A simple approach to determine reactive solute transport using time domain reflectometry. *Soil Sci. Soc. Am. J.*, 64, 12–18.

Vogeler, I., Green, S., Nadler, A., and Duwig, C., 2002. Measuring transient solute transport through the vadose zone using time domain reflectometry. *Aust. J. Soil Res.*, 39, 1359–1369.

Ward, A.L., Kachanoski, R.G., and Elrick, D.E., 1994. Laboratory measurements of solute transport using time domain reflectometry. *Soil Sci. Soc. Am. J.*, 58, 1031–1039.

Waxman, M.H., and Smits, L.J.M., 1968. Electrical conductivity in oil bearing shaly sands. *Soc. of Petroleum Eng. J.*, 243, 107–122.

Weast, R.C. (Ed.) 1965. *Handbook of Chemistry and Physics*, 46th ed., P.D-81. The Chemical Rubber Co., Ohio.

Wraith, J.M., Comfort, S.D., Woodbury, B.L. and Inskeep, W.P., 1993. A simplified analysis approach for monitoring solute transport using time-domain reflectometry. *Soil Sci. Soc. Am. J.*, 57, 637–642.

Wyseure, G.C.L., Mojid, M.A., and Malicki, M.A., 1997. Measurement of volumetric water content by TDR in saline soils. *Eur. J. Soil Sci.*, 48, 347–354.

Yanuka, M., Topp, G.C., Zegelin, S., and Zebchuk, W.D., 1988. Multiple reflection and attenuation of time domain reflectometry pulses: theoretical considerations for applications to soil and water. *Wat. Resour. Res.*, 24, 939–944.

Zegelin, S.J., White, I., and Jenkins, D.R., 1989. Improved field probes for soil water content and electrical conductivity measurement using time-domain reflectometry. *Wat. Resour. Res.*, 25, 2367–2376.

Zegelin, S.J., White, I., and Russell, G.F., 1992. A critique of the time domain reflectometry technique for determining field soil-water content. In: *Advances in Measurement of Soil Properties: Bringing Theory into Practice*. SSSA Special Publication No. 30.

11   Characterization of Solute Transport Through Miscible Displacement Experiments

J. *Álvarez-Benedí*
Instituto Tecnológico Agrario de Castilla y León,
Valladolid, Spain

C. M. Regalado and A. Ritter
Departamento de Suelos y Riegos, Instituto Canario de
Investigaciones Agrarias (ICIA), Ctra. del Boquerón s/n,
Valleguerra, Tenerife, Spain

S. Bolado
Departamento de Ingeniería Química
Universidad de Valladolid, Valladolid, Spain

CONTENTS

1-5667-0657-2/05/$0.00 + $1.50
© 2005 by CRC Press

11.1   CHARACTERIZATION OF SOLUTE TRANSPORT

The characterization of solute transport processes in subsurface systems has received increased attention by environmental science researchers in recent decades, especially concerning the fate of pollutants in soils. There are numerous reasons for the increased attention to the vadose zone processes, as they play a key role in the behavior of subsurface contaminants (see Chapter 1). Characterization of water and solute transport processes in soils is a complex task, which requires the coupling of mechanistic models and corresponding appropriate data generation on water contents and solute concentration (and, eventually, heat fluxes). Mathematical models describing the most relevant processes that govern solute transport, including the advection-dispersion equation and nonequilibrium during transport, were introduced in Chapter 3. In addition, *ad hoc* experimental methodologies for monitoring water and solute concentration have been described in Chapters 5, 9, and 10. Here, the development of methodological approaches for the characterization of solute transport processes is described. Thus, experimental strategies to elucidate transport mechanisms are discussed, without description of the monitoring techniques that can be used to generate the data.

Miscible displacement experiments are perhaps the most important among the available methodological approaches for characterizing solute transport. When such experiments are properly designed, they can provide valuable information about processes that affect solute movement, such as hydro-dynamic dispersion, adsorption, degradation, and transformation phenomena (Ersahin et al., 2002). These consist in applying a solute at a specific point of the soil and monitoring the evolution of solute concentration with time at a certain distance. This is a general definition, which can be applied to tracer tests for a wide range of scales. For example, tracers have been used for characterizing regional velocity, hydraulic conductivity distributions, or dispersivity (Boulding and Ginn, 2004). The main constraints on the application of this methodology are spatial variability of soils and the scale dependence at which transport studies are performed (Winton and Weber, 1996). The challenge for studies at large scale is to obtain adequate data collection in order to characterize the temporal and spatial distribution of model parameters and variable inputs. Therefore, the critical aspect of these studies is to develop a sampling strategy that will reflect the spatial heterogeneity of the physicochemical parameters and variables used in the functional solute transport model. Chapter 17 is devoted to these studies.

Field studies provide useful information under real, but specific, non-reproducible conditions. In the field, it is difficult to control the boundary conditions and to obtain detailed and reliable data on water content and solute concentrations at different depths within the soil profile. On the other hand, batch equilibration experiments at the laboratory scale can be carried out under very controlled conditions, and they permit one to isolate different mechanisms such as sorption, distribution, volatilization, or degradation. The main concern with these laboratory data is that, in general, results are scale-dependent and therefore difficult to extrapolate to field scenarios. An intermediate scale that combines several aspects of both field and laboratory scales can be obtained with soil column lysimeters. These are devices for measuring water and solute transport from a column of soil under controlled conditions (Führ et al., 1997). The dimensions of soil columns found in the literature can range from diameters of 2–3 cm up to 1 m, and from 10 cm to more than 3 m in height (Coltman et al., 1991). This gives an idea of the wide variety of experimental situations that can be covered, from situations close to laboratory scale to others more related to field scenarios (Figure 11.1). Experiments at these scales are described in this book (see, e.g., Chapters 9, 10, and 18 for studies at microplot, lysimeter, and field scales, respectively. Thus, the advantages of soil columns are that several aspects of the field and laboratory scales can be combined.

Soil column lysimeters provide a fundamental understanding of solute transport processes by combining the ability to imitate a natural environment and the possibility of developing different experimental strategies (e.g., water flow or boundary conditions). Through different experimental configurations, lysimeters allow the control of boundary conditions, for example, a constant flux or hydraulic pressure at the inlet (top) and free drainage or a specific pressure at the outlet (bottom). By using this control of boundary conditions, lysimeters can generate flow conditions close to the real environment. Alternatively, situations far from field conditions can be also achieved to elucidate the role of transport mechanisms. For example, steady-state experiments can be easily performed under saturated or unsaturated moisture conditions.

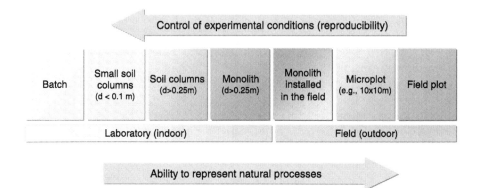

FIGURE 11.1 Basic scales for soil-solute studies. Reproducibility increases as the control of the experimental conditions increases; however, the degree of representation of natural environments demands studies at larger scale.

Considering the degree of disturbance of the soil column contained in the lysimeter, two basic types of lysimeters may be considered: filled-in disturbed soil matrix and natural undisturbed lysimeters (soil profile). In a disturbed soil column, several soil properties can be handled during the installation of the experimental setup. Among them, the most interesting properties are soil bulk density, soil aggregation, organic matter content, or the presence of stones. The control of soil properties allows obtaining an approximately homogeneous medium or, alternatively, the variation of soil properties with depth (e.g., to generate soil horizons). Natural undisturbed lysimeters (monoliths) are composed of a vertical section of a soil profile removed from the soil. They are costly to extract but, when they have appropriate dimensions, can represent accurately the transport processes in a natural scenario. A comparison of disturbed and undisturbed lysimeters is given in Table 11.1.

Soil lysimeters have been often criticized because they may suffer from sidewall flow, i.e., artificial flow paths along the soil-wall interface that allow rapid preferential flow and solute movement, which are not representative of the field conditions. This problem is reduced when using large lysimeters with big cross-sectional area (Cassel et al., 1973; Bergström, 1990; Schneider and Howell, 1991). Also, alternative designs such as annular rings have been developed to minimize sidewall flow, which are easy to implement in disturbed soil columns (Corwin, 2000).

The objective of this chapter is to describe experimental methodologies to characterize soil-water-solute processes through transport experiments and application of related conceptual models. Based on the above-mentioned advantages, the development and analysis of miscible displacement experiments are described in Sec. 11.2, with emphasis on the application at the lysimeter scale. Section 11.3 will focus on the description of methodologies for characterizing nonequilibrium transport processes. This chapter uses most of

TABLE 11.1
Comparison Between Disturbed and Undisturbed Soil Lysimeters

Description	Disturbed	Undisturbed
Soil properties	Homogeneous	Heterogeneous
Number of replicates needed	Reduced	Large
Construction/obtaining	Easier	Difficult
Representative of real conditions	No	Yes
Soil macrostructure	Destroyed	Intact
Applicability to field	Questionable	Relatively applicable
Mathematical description	Easier to model	More difficult to model
Capability for characterizing soil processes	High	Low

Source: Coltman et al., 1991.

the conceptual models (soil-water and soil-solute transport processes) presented in Chapter 3. We therefore strongly recommend to the reader a previous analysis of those concepts that will be used in this chapter.

11.2 THE BREAKTHROUGH CURVE

11.2.1 THE MISCIBLE DISPLACEMENT EXPERIMENT AND ITS MATHEMATICAL DESCRIPTION

A miscible displacement experiment consists of the mutual mixing and movement of two fluids that are soluble in each other. Miscibility implies that the displacing solution mixes with the displaced solution, producing a dispersive front. As a particular example, consider an experiment where water is flowing through a soil column at a certain stationary rate. Consider a tracer consisting of a conservative solute (i.e., which does not undergo any chemical reaction), and sorption or phase changes such as volatilization are also not occurring. Consider also that a pulse (step change) of such miscible tracer is applied to the inlet (e.g., top of the soil column) and the effluent flowing through the system is analyzed at the outlet (bottom). This is schematically illustrated in Figure 11.2. We will refer to the volume of water in the soil column, in either saturated or unsaturated conditions, as the pore (water) volume, V. Now assume that initially the soil was solute-free. Ideally, we would expect that a sharp piston-like solute pulse applied at the top and flowing into the column is transported and exits the column without modifying its original shape. However, because of the mixing of the fluids and molecular diffusion of the solute, as the solute moves down the pulse the initial shape smoothes out. Other contribution to the smoothing of the original shape, due to the porous nature of the soil media, is the mechanical dispersion. It takes place because the solute must take different tortuous paths, with various

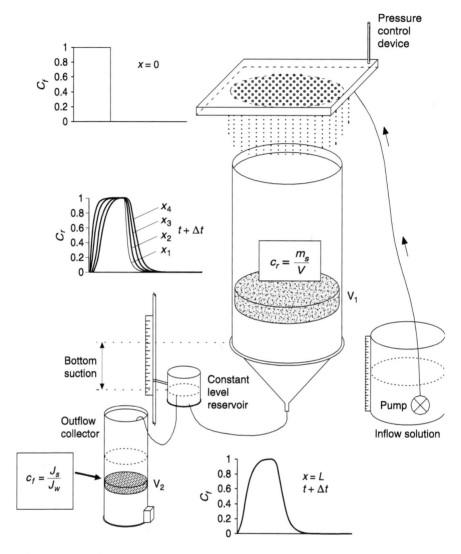

FIGURE 11.2 Schematic illustration of a miscible displacement experiment.

geometries and sizes, moving faster in some and slower in others. These two contributions to the overall effect of hydrodynamic dispersion were reviewed in Chapter 3.

Usually this results in different solute molecules (or ions) arriving earlier or later at the outlet. The time taken by each particle to cross through the soil column is known as the residence or breakthrough time. An illustrative graph of the process taking place is the so-called breakthrough curve (BTC). A BTC is a plot of the relative solute concentration in the outflow from a column of soil or porous material against the volume of outflow, after a step change in solute concentration has been applied to the inlet end of the column (S.S.S.A., 2001). A more general description of BTC includes any plot of solute concentration versus time at a fixed location in the soil (Skaggs and Leij, 2002) (Figure 11.2). The BTC describes the distribution of resident times in a column, and as such, its shape may be characterized by the moments of the distribution (see Sec. 11.2.2.2), as is generally done with statistical probability distributions. The shape adopted by the BTC curve can render useful information about whether the solute is being adsorbed or degraded, or about how it is being preferentially transported. Thus, BTCs can provide a fundamental understanding of transport processes for characterization purposes.

Following the example depicted in Figure 11.2, consider the shaded area representing a solute volume added to a soil column at time $t = t_0$. As we have discussed above, because of the different flow velocities within the soil profile, at time $t_0 + \Delta t$ we would expect different solute concentrations depending on whether we sample a volume V_1 by coring (destructive sampling) or we measure the soil column's effluent solute concentration in a volume V_2. This leads us to distinguish between the following two types of average solute concentrations: flux and resident concentrations.

11.2.1.1   Flux, Resident, and Time-Averaged Concentrations

11.2.1.1.1   The Transport Equation

With regard to the above-referred solute transport processes, so far our description of the mechanisms responsible for the smoothing of the initial solute sharp peak has been phenomenological. However, we can go further into a mathematical description of solute transport in the liquid phase by specifying the advective and dispersive components of the solute flux, J_s [$ML^{-3}T^{-1}$] (see Chapter 3):

$$J_s = J_w c \mid J_D = J_w c - \theta D \nabla c \qquad (11.1)$$

In Eq. (11.1) J_w [LT^{-1}] stems for flux of water, and θ [$L^3 L^{-3}$] is the volumetric water content, such that we can define the ratio J_w/θ as the mean pore water velocity, v [LT^{-1}]. Dividing Eq. (11.1) by θ, the first term in the right-hand side would then becomes a velocity, v, times the dissolved solute concentration, c [ML^{-3}]. Therefore, this first term represents an advective (convective) transport of solute, or the passive movement of solute that is transported by the flowing water. By contrast, the second term in the right-hand side corresponds to the dispersive flux, J_D [$ML^{-3}T^{-1}$], characterized by the hydrodynamic dispersion coefficient D [L^2T^{-1}]. If we now consider a unitary soil volume within our original soil column and invoke the mass conservation principle within such a unit volume,

$$\frac{\partial c_T}{\partial t} + \nabla J_s = 0 \qquad (11.2)$$

With $c_T = \theta \cdot c$, c_T being the total solute concentration, we obtain the advection-dispersion equation (ADE) for a non sorbed solute. This equation was used previously for a constant hydrodynamic dispersion coefficient, D, in Eqs. (3.19) and (10.20):

$$\frac{\partial c}{\partial t} = \frac{\partial}{\partial x}\left[D\frac{\partial c}{\partial x}\right] - v\frac{\partial c}{\partial x} \qquad (11.3$$

If the solute is subject to a distribution into different phases, the total solute concentration will be given by a mass balance taking into account all the possible phases involved in the solute transport. For example, for a sorbed, nonvolatile compound, the total solute concentration, c_T [ML^{-3}], can be written in terms of the soil bulk density, ρ_b [ML^{-3}], and sorbed solute concentration, S (mass of sorbed solute per unit mass of dry soil [MM^{-1}]), as

$$c_T = \rho_b S + \theta c \qquad (11.4$$

In case of a sorbed solute, under steady water flow ($\partial J_w/\partial t = 0$ and $\partial\theta/\partial t = 0$) and a uniform water content ($\partial\theta/\partial x = 0$), and considering a linear sorption isotherm given by $S = K_D c$ (see Chapter 3), the ADE equation takes the form of:

$$R\frac{\partial c}{\partial t} = D\frac{\partial^2 c}{\partial x^2} - v\frac{\partial c}{\partial x} \qquad (11.5$$

where R is the retardation factor, which accounts for the relative travel velocity of a sorbed solute ($R > 1$) compared to a nonsorbed tracer ($R = 1$). See Chapter 3 for a detailed description of soil-solute transport processes.

11.2.1.1.2 Flux Averaged, and Time Resident Concentrations

Since solute concentration is a microscopic magnitude and its observed macroscopic values are averaged over a volume or a flux, different types of averaged concentrations can be defined. The following discussion is based on papers by Kreft and Zuber (1978), van Genuchten and Parker (1984), Parker and van Genuchten (1984), and Toride et al. (1993). Concentrations in a macroscopic representative elementary volume (REV) can be defined as the mass of solute in the liquid phase per volume of liquid phase. Since this type of volume-averaged solute concentration is a measure of the total solute mass residing at a particular time within our REV, this is known as resident concentration. The solved solute concentration, c_r, is a resident concentration (used in Chapter 10).

A useful second type of concentration may be also defined. Consider a cross section of our original monolith. If we look at the mass of water and solute crossing this soil section during an elementary time interval, we can define the solute to water flux ratio as the flux concentration, c_f:

$$c_f = \frac{J_s}{J_w} \qquad (11.6)$$

The effluent solute concentration is a clear example of such a flux concentration.

By substituting Eq. (11.6) into Eq. (11.1) for a solute flux with advective and dispersive components, it can be shown that resident and flux concentrations are related for an advective and dispersive transport:

$$c_f = c_r - \frac{D}{v}\nabla c_r \qquad (11.7)$$

When no solute sorption occurs, i.e., $S=0$ in Eq. (11.4), and under steady water flow ($\partial\theta/\partial t = \nabla \cdot J_w = 0$) resident and flux concentrations are related through the mean pore water velocity, $v=J_w/\theta$, by substituting Eq. (11.4) into the mass conservation equation (11.2):

$$\frac{\partial c_r}{\partial t} = -\nabla J_s = -v\nabla c_f \qquad (11.8)$$

Further expressions similar to Eq. (11.7) and Eq. (11.8), relating both types of concentrations, can be found in Parker and van Genuchten (1984).

When water transport is slow and soil heterogeneity is neglected, the flux concentration, c_f, will approach the resident concentration, c_r. However, in general, these are not coincident (Figure 11.2).

The distinction between resident and flux concentrations thus has practical implications. When measuring soil concentrations in saturation extract of a disturbed soil sample or using non-destructive techniques such as time domain reflectometry (TDR) (see Chapter 10), we will obtain resident solute concentrations. By contrast, when interpreting breakthrough curves from the effluent at the outlet of the column, we will be considering flux concentrations. When probing soil water by suction extractors and pore cups, we will be approaching the flux type concentration. However, in this last case, the type of concentration is not so evident, since this will depend on the relation between flow velocity and suction time employed. Van Genuchten and Wierenga (1977) pointed out that soil solution sampled by porous cups may likely represent mobile water (see nonequilibrium models in Chapter 3). Disturbance of the solute flow path when relatively high suctions are applied may also take place (van der Ploeg and Beese, 1977). Comparison of solute concentration from the suction extractors with that of soil coring may thus lead to contradictory results. For example, Alberts et al., (1977) reported significant differences between soil nitrate concentrations determined by either soil coring or solution extractors.

Finally, macroscopic averaged concentrations can be also defined with respect to a temporal observation scale, such that the time-averaged concentration (c_t) over a time increment (Δt) around a discrete time (t_0) is (Fischer et al., 1979):

$$c_t = \frac{1}{\Delta t} \int\limits_{t_0-\Delta t/2}^{t_0+\Delta t/2} c\,dt \qquad (11.9$$

Where c is a solution of Eqs. (11.1) and (11.2). Volumetric liquid content in multifluid flow experiments may be determined by gamma ray attenuation. A major drawback of gamma ray attenuation is that long counting times (usually > 1 min) are needed. Thus time-averaged concentrations are a useful way of expressing such volumetric contents.

11.2.1.2 Boundary Conditions

The advective-dispersive equation (ADE) [Eq. (11.3)] is a partial differential equation. As such, solutions of the ADE exist and are unique if proper initial' and boundary conditions are set up. In general, such conditions will have a different form whether written in terms of flux or resident concentrations. We will define boundary conditions in terms of resident concentrations, and by invoking relation (11.7), one can write those as flux concentrations. For simplicity we will consider only finite or (semi)infinite one-dimensional domains, which are close to situations in laboratory experiments. More general conditions in complex domains, such as those often found in field scenarios, may be set up by defining a proper domain.

11.2.1.2.1 Inlet Boundary Conditions

A straightforward way of mathematically formulating boundary conditions is by fixing the solute flux [Eq. (11.1)] to some prescribed time-dependent concentration input, $c_0(t)$. For a one-dimensional system we have

$$vc_r - D\frac{\partial c_r}{\partial x}\bigg|_{x=0} = vc_0(t) \qquad (11.10$$

The way Eq. (11.10) has been formulated ensures that mass is conserved across $x=0$. Equation (11.10) is known as Robin, third-type, or flux-type inlet boundary condition. From a practical point of view, this is the kind of boundary condition we will encounter when pumping, ponding, or sprinkling a solution onto a soil column (Figures 11.2 and 11.3). Alternatively, when the concentration rather than the flux at the inlet boundary is prescribed, Dirichlet, first-type, or concentration-type boundary conditions are set up:

$$c_r(0, t) = c_0(t) \qquad (11.11$$

FIGURE 11.3 Soil monoliths under controlled boundary conditions for solute transport studies developed at ICIA, Spain. Sensors and monitoring devices include TDR, tensiometers, and suction samplers.

As opposed to the third-type boundary conditions [Eq. (11.10)], in using first-type boundary conditions [Eq. (11.11)] mass conservation principles may not be considered. Thus, mass balance errors may appear when solving the ADE subject to Eq. (11.11), especially at early times in high dispersive systems with low pore water velocity (see van Genuchten and Parker, 1984). Additionally, in practice, it is easier to control the rate of solute application than the concentration at the soil surface; hence the third-type boundary condition [Eq. (11.10)] is preferred when the ADE is written in terms of resident concentrations. However, substitution of the third-type boundary condition [Eq. (11.10)] into Eq. (11.7) yields:

$$c_f(0, t) = c_0(t) \tag{11.12}$$

which is equivalent to a first-type boundary condition in terms of flux-averaged concentrations. Hence, when written in terms of c_f inlet first-type boundary conditions are recommended for solving the ADE.

11.2.1.2.2 Outlet Boundary Conditions

As for the inlet, both first- and third-type boundary conditions may be prescribed across the exit boundary. We may, however, here distinguish between finite or semi-infinite systems. For a finite system of length L, a similar third-type boundary condition as the one in Eq. (11.10) may be fixed across the exit boundary:

$$vc_r - D\frac{\partial c_r}{\partial x}\bigg|_{x=L} = vc_e(t) \tag{11.13}$$

Because of the addition of an extra unknown, the effluent concentration, $c_e(t)$, an additional condition is necessary to determine $c_e(t)$. By assuming continuity at $x = L$ for the solute concentration, i.e.:

$$c_r(L, t) = c_e(t) \tag{11.14}$$

substitution of Eq. (11.14) into Eq. (11.13) yields the zero-gradient condition necessary for the system of Eqs. (11.3) and (11.13) to be determinate in a finite soil column, i.e.:

$$\frac{\partial c_r}{\partial x}\bigg|_{x=L} = 0 \tag{11.15}$$

Alternatively, for a semi-infinite soil system:

$$\frac{\partial c_r}{\partial x}\bigg|_{x=\infty} = 0 \tag{11.16}$$

As with Eq. (11.10), Eq. (11.13) requires the exit reservoir to be physically disconnected from the soil surface or, when connected, that diffusion–dispersion phenomena are negligible i.e., the exit boundary does not affect solute concentrations inside the column. This is, for example, the case of free dripping into a fraction collector.

Free drainage is, however, a rather limiting boundary condition as the water flux and soil water content cannot be controlled. A more interesting situation consists of applying a prescribed suction (pressure-head increments) at the bottom of the column (Figure 11.3). This permits isolation of the contribution of different pore classes to the bulk flow. In the laboratory, pressure heads are imposed by depositing the soil on a bed of fine sand and/or porous material and applying a passive (hanging water column) or active (vacuum pump) suction. With a sand-kaolin porous bed, suctions as high as 5 m can be achieved, although the usual suction range for sands is 0.15–1 m.

For suctions up to 2 m, a glass microfiber membrane covered with a layer of silica flour can be used (Ball and Hunter, 1988). For suctions higher than 5 m, porous ceramic endplates and active vacuum pumping are compulsory. Porous ceramics may pose problems with large monoliths, because these are unusually available in large diameters (standard diameter 28 cm), and they may crack. Alternatively, porous metal plates (www.mottcorp.com) may be attached to the end of the soil column.

11.2.1.3 Tracers

The requirements for a chemical tracer were summarized by Davis et al. (1980):

Chemical stability, i.e., they must not be degraded chemically or biologically during the course of the experiment.

Minimum sorption, i.e., the tracer should not be sorbed or retarded by the soil.

Low or nonexistent background concentrations, i.e., the tracer should be exotic to the soil environment or be present naturally in low concentrations.

Chemical substances that generally satisfy these three conditions are derivatives of the benzoic acid (fluorobenzoate), anions such as iodide (I^-) or thiocyanate (SCN^-), low molecular weight anions such as bromide (Br^-), labeled (deuterated or tritiated) water, and dissolved gases such as He and Ne (Davis et al., 1980; Bowman, 1984; Brusseau, 1993b; Benson and Bowman, 1994; Jardine et al., 1998). Other possible candidates are nitrate (NO_3^-) and chloride (Cl^-). However, biological and chemical degradability in the former, and high concentrations present in irrigation water and soil, together with the possibility of anion exchange in the latter, make them unsuitable as reliable tracers (Bowman, 1984). Bromide is usually employed as a tracer and has become widely considered as virtually nonsorbing. However, several studies reported a linear sorbing behavior of bromide in soils or sediments that contain significant amount of variable-charge minerals (Brooks et al., 1998; Seaman et al., 1995; Katou et al., 1996; Vogeler et al., 2000). Variable charge soils, such as Andisols, are exceptional in their response with ionic tracers because of their large surface area (as high as $750 \, m^2 \, g^{-1}$) and charge dependence on the pH and ionic strength of the media (Wada, 1980; Ishiguro et al., 1992; Vogeler et al., 2000). Other desirable properties of tracers are minimal environmental impact and economic viability, two characteristics that are obviously not satisfied by labeled water (Bowman, 1984). Additionally some sorption of H_2O has been previously reported in an aggregated tropical soil of volcanic origin (Seyfried and Rao, 1987).

An interesting property of a tracer is the possibility of using an on-line measurement of concentration. For example, fluorobenzoates can be analyzed by UV absorbance or bromide through TDR (see Chapters 9 and 10). These techniques have been extensively used in experiments with a single tracer.

However, tracers can be applied not only in solitary but also in multiple-tracing experiments (Maloszewski and Zuber, 1993). Although this requires an off-line determination of the solute concentration, additional information can be obtained. For example, taking advantage of the different diffusion coefficients of several tracers, miscible displacement experiments may elucidate the role of diffusive mechanisms in solute transport. Also, under nonequilibrium conditions, different tracers will render distinct BTCs. These effects can be exploited to elucidate transport mechanisms, as discussed in Sec. 11.3.

11.2.2 ANALYSIS OF THE BREAKTHROUGH CURVE

11.2.2.1 The Effect of Transport Mechanisms on the BTC

Different initial and boundary conditions can be used to generate data on concentration versus depth or time. Further analysis can be carried out through inverse modeling for solute transport characterization. However, a pulse in the inlet concentration and the subsequent effluent analysis is by far the most used strategy in transport experiments under controlled conditions. Analysis of the generated BTCs offers valuable information for characterizing the solute transport mechanisms. Such analysis is usually performed on the basis of dimensionless units. The corresponding dimensionless formulation was introduced in Chapter 3. These variables are: $C(x, t) = c(x, t)/c_0$; $X = x/L$ $T = vt/L$; and $P = vL/D$; where $c(x, t)$ represents solute concentration at any time (t) and location (x), c_0 is the pulse concentration at the inlet, and L is the column length.

A rough analysis of the BTC shape gives an indication of the mechanism governing solute transport. Figure 11.4 represents BTCs generated for transport experiments under different dominating transport mechanisms (Nielsen and Biggar, 1962). In fact, any experimentally measured BTC may be considered one or a combination of any of the curves shown there.

Curve a in Figure 11.4 represents nonreactive transport of a tracer with only an advective component. This results from piston-like flow (i.e., narrow range of pore water velocities), so that the solute front breaks through exactly when one pore volume is displaced. This type of BTC can be considered as an ideal case and would rarely occur in soils because they contain a distribution of pore size.

Curves b and c in Figure 11.4 describes situations where the conservative nonsorbed tracer is spread as a result of hydrodynamic dispersion, its main components being molecular diffusion and mechanical dispersion (see Chapter 3,). Figure 11.4b shows a BTC obtained under a nonreactive, advective-dispersive transport. Here, solute dispersion causes curve smoothness, and when one pore volume is displaced, soil solute concentration is exactly half of the inflow concentration (i.e., $c(x, T) = 0.5c_0$). Curve c in Figure 11.4 represents a similar case, but with a wider range in the distribution of pore water velocities and, hence, of residence times (note that BTC is not symmetrical). This is typical of soils with equal-sized aggregates.

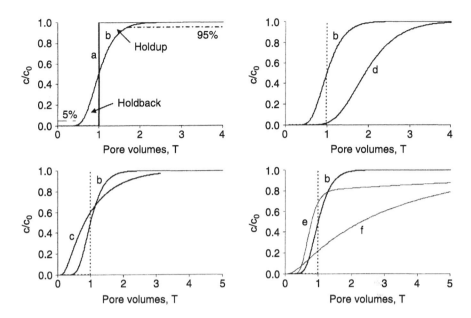

FIGURE 11.4 Basic BTC types for miscible displacement experiments.

Notice how, independently of the curve shape, in the three cases (a, b and c) the area below the BTC up to one pore volume is equal to the area above the BTC for T (number of pore volumes) greater than one. These areas are useful BTC attributes. The holdback is a measure of the efficiency of fluid displacement along the column (Danckwerts, 1953). A large holdback indicates difficult displacement. A similar characteristic, the holdup, is related to the amount of solutes that can be stored within the soil. Mathematically, both attributes are defined as:

$$holdback = \int_{0}^{1} C(x, T)dT \qquad (11.17)$$

$$holdup = \int_{1}^{\infty} [1 - C(x, T)]dT \qquad (11.18)$$

In addition, according to Biswas et al. (1991), two main points can be distinguished from the BTC: breakthrough $c(x, T) = 0.05c_0$, and exhaustion $c(x, T) = 0.95c_0$. Figure 11.4 shows the graphical representation of these points.

Curves d–f in Figure 11.4 represents situations where, in addition to advective-dispersive processes, solute transport is also affected by chemical or physical interactions. In these cases the areas under and above the BTC (as described before) are different. Curve d in Figure 11.4 describes solute transport, where in addition to the processes already mentioned, solutes interact and are retained by the soil matrix, so that the BTC is shifted to the right. This is the case of cations and organic chemicals that can be sorbed to soil constituents. Note that the shape of the sorption isotherm plays a decisive role in the symmetry of the BTC. On the other hand, a left-shifted BTC would be observed as a consequence of processes that produce a solute concentration increase (anion exclusion) or an incomplete mixing throughout the entire soil solution (stagnant pores).

When nonequilibrium exists (e.g., inhomogeneity and imperfect solute mixing or kinetics sorption), the resulting BTC is skewed (curve e in Figure 11.4). This occurs, for example, in soils with aggregates or large macropores, where rapid solute transport through the more permeable region causes early breakthrough. On the other hand, the slow solute diffusion into and out of a less permeable (stagnant or immobile) region produces early breakthrough and tailing (Mallants et al., 1994; Jardine et al., 1998). BTC symmetry is also an interesting feature to look at from a descriptive point of view. Figure 11.4 shows that BTC asymmetry may be due to a very large dispersive transport, low P values (Figure 11.4c), nonequilibrium during transport (Figure 11.4e), or a nonlinear sorption-isotherm (curve f in Figure 11.4).

11.2.2.2 Moment Analysis

Moment analysis is a statistical technique that has been successfully applied in chemical engineering, chromatography, and soil solute transport studies. Time moments provide useful and physically meaningful descriptors of the concentration breakthrough (Valocchi, 1985). The absolute moments are defined as:

$$m_n(x) = \int_0^\infty t^n C(x, t)\,dt \qquad (11.19)$$

The experimental n^{th} *normalized absolute* moment of the distribution of solute concentration versus time, t, at any location, x, is given by:

$$\mu'_n(x) = \frac{m_n(x)}{m_0(x)} = \frac{\int_0^\infty t^n C(x, t)\,dt}{\int_0^\infty C(x, t)\,dt} \qquad (11.20$$

where m_n is the nth absolute moment and m_0 is the zero moment, that corresponds to the total mass added and passing through location, x (for a conservative nonsorbed tracer, $m_0 = C_0 t_0$, with C_0 the inflow concentration and $_0$ the solute pulse time). In analogous way, the nth normalized central moment is defined as:

$$\mu_n(x) = \frac{\int\limits_0^\infty \left[t - \mu_1'(x)\right]^n C(x, t)dt}{m_0(x)} \tag{11.21}$$

The first three temporal moments of a BTC are often used to describe BTC shape parameters. The first moment (μ_1') corresponds to the mean residence or breakthrough time. The second (μ_2) is the variance or the average pulse spread relative to μ_1'. The coefficient of skewness (SK) or degree of BTC asymmetry is given by $\mu_3 \cdot \mu_2^{-3/2}$ (Mallants et al., 1994). Symmetrical BTCs ($SK = 0$) indicate local equilibrium during solute transport or a linear sorption isotherm in the case of sorbed solutes.

It should be mentioned that when calculating the experimental moments it is important to use a large data set of reliable measurements of the BTC (Leij and Dane, 1991). If such a data set is available, important information on the transport parameters can be estimated using the method of moments. Among them, the parameters commonly estimated include: (1) mean residence time (see example below); (2) dispersion coefficient (D); (3) pore water velocity (v) for conservative tracers (e.g., Maloszewski et al., 1994; Pang et al., 1998); (4) retardation factor (R) for sorbing solutes (Jacobsen et al., 1992; Rubin et al., 1997); and (5) degradation rate for reactive solutes (Pang et al., 2003). To achieve these estimations, the moments calculated from BTC data are compared to theoretical moments from a given model. Concerning the theoretical moments for a particular model, these are obtained by substituting the analytical solution of the model, $C(x, t)$, into Eqs. (11.19), (11.20), and (11.21), and performing the corresponding integration. However, these calculations are only possible for a limited number of cases subject to specific boundary conditions. Such theoretical moments may be derived using a Laplace transformed version of the transport model (Kucera, 1965; Valocchi, 1985; Pang et al., 2003); therefore, a solution of the transport equations is not required.

Table 11.2 shows theoretical moments published by Valocchi (1985) and Vanderborght et al. (1997) for local equilibrium and nonequilibrium models with instantaneous solute application (Dirac delta function). Further expressions of theoretical moments for other models and boundary conditions can be found in Jury and Sposito (1985), Valocchi (1986), Leij and Dane (1991, 1992), Das and Kluitenberg (1996), Toride and Leij (1996), Espinoza and Valocchi (1998), and Young and Ball (2000), among others.

An example of practical application of moments to estimate tortuosity, mean residence time, and deformation of tritium breakthroughs from soil columns was presented by Ma and Selim (1994). Solute residence time t_m (defined as the time required for a solute molecule or ion to traverse the column) can be calculated as the first normalized absolute moment (Leij and Dane, 1991). A first interesting conclusion of the results presented in Table 11.2

TABLE 11.2
Theoretical Temporal Moments for Equilibrium and Nonequilibrium Models for Instantaneous Solute Application

Moment	Local equilibrium	Physical nonequilibrium
μ_1'	$\dfrac{xR}{v}$	$\dfrac{xR}{v}$
μ_2	$\dfrac{2xDR^2}{v^3}$	$\dfrac{2x\beta D_m R^2}{\theta v^3} + \dfrac{2x(1-\beta)^2 R^2 \theta}{\alpha v}$
μ_3	$\dfrac{12xD^2 R^3}{v^5}$	$\dfrac{6x(1-\beta)^3 R^3 \theta^2}{\alpha^2 \theta v} + \dfrac{12xD_m(1-\beta)^2 R^3 \beta \theta}{\alpha v^3} + \dfrac{12xD_m^2 R^3 \beta}{v^5}$

J_w: water flux density; D: dispersion coefficient; D_m: dispersion coefficient for the mobile region; β: ratio of mobile water respect to the total water content, θ; α: first-order mass exchange rate; v: average pore water velocity.

Source: Valocchi, 1985; Vanderborght et al., 1997.

is that non-equilibrium does not affect t_m (the first moment is equivalent for equilibrium and nonequilibrium models). Also, if the pore water velocity v is measured from controlled water flux, an equivalent column length, L_e, may be defined such that $L_e = \mu_1' v$. L_e is a measure of the effective solute transport length (Ma and Selim, 1994). In addition, an apparent tortuosity factor may be obtained from $\tau = L/L_e$. Several authors (Porter et al., 1960; van Schaik and Kemper, 1966; Gillham et al., 1984; Sadghi et al., 1989; Ma and Selim, 1994) incorporated τ into an effective diffusion coefficient (D_e) and performed τ-estimations based on measured D_e (see Chapter 3).

Spatial moments may be also useful for describing spatial distributions of solute concentration during transport (Parker and van Genuchten, 1984). Goltz and Roberts (1987) presented analytical solutions for the spatial moments of a sorbing solute in an infinite porous media. However, these solutions are given in terms of integrals that are difficult to evaluate. In this context, Cunningham et al. (1999) found expressions that greatly simplify the numerical evaluation of the spatial moments.

11.2.2.3 Characterizing Transport Mechanisms Through Inverse Modeling

Alternative techniques to estimate transport parameters from miscible displacement experiments are the so-called "indirect methods." These techniques treat some of the model parameters as unknowns and are estimated based on one or more experimentally measurable variables. Inverse modeling is a particular and more complex technique (since it requires the development of sophisticated algorithms), but it has become an increasingly attractive form of indirect parameter estimation. Basically, the process searches for the best set of parameters in an iterative procedure. Thus, by varying the parameters and comparing the response of the variables measured during an experiment with the numerical solution given by the model, BTC parameters are optimized. In the context of solute transport, this is performed by minimizing a suitable objective function that expresses the discrepancy between the output of the transport model and solute concentration measurements. The advantage of the method is its versatility in reflecting several experimental situations and thus, being applied to BTCs, field solute concentration profiles, etc. This approach requires the coupling of the transport model with an optimization algorithm. Chapter 20 discusses in detail the applications of inverse modeling. Here we will briefly comment on the inverse techniques for characterizing solute transport from BTC data.

The successful identification of the transport parameters depends on several issues that should be taken into account:

1. Inverse modeling requires a reliable and detailed (in time and/or space) data set of solute transport, which is often difficult to obtain (Jacques et al., 2002).
2. The inversion algorithm must be efficient and robust (Kool et al., 1987). Algorithms that greatly depend on initial conditions may not be adequate.
3. The model must be appropriate for the transport conditions, e.g., a model based on the local-equlibrium assumption (see Chapter 3) should not be used for describing asymmetrical BTCs.
4. When choosing the parameters to optimize, parameter correlation should be studied. When correlation exits, small variations of a certain parameter may be compensated with changes in others (Koch and Flühler, 1993).

In this context, before performing an inverse procedure it is important to obtain additional information about the transport processes involved.

One of the most popular codes for analysis of BTCs through inverse modeling is CXTFIT, which uses least squares for parameter estimation. The algorithm allows the estimation of transport parameters from a wide variety of experimental situations, including stochastic models which may be applied at field scale. CXTFIT was developed by Parker and van Genuchten (1984), and it has been updated by Toride et al. (1995) and Šimůnek et al. (2000). Chapter 20 discusses the use of more sophisticated optimization algorithms such as the global multilevel search coordinates (GMCS) of Huyer and Neumaier (1999). This algorithm combines global and local search capabilities with a multilevel approach. The GMCS is a good alternative to other existing optimization routines, because it can deal with objective functions with complex topography, it does not require powerful computing resources, and initial values of the parameters to be optimized are not needed.

11.2.2.4 Application for Sorbed Solutes: The Estimation of the Retardation Factor

Transport of sorbing solutes through a porous media is characterized by a delayed breakthrough (curve d in Figure 11.4). This delay is usually quantified by means of a retardation factor, R. The retardation factor is perhaps one of the most interesting parameters, as it is closely related to the relative travel speed of a particular solute, when compared to water flow. Determination of R is essential to successfully describe solute transport with mathematical models. For sorbed solutes, R values are greater than unity (i.e., solutes move at slower velocity than bulk water). For nonsorbing solutes $R = 1$, whereas values of $R < 1$ are an indication of solutes traveling faster than the bulk water (due to a size or anion exclusion).

Mathematical description of the retardation factor depends on the adsorption isotherm equation. For a linear isotherm ($S = K_d C$), R is given by

$$R = 1 + \frac{\rho_b}{\theta}\frac{dS}{dC} = 1 + \frac{\rho_b K_d}{\theta} \qquad (11.22$$

where θ is the volumetric soil water content $[L^3 L^{-3}]$, ρ_b is the soil bulk density $[ML^{-3}]$, and K_d is the distribution coefficient $[L^3 M^{-1}]$. For nonlinear adsorption isotherms, R is a concentration-dependent factor. For example, for a Freundlich isotherm ($S = K_f C^n$), with K_f and n the isotherm coefficients, the retardation factor is given by

$$R = 1 + \frac{\rho_b}{\theta}\frac{dS}{dC} = 1 + \frac{\rho_b}{\theta} n K_f C^{n-1} \qquad (11.23$$

Several methods have been used to estimate the retardation factor from BTC data (see, for example, Jacobson et al., 1984; Bouchard et al., 1988; Maraqa et al., 1998):

R₁ R is obtained as the eluted pore volume (T) when the solute concentration is half the inflow concentration ($c = 0.5 c_0$, curve a in Figure 11.5). This procedure can be affected by BTC asymmetry, though it is a fast approach.

R₂ R is estimated from the area above the BTC equivalent to the area between the elution curve and the step input curve divided by the area above the BTC for a conservative nonsorbed tracer (curve b in Figure 11.5). This procedure requires two BTCs under the same conditions (the tracer and solute curves, respectively).

R₃ R is calculated with the method of moments (MOM), using the mean residence time calculated as the first normalized absolute moment (μ_1') according to Eq. (11.20) and the pulse duration (t_0) (curve c in Figure 11.5) (Leij and Dane, 1991):

$$R = \frac{\mu_1' - 0.5 t_0}{x/v} \qquad (11.24$$

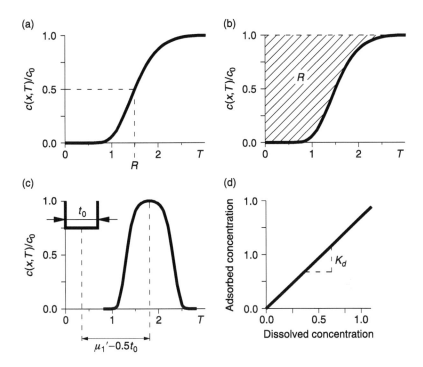

FIGURE 11.5 Estimations of retardation factor: (a) eluted pore volume when $C(x, T) = 0.5$; (b) area above the BTC; (c) from mean residence time (μ_1'); (d) using Eqs. (11.21) or (11.22) and a sorption isotherm.

R_4 Curve fitting technique, by means of inverse simulation techniques. Although this procedure needs the appropriate codes for the model and the optimization algorithm, it can analyze, with the same accuracy, different experimental situations (e.g., symmetrical and asymmetrical BTCs).

Alternatively, an independent estimation of the retardation factor can be achieved using batch equilibration data and Eqs. (11.22) or (11.23), where θ and ρ_b are measured in the transport experiment and K_d or K_f are obtained from the slope of the sorption isotherm (Figure 11.5d).

As stated above, symmetrical BTCs are indicative of nonsorbed or linear sorption isotherms and sorption equilibrium during transport. Bouchard et al. (1988) compared methods R_1 and R_2 for estimating the retardation factor. For symmetrical BTCs both methods yielded similar results (in fact, for symmetrical BTCs the four methods mentioned above yield comparable results). When the obtained BTC is asymmetrical, these same authors suggested an interesting measure of BTC asymmetry, which consists of estimating the ratio between R calculated by these two methods (R_1/R_2).

They also found agreement between batch (R_4) and column methods for symmetrical BTCs. For asymmetrical BTCs, the estimation of R from independent batch equilibration and area calculation methods R_2 were also comparable, whereas the R_1 method did not give reliable results (as expected from Figure 11.4). Maraqa et al. (1998) compared the values of the retardation factor of benzene and dimethylphthalate, estimated by MOM and inverse modeling from nine BTCs. In spite of the lack of information on confidence intervals for results obtained with the MOM, the two methods gave comparable results. However, the values of independent estimations from batch equilibration experiments were not comparable. In these experiments, BTCs of benzene and dimethylphthalate presented asymmetry.

Considering independent parameter estimations, batch experiments can provide rough estimations of the retardation factor. However, experimental variables, such as the soil-to-solution ratio, have a direct effect on the results obtained (Chapter 12). Thus, the predictive capacity of the batch method is limited. Considering BTC analysis based on $c = 0.5c_0$ (R_1) and area above the BTC (R_2) can be used as a fast estimation of BTC asymmetry, as proposed by Bouchard et al. (1998). The former is only valid for conservative nonsorbed tracers. Finally, the MOM (R_3) and inverse techniques (R_4) are the more appropriate methods when working with asymmetrical BTCs. The main advantage of the curve fitting is that a large set of experimental data is not needed. On the other hand, the accuracy of the MOM will be low with small data sets. However, MOM gives no ambiguous parameters and does not require any prior information, such as the initial values for the fitted parameters in inverse methods.

11.2.3 BEYOND THE BTC

Breakthrough curves can provide useful information about miscible solute transport in the soil environment, but either solely or in combination with BTC analysis, alternative procedures can be also employed. Transport experiments in the field scenario are technically more difficult to perform. Also, due to field spatial variability, temporal fluctuation of soil parameters, and uncontrolled environmental conditions, such experiments may be too simplistic. One of such alternatives to characterize solute transport in soil, also frequently used with soil monoliths, is preferential flow and macropores mapping by application of dyes (Flury and Flühler, 1995).*

Dying does not exclude other techniques, and in fact dye staining is recommended in combination with miscible displacement experiments (Wildenschild et al., 1994). Burkhardt et al. (2003) combined multitracing experiments with fluorescent microspheres to identify preferential flow paths in a field experiment. Instead of funneling the leaching into a single reservoir,

---

* Some pictures and animated video sequences may be found in www.bee.cornell.edu/swlab/pfweb/educators/image.htm

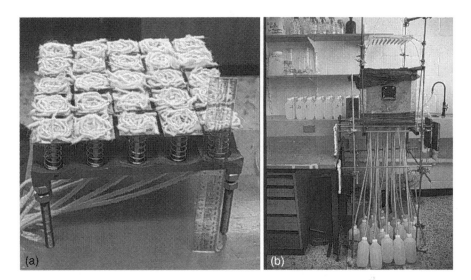

FIGURE 11.6 (a) Three-dimensional view of the alloy-cast base plate installed with spring-loaded wick lysimeters prior to mounting on soil column. (b) Multisegment wick percolation system preferential flow apparatus used in solute transport studies. (Images courtesy of F. Stagnitti)

Wildenschild et al. (1994) used a 9 × 9 cell collection grid to evaluate the spatial distribution of the outflow from an undisturbed soil monolith. Outflow at the base of the monolith was by free drainage, and so their system is not feasible for low-moisture conditions. Some devices have, however been designed for collecting the outflow solution under unsaturated conditions. The Alundum tension plate sampler is an example of these. Fiberglass multisegment wick lysimeters can also be used to sample unsaturated flow in field sites or alternatively in undisturbed soil cores (Boll et al. 1992; de Rooij and Stagnitti, 2002). Figure 11.6 shows the use of the preferential flow apparatus used in solute transport studies in Australia (Stagnitti et al., 1999, 2003). The apparatus was originally developed by Tammo Steenhuis, John Selker, and Jan Boll at Cornell Univeristy.

In the field, dye application is necessarily followed by either soil coring or trench excavation and photographing. Image analysis of digitalized photographs can provide quantitative information of size, shape, and proportion of preferential paths, and whether these are associated to plant roots or wormholes (Kulli et al., 2003). In the lab, dying is normally followed by some destructive technique as soil slicing and photographing or analysis of the solution extracts after destructive soil sampling. Binley et al. (1996) used dyes to verify tomographic images of a large monolith in combination with breakthrough curve analysis. X-ray tomography is a nondestructive technique that can provide 3-D images of macropore networks (see Chapter 7). However, its low resolution (< 0.5 mm) and high cost may limit their extensive use (Peyton et al., 1992).

11.3 TECHNIQUES FOR CHARACTERIZING NONEQUILIBRIUM DURING SOLUTE TRANSPORT IN SOILS

11.3.1 TECHNIQUES BASED ON BREAKTHROUGH CURVES

From the previous discussion, it is seen that analysis of a single BTC can provide information on the role of several transport mechanisms such as dispersion, sorption, or degradation. If a BTC is symmetrical and $c = 0.5c_0$ at one pore volume, we can conclude that sorption and nonequilibrium are negligible and dispersion is moderate. The same degree of symmetry in a BTC shifted to higher pore volumes reveals sorption under local equilibrium. However, BTC asymmetry can be caused by nonequilibrium (from a physical or chemical nature), isotherm nonlinearity, or high hydrodynamic dispersion. Among them, isotherm nonlinearity is only found for sorbed solutes at relatively high solute concentrations, and high hydrodynamic dispersion is evidenced only at high Peclet numbers (P). For this reason, under natural scenarios, and experiments in soil columns, which intend to be close to those conditions, nonequilibrium is by far the most important source of BTC asymmetry.

Nonequilibrium during solute transport in soils has become apparent from the earlier studies on characterization of solute transport. Laboratory and field studies that document the preferential movement of solutes coupled with kinetics mass transfer within different flow domains have been increasingly reported in the scientific literature (see Chapters 3 and 8). This physical non-equilibrium affects all solutes, including conservative nonsorbed tracers. The simplest conceptualization of physical nonequilibrium, which consists of dividing the soil flow domain into mobile and immobile regions, is described in Chapter 3.

The sorbed solutes can undergo an additional source of nonequilibrium if the sorption rate is not rapid when compared to other processes (e.g., advection or dispersion). As reported by Brusseau and Rao (1989), it was initially thought that, because of the generally slow movement of water in the subsurface, equilibrium conditions should prevail. However, detailed laboratory and field investigations have revealed that in many cases the so-called local equilibrium assumption is invalid. The effects of both physical and chemical types of nonequilibrium on the BTC are identical, producing asymmetrical BTCs, with earlier breakthrough, increased time to complete breakthrough, and elution-front tailing. This coincidence of the shape of BTCs for the two different nonequilibrium sources is not surprising, as the equivalence between the mathematical formulations of the two-region (i.e., physical nonequilibrium) and two-site (i.e., chemical nonequilibrium) models was presented in Chapter 3. In addition, it is mentioned above that isotherm nonlinearity also influences the shape of the BTC of a sorbed solute. For example, for $n < 1$ in the Freunlich isotherm [used in Eq. (11.23)], an elution-front tailing effect is generated.

As shown by Jardine et al. (1998), experiments involving nonequilibrium are often performed under a single set of conditions, making it difficult to accurately describe observations in mathematical terms. However, based on the above considerations, a single experimental condition by itself is not usually sensitive enough to accurately describe the processes present under nonequilibrium. Several experimental strategies can be developed for characterizing the origin of BTC asymmetry.

For example, as mentioned above, if a sorbed solute exhibits an asymmetrical BTC, this behavior can be attributed to physical nonequilibrium, chemical nonequilibrium, or sorption isotherm nonlinearity. However, using a nonsorbed tracer under the same experimental conditions, the role of physical nonequilibrium could be determined, as chemical non-equilibrium is not present if sorption does not exist. Similarly, using different solute concentrations in the pulse for generating the BTC, isotherm nonlinearity could be identified as the responsible for BTC asymmetry.

Based on a classification given by Jardine et al. (1998), experimental techniques for confirming and quantifying transport mechanisms can be divided among the following categories:

Controlling flow-path dynamics:
   Variation in pore-water flux in miscible displacement experiments: the objective is to alter the rate of approach to equilibrium by variation of the experimental flux. Also, the relative role of molecular diffusion and mechanical dispersion in the total hydrodynamic dispersion can be determined as mechanical dispersion depends on pore water velocity (see Chapter 3).
   Variation in pressure-head: the purpose of this technique is to collect water and solutes from selected sets of pore classes in order to determine how each set contributes to the bulk flow and transport processes (see boundary conditions in Sec. 11.2.1.2).
Tracing:
   Multiple conservative tracers with different mass transfer diffusion coefficients: the basic principle of this technique is that tracers with larger molecular diffusion will travel faster through the matrix micropores. Thus, comparing the BTCs of several conservative tracers, the differences in the observed BTCs will be attributed to differences in the diffusive (versus advective) transport.
   Multiple conservative tracers with grossly different sizes: this technique usually uses dissolved solutes and nonsorbed colloidal tracers. The differences in BTCs (i.e., an early arrival of the colloidal tracers) will reveal a size exclusion process.
   Tracers for flow-path mapping and imaging: the objective is to label the flow paths for evaluating nonequilibrium by means of application of dyes (see Sec. 11.2.3).
Isolating diffusion processes with flow interruption: during a miscible displacement experiment, the flow is interrupted and the hydraulic

nonequilibrium (e.g., the difference between mobile and immobile regions) is eliminated. During the flow interruption, the system approaches an equilibrium state as solute mass transfer continues between mobile and immobile regions. When flow is resumed, the concentration of the BTC is perturbed proportionally to the magnitude of the existing nonequilibrium.

Each of the above techniques provides different additional information on solute transport mechanisms, including nonequilibrium. As suggested by Jaynes et al. (1998) the best results are obtained when the techniques are combined (e.g., flow interruption + multiple tracers, or variation in pore-water flux + tracers with different sizes). Although the above classification was initially developed for physical nonequilibrium, the proposed experimental strategies can be valid for determining all transport mechanisms responsible for asymmetry in BTCs.

A major limitation of the above-mentioned techniques is their difficult application at field scale. Very few attempts have been reported to quantify physical nonequilibrium processes at the field scale, and to the authors' knowledge, there have not been any application of flow interruption at large scale. This is, in fact, not surprising, as these techniques require an exhaustive control of the soil boundary conditions. For this reason, simple alternative methods for investigating nonequilibrium are very interesting for application at field scale. Among them, the estimation of nonequilibrium parameters from simple experiments has given results comparable to other BTC analysis. The two more extended procedures will be presented in Sec. 11.3.2.

11.3.1.1   Effect of Variation of the Pore Water Velocity

The effect of variation of the pore water velocity on the BTCs can be anticipated from the theoretical description of solute transport given in Chapter 3. For a nonsorbed conservative tracer, the hydrodynamic dispersion coefficient, D, plotted against the pore water velocity will give a linear relationship if mechanical dispersion is the dominant mechanism [Eq. (3.25)]. A dominating effect of mechanical dispersion will also produce identical BTCs for different tracers if other mechanisms (such as size exclusion) are not present. However, differences can appear at low flow rates with solutes of different molecular diffusion, as under these circumstances mechanical dispersion is not the dominating mechanism (Hu and Brusseau, 1994).

With regard to the effect of pore water velocity in nonequilibrium parameters, De Smedt and Wierenga (1984) compared several transport experiments in glass beads under saturated and unsaturated conditions. Considering the unsaturated experiments, they found a linear relationship between the mobile volumetric water content, θ_m, and the water flux. Since under unsaturated conditions the water content increases with flux, they concluded that the mobile fraction was related to the total water content (θ

through a proportionality constant ($\theta_m = 0.853\theta$). Note that the immobile water content also must follow a proportional increase with the pore water velocity (as $\theta = \theta_m + \theta_{im}$). The mass transfer coefficient α [Eqs (3.8) and (3.21)] also varied with the average pore water velocity in the mobile zone, v_m (see Chapter 8). An increase in α with pore water velocity was explained by the increase of water content and thus cross-sectional area for mass transfer, which is greater at larger pore water velocities (De Smedt and Wierenga, 1984).

11.3.1.2 Single and Multiple Tracers

BTCs obtained under the same hydrodynamic conditions, with solutes that are subjected to different transport mechanisms, are an important source of information for characterization purposes. BTCs can be obtained sequentially for different solutes, or tracers can be applied simultaneously in multiple-tracing experiments. The simultaneous multiple tracing experiments are less costly and time consuming than sequential tracing, although they may require reliable methods of chemical analysis for solute concentration. Sequential tracing allows on-line detection, whereas simultaneous tracing analysis must normally be performed off-line (Maloszewski and Zuber, 1993).

The comparison of BTCs from conservative nonsorbed tracers with the BTCs of sorbed solutes can provide useful information on the role of sorption processes and the possible existence of nonequilibrium sorption. Also, taking advantage of the different diffusion coefficients of several tracers, multiple-tracing displacement experiments may distinguish between diffusive and advective dominant mechanisms in structured soils (Jardine et al., 1998). When advective transport is prevalent, BTCs will be similar. By contrast, if diffusion is the dominant transport mechanism, BTCs will be different. Tracers with a larger diffusion coefficient will exhibit a longer end-tailing in their breakthrough curve. For example, the PFBA diffusion constant is 40% smaller than that of bromide. When applied to fractured weathered shale, since the PFBA diffused more slowly into the soil matrix, the PFBA breakthrough at the column outlet was initially faster than bromide. Conversely, the PFBA BTC took longer to approach exhaustion time (Jardine et al., 1998).

Multiple tracing can also use tracers of different sizes, such as solvable solutes and colloids. Because of their large size, colloids are excluded from specific pore sizes of the soil matrix, and therefore they serve as an excellent tracer to quantify advective transport. Viruses, bacteria, flurorescent microspheres, synthetic polymers and labeled Fe-oxide particles have all been used as colloidal tracers (Jardine et al., 1998).

11.3.1.3 The Flow-Interruption Technique

The flow-interruption technique is perhaps the most promising approach for characterizing transport mechanisms. This technique is based on similar methodological approaches used in chemical engineering (Kunin and

Myers, 1947), chromatography (Knox and Mclaren, 1964), and soil science (Muraly and Aylmore, 1980; Brusseau et al., 1997) to evaluate the role of different transport mechanisms.

The physical nonequilibrium of solute concentration resulting from preferential flow of soil water has led to models where the soil is partitioned into two regions (see Chapters 3 and 8). In the mobile region, the solute transport occurs mainly by advection. In the remaining immobile region, the solute transport occurs through diffusive exchange with the mobile region. During miscible displacement experiments, variations in the inlet concentration will rapidly affect the advective (mobile) domain. Then, the diffusive barrier between the two regions produces a nonequilibrium effect. Under this situation, if the flow is stopped, the two different flow regions disappear and the system will tend to a new equilibrium state by means of diffusive transport. Once the flow is resumed, the advection is restored but a perturbation on the effluent concentration will be observed. This perturbation can consist of an increment if, during the interruption, the solute has moved from the immobile to the mobile region, or a decrease if the solute transfer is in the inverse sense (Figure 11.7). Thus, by using this method the contributions of macro- and micropore transport mechanisms can be separated (Koch and Flühler, 1993).

In addition to the diffusive mass transfer, the transport and fate of many contaminants in subsurface systems can be influenced by several rate-limited processes, such as rate-limited sorption, and transformation reactions. Flow-interruption techniques can provide valuable information in the study of the different solute transport processes. Considering the analysis of BTCs, both physical nonequilibrium and hydrodynamic dispersion can cause asymmetry (curves c, e in Figure 11.4). However, only the former produces a noticeable perturbation concentration in upon interruption of flow under typical conditions. In addition, both rate-limited and nonlinear sorption can cause breakthrough curves to exhibit tailing. Nevertheless, only rate-limited

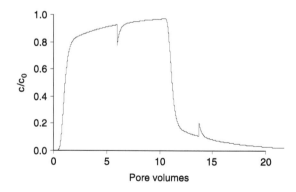

FIGURE 11.7 Breakthrough curve with flow interruption for a conservative tracer under nonequilibrium transport. Flow interruption during tracer elution produces a positive perturbation. Note that abscissa does not represent time but dimensionless eluted volume.

sorption induces a perturbation of concentration in the BTC when the flow is interrupted. There are other processes that can cause a perturbation in effluent concentration after a flow interruption, such as solute losses by degradation and volatilization. Therefore, these processes must be controlled when readily degradable or volatile compounds are used.

Both changes in the effluent concentrations or concentration profiles within the column can be analyzed after a flow interruption. As stated by Brusseau et al. (1989), the former are experimentally more feasible and theoretically more appealing, since the temporal first moment (i.e., center of mass equivalent to the retardation factor) is independent of the kinetics constants, whereas the corresponding spatial moment is not. Table 11.3 shows the experimental approach proposed by Brusseau et al. (1989) to investigate processes under nonequilibrium transport. Brusseau et al. (1997) further presented a method that can differentiate among several transport mechanisms such as film and intraparticle diffusion, intraorganic matter diffusion, and rate-limited interactions between sorbate and mineral components of the sorbent.

Figure 11.7 shows a simulation of a flow-interruption experiment. Resuming the flow after an interruption during tracer injection (curve a in Figure 11.7) results in decreased tracer concentration, as part of the solute in the mobile region had enough time during the interruption to be transferred to the immobile region. Conversely, during tracer displacement, the concentration in the immobile region is higher than in the mobile region. Thus, resuming the flow after an interruption results in increased tracer concentration (curve b in Figure 11.7).

As described by Koch and Flühler (1993), both diffusion in macropores (i.e., vertical diffusion in the direction of flow) and diffusion into micropores (i.e., into the soil matrix) may introduce a change of outflow concentration after a flow interruption. The two processes may compensate each other. For example, considering the breakthrough front (curve a in Figure 11.5); diffusion in macropores increases the outflow concentration, whereas diffusion into micropores decreases it. The influence of the two diffusion processes depends on the concentration gradients in the direction of flow, and between macro- and microporosities. In general, concentration gradients between mobile and immobile regions are greater than gradients in the direction of flow due to the effect of advective transport. The maximum magnitude of perturbation in the BTC will be obtained when differences between solute concentrations in the mobile and immobile regions are higher (i.e., higher slope of the BTC).

Concerning the time extent of the interruption, we will follow with the previous example during the breakthrough front (curve a in Figure 11.5). Short times will not allow mass transfer between macro- and microporosity, but long intervals of interrupted flow will allow diffusion in the macropores (i.e., vertical diffusion in the direction of flow) and the solute concentration in macropores will decrease. Thus, an optimum in sensitivity of the technique exists, which can be estimated through numerical simulations for a particular system (Koch and Flühler, 1993). The compensating effect of macropore

TABLE 11.3
Techniques for Investigating Transport Processes Through Miscible Displacement Experiments

Step[a]	Technique	Use	Explanation
		A. Miscible displacement experiment, single/multiple tracers	
1	Miscible displacement with a single conservative nonsorbed tracer	Physical nonequilibrium	BTC analysis (Sec. 11.2.2) or inverse modeling can reveal nonequilibrium during transport.
2	Miscible displacement with a conservative sorbed solute	Sorption-related nonequilibrium	If parameters from step 1 can reproduce accurately the BTC of the sorbed solute, the sorption-related nonequilibrium is negligible (Brusseau, 1993; Hu and Brusseau, 1995)
2	Miscible displacement with different conservative sorbed solutes	Sorption-related nonequilibrium	Nonequilibrium (i.e., BTCs asymmetry) increases as the sorption increases (Bouchard et al., 1988)
3	Miscible displacement with multiple tracers	Physical nonequilibrium	If BTCs are comparable for tracers with different molecular diffusion coefficient, transport is not affected by physical nonequilibrium (the role of diffusive mass transfer between the eventual mobile-immobile regions is negligible)
		B. Miscible displacement controlling flow path dynamics	
2	Miscible displacement under different pore water velocities with a conservative nonsorbed tracer	Study of dispersivity	When hydrodynamic dispersion is the predominant source of dispersion, the shape of the BTCs measured for nonreactive tracers should be independent of velocity (Brusseau, 1993; Bouchard et al., 1988). Differences can appear at low flow rates with solutes of different molecular diffusion (Hu and Brusseau, 1994).

No.	Experimental conditions	Process	Interpretation
2	Miscible displacement under different pore water velocities with different conservative nonsorbed tracers	Physical nonequilibrium	Rapid initial breakthrough and extended tailing of a conservative nonsorbed tracer becomes more evident as pore-water velocity increases and residence time of tracer in the mobile region decreases (Nkedi-Kizza et al., 1982, 1983).
		Role of molecular diffusion and mechanical dispersion in the hydrodynamic dispersion	The relative importance of mechanical dispersion is a function of the pore-water velocity, whereas the molecular diffusion is a constant that differs from solute to solute [see Chapter 3, Eq. (3.26)]. At low pore-water velocities, BTCs will differ. However, increasing the pore water velocity, mechanical dispersion will become the main component of hydrodynamic dispersion and BTCs of different solutes will be comparable (Hu and Brusseau, 1994).
3	Miscible displacement under unsaturated conditions	Physical non-equilibrium and preferential flow	If asymmetry of the BTC increases with water content, enhanced preferential flow is present as large pores are involved in the transport process (Seyfried and Rao, 1987).
3	Miscible displacement under different pore water velocities with a conservative sorbed solute	Sorption related nonequilibrium	As pore-water velocity increases, the attainment of sorption equilibrium is limited temporally by solute column residence time. BTC asymmetry increases when compared to a nonsorbed tracer (Bouchard et al., 1988).

(Continued)

TABLE 11.3
Continued

Step[a]	Technique	Use	Explanation
		C. Flow interruption (Brusseau et al., 1989)	
1	Flow interruption with a nonsorbing conservative tracer	Physical nonequilibrium	If the BTC exhibits a perturbation after the flow interruption, physical nonequilibrium (mass transfer between mobile and immobile regions) is present.
2	Flow interruption with a conservative sorbed solute (volatilization and degradation must be negligible)	Nonequilibrium related to the sorption process	If the parameters used for the nonsorbed tracer can describe the BTC, sorption-related nonequilibrium is negligible.
3	Flow interruption with a conservative sorbed solute with sorbent organic matter reduced (or eliminated)	Nonequilibrium related to intra organic matter diffusion	If the parameters used for the sorbed solute can describe the BTC, the effect of organic matter on nonequilibrium is negligible.
4	Miscible displacement experiment without flow interruption	Significant physical nonequilibrium of film diffusion	If parameters from step 2 can reproduce the BTC, film diffusion is negligible, as film diffusion is not present during flow interruption.

[a] Steps denote alternative experimental routes.

diffusion is, however, not so evident in all the flow-interruption experiments (Reedy et al., 1996).

A variation of the flow-interruption technique more applicable at the field scale consists of intermittent leaching, where flow and nonflow periods are sequentially established. Alternating wet and dry periods in experiments performed at laboratory scale, Cote et al. (1999) enhanced leaching (i.e., solute removal from soil matrix) by up to 20%. Extrapolating these results to other soils, the use of alternating wet and dry periods can improve management of solutes, for example, in salinity control and soil remediation. Intermittent flow takes advantage of the same physical principles in remediation as pump-and-treat technologies. Here, advective flow is used to perform solute or contaminant withdrawal from the mobile region, whereas the solute from the immobile region is subject to a mass transfer kinetics barrier. If a sufficiently large nonflow period takes place, mass transfer from the immobile to the mobile region will tend to equilibrium under this new scenario. In the subsequent advective period, the solute diffused from the immobile region in the previous nonflow stage will be removed, and the overall efficiency will be improved. Another application at field scale is given by Al-Sibai et al. (1997) for removing excess salts from soil. During leaching, water flows preferentially through macropores between aggregates. Consequently, the solute within aggregates is removed much more slowly. For this reason intermittent flow can be more efficient, because it allows time for solute to diffuse to the surfaces of aggregates during the rest period and subsequently be removed in macropore flow.

11.3.2 ESTIMATION OF NONEQUILIBRIUM PARAMETERS FROM SIMPLE EXPERIMENTS

An obstacle to the practical application of the two-region models at field scale is the estimation of the two key parameters accounting for nonequilibrium: the volumetric water content in the mobile region (θ_m) and the solute mass transfer coefficient between the two regions (α) [see Eq. (3.8)]. Usually these key parameters are found by inverse simulation techniques, optimizing non-equilibrium model response to fit the experimental data. This requires the generation of, at least, a complete BTC, which can require an intensive field experimental effort. Also, as mentioned above, several sets of model parameters can describe a given experimental BTC equally well (Koch and Flühler, 1993), and thus, the inverse modeling procedure does not necessarily produce a unique solution. For such reasons, it is interesting to evaluate methods for independent estimation of nonequilibrium parameters at field and laboratory scales.

Several procedures have been presented to determine θ_m and α from independent experiments. Ma and Selim (1998) reviewed the most relevant approaches based on soil geometry, water content monitoring during infiltration, air permeability, and macropore tracing. Clothier et al. (1992, 1995), Jaynes et al. (1995), and Jaynes and Horton (1998) have proposed attractive simplified approaches for estimating the mobile fraction and the solute mass transfer coefficient. These approaches rely on the assumption that as infiltration exceeds one pore volume, the effect of hydrodynamic dispersion is negligible. Therefore, the resident solute concentration can be determined predominantly by the mobile water fraction and the solute mass transfer rate between mobile and immobile regions.

11.3.2.1 Single Tracer

Clothier et al. (1992) proposed a single-tracer technique using a disk permeameter (Perroux and White, 1988) for measuring the soil apparent water mobile fraction during near-saturated flow. Initial charging of the mobile and immobile fractions (θ_m and θ_{im}) is achieved by first wetting the soil using pure water with a disk permeameter until steady state. The disk is then removed and the pure water quickly replaced with a solution containing a tracer at concentration c_0. The permeameter is immediately replaced and the tracer infiltrates the soil beneath the disk. Finally, the disk is removed and a vertical face is quickly excavated in the soil, taking soil samples from underneath the disk. The total volumetric water content (θ) and tracer concentration (c) is then measured in the soil samples. The solute mass balance is thus expressed as:

$$\theta c = \theta_m c_m + \theta_{im} c_{im} \tag{11.25}$$

The tracer concentration in the mobile region is assumed to be that supplied by the disk ($c_m = c_0$). In addition, the tracer is chosen so that the initial concentration in the soil is negligible, and the mass transfer coefficient, α, is considered sufficiently small so that the immobile water remains free of tracer at the sampling time ($c_{im} = 0$). Then, the immobile water content can be written as:

$$\theta_{im} = \theta - \theta_m = \theta\left[1 - \frac{c}{c_m}\right] \tag{11.26}$$

This technique is very simple to apply at laboratory and field scale.

Sampling times must meet the two assumptions mentioned above: the tracer concentration in the immobile region should remain zero ($c_{im} = 0$), and the mobile region should be free of dispersive effects with $c_m = c_0$. Snow (1999) described that proper application of the single-tracer technique requires that sampling takes place between two limiting times t_{min} (required for $c_m = c_0$) and $_{max}$ (length of time for which $c_{im} \approx 0$). These times can be estimated from the soil's dispersivity (λ), the water flux density (J_w), and the solute mass transfer coefficient α (see Chapter 3). High values of λ or low values of J_w will increase $_{min}$; also, as α increases, t_{max} decreases.

Using a numerical simulation of the transport equation for different λ, J_w and α values, Snow (1999) showed that the application of the single-tracer technique to measure θ_m directly has limitations under certain soil and experimental conditions.

11.3.2.2   Sequential Tracer Technique

Jaynes et al. (1995) presented a method to estimate θ_{im} and α using a sequence of conservative, noninteracting tracers with similar size and diffusive properties. This method assumes mass transfer taking place between mobile and immobile regions ($\alpha \neq 0$, $c_{im} \neq 0$), which is used to obtain simultaneous estimates of the nonequilibrium parameters (θ_{im} and α). The solution in the disk is replaced several times. Each time an additional, different tracer is added to the solution. The tracers are chosen so that they can be analyzed easily, although they must have the same values of the transport parameters. A single soil sample is taken an analyzed for the concentration of each tracer (c). As in the single tracer case (Sec. 11.3.2.1), the method assumes that the initial tracer concentration in the soil is zero. Here also it is assumed that the tracer concentration in the mobile region is constant and equal to the input solution $c_m = c_0$), while $c_m \approx c_0$ as $c_{im} \neq 0$. After separation of variables, the solution of the equation accounting for mass transfer between mobile and immobile regions ($\theta_{im}\, \partial c_{im}/\partial t = \alpha(c_m - c_{im})$) [see Eq. (3.22)] is:

$$\ln\left(1 - \frac{c}{c_0}\right) = \ln(1 - \beta) - \frac{\alpha}{\theta_{im}} t^* \qquad (11.27)$$

where β is the ratio of mobile water (θ_m) respect to the total water content (θ); $^* = t - (x/v)$ is the time for each tracer front to reach the sampling depth (x); t is the time since each tracer was applied; and v is the average pore water velocity Thus, by regression of $\ln(1 - c/c_0)$ vs. t^*, both θ_{im} and α can be obtained. Although only two tracers are theoretically needed, the use of additional tracers allows the confirmation of the log-linear behavior in Eq. (11.27) (Lee et al., 2000). Nonlinearity of $\ln(1 - c/c_0)$ can be found at relatively short sampling times, because c_m does not instantaneously rise to the input concentration c_0 (Snow, 1999).

The assumption of $c_m = c_0$ could be questionable and may not be correct for $\alpha > 0$. However, after a series of numerical simulations, Snow (1999) concluded that, assuming that c_m rise instantaneously to c_0 at the sampling depth and that $c < c_0$, produces only negligible errors. Sampling time t^* must be choosen between two limits so that $t_{min} < t^* < t_{max}$. The value of the limiting sampling time t_{min} for the sequential-tracer method requires that $c_m \approx c_0$ and that $c_{im} > 0$ for the last applied tracer. As in the case of the single trace method, t_{max} is given by the diffusion of the first tracer into θ_{im}. At t_{max} $c \rightarrow c_0$, and $\ln(1 - c/c_0)$ is undefined.

Although it is not as simple as the single tracer method, it only needs a single soil sample, and total experimental time is relatively short, giving estimates of the additional parameter α. The method is appropriate for field studies, as it is not necessary to use intensive sampling for curve analysis.

Snow (1999) observed that for the sequential tracer technique, at low values of β, the sampling opportunities are more generous than those of the single tracer technique. However, as β increases, the sampling opportunities become increasingly restrictive, which makes appropriate sampling difficult. A similar study was carried out by Jaynes and Shao (1999) for evaluating the range of working conditions for which the assumption of negligible dispersion was appropriate.

The application of single-tracer and sequential-tracer techniques should, therefore, be combined with numerical analysis based on the solution of the ADE equation to assess the range of soil and sampling conditions.

Lee et al. (2000) compared the values of the sequential tracer techniques to those obtained after inverse simulation of the BTCs in soil column studies under controlled laboratory conditions. Most of the estimated α values using the sequential-tracer method were within the 95% confidence interval (CI) of the BTC estimates. For 7 out of 10 experiments, the estimates of $(1 - \beta)$ were within the 95% CI of the estimates obtained from the BTC data. BTCs generated using the sequential tracer method were similar to observed BTCs.

11.4   RECOMMENDATIONS AND FUTURE RESEARCH

Solute transport processes involve a large variety of transport mechanisms related to soil, water, and solute interactions. Thus, the investigation of these mechanisms is a challenge. Miscible displacement experiments can provide useful information, especially if several experiments can be performed varying hydrodynamic flow conditions and tracers under the same scenario. This is feasible at the lysimeter scale, but many limitations are present at field scale. Also, experiments at large scale are costly and introduce other variables such as spatial variability. Thus, the lysimeter scale can be the best option for determining the role of the different transport mechanisms, while field experiments could be used for validation purposes. The development of on-site methods for estimating resident or flux averaged solute concentrations will give new opportunities to improve the characterization of solute transport processes. Among them, the two more relevant alternatives would be simultaneous on-line monitoring of solute concentration of multiple tracers and the generation of simultaneous BTCs at various depths. Experimental setups at the lysimeter scale constitute a trade-off between the degree of description and the ease of control of experimental conditions. However, several limitations such as sidewall flow and possible scale effects must be considered. Future efforts should be directed to the construction of lysimeters based on large soil monoliths with accurate control of boundary conditions.

The use of new tracers may offer alternative opportunities for characterizing solute transport mechanisms. Quantitative structure-activity relationships (QSARs) can be used to identify chemicals with distinct potential capabilities for promoting different transport mechanisms (Brusseau, 1993a). For example, values for distribution coefficients can be estimated by QSAR developed through correlation with a variety of physical or chemical properties. Octanol-water partition coefficients, aqueous solubilities, molecular connectivity indices, molecular weight, molecular surface area, and reverse-phase high-performance liquid chromatography retention times can be used for such correlation. These studies can provide insights for consolidating experimental strategies in solute transport studies.

Problems with inverse simulation methods to estimate nonequilibrium model parameters from BTC arise from two main areas (Brusseau et al., 1989). (1) The curve fitting is weighted towards the central part of the BTC (i.e., higher effluent concentrations, after the holdup in Figure 11.4). Here, the minimized sum of the squares is high while sensitivity to model parameters is low. (2) The sensitivity analysis of model parameters (Koch and Flühler, 1993) shows that different combinations of parameter values can produce similar simulations. An interesting alternative to overcome these problems is the flow-interruption technique. Perturbations in the BTCs supply additional information on transport nonequilibrium, thus avoiding ambiguous parameter estimations. Two sets of parameters producing similar noninterrupted BTCs will produce significant differences in the BTCs obtained with the flow-interruption method. Thus, this technique enables more reliable determination by inverse modeling of parameters characterizing solute transport.

The application of the method of moments (MOM) may solve ambiguity of inverse modeling techniques. However, this methodology requires the generation of a large experimental data set, which can be difficult and costly to obtain. The main concern related to the MOM is that a transport model is assumed to provide information on physical parameters from statistic analysis. The MOM does not provide information on the quality of model response to an experimental situation. On the other hand, inverse simulation can provide information when the used model is not adequate. For example, using a nonequilibrium two-region transport model, a value for $\beta \approx 1$ indicates that local equilibrium can be assumed [see Eq. (3.21)].

Independent estimation of model parameters is always a desirable counterpart to inverse modeling. Several methodologies can provide independent estimates for the retardation factor, as described in this chapter. In addition, simple methods can provide estimations for the physical nonequilibrium parameters.

Each system requires a specific experimental strategy. Table 11.3 summarizes several experimental alternatives, that can offer relevant information of solute transport processes, especially if they are combined. Also, complementary valuable information can be generated from the independent estimation methods described in this chapter.

ACKNOWLEDGMENTS

The authors would like to thank Ms. Aoife Doran for her valuable advice concerning the language revision of this chapter.

REFERENCES

Al-Sibai, M., M.A. Adey, and D.A. Rose, 1997. Movement of solute through a porous medium under intermittent leaching. Eur. J. Soil Sci. 48(4): 711–725.

Alberts, E.E., R.E. Burnell, and G.E. Schuman, 1977. Soil nitrate-nitrogen determined by coring and solution extraction techniques. Soil Sci. Soc. Am. J. 41: 90–92.

Ball, B.C., and R. Hunter, 1988. The determination of water release characteristics of soil cores at low suctions. Geoderma 43: 195–212.

Boll, J., J.S. Selker, B.M.J.S. Nijssen, T.S.B.M. Steenhuis, J. vanT. S. Van Winkle, and E. Jolles, 1991. Water quality sampling under preferential flow conditions. In: R.G. Allen, T.A. Howell, W.O. Pruitt, I.A. Walter, and M.E. Jensens (eds.), Lysimeters for Evapotranspiration and Environmental Measures, Proceedings American Society of Civil Engineers, International Symposium on Lysimetry, New York. American Society of Civil Engineers, 290–298.

Benson, C.F., and R.S. Bowman, 1994. Tri- and tetrafluorobenzoates as nonreactive tracers in soil and groundwater. Soil Sci. Soc. Am. J. 58: 1123–1129.

Bergström, L., 1990. Use of lysimeters to estimate leaching of pesticides in agricultural soils. Environ. Pollut. 67: 325–347.

Biswas, N., R.G. Zytner, J.A. McCorquodale, and J.K. Bewtra, 1991. Prediction of the movement of perchloroethylene in soil columns. Water Air Soil Pollut. 60: 361–380.

Binley A., S. Henry-Poulter, and B. Shaw, 1996. Examination of solute transport in an undisturbed soil column using electrical resistance tomography. Water Resour. Res. 32: 763–769.

Boll, J., T.S. Steenhuis, and J.S. Selker, 1992. Fiberglass wicks for sampling water and solutes in the vadose zone. Soil Sci. Soc. Am. J. 56: 701–707.

Bouchard D.C., A.L. Wood, M.L. Campbell, P. Nkedi-Kizza, and P.S.C. Rao, 1988. Sorption nonequilibrium during solute transport. J. Contam. Hydrol. 2: 209–223.

Boulding R.J., and J.S. Ginn (eds.), 2004. Soil and Ground Water Tracers. In: Boulding R.J., and J.S. Ginn (eds.). Practical Handbook of Soil, Vadose zone, and Ground Water Contamination, 2nd ed, 305–343. Lewis Publishers, Boca Raton, FL.

Bowman, R. S., 1984. Evaluation of some new tracers for soil water studies. Soil Sci. Soc. Am. J. 48: 987–99.

Brooks, S.C., D.L. Taylor, and P.M. Jardine, 1998. Thermodynamics of bromide exchange on ferrihydrite: implications for bromide transport. Soil Sci. Soc. Am. J. 62: 1275–1279.

Brusseau, M.L., 1993a. Using QSAR to evaluate phenomenological models for sorption of organic compounds by soil. Environ. Toxicol. Chem. 12(10), 1835–1846, 1993.

Brusseau, M.L., 1993b. The influence of solute size, pore water solute size, pore water velocity, and intraparticle porosity on solute dispersion and transport in soil. Water Resour. Res. 29: 1071–1080.

Brusseau M.L., Q.H. Hu, R. Srivastava, 1997. Using flow interruption to identify factors causing nonideal contaminant transport. J. Contam. Hydrol. 24: 205–219.

Brusseau, M.L., and P.S.C. Rao, 1989. Sorption nonideality during organic contaminant transport in porous media. CRC Crit. Rev. Environ. Cont. 19: 33–99.

Burkhardt, M., R. Kasteel, and H. Vereecken, 2003. Multi-tracing experiments with polymeric microspheres in undisturbed soil. Geophys. Res. Abstr. 5: 09304.

Cassel, D.K., T.H. Krueger, F.W. Schroer, and E.B. Norum, 1973. Solute movement through disturberd and undisturbed soil cores. Soil Sci. Soc. Am. Proc. 38: 36–40.

Clothier B.E., Kirkham M.B., and J.E. McLean, 1992. In situ measurement of the effective transport volume for solute moving through soil. Soil Sci. Soc. Am. J. 56: 733–736.

Clothier, B.E., L. Heng, G.N. Magesan, and I. Vogeler, 1995. The measured mobile-water content of an unsaturated soil as a function of hydraulic regime. Aust. J. Soil. Res. 33: 397–414.

Coltman K.M., N.R. Fausey, A.D. Ward, and T.J. Logan, 1991. Soil columns for solute transport studies. A review. ASAE Paper 91–2150, Presented at 1991 ASAE International Summer Meeting. Alburquerque, NM.

Corwin, D.L., 2000. Evaluation of a simple lysimeter-design modification to minimize sidewall flow. J. Contam. Hidrol. 42: 35–49.

Cote, C.M., K.L. Bristow, and P.J. Ross, 1999. Quantifying the influence of intra-aggregate concentration gradients on solute transport. Soil Sci. Soc. Am. J. 63(4): 759–767.

Cunningham, J.A., M.N. Goltz, and P.V. Roberts, 1999. Simplified expressions for spatial moments of ground-water contaminant plumes. J. Hydrol. Eng. 4: 377–380.

Danckwerts, P.V., 1953. Continuous flow systems. Distribution of residence times. Chem. Eng. Sci. 2: 1–13.

Dane, J.H, and G.C. Topp (eds.), 2002. Methods of Soils Analysis, Part 4. SSSA Book Series, 5. Madison, WI.

Das, B.S., and G.J. Kluitenberg, 1996. Moment analysis to estimate degradation rate constants from leaching experiments. Soil Sci. Am. J. 60: 1724–1731.

Davis, S.N., G.M. Thompson, H.W. Bentley, and G. Stiles, 1980. Groundwater tracers—a short review. Ground Water 18: 14–23.

de Rooij, G.H., and F. Stagnitti, 2002. Spatial and temporal distribution of solute leaching in heterogeneous soils: analysis and application to multisampler lysimeter data. J. Contam. Hidrol. 54: 329–346.

De Smedt F., and P.J. Wierenga, 1984. Solute transfer through columns of glass beads. Water Resour. Res. 20: 225–232.

Ersahin, S., R.I. Papendick, J.L. Smith, C.K. Keller, and V.S. Manoranjan, 2002. Macropore transport of bromide as influenced by soil structure differences. Geoderma 108: 207–223.

Espinoza, C., and A.J. Valocchi, 1998. Temporal moments analysis of transport in chemically heterogeneous porous media. J. Hydrol. Eng. 3(4): 276–284.

Fischer, H.B., E. List, R.C.Y. Koh, J. Imberger, and N.H. Brooks, 1979. Mixing in inland and coastal waters. Academic Press, New York, 483 pp.

Flury, M., and H. Flühler, 1995. Tracer characteristics of brilliant blue FCF. Soil Sci. Soc. Am. J. 59: 22–27.

Führ, F., R.J. Hance, J.R. Plimmer, and J.O. Nelson, 1997. The Lysimeter Concept. ACS Symposium Series 699. American Chemical Society, Washington, DC.

Gillham, R.W., M.J.L. Robin, D.J. Dytynshyn, and H.M. Jonhnston, 1984. Diffusion of nonreactive and reactive solutes through fine-grained barrier materials. Can. Geotech. J. 21: 541–550.

Goltz, M.N., and P.V. Roberts, 1987. Using the method of moments to analyze three-dimensional diffusion-limited solute transport from spatial and temporal perspectives. Water Resour. Res. 23: 1575–1585.

Hu, Q., and M.L. Brusseau, 1994. The effect of solute size on diffusive-dispersive transport in porous media. J. Hydrol. 158: 305–317.

Hu, Q., and M.L. Brusseau, 1995. Effect of solute size on transport in structured porous media. Water Resour. Res. 31: 1637–1646.

Huyer, W., and A. Neumaier, 1999. Global optimization by multilevel coordinate search. J. Global Optimization 14: 331–355.

Ishiguro, M., K.C. Song, and K. Yuita, 1992. Ion transport in an allophanic Andisol under the influence of variable charge. Soil Sci. Soc. Am. J. 56: 1789–1793.

Jacobsen, O.H., F.J. Leij, and M.Th. van Genuchten, 1992. Parameter determination for chloride and tritium transport in undisturbed lysimeters during steady flow. Nord. Hydrol. 23: 89–104.

Jacobson, D.R., B.C. Garrett, and T.A. Bishop, 1984. Comparison of batch and column methods for assessing the leachability of hazardous waste. Environ. Sci. Technol. 18: 666–673.

Jacques, D., J. Šimůnek, A. Timmerman, and J. Feyen, 2002. Calibration of Richards' and convection-dispersion equations to field-scale water flow and solute transport under rainfall conditions. J. Hydrol. 259: 15–31.

Jardine, P.M., R. O'Brien, G.V. Wilson, and J.P. Gwo, 1998. Experimental techniques for confirming and quantifying physical nonequilibrium processes in soils. In: Selim, H.M., and L. Ma (eds.), Physical Nonequilibrium in Soils, Modeling and Application. Ann Arbor Press, Chelsea, MI, pp. 243–271.

Jaynes, D.B., and M. Shao, 1999. Evaluation of a simple technique for estimating two-domain transport parameters. Soil Sci. 164(2): 82–91.

Jaynes, D.B., and R. Horton, 1998. Field parametrization of the mobile/immobile domain model. In: H.M. Selim, and L. Ma (eds.), Physical Nonequilibrium in Soils, Modelling and Application. Ann Arbor Press, Chelsea, MI, pp. 297–310.

Jaynes, D.B., S.D. Logsdon, and R. Horton, 1995. Field method for measuring mobile/immobile water content and solute transfer rate coefficient. Soil Sci. Soc. Am. J. 59: 352–356.

Jury, W.A., and G. Sposito, 1985. Field calibration and validation of solute transport models for the unsaturated zone. Soil Sci. Soc. Am. J. 49: 1331–1341.

Katou, H., B.E. Clothier, and S.R. Green, 1996. Anion transport involving competitive adsorption during transient water flow in an Andisol. Soil Sci. Soc. Am. J. 60: 1368–1375.

Knox, J.H., and L. McLaren, 1964. A new gas chromatographic method for measuring gaseous diffusion coefficients and obstructive factors. Anal. Chem., 36: 1477–1485.

Koch, S., and H. Flühler, 1993. Nonreactive solute tansport with micropore diffusion in aggregated porous-media determined by a flow-interruption method. J. Contam. Hydrol. 14(1): 39–54.

Kool, J.B., Parker, J.C., and M.Th. van Genuchten, 1987. Parameter estimation for unsaturated flow and transport models — a review. J. Hydrol. 91: 255–293.

Kreft, A., and A. Zuber, 1978. On the physical meaning of the dispersion equation and its solutions for different initial and boundary conditions. Chem. Eng. Sci. 33: 1471–1480.

Kucera, E., 1965. Contribution to the theory of chromatography: linear nonequilibrium elution chromatography. J. Chromatogr. 19: 237–248.

Kulli, B., C. Stamm, A. Papritz, and H. Flühler, 2003. Discrimination of flow regions on the basis of stained infiltration patterns in soil profiles. Vadose Zone J. 2: 338–348.

Kunin R., and R.J. Myers, 1947. Rates of anion exchange in ion-exchange resins. J. Phys. Colloid Chem. 51: 1111–1115.

Lee, J., C.B. Jaynes, and R. Horton, 2000. Evaluation of a simple method for estimating solute transport parameters: laboratory studies. Soil Sci. Soc. Am. J. 64: 492–498.

Leij, F.J., and J.H. Dane, 1991. Solute transport in a two-layer medium investigated with time moments. Soil Sci. Soc. Am. J. 55: 1529–1535.

Leij, F.J., and J.H. Dane, 1992. Moment method applied to solute transport with binary and ternary exchange. Soil Sci. Soc. Am. J. 56(3): 667–674.

Ma, L., and H.M. Selim, 1994. Tortuosity, mean residence time, and deformation of tritium breakthroughs from soil columns. Soil Sci. Soc. Am. J. 58: 1076–1085.

Ma, L., and H.M. Selim, 1998. Coupling of retention approaches to physical nonequilibrium models. In: Selim, H.M., and L. Ma (eds.), Physical Non-equilibrium in Soils, Modeling and Application. Ann Arbor Press, Chelsea, MI, pp. 83–115.

Mallants, D., M. Vanclooster, M. Meddahi, and J. Feyen, 1994. Estimating solute transport in undisturbed soil columns using time domain reflectometry. J. Contam. Hydrol. 17: 91–109.

Maloszewski, P., and A. Zuber, 1993. Tracer experiments in fractured rocks: matrix diffusion and the validity of models. Water Resour. Res. 29: 2723–2735.

Maloszewski, P., R. Benischke, T. Harum, and H. Zojer, 1994. Estimation of solute transport parameters in heterogeneous groundwater system of a karstic aquifer using artificial tracer experiments. Water Down Under 94, vol. 2. The Institute of Engineers, Australia, Barton, ACT, pp. 105–111, part A.

Maraqa, M.A., X. Zhao, R.B. Wallace, and T.C. Voice, 1998. Retardation coefficients of nonionic organic compounds determined by batch and column techniques. Soil Sci. Soc. Am. J., 62: 142–152.

Nielsen, D.R., and J.W. Biggar, 1962. Miscible displacement: iii. Theoretical considerations. Soil Sci. Soc. Am. Proc. 26: 216–221.

Nkedi-Kizza, P., Rao, P.S.C., Jessup, R.E., and J.M. Davidson, 1982. Ion exchange and diffusive mass transfer during miscible displacement through an aggregated oxisol. Soil Sci. Soc. Am. J. 46: 471–476.

Nkedi-Kizza, P., J.W. Biggar, and M.Th. Genuchten, P.J. Wierega, H.M. Selim, J.D. Davidson, and D.R. Nielsen, 1983. Modeling tritium and chloride 36 transport through an aggregated oxisol. Water Resour. Res. 19: 691–700.

Pang, L., M.E. Close, and M. Noonan, 1998. Rhodamine WT and *Bacillus subtilis* transport through an alluvial gravel aquifer. Ground Water 36(1): 112–122.

Pang, L., M. Goltz, and M. Close, 2003. Application of the method of temporal moments to interpret solute transport with sorption and degradation. J. Contam. Hydrol. 60: 123–134.

Parker, J.C., and M.Th. van Genuchten, 1984. Flux-averaged and volume-averaged concentrations in continuum approaches to solute transport. Water Resour. Res. 20: 866–872.

Perroux, K.M., and I. White, 1988. Designs for disc permeameters. Soil Sci. Sic. Am. J. 52: 1205–1214.

Peyton, R.L., B.A. Haeffner, S.H. Anderson, and C.J. Gantzer, 1992. Applying x-ray to measure macropore diameters in undisturbed soil cores. Geoderma 53: 329–340.

Porter, L.K., W.D. Kemper, R.D. Jackson, and B.A. Stewart, 1960. Chloride diffusion in soils as influenced by moisture content. Soil Sci. Soc. Am. Proc. 24: 460–463.

Reedy, O.C., P.M. Jardine, G.V. Wilson, H.M. Selim, 1996. Quantifying the diffusive mass transfer of nonreactive solutes in columns of fractured saprolite using flow interruption. Soil Sci. Soc. Am. J. 60: 1376–1384.

Rubin, Y., M.A. Cushey, and A. Wilson, 1997. The moments of the breakthrough curves of instantaneously and kinetically sorbing solutes in heterogeneous geologic media: prediction and parameter inference from field measurements. Water Resour. Res. 33(11): 2465–2481.

Sadghi, A.M., D.E. Kissel, and M.L. Cabrera, 1989. Estimating molecular diffusion coefficients of urea in unsaturated soil. Soil Sci. Soc. Am. J. 53: 15–18.

Schneider, A.D., and T.A. Howell, 1991. Large, monolithic, weighing lysimeters. In: R.G. Allen, T.A. Howell, W.O. Pruitt, I.A. Walter, and M.E. Jensen (eds.), Lysimeters for Evapotranspiration and Environmental Measurements. Proceedings of the International Symposium on Lysimetry. ASCE: 37–45. New York.

Seaman, J.C., P.M. Bertsch, and W.P. Miller, 1995. Ionic tracer movement through highly weathered sediments. J. Contam. Hydrol. 20: 127–143.

Seyfried, M.S., and P.S.C. Rao, 1987. Solute transport in undisturbed columns of an aggregated tropical soil: preferential flow effects. Soil Sci. Soc. Am. J. 51: 1434–1444.

Šimůnek, J., M.Th. van Genuchten, M. Sejna, N. Toride, and F.J. Leij, 2000. STANMOD studio of analytical models for solving the convection-dispersion equation. Version 1.0. International Ground Water Modeling Center. Colorado School of Mines, Golden, CO.

Skaggs, T.H., and F.J. Leij, 2002. Solute transport: theoretical background. In: Dane J.H., and G.C. Topp (eds.), Methods of Soil Analysis. Part 4. Physical Methods. SSSA Book Series 5, American Society of Agronomy, Madison, WI, pp. 1353–1380.

Snow, V.O., 1999. In situ measurement of solute transport coefficients: assumptions and errors. Soil Sci. Soc. Am. J. 63: 255–263.

S.S.S.A., 2001. Glossary of Soil Science Terms. Soil Science Society of America, Madison, WI. [Online: http://www.soils.org/sssagloss/intro.html]

Stagnitti, F., Villiers, N., Parlange, J.-Y., Steenhuis, T.S., de Rooij, G., Li, L., Barry, D.A., Xiong, X., and Li, P. 2003. Solute and contaminant transport in heterogeneous soils. Bull. Environ. Contam. Toxicol. 71(4): 737–745.

Stagnitti, F., Allinson, G., Sherwood, J., Graymore, M., Allinson, M., Turoczy, N., Li, L., and Phillips, I. 1999. Preferential leaching of nitrate, chloride and phosphate in an Australian clay soil. Toxicol. Environ. Chem. 70: 415–425.

Toride N., F.J. Leij, and M.Th. van Genuchten, 1993. Flux-averaged concentrations for transport in soils having nonuniform initial solute distributions. Soil Sci. Soc. Am. J. 57: 1406–1409.

Toride, N., F.J. Leij, and M.Th. van Genuchten, 1995. The CXTFIT code for estimating transport parameters from laboratory or field tracer experiments. U.S. Salinity Laboratory Research Report 137. Riverside, CA.

Toride, N., and F.J. Leij, 1996. Convective-dispersive stream tube model for field-scale solute transport: 1. Moment analysis. Soil Sci. Soc. Am. J. 60: 342–352.

Valocchi, A., 1985. Validity of the local equilibrium assumption for modelling sorbing solute transport through homogeneous soils. Water Resour. Res. 21(6): 808–820.

Valocchi, A., 1986. Effect of radial flow on deviations from local equilibrium during sorbing solute transport through homogeneous soils. Water Resour. Res. 22(12): 1693–1701.

van der Ploeg, R.R., and F. Beese, 1977. Model calculations for the extraction of soil water by ceramic cups and plates. Soil Sci. Soc. Am. J. 41: 466–470.

van Genuchten, M.Th., and J.C. Parker, 1984. Boundary conditions for displacement experiments through short laboratory soil columns. Soil Sci. Soc. Am. J. 48: 703–708.

van Genuchten, M.Th., and P.J. Wierenga, 1977. Mass transfer studies in sorbing porous media: II. Experimental evaluation with tritium ($3H_2O$). Soil Sci. Soc. Am. J. 41: 272–278.

van Schaik, J.C., and W.D. Kemper, 1966. Chloride diffusion in clay-water system. Soil Sci. Soc. Am. Proc. 30: 22–25.

Vanderborght, J., D. Mallants, M. Vanclooster, and J. Feyen, 1997. Parameter uncertainty in the mobile-immobile solute transport model. J. Hydrol. 190: 75–101.

Vogeler, I., C. Duwig, B.E. Clothier, and S.R. Green, 2000. A simple approach to determine reactive solute transport using time domain reflectometry. Soil Sci. Soc. Am. J. 64: 12–18.

Wada, K., 1980. Mineralogical characteristics of Andisols. In: B.K.G. Theng (ed.), Soils with Variable Charge, pp. 281–299. New Zealand Society of Soil Science, Offset Publications, New Zealand.

Wildenschild, D., K.H. Jensen, K. Villholth, and T.H. Illangasekare, 1994. A laboratory analysis of the effect of macropores on solute transport. Ground Water 32: 381–389.

Winton, K., and J.B. Weber, 1996. A review of field lysimeter studies to describe the environmental fate of pesticides. Weed Technol. 10: 202–209.

Young, D.F., and W.P. Ball, 2000. Column experimental design requirements for estimating model parameters from temporal moments under nonequilibrium conditions. Adv. Water Resour. 23: 449–460.

12 Methods to Determine Sorption of Pesticides and Other Organic Compounds

*Juan Cornejo, Mª Carmen Hermosín,
Rafael Celis, and Lucía Cox*
Instituto de Recursos Naturales y Agrobiología de Sevilla,
CSIC, Sevilla, Spain

CONTENTS

1-5667-0657-2/05/$0.00 + $1.50
© 2005 by CRC Press

12.1   INTRODUCTION

From the beginning of the second half of the last century, thousands of tons of xenobiotics such as agrochemicals and many other chemicals have reached the soil in a voluntary or accidental way. As discussed in Chapter 1, most of those chemicals are organic compounds, mainly pesticides and organic solvents used as part of pesticides formulations and others forming part of municipal, industrial wastes and composts of different origin applied to the soil as fertilizers and/or amendments. The use of pesticides has coincided with a tremendous increase in agricultural productivity. Pesticides are used worldwide in plant protection to control or destroy weeds, insects, fungi, and other pests. The recent trend toward conservation-tillage systems has also led to increasing reliance on chemical pesticide use (Logan et al., 1987). Most pesticides reach the soil during or immediately after treatment. The soil, which is the main recipient of all pesticides, plays a leading role in the environmental fate of these chemicals and in the protection of surface and ground waters.

The environmental fate of pesticides in soil is viewed with great concern today due to the problems resulting from the repetitive application of persistent and mobile molecules affecting surface and ground water quality, the food chain, and finally the human being. Along with increasing concern about chemical contamination of various ecosystems, much emphasis has been put on designing suitable methods to characterize the different processes affecting the fate of pesticides in soil (Cornejo and Jamet, 2000) and to the development of predictive models, as discussed in Chapters 3 and 21.

Besides its function as a terrestrial ecosystem and food producer, the soil plays a unique role as a filtering, buffering, and transforming system for contaminants of anthropogenic origin. The soil behaves as an active filter, where the chemicals are degraded by biological and nonbiological processes, and as a selective filter because it is able to retain some chemicals and to prevent their leaching to ground water or losses by volatilization. However, the soil does not always act as an environmental "sink," and many xenobiotics move from soil to other weak ecosystem compartments. Both the fate of the agrochemicals in the soil and their dispersion in the environment depend on the characteristics and the overall functioning of this ecosystem (Figure 12.1). In the soil, pesticides, like other organic chemicals, are affected by the simultaneous influence of sorption-desorption, leaching, volatilization, and biotic and abiotic degradation phenomena. All these processes are dynamic and nonlinear.

To understand the physical, chemical, and biological processes that take place in soil, it is necessary to know the soil characteristics for any given study. It is not possible to generalize and/or extrapolate the behavior of a given chemical in one soil to another. Soil is a very heterogeneous, three-dimensional matrix in which solid, liquid, and gas phases form a dynamic system in which biota is embedded.

Soil activity is a very complex combination of biotic and abiotic processes taking place in the bulk and/or internal (pore) or external surface of particles

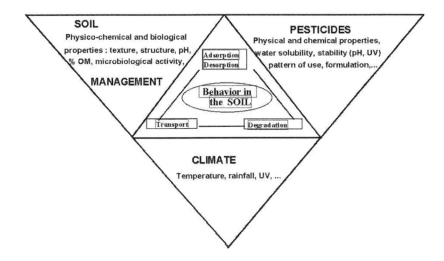

FIGURE 12.1 Factors and processes affecting the behavior of pesticides in soil.

where chemicals from liquid and gas phases interact with them and with microorganisms. The recent recognition of soil as a fragmented porous medium appropriate for fractal representation (Anderson et al., 1980) has resulted in increasing interest in the development of models relating to soil fractal parameters, such as the soil-surface fractal dimension or structure-dependent soil properties such as solute movement. Very recently, Celis et al. (2001) showed a link between soil porosity, as described by the soil-surface fractal dimension (D_s), and the behavior of polar herbicides in soil.

The colloidal soil constituents, such as metal oxides and oxyhydroxides (Al, Fe, Mn, etc.), $CaCO_3$, clay minerals, organic matter, and polysaccharides, can be present as separate components, or they can be more or less associated. These particles are included in the clay-size fraction of soils, considered as the fraction with an equivalent spherical diameter (e.s.d.) of $< 2\,\mu m$. Colloidal soil constituents are able to establish a great range of interparticle forces contributing to the formation of aggregates with different degrees of stability. The soil materials are made up of these particle associations ranging in size from nanometers (molecules) to rocks. However, from an agronomic point of view, it is believed that clay particles of $< 20\,\mu m$ form domains by a mixture of clay microstructures, biopolymers, and microorganisms like bacteria, more or less bound by cementing agents; domains together with plant debris form the microaggregates of $< 250\,\mu m$, and macroaggregates of $> 250\,\mu m$ by fungal hyphae in roots (Oades and Waters, 1991) (see also Chapter 7). Within particles and aggregates, pores of different shapes and sizes are generally present, greatly affecting the transport of soil solutes (nutrients, pollutants, etc.). The pore size classification depends on the subject being studied and the techniques used for its determination. The International Union of Pure and Applied Chemistry (IUPAC) classifies pore size according to aperture (d): macropores ($d > 50\,nm$), mesopores ($50\,nm > d > 2\,nm$), and micropores

$d < 2$ nm). According to the occurrence and availability of soil water, the largest category of pore size is the macropore (MP, > 30 μm), normally originated by roots, cracking effects, and earthworms. These macropores can be connected, forming voids and channels (> 750 μm), and are always involved in the preferential flow of water and chemicals through the soil profile. The second category of pore size is the mesopore (mP, $0.2–30$ μm), considered a water reservoir for plants holding water strongly enough to prevent drainage. The micropore (μP, < 0.2 μm) contains water not accessible to plants, making water-solute interchanges very difficult. These internal surfaces are to some extent responsible for the slow sorption processes observed in soils and account for much of the hysteretic effects.

The movement and retention of soil organisms depend on their own size and the size and shape of the pores. Organisms in soil consist of microflora such as bacteria and fungi, microfauna such as protozoa and nematodes, mesofauna such as enchytreids, acari, and collembella, and macrofauna such as earthworms and slugs. Soil microorganisms are found generally in the pores within aggregates, especially in the storage pores ($0.5–50$ μm e.s.d.) and on the soil colloidal surfaces. It has been suggested by Bergstrom and Stenstrom (1998) that much of the soil pore volume (< 2 μm) is physically inaccessible to soil microorganisms. Organic chemicals diffusing into such small pores are less available for degradation, with the diffusion process and sorption being time-dependent processes — generally slow. The increased soil-compound contact time results in the chemical becoming less extractable and potentially less bioavailable (Jensen, 2002). This aging process can lead to a situation where the compound becomes more and more difficult to remove from the soil matrix by traditional chemical extraction techniques. A general scheme of transport of organics through the soil matrix is represented in Figure 12.2. Transport processes are discussed in detail in Chapters 3, 9 and 11.

The behavior of organic chemicals in soils as a combination of interactions with soil constituents has been used in the past to help to understand their role in soil as a whole (Calvet, 1980; Cox et al., 1997a). However, it is well known that mineral and organic colloids in soils are generally associated, forming active organo-mineral complexes (Cornejo and Hermosín, 1996), and are present in different soil size fractions, showing diverse surface reactivity depending on the association (Celis et al., 1997a, 1997b). Many processes take place in soils, but sorption is probably the controlling step in the transport of nutrients and pollutants along the soil profile and hence the process responsible for their availability (Cornejo and Jamet, 2000; Pignatello, 2000).

12.2   SORPTION AND OTHER SOIL PROCESSES

Sorption affects every other aspect of organic chemical behavior in the soil. Sorption will determine movement, biological activity, and persistence. Sorption of pesticides and other organic compounds on soil surfaces makes these molecules less available for leaching, volatilization, and biodegradation, whereas desorption means that pesticide and organic chemicals are available

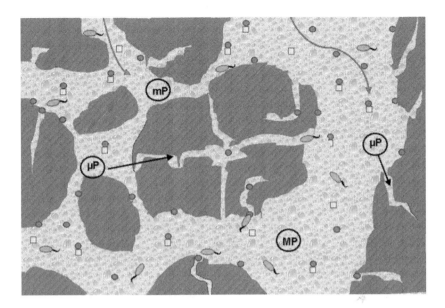

FIGURE 12.2 Transport of organics through the soil profile. •, Organic; □, particle; ∾, microorganism; μP, micropore; mP, mesopore; MP, macropore.

for transport and degradation processes. Variation in magnitude of sorption will give rise to variation in degradation and transport rates even in soils with similar microbial communities and activities or soils with similar hydraulic conductivities. Also the biological activity of many soil-applied pesticides has been related to sorption and its reversibility. It is well known that soils with high organic matter content require higher doses of pesticides than those with low organic matter content. Although most laboratory experiments are designed for investigating sorption, leaching, and degradation of pesticides and other organic chemicals in isolation, an understanding of the interactions among simultaneous processes is essential for explaining field observations and for the development of predictive models coupling sorption, degradation, and leaching in soils (Rao et al., 1993). In this section we will review how sorption directly or indirectly affects leaching and degradation. Figure 12.3 summarizes the linkages between sorption and leaching and degradation discussed here. Linkages between sorption and volatilization can be found in Chapter 13.

12.2.1 SORPTION-LEACHING

Experimental data for many pesticides and organic compounds show an inverse relation between sorption and leaching. Kookana et al. (1995), in a field study with nine different pesticides, observed that variations in leaching depth corresponded inversely with sorption coefficients. Also under field conditions, Laabs et al. (2002) observed that cumulative efflux of ten pesticides in percolates of two different soils were closely related with their groundwater ubiquity score (GUS), a leachability index defined by Gustafson (1989) which

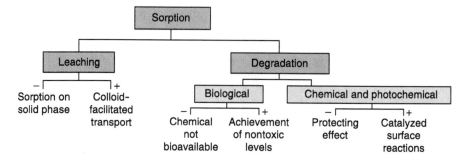

FIGURE 12.3 Sorption as affecting leaching and degradation of pesticides and other organic contaminants.

classifies a chemical's potential to leach to groundwater according to the combined effect of degradation and sorption processes:

$$GUS = \log(t_{1/2}\text{-soil}) \times [4 \log(K_{oc})] \tag{12.1}$$

where $t_{1/2}$-soil is the soil degradation half-life (see Chapter 3) assuming first-order kinetics, and K_{oc} sorption coefficient normalized to soil organic carbon content.

One exception to this direct relationship between sorption and leaching is transport through preferential flow pathways or macropores, which allow pesticide and other organic molecules to leach high depths through the soil profile without contact with the soil matrix (Brusseau et al., 1990; Flury, 1996). (This phenomenon is discussed in detail in Chapter 8.) De Jonge et al. (2000) observed that transport of the highly sorbable glyphosate in undisturbed soil columns was strongly governed by macropore flow; the amounts leached from a sandy loam soil were 50–150 times larger than from a coarse sandy soil, despite the lower sorption coefficients measured for the sandy soil when compared with the sandy loam soil. Also Cox et al. (1999a) attributed to preferential flow through macropores the higher leaching of polar herbicides in soils under reduced tillage systems versus conventional tillage, despite the similar sorption coefficients measured in both soils. Preferential pathways are also one reason why leaching of pesticides is often faster than predicted using simple models (Turin and Bowman, 1997; Larsson and Jarvis, 2000).

Sorption of organic compounds on mobile soil colloids also affects leaching through facilitated transport mechanisms: sorbed on soil colloids, pesticides and other organic contaminants can be transported through the soil to surface water or ground water (Figure 12.2). In many studies of organic chemical transport, a two-phase approach has been used in which chemicals are partitioned between an immobile solid phase and a mobile aqueous phase (see Chapter 11). Colloids in water act as a third mobile solid phase that can sorb organic molecules in the same manner as to the immobile solid phase enhancing transport of contaminants (McCarthy and Zachara, 1989; McGechan and Lewis, 2002). McCarthy and Zachara (1989) suggest a size range of 0.1–10 µm for various soil mineral–derived colloids, including clays, iron oxides, silica, and lime. Soil organic matter and organic particles from manure, slurry, and other organic wastes have a wide range of physical sizes, but the smaller ones provide the larger number of sites for sorption of pollutants (McGechan and Lewis, 2002). According to Kretzschmar et al. (1999), colloid-facilitated transport is environmentally relevant when (1) mobile colloids are present in large concentrations, (2) the particles are transported over significant distances through uncontaminated zones of the porous medium, and (3) the contaminant sorbs strongly to the mobile particles and desorbs only slowly. There is very little published information on sorption to mobile colloids. O'Connor and Connolly (1980) studied the effect of organic matter and clay content on the partition of contaminants between the dissolved phase and the sorbed phase for a number of pesticides and heavy metals and observed an inverse relationship between the concentration of sorbing colloids and the partition coefficient. The same was observed by Voice et al. (1983) using different sediments and hydrophobic pollutants. In a study of the movement of glyphosate in undisturbed soil columns, de Jonge et al. (2000) reported that the contribution of colloid transport to total glyphosate leaching varied, depending on soil treatment, from < 1 to 52% of the total herbicide leached.

Interaction of pesticide and other organic molecules with dissolved organic matter (DOM) also affects leaching through facilitated transport mechanisms. These interactions in solution have been shown to increase the solubility of the compound, resulting in an increase in leaching (Chiou et al., 1987; Graber et al., 1995; Nelson et al., 1998; Cox et al., 2000; Graber et al., 2001). Equilibrium dialysis methods are generally used to confirm these interactions (Nelson et al., 1998; Celis et al., 1998; Cox et al., 2001). According to Lee and Farmer (1989), the impact of DOM on the fate of an organic chemical depends on the fraction of chemical associated with DOM, which is determined by the association coefficient of DOM and the concentration of DOM in solution. Organic chemical interactions in solution have been shown to depend on the nature and source of the DOM. Lee and Farmer (1989) found that the herbicide napropamide had higher affinity for dissolved humic acids than for dissolved fulvic acids, and Clapp et al. (1996) found that this interaction was higher for soil-derived humic acid than for humic acids from peat or water sources.

Pesticide interactions in solution are of particular interest when organic amendments are applied to soils, since the incorporation of organic amendments introduces both solid organic matter and DOM. Although in most cases the use of organic amendments produces a reduction in leaching, in some cases the contrary is observed. The reasons for reduction in leaching are an increase in sorption as a consequence of the generation of new organic sorption sites on soil surfaces (Zsolnay, 1992; Guo et al., 1993; Barriuso and Koskinen, 1996; Cox et al., 1997b, 1999b) and a significant increase in degradation rates due to the increase in organic matter content and subsequent increase in microbial populations and microbial activity in organic amended soils (Felsot and Dzantor, 1995; Topp et al., 1996; Moorman et al., 2001). On the other hand, organic amendments can also enhance transport of pesticides and organic compounds through two different mechanisms: interactions in solution between DOM and organic compound molecules (Lee and Farmer, 1989; Nelson et al., 1998; Cox et al., 2000, 2001; Graber et al., 2001; Flores-Céspedes et al., 2002) and competition between DOM molecules and pesticide or other organic molecules for sorption sites, giving rise to a reduction in sorption of the pesticide (Businelli, 1997; Celis et al., 1998).

Another soil management practice that affects sorption and leaching is tillage system. While reduced or no-tillage systems have been shown to increase sorption of pesticides due to increase in organic matter of surface soil (Levanon et al., 1993; Dao, 1995; Fermanich et al., 1996), the abundance of macropores in these systems has been reported to increase leaching (Cox et al., 1999a; Isenssee et al., 1990; Gaston and Locke, 1996). Consequently, as pointed out by Cox et al. (1999a), changes in natural field conditions, soil spatial variability, experimental scale, soil characteristics, and even the nature of the chemical under study determine the effect of tillage system on sorption of pesticides and how this affects leaching in soil.

12.2.2 SORPTION-DEGRADATION

Pesticides and other organic chemicals in soil degrade through microbiological, chemical and/or photochemical processes, as will be discussed in Chapter 15. The effect of sorption on degradation depends on the nature of the transformation.

Chemical degradation can take place either in the soil solution or in the sorbed phase (Calvet et al., 1989). Sorption can protect organic chemicals from degradation in the soil solution, as observed for the herbicide dichlobenil by Briggs and Dawson (1970), who found an inverse relation between nonbiological hydrolysis rate and sorption in a study with 34 surface and subsurface soils. Degradation of triazine herbicides in soil is by chemical hydrolysis and, potentially, by microbial metabolism (Kaufman and Kearney, 1970; Walker and Blacklow, 1994). Walker and Blacklow (1994) confirmed chemical hydrolysis of simazine and atrazine by soil sterilization, and observed a positive relationship between triazine $t_{1/2}$ and the proportion of applied herbicide sorbed by the soils.

The importance of nonbiochemical reactions in decomposition of organic molecules sorbed on *clay surfaces* has been emphasized in different reviews and studies (Mortland, 1970; Cornejo et al., 1983; Pusino et al., 1988; Sánchez-Camazano and Sánchez-Martín, 1991; Cox et al., 1994). Cox et al. (1994) suggested that the insecticide methomyl degrades through acid hydrolysis catalyzed by the clay mineral surface where the pesticide was sorbed, since no significant degradation of methomyl was observed in 24 hours in aqueous solution at pH 2. A similar mechanism was proposed for triazine herbicides.

Following sorption, surface acidity of smectite surface catalyzes protonation and hydrolysis of sorbed atrazine (Laird, 1996; Celis et al., 1997a).

Soil humic substances have been also shown to catalyze hydrolysis of pesticides such as atrazine (Gamble and Khan, 1985). Sorption of atrazine into the undissolved phase through hydrogen bonding between undissociated carboxyl groups leads directly to catalyzed hydrolysis of the sorbed herbicide.

Sorption reduces biodegradation, since sorbed molecules are unavailable to soil microorganisms. According to Scow and Hutson (1992), sorption influences kinetics of biodegradation not only by modifying the concentration of the chemical in the soil solution, but also directly by inducing concentration-dependent regulatory responses of microorganisms. Allen and Walker (1987) found that metamitron degradation in soil was controlled primarily by its availability in the soil solution and suggest that it occurs most rapidly in light-textured soils with high sand, low clay, and low organic matter contents. Tests with diallate added to mixtures of activated charcoal and soil to give varying percentages of binding of the herbicide revealed a direct relationship between the amount of herbicide sorbed and degradation rate (Anderson, 1981).

However, in some cases toxicity has been reported to be the main factor controlling biodegradation (Martins and Mermoud, 1998; Moorman et al., 2001). In these cases, sorption would reduce concentration of the chemical in the soil solution to nontoxic levels and would indirectly favor biodegradation (van Loosdrecht et al., 1990).

In addition to removing pesticide and other organic compounds from soil solution, sorption affects biodegradation in a different way. Microorganisms are also sorbed to soil surfaces, and, once sorbed, cells may alter their metabolic activity, which can be greater, less than, or similar to cells free in solution (Alexander, 1999). Sorption of nutrients and growth factors may also reduce the rate of microbial growth.

12.3 CHARACTERIZING SORPTION-DESORPTION PROCESSES

12.3.1 MEASURING SORPTION

12.3.1.1 Sorption Equilibrium

Organic chemical sorption on soils can be easily measured in the laboratory, although such measurements have been considered estimates because laboratory-derived sorption parameters constitute estimates of sorption in field soils (Green and Karickhoff, 1990). The most common methodology to characterize sorption equilibrium is indirect batch-suspension measurement, consisting of agitating soil and an organic chemical solution in a closed container for a sufficient time to achieve apparent equilibrium in the system, typically 24 hours. After equilibration, the suspension is centrifuged and concentration of the organic compound is determined in the supernatant. Differences between the initial and final solution concentration of the organic chemical are assumed to be sorbed to the soil. By using different initial concentrations, a sorption isotherm, which represents the relation between the amount of chemical sorbed and the chemical concentration in the solution at equilibrium in a range of concentrations of interest, can be obtained. An inherent problem that may interfere with this procedure is the loss of chemical due to processes other than sorption to soil particles, such as degradation, volatilization, or sorption to the centrifuge tubes. Use of control chemical solutions (without soil) can help identify pesticide and other organic chemical losses due to volatilization or sorption to the soil container, whereas use of initially air-dry soil, a brief equilibration period, and low temperatures have been suggested to minimize degradation during the sorption measurement. Nevertheless, since the antecedent soil conditions, equilibration period, and temperature are all experimental variables that can affect sorption results, the choice of the experimental protocol requires careful consideration of the organic chemical and soil characteristics (Green and Karickhoff, 1990).

Although other sorption models have been used, because of its simplicity and general utility, the Freundlich equation has been a preferred model to describe organic chemical sorption isotherms:

$$C_s = K_f C_e^{N_f} \tag{12.2}$$

where C_s is the amount of chemical sorbed at the equilibrium concentration C_e and K_f and N_f are empirical constants.

For many modeling purposes, sorption has been assumed to be independent of the organic chemical concentration, that is, N_f has been assumed to be 1, resulting in a simplified equation, $C_s = K_d C_e$, where K_d is the linear distribution coefficient. For many environmental contexts, this approximation is probably acceptable considering errors associated to additional approaches. The error induced by the assumption of linearity depends on the value of N_f and the solution concentration (Hamaker and Thompson, 1972).

12.3.1.2 Desorption

Most soil-organic chemical systems are seldom, if ever, at equilibrium, so that continuous solute transfer between the solid and solution phases occurs in field conditions. As the organic compound dissipates by degradation or transport processes, desorption replenishes organic chemicals into the soil solution. Therefore, the release (desorption) of the sorbed compound from soil particles is of fundamental importance to predict the distribution of the chemical in the soil, becoming particularly important in predicting the fate and mobility of contaminants in already contaminated soils and developing remediation strategies (Scheidegger and Sparks, 1996).

While organic chemical sorption by soil and its constituents has been extensively documented, desorption is much less understood. Many questions remain, such as the causes of the nonsingularity (hysteresis) between the sorption and desorption isotherms frequently observed in batch laboratory experiments, such as those based on successive dilution methodology (Koskinen et al., 1979; Clay et al., 1988; Clay and Koskinen, 1990; Barriuso et al., 1994; Carton et al., 1997; Delle Site, 2001). The presence of non–single-valued sorption-desorption relationships has been attributed to a number of experimental artifacts, such as nonattainment of sorption equilibrium, removal of soil particles during desorption, formation of precipitates or loss of chemical due to volatilization, degradation, or both (Koskinen et al., 1979; Calvet, 1980; Altfelder et al., 2000; Lesan and Bhandari, 2003). Changes in solution composition during the desorption experiment (i.e., pH changes or removal of dissolved organic matter) may also contribute to hysteresis (Clay et al., 1988). However, besides experimental artifacts, sufficient evidence exists to suggest that hysteretic behavior can be due to a portion of organic chemical that is strongly or irreversibly bound to soil, where desorption is kinetically so slow that it would require a prohibitive experimental time to be observed (Karickhoff, 1980; Di Toro and Horzempa, 1982; Wauchope and Myers, 1985; Clay and Koskinen, 1990; Gu et al., 1994; Kan et al., 1994; Celis and Koskinen, 1999a). It has also been proposed that sorption-desorption nonsingularity is due to deformation of the sorbent during sorption, which results in the pathway of sorption being different from the pathway of desorption and solute entrapment after sorption (Adamson, 1990; Kan et al., 1994; Schlebaum et al., 1998; Braida et al., 2003). Thus, there is a need to account for sorption-desorption hysteresis in a quantitative and consistent way to better incorporate laboratory sorption-desorption data into models of organic chemical behavior in soil.

On the basis of the assumption that strongly bound pesticide would not be available for desorption, some authors have fit desorption isotherm data to equations based on two-compartment models that attributed some of the observed hysteresis to nondesorbable molecules (Di Toro and Horzempa, 1982; Barriuso et al., 1992; Benoit et al., 1996). The two-compartment model assumes that in one compartment the retention force is weak and allows easy desorption, while in the other compartment the molecules are strongly retained and are either nondesorbable (Di Toro and Horzempa, 1982) or desorbable only at high dilutions (Barriuso et al., 1982). From such models, indirect estimates of the fraction of organic chemical irreversibly bound to soil, as explaining sorption-desorption hysteresis, have been made.

Direct estimates of strongly bound pesticide and other organic compound residues can also be found in the literature, although attempts to use these estimates to explain the hysteretic behavior of desorption isotherms are more scarce (Clay and Koskinen, 1990). In these studies, organic chemical residues are considered bound when the chemical species cannot be extracted by methods commonly used for residue analyses or after exhaustive extraction (Khan, 1982; Roberts, 1984; Gilchrist et al., 1993). However, it is important to bear in mind that "extractability" of a compound will be operationally defined by the nature of the extractant and the experimental conditions under which an extraction is carried out (Gevao et al., 2000). In fact, differences in the ability of specific solvents or experimental conditions to extract compounds from environmental solids can provide information regarding the sorption strength (Northcott and Jones, 2000). The extraction protocols used range from classical solvent extraction techniques, including batch shaking and Soxhlet extractions, to newly developed methodologies, such as microwave-assisted extraction, ultrasonic extraction, accelerated solvent extraction, supercritical fluid extraction, and subcritical water extraction (Dean, 1996; Northcott and Jones, 2000; Loibner et al., 2000; Macutkiewicz et al., 2003). In particular, supercritical fluid extraction has become a promising technique due to its advantages over classical solvent extractions and the almost unlimited extraction options resulting from the range of available temperature and pressure combinations that can be further expanded by the choice of the modifying agent (Langenfeld et al., 1994; Camel, 1997; Northcott and Jones, 2000; Loibner et al., 2000; Motohashi et al., 2000). The use of radiolabeled compounds, pyrolisis, thermal desorption, and spectroscopic methods have also provided essential information on the interactions of bound residues in soil, including residue distribution in soil fractions and mechanistic information at molecular level (Khan and Hamilton, 1980; Koskinen and Harper, 1990; Northcott and Jones, 2000).

As mentioned above, even though direct estimates of strongly bound organic chemical residues are abundant in the literature, attempts to use these estimates to explain the hysteretic behavior of the desorption isotherms are very scarce. One exception is the recent work of Celis and Koskinen (1999a), where a simple batch methodology based on isotopic exchange was used to quantitatively estimate the irreversible and reversible components of sorption (Figure 12.4). Monitoring of the exchange between ^{12}C-pesticide molecules and ^{14}C-labeled pesticide molecules in preequilibrated soil suspensions allowed characterization of pesticide exchange kinetics and estimation of amounts of pesticide that did not participate in reversible sorption-desorption equilibria. A two-compartment model was then applied to describe the experimental data point of the sorption isotherms as the sum of a reversible component and a nondesorbable, irreversible component, which allowed accurate prediction of sorption-desorption hysteresis during successive desorption cycles. Application of this technique to various soil-pesticide combinations and different experimental conditions suggested an increase in the irreversible component of sorption at lower pesticide concentrations and longer equilibration times (Celis and Koskinen, 1999b).

12.3.1.3 Sorption Kinetics

Although often regarded as instantaneous for modeling purposes, the sorption process in soil may require long time periods to reach equilibrium (Pignatello and Xing, 1996; Altfelder et al., 2000). A common observation is that sorption and desorption of organic compounds by soils and soil constituents is characterized by a rapid rate followed by a much slower approach to an apparent equilibrium (Karickhoff, 1980; Sparks, 1989), the division between

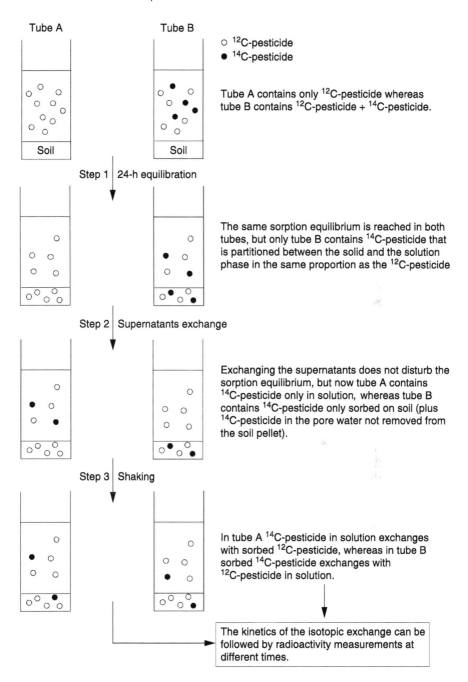

FIGURE 12.4 Scheme of the isotopic exchange experiment. (From Celis and Koskinen, 1999a.)

these two stages being inherently arbitrary. Because changes in the solution-phase concentration can be very small over long periods, in many routine sorption-desorption experiments an agitation time of 24 hours has been considered sufficient to establish equilibrium. However, there is evidence that the magnitude of the slow fraction may not be insignificant and that increases of as much as 10-fold in the sorption of hydrophobic organic compounds can take place by increasing the contact time from a few days to several weeks or months (Boesten and van der Pas, 1983; Pignatello and Xing, 1996; Cox et al., 1998; Altfelder et al., 2000).

Understanding the mechanisms responsible for slow sorption processes is fundamental for the development of valid kinetics of sorption-desorption in soil. These mechanisms have been grouped into two general categories: transport-related processes (or physical nonequilibrium) and sorption-related processes (Brusseau et al., 1991) (Figure 12.5). Transport-related nonequilibrium affects both sorbing and non-sorbing solutes and is a result of the existence of a heterogeneous flow domain (Brusseau et al., 1991). The solute moves from mobile pore water to less mobile water, such as water surrounding the sorbent surface or that existing between soil particles, resulting in slow exchange (Brusseau et al., 1991; Kearney et al., 1997). Sorption-related processes affect only sorbing solutes and may result from chemical nonequilibrium or from rate-limited diffusive mass transfer. Chemical nonequilibrium occurs when the activation energy for sorption or desorption is high. Behavior attributable to transport-related nonequilibrium has been observed in aggregated, macroporous, heterogeneous (with respect to hydraulic conductivity), and fractured porous media (Brusseau and Rao, 1991).

Although it has been ruled out as a probable cause of nonequilibrium for nonpolar organic compounds, chemical nonequilibrium may become important for organic chemicals containing polar functional groups, such as

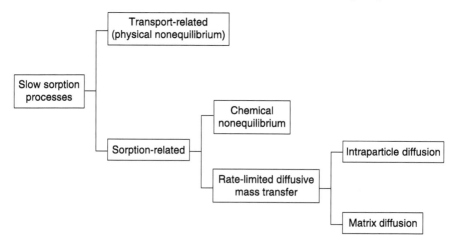

FIGURE 12.5 Mechanisms of slow sorption of organic compounds in soil.

pesticides (Brusseau et al., 1991). For these compounds, high activation energy for sorption/desorption can be a consequence, for example, of simultaneous interaction of the pesticide molecule with multiple points of the sorbent surface or steric hindrance to the sorption or desorption reaction (Pignatello and Xing, 1996).

Rate-limited diffusive mass transfer is considered the major cause of slow kinetics by many researchers. To reach sorption sites, diffusing molecules must traverse bulk liquid, the relative stagnant liquid film at the soil surface (film diffusion), pores within the particles (intraparticle diffusion), and penetrable soil phases (matrix diffusion) (Pignatello and Xing, 1996). Slow intraparticle diffusion has been related to movement through rigid micropores of the same order of magnitude (nanometer size) as the diffusing compound (Cornelissen et al., 1998), whereas slow matrix diffusion has been linked to penetration of the solute within the polymeric-type structure of soil organic matter. In this regard, soil organic matter has been proposed to consist of two different domains: a rubbery domain that is flexible due to relatively weak forces between structural units, and a glassy domain that is more condensed and rigid and is responsible for slow matrix diffusion of hydrophobic organic compounds (Pignatello, 1998; Altfelder et al., 2000).

Batch slurry and incubation experiments have been used to characterize the kinetics of organic chemical sorption-desorption by soil. Cox et al. (1998) found an increase in the sorption coefficient (K_d) values for the insecticide imidacloprid in three different soils by an average factor of 2.8 during an incubation period of 16 weeks. Similar increases in sorption with incubation time have been reported for the insecticide carbofuran (Shelton and Parkin, 1991) and soil-applied herbicides, such as cyanazine and metribuzin (Boesten et al., 1989), alachlor (Xue and Selim, 1995), and isoproturon and linuron (Gaillardon and Sabar, 1994; Cox and Walker, 1999). This increase in sorption coefficients has often been attributed to two types of sorption sites in the soil: external sites, on which sorption is relatively fast, and internal sites, which are less accessible and reach equilibrium slowly. In relation to this, there is evidence that desorption becomes more difficult with increasing residence time in the soil, which has been shown to be particularly important in well-structured soils (Loehr and Webster, 1996; Walker et al., 1999).

Soil column experiments are more time-consuming than simple batch mixing, but can provide detailed information on sorption and transport phenomena under dynamic conditions and, therefore, are closer to field conditions. Column experiments have given insight into both sorption kinetics and soil water flow dispersion. Thus, Kay and Elrick (1967) demonstrated very early that a simple chromatographic sorption equilibrium model, in which K_d determined retardation, applies only at very slow flows, which was one of the earliest demonstrations that sorption kinetics are not instantaneous (Wauchope et al., 2002). In addition, columns may also be used for measuring the mobility of aged pesticide residues in soil, which may be important for purposes of risk assessment of already contaminated soils and remediation designs. Pignatello et al. (1993) compared soil column elution of aged atrazine and metolachlor with elution of "fresh" herbicide, concluding that batch partition coefficients greatly underestimate the apparent true values, since the time scale for sorption is many months.

Batch and column sorption data generated under laboratory conditions have been used to determine sorption parameters for kinetic sorption models, but progress in incorporating sorption kinetic data into chemical transport modeling has been slow (Cox et al., 1994). Wauchope and Myers (1985) found that their sorption-desorption data fit a sequential-equilibria model that assumed that two reversible equilibria occurred sequentially. The first equilibrium involved labile sites and was assumed to be rapid, whereas the second equilibrium was assumed to result from a much slower reversible interchange of pesticide between labile and restricted soil sites. The two-site model has been shown to adequately describe nonequilibrium transport of a wide range of organic compounds. Extensions, such as assuming a three-compartment model or even a continuous distribution of sorption kinetic constants, have also been proposed (Boesten et al., 1989; Deitsch et al., 2000; Wauchope et al., 2002). As pointed out by Green and Karickhoff (1990), it is important to note that sorption modeling should be designed to accommodate sorption-desorption behavior in specific environmental systems. A kinetic approach may be most appropriate in some cases and equilibrium sorption may be adequate in others. The methods of measuring or estimating sorption parameters must be consistent with the intended modeling use.

12.3.2 ESTIMATING SORPTION

12.3.2.1 Characterizing Sorption at Field Scale

Despite the practical objective of any sorption characterization effort to predict sorption at the field scale, the complexity of natural environments make the attainment of representative conditions in the laboratory very difficult. In fact, design of experimental procedures for sorption characterization is usually driven by practical concerns (Wauchope et al., 2002). Soil structure, for example, is often destroyed during sample collection and preparation prior to sorption analysis, and a 24-hour mixing experiment may cause further aggregate breakdown (Wauchope et al., 2002). It has also been suggested that batch slurry experiments, performed under low soil : solution ratios, may not give a correct picture of the sorptive capacity of a soil under field conditions where water is less abundant (Grover and Hance, 1970; Koskinen and Cheng, 1983). To circumvent this limitation, new methods based on supercritical fluid extraction and modified centrifugation techniques have been proposed in an attempt to characterize sorption-desorption at water contents closer to those occurring in the field (Koskinen and Rochette, 1996; Walker and Jurado-Exposito, 1998). Similarly, temperature and solution composition are usually not strictly controlled in routine sorption experiments, but their effect on organic chemical sorption can be significant (Wauchope et al., 2002).

In particular, the amount and nature of dissolved organic matter in the soil solution have been shown to strongly influence sorption and desorption of organic chemicals by soil (Lee and Farmer, 1989; Lee et al., 1990; Celis et al., 1998; Cox et al., 2000). Dissolved organic matter–facilitated transport can represent the dominant mechanism for the migration of strongly sorbing contaminants, such as hydrophobic pesticides and polycyclic aromatic hydrocarbons (Prechtel et al., 2002).

An additional obstacle that can make sorption measurements in the laboratory different from sorption at the field scale is the spatial and temporal variability of soil properties, especially over large field areas. Oliveira et al. (1999) reported changes in the K_d value for the herbicide imazethapyr from 0.18 to 3.78 L kg^{-1} across a 31.4-ha field plot. Similar variations of the distribution coefficient, K_d, have been observed for atrazine (0.47–1.70 L kg^{-1}), isoproturon (0.47–1.81 L kg^{-1}), and metamitron (0.55–2.21 L kg^{-1}) (Coquet, 2003). Much of the observed variations have been related to changes in pH, composition, and porosity characteristics of the soil across the investigated field areas.

Changes in sorption coefficients with depth have been reported to be even more pronounced than those due to horizontal variability of soil, due to greater variation of soil properties. Thus, Njoroge et al. (1998) reported a 100-fold decrease in the K_d value for the sorption of 1,2,4-trichlorobenzene and tetrachlorobenzene within a soil profile extending 2.5 m below ground surface. Interestingly, this decrease in K_d with depth was greater than could be predicted on the basis of a 10-fold decrease in organic matter content and despite significantly greater surface area in the lower horizons. The authors attributed the lower K_d values of the deeper soil horizons to a combination of lower organic matter content, a more hydrophilic nature of the organic matter, and a more intimate association of the organic matter with the mineral fraction, affecting its accessibility, sorptivity, or both.

Environmental variation of soil properties can considerably reduce the value of accurately characterizing sorption using expensive or time-consuming studies. According to Cleveland (1996), better balance between study detail and natural environmental variation is needed in the current regulatory requirements for pesticide sorption/desorption studies. In her work on the current state of registration requirements for mobility assessments of pesticides, Cleveland (1996) suggests that the information provided by single concentration batch K_d measurements on a large suite of soils (i.e., 30 soils) would be more useful than that provided by a detailed isotherm analysis on fewer soil samples, because these data would provide a probability distribution for K_d across the intended market at environmentally relevant concentrations. This distribution could be used to determine the dominant soil properties that influence sorption for a given test substance and would also provide input to construct probabilistic modeling. Further characterization of sorption kinetics under specific conditions would provide temporal information of the sorption process and allow temporal models to become more realistic (Cleveland, 1996).

12.3.2.2   Estimating Sorption from Easily Measurable Soil Properties

An alternative to experimentally characterizing sorption for a given organic chemical-soil combination is to use estimation procedures based on previously observed correlations of sorption with easily measurable soil properties. These sorption estimates can also provide valuable information to predict changes in sorption as a function of spatial and temporal variations of soil properties. For example, Oliveira et al. (1999) identified two distinct patterns for imazethapyr sorption on a field area: areas in which pH > 6.25 and K_d < 1.5, where K_d variation was based primarily on pH, and areas in which pH < 6.25 and K_d > 1.5, where other soil properties such as organic carbon content had a significant influence on K_d variation. Accordingly, based on easily measurable soil properties, they divided the field into two potential management areas with different leaching potential, which provided a rationale for site-specific imazethapyr application. Examples of important physico-chemical properties that have been correlated with the sorption of organic chemicals in soil are described below.

12.3.2.2.1   Organic Carbon Content

The most widely accepted approach for estimation of sorption from soil properties is that which assumes that sorption of nonpolar organic compounds is related directly to the organic carbon content of the soil. This approach is based on the fact that unless soil organic carbon is very low, sorption of many nonpolar organic compounds occurs mainly on soil organic matter (Chiou, 1989; Koskinen and Harper, 1990). A relationship based on the soil organic carbon content was introduced by Hamaker and Thompson (1972) for the analysis of sorption constants:

$$K_{oc} = K_d/f_{oc} \qquad (12.3$$

where an organic-carbon normalized sorption coefficient (K_{oc}) is calculated by dividing the sorption distribution coefficient (K_d) by the organic carbon content of the soil (f_{oc}).

According to Green and Karickhoff (1990), the use of K_{oc} has been debated since its introduction relative to pesticides in the 1960s. The use of K_{oc} assumes that soil mineral components are not significantly active in the sorption process, and its utility will depend heavily on the compound, the level of organic carbon, and the nature of the organic matter (Hamaker and Thompson, 1972; Pusino et al., 1992; Cleveland, 1996; Ahmad et al., 2001). Pusino et al. (1992) reported a reduction in variation of sorption constants for metolachlor from 17-fold to 4-fold when relating sorption to the organic carbon content of the soil. They also observed a reduction in K_{oc} values after reduction of soil organic matter content by H_2O_2 treatment, substantiating the role of organic matter in the sorption of the herbicide metolachlor. On the contrary, Cox et al. (1997a) and Celis et al. (1997b) observed an increase in sorption coefficients of polar herbicides with peroxidation indicating a greater contribution of mineral surfaces when organic matter is removed. The high contribution of mineral surfaces in soil sorption of pesticides is a reason for variability in K_{oc} values.

Variations in K_{oc} among soils can also indicate that the nature of soil organic matter affects its sorption capacity (Hayes et al., 1989; Gerstl and Kliger, 1990). Soil organic matter is heterogeneous in nature (Xing, 1997). Ahmad et al. (2001) studied the structural composition of soil organic matter from 27 soils with different vegetation and ecological zones of Australia and Pakistan using CPMAS ^{13}C NMR and observed that the variations in K_{oc} values of the pesticides carbaryl and phosalone could be explained only when variations in the aromatic component of soil organic matter were taken into account. Another drawback of using K_{oc} is that when low organic carbon soil is used to generate K_{oc} values, which are then applied to soils with higher organic carbon contents, the sorption may be either overestimated (Sukop and Cogger, 1992) or underestimated (Njoroge et al., 1998).

12.3.2.2.2 Clay Content

Because for highly polar or ionic organic compounds and for soils with low organic carbon content, the soil mineral fraction can significantly contribute to the sorption process (Mingelgrin and Gerstl, 1983), attempts have been made to incorporate this contribution in estimation procedures. Thus, Hermosín et al. (2000a, 2000b) suggest the use of a clay-normalized sorption coefficient, $K_{clay} = K_d/f_{clay}$, where f_{clay} represents the clay fraction of the soil, i.e., the sum of inorganic and organic components present in the $< 2\,\mu$m fraction of the soil. They showed that a decrease in the variation coefficient from K_d (or K_f) to K_{clay} occurred for many polar, ionic, and ionizable pesticides (Figure 12.6). This fact indicates more homogeneous behavior of the clay fraction in pesticide sorption as compared with the whole soil. Similarly, Weidenhaupt et al. (1997) suggest that the overall sorption of triorganotin biocides (TOTs) by natural sediment, $K_{d, tot}$, can be roughly estimated considering only sorption of TOT^{+} cations to mineral surfaces, $K_{d, min}$, and hydrophobic partitioning of neutral TOT-OH species into organic matter, $K_{d, oc}$:

$$K_{d, tot} = K_{d, min} + K_{d,oc} = f_{min}K_{min} + f_{oc}K_{oc} \qquad (12.4)$$

where f_{min} represents the weight fraction of minerals bearing negatively charged sites, K_{min} is the sorption coefficient of TOTs at such minerals, f_{oc} is the weight fraction of particulate organic carbon, and K_{oc} is the partitioning constant of TOT-OH species into natural organic carbon. The authors further suggest that K_{oc} and K_{min} can be calculated from the octanol/water partition coefficient of TOT-OH and the sorption of TOTs on model minerals, respectively. As pointed out by Hermosín et al. (2000a), it should be noted that accurate estimation procedures that take into account the sorptive capacity of the soil mineral fraction require knowledge of the nature of the

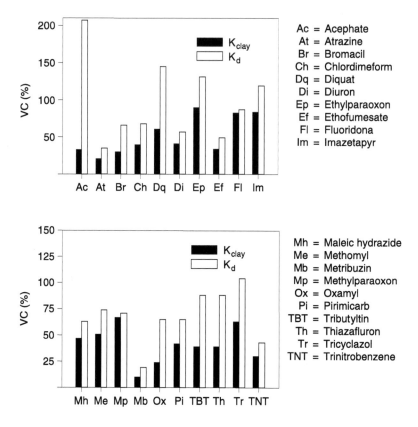

FIGURE 12.6 Variation coefficients (VC) for pesticide K_d and K_{clay} sorption coefficients in soils. (From Hermosín et al., 2000a.)

minerals present in soil (e.g., expandable/nonexpandable phyllosilicates, metal oxides), because some mineral components can be considerably more sorptive than others.

12.3.2.2.3 Other Soil Properties

Although their use has been limited, alternatives to the organic carbon- and clay-referenced sorption approaches have been proposed as being desirable for some organic chemical-soil combinations. Based on linear regressions between measured K_d values for 35 pesticides and the soil specific surface area (SSA), Pionke and DeAngelis (1980) developed an estimation methodology to predict K_d from SSA. For a diverse group of pesticides the linear regression of K_d with SSA gave consistently higher coefficients of determination than did the linear regression of K_d with organic carbon content, indicating the usefulness of the specific surface area approach (Green and Karickhoff, 1990). Similar estimation procedures have been proposed to include other easily measurable soil properties such as pH, textural variables, or cation exchange capacity (Oliveira et al., 1999; Jacques et al., 1999; Dubus et al., 2001).

12.3.3 STRENGTHS AND WEAKNESSES

Among the different methodologies for sorption-desorption characterization discussed in this chapter, batch equilibration technology remains most widely used at this time. The resulting simple equilibrium constant, K_d, frequently obtained from these experiments, implies ideal conditions of instantaneous equilibrium, isotherm linearity, and desorption reversibility. Even though under real situations organic chemicals are seldom, if ever, at these conditions, batch K_d values are believed to be adequate for discriminating between broad leaching classes of chemicals and for use in early-tier risk assessment (Wauchope et al., 2002). Attempts to develop more precise sorption descriptions have been made by considering nonlinear sorption equations, kinetic equations, and hysteretic behavior. These attempts, however, have resulted in little improvement of sorption predictions under many field situations, mainly because the large spatial and temporal variability of soil properties occurring in the field limits the value of accurately characterizing sorption under specific conditions in the laboratory using expensive or time-consuming studies. An additional obstacle is the complexity to represent realistic soil conditions, such as pore structure, water content, or water flow, in laboratory sorption studies.

Although indirect methods of estimating sorption have been used when actual measurement of sorption is impossible (Green and Karickhoff, 1990; Kearney et al., 1997), attempts to predict the sorption behavior of organic chemicals in soil as a combination of interactions with separated soil constituents have had only limited success (Wauchope et al., 2002). For uncharged, highly hydrophobic compounds, sorption has been successfully related to the soil organic matter content. In contrast, the range of soil components with affinity for organic compounds containing polar function-alities is large, and the components interact with each other in complex ways, altering the amount and nature of the surface ultimately exposed for sorption processes (Celis et al., 1997b). Thus, according to Wauchope et al. (2002), if one wishes to know accurately the degree of sorption of a specific organic compound in a specific soil, the approach is still empirical.

12.4 RECOMMENDATIONS AND FUTURE RESEARCH

The degree of rigor required in the characterization of sorption depends on the accuracy required for the intended use (Kearney et al., 1997). To know how the sorption process interacts with other processes, the sorption mechanisms and factors influencing sorption must be known. For this purpose, rigorous sorption studies under specific experimental conditions are valuable for understanding sorption at the field scale. In this regard, research directed to elucidate the relative contribution of the various soil constituents (not just as isolated components, but when associated into organomineral complex particles) to the sorptive properties of soils is needed. Sorption characterization experiments directed to improve our understanding of hysteretic and kinetic behavior of organic compounds in soil, in particular for aged residues, are also needed. Nevertheless, many environmental quality applications require only approximate estimates of sorption. In these cases, simplifying assumptions in sorption characterization or indirect estimates of sorption may be acceptable (Green and Karickhoff, 1990).

ACKNOWLEDGMENT

This work has been partially financed by Junta de Andalucía through Research Group RNM124, the MCYT projects AGL-2001-1554, REN-2001-1700-C02-01/TECNO, and the FP5 EU Project EVK1-CT-2001-00105 [http://www.liberation.dk].

REFERENCES

Adamson, A.W. 1990. Physical Chemistry of Surfaces. John Wiley, New York, 800 pp.
Ahmad, R., R.S. Kookana, A.M. Alston, and J.O. Skjemstad. 2001. The nature of soil organic matter affects sorption of pesticides.1. Relationships with carbon chemistry as determined by ^{13}C CPMAS NMR spectroscopy. Environ. Sci. Technol. 35:878–889.
Alexander, M. 1999. Sorption. p. 117. In Biodegradation and Bioremediation. Academic Press, San Diego, CA.
Allen, R., and A. Walker. 1987. The influence of soil properties on the rates of degradation of metamitron, metazachlor and metribuzin. Pestic. Sci. 18:95–111.
Altfelder, S., T. Streck, and J. Richter. 2000. Nonsingular sorption of organic compounds in soil: the role of slow kinetics. J. Environ. Qual. 29:917–925.
Anderson, A.W., A.B. McBratney, and J.W. Crawford. 1998. Application of fractals to soil studies. Adv. Agron. 63:1–76.
Anderson, J.P.E. 1981. Soil moisture and the rates of biodegradation of diallate and triallate. Soil Biol. Biochem. 13:155–161.
Barriuso, E., and W.C. Koskinen. 1996. Incorporating nonextractable atrazine residues into soil size fractions as a function of time. Soil Sci. Soc. Am. J. 60:150–157.
Barriuso, E., D.A. Laird, W.C. Koskinen, and R.H. Dowdy. 1994. Atrazine desorption from smectites. Soil Sci. Soc. Am. J. 58:1632–1638.
Barriuso, E., U. Baer, and R. Calvet. 1992. Dissolved organic matter and adsorption-desorption of dimefuron, atrazine, and carbetamide by soils. J. Environ. Qual. 21:359–367.
Benoit, P., E. Barriuso, S. Houot, and R. Calvet. 1996. Influence of the nature of soil organic matter on the sorption-desorption of 4-chlorophenol, 2,4-dichlorophenol and the herbicide 2,4-dichlorophenoxyacetic (2,4-D). Eur. J. Soil Sci. 47:567–578.
Bergstrom, L., and J. Stenstrom. 1998. Environmental fate of chemicals in soils. Ambio 27:16–23.
Boesten, J.J.T.I., and L.J.T. van der Pas. 1983. Test of some aspects of a model for the adsorption/desorption of herbicides in soil. Aspects Appl. Biol. 4:495–501.

Boesten, J.J.T.I., L.J.T. van der Pas, and J.H. Smelt. 1989. Field test of a mathematical model for non-equilibrium transport of pesticides in soil. Pestic. Sci. 25:187–203.

Braida, W.J., J.J. Pignatello, Y.F. Lu, P.I. Ravikovitch, A.V. Neimark, and B.S. Xing. 2003. Sorption hysteresis of benzene in charcoal particles. Environ. Sci. Technol. 37:409–417.

Briggs, G.G., and J.E. Dawson. 1970. Hydrolysis of 2,6-dichlorobenzonitrile in soils. J. Agric. Food Chem. 18:97–99.

Brusseau, M.L., and P.S.C. Rao. 1990. Modelling solute transport in structured soils: a review. Geoderma 46:169–192.

Brusseau, M.L., R.E. Jessup, and P.S.C. Rao. 1991. Nonequilibrium sorption of organic chemicals: elucidation of rate-limiting processes. Environ. Sci. Technol. 25:134–142.

Businelli, D. 1997. Pig slurry amendment and herbicide coapplication effects on s-triazine mobility in soil: an adsorption-desorption study. J. Environ. Qual. 26:102–108.

Calvet, R. 1980. Adsorption-desorption phenomena. p. 1–30. *In* R.J. Hance (ed.), Interactions Between Herbicides and the Soil. Academic Press, New York.

Calvet, R. 1989. Adsorption of organic chemicals in soils. Environ. Health Persp. 83:145–177.

Camel, V. 1997. The determination of pesticide residues and metabolites using supercritical fluid extraction. Trends Anal. Chem. 16:351–369.

Carton, A., T. Isla, and J. Alvárez-Benedi. 1997. Sorption-desorption of imazamethabenz on three Spanish soils. J. Agric. Food Chem. 45:1454–1458.

Celis, R., J. Cornejo, M.C. Hermosín, and W.C. Koskinen. 1997a. Sorption-desorption of atrazine and simazine by model soil colloidal components. Soil. Sci. Soc. Am. J. 61:436–443.

Celis, R., L. Cox, M.C. Hermosín, and J. Cornejo. 1997b. Sorption of thiazafluron on iron- and humic acid-coated montmorillonite. J. Environ. Qual. 26:472–479.

Celis, R., E. Barriuso, and S. Houot. 1998. Sorption and desorption of atrazine by sludge-amended soil: dissolved organic matter effects. J. Environ. Qual. 27:1348–1356.

Celis, R., and W.C. Koskinen. 1999a. An isotopic exchange method for the characterization of the irreversibility of pesticide sorption-desorption in soil. J. Agric. Food Chem. 47:782–790.

Celis, R., and W.C. Koskinen. 1999b. Characterization of pesticide desorption from soil by the isotopic exchange technique. Soil Sci. Soc. Am. J. 63:1659–1666.

Celis, R., L. Cox, M.C. Hermosín, and J. Cornejo. 2001. Porosity and surface fractal dimension of soils as affecting sorption, degradation and mobility of polar herbicides. p. 77–82. *In* A. Walker (ed.), Proc. BCPC Symp. on Pesticide Behaviour in Soils and Water, Brighton, U.K. 13–15 November 2001. BCPC, Brighton.

Chiou, C.T. 1989. Theoretical considerations of the partition uptake of nonionic organic compounds by soil organic matter. p. 1–29. *In* B.L. Sawhney and K. Brown (eds.), Reactions and Movement of Organic Chemicals in Soils. SSSA Special Publication. No. 22. SSSA, Madison, WI.

Chiou, C.T., D.E. Kile, T.I. Brinton, R.L. Malcom, J.A. Leenheer, and P. MacCarthy. 1987. A comparison of water solubility enhancements of organic solutes by aquatic humic materials and commercial humic acids. Environ. Sci. Technol. 21:1231–1234.

Clapp, C.E., R. Lio, M.H.B. Hayes, and U. Mingelgrin. 1996. Stability of complexes formed by the herbicide napropamide and soluble humic acids. p. 305–315.

In U. Mingelgrin (ed.), Humic Substances and Organic Matter in Soil and Water Environments. Characterization, Transformations and Interactions. IHSS, Birmingham, UK.

Clay, S.A., and W.C. Koskinen. 1990. Characterization of alachlor and atrazine desorption from soils. Weed Sci. 38:74–80.

Clay, S.A., R.R. Allmaras, W.C. Koskinen, and D.L. Wyse. 1988. Desorption of atrazine and cyanazine from soil. J. Environ. Qual. 17:719–723.

Cleveland, C.B. 1996. Mobility assessment of agrichemicals: current laboratory methodology and suggestions for future directions. Weed Technol. 10:157–168.

Coquet, Y. 2003. Variation of pesticide sorption isotherm in soil at the catchment scale. Pest Manag. Sci. 59:69–78.

Cornejo, J., and M.C. Hermosín. 1996. Interaction of humic substances and soil clays. p. 595–624. *In* A. Piccolo (ed.), Humic Substances in Terrestrial Ecosystems. Elsevier, Amsterdam.

Cornejo, J., and P. Jamet. 2000. Pesticide/Soil Interactions: Some Current Research Methods. INRA Editions, Paris.

Cornejo, J., M.C. Hermosín, J.L. White, J.R. Barnes, and S.L. Hem. 1983. Role of ferric iron in the oxidation of hydrocortisone by sepiolite and palygorskite. Clays Clay Miner. 31:109–112.

Cornelissen, G., P.C.M. Van Noort, and H.A.J. Govers. 1998. Mechanism of slow desorption of organic compounds from sediments: a study using model sorbents. Environ. Sci. Technol. 32:3124–3131.

Cox, L., M.C. Hermosín, and J. Cornejo. 1994. Interactions of methomyl with montmorillonites. Clay Miner. 29:767–774.

Cox, L., M.C. Hermosín, R. Celis, and J. Cornejo. 1997a. Sorption of two polar herbicides in soils and soil clays suspensions. Water Res. 31:1309–1316.

Cox, L., R. Celis, M.C. Hermosín, A. Becker, and J. Cornejo. 1997b. Porosity and herbicide leaching in soils amended with olive-mill wastewater. Agric. Ecosys. Environ. 65:151–161.

Cox, L., W.C. Koskinen, and P.Y. Yen. 1998. Changes in sorption of imidacloprid with incubation time. Soil Sci. Soc. Am. J. 62:342–347.

Cox, L., and A. Walker. 1999. Studies of time-dependent sorption of linuron and isoproturon in soils. Chemosphere 38:2707–2718.

Cox, L., M.J. Calderón, M.C. Hermosín, and J. Cornejo. 1999a. Leaching of clopyralid and metamitron under conventional and reduced tillage systems. J. Environ. Qual. 28:605–610.

Cox, L., M.C. Hermosín, and J. Cornejo. 1999b. Leaching of simazine in organic amended soils. Comm. Soil Sci. Plant Anal. 30:1697–1706.

Cox, L., R. Celis, M.C. Hermosín, J. Cornejo, A. Zsolnay, and K. Zeller. 2000. Effect of organic amendments on herbicide sorption as related to the nature of the dissolved organic matter. Environ. Sci. Technol. 34:4600–4605.

Cox, L., A. Cecchi, R. Celis, M.C. Hermosín, W.C. Koskinen, and J. Cornejo. 2001. Effect of exogenous carbon on movement of simazine and 2,4-D in soils. Soil Sci. Soc. Am. J. 65:1688–1695.

Dao, T.H. 1995. Subsurface mobility of metribuzin as affected by crop residue placement and tillage method. J. Environ. Qual. 24:1193–1198.

de Jonge, H., L.W. de Jonge, and O.H. Jacobsen. 2000. [C-14]glyphosate transport in undisturbed topsoil columns. Pest Manag. Sci. 56:909–915.

Dean, J.R. 1996. Effect of soil-pesticide interactions on the efficiency of supercritical fluid extraction. J. Chromatography A 754:221–233.

Deitsch, J.J., J.A. Smith, T.B. Culver, R.A. Brown, and S.A. Riddle. 2000. Distributed-rate model analysis of 1,2-dichlorobenzene batch sorption. Environ. Sci. Technol. 34:1469–1476.

Delle Site, A. 2001. Factors affecting sorption of organic compound in natural sorbent/water systems and sorption coefficients for selected pollutants: a review. J. Phys. Chem. Ref. Data 30:187–439.

Di Toro, D.M., and L.M. Horzempa. 1982. Reversible and resistant components of PCB adsorption-desorption isotherms. Environ. Sci. Technol. 16:594–602.

Dubus, I.G., E. Barriuso, and R. Calvet. 2001. Sorption of weak organic acids in soils: clofencet, 2,4-D and salicylic acid. Chemosphere 45:767–774.

Felsot, A.S., and E.K. Dzantor. 1995. Effect of alachlor concentration and an organic amendment on soil dehydrogenase activity and pesticide degradation rate. Environ. Toxicol. Chem. 14:23–28.

Fermanich, K.J., W.L. Bland, B. Lowerg, K. McSweeney. 1996. Irrigation and tillage effects on atrazine and metabolite leaching from a sandy soil. J. Environ. Qual. 25:1291–1299.

Flores-Céspedes, F., E. González-Pradas, M. Fernández-Pérez, M. Villafranca-Sánchez, M. Socias-Viciana, and M.D. Ureña-Amate. 2002. Effects of dissolved organic carbon on sorption and mobility of imidacloprd in soil. J. Environ. Qual. 31:880–888.

Flury, M. 1996. Experimental evidence of transport of pesticides through field soils — a review. J. Environ. Qual. 25:25–45.

Gaillardon, P., and M. Sabar. 1994. Changes in the concentrations of isoproturon and its degradation products in soil and soil solution during incubation at two temperatures. Weed Res. 34:243–251.

Gamble, D.S., and S.U. Khan. 1985. Atrazine hydrolysis in soils: catalysis by the acidic functional groups of fulvic acid. Can. J. Soil Sci. 65:435–443.

Gaston, L.A., and M.A. Locke. 1996. Bentazon mobility through intact unsaturated columns of conventional and no-till Dundee soil. J. Environ. Qual. 25:1350–1356.

Gerstl, Z., and L. Kliger. 1990. Fractionation of the organic matter in soils and sediments and their contribution to the sorption of pesticides. J. Environ. Sci. Health B 25:729–741.

Gevao, B., K.T. Semple, and K.C. Jones. 2000. Bound residues in soils: a review. Environ. Pollut. 108:3–14.

Gilchrist, G.F.R., D.S. Gamble, H. Kodama, and S.U. Khan. 1993. Atrazine interactions with clay minerals: kinetics and equilibria of sorption. J. Agric. Food Chem. 41:1748–1755.

Graber, E.R., I. Dror, F.C. Bercovich, and M. Rosner. 2001. Enhanced transport of pesticides in a field trial with treated sewage sludge. Chemosphere 44:805–811.

Graber, E.R., Z. Gerstl, E. Fischer, and U. Mingelgrin. 1995. Enhanced transport of atrazine under irrigation with effluent. Soils Sci. Soc. Am. J. 59:1513–1519.

Green, R.E., and S.W. Karickhoff. 1990. Sorption estimates for modeling. p. 79–101. In H.H. Cheng (ed.), Pesticides in the Soil Environment: Processes, Impacts, and Modeling. SSSA Book Series 2, Madison, WI.

Grover, R., and R.J. Hance. 1970. Effect of ratio of soil to water on adsorption of linuron and atrazine. Soil Sci. 109:136–138.

Gu, B., J. Schmitt, Z. Chen, L. Liang, and J.F. McCarthy. 1995. Adsorption and desorption of natural organic matter on iron oxide: mechanisms and models. Environ. Sci. Technol. 28:38–46.

Guo, L., T.J. Bicki, A.S. Felsot, and T.D. Hinesly. 1993. Sorption and movement of alachlor in soil modified by carbon-rich wastes. J. Environ. Qual. 22:186–194.

Gustafson, D.I. 1989. Groundwater-Ubiquity Score, a simple method for assessing pesticide leachability. Environ. Toxicol. Chem. 8:339–357.

Hamaker, J.W., and J.M. Thompson. 1972. Adsorption. p. 49–143. *In* C.A.I. Goring and J.W. Hamaker (eds.), Organic Chemicals in the Soil Environment. Marcel Dekker, New York.

Hayes, M.H.B., P. MacCarthy, R.L. Malcom, and R.S. Swift. 1989. Humic Substances II. In Search for Structure. John Wiley & Sons, New York, p. 689.

Hermosín, M.C., J. Cornejo, and L. Cox. 2000a. Calculation and validation of Kclay as predictor for polar or ionic pesticide adsorption by soils. p. 131–140. *In* J. Cornejo and P. Jamet (eds.), Pesticide/Soil Interactions: Some Current Research Methods. INRA, Paris.

Hermosín, M.C., J. Cornejo, and L. Cox. 2000b. Suggesting a Kclay as prediction factor for polar or ionic pesticides adsorption by soil or sediment. p. 18. *In* Abstracts, Bouyoucos Conference on Environmental Chemistry at the Clay-Water Interface, Honolulu, Hawaii. 6–9 March, 2000.

Isensee, A.R., R.G. Nash, and C.S. Helling. 1990. Effect of conventional vs. no-tillage on pesticide leaching to shallow groundwater. J. Environ. Qual. 19:434–440.

Jacques, D., C. Mouvet, B. Mohanty, H. Vereecken, and J. Feyen. 1999. Spatial variability of atrazine sorption parameters and other soil properties in a podzoluvizol. J. Contam. Hydrol. 36:31–52.

Jensen, J.J. 2002. Development of a decision support system for sustainable management of contaminated land by linking bioavailability, ecological risk and groundwater pollution of organic pollutants. FP5 EU Project EVK1-CT-2001-00105. http://www.liberation.dk (accessed April 2004).

Kan, A.T., G.M. Fu, and M.B. Tomson. Adsorption-desorption hysteresis in organic pollutant and soil sediment interaction. Environ. Sci. Technol. 28:859–867.

Karickhoff, S.W. 1980. Sorption kinetics of hydrophobic pollutants in natural sediments. p. 193. *In* R.A. Baker (ed.), Contaminants and Sediments. Ann Arbor Science Publishers, Ann Arbor, MI.

Kaufman, D.D., and P.C. Kearney. 1970. Microbial degradation of triazine herbicides. Residue Rev. 32:235–265.

Kay, B.D., and D.E. Elrick. 1967. Adsorption and movement of lindane in soils. Soil Sci. 104:314–322.

Kearney, P.C., D.R. Shelton, and W.C. Koskinen. 1997. Soil chemistry of pesticides. p. 419–451. *In* Encyclopedia of Chemical Technology, 4th ed., Vol. 22. John Wiley & Sons, New York.

Khan, S.U. 1982. Bound pesticide residue in soil and plants. Residue Rev. 84:1–25.

Khan, S.U., and H.A. Hamilton. 1980. Extractable and bound (unextractable) residues of prometryn and its metabolites in an organic soil. J. Agric. Food Chem. 28:126–132.

Kookana, R.S., H.J. Di, and L.A.G. Aylmore. 1995. A field study of leaching and degradation of nine pesticides in a sandy soil. Aust. J. Soil Res. 33:1019–1030.

Koskinen, W.C., and E.A. Rochette. 1996. Atrazine sorption-desorption in field-moist soils. Intern. J. Environ. Anal. Chem. 65:223–230.

Koskinen, W.C., and H.H. Cheng. 1983. Effects of experimental variables on 2,4,5-T adsorption-desorption in soil. J. Environ. Qual. 12:325–330.

Koskinen, W.C., and S.S. Harper. 1990. The retention process: mechanisms. p. 51–77. *In* H.H. Cheng (ed.), Pesticides in the Soil Environment: Processes, Impacts, and Modeling. SSSA Book Series 2, Madison, WI.

Koskinen, W.C., G.A. O'Connor, and H.H. Cheng. 1979. Characterization of hysteresis in the desorption of 2,4,5-T from soil. Soil Sci. Soc. Am. J. 43:871–874.

Kretzchmar, R., M. Borkovec, D. Grolimund, and M. Elimelech. 1999. Mobile subsurface colloids and their role in contaminant transport. Adv. Agron. 66:121–193.

Laabs, V., W. Amelung, A. Pinto, and W. Zech. 2002. Fate of pesticides in tropical soils of Brazil under field conditions. J. Environ. Qual. 31:256–268.

Laird, D.A. 1996. Interactions between atrazine and smectite surfaces. ACS Symp. Ser. 630:86–100.

Langenfeld, J.J., S.B. Hawthorne, D.J. Miller, and J. Pawliszkyn. 1994. Role of modifiers for analytical-scale supercritical fluid extraction of environmental samples. Anal. Chem. 66:909–916.

Larsson, M.H., and N.J. Jarvis. 2000. Quantifying interactions between compound properties and macropore flow effects on pesticide leaching. Pest Manag. Sci. 56:133–141.

Lee, D.J., and W.J. Farmer. 1989. Dissolved organic matter interactions with napropamide and four other nonionic pesticides. J. Environ. Qual. 18:468–474.

Lee, D.-Y., W.J. Farmer, and Y. Aochi. 1990. Sorption of napropamide on clay and soil in the presence of dissolved organic matter. J. Environ. Qual. 19:567–573.

Lesan, H.M., and A. Bhandari. 2003. Atrazine sorption on surface soils: time-dependent phase distribution and apparent desorption hysteresis. Water Res. 37:1644–1654.

Levanon, D., E.E. Codlin, J.J. Meisinger, and J.L. Starr. 1993. Mobility of agrochemicals through soil from two tillage systems. J. Environ. Qual. 22:155–161.

Loehr, R.C., and M.T. Webster. 1996. Behavior of fresh vs aged chemicals in soil. J. Soil Contam. 5:361–383.

Logan, T.J., J.M. Davidson, J.L. Baker, and M.R. Overcash (ed.). 1987. Effects of Conventional Tillage on Groundwater Quality. Lewis Publ., Chelsea, p. 292.

Loibner, A.P., M. Holzer, M. Gartner, O.H.J. Szolar, and R. Braun. 2000. The use of sequential supercritical fluid extraction for bioavailability investigations of PAH in soil. Bodenkultur 51:225–233.

Macutkiewicz, E., M. Rompa, and B. Zygmunt. 2003. Sample preparation and chromatographic analysis of acidic herbicides in soils and sediments. Crit. Rev. Anal. Chem. 33:1–17.

Martins, J.M., and A. Mermoud. 1998. Sorption and degradation of four nitroaromatic herbicides in mono and multi-solute saturated/unsaturated soil batch systems. J. Contam. Hydrol. 33:187–210.

McCarthy, J.F., and J.M. Zachara. 1989. Subsurface transport of contaminants. Environ. Sci. Technol. 23:496–502.

McGechan, M.B., and D.R. Lewis. 2002. Transport of particulate and colloid-sorbed contaminants through soil, Part 1: general principles. Biosyst. Eng. 83:255–273.

Mingelgrin, U., and Z. Gerstl. 1983. Reevaluation of partitioning as a mechanism of nonionic chemicals adsorption in soils. J. Environ. Qual. 12:1–11.

Moorman, T.B., J.K. Cowan, E.L. Arthur, and J.R. Coats. 2001. Organic amendments to enhance herbicide biodegradation in contaminated soils. Biol. Fertil. Soils 33:541–545.

Mortland, M.M. 1970. Clay organic complexes and interactions. Adv. Agron. 22:75–117.

Motohashi, N., H. Nagashima, and C. Parkanyi. 2000. Supercritical fluid extraction for the analysis of pesticide residues in miscellaneous samples. J. Biochem. Biophys. Methods 43:313–328.

Nelson, S.D., J. Letey, W.J. Farmer, C.F. Williams, and M. Ben-Hur. 1998. Facilitated transport of napropamide by dissolved organic matter in sewage sludge amended soil. J. Environ. Qual. 27:1194–1200.

Njoroge, B.N.K., W.P. Ball, and R.S. Cherry. 1998. Sorption of 1,2,4-trichlorobenzene and tetrachloroethene within an authigenic soil profile: changes in K_{oc} with soil depth. J. Contam. Hydrol. 29:347–377.

Northcott, G.L., and K.C. Jones. 2000. Experimental approaches and analytical techniques for determining organic compound bound residues in soil and sediment. Environ. Pollut. 108:19–43.

O'Connor, D.J., and J.P. Connolly. 1980. The effect of concentration of adsorbing solids on the partition coefficient. Water Res. 14:1517–1523.

Oades, J.M., and A.G. Waters. 1991. Aggregate hierarchy in soils. Aust. J. Soil Res. 29:815–828.

Oliveira, R.S., W.C. Koskinen, F.A. Ferreira, B.R. Khakural, D.J. Mulla, and P.J. Robert. 1999. Spatial variability of imazethapyr sorption in soil. Weed Sci. 47:243–248.

Pignatello, J.J. 1998. Soil organic matter as a nanoporous sorbent of organic pollutants. Adv. Colloid Interface Sci. 77:445–467.

Pignatello, J.J. 2000. The measurement and interpretation of sorption and desorption rates for organic compounds in soil media. Adv. Agron. 69:1–73.

Pignatello, J.J., and B. Xing. 1996. Mechanisms of slow sorption of organic chemicals to natural particles. Environ. Sci. Technol. 30:1–11.

Pignatello, J.J., F.J. Ferrandino, and L.Q. Huang. 1993. Elution of aged and freshly added herbicides from a soil. Environ. Sci. Technol. 27:1563–1571.

Pionke, H.B., and R.J. DeAngelis. 1980. Method for distributing pesticide loss in field runoff between the solution and adsorbed phase. p. 607. In CREAMS, A Field Scale Model for Chemicals, Runoff, and Erosion from Agricultural Management Systems, USDA Conservation Res. Rep. 26. USDA, SEA, Washington, DC.

Prechtel, A., P. Knabner, E. Schneid, and K.U. Totsche. 2002. Simulation of carrier-facilitated transport of phenanthrene in a layered soil profile. J. Contam. Hydrol. 56:209–225.

Pusino, A., C. Gessa, and H. Kozlowsi. 1988. Catalytic hydrolysis of quinalphos on homoionic clays. Pestic. Sci. 24:1–8.

Pusino, A., W. Liu, and C. Gessa. 1992. Influence of organic matter and its clay complexes on metolachlor adsorption on soil. Pestic. Sci. 36:283–286.

Rao, P.S.C., C.A. Bellin, and M.L. Brusseau. 1993. Coupling biodegradation of organic chemicals to sorption and transport in soils and aquifers: paradigms and paradoxes. p. 1–26. In Sorption and Degradation of Pesticides and Organic Chemicals in Soil. SSSA Special Publication no. 22, Madison, WI.

Roberts, T.R. 1984. Non-extractable pesticide residues in soil and plants. Pure Appl. Chem. 56:945–956.

Sánchez-Camazano, M., and M.J. Sánchez-Martin. 1991. Hydrolysis of azinphosmethyl induced by the surface of smectites. Clays Clay Miner. 39:609–613.

Scheidegger, A., and D.L. Sparks. 1996. A critical assessment of sorption-desorption mechanisms at the soil mineral/water interface. Soil Sci. 161:813–831.

Schlebaum, W., A. Badora, G. Schraa, and W.H. Van Riemsdijk. 1998. Interactions between a hydrophobic organic chemical and natural organic matter: equilibrium and kinetic studies. Environ. Sci. Technol. 32:2273–2277.

Scow, K.M., and J. Hutson. 1992. Effect of diffusion and sorption on the kinetics of biodegradation: theoretical considerations. Soil Sci. Soc. Am. J. 56:119–127.

Shelton, D.R., and T.B. Parkin. 1991. Effect of moisture on sorption and biodegradation of carbofuran in soil. J. Agric. Food Chem. 39:2063–2068.

Sparks, D.L. 1989. Kinetics of pesticide and organic pollutant reactions. p. 128. *In* Kinetics of soil Chemical Processes. Academic Press, San Diego, CA.

Sukop, M., and C.G. Cogger. 1992. Adsorption of carbofuran, metalaxyl, and simazine: K_{oc} evaluation and relation to soil transport, J. Environ. Sci. Health 27:565–590.

Topp, E., L. Tessier, and e.g., Gregorich. 1996. Dairy manure incorporation stimulates rapid atrazine mineralization in an agricultural soil. Can. J. Soil Sci. 76:403–409.

Turin, H.J., and R.S. Bowman. 1997. Sorption behaviour and competition of bromacil, napropamide, and prometryn. J. Environ. Qual. 26:1282–1287.

Van Loosdrecht, M.C.M., J. Lyklema, W. Norde, and A.J.B. Zehnder. 1990. Influence of interfaces on microbial activity. Microbiol. Rev. 54:75–87.

Voice, T.C., C.P. Rice, and W.J. Weber. 1983. Effect of solids concentration on the sorptive partitioning of hydrophobic pollutants in aquatic systems. Environ. Sci. Technol. 17:513–518.

Walker, A., and M. Jurado-Exposito. 1998. Adsorption of isoproturon, diuron and metsulfuron-methyl in two soils at high soil-solution ratios. Weed Res. 38:229–238.

Walker, A., I.J. Turner, J.E. Cullington, and S.J. Welch. 1999. Aspects of the adsorption and degradation of isoproturon in a heavy clay soil. Soil Use Manag. 15:9–13.

Walker, S.R., and W.M. Blacklow. 1994. Adsorption and degradation of triazine herbicides in soils used for lupin production in Western Australia: laboratory studies and simulation model. Aust. J. Soil Res. 32:1189–1205.

Wauchope, R.D., and R.S. Myers. 1985. Adsorption-desorption kinetics of atrazine and linuron in freshwater-sediment aqueous slurries. J. Environ. Qual. 14:132–136.

Wauchope, R.D., S. Yeh, J.B.H.J. Linders, R. Kloskowski, K. Tanaka, B. Rubin, A. Katayama, W. Kordel, Z. Gerstl, M. Lane, and J.B. Unsworth. 2002. Pesticide soil sorption parameters: theory, measurement, uses, limitations and reliability. Pest Manag. Sci. 58:419–445.

Weidenhaupt, A., C. Arnold, S.R. Muller, S.B. Haderlein, and R.P. Schwarzenbach. 1997. Sorption of organotin biocides to mineral surfaces. Environ. Sci. Technol. 31:2603–2609.

Xing, B. 1997. The effect of the quality of soil organic matter on sorption of naphtalene. Chemosphere 35:633–642.

Xue, S.K., and H.M. Selim. 1995. Modeling adsorption-desorption kinetics of alachlor in a Typic Fragiudalf. J. Environ. Qual. 24:896–903.

Zsolnay, A. 1992. Effect of an organic fertilizer on the transport of the herbicide atrazine in soil. Chemosphere 24:663–669.

13 Methods for Measuring Soil-Surface Gas Fluxes

Philippe Rochette
Agriculture and Agri-Food Canada, Sainte-Foy,
QC, Canada

Sean M. McGinn
Agriculture and Agri-Food Canada,
Lethbridge, AB, Canada

CONTENTS

1-5667-0657-2/05/$0.00 + $1.50
© 2005 by CRC Press

13.1   INTRODUCTION

A variety of gases, some in very large quantities, are exchanged at the soil surface in terrestrial ecosystems. Measurement of the rates at which these exchanges take place can serve several goals. Gaseous emissions are often used to assess the impacts of agricultural and forest management on the atmospheric environment (e.g., ammonia, greenhouse gases, and pesticides). Also, several soil biological or chemical reactions can be more easily quantified by measuring the emission rate of their gaseous products than by monitoring the rate of change in the amounts of soil substrates (e.g., soil carbon and nitrogen) (see Chapter 15). Finally, real-time measurements of evaporation rates from agricultural fields can be used in support to irrigation scheduling (see Chapter 4).

An exchange of gas between soil and the atmosphere (Fg) occurs when the gas concentration in the air near the soil surface is greater or smaller than in the air above. The creation and maintenance of such a vertical concentration gradient implies gas production or consumption and its transfer in the soil profile at depths ranging from zero to several meters. Gas can be produced in soils by chemical reactions and biological processes. Several factors influence the rate of these processes, including the nature and the amounts of substrates, redox potential, pH, temperature, water content, and aeration.

Gases that are emitted at the soil surface arise from a variety of sources. Natural cycles of carbon and nitrogen both involve a large atmospheric pool. For example, the carbon present as CO_2 in the atmosphere is about 30% of the total carbon found in the vegetation and soil organic matter. Other carbon gases include methane and volatile organic substances that have important environmental impacts but are less abundant than CO_2. Similarly, atmospheric dinitrogen is by far the most abundant form of nitrogen on earth. Other forms of gaseous nitrogen such as ammonia, nitric oxide, and nitrous oxide impact atmospheric chemistry or greenhouse effect.

The objective of this chapter is to describe methods that can be used for measuring gas fluxes at the soil surface. The best protection against selecting an inadequate gas flux–measuring technique for a given situation remains a good understanding of the scientific basis of the available methods. Only with such knowledge will a scientist know when a combination of site characteristics and scientific objectives is suitable for a given technique. For this reason, our first objective has been to provide a detailed technical and scientific description of each technique.

Most techniques for estimating the gas exchange at the soil surface can be categorized based on using a soil mass balance, chambers, or micrometeorological theory and measurement. Within the last two categories are a variety of techniques that can be selected depending on the study requirements. For example, Harper (1988) reported on several techniques used to calculate the exchange rate of ammonia at the soil surface. In selecting an appropriate technique, consideration must be made regarding the availability of equipment and labor, sample analysis capacity, resolution of sensors over the sampling period, treatment plot size, number of treatments and replications, sampling interval and ease of operation.

13.2   SOIL MASS BALANCE APPROACH

A mass balance can be utilized to estimate gaseous emissions where there is a significant exchange of a substance at the soil surface over the sampling period, relative to the abundance within the soil profile. For example, the evaporation of water from a soil over several days may be large enough to resolve this loss as a decrease in soil water content in the soil profile if other terms in the water balance are known, e.g., surface runoff and deep percolation. The mass balance technique can also be successful to monitor gas deposition rate using methods that avoid problems associated with soil sampling, e.g., spatial variability. Here, the change in mass of the compound of interest is calculated using a prepared substrate exposed to atmosphere. For example, McGinn et al. (2003) used open Petri dishes (under a rain shelter) that contained oven-dried mixed soil to assess the dry deposition of ammonia to fields in relation to distance from a livestock facility. In such studies, unexposed substrates are used as a control treatment.

The mass balance approach is not appropriate for situations where the gas exchange rate is small over the sampling period. For example, daily soil respiration losses typically represent approximately 0.05% of total soil organic carbon. In this case, this change would be impossible to resolve owing to the spatial variability in carbon content of soil in the field. Denmead et al. (1977) sampled soil after the application of anhydrous ammonia fertilizer and found a sampling standard error approximately 15 times larger than the ammonia emission. Problems with balancing the loss of ammonia by sampling changes in the soil nitrogen were also noted by Beauchamp et al. (1978). In these cases, ammonia loss was small relative to the size of the soil nitrogen pool, such that the soil balance approach had insufficient resolution to reliably estimate ammonia loss over the sampling period. Although a soil mass balance may estimate total losses of a substance, it gives little detail on the transformations. For example, Smith et al. (1997) conducted a soil mass balance for herbicides and measured volatilised amounts over the same time period. Between 40 and 60% of the applied herbicide was unaccounted for and attributed to photochemical or biological losses that were not measured.

The major advantage of using a soil mass balance is that it is a direct measurement of the amount of substance lost or gained at the soil surface and requires minimal equipment. However, it requires intensive sampling to overcome spatial variability and cannot be used for emission/deposition rates that are too small to resolve for a given sampling protocol.

13.3   CHAMBER TECHNIQUES

Chamber techniques estimate soil-surface gas emissions by measuring the mass balance of the target species in an enclosure placed on the soil. They were first used several decades ago (Bornemann, 1920) and are still the most commonly used method in many situations. Chambers are simple and relatively inexpensive to build and have been used under a wide range of laboratory and field conditions. Their estimates of Fg are reliable when appropriate precautions are taken. However, they are an intrusive method, and the placement of any chamber types on the soil surface can modify the flux that was emitted prior to chamber deployment. In this section we will discuss how their design and operation can minimize measurement biases associated with their placement on the soil surface.

13.3.1   CHAMBER IMPACTS ON GAS FLUXES

13.3.1.1   Soil and Air Temperature and Humidity

Soil biological activity largely occurs in the top 25 cm of soil, and changes in near-surface soil temperature and water content during deployment affect gas production and Fg. Soil temperature (Ts) inside chambers depends on the net energy balance at the soil surface and how that energy is partitioned into sensible, latent, and soil heat fluxes. Chamber wall optical properties can be changed to reproduce energy fluxes inside chambers similar to the outside. However, the desirable optical properties differ depending on soil and meteorological conditions (Matthias and Peralta-Hernández, 1998), and the same chamber design cannot perform equally well under contrasting experimental conditions. Considering the difficulties in maintaining temperature and humidity constant in the chamber headspace during deployment, most chamber users opt for insulated and reflective chambers that are usually adequate to prevent large differences in both air temperature (Ta) and Ts for short deployments (≤ 1 h) (Matthias et al., 1980). Such chambers inhibit energy exchange with the surroundings and rely on the thermal inertia of the system to limit temperature variations. It is then assumed that remaining small variations in Ts and Ta have negligible effects on Fg during the short deployment period. For long deployment periods (hours to days), Matthias et al. (1980) measured large variations in Ts for several types of chamber materials and significant biases in Fg should be expected when most chambers are left in place for > 1 hour.

Changes in soil water content (Hs) result in variations in gas flux across the soil surface because of the impact of Hs on biological gas production rate (Linn and Doran, 1984) and its strong influence on gas transport rates. Soil moisture inside chambers can be modified by excluding rainfall water and reducing evaporation (Edwards, 1974). Accordingly, chambers or collars should be relocated when Hs inside the collar differs from the outside. Also, chamber measurements during rainfall can be biased by the mass flow of gases into the chamber associated with the infiltrating wetting front outside the chamber.

Changes in air temperature and humidity (Ha) inside nonvented chambers induce variations in air pressure (P) inside the chamber. The resulting expansion or contraction of chamber air pushes the air into or pulls it out of the soil and substantially alters the measured gas flux across the soil surface (Hutchinson and Livingston, 2001). The use of insulated and reflective chambers reduces both warming/cooling and evaporation rates inside the chamber but rarely eliminates them (Matthias et al., 1978; Rochette et al., 1997; Goulden and Crill, 1997). A properly designed venting tube (Hutchinson and Mosier, 1981) eliminates biases associated with P changes but results in leakage or contamination problems. Procedures for correcting Fg estimates for leakage or contamination related to temperature and humidity changes are specific to chamber types and are detailed in Rochette and Hutchinson (2004). In summary, it not recommended to use chamber designs that result in cooling during deployment, and the effects of increases in temperature can be corrected for by using the perfect gas law. Such corrections differ between chamber types but are all easily applied by expressing gas concentrations at the appropriate temperature. Changes in Ha during deployment are almost always positive, and their diluting effect on gas concentration ($[G]$) can be corrected for by expressing $[G]$ at a constant Ha. Drying all air samples prior to analysis followed by a correction for ambient Ha is a simple means to account for most of the effects of changes in Ha during deployment (Rochette and Hutchinson, 2004).

13.3.1.2   Chamber Headspace Gas Concentration

In any chamber types, Fg is rarely performed at a $[G]$ inside the chamber ($[G]_{ch}$) equal to that outside the chamber ($[G]_{amb}$). This has important implications since headspace $[G]$ influences gas vertical movement in soil. Under steady-state conditions, the rate of soil-surface gas emission or uptake is equal to its rate of production or consumption at depth, and the vertical $[G]$ profile is constant. Changes in $[G]$ above the soil surface following chamber deployment affect the gas flux into the chamber by altering the vertical $[G]$ profile (Matthias et al., 1978; Healy et al., 1996). Optimum chamber deployment strategies and flux calculation procedures for minimizing the effects of changing $[G]_{ch}$ on Fg are specific to the chamber types and will be presented in Sec. 13.3.4.

13.3.2 CHAMBER DESIGN

Stainless steel, aluminum, acrylic plastic, polyvinyl chloride, and rigid materials have been used to build chambers, and the chamber geometry can be adapted to specific experimental conditions or objectives (Rochette et al., 1997; Norman et al., 1997; Rochette and Hutchinson, 2004), as long as adequate air mixing is achieved (Hutchinson and Livingston, 2001). Most chambers should also have a venting tube to account for the effects of chamber volume changes induced by variations in Ta and Ha during deployment and at the time of chamber placement on the soil surface and headspace air sampling (Hutchinson and Livingston, 2001; Rochette and Hutchinson, 2004). There is empirical evidence that mechanical mixing and barometric P fluctuations above soil surface influence gas emissions (Kimball, 1983), and a properly designed venting tube should also transmit changes in external atmospheric P to the chamber headspace (Livingston and Hutchinson, 1995; Hutchinson and Livingston, 2001).

Molecular diffusion and thermally driven convective movements are often sufficient to ensure that air samples are representative of the mean headspace gas concentration (Matthias et al., 1980; Hutchinson et al., 2000). However, fans can be used to more rapidly achieve steady-state conditions following the deployment of flow-through chambers and to help generate air mixing intensities that match predeployment levels. Indeed, deployment of a chamber in which the turbulence intensity differs from predeployment conditions may result in transient effects on gas transfer and Fg estimates (Matthias et al., 1980; Hanson et al., 1993; Livingston and Hutchinson, 1995; Healy et al., 1996; Reicosky et al., 1997; Le Dantec et al., 1999; Hutchinson et al., 2000; Janssens et al., 2000; Rochette and Hutchinson, 2004).

Flux calculations using chamber techniques are based on the assumption that only vertical gas transport occurs in the soil-chamber system or that other gas exchange is measured. Inadequate sealing of the soil-chamber system induces biases in Fg estimates if $[G]_{ch} \neq [G]_{amb}$. Leakage or contamination of chamber air can occur through openings such as the venting tube and imperfect seals at the chamber-collar and chamber- or collar-soil joints. A properly designed venting tube allows transmission of barometric pressure fluctuations without causing significant leakage or contamination (Hutchinson and Livingston, 2001). A good seal between the collar and the chamber can be provided by water or gaskets made of rubber or closed-cell foam. Chambers or collars need to be inserted to a depth that prevents lateral diffusion of gas in response to the deformation of the vertical soil gas concentration gradient during deployment (Healy et al., 1996). The appropriate insertion depth increases with increasing deployment time and soil gas diffusivity. Model simulations suggest that a 10–15 cm insertion depth limits biases in Fg estimates to $<1\%$ for deployments ≤ 60 min in a soil having an air-filled porosity of 0.3 m^3 m^{-3} (Hutchinson and Livingston, 2001). Finally, most chamber users have solved the soil disturbance problems related to chamber insertion by using collars that are installed prior to the measurements.

13.3.3 AIR SAMPLING AND GAS CONCENTRATION ANALYSIS

Gas concentration can be measured in real time during chamber deployment or later in the laboratory on samples taken during deployment. Gas analyzers can be used for near-continuous determination of the concentration of certain gases. Differences between the absorbance of samples and a reference of known concentration have been determined using acoustic or radiation detectors, both of which are portable and adaptable to flow-through chamber systems (Rochette et al., 1991; Zibilske, 1994; Ambus and Robertson, 1998; Welles and McDermitt, 2004). When analysis is made *a posteriori* in the laboratory, air samples are usually stored in syringes or in preevacuated glass vials, depending on the anticipated delay before analysis.

There are few reports of the efficiency of containers for air sample storage. Rochette and Bertrand (2003) have shown that the N_2O concentration of air samples kept in certain types of polypropylene syringes can change rapidly with time. Consequently, it is highly recommended to test these syringes before using them for storing air samples even for periods as short as a one hour. Scott et al. (1999) recovered 99.86% of a 10 μmol N_2O mol^{-1} gas standard kept in aluminium sealed tubes after 7 days. The N_2O losses were greater from tubes made of tedlar ($\approx$3.8%), butyl rubber $\approx$8.8%), teflon ($\approx$25%), tygon ($\approx$93%), and nylon-silicon ($\approx$97%). After 14 days, Segschneider et al. (1997) reported a 9% decrease in a 1.5 μmol N_2O mol^{-1} standard stored in crimp-top 20 mL glass vials fitted with butyl rubber stoppers. When the same vials were flushed with pure hydrogen, N_2O concentration increased to approximately 0.125 μmol mol^{-1}, or a 33% contamination, after 14 days. Stoppers or septa made of other materials including ethylene propylene diene monomer (EPDM) or silicone were much less efficient than butyl rubber stoppers. Laughlin and Stevens (2003) tested glass vials (12 mL Exetainers, Labco, High Wycombe, UK) for storage of a gas mixture of known N_2O concentration and isotopic ratio. The N_2O concentration of the gas samples stored at ambient pressure (100 kPa) decreased by 30% during a period of 50 weeks while the change in ^{15}N enrichment was not detectable.

Pressurizing samples (200 kPa) into the preevacuated Exetainers using a syringe allows for travel over long distances and analysis on a gas chromatograph equipped with a headspace autosampler. Handling samples at above-ambient pressure minimizes contamination during storage and when a subsample is taken for analysis, and residual positive P after analysis can be used to detect defective vials. For maximum performance, vials should be evacuated, flushed with an inert gas (e.g., N_2 or He), and evacuated again to avoid problems associated with dilution and contamination of air samples. Rochette and Bertrand (2003) showed that double septa (rubber and silicone) twist-on caps provide an excellent seal for glass vials. Vacuum level 135 days after evacuation was 98 and 96% for vials used for the first and seventh time, respectively. The average contamination rate of N_2O samples during the first 129 days of storage was 0.20% d^{-1} with a rubber septum alone and 0.13% d^{-1} when a silicone septa was added. In comparison, nitrous oxide concentration in a gas mixture of known concentration stored in polypropylene syringes (20 mL, Becton Dickinson, Rutherford, NJ) decreased by 11 and 47% during the first 8 and 166 hours, respectively (Rochette and Bertrand, 2003). Airtight glass syringes provide a better seal but are expensive, cumbersome, and cannot be easily adapted to automated analysis. They are also subject to contamination when temperature at the time of sampling is higher than during storage or analysis. Whenever chambers are used, samples should be taken from a source of known [G] following the same procedure as for chamber air samples. These samples should be stored and analyzed in the same way as the unknown samples to assess sample handling efficiency.

Gas chromatographs are extensively used for [G] determination in air samples. They are most often used in the laboratory, but can be adapted for field operation (Christensen, 1983; Lotfield et al., 1992). General information for determination of [G] using a gas chromatograph can be found in Lodge (1988), Smith and Conen (2003), and Rochette and Hutchinson (2004). Several combinations of column, detector, and analysis conditions can be used and the resolution of the analysis can be modified by changing sample volume, detector T, etc. Gas chromatographs can be equipped with headspace autoinjectors for automation and sampling loops to keep the injected volume constant. Errors in the reported concentration of commercially available standard gases were often found to exceed supplier's claim. It is therefore highly recommended to verify secondary standard gases against primary standards prepared by qualified laboratories. Vials used to handle standard should be prepared using the same procedures as for those used to handle air samples. For some gases such as CO_2, IRGA analyzers can also be used in the laboratory for small air samples (Parkinson, 1981; Bekku et al., 1995) or in the field for flow-through chambers (Rochette et al., 1991; Norman et al., 1992). The property of some gases to react with alkali or acidic substrates has also been used to quantify gas exchange in chambers. Aqueous solutions of NaOH and KOH have been utilized for CO_2, and weak acids have been used to trap ammonia.

13.3.4 CHAMBER TYPES

A wide variety of chamber systems are used to measure Fg. In the literature, chamber types have been referred to as being open or closed, dynamic or static, or more simply as inverted boxes. Livingston and Hutchinson (1995) proposed a more rigorous classification of chamber types. According to this system, all chambers can be grouped into steady-state (SS) or non–steady-state (NSS) types depending on if Fg is calculated under constant or changing $[G]_{ch}$

and into flow-through (FT) or non–flow-through (NFT) sub-types depending on if air is circulated through the chamber or not.

13.3.4.1 Steady-State Chambers

Many problems related to chamber measurements of soil-surface gas fluxes are caused by increases or decreases of [G] inside the chamber during deployment. Steady-state (SS) chambers reduce these impacts by operating at constant $[G]_{ch}$.

13.3.4.1.1 *Flow-Through Chambers*

In flow-through (FT)-SS chambers, $[G]_{ch}$ is controlled by passing air through the chamber at a known constant flow rate (f; $m^3 s^{-1}$), and Fg ($mol\,m^{-2}\,s^{-1}$) is calculated as follows:

$$Fg = \frac{(f/A)(G_O - G_i)}{Mv} \tag{13.1}$$

where A is the soil area covered by the chamber, Mv ($m^3\,mol^{-1}$) is the molar volume of air at chamber air temperature and pressure, and Gi and Go are the [G] ($mol\,mol^{-1}$) of air entering and leaving the chamber, respectively. The performance of FT-SS chamber systems depends on accurate measurements of f and [G]. Ideally, f should be measured using electronic mass flow meters that provide continuous real-time Ta- and P-corrected data. At remote field sites where electrical power is not available, rotameters can provide adequate data if they are calibrated carefully and corrected for changing Ta and P. Gas concentration of the incoming and outgoing air can be measured using portable gas analyzers (Rayment and Jarvis, 1997) or by using chemical traps (Lockyer, 1984) (Figure 13.1).

Estimates of Fg by FT-SS chambers can be significantly biased by pressure gradients induced by the air circulation system. Kanemasu et al. (1974) first showed that drawing air into rather than pushing it through a FT-SS chamber resulted in large increases in Fg. For this reason, most recent FT-SS chamber systems are designed to maintain negligible pressure gradients between the inside and the outside of the chamber during deployment (Denmead, 1979; Rayment and Jarvis, 1997; Fang and Moncrieff, 1998).

Many problems associated with long deployment of chambers are less important in FT-SS than in NSS chambers. For example, the air flow transports excess energy and accumulated gases outside of FT-SS chambers and helps maintain [G], Ha, Ta and Ts in the headspace close to the outside conditions (Matthias et al., 1980). Designing FT-SS systems with a retractable lid that is closed only at the time of Fc measurement can also reduce the problems associated with long deployments (McGinn et al., 1998). Correcting measured [G] in FT-SS chambers for differences in Ta and Ha between

FIGURE 13.1 Wind tunnels are an example of flow-through steady-state chambers. The acrylic plastic dome is covering an area of $1\,m^2$ ($0.5 \times 2\,m$) and is connected to the fan housing. Wind tunnels are extensively used for measuring ammonia volatilization.

ingoing and outgoing air may be required if gas analyzers and flowmeters do not have built-in correction (Rochette and Hutchinson, 2004).

The smaller differences between $[G]_{ch}$ and $[G]_{amb}$ in SS and in NSS chambers reduce the errors associated with leakage through imperfect chamber seals and lateral diffusion beneath the chamber walls (Hutchinson and Livingston, 2001). However, maintaining $[G]_{ch}$ equal to $[G]_{amb}$ is not possible in FT-SS chambers when they are swept with a constant flow of ambient air. The concentration at equilibrium may be substantially lower $Fg < 0$) or higher ($Fg > 0$) than $[G]_{amb}$ (Rayment, 2000) and the time required to reach it can vary from as little as 2 minutes (Matthias et al., 1978) to as long as 60 min (Denmead, 1979). When zero air is used instead of ambient air, very accurate measurement of f and $[G]_{ch}$ are required because a small change in f can have a large effect on $[G]_{ch}$ and result in significant measurement errors (Hutchinson et al., 2000).

13.3.4.1.2 Non–Flow-Through Chambers

A near-constant $[G]_{ch}$ during deployment of a SS chamber can also be achieved without air flow through the chamber when $Fg > 0$. In a NFT-SS chamber, the emitted gas is removed from the chamber headspace by reaction with a chemical placed in a vessel inside the chamber. Such chambers are typically deployed for long periods (p; s), often 12 or 24 h, and the amount of gas that has reacted with the chemical trap ($C_{g,a}$; mol) is determined by *a posteriori* laboratory analyses. Soil-surface gas emissions (mol m^{-2} s^{-1}) in NFT-SS chambers are calculated as follows:

$$Fg = \frac{C_{g,a} - C_{g,b}}{Ap} \qquad (13.2)$$

where A is the enclosed soil area (m^2) and $C_{g,b}$ (mol) is the amount of gas absorbed by an identical (blank) trap, handled the same as other traps, except that it has been placed in a chamber deployed over a non-emitting surface.

NFT-SS chambers were first described by Bornemann (1920) and Lundegårdh (1921) to quantify soil respiration *in situ* by reaction of CO_2 with an alkali substrate. Attempts have been made to adapt this technique to various gases, but NFT-SS chambers have been and remain almost exclusively used to measure soil respiration. When compared to FT-SS and NSS systems, the NFT-SS chamber has been reported to underestimate (Ewel et al., 1987; Norman et al., 1992; Rochette et al., 1992) or overestimate (Bekku et al., 1997) CO_2 fluxes. Consequently, NFT-SS chambers are often considered inaccurate and have been largely replaced by more sophisticated chamber systems. However, more recent empirical data (Rochette et al., 1997) and simulation studies (Hutchinson and Rochette, 2003) have provided evidence that properly designed NFT-SS chambers can provide a valuable assessment of soil respiration *in situ*.

Hutchinson and Rochette (2003) and Rochette and Hutchinson (2004) recently reviewed the factors influencing the performance of NFT-SS chambers for soil respiration measurements. Two main problems were identified: large deviations in Ts inside the chamber associated with long deployment times and large differences between $[G]_{ch}$ and $[G]_{amb}$ due to an imperfect design of the chemical traps. Over the typically long deployment times of NFT-SS chambers, there are no unique chamber wall properties that would result in Ts inside chambers equal to the outside under the range of soil and meteorological conditions encountered under field conditions. Insulated and reflective chambers that minimize changes in Ts during short deployments are not efficient for longer deployments, and NFT-SS chambers will therefore experience significant deviations in Ts between the inside and the outside of the chambers. The changes in Ts are an important problem with NFT-SS chambers because of the associated changes in biological and physico-chemical equilibrium that control gas production. In opaque chambers, Ts is lower than outside values during daytime but greater at night (Sharkov, 1984). Therefore, we recommend 24 hour-deployments of opaque NFT-SS chambers to reduce the overall Ts bias in soil-surface gas flux measurements. Of course, problems associated with changes in Ts are minimal where diurnal changes in energy balance are small, such as in locations shaded from direct sunlight (under dense canopies) or in the laboratory.

Inside chambers, Ta at any given time is usually close to Ts (Matthias and Peralta-Hernández, 1998), and significant changes in Ta and Pa should be expected during long NFT-SS chamber deployments. The impact of the changes in P on Fg during deployment is difficult to account for because both warming and cooling usually occur during their typically long deployment and because Ta is not usually monitored in these chambers. Hutchinson and Rochette (2003) used model simulations to assess the impact of variations in P inside NFT-SS chambers on Fg estimates. They concluded that the net effect of short-term changes in chamber pressure on NFT-SS chamber performance should be small if the system is closed and the deployment period is long. Consequently, they concluded that NFT-SS chambers should be nonvented.

A perfectly designed NFT-SS chamber has a trap that reacts with the target gas at a rate equal to Fg, and $[G]_{ch}$ is equal to $[G]_{amb}$ during deployment. Such a perfect match between the chemical sink and the soil gas sources is rarely achieved, and nearly all NFT-SS measurements of Fg are made at $[G]_{ch} \neq [G]_{amb}$. The accumulation or depletion of gas inside the chamber represents a bias in Fg estimates because of the differences between the amounts of gas that were emitted at the soil surface and the amounts that are trapped. Traps should therefore be designed to avoid large deviations in $G]_{ch}$. Rochette and Hutchinson (2004) reviewed the literature and concluded that a well-designed alkali trap for CO_2 (NaOH) should have alkali/soil ratio of approximately 0.2 and have a total absorption capacity that is three times greater that the expected cumulated flux (Haber, 1958; Kirita, 1971; Gupta and Singh, 1977; Sharkov, 1984). In such chambers, the relative impact of a change in $[G]$ on Fg estimates decreases with increasing deployment time, and the estimated bias on Fg should be $< 5\%$ for 24 hour-deployments (Hutchinson and Rochette, 2003). Risks of leakage or contamination of chamber air are more important for NFT-SS than for other chamber types because of the typically longer deployment times. It is therefore recommended to increase the tightness of the seal by inserting the chamber walls or the frames to greater depths than for other chamber types (Rochette and Hutchinson, 2004).

13.3.4.2 Non–Steady-State Chambers

Non–steady-state chambers are the most simple and the most often used chamber technique for measuring soil-surface gas emissions (Figure 13.2). In NSS chambers, emitted gas accumulates during deployment and Fg ($mol\, m^{-2}\, s^{-1}$) is estimated using that rate of accumulation ($\partial G / \partial t$; $mol\, mol^{-1}\, s^{-1}$):

$$Fg = \frac{(\partial G / \partial t)(V_{ch}/A)(1 - e/P)}{Mv} \qquad (13.3$$

where $\partial G / \partial t$ is determined on dry air samples, V_{ch} (m^3) is the chamber volume, A is the chamber area (m^2), Mv is determined at chamber air temperature at

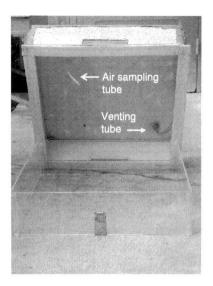

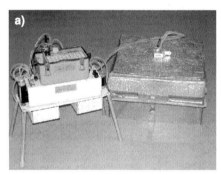

FIGURE 13.2 Non–flow-through chambers are the simplest and the most common type of chamber techniques used to measure gas fluxes at the soil surface. They can be used (a) with (flow-through) or (b) without (non–flow-through) an on-site gas analyzer.

deployment time $= 0$ (m^3 mol^{-1}), e is the partial pressure of water vapor inside chamber at deployment time $= 0$ (kPa), and P is barometric pressure (kPa).

Changes in $[G]_{ch}$ during deployment impact Fg by decreasing or increasing the gas concentration gradient between the soil and the chamber headspace. As a result, Fg at any given time after the placement of the chamber on the soil surface is an underestimate of the flux that was occurring before the chamber was installed. Furthermore, the underestimation becomes larger with increasing deployment time as the depth at which soil concentration in G is being affected increases (Makarov, 1959; Naganawa and Kyuma, 1991; Healy et al., 1996; Rochette et al., 1997; Rayment, 2000). Accurate estimates of

Fg using NFT chambers require projecting a $\partial G/\partial t$ representative of the predeployment soil [G] profile using the measured values of $[G]_{ch}$ during deployment.

The most common approach to minimize the impact of changing [G] on *Fg* within a NSS chamber is to estimate $\partial G/\partial t$ as early as possible during deployment. A simple mathematical model is used to describe the temporal changes in $[G]_{ch}$, and $\partial G/\partial t$ is estimated from the slope of that curve extrapolated to the moment of chamber deployment. Diffusion theory predicts that the rate of increase in $[G]_{ch}$ decreases with time after deployment. For this reason, quadratic, cubic, or exponential models usually yield better estimates of $\partial G/\partial t$ than linear models (Hutchinson and Mosier, 1981; Anthony et al., 1995; Pedersen, 2000).

The changing $[G]_{ch}$ during deployment makes NFT chambers more sensitive than SS chambers to errors related to leakage. Therefore, NFT chamber performance increases with decreasing soil gas diffusivity (Livingston and Hutchinson, 1995; Hutchinson and Livingston, 2002). Measured values of $[G]_{ch}$ must be corrected to account for the effect of *Ta* and *Ha* variations during deployment. Detailed description of the procedures specific to FT- and NFT-NSS chambers were presented by Rochette and Hutchinson (2004).

13.3.4.2.1   Non–Flow-Through Chambers

Among all chamber types, the NFT-NSS chamber is the most commonly used because of its low cost, simplicity, and adaptability to remote sites. Consequently, it has been used to measure the flux of several gases under a wide range of conditions (Lundegårdh, 1926; Makarov, 1959; Gupta and Singh, 1977; Matthias et al., 1980; Hutchinson and Mosier, 1981; Parkinson, 1981; Crill, 1991; Striegl et al., 1992). Estimates of $\partial G/\partial t$ in NFT-NSS chambers using a nonlinear model is made difficult by the limited number of air samples than can be taken during deployment. Nonlinear models require at least three, but preferably four or more samples, and sampling strategy should provide adequate description of the rapid changes in $\partial G/\partial t$ shortly after deployment (Healy et al., 1996).

There is no advantage to deploying NFT-NSS chambers longer than the period required to adequately describe $\partial G/\partial t$. This optimum duration decreases with increasing *Fg*, precision of gas analysis, headspace air mixing intensity, and decreasing volume/area ratio. Chamber users often opt for relatively long deployment for various reasons. However, the advantages of longer deployment periods must be weighed against the risk of increasing the underestimation of *Fg* (Healy et al., 1996). For example, Rochette (unpublished data) found that reducing the deployment duration of NFT-NSS chambers (with four air samples) from 60 to 24 minutes greatly improved the agreement between CO_2 flux estimates obtained by NFT- and FT-NSS chambers.

13.3.4.2.2   Flow-Through Chambers

The only difference between FT- and NFT-NSS chambers is that $[G]_{ch}$ inside the FT-NSS chamber is monitored by an on-site gas analyzer. The continuous monitoring of $[G]_{ch}$ during deployment of a FT-NSS chamber results in several advantages, including the early detection of experimental problems, a better description of the temporal pattern of $[G]_{ch}$, the determination of $\partial G / \partial t$ over shorter deployments, smaller gas leakage and changes in air and soil temperature and humidity, and the possibility of near-continuous monitoring of *Fg* over long periods (Lotfield et al., 1992; Goulden and Crill, 1997; Ambus and Robertson, 1998). However, the advantages of shorter deployment periods come at a price: FT-NSS chambers rarely permit simultaneous measurement of several gases and are more susceptible to be affected by soil disturbance, changes in the air mixing regime at the soil surface, and pressure effects.

13.3.5   STRENGTHS AND WEAKNESSES OF CHAMBER TECHNIQUES

The choice of the most appropriate technique for measuring soil-surface gas exchange is influenced by several factors, including scientific objectives, site characteristics, and budget. Chambers are not well suited for situations where spatial and temporal integration of gas fluxes are required. Their deployment on the soil usually influences the flux being measured. However, they can provide reliable measurements when adequate procedures are followed. Chambers can measure small fluxes, sample small-scale spatial variability, discriminate between several sources within an ecosystem, be used at remote locations, and be operated at relatively low cost. Also, because of their small dimension they can be used by soil and crop scientists in typical agronomical plots. Furthermore, chambers require minimal expertise, and even an inexperienced operator can make high-quality measurements with relatively little training (Table 13.1).

13.4.   MASS EXCHANGE USING MICROMETEOROLOGICAL TECHNIQUES

Micrometeorological techniques employ a combination of atmospheric turbulence theory and measurement to estimate gas flux to or from a surface. Ideally, these techniques require emission or deposition rates that are uniform across the surface of interest. In addition, the sampled air must be characteristic of the surface flux of the surface in question. To achieve this condition, measurements are made within the adapted layer (Reitsma, 1978), also known as the internal boundary layer (Rosenberg et al., 1983). As air passes over a newly encountered uniform surface, the properties of the airflow in the layer adjacent to the surface adjust to that particular surface. The height of the adapted layer increases with distance from the leading edge of the homogeneous surface (fetch). The relationship between fetch and height of the adaptive layer depends on the mixing capacity of air, referred

TABLE 13.1
Chamber Technique Characteristics

Chamber type	Limitations	Strengths
Flow-through steady-state	Requires on-site gas analyzer Requires electrical power Usually one gas at a time Low sensitivity	Low feedback effects on Fg Can be automated Possibility for continuous measurements of Fg Possibility to control temperature and humidity
Non–flow-through steady-state	Usually one gas at a time Cannot be automated Potentially high feedback effects on Fg	Very high sensitivity Temporal integration of Fg No electrical power needed Low costs
Flow-through non–steady-state	Requires on-site gas analyzer Requires electrical power Usually one gas at a time Low sensitivity	Low feedback effects on Fg Low costs Requires short deployment Can be automated
Non–flow-through non–steady-state	Cannot be easily automated High feedback effects on Fg	High sensitivity Several gases simultaneously Low costs No electrical power needed

to as the atmospheric stability (Leclerc and Thurtell, 1990). In general, the fetch requirement is between 50–100 times the upper height of the measurements.

13.4.1. Aerodynamic Technique

One of the first field studies to employ micrometeorological theory and measurement was conducted by Thornthwaite and Holtzman (1939), who used an aerodynamic technique to measure evapotranspiration over grass. Their work was based on a collection of publications on fluid dynamics in the 1930s by Kármán, Prandtl, Rossby, and Sverdrup.

The aerodynamic technique as employed by Thornthwaite and Holtzman (1939) required measurements of gas concentration and wind speed at two heights. It is now understood that accuracy is greatly improved using several measurement heights. In the original paper by Thornthwaite and Holtzman (1939), no correction was made for atmospheric stability to account for variations in the mixing of the air above the surface, although they did recognize its importance. Today, an atmospheric stability correction is a necessary component in using this technique.

The aerodynamic technique is highly dependent on accurate measurements of the wind speed and concentration with height. As such, care is needed when selecting a set of anemometers to ensure sensitivity at low wind speeds where anemometer characteristics must be well matched, e.g., the start-up threshold.

This technique uses the logarithmic wind speed profile (to estimate the friction velocity), $\mu*$ (m s^{-1}):

$$\mu = \frac{\mu*}{k}\left[ln\left(\frac{z-d}{z_0}\right) - \psi_m\right] \tag{13.4}$$

where μ is the wind speed (m s^{-1}), k is von Karman's constant (0.41), z is the measurement height (m), d is the zero plane displacement height (m), z_0 is aerodynamic roughness (m), and Ψ_m is a profile correction for atmospheric stability (subscript refers to momentum). Estimating $\mu*$ is done using a least-squares fit of the regress of μ against (ln $((z-d)/z_0) - \Psi_m$), where the slope of the line is $\mu*/k$. The value of d and z_0 can be estimated from the characteristics of the surface. For a smooth soil surface, d is generally set to zero since it is often very small relative to the measurement height z. However, Abtew et al. (1989) did estimate d for uniform sized clods mixed with soil (half exposed) as 0.35 times the clod diameter. The aerodynamic roughness z_0 (also called the momentum roughness or roughness length) is often determined as the y intercept on the least-squares regression of ln(z) or ln$(z-d)$ against wind speed under neutral atmospheric stability $\Psi_m \approx 0$). It has also been estimated from the diameter of the surface clods (for soil surfaces) or 0.5 times the soil ridge height assuming perpendicular wind direction (Abtew et al., 1989). For crop surfaces Campbell (1977) estimated d and z_0 as 0.64 and 0.13 times the crop height, respectively. A method for estimating Ψ_m is reported by Paulson (1970). This term is estimated using the Monin-Obukhov stability parameter (z/L), a measure of the buoyant (heating) to mechanical (wind) produced turbulence. The value of the Monin-Obukhov length (L) is calculated from measurements made using an ultrasonic anemometer/thermometer sensor that estimates friction velocity and sensible heat flux. It can also be estimated from vertical profile measurements of wind speed and air temperature, employing a relationship between z/L and another stability parameter, the Richardson number (Saugier and Ripley, 1978; Sutton et al., 1992).

The second step is to estimate concentration scale (also known as the friction concentration), $c*$ (g m^{-3}), using the following equation:

$$c_a = \frac{c*}{k}\left[\ln\left(\frac{z-d}{z_g}\right) - \psi_g\right] \tag{13.5}$$

where c is the gas concentration (g m^{-3}) at $z-d$, and z_g is the height (m) where the concentration is zero. As with $\mu*$, the concentration scale $c*$ is estimated as the slope of c against (ln $((z-d)/z_g) - \Psi_g$) that gives the value of $c*/k$. The final step is to calculate the flux density (g m^{-2} s^{-1}) as the product of μ* and c*. Generally, the flux density is determined using a 20- to 30-minute averaging period.

One advantage of the aerodynamic technique is its flexibility. It has been used to estimate heat, water vapor, and carbon dioxide above native grassland (Saugier and Ripley, 1978) and to measure ammonia exchange over a pasture (Harper et al., 1983), over moorland (Sutton et al., 1992), and over a spruce stand (Andersen et al., 1993). This technique is also capable of estimating volatile herbicides (Smith et al., 1997) and pesticide emissions, e.g., methyl bromide (Yates et al., 1996). Saugier and Ripley (1978) recommended that the aerodynamic technique be used for relatively smooth surfaces in the daytime or during the night when atmospheric stability is weak to moderate. One major disadvantage in using the aerodynamic technique occurs when wind speeds are low. This condition leads to inaccurate estimates of wind speed gradients and therefore large errors in the flux density.

13.4.2   Bowen Ratio–Energy Balance Technique

The Bowen ratio–energy balance technique, like the aerodynamic technique, requires large plot treatments to ensure that the air at the measurement heights is characteristic of the surface flux (see Chapter 4, Sec. 4.3.2.2). However, unlike the aerodynamic technique, there is no requirement for a wind speed profile that is difficult to measure accurately under low wind speed conditions.

The Bowen ratio–energy balance technique estimates fluxes from measurements of the gradient concentrations (two heights) of temperature, water vapor, and the gas of interest (Figure 13.3). The first step is to examine the surface energy balance (Tanner, 1960) and eliminate the terms that are insignificant given the surface and fetch conditions and sampling duration. For a large homogenous bare soil surface, the energy balance [Eq. (13.6)] includes radiant energy lost or gained at the soil surface (net radiation flux density, Rn; $W m^{-2}$), energy used to evaporate water from the soil surface (latent heat flux density, LE; $W m^{-2}$), and energy at the soil surface that is exchanged with the subsoil (soil heat flux density, G; $W m^{-2}$ or atmosphere (sensible heat flux density, H; $W m^{-2}$). Terms referring to storage of energy by vegetation cover and horizontal/vertical energy divergence are assumed to be zero. Where the measurements are made close to the leading edge of a transition in surfaces, e.g., irrigated/nonirrigated surfaces, the value of the divergence terms can be more significant.

$$R_n - G = LE + H$$
(13.6

Over a 24-hour period, the net value of G can be small and therefore is often ignored. However, G is required for flux estimates made within 24 hours. Both Rn and G are usually measured directly, using a net radiometer and buried soil heat flux plates (in conjunction with soil temperature near the surface), respectively. Since exposing the soil heat flux plates to direct sunlight would bias the measurement of soil heat flux, the plates are buried well

FIGURE 13.3 A Bowen ratio–energy balance field unit monitoring air temperature, water vapor, and carbon dioxide gradients (at two levels shown) that are used to determine the exchange of heat, water, and carbon dioxide between the surface and atmosphere.

below the surface, e.g., at 5 cm depth. The soil temperature change over the sampling duration in the 0–5 cm layer is used to determine heat storage. This in turn requires knowledge of the specific heat capacity and hence the soil moisture content. The soil surface heat flux is the difference between the stored heat (0–5 cm) and heat flux measured by the plates (at 5 cm). The second step [Equation (13.7)] is to estimate the ratio H/LE (β); (Bowen, 1926) from the psychrometric constant (γ); (Pa °C^{-1}), ratio of eddy diffusivities for heat and water vapor (K_h/K_v), gradient measurements of potential temperature ($\triangle\theta/\triangle z$; °C m^{-1}), and water vapor pressure ($\triangle e/\triangle z$; Pa m^{-1}). The gradient measurements are made at the same heights above the surface. The derivations of H and LE come from the flux gradient relationships, where $H = (\rho C_p K_h \triangle\theta/\triangle z)$, $LE = (L_v \rho K_v \varepsilon P \triangle e/\triangle z)$, ρ is the air density (g m^{-3}), C_p is the specific heat of air (J g^{-1} C^{-1}), L_v is the latent heat of vaporization of water (J g^{-1}), ε is the ratio of the molecular weight of water to air (0.622), and P is the atmospheric pressure (Pa).

$$\beta = \frac{H}{LE} = \gamma = \frac{K_h}{K_v}\frac{\triangle\theta}{\triangle e} \tag{13.7}$$

In calculating β, the assumption is made that the ratio of eddy diffusivities (also called the turbulent transfer coefficients) is equal to one. The assumption of equal eddy diffusivities between conservative scalars arises from similarity theory. There are some practical situations where the assumption of similarity of eddy diffusivities is problematic. For example, Rosenberg et al. (1983) reported that local advection conditions may cause fluxes to be underestimated since eddy diffusivities are not similar in this situation.

The third step [Eq. (13.8)] is to solve for LE by rearranging the energy balance equation where $LE + H$ is rewritten as $LE(1 + \beta)$:

$$LE = \frac{Rn - G}{1 + \beta} \qquad (13.8$$

Once Rn, G, and LE are known, H is estimated as the residual term in the energy balance equation ($H = Rn - G - LE$). This technique is widely used to determine CO_2 flux, but application to other gases is possible where concentration gradients can be determined over a few hours. In this procedure, the eddy diffusivity of the gas of interest and that of heat are equal such that the ratio of gradient measurements is a function of the ratio of flux densities:

$$Fg = bH\frac{\Delta c}{\Delta \theta} \qquad (13.9$$

In deriving the flux density in this manner, care is needed to ensure that the correct conversion constant (b) is used. For CO_2 gradients measured in ppm and Fg expressed as $g\,m^{-2}h^{-1}$, the value of b is 0.005408 as determined from $(Mw_{CO_2}/Mw_{AIR})\,10^{-6}\,C_p^{-1}\,3600\}$ where Mw is the molecular weight, 10^{-6} is the conversion for ppm, C_p is the specific heat of air ($J\,g^{-1}\,°C^{-1}$), and 3600 is the conversion for seconds in an hour.

The major advantage of the Bowen ratio–energy balance technique is that the instrumentation is relatively simple and fluxes can be averaged every 20–30 minutes to give the diurnal flux pattern. Care must be taken to ensure that gradient measurements contain no instrument bias, especially when gradients are small. For gas concentration gradients, this is commonly accomplished by using a single detector (e.g., a flow through gas analyzer) that is switched between sampling heights and assuming instrument drift in minimal. Equally effective is the use of two detectors that allows continuous gradient measurements, where the detectors are switched between sampling heights. The gradient gas concentration must also be corrected for density differences due to the difference in water vapor density between the two heights (Webb et al., 1980). Alternately, the water vapor in the air stream can be removed prior to being analyzed for gas concentration provided the desiccant placed in the air stream does not interact with the gas of interest. For example, for concentration measurements of CO_2 the desiccant that has been used is magnesium perchlorate (McGinn and Akinremi, 2001). Some of the newer infrared gas analyzers, which monitor water vapor pressure and CO_2 concentration simultaneously, allow autocorrection of CO_2 concentration. Sinclair et al. (1975) reported that the error in determining CO_2 flux ranged from 15% to more than 40% when sensible heat flux was small.

A common disadvantage of this technique, aside from the instrumentation requirement and large plot size, is the assumption of similarity in transfer coefficients. As well, during times in the day when β approaches -1 (when $LE = H$) and when $Rn + G$ should also approach zero, the calculation of LE is more susceptible to measurement errors in either the denominator or numerator terms in Eq. (13.8). One solution has been to identify when β, over the 20- to 30-minute averaging period, is between -0.8 and -1.2 and to calculate LE from adjacent time series values (McGinn and Akinremi, 2001).

13.4.3   EDDY COVARIANCE TECHNIQUE

The eddy covariance (also known as the eddy correlation or eddy flux) technique is based on the concept that the transfer of a gas from or to a surface is predominantly governed by the movement of small parcels of air, known as eddies. In determining the vertical flux from a surface, an understanding of the vertical movement of a gas in eddies can be characterized as the product of the fluctuations in gas concentration, c', and vertical wind speed, w' which is averaged over a short period:

$$Fg = \overline{c'w'} \qquad (13.10)$$

The protocol for the eddy covariance technique is to measure an average (20–30 min) of the instantaneous cross products of c' and w'. The determination of w' is usually made using an ultrasonic anemometer. Typically, a fast-response (fine junction) thermocouple is mounted close to the wind speed measurement to allow determination of H, i.e., $H = \rho C_p \overline{w'T'}$ where T' is the fluctuation in air temperature (°C). Commercially available fast-response open-path sensors for water vapor and CO_2 can be mounted in the vicinity of the ultrasonic anemometer to calculate these fluxes, e.g., $LE = \overline{w'\rho_v'}$ where ρ_v' is the water vapor density (g m^{-3}).

The advantage of this technique is that it is a direct measurement in that it requires no assumption about the characteristics of turbulence transfer. This necessitates the use of fast-response sensors and recorders in the order of 10–20 Hz (10–20 samples per second), depending on the measurement height (Kaimal and Finnigan, 1994). It has been used mainly for determining fluxes of heat, water vapour, and carbon dioxide because of the availability of fast-response sensors for these gases. Civerolo and Dickerson (1998) measured nitric oxide emissions using eddy covariance above tilled and untilled cornfields.

A major disadvantage of this technique is that the determination of c for many trace gases can be difficult to obtain from analyzers at a fast enough frequency to characterize eddy transfer. The use of closed-path analyzers (where detection is made inside a flow-through sample tube) is possible where the time response of the analyzer is sufficient for use with eddy covariance. Where such an analyzer is used to measure c', there must be a realignment of c time series with the w' time series prior to doing the cross product to compensate for the time it takes for the gas to travel to (via tubing) and through the analyzer relative to the more instantaneous w' measurement made using a ultrasonic anemometer. For some gases, close-path analyzers may also be a poor choice if filtering of high-frequency detection exists due to absorption characteristics, e.g., the high absorption affinity of ammonia by many different materials. Aside from instrument limitations, other factors that impact eddy covariance flux calculations include nature of airflow at the site and assumptions regarding the mean vertical wind speed equaling zero, as well as limitations during flux measurements at night (Massman and Lee, 2002).

13.4.4 RELAXED EDDY ACCUMULATION TECHNIQUE

The relaxed eddy accumulation technique (also called condition sampling) (Businger and Oncley, 1990; Pattey et al., 1993; Baker, 2000) is an attractive modification of the eddy accumulation (EA) technique. This latter technique, first discussed by Desjardins (1972), operates by accumulating air into two samples, the upward and downward air that is based on the measurement of w' The rate at which the air is sampled is proportional to the magnitude of w' The eddy accumulation technique eliminated the need for high-resolution and fast-response detection of trace gases used in the eddy covariance technique. The more recent relaxed eddy accumulation technique is a simplification of the eddy accumulation technique where air is sampled at the same rate for both upward and downward moving eddies, regardless of the magnitude of w'.

In the relaxed eddy accumulation technique, the upward and downward air samples are later analyzed for the difference in concentration ($\triangle c$; $g\,m^{-3}$). The standard deviation of the vertical wind speed (σ_w; $m\,s^{-1}$) during the sampling period is measured using an ultrasonic anemometer. The flux density [Eq. (13.11)] is the product of these measurements and an empirical coefficient (β') that can vary between 0.4 and 0.63 (Milne et al., 1999). Katul et al. (1996) report the value of β' for several studies. They concluded that a β' value of 0.58 gave the best fit against eddy covariance flux data:

$$Fg = \beta'\triangle c\sigma_w \qquad (13.11$$

This approach has been used for gases such as carbon dioxide (Pattey et al., 1993), volatilized pesticides (Majewski, 1999), and ozone and water vapor (Katul et al., 1996). The principal advantage of the technique is that it allows for analysis of concentration using high-resolution laboratory methods, e.g., gas chromatography, and as such is applicable to a wide variety of trace gases. However, it also demands a specialized instrumentation package that is not commercially available at this time.

13.4.5 COMBINED TECHNIQUES

The use of energy balance and aerodynamic techniques has given way to combination models utilized for estimating potential evaporation from a surface. A well-known example is the Penman equation (Penman, 1948) used to determine potential evaporation from a wet surface. Further modifications have been developed for water-limiting conditions of the surface. In general, the water loss is determined as a residual of the energy terms [Equation (13.6)], where G is ignored and H is estimated using assumptions based on the surface characteristics.

Tanner (1960) suggested that the profile (aerodynamic) or eddy fluctuation (covariance) technique could be used to estimate H directly and then used in combination with the energy balance to estimate LE. Itier et al. (1985) used a simplified aerodynamic technique to estimate H using measured wind speed and air temperature at only two heights. The determination of H is based on formulation made using four boundary layer stability classes. The sensible heat flux can then be used to estimate LE from a surface as a residual in the energy balance equation where Rn and G are measured.

Gases other than water vapor could be estimated once H is known and where gradient temperature and gas measurements are measured. Combining the eddy covariance technique to determine eddy diffusivities along with the flux gradient relations has been used for nitrous oxide (Wagner-Riddle et al., 1996; Laville et al., 1999), methane (Miyata et al., 2000), and total hydrocarbon emissions (Ausma et al., 2001).

The advantage of this technique is that additional measurements of the energy balance can be useful in understanding the surface processes at the study site. Some of the disadvantages include the technically demanding instrumentation, large plot size, and the accumulation of error in the residual term or the assumption of similarity in eddy diffusivities.

13.4.6 INTEGRATED HORIZONTAL FLUX TECHNIQUE

The integrated horizontal flux technique is essentially a simplified mass accounting of a gas emitted from a well-defined surface area into the air above that surface. It requires the surface to be a significant source of the gas of interest relative to the upwind surface. The technique can be applied to smaller plots than those used in the aerodynamic, Bowen ratio–energy balance, or eddy covariance techniques. As such, there is greater opportunity in using the integrated horizontal flux technique for simultaneous replication and treatment sampling.

The premise of this technique is that the flux density from a plot can be resolved from wind speed and concentration measurements in two planes: the axis along the wind direction (fetch) and the height (profile). Where the wind speed profile is the same for the upwind and downwind perimeter of the plot, as is the case for horizontally uniform surface, only one wind speed profile measurement is needed. Generally, where there is no significant source or sink upwind of the plot for the gas in question, the concentration profile can be characterized at one upwind height. In addition, the top height of the downwind measurements must be higher than the dispersion plume from the treated plot or be high enough to allow extrapolation of the concentration measurements to this height (where upwind concentrations exist).

The amount of gas entering through the upwind or downwind perimeter of the plot, at a specific location and height, is the product of wind speed and gas concentration (known as the horizontal flux). The vertical flux density from the plot is estimated from the vertical integration (z_1 to z_x) of the difference between the upwind (up) and downwind (dw) horizontal fluxes $(u\,c)_{dw} - (u\,c)_{up}$ and the distance the air travels across the plot (x) [Eq. (13.12)]. For noncircular plots, x must be determined from the mean vector wind direction. It is assumed in Eq. (13.12) that wind speed does not change along distance x (for any given height), but where homogenous surface conditions do not exist, both the upwind and downwind wind speeds must be used:

$$Fg = \frac{1}{x} \int_{z_1}^{z_x} \bar{u}\left(\bar{c}_{dw} - \bar{c}_{up}\right)dz \qquad (13.12$$

In addition, when c_{up} is not significantly changed with height, i.e., no significant source or sink is in the upwind vicinity, only a single measurement of upwind gas concentration is required. Leoning et al. (1985) reported that the cross product of the averages of u and c (as opposed to the instantaneous cross products) caused an overestimation in emission (Eq. 13.12) of 15%.

The integrated horizontal flux technique allows for measuring gas concentration by near instantaneous detection of concentration (e.g., using laser or infrared analyzer) or by accumulating gaseous samples and analyzing in the laboratory for an average concentration over the period. For example, samplers that accumulate ammonia include simple acidic mediums (e.g., phosphoric acid) (Bussink, 1994) or denuders such as the Ferm tube (Ferm, 1991) or Leuning sampler (Leuning et al., 1985). For passively ventilated samplers (vented by the wind) with fixed orientation, the wind speed component parallel to the axis of the sampler is used.

This method is commonly used for small circular plots where measurements are located in the centre and x is constant (radius). Wilson and Shum (1992) indicated that the accuracy of this method deteriorates with decreasing plot radius and recommended that a radius of 20 m or more be used. Where measurements are made on the downwind perimeter of the plot, the adjacent edges of the plot delineate the effective wind directions that can be used.

As an alternative approach, Ryden and McNeill (1984) calculated the mean horizontal flux after regressing u and c against $\ln(z)$:

$$F_g = \frac{1}{x}\Big[AD \{z(\ln z)^2 - 2z\ln z + 2z\} + (BD - AE)\{z(\ln z - 1)\}$$
$$+ EBz - \bar{c}_b D\{z(\ln z - 1)\} - \bar{c}_b Ez \Big]_{z_1}^{z_2} \qquad (13.3$$

Since the vertical profile of gas changes with atmospheric stability, it is necessary to sample at five or more heights to correctly define the profile (e.g., Harper et al., 1983).

A major advantage of the integrated horizontal flux is that it is a relatively straightforward balance, requiring only concentration and wind speed profile. There is no requirement for knowing atmospheric stability. One disadvantage relates to the measurement at low wind speeds where many of the commonly used cup-type anemometers become unreliable below 1 m/s. This technique also requires a well-defined surface source.

13.4.7 MASS DIFFERENCE TECHNIQUE

The mass difference technique (Denmead et al., 1998) is similar to the integrated horizontal flux technique theory in that it integrates perimeter measurements made at different heights. However, instead of a single point measurement (at specific heights) along the upwind and downwind perimeter, the mass difference technique uses a line average measurement of the concentration along each perimeter of a square plot at each height. In this configuration, the flux density [Eq. (13.14)] is calculated from the distance along the perimeter (X; m), average concentration (c_1, c_2, c_3, c_4 subscripts refer to each whole perimeter concentrations, where 2 and 4 are opposite sides; $g\,m^{-3}$) (Figure 13.4) measured along the perimeter, and the perpendicular wind speed component to the perimeters (U and V):

$$Fg = X \int_{z_1}^{z_x} \left[\overline{U}(\overline{c_4} - \overline{c_2}) + \overline{V}(\overline{c_3} - \overline{c_1}) \right] dz \qquad (13.14)$$

One advantage of the mass difference technique is its flexibility, e.g., it is reported to be sensitive enough for measuring fluxes of carbon dioxide, methane, and nitrous oxide for various scenarios (Denmead et al., 1998). The technique has also been useful for quantifying methane flux from cattle (Harper et al., 1999). Disadvantages of the technique are the need for accurate wind speed profiles and the specialized setup for monitoring line concentration along the perimeters at different heights.

13.4.8 THEORETICAL PROFILE SHAPE TECHNIQUE

The trajectory simulation model of Wilson et al. (1982) is used to relate surface fluxes from a small plot to concentration and wind speed (the product being the horizontal flux) in the dispersion plume above the plot where surface roughness and atmospheric stability is known. A simplification of the trajectory simulation model is the theoretical profile shape technique (Wilson et al., 1983).

In the theoretical profile shape technique, the trajectory simulation is used to identify a height above a surface source, denoted as ZINST, where the

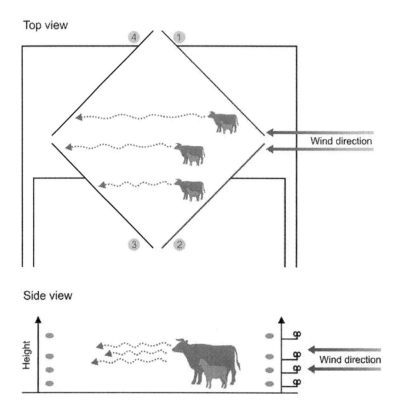

FIGURE 13.4 Schematic of setup to monitor gas emissions using the mass difference technique. Gas concentrations correspond to perimeters 1 to 4. (Adapted from Harper et al. (1999).)

vertical flux is proportional to the normalized horizontal flux regardless of atmospheric stability [Eq. (13.15)]. In this manner, for a given surface condition and ZINST, the measurements needed to estimate flux density are simply the concentration and wind speed measurement made at height ZINST:

$$Fg = \frac{u\,c}{\overline{uc}/F_0} \qquad (13.15$$

where the denominator $\overline{uc}/F_0$ is a constant simulated based on a given plot size and roughness length. In practice, this approach is used where the measurements are located in the center of a circular plot.

The major advantage of using this technique is the minimum amount of sampling locations, i.e., only one concentration and one wind speed measurement. In addition, it can be applied to a variety of emissions: for example, the technique has been used to study the emission of additives to a surface such as pesticides (Rice et al., 2002) and ammonia from organic fertilizer application (Gordon et al., 2001). However, the theoretical profile shape technique also relies on knowing the height of ZINST before conducting the measurements. These values have been reported for circular plots of 20 and 50 m radius (Wilson et al., 1982).

13.4.9 BACKWARD LAGRANGIAN STOCHASTIC TECHNIQUE

Another variation of the trajectory model of Wilson et al. (1982), was introduced by Flesch et al. (1995), is known as the backward Lagrangian stochastic technique. The simulation model allows a user to determine the backward trajectories of a gas, starting at the sensor and moving to the surface source (Figure 13.5). In this technique, the denominator in Eq. (13.15) is calculated from a simulation of gas trajectories between the sampling location in the dispersion plume and the emitting surface. Unlike the theoretical profile shape technique, however, the measurements can be made anywhere in the dispersion plume (not just at ZINST in a circular plot). The backward Lagrangian stochastic technique requires knowledge of the sensor height, geometry of the sensor/source, surface roughness, and atmospheric stability. Ideally, stability is estimated by calculating the Monin-Obukhov stability length, but, the commercially available software for the

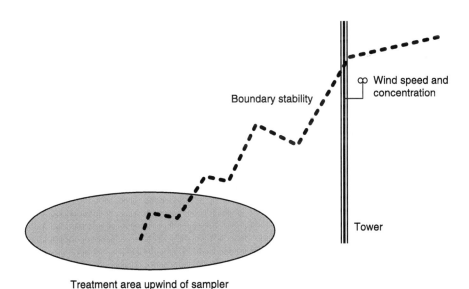

Treatment area upwind of sampler

FIGURE 13.5 Configuration of the backward Lagrangian stochastic technique showing upwind source and measurement of atmospheric stability, target gas concentration, and wind speed within the dispersion plume. Dashed line is a simulated trajectory through sample point.

technique (WINDTRAX; Thunder Beach Scientific, Halifax, Canada) allows for several options for atmospheric stability. The backward Lagrangian stochastic technique is perhaps the most flexible micrometeorological technique available, as it allows for flux estimates using remote measurements downwind of different shaped treatment surfaces of different shapes. One of the limiting features of this technique is the need for homogeneous surface conditions. Flesch et al. (1995) showed that volatile chemical fluxes calculated with the backward Lagrangian stochastic technique compared to within 10% (bias error) and 20% (random error) of that calculated with the mass balance technique.

The major advantage of this technique is it can be applied to a wide variety of sources by measuring some distance downwind of the source (within the plume) to minimize turbulent effects at the source. Both line average and single point concentration can be used. Commercial software is also available that makes the modeling of dispersion very easy. One disadvantage is that, although simple alternatives to characterizing stability within commercial software exist, the most reliable means is to measure it. In addition, it requires an isolated source of the emission.

13.4.10. STRENGTHS AND WEAKNESSES OF MICROMETEOROLOGICAL TECHNIQUES

In comparison to chamber techniques, the general advantages of micrometeorological techniques are that they are non-intrusive and can provide spatial averaging that negates the need for many sampling locations (Table 13.2). In addition, a few of these techniques allow near continuous flux estimates, while others allow temporal averaging through sampling accumulation. Complete instrumentation packages are available commercially for several of the micrometeorological techniques, e.g., the Bowen ratio–energy balance and eddy covariance for some gases. Micrometeorological techniques can also accommodate large plots (e.g., aerodynamic, Bowen ratio–energy balance, eddy covariance, eddy relaxation techniques) as well as estimating fluxes from small plots (e.g., mass difference, integrated horizontal flux, theoretical profile shape, dispersion modeling techniques).

One disadvantage of micrometeorological techniques, relative to some chamber approaches, is that many of these techniques rely on an understanding of micrometeorological instrumentation and processes. The number of treatments and replication is generally less for micrometeorological techniques owing to larger plot size requirements and equipment costs.

13.5. RECOMMENDATIONS AND FUTURE RESEARCH

Selection of the most appropriate technique to measure soil-surface gas exchange requires an understanding of both the principles of operation of the techniques and the theory of gas transfer from soil to the atmosphere. The selection is usually confounded further by the availability of time,

TABLE 13.2
Characteristics of Micrometeorological Techniques

Technique	Stability	Number of sample points	Detection response type	Plot size	Major limitations
Aerodynamic	Yes	5	Slow	Large	Large error at low wind speed, e.g., calm night
Bowen ratio–energy balance	No	2	Slow	Large	Assumption of similarity in transfer coefficients
Eddy covariance	No	1	Fast	Large	Specialized gas detectors
Integrated horizontal flux	No	5 downwind plus upwind	Slow or accumulation	Small or large	Isolated source
Eddy accumulation	No	1 intake and 2 accumulations	Accumulation	Large	Fast response valves and variable pump flow
Relaxed eddy accumulation	No	1 intake and 2 accumulations	Accumulation	Large	Fast response valve system
Theoretical shape profile	No	1	Slow or accumulation	Small	Prior knowledge of ZINST
Trajectory modeling	Yes	1	Slow or accumulation	Small or large	Indirect measurement, isolated source, uniform air flow

labor and equipment, measurement sites restrictions, and experimental objectives. As a result, no single technique is ideal for all situations. However, we recommend whenever possible to use more than one technique to increase confidence in the flux estimates. For example, measurements made with the more traditional micrometeorological techniques, e.g., Bowen ratio, could be used to also generate fluxes using a dispersion model technique with a few extra measurements.

Chamber estimates of Fg are reliable when appropriate principles of operation are used. However, the intrusive nature of chamber techniques makes them vulnerable to measurement biases. The accuracy of the various chamber techniques is difficult to assess under field conditions because the true Fg is unknown. There is a need for adequate laboratory calibration systems to measure the accuracy of the various chamber systems and to compare their estimates against a known source (Nay et al., 1994; Nay and Bormann, 2000; Widén and Lindroth, 2003). Moreover, integration of chamber estimates of Fg over space and time are often required to satisfy research objectives. This integration exercise is made difficult by the usually large variability of Fg Additional research is needed to guide selection of an adequate chamber type, chamber design, and sampling methodology that minimize errors when individual chamber measurements are integrated over time and space.

Limitations on the use of micrometeorological techniques come from both the measurement and micrometeorological theory components of these techniques. There remains research to be done on developing fast-response analyzers for deployment with the eddy covariance technique to expand the application beyond that for water vapor and carbon dioxide gases. We also recognize the need to develop open-path laser technology for a variety of gases to simplify measurements needed in utilizing the integrated horizontal flux technique. For example, recent development of open-path lasers for ammonia and methane gas has greatly simplified the measurement requirements of the integrated horizontal flux technique for collecting flux data from manure amended soil plots. The availability of an open-path laser to measure concentration of nitrous oxide would accelerate research on this greenhouse gases.

The application of the dispersion models discussed in this paper has been limited by the lack of data on their accuracy. Research is needed to provide a better understanding of the sensitivity of dispersion techniques relative to more standard micrometeorological techniques. Broadening the conditions suitable for employing dispersion models would increase their utilization. For example, the impact of turbulent flow around the source would be useful to incorporate into the dispersion model to allow measurements close to the source to be used.

NOTATION

A	soil area covered by the chamber (m^2)
b	conversion constant

c	gas concentration ($g\,m^{-3}$)
c_b	background gas concentration ($g\,m^{-3}$)
c_{dw}	downwind moving gas concentration ($g\,m^{-3}$)
c_{up}	upwind moving gas concentration ($g\,m^{-3}$)
c'	instantaneous gas concentration ($g\,m^{-3}$)
$c*$	concentration scale ($g\,m^{-3}$)
$C_{g,a}$	amount of gas in chemical trap (mol)
$C_{g.b}$	amount of gas in blank trap (mol)
C_p	specific heat of air ($J\,kg^{-1}\,C^{-1}$)
d	zero plane displacement height (m)
e	partial pressure of water vapor inside chamber
f	flow rate through chamber ($m^3\,s^{-1}$)
Fg	soil-air gas exchange ($mol\,m^{-2}\,s^{-1}$) or ($g\,m^{-2}\,s^{-1}$)
G	soil heat flux density ($W\,m^{-2}$)
G_i	chamber intake gas concentration ($mol\,mol^{-1}$)
G_0	chamber exhaust gas concentration ($mol\,mol^{-1}$)
$G]$	gas concentration ($mol\,mol^{-1}$)
$G]_{ch}$	gas concentration inside chamber ($mol\,mol^{-1}$)
$G]_{amb}$	gas concentration outside chamber ($mol\,mol^{-1}$)
H	sensible heat flux density ($W\,m^{-2}$)
Ha	humidity inside chamber (%)
Hs	soil water content (%)
k	von Karman constant (0.41)
K_h	eddy diffusivity for heat ($m^2\,s^{-1}$)
K_v	eddy diffusivity for water vapor ($m^2\,s^{-1}$)
L	Monin-Obukhov length (m)
L_v	latent heat of vaporization ($J\,g^{-1}$)
LE	latent heat flux density ($W\,m^{-2}$)
Mv	molar volume ($m^3\,mol^{-1}$)
Mw	molecular weight
p	time period (s)
P	barometric air pressure (kPa)
Rn	net radiation flux density ($W\,m^{-2}$)
T'	instantaneous air temperature (°C)
Ts	soil temperature (°C)
Ta	air temperature (°C)
μ	horizontal wind speed ($m\,s^{-1}$)
$\mu*$	friction velocity ($m\,s^{-1}$)
U, V	wind speed vectors ($m\,s^{-1}$)
V_{ch}	chamber volume (m^3)
w'	instantaneous vertical wind speed ($m\,s^{-1}$)
x	windward distance across emission source to measurement (m)
X	perimeter length (m)
z	measurement height (m)
z_o	aerodynamic roughness (m)
β	Bowen ratio

β'	empirical coefficient
ε	ratio of molecular weight of water to air (0.622)
ρ	air density (g m^{-3})
ρ_v	water vapor density (g m^{-3})
ρ'_v	instantaneous water vapor density (g m^{-3})
$\delta G/\delta t$	accumulation rate of gas concentration $(\text{mol mol}^{-1}\,\text{s}^{-1})$
$\triangle c$	difference in up- and down-going gas concentration (g m^{-3})
θ	potential temperature $(^\circ\text{C})$
e	water vapor pressure (Pa)
σ_w	standard deviation of the vertical wind speed (m s^{-1})
γ	psychrometric constant (Pa C^{-1})
Ψ_g	atmospheric stability for target gas
Ψ_m	atmospheric stability for momentum

REFERENCES

Abtew, W., J.M. Gregory, and J. Borrelli. 1989. Wind profile: estimation of displacement height and aerodynamic roughness. Trans. ASAE 32:521–527.

Ambus, P., and G.P. Robertson. 1998. Automated near-continuous measurement of carbon dioxide and nitrous oxide fluxes with a photoacoustic infra-red spectrometer and flow-through soil cover boxes. Soil Sci. Soc. Am. J. 62:394–400.

Andersen, H.V., M.F. Hovmand, P. Hummelshøj, and N.O. Jensen. 1993. Measurements of ammonia flux to a spruce stand in Denmark. Atmos. Environ. 27A:189–202.

Anthony, W.H., G.L. Hutchinson, and G.P. Livingston. 1995. Chamber measurement of soil-atmosphere gas exchange: Linear vs. diffusion-based flux models. Soil Sci. Soc. Am. J. 59:1308–1310.

Ausma, S., G.C. Edwards, E.K. Wong, T.J. Gillespie, C.R. Fitzgerald-Hubble, L. Halfpenny-Mitchell, and W.P. Mortimer. 2001. A micrometeorological technique to monitor total hydrocarbon emissions from landfarms to the atmosphere. J. Environ. Qual. 30:776–785.

Baker, J.M. 2000. Conditional sampling revisited. Agric. For. Meteor. 24:59–65.

Beauchamp, e.g., G.E. Kidd, and G.W. Thurtell. 1978. Ammonia volatilization from sewage sludge applied to the field. J. Environ. Qual. 7:141–146.

Bekku, Y., H. Koizumi, T. Nakadai, and H. Iwaki. 1995. Measurement of soil respiration using closed chamber method: an IRGA technique. Ecol. Res. 10:369–373.

Bekku, Y., H. Koizumi, T. Nakadai, and H. Iwaki. 1997. Examination of four methods for measuring soil respiration. Appl. Soil Ecology. 5:247–254.

Bornemann, F. 1920. Kohlensaure und Pflanzenwachstum. Mitt. Dtsch. Landwirtsch.-Ges. 35:363.

Bowen, I.S. 1926. The ratio of heat losses by conduction and by evaporation from any water surface. Phys. Rev. 27:779–787.

Businger, J.A., and S.P. Oncley. 1990. Flux measurement with conditional sampling. J. Atmos. Ocean. Tech. 7:349–352.

Bussink, D.W. 1994. Relationships between ammonia volatilization and nitrogen fertilizer application, rate, intake and excretion of herbage nitrogen by cattle on grazed swards. Fertilizer Res. 38:111–121.

Campbell, G.S. 1977. An Introduction to Environmental Biophysics. Springer-Verlag, New York.

Christensen, S. 1983. Nitrous oxide emission from the soil surface: continuous measurement by gas chromatography. Soil Biol. Biochem. 15:481–483.

Civerolo, K.L., and R.R. Dickerson. 1998. Nitric oxide soil emissions from tilled and untilled cornfields. Agric. For. Meteor. 30:307–311.

Crill, P.M. 1991. Seasonal patterns of methane uptake and carbon dioxide release by a temperate woodland. Global Biogeochem. Cycles 5:319–334.

Denmead, O.T. 1979. Chamber systems for measuring nitrous oxide emissions from soils in the field. Soil Sci. Soc. Am. J. 43:89–95.

Denmead, O.T., J.R. Simpson, and Freney, J.R. 1977. A direct field measurement of ammonia emission after injection of anhydrous ammonia. Soil Sci. Soc. Am. J. 41:1001–1004.

Denmead, O.T., L.A. Harper, J.R. Freney, D.W.T. Griffith, R. Leuning, and R.R. Sharpe. 1998. A mass balance method for non-intrusive measurements of surface-air trace gas exchange. Atmos. Environ. 32:3679–3688.

Desjardins, R.L. 1972. A study of carbon-dioxide and sensible heat fluxes using the eddy correlation technique. Ph.D. dissertation. Cornell University, Ithaca, NY.

Edwards, N.T. 1974. A moving chamber design for measuring soil respiration rates. Oikos 25:97–101.

Ewel, K.C., W.P. Cropper, and H.L. Gholz. 1987. Soil CO_2 evolution in Florida slash pine plantations. I. Changes through time. Can. J. For. Res. 17:325–329.

Fang, C., and J.B. Moncrieff. 1998. An open-top chamber technique for measuring soil respiration and the influence of pressure difference on CO_2 flux measurements. Func. Ecol. 12:319–326.

Ferm, M. 1991. A sensitive diffusional sampler. Report B-1020, Swedish Environmental Research Institute, Gothenburg, Sweden.

Flesch, T., J.D. Wilson, and E. Yee. 1995. Backward-time Lagrangian stochastic dispersion models, and their application to estimate gaseous emissions. J. Appl. Meteor. 34:1320–1332.

Gordon, R., R. Jamieson, V. Rodd, G. Patterson, and T. Harz. 2001. Effects of surface manure application timing on ammonia volatilization. Can. J. Soil Sci. 81:525–533.

Goulden, M.L., and P.M. Crill. 1997. Automated measurements of CO_2 exchange at the moss surface of a black spruce forest. Tree Physiol. 17:537–542.

Gupta, S.R., and J.S. Singh. 1977. Effect of alkali volume and absorption area on the measurement of soil respiration in a tropical sward. Pedobiologia 17:223–239.

Haber, W. 1958. Oekologische Untersuchungen der Bodenatmung. [Ecological analysis of soil respiration (review of methods)]. Flora 146:109–157.

Hanson, P.J., S.D. Wullschleger, S.A. Bohlman, and D.E. Todd. 1993. Seasonal and topographic patterns of forest floor CO_2 efflux from an upland oak forest. Tree Physiol. 13:1–15.

Harper, L.A. 1988.Comparisons of methods to measure ammonia volatilization in the field. *In*: B. R. Bock and D. E. Kissel (eds.), Ammonia Volatilization from Urea Fertilizers. National Fertilizer Development Centre, Bulletin Y-206, Tennessee Valley Authority, Muscle Shoals, AL.

Harper, L.A., V.R. Catchpoole, R. Davis, and K.L. Weir. 1983. Ammonia volatilization: soil, plant and microclimate effects on diurnal and seasonal fluctuations. Agron. J. 75:212–218.

Harper, L.A., O.T. Denmead, J.R. Freney, and F.M. Byers. 1999. Direct measurements of methane emissions from grazing and feedlot cattle. J. Anim. Sci. 77:1392–1401.

Healy, R.W., R.G. Striegl, T.F. Russell, G.L. Hutchinson, and G.P. Livingston. 1996. Numerical evaluation of static-chamber measurements of soil-atmosphere gas exchange: Identification of physical processes. Soil Sci. Soc. Am. J. 60:740–747.

Hutchinson, G.L., and G.P. Livingston. 2001. Vents and seals in non-steady-state chambers for measuring gas exchange between soil and the atmosphere. Eur. J. Soil Sci. 52:675–682.

Hutchinson, G.L., and G.P. Livingston. 2002. Soil-atmosphere gas exchange. In: J. H. Dane and G. C. Topp (eds.), Methods of Soil Analysis. Part 4, SSSA Book Ser. 5 SSSA, Madison, WI. p. 1159–1182.

Hutchinson, G.L., G.P. Livingston, R.W. Healy, and R.G. Striegl. 2000. Chamber measurement of surface-atmosphere trace gas exchange: dependence on soil, interfacial layer, and source/sink properties. J. Geophys. Res. 105:8865–8876.

Hutchinson, G.L., and A.R. Mosier. 1981. Improved soil cover method for field measurement of nitrous oxide fluxes. Soil Sci. Soc. Am. J. 45:311–316.

Huthchinson, G.L., and P. Rochette. 2003. Non-flow-through steady-state chambers for in situ measurement of soil respiration: numerical evaluation of their accuracy and performance. Soil Sci. Soc. Am. J. 67:166–180.

Itier, B., P. Cellier, and C. Riou. 1985. Actual evapotranspiration measurement by a simplified aerodynamic system. In: Advances in Evapotranspiration, Proceedings of the National Conference on Advances in Evapotranspiration, ASAE, Dec 16–17 Chicago IL.

Janssens, I.A., A.S. Kowalski, B. Longdoz, and R. Ceulemans. 2000.Assessing forest soil CO_2 efflux: an in situ comparison of four techniques. Tree Physiol. 20:23–32.

Kaimal, J.C., and J.J. Finnigan. 1994. Atmospheric Boundary Layer Flows: Their Structure and Measurement. Oxford University Press, New York.

Kanemasu, E.T., W.L. Powers, and J.W. Sij. 1974. Field chamber measurements of CO_2 flux from soil surface. Soil Sci. 118:233–237.

Katul, G.G., P.L. Finkelstein, J.F. Clarke, and T.G. Ellestad. 1996. An investigation of the conditional sampling method used to estimate fluxes of active, reactive, and passive scalars. J. Appl. Meteor. 35:1835–1845.

Kimball, B.A. 1983. Canopy gas exchange: gas exchange with soil. p. 215–226. In: Taylor, H.M., W.R. Jordan and T.S. Sinclair (eds.), Limitations to Efficient Water Use in Crop Production. ASA-CSSA-SSSA Publication, Madison, WI.

Kirita, H. 1971. Re-examination of the absorption method of measuring soil respiration under field conditions. III. Combined effect of the covered ground area and the surface area of KOH solution on CO_2 absorption rates. Jpn. J. Ecol. 21:43–47.

Laughlin R. J., and Stevens, R. J. 2003. Changes in composition of nitrogen-15-labeled gases during storage in septum-capped vials. Soil Sci. Soc. Am. J. 67:540–543.

Laville, P., C. Jambert, P. Cellier, and R. Delmas. 1999. Nitrous oxide fluxes from a fertilised maize crop using micrometeorological and chamber methods. Agric. For. Meteor. 30:19–38.

Le Dantec, V., D. Epron, and E. Dufrêne. 1999. Soil CO_2 efflux in a beech forest: comparison of two closed dynamic systems. Plant Soil 214:125–132.

Leclerc, M.Y., and G.W. Thurtell. 1990. Footprint prediction of scalar flux using a Markovian analysis. Bound.-Layer Meteor. 52:247–258.

Leuning, R., J.R. Freney, O.T. Denmead, and J.R. Simpson. 1985. A sampler for measuring atmospheric ammonia flux. Atm. Environ. 19:1117–1124.

Linn, D.M., and J.W. Doran. 1984. Effect of water-filled pore space on carbon dioxide and nitrous oxide production in tilled and non-tilled soils. Soil Sci. Soc. Am. J. 48:1267–1272.

Livingston, G.P., and G.L. Hutchinson. 1995. Enclosure-based measurement of trace gas exchange: Applications and sources of error. p. 14–51. *In*: P.A. Matson and R.C. Harriss (eds.), Biogenic Trace Gases: Measuring Emissions from Soil and Water. Blackwell Science Ltd., Oxford, UK.

Lockyer, D.R. 1984. A system for the measurement in the field of losses of ammonia through volatilization. J. Sci. Food Agric. 35, 837–848.

Lodge, J.P. 1988. Methods of Air Sampling and Analysis. 3rd ed. Lewis Publishers, CRC Press, Boca Raton, FL.

Loftfield, N.S., R. Brumme, and F. Beese. 1992. Automated monitoring of nitrous oxide and carbon dioxide flux from forest soils. Soil Sci. Soc. Am. J. 56:1147–1150.

Lundegårdh, H. 1921. Ecological studies in the assimilation of certain forest plants and shore plants. Sven. Bot. Tidskr. 15:46–94.

Lundegårdh, H. 1926. Carbon dioxide evolution of soil and crop growth. Soil Sci. 23:417–454.

Mosier, A.R., and G.L. Hutchinson. 1981. Nitrous oxide emissions from cropped fields. J. Environ. Qual. 10:169–173.

Majewski, M.S. 1999. Micrometeorological methods for measuring the post-application volatilisation of pesticides. Water Air Soil Pollut. 115:83–113.

Makarov, B.N. 1959. A simple method of determining soil respiration. Pochvovedeniye 9:119–122.

Massman, W.J., and X. Lee. 2002. Eddy covariance flux corrections and uncertainties in long-term studies of carbon and energy exchanges. Agric. For. Meteor. 113:121–144.

Matthias, A.D., A.M. Blackmer, and J.M. Bremner. 1980. A simple chamber technique for field measurement of emissions of nitrous oxide from soils. J. Environ. Qual. 9:251–256.

Matthias, A.D., and A.R. Peralta-Hernández. 1998. Modelling temperatures in soil under an opaque cylindrical enclosure. Agric. Forest Meteorol. 90:27–38.

Matthias, A.D., D.N. Yarger, and R.S. Weinbeck. 1978. A numerical evaluation of chamber methods for determining gas fluxes. Geophys. Res. Lett., 5:765–768.

McGinn, S.M., and O.O. Akinremi. 2001. Carbon dioxide balance pf a crop-fallow rotation in western Canada. Can. J. Soil Sci. 81:121–127.

McGinn, S.M., O.O. Akinremi, H.D.J. McLean, and B.H. Ellert. 1998. An automated chamber system for measuring soil respiration. Can. J. Soil Sci. 78:573–579.

McGinn, S.M., H.H. Janzen, and T. Coates. 2003. Atmospheric ammonia, volatile fatty acids, and other odorants near beef feedlots. J. Environ. Qual. 32:1173–1182.

Milne, R., I.J. Beverland, K. Hargreaves, and J.B. Moncrieff. 1999. Variation of the (coefficient in the relaxed eddy accumulation method. Bound.-Layer Meteor. 93:211–225.

Miyata, A., R. Leuning, O.T. Denmead, J. Kim, and Y. Harazona. 2000. Carbon dioxide and methane fluxes from an intermittently flooded paddy field. Agric. For. Meteor. 24:287–303.

Naganawa, T., and K. Kyuma. 1991. Concentration dependence of CO_2 evolution from soil in chamber with low CO_2 concentration (< 2000 ppm) and CO_2 diffusion/ sorption modelling in soil. Soil Sci. Plant Nutr. 37:381–386.

Nay, S.M., and B.T. Bormann. 2000. Soil carbon changes: comparing flux monitoring and mass balance in a box lysimeter experiment. Soil Sci. Soc. Am. J. 64:943–948.

Nay, S.M., K.G. Mattson, and B.T. Bormann. 1994. Biases of chamber methods for measuring soil CO_2 efflux demonstrated with a laboratory apparatus. Ecology 75:2460–2463.

Norman, J.M., R. Garcia, and S.B. Verma. 1992. Soil surface CO_2 fluxes and the carbon budget of a grassland. J. Geophys. Res. 97:18,845–18,853.

Norman, J.M., C.J. Kucharik, S.T. Gower, D.D. Baldocchi, P.M. Crill, M.B. Rayment, K. Savage, and R.G. Striegl. 1997. A comparison of six methods for measuring soil-surface carbon dioxide fluxes. J. Geophys. Res. 102:28,771–28,777.

Parkinson, D. 1981. An improved method for measuring soil respiration in the field. J. Appl. Ecol. 18:221–228.

Pattey, E., R.L. Desjardins, and P. Rochette. 1993. Accuracy of the relaxed eddy-accumulation technique, evaluated using CO_2 flux measurements. Boundary-Layer Meteorol. 66:341–356.

Paulson, C.A. 1970. The mathematical representation of wind speed and temperature profiles in the unstable atmospheric surface layer. J. Appl. Meteor. 9:857–861.

Pedersen, A.R. 2000. Estimating the nitrous oxide emission rate from the soil surface by means of a diffusion model. Scand. J. Stat. 27: 385–403.

Penman, H.L. 1948. Natural evapotranspiration from open water, bare soil and grass. Proc. R. Soc. London Ser. A. 193:120–145.

Rayment, M.B. 2000. Closed chamber systems underestimate soil CO_2 efflux. Eur. J. Soil Sci. 51:107–110.

Rayment, M.B., and P.G. Jarvis. 1997. An improved open chamber system for measuring soil CO_2 effluxes in the field. J. Geophys. Res. 102:28,779–28,784.

Reicosky, D.C., W.A., Dugas, and H.A. Torbert. 1997. Tillage-induced soil carbon dioxide loss from different cropping systems. Soil Till. Res. 41:105–118.

Reitsma, T. 1978. Wind-Profile Measurements Above a Maize Crop. Agric. Res. Report, Wageningen, The Netherlands.

Rice, C.P., C.B. Nochetto, and P. Zara. 2002. Volatilization of trifluralin, atrazine, metolachlor, chlorpyifos, alpha-endosulgan, and beta-endosulfan from freshly tilled soil. J. Agric. Food Chem. 50:4009–4017.

Rochette, P., and N. Bertrand. 2003. Soil air sample storage and handling using polypropylene syringes and glass vials. Can. J. Soil Sci. 83:631–637.

Rochette, P., R.L. Desjardins, and E. Pattey. 1991. Spatial and temporal variability of soil respiration in agricultural fields. Can. J. Soil Sci. 71:189–196.

Rochette, P., B. Ellert, E.G. Gregorich, R.L. Desjardins, E. Pattey, R. Lessard, and B.G. Johnson. 1997. Description of a dynamic closed chamber for measuring soil respiration and its comparison with other techniques. Can. J. Soil Sci. 77:195–203.

Rochette, P., E.G. Gregorich, and R.L. Desjardins. 1992. Comparison of static and dynamic closed chambers for measurement of soil respiration under field conditions. Can. J. Soil Sci. 72:605–609.

Rochette, P., and G.L. Hutchinson. 2004. Measurement of soil respiration *in situ* chamber techniques. In J.L. Hatfield (ed.), Micrometeorology in Agriculture. ASA, CSSA and SSSA, Madison, WI.

Rosenberg, N.J., B.L. Blad, and S.B. Verma. 1983. Microclimate: The Biological Environment. John Wiley & Sons, New York.

Ryden, J.C., and J.E. McNeill. 1984. Application of the micrometeorological mass balance method to the determination of ammonia loss from a grazed sward. J. Sci. Food Agric. 35:1297–1310.

Saugier, B., and E.A. Ripley. 1978. Evaluation of the aerodynamic method of determining fluxes over natural grassland. Quart. J. R. Met. Soc. 104:257–270.

Scott, A., I. Crichton, and B.C. Ball. 1999. Long-term monitoring of soil gas fluxes with closed chambers using automated and manual systems. J. Environ. Qual. 28:1637–1643.

Segschneider, H.J., Sich, I., and Russow, R. 1997. Use of a specifically configured gas chromatographic system for the simultaneous determination of methane, nitrous oxide and carbon dioxide in ambient air and soil atmosphere. p. 211–218. *In*: Jarvis, S.C. and Pain, B.F. (eds.), Gaseous Nitrogen Emissions from Grasslands. CAB international, Wellingford.

Sharkov, I.N. 1984. Determination of the rate of soil CO_2 production by the absorption method. Sov. Soil Sci. 16:102–111.

Sinclair, T.R., L.H. Allen, and E.R. Lemon. 1975. An analysis of errors in the calculation of energy flux densities above vegetation by a Bowen-ratio profile method. Bound.-Layer Meteor. 8:129–139.

Smith, A.E., L.A. Kerr, and B. Caldwell. 1997. Volatility of ethalfluralin, trifluralin, and triallate from a field following surface treatments with granular formulations. J. Agric. Food Chem. 45:1473–1478.

Smith, K.A., and F. Conen. 2003. Measurement of trace gases, I: gas analysis, chamber methods, and related procedures. p. 433–476. *In*: K.A. Smith and M.S. Cresser (eds.), Soil and Environment Analysis; Modern Instrumental Techniques. Marcel Dekker, New York.

Striegl, R.G., T.A. McConnaughey, D.C. Thorstenson, E.P. Weeks, and J.C. Woodward. 1992. Consumption of atmospheric methane by desert soils. Nature 357:145–147.

Sutton, M.A., J.B. Moncrieff, and D. Fowler. 1992. Deposition of atmospheric ammonia to moorlands. Environ. Pollut. 75:15–24.

Tanner, C.B. 1960. Energy balance approach to evapotranspiration from crops. Soil Sci. Soc. Am. Proc. 24:1–9.

Thornthwaite, C.W., and B. Holtzman. 1939. The determination of evaporation from land and water surfaces. U.S. Monthly Weather Rev. 67: 4–11.

Wagner-Riddle, C., G.W. Thurtell, K.M. King, G.E. Kidd, and E.G. Beauchamp. 1996. Nitrous oxide and carbon dioxide fluxes from a bare soil using a micrometeorological approach. J. Environ. Qual. 25:898–907.

Webb, E.K., G.I. Pearman, and R. Leuning. 1980. Correction of flux measurements for density effects due to heat and water vapour transfer. Quart. J. Roy. Meteorol. Soc. 106:85–100.

Welles, J.M., and D.K. McDermitt. 2004. Measuring carbon dioxide in the atmosphere. *In*: J.L. Hatfield (ed.), Micrometeorology in Agriculture. ASA, CSSA and SSSA, Madison, WI.

Widén, B., and A. Lindroth. 2003. A calibration system for soil carbon dioxide efflux measurement chambers: Description and application. Soil Sci. Soc. Am. J., 67:327–334.

Wilson, J.D., and V.K.N. Shum. 1992. A re-examination of the integrated horizontal flux method for estimating volatilization from circular plots. Agric. For. Meteor. 57:281–295.

Wilson, J.D., G.W. Thurtell, G.E. Kidd, and E.G. Beauchemin. 1982. Estimation of the rate of gaseous mass transfer from a surface source plot to the atmosphere. Atmos. Environ. 16:181–1867.

Wilson, J.D., V.R. Catchpoole, O.T. Denmead, and G.W. Thurtell. 1983. Verification of a simple micrometeorological method for estimating the rate of gaseous mass transfer from the ground to the atmosphere. Agric. Meteor. 29:183–189.

Yamulki, S., and Jarvis, S. C. 1999. Automated chamber technique for gaseous flux measurements: evaluation of a photoacoustic infrared spectrometer-trace gas analyzer. J. Geophys. Res. 104: 5463–5469.

Yates, S.R., F.F. Ernst, J. Gan, F. Gao, and M.V. Yates. 1996. Methyl bromide ,emissions from a covered field: II. Volatilization. J. Environ. Qual. 25:192–202.

Zibilske, L.M. 1994. Carbon mineralisation. p. 836–864. *In*: Methods of Soil Analysis. Part 2. Microbiological and Biochemical Properties. SSSA Book Ser. 5 SSSA. Madison, WI.

14 Chemical Methods for Soil and Water Characterization

Yuncong Li and Meifang Zhou
University of Florida, Homestead, Florida, U.S.A.

Jianqiang Zhao
Florida Department of Agriculture and Consumer Services,
Tallahassee, Florida, U.S.A.

CONTENTS

1-5667-0657-2/05/$0.00 + $1.50
© 2005 by CRC Press

14.1 INTRODUCTION

Chemical analysis is one of the most important tools for soil and water characterization. Soil testing began in the nineteenth century and was primarily used for soil classification and fertilizer recommendations. Recently the purpose of soil testing has been extended beyond improving crop production to encompass environmental evaluation. Before Daubeny (1845) proposed the concept of active (plant-available) and dormant (plant-unavailable or slowly available) forms of plant nutrients in soils, only total nutrient analysis was conducted. In the century and a half since that time, many chemical solutions ("extractants") have been tested and used for the determination of plant-available nutrients in soils. In addition, emphasis has shifted from the simple determination of total concentrations to determining the various fractions of the forms of each element in soil. Often the most commonly used extractants are mixtures of two or more chemicals. Each of the different extractants requires a corresponding specific set of analytical procedures for quantifying the elements of interest in the extraction solution. These analyses have been widely used for agronomic purposes, contaminated soil remediation, ecosystem restoration, and solute transport modeling. Some methods were developed to be universally applicable procedures. However, because of complexity of soils, no method has been accepted as universally applicable. Soil testing laboratories in each region of the United States have established standard extractants and analytical methods for selected chemicals for agronomic purposes only. In 1980 the U.S. Environmental Protection Agency (USEPA) published a document, Test Methods for Evaluating Solid Waste, Physical/Chemical Methods (SW-846), which includes testing methods for soils and other solids. The third edition of this document, released recently, lists 127 standard methods (USEPA, 2004).

The first edition of Standard Methods for the Examination of Water and Wastewater was published in 1905 by the American Public Health Association (APHA). The 20th edition, released in 1998, includes more than 350 testing methods (APHA, 1998). The USEPA evaluates and certifies drinking water and wastewater analytical methods. The EPA's Office of Water recently released a CD-ROM that contains 330 drinking water and wastewater methods (USEPA, 1999). All of analytical laboratories certified by the National Laboratory Environmental Accreditation Conference (NELAC) must comply with the use of USEPA-approved or other standard methods for water analyses.

Although several hundred methods for soil and water analysis are currently available, many new analytical methods are being developed because analytical technology is being improved continually. Often the selection of an analytical method out of the plethora of new and existing methods is difficult, but crucially important, since use of an improper method will lead to false determinations. The purpose of this chapter is to aid researchers and analysts to understand significant differences between various methods and to select the appropriate analytical methods for soil and water characterization.

14.1.1 CRITERIA FOR METHOD SELECTION

14.1.1.1 Using Standard Methods

Analytical methods that have been reviewed, validated, and approved by various societies and organizations are designated as standard methods. If available, a method that has been published as a certified international, regional, or national standard should be used for the desired analysis. Such methods generate strongly defensible analytical data for scientific or legal purposes. Any modifications of these authoritative standard methods should be validated, documented, and, if possible, reviewed and approved by an accreditation organization. If a nonstandard method is used, it should be validated appropriately before it is used, and the results from the modified method should be comparable to those obtained with a standard method. The International Organization for Standardization (ISO) certifies standard methods for universal use. In the United States, most standard methods for soil and water analysis have been developed by scientists affiliated with the USEPA, APHA, Soil Science Society of America (SSSA), American Society for Testing and Materials (ASTM), and Association of Official Analytical Chemists–International (AOAC). Selected standard methods certified by the USEPA are listed in Table 14.1.

14.1.1.2 Fitting to Analytical Purposes

The analytical method selected must be suitable for solving the analytical problems of concern. Many methods are developed specifically for particular analytical problems. For example, the Mehlich-1 (double acid) (Mehlich, 1953) extraction method is used for measuring plant available nutrients in acid soils, while the Olsen method ($NaHCO_3$) (Olsen et al., 1954) is used for measuring plant available nutrients in calcareous soils. USEPA methods (SW-846) have been approved for chemical analyses of soils and other solids in order to achieve compliance with RCRA (Resource Conservation and Recovery Act) regulations. Analytical methods for drinking water quality are well established, and some of these methods are listed in Table 14.1. If no existing method is available for a specific analysis, a new method should be developed, validated, and, if possible, certified by an accreditation agency, such as NELAC. Many *in situ* methods are well developed and serve special purposes, but most of them are non-standard methods. Validating these methods with a standard method before use is important and often essential.

14.1.1.3 Meeting the Method Detection Limit

The method detection limit (MDL) is a one of the critical criteria for selecting analytical methods. In order to achieve low detection limits, expensive instruments, well-trained personnel, and strictly controlled environmental conditions are often required. In addition great care must be exercised in sample preparation. Often these stringent requirements increase both the analytical costs and the time required to obtain results. Therefore, clients frequently request simple, cheap, and quick methods or devices for soil and water analyses. However, skimpiness almost always undermines the quality of analyses, especially the minimum detection limits. For example, soil pH is commonly and readily measured in the field. However, most portable pH meters are inaccurate. Generally, these meters either lack or have a deficient calibration function. Of course, both portable pH meters and pH

TABLE 14.1
USEPA Methods and APHA Standard Methods for Selected Analyses

Analysis	MCL[a] (mg/L)	MDL[b] (mg/L)	Standard method	Recommended method	Alternative method
Aluminum	0.05–0.2		Water: EPA 200.7, 200.8, 202.1, 202.2, 200.9; SM 3111D and E, 3113B, 3120B, 3125B Soil: 6110B, 6020, 7020	SM 3120B, EPA 200.7 or 6010B (ICP)	SM 3113B, EPA 202.2 (GFAA) or EPA 7020 (FLAA)
Ammonia		0.01–2	Water/Soil: EPA 350.1, 350.2, 350.3, SM 4500-NH3B, C, D, E, F, G, and H	EPA 350.1, SM 4500-NH3G, H (Autoanalyzer)	EPA 350.3, SM-NH3D (ion selective electrode)
Antimony	0.006	0.0008–0.003	Water: EPA 200.7, 200.8, 200.9, 204.2; SM 3113B, 3120B, 3125B. Soil: EPA 6010B, 6020, 7040, 7041, 7062	SM 3120B, EPA 200.7 or 6010B (ICP)	SM 3113B, EPA 204.2 or 7041 (GFAA)
Arsenic	0.01		Water: EPA 200.7A, 200.8, 200.9, 206.2; SM 3113B, 3114B, 3120B, 3125B Soil: EPA 6010B, 6020, 7060A, 7061A, 7062	SM 3120B, EPA 200.7, or 6010B (ICP)	SM 3113B EPA 206.2 or 7060A (GFAA), SM 3114B (hydride generation-AA)
Barium	2	0.001–0.1	Water: EPA 200.7, 208.1, 200.8, 208.2; SM 3113B, 3120B, 3125B Soil: EPA 6010B, 6020, 7080A, 7081	EPA 200.7, SM 3120B or 6010B (ICP)	SM 3113B, EPA 208.2, or 7081 (GFAA)
Beryllium	0.004	0.00002–0.0003	Water: EPA 200.7, 200.8, 200.9, 210.2; SM 3111D and E, 3113B, 3120B, 3125B Soil: EPA 6010B, 6020, 7090, 7091	EPA 200.7, SM 3120B or 6010B (ICP)	SM 3113B, EPA 210.2 or 7091 (GFAA)

Analyte			Water/Soil methods	(ICP)/(IC)	
Bromide			Water/soil: EPA212.3 Water/soil: SM 4110B, 4500BrB, EPA 300.0, 320.1, 9211	(ICP) SM 4110B, EPA 300.0 (IC)	(spectrometer) EPA9211 (bromide-specific electrode)
Cadmium	0.005	0.0001–0.001	Water: EPA 200.7, 200.8, 200.9, 213.2; SM 3120B, 3113B, 3125B, 3111 B and C. Soil: EPA 6010B, 6020, 7030, 7031A	SM 3120B, EPA 200.7 or 6010B (ICP)	SM 3113B, EPA 213.2 or 7131A (GFAA)
Calcium			Water: EPA 200.7, 215.2, 300.7; SM 3120B, 3500CaB, 3111B, D and E Soil: EPA 6010B, 6020, 7140	SM 3120B, EPA 200.7 or 6010B (ICP)	EPA215.2 (Titration) EPA 7140, SM 3111B (FLAA)
Chloride	250		Water: EPA300.0, 325.1, 325.2, 325.3, 9250, 9252, 9253; SM 4110B, 4500B, C, D, and E.	EPA300.0, SM 4110B (IC)	EPA9212 (chloride-specific electrode)
Chromium	0.1	0.001–0.007	Water: EPA 200.7, 200.8, 200.9, 218.4, 218.6; SM 3120B, 3113B, 3125B, 3111B and C, 3500CrB, and C. Soil: EPA 6010B, 6020, 7190, 7191, 7197	SM 3120B, EPA 200.7 or 6010B (ICP)	SM 3113B, EPA 218.2 or 7191 (GFAA)
Cobalt	0.1		Water: SM 3120B, 3113B, 3125B Soil: EPA 6010B, 6020, 7200, 7201	SM 3120B, EPA 6010B (ICP)	SM 3113B, EPA 7201 (GFAA)
Copper	0.1	0.001–0.05	Water: EPA 200.7, 200.8, 200.9, 220.1, 220.2; SM 3111B and C, 3113B, 3120B, 3125B, 3500CuB and C Soil: EPA 6010B, 6020, 7210, 7211	SM 3120B, EPA 200.7 or 6010B (ICP)	SM 3113B, EPA 220.2 or 7211 (GFAA)

(Continued)

TABLE 14.1
Continued

Analysis	MCL[a] (mg/L)	MDL[b] (mg/L)	Standard method	Recommended method	Alternative method
Cyanide	0.2	0.005–0.02	Water: EPA 335.2, 335.3, 335.4, 9010, 9012, 9013, 9014; SM 4500CNC, D, E, F, G, H, I, J, K, L, M, N and O	SM 4500CNO, EPA335.3, 335.4 (Autoanalyzer)	EPA9213 (specific electrode)
Fluoride	4		Water: EPA 300.0, 340.1, 340.2, 340.3, 9214; SM 4110B, 4500FC, D, E and G	EPA300.0, SM 4110B (IC)	EPA9214, SM 4500FC (specific electrode)
Iron	0.3		Water: EPA 200.7, 236.1, 236.2; SM 3111B and C, 3113B, 3120B, 3500FeB	SM 3120B, EPA 200.7 or 6010B	SM 3113B, EPA 236.2 or 7381 (GFAA)
			Soil: EPA 6010B, 6020, 7380, 7381	(ICP)	
Kjeldahl cnitrogen			Water/soil: EPA 351.1, 351.2, 351.3; SM 4500NorgB, C and D	SM 4500NorgD, EPA351.1, or 351.2 (Digestion/Autoanalyzer)	EPA351.3 (titration)
Lead	0.015	0.001	Water: EPA 200.7, 200.8, 200.9, 239.1, 239.2; SM 3111 B and C, 3113B, 3120B, 3125B, 3130B	SM 3120B, EPA 200.7 or 6010B	SM 3113B, EPA 239.2 or 7420 (GFAA)
			Soil: EPA 6010B, 6020, 7420, 7421	(ICP)	
Lithium			Water: SM 3111B, 3120B, 3125B, 3500LiB	SM 3120B, EPA 6010B	SM 3113B, EPA 7430 (FLAA)
			Soil: EPA 6010B, 6020, 7430	(ICP)	
Magnesium			Water: EPA 200.7, 242.1; SM 3111B, 3120B, 3125B	SM 3120B, EPA 200.7 or 6010B	SM 3111B, EPA 242.1 or 7450 (FLAA)
			Soil: EPA 6010B, 6020, 7450	(ICP)	

Manganese	0.05	Water: EPA 200.7, 243.1, 243.2; SM 3111B and C, 3120B, 3125B, 3500MnB Soil: EPA 6010B, 6020, 7460, 7461	SM 3120B, EPA 200.7 or 6010B (ICP)	SM 3133B, EPA 243.2 or 7461 (GFAA)
Mercury	0.002	EPA 245.1, 245.2, 245.5, 20C.8, 1630, 1631	EPA1630 (methyl), 1631 (total)	EPA 245.2, 245.5
Molybdenum	0.0002	Water: SM 3111D and E, 3113B, 3120B, 3125B Soil: EPA 6010B, 6020, 7480, 7481	SM 3120B, EPA 6010B (ICP)	SM 3113B, EPA 7481 (GFAA)
Nickel		Water: EPA 200.7, 200.8, 200.9, 249.1, 249.2, SM 3111B and C, 3113B, 3120B, 3125B Soil: EPA 6010B, 6020, 7520, 7521	SM 3120B, EPA 200.7 or 6010B (ICP)	SM 3113B, EPA 249.2 or 7521 (GFAA)
Nitrate	0.01–1	Water/soil: EPA 300.0, 352.1, 353.1, 353.2, 353.3; SM 4110B, 4500NO3B, D, E, F, H, I	SM 4500NO3F and I, EPA 353.2 (Autoanalyzer)	SM 4110, EPA 300.0 (IC)
Nitrite	0.004–0.05	Water/soil: EPA 300.0, 353.2, 354.1; SM 4110B, 4500NO2B, 4500NO3F	SM 4500NO2B, 4500NO3F and I, EPA 353.2 (Autoanalyzer)	SM 4110, EPA 300.0 (IC)
pH	6.5–8.5	Water/soil: EPA 150.1; SM 4500HB	SM 4500HB, EPA 150.1 (pH electrode)	
Phosphorus (total)		Water/soil: EPA365.1, 365.2, 365.3, 365.4; SM 4500PC, D, E and F	SM 4500PF, EPA 365.2 (Autoanalyzer)	SM 4500PE, EPA365.3 (manuel method)
Phosphorus (ortho-)		Water/soil: EPA300.0, 365.1, 365.2, 365.3, 365.4, SM 4500PC, D, E and F	SM 4500PF, EPA 365.1 (Autoanalyzer)	SM 4500PE, PA365.3 (manuel method) EPA300.0 (IC)

(Continued)

TABLE 14.1
Continued

Analysis	MCL[a] (mg/L)	MDL[b] (mg/L)	Standard method	Recommended method	Alternative method
Potassium			Water: EPA200.7, 258.1; SM 3111B, 3120B, 3125B, 3500KB and C. Soil: EPA 6010B, 6020, 7610	SM 3120B, EPA 200.7 (ICP)	SM 3111B, EPA 258.1 or 7610 (FLAA)
Selenium	0.05	0.002	Water: EPA 200.7, 200.8, 200.9, 270.2; SM 3113B, 3114B and C, 3120B, 3125B, 3500SeC, D and E Soil: EPA 6010B, 6020, 740, 7741A, 7742	SM 3120B, EPA 200.7 or 6010B (ICP)	SM 3113B, EPA 270.2 or 7740 (GFAA)
Silica			Water: SM 3111D, 3120B, 4500SiC, D, E and F Soil: EPA 6010B, 6020	SM 3120B, EPA 6010B (ICP)	SM4500SiF (Autoanalyzer)
Silver	0.1		Water: EPA 200.7, 200.8, 200.9, 272.1, 272.2; SM 3111B and C, 3113B, 3120B, 3125B Soil: EPA 6010B, 6020, 7760A, 7761	3120B, EPA 200.7 or 6010B (ICP)	SM 3113B, EPA 272.2 or 7761 (GFAA)
Sodium	20		Water: EPA 200.7, 273.1; SM 3111B, 3120B, 3125B, 3500NaB Soil: EPA 6010B, 6020, 7770	SM 3120B, EPA 200.7 or 6010B (ICP)	SM 3111B, EPA 273.1 or 7770 (FLAA)
Specific conductance (EC)			Water/soil: EPA 120.1; SM 2510B	SM 2510B, EPA120.1 (probe)	
Sulfate	250		Water: EPA 300.0, 375.1, 375.2, 375.3, 375.4; SM 4110B, 4500SO42C, D, E and F	SM 4110B, EPA300.0 (IC)	SM4500SO42F (Autoanalyzer) EPA 375.1 (spectrometer)
Strontium			Water: SM 3111B, 3120B, 3125B, 3500SrB	SM 3120B, EPA 6010B	SM 3111B, EPA 7780

Thallium	0.002	0.0007–0.001	Water: EPA 200.7, 200.8, 200.9, 279.1, 279.2; SM 3111B, 3113B, 3120B, 3125B Soil: EPA 6010B, 6020, 7840, 7841	SM 3120B, EPA 200.7 or 6010B (ICP)	SM 3113B, EPA 279.2 or 7841 (GFAA)
Tin			Water: SM 3111B, 3113B, 3120B, 3125B Soil: EPA 6010B, 6020, 7870	3120B, EPA 6010B (ICP)	3113B, EPA 7770 (FLAA)
Vanadium			Water: SM 3111D and E, 3113B, 3120B, 3125B, 3500VB Soil: EPA 6010B, 6020, 7910, 7911	SM 3120B, EPA 6010B (ICP)	SM 3113B, EPA 7911 (GFAA)
Zinc	5		Water: EPA 200.7, 289.1, 289.2; SM 3111B and C, 3120B, 3125B, 3500ZnB Soil: EPA 6010B, 6020, 7950, 7951	SM 3120B, EPA 200.7 or 6010B (ICP)	SM 3113B, EPA 289.2 or 7951 (GFAA)
Total organic carbon (TOC)			Water: EPA415.1; SM 5310B, C and D Soil: EPA 9060	SM 5310B, EPA415.1 or 9060 (carbon analyzer) (ICP)	
Alachlor	0.002	0.0002	EPA 505, 507, 525.2, 508.1	EPA525.2 (GC-MS)	EPA507 (GC-NPD)
Aldicarb		0.0001	EPA 531.1	EPA 531.1 (HPLC-fluorescence)	
Atrazine	0.003	0.0001	EPA 505, 507, 525.2, 508.1	EPA525.2 (GC-MS)	EPA507 (GC-NPD)
Carbaryl		0.0001	EPA 531.1	EPA 531.1 (HPLC-fluorescence)	
Carbofuran	0.04	0.00009	EPA 531.1	EPA 531.1 (HPLC-fluorescence)	
Chlordane	0.002	0.0002	EPA 505, 508, 525.2, 508.1	EPA525.2 (GC-MS)	EPA507 (GC-NPD)

(Continued)

TABLE 14.1
Continued

Analysis	MCL[a] (mg/L)	MDL[b] (mg/L)	Standard method	Recommended method	Alternative method
Dalapon	0.2	0.001	EPA 515.1, 552.1	EPA515.1 (GC-ECD)	EPA552.1 (GC-ECD)
Dibromochloropropane (DBCP)	0.0002	0.00002	EPA 504.1, 551.1	EPA551.1 (GC-ECD)	
2,4-D	0.07	0.0001	EPA 515.2, 515.1, 555	EPA515.1 (GC-ECD)	EPA555 (HPLC-UV)
Dinoseb	0.007	0.0002	EPA 515.1, 515.2, 555	EPA515.1 (GC-ECD)	EPA555 (HPLC-UV)
Diquat	0.02	0.0004	EPA 549.1	EPA549.1 (HPLC-UV)	
Endothall	0.1	0.009	EPA 548.1	EPA548.1 (GC-FID or MS)	
Endrin	0.002	0.00001	EPA 505, 508.1, 509, 525.2	EPA 508.1 (GC-ECD)	EPA525.2 (GC-MS)
Ethylene dibromide (EDB)	0.00005	0.00001	EPA 504.1, 551	EPA551.1 (GC-ECD)	
Glyphosate	0.7	0.006	EPA 547	EPA 547 (HPLC-fluorescence)	
Heptachlor	0.0004	0.00004	EPA 505, 508, 525.2	EPA 508 (GC-ECD)	EPA 525.2 (GC-MS)
Hepatchlorepoxide	0.0002	0.00002	EPA 505, 508, 525.2	EPA 508 (GC-ECD)	EPA 525.2 (GC-MS)
Hexachlorobenzene	0.001	0.0001	EPA 505, 508, 525.2	EPA 508 (GC-ECD)	EPA 525.2 (GC-MS)

Hexachlorocyclo-pentadiene	0.05	0.0001	EPA 505, 508, 525.2	EPA 508 (GC-ECD)	EPA 525.2 (GC-MS)
Lindane	0.0002	0.0002	EPA 505, 508, 525.2	EPA 508 (GC-ECD)	EPA 525.2 (GC-MS)
Methiocarb		0.0001	EPA 531.1	EPA 531 (HPLC-fluorescence)	
Methomyl		0.0001	EPA 531.1	EPA 531 (HPLC-fluorescence)	
Methoxychlor	0.04	0.0001	EPA 505, 508, 525.2	EPA 508 (GC-ECD)	EPA 525.2 (GC-MS)
Propoxur		0.0001	EPA 531.1	EPA 531 (HPLC-fluorescence)	
Oxamyl (Vydate)	0.2	0.002	EPA 531.1	EPA 531.1 (HPLC-fluorescence)	

[a]Maximum contaminant level (MCL) is the higest level of a contaminant that is allowed in primary or secondary drinking water standards in the United States.
[b]Method detection limit (MDL) is the lowest concentration of a substance that can be measured with 99% confidence that the analyte concentration is greater than zero.

paper will serve the purpose if the need is merely to determine whether the soil is either acidic or alkaline. A portable ion selective electrode for nitrate, such as in the Horiba Cardy meter, has been reported to be useful for measuring nitrate in plant tissues in the field (Davenport and Jabro, 2001). However, an instrument with such an electrode cannot be used for water quality analyses, because its guaranteed lower detection limit is 62 mg/L, whereas the drinking water standard for NO_3-N in the United States is 10 mg/L. Maximum contamination levels (MCLs) for drinking water are included in Table 14.1.

14.1.2 ASSESSMENT OF UNCERTAINTY

According to ISO/IEC (International Electrotechnical Commission) 17025, analytical laboratories must follow procedures for estimating analytical uncertainties (Ellison et al., 2000). All relevant uncertainty components must be documented and reported. Typical sources of uncertainty are 1) sampling; 2) storage conditions; 3) instrument effects and reagent purity; 4) assumed stoichiometry; 5) measurement conditions; 6) sample effects; 7) computational effects; 8) blank correction; 9) operator effects; and 10) random effects. Uncertainty from sampling has always been of paramount concern. No matter how excellent the analytical method or how carefully the analysis is performed, the analytical result will be false if the samples are not representative of the original material. As analytical technology improves, new methods require smaller sample sizes, and therefore the uncertainties associated with sampling become increasingly important. Sample sizes used by various carbon and nitrogen analyzers range from 0.01 to 5 g for soil analysis (see Table 14.4). Because of the heterogeneous nature of soils, measurements of the composition of soil based on subsamples of less than 0.1 g soil are highly uncertain. Reduction of the particle size by grinding or milling can improve analyses made with instruments and methods, which use small-sized samples. Sample storage conditions before and after samples arrive at the analytical laboratory may also affect the analytical results. A cooler with ice should be used to transport water samples. A recent survey of pipettes used in a typical laboratory showed that 36% of pipettes failed accuracy and precision tests (Pavlis, 2004). Frequent calibration is needed to improve the performance of pipettes. Ellison et al. (2000) provide detailed descriptions of other sources of uncertainty, as well as how to estimate and report uncertainty based on ISO requirements. In the United States, NELAC has incorporated uncertainty measurements in its quality system manual, but it lacks detailed descriptions of uncertainty sources and corresponding methods of estimation. However, NELAC requires all certified laboratories to compile strict quality assurance/quality control procedures (QA/QC) (NELAC, 2002).

The following discussion will focus on analytical methods for major chemicals in soil and water. These methods have been used for agronomic and environmental analyses.

TABLE 14.2
Nitrogen Forms in Soil and Water and Analytical Methods

Form	Phase	Agricultural/environmental impact	Analytical method
Ammonium (NH_4^+)	Aqueous, adsorbed	Plant uptake Surface water/groundwater contamination	*In situ*, laboratory
Nitrate (NO_3^-)	Aqueous	Plant uptake Surface water/groundwater contamination	*In situ*, laboratory
Nitrite (NO_2^-)	Aqueous	Plant uptake Surface water/groundwater contamination	Laboratory
Organic N ($R–NH_2$)	Solid	Source of mineralization Source of N in sediment	Laboratory
Ammonia (NH_3)	Gaseous	Product of volatilization Atmospheric contamination	No reliable method
Nitrous oxide/nitric oxide (NO_x)	Gaseous	Product of denitrification Greenhouse effects	No reliable method

14.2 CRITICAL DISCUSSION OF ANALYTICAL METHODS OF SOIL AND WATER

14.2.1 NITROGEN

14.2.1.1 Nitrogen in Soil and Water

Nitrogen (N), an essential nutrient for crop production, is also a major pollutant in groundwater and surface water. Nitrogen is introduced into soils largely through applications of chemical fertilizers and soil amendments (manures, biosolids, compost, etc.), symbiotic N fixation, and atmospheric deposition. Synthetic inorganic fertilizer is the main N source for agricultural soils; application of inorganic N can be as high as 300 kg N/ha/year. For natural areas with sparse legume populations (uncultivated land), atmospheric deposition is the main source of N, normally less than 15 kg N/ha/year. However, N in groundwater or surface runoff from adjacent farmland often affects the N balance in natural areas.

Nitrogen in soils generally is classified as inorganic (ammonia, nitrate, nitrite, nitrous oxide, nitric oxide, and elemental nitrogen) or organic (amino acids or proteins and numerous other unidentified organic compounds) (Table 14.2). Ninety-five percent or more of N in surface soils usually is in organic forms. However, inorganic forms of N are of special concern because they directly impact plant growth and the environment. Ammonia and nitrate are major inorganic forms of N in soils that are readily taken up by plants and that have the potential to leach into groundwater or enter surface water through runoff. Nitrite, an intermediate product of nitrification, may also be present, often in amounts below detectable levels. Most soil-testing laboratories only determine and report ammonia and nitrate for the inorganic N category. Ammonia gas, the product of volatilization, accounts for up to 30% loss of nitrogen fertilizer. A fairly reliable approach for determining ammonia volatilization is laboratory incubation (He et al., 1999). Denitrification reactions produce nitrous oxide (N_2O), nitric oxide (NO), and nitrogen gas (N_2). There are no reliable quantitative analytical methods for these nitrogen gases. Two field methods have been reported for measuring denitrification: acetylene inhibition and use of the stable isotope [15]N (Mosier and Klemedtsson, 1994). Acetylene is used to inhibit the reaction of N_2O to N_2 followed by the determination of N_2O by gas chromatography (GC). The efficiency of acetylene inhibition is questionable (Bernot et al., 2003), and the procedure normally requires 24 hours or less (Beauchamp and Bergstrom, 1993). The stable isotope ^{15}N method involves the use of mass spectrometry (MS) to measure the ratio of the N isotopes in N gas emitted after ^{15}N-labeled fertilizer has been applied. The major weakness of the method is that it does not account for N denitrification in native soil. [15]N analysis is costly and time consuming.

A number of solvents including water have been used to extract inorganic N from soils (Maynard and Kalra, 1993). Solvent solutions that have been reported for nitrate extraction include 0.01 M $CaCl_2$, 0.5 M $NaHCO_3$, 0.01 M $CuSO_4$, saturated $CaSO_4 + 0.03$ M NH_4F, 0.015 M H_2SO_4 and 2 M KCl. Potassium salt solutions (0.05 M K_2SO_4, 0.1, 1, or 2 M KCl) have been used to extract NH_4. Nitrate is water soluble and can be extracted with water. However, NH_4 is strongly bound to soil particles and is not readily extracted with water. The most common extracting solution for soil inorganic N is 2 M KCl. Determination of organic N in soils will be discussed later.

Nitrogen concentration in groundwater and surface water is usually very low. However, excessive N in surface and groundwater systems used for drinking water is dangerous to human health, especially to infants less than 6 months old. The U.S. drinking water standard for nitrate N is not to exceed 10 mg/L. High N concentrations in surface water increase the growth or "bloom" of algae and other aquatic plants. These algal blooms lead to oxygen depletion, which in turn kills fish. This process is called eutrophication. Most studies of eutrophication have focused on inorganic N. However, organic N may also play an important role (Seitzinger and Sanders, 1997). Analysis of both inorganic N and organic N in water will be discussed.

14.2.1.2   Laboratory Methods for Ammonia Determination

The Nesslerization method is the classic colorimetric method for ammonia analysis. It has been the most widely applied method for more than a century. This method had been listed as the standard method in 19th edition of APHA (APHA, 1995). However, this method has been discontinued as a standard method in latest edition of Standard Methods because it involves the use of mercury (APHA, 1998). Currently the phenate colorimetry method is the most widely used for ammonia determination in water. The distillation-titration method is used for samples with the high ammonia concentrations, such as wastewater. The ammonia-selective electrode method is applicable over a wide range of ammonia concentrations. Currently the indophenol blue method is the most popular for ammonia analysis in water.

14.2.1.2.1   Indophenol Blue Colorimetry

The first report on determining ammonia by the indophenol blue reaction involved the mixing of ammonia, phenol, and hypochlorite in 1859 (Grasshoff, 1976). Since the reaction is very slow, different catalysts such as manganese, acetone, and nitroprusside have been used to accelerate the reaction (Crowther and Lage, 1956; Horn and Squire, 1967). Most methods now use nitroprusside, since it improves the sensitivity and stability of the indophenol blue color complex. The optimum final pH for the reaction is between 10.5 and 11.5. The most widely used phenolic reagent is phenol and sodium salicylate. Sodium salicylate is nontoxic and can be used in open containers and reaction cells, while phenol is highly toxic, and waste must be disposed of appropriately. The advantage of phenate chemistry is that the optimum pH (10.5–11.5) is lower than that of salicylate chemistry ($\sim$12.6). The high pH promotes the precipitation of Mg^{2+}, and this interferes with the colorimetric measurement. Since saline waters and wastewaters have relative high Mg^{2+} contents, the phenate method is preferred for saline water and waste analyses. Chelating agents, such as EDTA citrate and sodium potassium tartrate, are often used to prevent the precipitation of Mg^{2+} at high pH condition. Sodium potassium tartrate and EDTA should be used for samples with high Mg^{2+} contents and/or high pH, especially when sodium salicylate is used as the phenolic agent.

14.2.1.2.2   Ion-Selective Electrode

The ion-selective electrode method is a relatively new but classic method and has received considerable attention over last 20 years due to its simplicity, convenience, and applicability to a wide range of ammonia concentrations in water. Ammonia is measured by diffusing of NH_3 gas through a gas-permeable membrane, thereby changing the internal pH that is sensed by a pH electrode. The unit is a complete electrochemical cell, and it is also referred to as a "gas-sensing probe." Diffusion of ammonia through the membrane is driven by the difference in the partial pressure between the water sample and internal solution of the probe. The equilibrium time is dependent on the ammonia concentration in the water, and, therefore, as the ammonia concentration decreases the equilibrium time increases.

Volatile amines interfere positively with ammonia analysis, i.e., inflate the measurements. The relative selectivity coefficients of amines are greater than unity, and the response times for ammines are nearly the same as for ammonia. However, since the concentrations of amines in water typically are much lower than those of ammonia, the interference of amines in the analysis of most water samples is insignificant. Mercury and silver interfere negatively with the analyses by complexing with ammonia, i.e., diminishing the measurements. Interference by mercury and silver can be prevented by the addition NaOH/EDTA to the samples.

14.2.1.2.3 Distillation-Titrimetric Method

The distillation step is used for analyzing samples with high ammonia concentrations and for removing interfering substances. Often the distillation step is necessary for the analysis of heavily polluted water and waste. Distillation can be used in conjunction with most ammonia analytical techniques, but it is a necessary pretreatment for use of the titrimetric method.

14.2.1.2.4 Nontraditional, New, or Advanced Methods

Fluorimetric method is a highly sensitive alternative to traditional colorimetric methods. This method involves a ternary reaction of o-phthaldiadehyde (OPA) under alkaline-buffered conditions (pH 9–10) in the presence of a reducing agent to produce intensely fluorescent isonodole. Mercapto-ethanol (ME) was used as the reducing agent when the method was initially developed. However, now sulfite is the preferred reducing agent, since it reacts so as to provide much greater sensitivity and selectivity for ammonia over amino acids. When ME is used as the reducing agent, primary amines, amino acids, and many inorganic species (such as sodium nitrate and calcium chloride) may interfere. In order to minimize interference, the gas diffusion fluorescence method has been developed, whereby NH is produced under high pH conditions and diffused across a gas-permeable membrane into a flowing stream of OPA/ME to produce a fluorescent adducts.

The flow injection fluorimetric method (Zhang and Dasgupta, 1989; Aminot et al., 2001), segmented flow fluorimetric method (Kerouel and Aminot, 1997), and manual fluorimetric method (Holmes et al., 1999) are all based on the reaction of ammonia with OPA and sulfite without significant interferences from amine and inorganic species. Phosphate buffer at pH near 11 is used for fresh water samples, while borate buffer at pH near 9 has been used for both the salt and fresh water samples. The relatively low pH of borate buffer prevents the precipitation of Mg^{2+} in seawater. The optimum pH window for the reaction is relatively small, and the control of the pH of the acidified samples is critical.

14.2.1.3 *In Situ* Methods for Ammonia Determination

14.2.1.3.1 Field Testing Kits

Many field "quick" testing kits for ammonia in soil and water are available commercially. However, little information is available to compare these methods with standard methods. Day (1996) and Burton and Pitt (2002) evaluated four test kits and concluded that these kits have acceptable MDLs. One kit requires refrigeration, and others contain mercury.

14.2.1.3.2 Field Monitoring Probes

Field-monitoring devices equipped with an ammonia-selective electrode have been used to continuously measure water quality parameters. Hydrolab-Hach Company (Loveland, CO, USA, http://www.hydrolab.com) and YSI Incorporated (Yellow Springs, OH, USA, http://www.ysi.com) are major manufacturers for these devices. The results of comparing two recent models are presented in Table 14.3. These devices accomplish ammonia analysis by means of an ammonia-selective electrode, and a data logger is used to record the measurements at a programmed interval. Measurements can be recorded at various intervals, such as daily or hourly, and continue for more than 2 months. However, frequent calibration is needed to assure the electrode is working properly. Weekly calibration is recommended. The probes are relatively sensitive for ammonia analysis with a resolution of 0.01 mg N/L for the Hydrolab series 4a and 0.001–1 mg N/L for the SYI 6600 Sonde. The initial cost of these instruments is high, and skilled personnel are needed to calibrate the probes.

14.2.1.3.3 Sophisticated Instruments for Field Analysis

Both the flow injection fluorescence method (FIA) (Aminot et al., 2001) and the gas diffusion fluorescence method (GDF) (Masserini and Fanning, 2000) have been used in the development of *in situ* ammonia analysis. FIA is more robust for *in situ* determination than air-segmented flow analysis, which is often used in the laboratory. Both analyzers are submersible systems that utilize the fluorescence-based chemistries described above, and each has a method detection limit of several nanomolars. The gas diffusion fluorescence FIA has been tested in field, and it can operate for many hours with relatively good accuracy and reproducibility.

14.2.1.4 Laboratory Methods for Nitrate and Nitrite Determination

Both the nitrate and nitrite ions can be directly determined by means of the ion chromatography method (IC). Nitrate also can be reduced to nitrite, and subsequently the nitrite is analyzed by the Griess assay. Many reducing agents have been used to facilitate the conversion of nitrate to nitrite, and the most common agents are copperized cadmium and hydrazine-copper.

TABLE 14.3

Comparison of *In Situ* Water Quality Monitoring Instruments Between Series 4a Multiprobe from Hydrolab-Hach Company (Loveland, CO, http://www.hydrolab.com) and YSI6600 Sonde from YSI Incorporated (Yellow Springs, OH, http://www.ysi.com)[a]

Parameter	Company	Range	Accuracy	Resolution
Ammonia	Hydrolab	0–100 mg/L-N	greater of ±5 of reading or ±2 mg/L-N	0.01 mg/L-N
	YSI	0–200 mg/L-N	greater of ±10% of reading or 2 mg/L	0.001–1 mg/L-N (range-dependent)
Nitrate	Hydrolab	0–100 mg/L-N	greater of ±5 of reading or ±2 mg/L-N	0.01 mg/L-N
	YSI	0–200 mg/L-N	greater of ±10% of reading or 2 mg/L	0.001–1 mg/L-N (range-dependent)
Chloride	Hydrolab	0.5–18,000 mg/L	greater of ±5 of reading or ±2 mg/l	4 digits
	YSI	0–1000 mg/L	greater of ±15% of reading or 5 mg/L	0.001–1 mg/L (range-dependent)
pH	Hydrolab	0–14 units	±0.2 units	0.01 units
	YSI	0–14 units	±0.2 units	0.01 units
Dissolved oxygen	Hydrolab	0 to 50 mg/L	±0.2 mg/L	0.01 mg/L
	YSI	0 to 50 mg/L	0 to 20 mg/L: greater of ±2% of reading or 0.2 mg/L; 20 to 50 mg/L: ±6% of reading	0.01 mg/L
Conductivity	Hydrolab	0–100 mS/cm	±1% of reading	4 digits
	YSI		±0.001 mS/cm	
Turbidity	Hydrolab	0–100 or 0–1000 NTU	±5% of reading	0.1 or 1 NTU
	YSI	0–1000 NTU	±0.5% of reading or 2 NTU	0.1 NTU

Please check websites for updated information on these instruments.

The UV spectroscopy method also has been used to directly determine the nitrate in unpolluted water with a low content of organic matter. The nitrate electrode method is another direct nitrate analysis method.

14.2.1.4.1 Griess Assay

Griess described the formation of diazo pigments in 1864, and he first used the reaction to measure the nitrite in 1879 (Griess and Bemerkungen, 1879). This method is based on nitrite, which results from the reduction of nitrate. Nitrite reacts with sulfanilamide to form a diazo compound that is then coupled with *n*-naftilethylenediamine hydrochloride in an acidic medium to form the azo dye. Maximum formation of the azo dye has been reported at pH 2.5–3 (Rider, 1946). However, pH near 2 may be the optimum for the nitrite determination, since the absorbance of the dye is constant, even though the absorbance at pH 2 is slightly lower than at pH 2.5 (Zhou, unpublished data).

Reduction of nitrate to nitrite. The most popular indirect method of nitrate analysis is based on the reduction of nitrate, to nitrite and subsequent determination of nitrite by Griess assay. The reduction can be accomplished by reaction with reducing reagent such as copper-cadmium, and copper-hydrazine, or through photochemical or enzymatic reactions. The photochemical and enzymatic reactions are relatively new methods that will be discussed in the nontraditional methods section. This indirect method is also called the NO_x method, since it determines both nitrate and nitrite. The nitrate concentration is calculated from the difference between NO_x and nitrite.

14.2.1.4.2 Using Copper-Cadmium

The use of cadmium as a reducing agent was reported in the 1960s. Since then, several kinds of cadmium reactors have been developed, such as the packed bed/column (Nydahl, 1976; Davison and Woof, 1978), wire-in-tube (Stainton, 1974; Willis, 1980), and open tubular (Willis and Gentry, 1987). When copper was introduced on the cadmium surface, the analytical results were improved (Okada et al., 1979). Copper-cadmium is preferred over zinc, hydrazine, and ultraviolet (UV) radiation because copper-cadmium provides near-quantitative reduction of nitrate to nitrite with negligible reduction of nitrite to lower oxidation species.

The cadmium reduction method that uses the copper-cadmium packed column has been adapted for continuous analysis: segmented flow or flow injection. Ammonium chloride and/or EDTA should be included in the buffer solution to prevent the precipitation of formed Cd^{2+} on the surface. The column must also be periodically regenerated to replace the metallic surface layer following the degradation of the reducing surface. This is caused by the gradual loss of copper, the precipitation of other species in the water, and adsorption of organic matter.

14.2.1.4.3 Use of Hydrazine Sulfate

The reduction of nitrate to nitrite by hydrazine sulfate in an alkaline medium and catalyzed by copper ions was first reported by Mullin and Riley in 1955. Since then several methods have been developed, and both the EPA and APHA Standard Methods documents have listed the hydrazine method as a reference method (USEPA, 1982; APHA, 1998). The main limitations of the method are that the rates of nitrate conversion to nitrite and the recovery of the nitrite originally present in the water both depend on the temperature, reaction time, pH, and hydrazine and copper concentrations. The use of a great excess of hydrazine and/or copper concentrations did not increases the conversion of nitrate to nitrite, but decreases the recovery of the nitrite originally present in the water, and eventually decreases the formation of the diazo dye from the nitrite that had been converted from the nitrate. Kahn and Brezenski (1967) reported that the hydrazine method gave low values. This was attributed to the consumption of the available hydrazine by environmental impurities, which resulted in the incomplete reduction of the nitrate to nitrite. It also been reported that the catalytic effect of copper is inhibited when natural chelating materials, such as humic or fulvic acids, exist in the water. The interference from humic and fulvic acids can be prevented by addition of zinc. Moreover the optimum hydrazine and copper concentrations in different methods with different instrument configurations are quite different (Downes, 1978; Hilton and Rigg, 1983, Oms et al., 1995; APHA, 1998); therefore, the optimum conditions in the reference method may not be directly adopted for use with a different analytical instrument.

14.2.1.4.4   Ion Chromatography

Nitrate and nitrite can be separated from other inorganic anions in aliquot samples by using ion chromatography (IC). Separation is mainly based on the ion exchange mechanism, which is the basis of ion pair/ion interaction chromatography, ion exclusion chromatography, chelation ion chromatography, and electrostatic chromatography. The main advantage of such IC techniques is their ability to simultaneously determine a range of different ions. However, the MDL of the IC method is higher than that of the continuous flow analyzer method.

Improvements of IC selectivity have mainly been obtained by developing new stationary phases. The stationary phases mainly are polystyrene divinylbenzene (PV-DVB) core particles surrounded by a monolayer of charged latex particles, which often have been used for the separation of nitrate and nitrite. Ethylvinylbenzene (EVB) cross-linked with 55% DVB used as core particles of the column makes the column 100% solvent compatible (Jackson et al., 2000). The core material is coated either permanently or dynamically with the molecules containing ionic functional groups such as alkyl quaternary ammonium or phosphonium, which act as fixed ion exchange sites of the stationary phases.

Nonsuppressed IC uses eluents of low equivalent conductance for separation. A device called a suppressor is added after the column in a suppressed IC instrument to modify the characteristics of the separated anions and of the eluent and therefore to improve the detection limit of conductometric detection. The suppressed IC with carbonate-bicarbonate as eluent has become a standard method for the analysis of nitrate, nitrite, and other anions by the EPA method (USEPA, 1982) and APHA's Standard method

(APHA, 1998). The use of hydroxide in the eluent has been studied recently for the hydroxide-selective column, and better method performance has been achieved (Jackson et al., 2000).

UV detection is also commonly used after conductometric detection. Because of the strong absorption of both the nitrate and nitrite in the UV region, a wavelength of 210 nm has been used to determine both ions after their separation via the column. Direct UV determination of nitrate and nitrite without using column separation will be discussed in the following section.

14.2.1.4.5   UV Method

The direct measurement of nitrate absorbance at 220 nm is used to screen water samples with low organic matter contents. Since nitrate, nitrite, and organic matter absorb between 200 and 230 nm, the separate determination of nitrate and nitrite by direct UV measurement is not possible at this region. Also, organic matter interferes in the determination of nitrate and nitrite. Since nitrite can be eliminated by addition of sulfamic acid and nitrate does not absorb at 275 nm, nitrate alone can be determined by addition of sulfamic acid, and the second measurement at 275 nm can be made to correct for the organic matter interference. Kishimoto et al. (2001) found that the coefficient of variation of relative absorbance of interfering substances at 215 nm was smaller and more stable than at 275 nm. In addition, most interfering substances adsorbed UV light at 215 and 220 nm. Kishimoto et al. (2001) proposed that absorbance at 215 nm instead of at 275 nm be used for interference correction. Wetters and Uglum (1970) used the spectral differences between nitrate and nitrite at longer wavelengths (300–360 nm) for simultaneous determination of both ions. It has been found that nitrate has negligible absorbance at 355 nm, whereas nitrite displays a characteristic absorption band at 302 nm. The adsorption ratio of nitrite at those two wavelengths is constant at 2.5, which is the correction factor for nitrate determination at 302 nm. Nitrate concentration in ground water has also been predicted using the UV/Vis spectroscopic data, processed by partial least-squares regression (Dahlen et al., 2000).

14.2.1.4.6   Nitrate Electrode

The main advantages of the nitrate electrode method are its relatively low cost, speed, and the possibility of the selective analyses of nitrate in complex matrices with little pretreatment. The main disadvantages are its relatively low selectivity and poor stability. Porous membrane electrodes have many applications in the analysis of in drinking water, natural water, residual water, and treated wastewater, and the nitrate electrode method has become a standard method (APHA, 1998). The interferences of chloride and bicarbonate are significant, but can be eliminated by addition of a buffer solution containing silver sulfate (Hulanick et al., 1974;

APHA, 1998.) The use of pH 3 buffer also maintains the constant ionic strength of the samples and of the standard. Nitrite interference can be masked by the addition of sulfamic acid to the samples. High humic substances also interfere. However, the disposable column packed with chemically bonded amine materials has been used to remove humic materials in water prior to nitrate measurement by means of the ion selective electrode (Csiky et al., 1985).

14.2.1.4.7 Nontraditional, New, or Advanced Methods Capillary Electrophoresis

Capillary electrophoresis (CE), often called capillary zone electrophoresis (CZE) in inorganic anion analysis, was developed for the simultaneous determination of nitrate and nitrite and other anions and has been proposed as a standard method in the latest edition of the APHA Standard Methods (APHA, 1998). The CZE method is based on differences in the mobilities of the various ions caused by differences in size and magnitude of charge when an electrical field is applied to the sample in a capillary. The attractive features of this method are its high efficiencies, tolerance to complex sample matrices, small sample requirements, small reagent consumption, and cost-effectiveness. Typically the method detection limit is similar to that of the IC. Sample stacking and on-capillary preconcentration techniques have been used to increase the sensitivity and decrease the detection limit, which is comparable with or superior to that of the Cu-Cd/ flow analysis method (Kaniansky et al., 1994; Okemgbo et al., 1999). The main disadvantages of this method are poorer precision and reproducibility compared to the IC method.

14.2.1.4.8 Photochemical and Enzymatic Nitrate Reductions

Photochemical and enzymatic nitrate reductions methods have been touted to be much more environmentally friendly than the Cu-Cd and Cu-hydrazine nitrate reduction methods. The photochemical reduction efficiency of nitrate to nitrite with a low-pressure lamp depends on the solution pH and the power of the lamp (Torro et al., 1998). These reduction efficiencies vary from 38.7% to 93%, according to different investigators (Takeda and Fujiwara, 1993; Motomizu and Masahiro, 1995; Torro et al., 1998; Mikuska and Vecera, 2003). Use of corn leaf nitrate reductase to catalyze the reduction of nitrate to nitrite has been reported to be more than 95% efficient (Pattonal et al., 2002). The reactivity of nitrate reductase, like that of other enzymes, is sensitive to environment conditions such as pH and temperature. The reactivity of nitrate reductase is maximal at pH ～ 7 and 30°C. The per sample cost of using nitrate reductase is about 20 times higher than that of the Cu-Cd reduction method without accounting for the expense of waste disposal in using the Cu-Cd reduction method.

14.2.1.5   *In Situ* Methods for Nitrate Determination

14.2.1.5.1   Field Testing Kits

There are more "quick" field-testing kits commercially available for nitrate and nitrite than for ammonia because the public has more concerns about nitrate in drinking water. USDA recommends using a nitrate/nitrite test strip to evaluate soil quality (USDA, 1999). In general, field-testing kits have low sensitivities and high MDLs. They are very helpful for qualitative analyses, but have limited roles for quantitative analyses.

14.2.1.5.2   Field Monitoring Probes

Both Hydrolab and YSI field monitoring instruments have a nitrate-selective electrode and can be used to continuously measure nitrate in water (Table 14.3). The probes are relatively sensitive for nitrate analyses, with a resolution of $0.01\,mg\,N/L$ for the Hydrolab series 4a and $0.001-1\,mg\,N/L$ for SYI 6600 Sonde. The accuracy of both instruments is $0.2\,mg\,N/L$.

14.2.1.5.3   Sophisticated Instruments for Field Analysis

Some *in situ* nitrate and nitrite methods have been developed that utilize a flow-injection analyzer (Blundell and Worsfold, 1995; Daniel et al., 1995; Petsul et al., 2001). In these instruments nitrate is reduced to nitrite using a cadmium column/redactor, and the resulting nitrite is determined using the Griess diazo-coupling reaction. The submersible analyzers are capable of *in situ* analysis of nitrate and/or nitrite in seawater with different depths or special variations (Daniel et al., 1995; Steimle et al., 2002). The short-term (several hours) nitrate and nitrite temporal variations in the water in a bay area were monitored using a submersible flow injection analyzer (Daniel et al., 1995). A battery-powered *in situ* FIA has been designed and tested to determine the long-term stability of the system for 23 days in laboratory (Blundell and Worsfold, 1995). An on-chip micro-flow injection analyzer (μFIA) has been developed for *in situ* nitrate and nitrite analyses with very low consumption of reagents and waste generation (Petsul et al., 2001). However, the μFIA requires further improvement, since its reproducibility is inadequate.

14.2.1.6   Organic N Determination

The digestion of soil and water samples to convert organic nitrogen compounds into inorganic forms is a necessary step to determine organic N. The Kjeldahl method is the classic and most popular method for organic N determination. Other methods such as persulfate, UV, microwave, and high-temperature oxidation methods will also be discussed.

14.2.1.6.1   Kjeldahl Method

The Kjeldahl method was first published in 1883 (Kjeldhal, 1883). The Kjeldahl method is the only standard in the APHA's Standard Methods for organic N analysis (APHA, 1998). Sample digestion is achieved by reaction of organic compounds with concentrated sulfuric acid at a high temperature (380°C) with the presence of potassium sulfate, and a catalyst to convert the organic N into ammonia. Potassium sulfate is used to increase the boiling point and, along with sulfuric acid, to produce the digestion temperature of 380°C. Copper, mercury, selenium, or titanium salts are used as catalysts. Mercury, which has highest efficiency, was previously the most popular catalyst. However, because of the high toxicity of mercury, a copper salt replaces it (Baethgen and Alley, 1989; Rohwedder and Pasquini, 1991), and the latter generally has good efficiency for determination of organic N in water samples. This method is used to determine the amino-nitrogen of most organic compounds in addition to free ammonia, but it fails to account for N in the form of azide, azine, azo, hydrazone, nitrate, nitrite, nitrile, nitro, nitroso, oxime, or semicarbazone. This method, referred to as the total Kjeldahl nitrogen method (TKN), provides the sum of organic N and ammonia. The ammonia in the digester is usually determined using the sodium salicylate or phenol indophenol blue colorimetry method and the distillation-titrimetric method. (see above).

14.2.1.6.2   Persulfate Method

The persulfate method was proposed as a standard in the 1995 edition of Standard Methods, and was officially adopted as such in the latest edition of Standard Methods (APHA, 1998). This digestion method uses a low concentration of nitrogen potassium persulfate as the oxidant to convert N compounds, such as organic N, ammonia, and nitrite into nitrate in an initial strongly alkaline medium (pH 12.5–13.2) under high temperature and pressure conditions (Koroleff, 1969; D'Elia et al., 1977). This method is relatively easer to use for total N analysis than the Kjeldahl procedure. The single digestion procedure for both total N (TN) and total P (TP) analysis has been developed (Langner and Hendrix, 1982; Ebina et al., 1983; Hoosmi and Sudo, 1986; Johnes and Heathwaite, 1992). This single digestion method requires that the initial high pH (12.6–12.8) needed for TN digestion then changes to the acidity (pH 2.0–2.1) required for TP digestion. This is accomplished during digestion due to the formation of potassium hydrogen sulfate from the potassium persulfate. However, when the water samples have high amounts of particulate P, the TP recovery may not adequate. Therefore, a separate acid persulfate digestion should be used regularly for TP analyses of samples with high amounts of particulate P.

14.2.1.6.3   High-Temperature Combustion Method

Very high-temperature (900–1100°C) combustion, or high-temperature (650–900°C) catalytic (Pt/Al2O3) combustion has been used to oxidize organic N compounds and inorganic N in water, soil, and sediment samples into nitric oxide. The resulting nitric oxide is directly measured with various detectors (e.g., chemiluminescent detector). The results from the high-temperature combustion (HTC) method were comparable with those of the UV method (Walsh, 1989), the persulfate method (Bronk, 2000), and the Kjeldahl digestion method (Clifford and McGaughey, 1982). Compared to other digestion methods, HTC is relatively easier and more convenient to apply, but the equipment required is more sophisticated (Table 14.4).

14.2.1.6.4 Nontraditional, New, or Advanced Methods

14.2.1.6.4.1 Microwave Method

The combined use of sulfuric acid and hydrogen peroxide for decomposition/ oxidation of organic substances and microwave irradiation has been applied as a time-saving alternative to the TKN digestion method for analyses of food, vegetation, and soils. A few reports have been published for this application to water samples (Collins et al., 1996). There are two fundamentally different methods of applying microwave energy in digestion. One approach uses the familiar cavity-type microwave; the other does not, but in it the sample is inserted directly into the wave-guide. The latter is referred to as focused or atmospheric pressure microwave and is often used in the TKN digestion method.

14.2.1.6.4.2 UV Method

The UV oxidation method, along with persulfate method and the high-temperature combustion method, is often used for seawater total nitrogen determination because the Kjeldahl digestion method is inappropriate for seawater analysis (D'Elia et al., 1977). The UV oxidation method is based on the fact that the decomposition of organic matter under UV irradiation takes place in the presence of small amounts of an oxidant, such as hydrogen peroxide, or persulfate. Hereby organic nitrogen and ammonia are converted into nitrite and nitrate. The digestion efficiency in some water samples using hydrogen peroxide as an oxidant is lower than when persulfate is used (Bronk et al., 2000). The UV digestion method often incorporates use of the automatic analyzer such as the FIA (McKelvie et al., 1994), or the segmented flow analyzer (Oleksy-Frenzel and Jekel, 1996) as an online digestion method. Online UV/persulfate digestion and oxidation with FIA has been proposed as a standard method in the latest edition of Standard Methods (APHA, 1998). This method may be suitable for dissolved organic nitrogen, but not for the particulate fraction that may require more drastic oxidation conditions for digestion.

14.2.2 PHOSPHORUS

14.2.2.1 Phosphorus in Soil and Water

Total phosphorus (P) concentration in surface soils varies from 200 to 5000 mg P/kg. Uncultivated soils generally have low P because the parent materials

TABLE 14.4
Selected Automated Carbon and Nitrogen Analyzers[a]

Brand names	Recent model	Sample types	Sample size	Element(s) analyzed	Analytical range	Analysis time (min)	Temperature	Company website
AnalytikJena	Multi EA3000	Soil, water	0.4–3 g	C, S, Cl	0.1–2000 mg/L C	2–4	1000	www.analytik-jena.com
Duratech	Carbon analyzer TC9300	Soil	0.5–5 g	C	30–500 µL	2	900	Home.ntelos.net/~duratech
Elemental	vario MAX Macro Elemental Analyzer	Soil, plant, water	5 mg–5 g	C, N, S	0.02–200 mg C 0.02–30 mg N 0.02–15 mg S	10	900	www.chnos.com
Hach	STIP-toc high temperature TOC analyzer	Water		C	2–50,000 mg/L	3–15	600–900	www.isco.com
Leco	CNS 2000 analyzer	Soil, plant	0.25–0.5 g	C, N, S	0.15 g	4.5	<1450	www.leco.com
Perkin-Elmer	2400 Series II CHNS/O Analyzer	Soil, plant, water	<0.5 g	C, N, S	0.001–3.6 mg C 0.001–6 mg N 0.001–2 mg S	6–8	1800	www.perkinelmer.com
Shimadzu	TOC-VCPH Analyzer	Soil, water	0.01–2 mL <0.5 g	C, N	0–25 g/L 0–4 g/L 0.1 mg–30 mg	3 4 6	680 900	www.shimadzu.com
Teledyne Tekmar	Apollo 9000 Analyzer	Water	>0.5 µL	C, N	4 ppb–2500 ppm	15	~680	www.teledynetekmar.com
Thermo	Flash EA. 1112 automatic Elemental Analyzers	Soil, Water	0.01–100 mg	C, H, N, S, O	0.01–100%	5–10		www.thermo.com
OI Cooperation	1020A TOC Analyzer	Water	0.025–0.4 mL	C	0.025–10,000 mg/L	5–7	900	www.oico.com

were low in P. Therefore, P is the most frequent limiting nutrient for agronomic crop productions in agricultural ecosystems and for plants in natural ecosystems. Using P fertilizers has significantly improved crop yields and quality worldwide. Application rates of P fertilizers can be as high as 300 kg P/ha/year. Organic soil amendments (animal manures, compost, and biosolids) also contribute large amounts of P in many regions. Phosphorus that builds up in the soils is susceptible to loss through surface runoff and leaching. Soil P is present both in inorganic and organic forms. In most soils 50–75% of total P exists as inorganic P. Organic P is mineralized to become inorganic P, mainly $H_2PO_4^{-1}$ in acid soils and HPO_4^{-2} in calcareous soils. Much of the inorganic P in the soil solution is either sorbed by soil particles or precipitated.

Traditional soil testing involves the use of an extractant to remove reproducibly and consistently a certain proportion of the inorganic P. Various extraction methods have been developed for various soils. Recommended extraction methods for P in U.S. soil are given in Table 14.5. For plant-available P, we recommend the use of Mehlich 3 extractant because it can be used for multinutrient extraction in a wide range of soils. However, the Mehlich 3 method has not been developed for soils with high concentrations of calcium carbonate.

The transport of P in soils is largely determined by the various forms of P that are present. Several chemical fractionation schemes have been developed to characterize P fractions in soils (Graetz and Nair, 1999). We recommend the method developed by Nair et al. (1995). The sequential extraction procedure distinguished six forms of P, with the extractant listed in parentheses: water soluble (deionized H_2O), exchangeable (1 M NH_4Cl), Fe- and Al-bound (0.1 M NaOH), Ca- and Mg-bound (0.5 M HCl), NaOH-extractable organic, and residual P.

Phosphorus that builds up in the agricultural soil from P fertilization and application of soil organic amendments is susceptible to loss in surface water through surface runoff, or in the groundwater through leaching. High concentrations of P in surface waters may also come from storm water runoff from urban areas, P mining, and industrial manufacturing or the use of P products.

Phosphorus is an important contaminant of surface water, since even low concentrations can lead to algal blooms, which diminish the recreational value of lakes and rivers. Restoration of the Everglades, a 6100 Km^2 natural wetland, required the adoption of an unusually low surface water P-concentration standard. This is so because levels of phosphorus above 10 μg/L are believed to alter the species composition of the natural plant and animal communities of the Everglades. The USEPA water quality criteria state that phosphates should not exceed 0.05 mg/L if streams discharge into lakes or reservoirs, 0.025 mg/L within a lake or reservoir, and 0.1 mg/L in streams or flowing waters not discharging into lakes or reservoirs. These limits are intended to control algal growth (USEPA, 1986).

There are several forms of P that can be measured. Total P (TP) is a measure of all the forms of phosphorus, dissolved or particulate. Soluble reactive

TABLE 14.5
Common Extractants Used for Plant Available Phosphorus in Soils in the United States

Soil tests	Chemical formula	Analysis	Specification	Ref.
AB-DTPA	1 M NH_4HCO_3 + 0.005 M DTPA (pH 7.5)	P, metals	Good for calcareous soils. The method also can be used for other nutrient extraction.	Soltanpour and Schwab, 1977
Bray P1	0.03 M NH_4F + 0.025 M HCl	P	Good for acidic or neutral soils. For calcareous soils, calcium carbonate can neutralize the acidity and precipitate the fluoride in the extractant and results low P results.	Bray and Kurtz, 1945
Mehlich 1 "double acid"	0.05 M HCl + 0.0125 M H_2SO_4	P, metals	Good for soils with pH < 7.0, low cation exchange capacity (CEC). For calcareous soils, calcium carbonate can neutralize the acidity in the extractant and results low results.	Mehlich, 1953
Mehlich 3	0.2 M CH_3COOH + 0.25 M NH_4NO_3 + 0.013 M HNO_3 + 0.015 M NH_4F + 0.001 M EDTA	P, metals	Good for wide range of soils. The method also can be used for multiple nutrient extraction.	Mehlich, 1984
Morgan	0.54 M CH_3COOH + 0.7 M CH_3COONa (pH 4.8)	P	Good for low CEC soils. For calcareous soils, calcium carbonate can neutralize the acidity in the extractant and results low P results.	Morgan, 1941
Olsen	0.5 M $NaHCO_3$ (pH 8.5)	P	Good for calcareous soils, but also performs well on moderately acid soils.	Olsen et al., 1954

P (SRP) is a measure of orthophosphate, the filterable (soluble, inorganic) fraction of P, and very small fraction of water-soluble organic P that is hydrolyzed into inorganic P in acid conditions. Both P and orthophosphate are often measured using a colorimetric method. If total phosphorus is being measured, all forms of P are converted to dissolved orthophosphate. The P analysis methods and digestion methods for converting organic, particulate, and condensed phosphates into orthophosphate will be discussed in this section.

14.2.2.2 Laboratory Methods for Phosphorus Determination

The Murphy and Riley (1962) method, also called the ascorbic acid method, is the most widely used. The tin (II) chloride method is some what more sensitive, but less stable, than the Murphy and Riley method. The vanadomolybdophosphoric acid method is useful for samples with high phosphorus concentrations. Ion chromatography and capillary ion electrophoresis can be used to determine orthophosphate with relatively high concentrations. The determination of TP in water involves the conversion of organic and condensed phosphates into orthophosphate. The use of heat, an acid, and an oxidizing agent are essential for the complete release of P and its conversion into orthophosphate. Autoclave digestion with acid persulfate is the most common and relatively simple TP digestion method. The perchloric acid digestion method is the most drastic and time-consuming method and is only recommended for particularly difficult samples such as sediments. The nitric acid–sulfuric acid digestion method is suitable for most samples.

14.2.2.2.1 Colorimetry Methods

The colorimetry methods listed in the APHA standard methods (APHA, 1998) are mostly based on the reaction of phosphate with acidic molybdate to form 12-molybdophosphate, which, when reduced, forms a highly colored phosphomolybdenum blue, using a variety of reductants.

14.2.2.2.1.1 The Ascorbic Acid Method

The ascorbic acid method is the most widely used method for batch and automated analyses. It uses ascorbic acid as a reductant with a catalyst of potassium antimony tartrate. Detection is achieved at either 660 or 880 nm, depending on the nature of the phosphomolybdenum blue species (Towns, 1986). The absorbance reading at 880 nm is about 1.4 times higher than that at 660 nm. The ascorbic acid method is less salt and temperature sensitive and produces a more stable phosphomolybdenum blue color than the tin (II) chloride method. Because both the color formation and selectivity of this method are pH and [H]/[Mo] dependent, it is very important that the final acid concentration is kept in the optimum range of 0.3–0.5 N (Zhou and Struve, 2004a). When the acid concentration and/or the [H]/[Mo] is low, the color formation is very rapid. However, molybdenum blue will form through the direct reduction of Mo (VI) even without phosphate being present, and silicate and arsenate will also interfere with phosphate analysis by forming the blue-colored silicate or arsenate molybdenum blue. When the acid concentration and/or the [H]/[Mo] are high, the rate of phosphomolybdenum blue formation becomes very slow, especially for samples with low phosphate concentrations. The method detection limit (MDL) of this method typically ranges from 2 to $10\,\mu g\,P\,L^{-1}$ The ascorbic acid method with low MDL has been developed using an autoanalyzer (Kennelley and Mylavarapu, 2002). However, this method has very limited application in natural water P analysis due to the high silicate interference. It is possible to develop a low MDL ($<1\,\mu g\,P\,L^{-1}$) P method with low silicate interference for an autoanalyzer such as flow injection analyzer, when acid concentration and/or [H]/[Mo] of a method are optimized base both on the maximum phosphomolybdenum blue formation without Mo(VI) reduction and the minimum silicate interference (Zhou and Struve, 2004b).

14.2.2.2.1.2   Tin (II) Chloride Method

The tin (II) chloride method can reduce 12-molybdophosphate very quickly to form phosphomolybdenum blue and gives very sensitive results. However, the colored product is not stable, and, therefore, it may only be used in the automated method in which the reaction time of each sample is strictly controlled. This method is also susceptible to salt interference. It is not suitable for marine and estuarine water samples, since chloride inhibits molybdophosphate reduction.

14.2.2.2.1.3   Vanadomolybdophosphoric Acid Method

Phosphate reacts with ammonium metavanadate and with ammonium molybdate under acid conditions to form unreduced yellow-colored vanado-molybdophosphoric acid. The latter has maximum absorbance at 470 nm. This method has relatively low sensitivity, and the interference and is suitable for analyzing wastewaters and highly polluted waters.

14.2.2.2.2   Chromatographic Techniques

IC and CE techniques (described above) have been used in orthophos-phate analyses. Usually these techniques are not suitable for total phosphorus analyses after digestion, since the acid and/or salt from the digestion reagents introduce high concentrations of anions that are likely to strongly interfere and affect the resolution and retention times of other anions. The sensitivities of the conductivity detection techniques for IC commonly used are generally inadequate for direct application to analyses of pristine waters, and preconcentration techniques may be required. Phosphate is relative difficult anion to determine by means of CE compared to nitrate and chloride (Van den Hoop and van Staden, 1997).

14.2.2.2.3 Digestion Method

The determination of TP in water samples involves the conversion of particulate, organic and condensed phosphates into orthophosphate under high temperature, high acidity, and in an oxidizing environment followed by determination of the released orthophosphate. After acid digestion, the digestate is usually neutralized to control the final the acid concentration (APHA, 1998). For the perchloric acid method and sulfuric acid–nitric acid method, neutralization may not be required. However, the neutralization step may not be required for the acid persulfate method, since the acid concentration from acid persulfate can be estimated. The acid persulfate autoclave method is the most common method for digestion of water samples in TP analysis. This method is able to completely recover the P from most organic P compounds, condensed P, and sediment P. However, when water samples have relatively high sediment and/or organics contents, it is important to make certain that the persulfate concentration is sufficiently high to oxidize the organics and that the acid concentration is sufficiently high to the release through hydrolysis of the phosphate from condensed phosphates and particles. At typical persulfate and acid concentrations, this method can completely recover P from water samples with organic carbon contents up to 100 mg/L and from the sediment in water samples with TP up to 200 µg P/L (Zhou and Struve, 2004a). The posttreatment procedures after digestion in EPA Methods 365.1, 365.2, and 365.3 (USEPA, 1982) are quite different from those in Standard Method (SM) 4500 (APHA, 1998). EPA Method 365.2 requires the neutralizing and filtering of the digestates before TP analysis, while EPA Method 365.1 requires cooling the digestates and EPA Method 365.3 requires both cooling and filtering of the digestates before TP analysis. In contrast, SM 4500 requires neutralizing and mixing of the digestates before TP analysis. When water samples have insoluble particulates, which is common in surface waters, the mixing step before taking an aliquot of digestate for TP analysis tends to result in overestimates of the TP concentrations in the samples (Zhou and Struve, 2004c). If the digestate is neutralized and allowed to stand before taking an aliquot of digestate for TP analysis, the TP concentration will be underestimated due to the adsorption/precipitation of phosphate on/with the Fe and Al hydroxides (Zhou and Struve, 2004c). Reacidification and settling after neutralization can prevent the P loss from solution due to precipitation/adsorption, as well as interference from the insoluble particulates. Since the acid concentration contributed by digestion can be estimated, and there is an optimum acid concentration plateau for the ascorbic acid method, neutralization and reacidification steps can be eliminated, and, thus, only the settling step is required before taking an aliquot digestate for TP analysis.

The Alkaline and Acid Persulfate Digestion Method can completely recover TP from sediment up to 100 µg P/L (Lambert and Maher, 1995) and 200 µg P/L (Zhou and Struve, 2004a), respectively. There were no significant differences in recovery of TP in sediment samples between the Persulfate Method and Nitric Acid-Sulfuric Acid Digestion Method. The Persulfate Digestion Method gives a somewhat better precision than the Sulfuric Acid-Nitric Acid Digestion Method.

14.2.2.2.4   Nontraditional, New, or Advanced Methods

14.2.2.2.4.1   Colorimetry Methods

Some cationic dyes such as malachite green, crystal violet, and quinaldine red react with phosphate at low pH to form the highly colored phosphomolybdate complex. The malachite green method typically uses two color reagents (ammonium molybdate/H_2SO_4 and polyvinyl alcohol/malachite green). It has been found to be the most sensitive method among the cationic dye methods (Cogan et al., 1999) and is also much more sensitive than the most commonly used ascorbic acid method. However, the MDL of the malachite green method is no better than that of the ascorbic acid method, and the malachite green method has not been as widely used as the ascorbic acid method due to fading of the color (McKelvie et al., 1995), and adsorption of the complex on the surface of the tubing of the automatic flow instrument (Munoz et al., 1997). A microplate reader has been devised for determination of absorbance of phosphomolybdate complex in the malachite green method (Cogan, et al., 1999; D'Angelo et al., 2001). It uses fewer reagents, is faster than manual reading, and eliminates the carryover problem associated with the automatic continuous flow analyzer. However, it is difficult to avoid the generation of air bubbles during the addition of either samples or reagents, or when mixing in the microplate wells. Bubbles are generated because most surface water samples contain natural surfactant, and the PVA polymer used to stabilize the color is also surface active. The complete mixing of the reagent and the sample inside a small diameter well without generation of the air bubbles is a most challenging task. Researchers using microplates in the two reagent malachite green method have concluded that special care must be taken to ensure complete mixing and to avoid formation of air bubbles in wells during sample and reagent addition and mixing (D'Angelo et al., 2001). A new single reagent malachite green method has been developed in one of the authors' laboratories for low-level phosphorus analysis. This method uses a repipettor to add the single color reagent to each sample in the microplate well to minimize the color instability problem, and it employs a sonicator for both mixing and debubbling (Zhou and Struve, 2004d).

14.2.2.2.4.2   Preconcentration Methods

Several solvent extraction techniques have been developed for colorimetric phosphorus analysis to improve detection limits and to minimize interference. Solvents such as pentanol, benzene, butanol, 2-butanol, chloroform, and butyl acetate have been used. However, the use of classical solvent extraction is very slow and in some cases involves toxic solvents. More recently, several methods using filter or solid phase preconcentration techniques have been developed in which phosphomolybdenum blue or phosphomolybdate-malachite green complex is adsorbed on the filter or column. The latter methods have MDLs of less than $1 \mu g L^{-1} P$ (Taguchi et al., 1985, Susanto et al., 1995; Heckemann, 2000; Zui and Birks, 2000). These methods were developed and tested for orthophosphate determination in water. A new method that utilizes solid phase extraction/preconcentration of a malachite green–phosphomolybdenum complex has been developed for the spectrophotometric determination of TP in water (Zhou and Struve, 2002). Water samples are digested using persulfate/sulfuric acid digestion in a microprocessor-controlled block digester inside a clean laminar-flow hood. The phosphorus in the sample is then collected as a malachite green–phosphomolybdenum complex on a cellulose nitrate membrane filter. The absorbance of the malachite green–phosphomolybdenum complex is determined using a microplate reader after the filter with the malachite green–phosphomolybdenum complex has been dissolved into methyl cellosolve. The preconcentration methods are able to determine trace level orthophosphate or total phosphorus. However the preconcentration step is quite time consuming, and some methods that include the preconcentration of phosphomolybdenum blue require large-volume water samples.

14.2.2.2.4.3 Digestion Methods for Total Phosphorus

The use of UV photo-oxidation of organic phosphorus compounds combined with thermal hydrolysis of condensed phosphates has been reported for analysis of water and wastewater (Benson et al., 1996). It was found necessary to use a mixture of acid and persulfate to form Caro's acid in order to obtain high recoveries of both organic and condensed phosphorus.

14.2.2.3 *In Situ* Methods for Phosphorus Determination

The major *in situ* phosphorus analytical instruments are designed for orthophosphate determination (Worsfold and Clinch, 1987; Blundell and Worsfold, 1995; Zheng et al., 1998; Hanrahan et al., 2001). Flow injection analysis (FIA) has been used for *in situ* orthophosphate analysis, and the instruments have been tested in the field for short periods (3 days) (Blundell and Worsfold, 1995; Hanrahan et al., 2001). The detection limit of the newly developed FIA field instrument ($\sim 6 \mu g$ P/L) was near the MDL of the laboratory instrument (Hanrahan et al., 2001). However, when using *in situ* high temporal resolution field orthophosphate data to refine the export coefficient model for total phosphorus loading, the assumption of a constant ratio of TP to orthophosphate must be made to estimate the TP concentration in water (Hanrahan et al., 2001). A remote P-analyzer has been tested and optimized in one of the coauthors' laboratory. It employs UV/thermal-induced persulfate digestion of total phosphorus in water and the ascorbic acid–phosphomolybdenum blue method for TP and total reactive phosphorus (TRP) determination. The remote P-analyzer were tested in the field for nearly 3 months (Struve et al., 2004).

The analytical parameters were remotely monitored, and the data were remotely processed. This remote P-analyzer successfully provided near real-time TP and TRP data of the water. The P concentration results from this analyzer were comparable to those obtained with the laboratory analyses, and the method detection limit was also similar ($4 \mu g/L$) to that of the laboratory method.

14.2.3 METALS

14.2.3.1 Metals in Soil and Water

When considering the possible impact of elements on soil and water quality, Wood (1974) classified elements into three categories: 1) non-critical (e.g., Na, K, Mg, Ca, H, O, N, C, P, Fe, S, Cl, Br, F, Li, Rb, Sr, Al, Si); 2) toxic and readily accessible (e.g., Be, Co, Ni, Cu, Zn, Sn, As, Se, Te, Pd, Ag, Cd, Pt, Av, Hg, Te, Pb, Ag, Cd, Pt, Av, Hg, Te, Pb, Sb, Bi); and 3) toxic but very insoluble or rare (e.g., Ti, Hf, Zr, W, Nb, Ta, Re, Ga, La, Os, Rh, Ir, Ru, Ba). Many of these elements are not of agronomic and environmental concern. It may be more useful to subdivide the metals routinely analyzed in soil testing laboratories into two groups, i.e., nutritional metals and toxic metals. Nutritional metals include K, Ca, Mg, Fe, Mn, Mo, Zn, and Cu. They are essential nutrients and sometimes are limiting factors for plant growth. Plants cannot grow well without adequate amounts of these elements in the soil. These elements are often applied to soils as fertilizers. High concentrations of these metals in surface water may increase algal growth and cause eutrophication. Common toxic metals, which include As, Pb, Cd, Cu, Cr, Ni, Zn, and Hg, are associated with anthropogenic activities such as disposal of wastes in landfills, land application of biosolids, pesticide applications, etc. Copper and Zn are both nutritional and toxic metals because they pose threats to water quality if the concentrations are high. The U.S. drinking water standard for Cu is $1300 \mu g/L$.

Metals exist in soils in various forms and can be placed into the following groups: water soluble, exchangeable, adsorbed, organic bound, associated with oxide, carbonate bound, and mineral (residual fraction). Many extraction (speciation) procedures were developed to characterize various forms of metals in soils. These procedures are not specific, but only selective. Shuman (1991) provides a detailed discussion of these extraction procedures. We recommend the Tessier method for calcareous soils and Shuman method for acidic soils (Tessier et al., 1979, Shuman, 1985).

Nutritional metals are often evaluated for their bioavailability with various soil extraction methods, which are different as sequential extraction produced, discussed. Sims and Johnson (1991) described these methods. The common extracting solutions for plant available metals in soils are listed in Table 14.5. These analyses are very important for the proper fertilization of crops.

14.2.3.2   Laboratory Methods for Metal Determination

The titrimetric (volumetric) method is the one of classical methods for metal analysis. The EDTA titrimetric method has been used to determine calcium and magnesium concentrations in water. Spectrophotometry, another classical method, is based on the formation of colored complex of the metal and specific reagents. Spectrophotometry is widely used for the determination of the major transition elements. Thus the ferrozine method is used for Fe (II) determination in water, especially in pore water. Ferrozine is a chelating agent that forms the pink ferrozine complex with Fe (II). Fe (III) also can be determined after addition of reducing agent such as ascorbic acid to convert Fe (III) to Fe (II).

Spectrometric methods that include atomic absorption (AA) spectrometry, inductively coupled plasma emission spectrometry (ICP), and inductively coupled plasma mass spectrometry (ICP-MS) are the most common instrumental methods for metal analysis. There are several AA methods that can be selected for metal analysis depending on the metal and/or concentration to be determined. Flame AA is a relatively inexpensive method that can be used to determine the major metals in most aquatic system. Electrothermal (graphite furnace) AA is about two orders of magnitude more sensitive than flame AA and is capable of analyzing most metals in water. The cold vapor AA method and the hydride generation–AA method are designed for mercury and arsenic/selenium analysis, respectively. The sensitivity of the ICP method is about one order of magnitude higher than that of the flame AA method. The ICP method has a very wide linear dynamic range (up to four to six orders of magnitude). Because most elements exist predominantly as singly charged ions in the plasma, ICP can be effectively used as an ionization source for mass spectrometry. ICP-MS is the one of the most sensitive, accurate, and reliable method for the metal analysis. The wide dynamic range of the ICP method and the potential for simultaneous analysis of various metals using the polychromator provides the possibility of analysis of both major and trace metals in single run.

14.2.3.3   *In Situ* Method for Metal Determination

A field-portable x-ray fluorescence (FPXRF) instrument is used by USEPA for analysis of metals in soils and sediments. It is a small, hand-held instrument ~1 kg) capable of rapid measurement (~2 min/sample) at low cost and suitable for the *in situ* analysis of metal contaminants (Raab et al., 1990). Results from FPXRF correlated extremely well with laboratory atomic spectrometry (Shefsky, 1997). However, the accuracy of the *in situ* method depends on site-specific conditions, especially contaminant particle size and distribution. EPA method 6200 discusses the limitations of the XRF and should be considered before using the data for anything more than screening.

14.2.4 ORGANIC MATTER/CARBON

14.2.4.1 Organic Carbon in Soils and Water

Soil organic matter (SOM) is the organic fraction of soils and consists primarily of plant, animal, and microbial residues in different stages of decomposition. It is restricted to organic materials that pass through a 2-mm sieve during soil preparation. Thus, large pieces of roots and other plant residues should be excluded from analysis. Soil organic matter contents range from less than 0.5% for desert soil to more than 15% for histosols (organic soils). Soil organic matter is only a small fraction of the total soil, but it is a very important component because it improves the physical, chemical, and biological properties of soil and has beneficial effects on soil quality.

Soil organic matter contains about 58% carbon and oxygen, hydrogen, nitrogen, sulfur, and other mineral nutrients. Organic matter in soils is usually estimated by multiplying the concentration of carbon by 1.724. However, the conversion factor used to relate organic C to soil organic matter could be as high as 2.5 (Broadbent, 1965). Because no reliable direct method for measuring SOM exists, it is probably best to use organic C as an index of SOM without conversion.

Soil organic carbon exists in soil in various forms or fractions, which have different functions with respect to soil quality. Several fractionation procedures have been developed to characterize soil organic mater fractions based on the solubility, molecular size, and electrostatic charge of the molecules (Swift, 1996). A detailed discussion of these fractionation methods was presented by Swift (1996).

The commonly used fractionation method for soil organic carbon is based on the solubility of humic substances and divides soil organic matter into fulvic acid, humic acid, and humin. The fulvic acid fraction is soluble in an alkaline solution (pH 10–13), and it is soluble in the acidified alkaline extract. The humic acid fraction precipitates in the acidified alkaline extract. The unextracted alkaline-insoluble residue is humin (Zinati et al., 2001). This fractionation procedure is just like those for P and for metals that are very objective.

Concentrations of organic carbon in surface and groundwater are very low. Organic carbon affects biological and chemical processes such as biochemical oxygen demand (BOD) and chemical oxygen demand (COD). Organic carbon in water can be divided into fractions of dissolved organic carbon (that passes through a 45 μm filter), particulate organic carbon (non dissolved organic carbon that is retained by a 0.45 μm filter), and volatile organic carbon. For most surface and groundwater, volatile organic carbon is negligible.

14.2.4.2 Organic Carbon Determination

The methods for soil organic matter range from a simple color comparison to the use of an expensive and precise automated carbon analyzer. Simple methods such as soil color are of little value because soil color is not related only to organic matter. Other methods are based on three principles: 1) wet oxidation, 2) dry combustion, and 3) weight loss. Three commonly used methods based on these principles for soil organic carbon determination are listed in Table 14.6. The Walkley-Black method is the widely accepted standard. The carbon analyzer has become the preferred instrument for determining total and organic carbon in soil and water. Loss-on-ignition (LOI) is a good alternative method because it is simple, quick, and cheap.

14.2.4.2.1 Walkley-Black Method (Wet Oxidation)

The Walkley-Black method uses potassium dichromate ($K_2Cr_2O_7$) with externally applied heat and back-titration to measure the amount of unreacted dichromate. Since the color changes during titration are not easily recognized, overtitration may occur. An alternative to titration is to measure Cr^{3+} (red color) directly using a spectrometer at 660 nm. The Walkley-Black method is widely used in the routine analysis in soil testing laboratories. The method is used primarily for mineral soils with organic carbon as high as 12%. It is not suitable for histosols (organic soils) because of incomplete digestion. The method is subject to interferences from chlorite, iron oxide, and manganese. The presence of large amounts of Cl^- and Fe^{2+} will lead to overestimation and that of Mn to underestimation of organic C.

14.2.4.2.2 Carbon Analyzers (Dry Combustion)

The carbon analyzer measures total C by converting all C in soil and water to CO_2 and then quantifying the evolved CO_2 by a detector [e.g., nodispersive infrared (NDIR) detector]. The principle of the carbon analyzer was described by Pansu et al. (2001). When using combustion methods, correction for inorganic C must be performed for soil with inorganic C materials, such as calcium carbonate, charcoal, plant materials, etc. The precision of the carbon analyzer is also directly related to the quality of the carrier gas. Oxygen purity of 99.9% is required. Sample size required by most analyzers is too small to obtain reliable estimates. It is impossible to obtain representative subsamples from samples of less than 0.1 g.

14.2.4.2.3 Loss-on-Ignition

The loss-on-ignition method measures the weight loss from soil caused by complete combustion. Briefly, a soil sample is dried at 105°C for 12 hours and ashed at 500°C for 16 hours (Zhang et al., 2004). The loss in weight between 105 and 500°C is calculated as organic matter content. Different ashing temperatures ranging from 360 to 600°C have been used (Schulte and Hopkins, 1996). Low ignition temperatures result in incomplete combustion and the underestimation of organic carbon, while high ignition temperatures cause losses of inorganic carbon and overestimation of organic carbon.

TABLE 14.6
Common Methods of Organic Matter/Carbon Determination for Soil and Water

Method	Principle	Strengths	Weakness	Ref.
Walkley-Black (wet oxidation)	Organic carbon is oxidized in an acid dichromate solution followed by back titration of the remaining dichromate with ferrous ammonium sulfate.	Widely accepted, no expensive instrument needed, relatively accurate	Need correction factor for incomplete oxidation, Produce hazardous waste, interference by Cl, Fe, Mn	Walkley-Black, 1934
Carbon analyzer (Dry combustion)	Organic carbon is oxidized in a furnace followed by direct determination of the evolved CO_2.	Accurate, quick	Expensive instrument and consumables, need pure O_2 gas	Pansu et al., 2001
Loss-on-ignition (LOI)	Organic carbon is calculated based on weight loss of a sample due to high temperature ignition.	Quick, cheap, simple	Not accurate, depends on many factors	Schulte and Hopkins, 1996

14.2.5 PESTICIDES

14.2.5.1 Pesticides in Soil and Water

Pesticides are widely used for crop production and public health by controlling insects, diseases, and weeds. Pesticides used to control insects have saved millions of people from diseases such as malaria, yellow fever, and typhus. Agricultural use of pesticides is one of the important practices for crop production. Since 1950, the use of pesticides in agriculture has played a major role in doubling the world food supply without increasing acreage used for food production. However, impacts of pesticides on the environment have increasingly raised the public's concerns. Pesticides applied to soil may be taken up by plants, volatilized into atmosphere, broken down by sunlight, or absorbed to soil particles. The fate of pesticide in soil largely depends on characteristics of the pesticide, soil type and character, and climate conditions. Pesticides in soils may be sorbed on soil particles, degraded into other chemical forms by physical/chemical processes and/or microbes, leached into ground water, or transferred into ground water through surface runoff. The physical/chemical properties of a pesticide have a great influence on its fate in the environment. For example, a pesticide with a high solubility and low degradation rate may have a great propensity to leach into ground water in an irrigated sandy soil. Conversely, if a pesticide is insoluble and strongly binds to soil particles, as when applied to a clay soil, it may potentially remain in surface soil or be lost through surface runoff.

In the survey conducted by USEPA in 1991, 10.4% of 94,600 community water system wells contained one or more pesticides and about 0.8% of these wells had pesticide concentrations above health-based limits (USEPA, 1991). Under the provisions of the Safe Drinking Water Act, the USEPA has MCLs for concentrations of 12 herbicides and 10 insecticides in drinking water (Table 14.1) (USEPA, 2003). Larson et al. (1997) reported organo-chlorine pesticides or their degradation products were commonly detected primarily in the 1960s and 1970s. In recent years the herbicides alachlor, atrazine, and simazine have frequently exceeded their MCLs.

The National Water-Quality Assessment Program of the U.S. Geological Survey collected water samples from 58 rivers and streams across the United States during 1992–1995 (Larson, 1999). Eleven herbicides, an herbicide degradation product, and three insecticides were detected in more than 10% of 2200 samples collected. The herbicides atrazine, metolachlor, prometon, and simazine were detected most frequently; among the insecticides, carbaryl, chlorpyrifos, and diazinon were detected the most frequently.

14.2.5.2 Sample Preparation

The analysis of pesticides in soil and water samples seldom can be performed directly because these compounds are usually present at trace levels (Nelson and Dowdy, 1988; Zhao and Page, 2001). Steps must be taken to efficiently separate these trace levels of pesticides in a matrix from the magnitude of interfering organic substances prior to instrumental analysis. A multistep process prior to instrumental analysis is often necessary, which would include extraction, cleanup, concentration, and derivatization:

1. Extraction is the quantitative removal of the pesticide from the matrix in which it is present. The extraction step must not coextract excessive amounts of materials that interfere with the analysis (Nelson and Dowdy, 1988). Liquid-liquid partitioning and solid-phase extraction are two commonly used techniques for pesticide extraction from water extraction. Extraction of pesticides from soil is far more complex than their recovery from water. The extraction efficiency of pesticides from soil can be affected by the analyst's choice of solvent, solvent to soil ratio, and extraction technique.
2. Cleanup is a term that usually connotes further separating incurred pesticides from matrix coextractants. To clean up the organic solvent extracts, liquid column chromatography is most commonly used. Florisil, alumina, silica gel, and charcoal are frequently used column packings. Gel permeation and ion-exchange columns have also been used for cleanup.
3. After extraction, the amounts of pesticides in water or organic solvents are often low enough that the sample must be concentrated prior to analysis in order to achieve reasonable detection limits. Solid-phase extraction can be used for concentration of pesticides directly out of water. Concentration of organic solvents resulting for liquid-liquid partitioning of water or organic solvent extraction of soil is performed by evaporation of the volatile solvent.
4. Derivatization is used to form a compound that is more readily chromatographed or more accurately or sensitively measured. Some pesticides such as substituted-urea herbicides are thermally labile and decompose at the high temperatures used in GC. These chemicals can be derivatized to form thermally stable species. Halogen-substituted derivatizing reagents are commonly used to increase volatility and to provide more selective, sensitive detection using an electron capture detector. For analysis by HPLC, pesticides can be reacted with UV chromophores to increase sensitivity to UV detectors. Some low-volatility, polar compounds such as acids, alcohols, amines, and aldehydes must be derivatized to increase volatility prior to analysis by GC. Organic acids and phenols are often converted to volatile methyl esters.

14.2.5.3 General Approach for Screening Pesticides in Soil and Water

Approximately 50,000 different pesticides composed of over 1000 active ingredients are registered in the United States. Many pesticides are mixtures of various active ingredients. It is very difficult to select analytical methods for pesticides. The following procedures generalize pesticide screening for soil and water samples: 1) sample collection (standard custody); 2) sample preparation; 3) extraction; 4) purification or cleanup; 5) concentration; 6) instrumental analysis; 7) data reduction (including library match); 8) result confirmation; and 9) report.

14.2.5.4   Laboratory Methods for Pesticide Determination

The most commonly used tools for quantitatively determining pesticides in water and soil matrix are GC and high-performance liquid chromatography (HPLC). Combination of GC or HPLC with mass spectrometry (MS) has become more common for qualitative determination or confirmation. Each method has its advantages and limitations. Details are given below.

14.2.5.4.1   Gas Chromatography

In general, organic compounds with relatively high volatility can be analyzed by GC. Similar to the HPLC separation principle, both stationary and mobile phases are applied in GC. But in gas chromatography, unlike in HPLC, the mobile phase does not interact with molecules of the analyte while it carries the analyte through the analytical column; separation is primarily achieved with the stationary phase only. The commonly used and chemically inert carrier gases in GC include helium, argon, hydrogen, and nitrogen. Selection of carrier gas is usually associated with the type of detector used. The ideal detector would have these characteristics: adequate sensitivity, good stability, and reproducibility; a linear response to solutes that extends over several orders of magnitude; a temperature range from room temperature to perhaps 400°C; a short response time that is independent of flow rate; high reliability and ease of use; similarity in response toward all solutes or alternatively a highly predictable and selective response toward one or more classes of solutes; and nondestruction of sample (Skoog, 1985). In reality, no "universal" detector exists that would meet all those ideal conditions. Three GC detectors among the most widely used ones in gas chromatography are described as the followings.

14.2.5.4.1.1   Electron-Capture Detectors

An electron from the emitter causes ionization of the carrier gas and the production of a burst of electrons. In the absence of organic species, a constant standing current between a pair of electrodes results from this ionization process. The current decreases, however, in the presence of those organic molecules that tend to capture electrons (Skoog, 1985). The electron-capture detector (ECD) is selective in its response, being highly sensitive toward molecules containing electronegative functional groups such as halogens, peroxides, quinones, and nitro groups. It can detect quantities in the pictogram (10^{-13} g) region (Table 14.7). It is insensitive toward functional

TABLE 14.7
Comparison of Instrumental Methods Commonly Used for Pesticide Analysis in Water and Soil

Method	Sensitivity (min sample, g)	Selectivity	Strength	Weakness
GC-ECD	10^{-13}	Halogens	Nondestructive to samples; highly sensitive with low detection limit	Insensitive toward functional groups such as amines, alcohols, and hydrocarbons
GC-FID	10^{-10}	None	The most widely used method for most organic compounds	Destructive to samples
GC-NPD	10^{-12}	N, P	Nondestructive to samples; highly sensitive	Only used for N- or P-containing herbicides
GC-MS	10^{-9}	Molecular ions	Strong tools for pesticide screen	High cost on maintaining instruments
HPLC-UV	10^{-9}	Wavelength	Nondestructive to samples; relatively easy method	Difficulty to separate similar compounds
HPLC-FLR	10^{-12}	Fluorescence	Highly sensitive; low detection limit	Derivatizing reagents must be made fresh; relatively high complexity in operation
HPLCMS	10^{-9}	Molecular ions	Useful tool for pesticide confirmation	High cost on maintaining instruments

groups such as amines, alcohols, and hydrocarbons. An important application of the ECD has been for the detection and determination of chlorinated insecticides, such as endosulfan, dicofol, and tetradifon. Electron capture detectors are highly sensitive and possess the advantage of not altering the sample significantly. On the other hand, their linear response range is usually limited to about two orders of magnitude (Skoog, 1985). Other pesticides that are typically analyzed with ECD include aldrin, captan, chlordane, chlorpyrifos, chlorpyrifos-methyl, diazinon, dieldrin, DDT, DDVP, endrin, heptachlor, marathon, methoxychlor, and parathion.

14.2.5.4.1.2   Flame Ionization Detector

Capillary columns are often employed in the FID analysis. The effluent from the column is mixed with hydrogen and air, and ignited. Organic compounds burning in the flame produce ions and electrons, which can conduct electricity through the flame. A large electrical potential is applied at the burner tip, and a collector electrode is located above the flame. An electrometer is used for measurement of the current. The FID is insensitive towards noncombustible gases such as H_2O, CO_2, SO_2, and NO_x. Therefore, the detector is a most useful one for the analysis of most organic samples, including those that are contaminated with water and the oxides of nitrogen and sulfur. The FID can detect quantities in the nanogram (10^{-10} g) region (Table 14.7). Some functional groups, such as carbonyl, alcohol, halogen, and amine, yield fewer ions or none at all and may not be detected by FID (Skoog, 1985).

14.2.5.4.1.3   Nitrogen-Phosphorus Detector

The NPD is a highly specific thermionic detector for organically bound nitrogen and phosphorus. The detector works by electrically heating a glass bead containing an alkali metal until electrons are emitted. These electrons are captured by stable intermediates to form a hydrogen plasma. The column effluent is ionized when directed into this plasma. A polarizing field directs these resulting ions to a collector anode creating a current. This detector's sensitivity is dependent upon the air flow, while its selectivity is affected by hydrogen flow. The NPD can detect quantities in the picogram (10^{-12} g) region (Table 14.7). This detector is most commonly used for EPA Method 614 and SW-846 Methods 8140 and 8141. Pesticides that are analyzed with NPD include atrazine, captan, chlorpyrifos, DDVP, diazinon, eradicane, malathion, naled, parathion, and pirimiphos-methyl.

14.2.5.4.2   High Performance Liquid Chromatography

HPLC has become popular in analytical laboratories. It is mostly used for analyzing materials that are thermally labile or nonvolatile. In contrast to GC, the mobile phase used in HPLC react with the molecules of analytes that also react with the stationary phase of the analytical column. Therefore, the separation or elution in the column is a process where the mobile phase, the stationary phase, and the compounds of interest are all interactive. The ability of the solute to interact selectively with both the stationary and mobile phases in HPLC provides additional parameters to achieve the desired separation. The most commonly used detectors in HPLC include UV and fluorescence detectors.

The UV absorption detectors commonly employed in HPLC can detect nanogram (10^{-9} g) quantities of a wide variety of materials (Table 14.7). Fluorescence and electrochemical detectors can detect quantities in the picogram (10^{-12} g) region (Table 14.7). Most of the detectors employed in HPLC are nondestructive so that sample components can be collected easily as they pass through the detector, which makes it possible to use online double detectors for multiresidue analysis. Pesticides that can be detected with a UV detector include bromacil, cyanazine, diuron, imidacloprid, metalaxyl, norflurazon, oryzalin, and simazine (Page and Ma, 1994). Pesticides of carbamate class are commonly analyzed with a fluorescence detector; these include aldicarb, aldicarb sulfoxide, aldicarb sulfone, carbaryl, carbfuran, methiocarb, methomyl, oxamyl, promecarbm, and propoxyl (Zhao and Page, 2001).

It is generally recognized that the greatest area for future development in HPLC is in the type of detectors that can be coupled to this technique. Optical detectors currently dominate the HPLC detector field, including UV absorbance and refractive index detectors and fluorescence detectors (Johnson and Stevenson, 1978). Electrochemical, flame ionization, atomic absorption, infrared, and vision-based optical detectors are available. The mass spectrometric detector has become popular in both HPLC and GC fields.

14.2.5.4.3 Mass Spectrometry

Mass spectrometry is perhaps the most generally applicable of all of the analytical tools available to the scientist in the sense that the technique is capable of providing qualitative and quantitative information about both the atomic and the molecular composition of inorganic and organic materials (Skoog, 1985). Mass spectroscopy evolved from studies at the beginning of this century of the behavior of positive ions in magnetic and electrostatic fields. In the 1940s, reliable mass spectrometers first became available from commercial sources. In the middle 1950s, commercial mass spectrometers were also developed for the qualitative and quantitative determination of the elements based upon the mass-to-charge ratio (m/z) of elementary ions formed in an electric spark. In the early 1960s, the major thrust in the development of mass spectrometry shifted towards the use for the identification and structural analysis of complex molecules. In the last three decades, mass spectroscopy became a strong technique for the identification and determination of both atomic and molecular species on surfaces and for studying the compositional changes of solids as a function of depth.

As mentioned earlier, the coupling of mass spectrometer as a useful detector with both GC and HPLC has becoming more and more popular.

All mass spectrometers consist of there distinct regions: ionizer, ion analyzer, and detector. An ionizer creates fragment ions that result from the ionization of organic molecules. Commonly used ionization techniques include electron spray, atmospheric pressure chemical ionization, electron impact, fast atom bombardment, and matrix-assisted laser desorption. The mass analyzer pumps away the uncharged fragments and molecules while it sorts the charged fragments (molecular ions) according to their m/z. The ever-increasing m/z of the charged fragments travels to the detector under high scanning voltage. The detector records the abundance of each fragment. Information derived from the detector and valid library spectrum matching technique is valuable in determining the molecular weight of the compound and its chemical structure. A combination of MS with GC or HPLC has become more common for confirmation since the technique is very valid when a total mass scan can be obtained from a single component. It also allows one to identify contaminants in the mixture if this becomes necessary. Another advantage is that multicomponents of the sample can be analyzed in a single analytical run. This may not be possible with some selective detectors, especially GC. There are many types of mass spectrometers and sample introduction techniques, which allow a wide range of analyses. GC/MS and LC/MS are common employed measures for pesticide determination.

14.2.5.5 *In Situ* Methods for Pesticide Determination

Because of the high cost of traditional laboratory analysis, many field-testing kits have been developed that are cheap, quick, and simple. Many of them provide no calibration with standard methods and do not have reliable results. However, testing kits based on immunoassays have been reported with detection limits for pesticides of 1–100 ppb for soil samples and 50–10 ppb for water samples. The cost-of-analysis range, depending on the target analytes and the manufacturer, is about $10–40/sample. Some immunoassays are not highly selective, and sample contaminants may interfere with the antigen-antibody reaction.

14.3   RECOMMENDATIONS AND FUTURE TRENDS

A number of analytical methods are available for characterization of soil and water. It is not easy to select an appropriate method for specific analytical problem. Comparison of results obtained with a standard method is an important criterion in the selection of an appropriate method. A new method developed for a specific purpose has to be validated and calibrated with a standard method or certified by an accreditation organization. Even when a standard method is used, many uncertainties may arise during chemical analysis. Calculating uncertainties and following QA/QC of the selected method are critically important for reliable chemical analyses. It is also important to verify the method parameters (such as reagent concentrations/volumes, carrier flow rate/pressure, time, temperature, wavelength, and others) of a recommended and/or standard method that are optimized for high sensitivity and low interference when an analytical instrument is introduced to the lab for low concentration analysis. For determination of inorganic N in soils, the preferred procedure includes sample extraction with 2 M KCl and analysis with an autoanalyzer such as air-segmented flow analyzer or flow-injection analyzer. A water-monitoring device with a selective electrode can be used for continuously measuring nitrate and ammonia. Total N and C can be analyzed using dry combustion or by means of CN analyzers. The total Kjeldahl method is an alternative for organic N analysis, while the Walkley-Black and LOI methods are suitable alternatives for organic C. Analysis of total P and metals in soil and water should follow USEPA or other standard digestion and determination procedures. There is no universal method for plant-available P or for metals and their fractionation. Each of the various methods has advantages and limitations. A method should be selected based on the analytical purpose. Field sampling of soil and water was not included in this chapter, but standardized sampling procedures should be followed in order to provide representative and uncontaminated samples to the laboratory.

Analytical technology is improving continuously. In the future many new analytical methods will be developed based on emerging technology and the need to address specific analytical problems. *In situ* water-monitoring devices will be improved dramatically during the next few years. New instruments using technologies other than selective electrodes will be used for field N and P monitoring. Field-portable x-ray fluorescence will be improved and become very useful for *in situ* analysis of metals. Carbon and N analyzers will become affordable for routine total C and N analysis once alternative consumable supplies for these instruments are available. Traditional methods, such as the Walkley-Black method for organic carbon and Kjeldahl method for total N, will be always useful as standard methods to validate and calibrate new methods. Combination of GC or HPLC with MS has become more common for qualitative determination of pesticides. Immunoassays have a promising future as *in situ* methods to screen pesticides quickly and cost-effectively. More and more analytical laboratories will acquire certification from accreditation organizations. This trend will improve the quality of analytical data and thereby benefit the scientists and clientele who use these results.

ACKNOWLEDGMENTS

The authors thank Waldemar Klassen and Michael Page for review of the manuscript. This research was supported by the Florida Agricultural Experiment Station and approved for publication as Journal Series No. N-02498.

REFERENCES

Aminot, A., R. Kérouel, and D. Birot. 2001. A flow injection-fluorometric method for the determination of ammonium in fresh and saline waters with a wiew to *in situ* analyses. Wat. Res. 35:1777–1785.

APHA (American Public Health Association). 1995. Standard Method for the Examination of Water and Wastewater, 19th ed. APHA, New York.

APHA. 1998. Standard Method for the Examination of Water and Wastewater, 20th ed. APHA, New York.

Baethgen, W.E. and M.M. Alley. 1989. A manual colorimetric procedure for measuring ammonium nitrogen in soil and plant Kjeldahl digests. Comm. Soil Sci. Plant Anal. 20:961–969.

Beauchamp, E.G. and D.W. Bergstrom. 1993. Denitrification. p. 351–357. In: M.R. Carter (ed.), Soil Sampling and Methods of Analysis. Lewis Publishers, Boca Raton, FL.

Benson, R.L., I.D. Mckelvie, B.T. Hart, Y.B. Truong, and I.C. Hamilton. 1996. Determination of total phosphorus in waters and wastewaters by on-line UV/ thermal induced digestion and flow injection analysis. Anal. Chim. Acta 326:29–39.

Bernot, M.J., W.K. Dodds, W.S. Gardner, M.J. McCarthy, D. Sobolev, and J.L. Tank. 2003. Comparing denitrification estimates for a Texas estuary by using acetylene inhibition and membrane inlet mass spectrometry. Appl. Envin. Micr. 69:5950–5956.

Blundell N.J. and P.J. Worsfold. 1995. The design and performance of a portable, automated flow injection monitor for the in-situ analysis of nutrients in natural waters. Environ. Industrial. 21:205–209.

Bray, R.H. and L.T. Kurtz. 1945. Determination of total, organic, and available forms of phosphorus in soils. Soil Sci. 59:39–45.

Broadbent, F.E. 1965. Organic matter. p. 1397–1400. In: C.A. Balck (ed.), Method of Soil Analysis. Agron. Monogr. 9, ASA, CSSA and SSSA, Madison, WI.

Bronk, D.A., M.W. Lomas, P.M. Glibert, K.J. Schukert, and M.P. Sanderson. 2000. Total dissolved nitrogen analysis: comparisons between the persulfate, UV and high temperature oxidation methods. Marine Chem. 69:163 178.

Burton, G.A. and R.E. Pitt. 2002. Stormwater Effects Handbook. Lewis Publishers, Boca Raton, FL.

Cogan, E.B., G.B. Birrell, and O.H. Griffith. 1999. A robotics-based automated assay for inorganic and organic phosphates. Anal. Biochem. 271:29–35.

Collins, L.W., S.J. Chalk, and H.M. Kingston. 1996. Atmospheric pressure microwave sample preparation procedure for the combined analysis of total phosphorus and Kjeldahl nitrogen. Anal. Chem. 68:2610–2614.

Clifford, D.A. and L.M. McGaughey. 1982. Simultaneous determination of total nitrogen and total oxygendemand in aqueous samples. Anal. Chem. 54:1345–1350.

Crowther, A.B. and R.S. Large. 1956. Improved conditions for the sodium phenoxide-sodium hypochlorite method for the determination of ammonia. Analyst 81:64–65.

Csiky, I., G. Markovarga, and J.A. Jonsson. 1985. Use of disposable cleanup columns for selective removal of humic substances prior to measurements with a nitrate ion-selective electrode. Anal. Chim. Acta 178:307–312.

Daubeny, C.G.B. 1845. Memoirs on the rotation of crops and on the quantity of inorganic matters abstracted from the soil by various plants under different circumstances. Roy. Soc. (London) Phil. Trans. 135:179–253.

D'Angelo, E., J. Crutchfield, and M. Vandiviere. 2001. Rapid, sensitive, microscale determination of phosphate in water and soil. J. Environ. Qual. 30:2206–2209.

Daniel, A., D. Birot, S. Blain, P. Tréguer, B. Lecïldé, and E. Menut. 1995. A submersible flow-injection analyzer for the in-situ determination of nitrite and nitrate in coastal waters. Marine Chem 51:67–77.

Dahlen, J., S. Karlsson, M. Backstrom, J. Hagberg, and H. Pettersson. 2000. Determination of nitrate and other water quality parameters in groundwater from UV/Vis spectra employing partial least squares regression. Chemosphere 40:71–77.

Davenport, J.R. and J.D. Jabro. 2001. Assessment of hand held ion selective electrode technology for direct measurement of soil chemical properties. Commun. Soil Sci. Plant Anal. 32:3077–3085.

Davison, w. and C. Woof. 1978. Comparison of different forms of cadmium as reducing agents for batch determination of nitrate. Analyst 103:403–406.

Day, J. 1996. Selection of appropriate analytical procedures for volunteer field monitoring of water quality. Thesis, Dept. Civil Eviron. Eng., University of Alabama, Birmingham, AL.

D'Elia, C.F., P.A. Steudler, and N. Corwin. 1977. Determination of total nitrogen in aqueous samples using persulfate digestion. Limnol. Oceanogr. 22:760–764.

Downes, M.T. 1978. Improved hydrazine reduction method for automated-determination of low nitrate levels in freshwater. Water Res. 12:673–675.

Ebina, J., T. Tsutsui, and T. Shirai. 1983. Simultaneous determination of total nitrogen and total phosphorus in water using peroxodisulfate oxidation. Water Res. 17:1721–1726.

Ellison, S.L.R., M. Rosslein, and A. Williams. 2000. Quantifying uncertainty in analytical measurement. EURACHEM/CITAC.

Graetz, D. and V.D. Nair. 1999. Inorganic forms of phosphorus in soils and sediments. p. 171–186. In: K.R. Reddy, G.A. O'Connor and C.L. Schelske (eds.), Phosphorus Biogeochemistry in Subtropical Ecosystems. Lewis Publishers, Boca Raton, FL.

Grasshoff, K. 1976. Methods of Seawater Analysis. Verlag Chemie, Weinheim. Dtsch.

Griess, J.P. and Z.A.H.H. Bemerkungen. 1879. Über einige Azoverbindungen. Ber. Dtsch. Chem. Ges. 12:426–428.

Hanrahan, G., M. Gledhill, P.J. Fletcher, and P.J. Worsfold. 2001. High temporal resolution field monitoring of phosphate in the River Frome using flow injection with diode array detection. Anal Chim. Acta 440:55–62.

He, Z.L., A.K. Alva, D.V. Calvert, and D.J. Banks. 1999. Ammonia volatilization from different nitrogen fertilizers and effects of temperature and soil pH. Soil Sci. 164:750–758.

Heckemann, H.J. 2000. Highly sensitive flow analysis determination of orthophosphate using solid phase enrichment of phosphomolybdenum blue without need for organic solvents in elution. Anal. Chim. Acta 410:177–184.

Hilton, J. and E. Rigg. 1983. Determination of nitrate in lake water by the adaptation of the hydrazine–copper reduction method for use on a discrete analyzer: performance statistics and an instrument-induced difference from segmented flow conditions. Water Res. 108:1026–1028.

Holmes, R.M., A. Aminot, R. Kerouel, B.A. Hooker, and B.J. Peterson. 1999. A simple and precise method for measuring ammonium in marine and freshwater ecosystems. Can. J. Fish. Aquat. Sci. 56:1801–1808.

Horn, D.B. and C.R. Squire. 1967. A improved method for estimation of ammonia in blood plasma. Clin. Chim. Acta 17:99.

Hosomi, M. and R. Sudo. 1986. Simultaneous determination of total nitrogen and total phosphorus in freshwater samples using persulfate digestion. Int. J. Environ. Studies 27:267–275.

Hulanick, A., R. Lewandow, and M. Maj. 1974. Determination of nitrate in water with a new construction of ion-selective electrode. Anal. Chim. Acta 69:409–414.

Jackson, P.E., C. Weigert, C.A. Pohl, and C. Saini. 2000. Determination of inorganic anions in environmental waters with a hydroxide-selective column. J. Chromatogr. 884:175–184.

Johnes, P.J. and A.L. Heathwaite. 1992. A procedure for the simultaneous determination of total nitrogen and total phosphorus in freshwater samples using persulfate microwave digestion. Water Res. 26:1281–1287.

Johnson, E.L. and R. Stevenson. 1978. Basic Liquid Chromatography. Varian Associates, Inc., Tempe, AZ.

Kahn, L. and F.T. Brezenski. 1967. Determination of nitrate in estuarine waters: comparison of a hydrazine reduction and a brucine procedure and modification of a brucine procedure. Environ. Sci. Tech. 1:488–491.

Kaniansky, D., I. Zelensky, A. Hybenova, and F.I. Onuska. 1994. Determination of chloride, nitrate, sulfate, nitrite, fluoride, and phosphate by online coupled capillary isotachophoresis-capillary zone electrophoresis with conductivity detection. Anal. Chem. 66:4258–4264.

Kennelley, E.D. and R.S. Mylavarapu. 2002. Low-level phosphorus analysis in the presence of silicate. Comm. Soil Sci. Plant Anal. 33:3189–3201.

Kerouel, R. and Aminot, A. 1997. Fluorometric determination of ammonia in sed and estuarine waters by direct segmented flow analysis. Mar. Chem. 57:265–275.

Kishimoto, N., I. Somiya, and R. Taniyama. 2001. Improved ultraviolet spectro-photometric method for determination of nitrate in natural waters. Second World Water Congress: Drinking Water Treatment. Water Sci. Technol. Water Supply 2:213–221.

Kjeldahl, J. 1883. A new method of determinating nitrogen in organic materials. Z. Anal. Chem. 22:366–382.

Koroleff, F. 1969. Determination of total nitrogen in natural waters by means of persulfate oxidation. Int. Counc. Explor. Sea (ICES) pap. C. M. C:8.

Lambert, D. and W. Maher. 1995. An evaluation of the efficiency of the alkaline persulphate digestion method for the determination of total phosphorus in turbid waters. Water Res. 29:7–9.

Langner, C.L. and P.F. Hendrix. 1982. Evaluation of a persulfate digestion method for particulate nitrigen and phosphorus. Water Res. 16:1451–1454.

Larson, S.J., R.J. Gilliom, and P.D. Capel. 1997. Pesticides in surface waters: current understanding of distribution and major influences. U.S. Geological Survey Fact Sheet FS-039-97, 4p. At URL http://ca.water.usgs.gov/pnsp/rep/fs97039/

Larson, S.J., R.J. Gilliom, and P.D. Capel. 1999, Pesticides in streams of the United States — initial results from the national water-quality assessment program. U.S. Geological Survey, Water-Resources Investigations Report 98-4222.

Masserini, R.T. and K.A. Fanning. 2000. A sensor package for the simultaneous determination of nanomolar concentrations of nitrite, nitrate, and ammonia in seawater by fluorescence detection. Mar. Chem. 68:323–333.

Maynard, D.G. and Y.P. Kalra. 1993. Nitrate and exchangeable ammonium nitrogen. p. 25–38. In: M.R. Carter (ed.), Soil Sampling and Methods of Analysis. Lewis Publishers, Boca Raton, FL.

McKelvie, I.D., D.M.W. Peat, and P.J. Worsfold. 1995. Techniques for the quantification and speciation of phosphorus in natural waters. Anal. Proc. Incl. Anal. Comm. 32:437–445.

McKelvie, I.D., M. Mitri, B.T. Hart, I.C. Hamilton, and A.D. Stuart. 1994. Analysis of total dissolved nitrogen in natural-waters by online photooxidation and flow-injection. Anal. Chim. Acta 293:155–162.

Mehlich, A. 1953. Determination of P, Ca, Mg, K, Na, NH_4. Soil Testing Division Publication No. 1–53. North Carolina Department of Agriculture, Agronomic Division. Raleigh, NC.

Mehlich, A. 1984. Mehlich 3 soil test extractant: a modification of Mehlich 2 extractant. Comm. Soil Sci. Plant An. 15:1409–1416.

Mikuska, P. and Z. Vecera. 2003. Simultaneous determination of nitrite and nitrate in water by chemiluinescent flow-injection analysis. Anal. Chim. Acta 495:225–232.

Morgan, M.F. 1941. Chemical soil diagnosis by the universal soil testing system. Conn. Agric. Exp. Stn. Bull. No. 450.

Mosier, A.R. and L. Klemedtsson. 1994. Measuring denitrification in the field. In: R.W. Weaver, S. Angle, P. Bottomly, D. Bezdicek, S. Smith, A. Tabatabai, and A. Wollum (eds.), Methods of Soil Analysis, Part 2. Microbiological and Biochemical Properties. SSSA Book Series, No 5. SSS, Madison, WI.

Motomizu, S. and S. Masahiro. 1995. Photo-induced reduction of nitrate to nitrite and its application to the sensitive determination of nitrate in natural waters. Anal. Chim. Acta 308:406–412.

Mullin, J.B. and J.P. Riley. 1955. The spectrophotometric determination of nitrate in natural waters, with particular reference to sea-water. Anal. Chim. Acta 12: 464–480.

Munoz, A., F.M. Torres, J.M. Estela, and V. Cerda. 1997. Evaluation of spectrophotometric methods for determination of orthophosphates by sequential injection analysis. Anal. Chim. Acta 50:21–29.

Murphy, J. and J.P. Riley. 1962. A modified single solution method for the determination of phosphate in natural waters. Anal. Chim. Acta 27:31–36.

Nair, V.D., D.A. Graetz, and K.M. Portier. 1995. Forms of phosphorus in soil profiles from dairies of south Florida. Soil Sci. Soc. Am. J. 59:1244–1249.

Nelson, D.W. and R.H. Dowdy. 1988. Methods for Ground Water Quality Studies. ARD-UNL, Lincoln, NB.

NELAC (National Environmental Laboratory Accreditation Conference). 2002. Quality systems. http://www.epa.gov/ttn/nelac/standard/chapter5.pdf

Nydahl, I.F. 1976. On the optimum conditions for the reduction of nitrate to nitrite by cadmium. Talanta 23:349–357.

Okada, M., H. Miyata, and K. Toei. 1979. Determination of nitrate and nitrite in river waters. Analyst 104:195–1197.

Okemgbo, A., H.H. Hill, W.F. Siems, and S.G. Metcalf. 1999. Reverse polarity capillary zone electrophoretic analysis of nitrate and nitrite in natural water samples. Anal. Chem. 71:2725–2731.

Oleksy-Frenzel, J. and M. Jekel. 1996. On-line multicomponent determination of organic compounds in water following gel permeation chromatographic separation. Anal. Chim. Acta 319:165–175.

Olsen, S.R., C.V. Cole, F.S. Watanabe, and L.A. Dean. 1954. Estimation of available phosphorus in soils by extraction with sodium bicarbonate. U.S. Dept. Agric. Circ. 939.

Oms, M.T., A. Cerda, and V. Cerda. 1995. Sequential injection-analysis of nitrites and nitrates. Anal. Chim. Acta 315:321–330.

Page, M. and Y. Ma. 1994. The analysis of water soluble herbicides and pesticides in well water or drinking water by high pressure liquid chromatography. In: House Method, PL-AES. Florida Dept. of Agric. Tallahassee, FL.

Pansu, M., J. Gautheyrou, and J. Loyer. 2001. Soil Analysis-Sampling, Instrumentation and Quality Control. A.A. Balkema Publishers, Exton, PA.

Patton, C.J., A.E. Fischer, and W.H. Campbell. 2002. Corn leaf nitrate reductase — a nontoxic alternative to cadmium for photometric nitrate determinations in water samples by air-segmented continuous-flow analysis. Environ. Sci. Tech. 36:729–735.

Pavlis, R. 2004. Surprising statistics on pipet performance. Am. Lab. 36:8–10.

Petsul, P.H., G.M. Greenway, and S.J. Haswell. 2001. The development of an on-chip micro-flow injection analysis of nitrate with a cadmium reductor. Anal. Chim. Acta 428:155–161.

Raab, G.A., R.E. Enwall, W.H. Cole, III, M.L. Faber, and L.A. Eccles. 1990. X-ray fluorescence field method for screening of inorganic contaminants at hazardous waste sites. In: M. Simmons (ed.), Hazardous Waste Measurements, Lewis Publishers, Chelsea, MI.

Rider, B.F. and M.G. Mellon. 1946. Colorimetric determination of nitrites. Ind. Eng. Chem. 18:96–99.

Rohwedder, J.J.R. and C. Pasquini. 1991. Differential conductimetry in flow-injection — determination of ammonia in Kjeldahl digests. Analyst 116:841–845.

Schulte, E.E. and B.G. Hopkins. 1996. Estimation of soil organic matter by weight loss-on-ignition. p. 21–31. In: F.R. Magdoff, M.A. Tatatabai, and E.A. Hanlon (eds.), Soil Organic Matter: Analysis and Interpretation. SSSA, Madison, WI.

Seitzinger, S. and R. Sanders. 1997. Contribution of dissolved organic nitrogen from rivers to estuarine eutrophication. Mar. Ecol. Prog. Ser. 159:1–12

Shefsky, S. 1997. Comparing field portable x-ray fluorescence (XRF) to laboratory analysis of heavy metals in soil. The International Symposium of Field Screening Methods for Hazardous Wastes and Toxic Chemicals, Las Vegas, NV, January 29–31, 1997.

Shuman, L.M. 1985. Fracnation method for microelements. Soil Sci. 140:11–22.

Shuman, L.M. 1991. Chemical forms of micronutrients in soils. p. 113–144. In: J.J. Mortvedt, F.P. Cox, L.M. Shuman, and R.M. Welch (eds.), Micronutrients in Agriculture, SSSA, Madison, WI.

Sims, J.T. and G.V. Johnson. 1991. Micronutrient soil tests. p. 427–476. In: J.J. Mortvedt et al. (eds.), Micronutrients in Agriculture, SSSA, Madison, WI.

Skoog, D.A. 1985. Principles of Instrumental Analysis. Holt, Rinehart and Winston, Austin, TX.

Soltanpour, P.N. and A.P. Schwab. 1977. A new soil test for simultaneous extraction of macro- and micronutrients in alkaline soils. Commun. Soil Sci. Plant Anal. 8:195–207.

Stainton, M.P. 1974. Simple, efficient reduction column for use in the automated determination of nitrate in water. Anal. Chem. 46:1616.

Steimle, E.T., E.A. Kaltenbacher, and R.H. Byrne. 2002. In situ nitrite measurements using a compact spectrophotometric analysis system. Marine Chem. 77:255–262.

Struve, D., M. Zhou, and T. Baber. 2004. Design and operation of a remote phosphorus analyzer. FWRC. Proc. (in press).

Susanto, J.P., M. Oshima, and S. Motomizu. 1995. Determination of micro amounts of phosphorus with malachite green using a filtration-dissolution preconcentration method and flow injection-spectrophotometric detection. Analyst 120:187–191.

Swift, W. 1996. Organic matter characterization. p. 1011–1069. In: D.L Sparks, A.L. Page, P.A. Hemke, R.H. Loeppert, P.V. Soltanpour, M.A. Tabatabai, C.T. Johnston, and M.E. Sumner (eds.), Methods of Soil Analysis, Part 3. Chemical methods, SSA Book Series, No 5. SSSA, Madison, WI.

Taguchi, S., E. Ito-Oka, K. Masuyama, I. Kasahara, and K. Goto. 1985. Application of organic solvent-soluble membrane filters in the perconcentration and determination of trace elements: spectrophotometric determination of phosphorus as phosphomolybdenum blue. Talanta 32:391–394.

Takeda, K. and Fujiwara, K. 1993 Determination of nitrate in natural waters with the photo-induced conversion of nitrate to nitrite. Anal. Chim. Acta 276:25–32.

Tessier, A., P.G.C. Campbell, and M. Bisson. 1979. Sequencial extraction procedure for the speciation of particulate trace metals. Anal. Chem. 51:844–851.

Torro, I.G., J.V.G. Mateo, and J.M. Calatayud. 1998. Flow-injection biamperometric determination of nitrate (by photoreduction) and nitrite with the $NO_2^-/1^-$ reaction. Anal. Chim. Acta 366:241–249.

Towns, T.G. 1986. Determination of aqueous phosphate by ascorbic-acid reduction of phosphomolybdic acid. Anal. Chem. 58:223–229.

USDA (United State Department of Agriculture). 1999. Soil quality test kit guide. The Soil Quality Institute, USDA/ARS/NRCS, Lincoln, NE.

USEPA (United State Environmental Protection Agency). 1982. Methods for chemical analysis of water and wastes. EPA 600/4-79-020. U.S. Gov. Print Office, Washington, DC.

USEPA. 1986. Quality of criteria for water. EPA 440/5-86-001. U.S. Gov. Print Office, Washington, DC.

USEPA. 1991. National pesticides survey. National Technical Information Service, Springfield, VA.

USEPA. 1999. Methods and guidance for the analysis of water, version 2, NTIS Order No. PB99-500206. National Technical Information Service, Springfield, VA.

USEPA. 2003. List of drinking water contaminants & MCLs. http://www.epa.gov/safewater/mcl.html#mcls

USEPA. 2004. Test methods for evaluating solid waste, physical/chemical methods (SW-846). Government Printing Office (GPO), Superintendent of Documents, Washington, DC.

Van den Hoop, M.A.G.T. and J.J. van Staden. 1997. Determination of phosphate in natural waters by capillary electrophoresis: influence of solution composition on migration time and response. J. Chromatogr. 770:321–328.

Walkley, A. and I.A. Black. 1934. An examination of the Degtjareff method for determining soil organic matter and a proposed modification of the chromic acid titration method. Soil Sci. 37:29–38.

Walsh, T.W. 1989. Total dissolved nitrogen in seawater — a new high-temperature combustion method and a comparison with photo-oxidation. Marine Chem. 26:295–311.

Wear, J.I. and C.E. Evens. 1968. Relationships of zinc uptake of corn and sorghum to soil zinc measured by three extractants. Soil Sci. Soc. Am. Proc. 32:543–546.

Wetters, J.H. and K.L. Uglum. 1970. Direct spectrophotometric simultaneous determination of nitrite and nitrate in ultraviolet. Anal. Chem. 42:335–340.

Willis, R.B. 1980. Reduction column for automated determination of nitrate and nitrite in water. Anal. Chem. 52:1376–1377.

Willis, R.B. and C.E. Gentry. 1987. Automated-method for determining nitrate and nitrite in water and soil extracts. Comm. Soil Sci. Plant Anal. 18:625–636.

Wood. J.M. 1974. Biological cycles for toxic elements in the environment. Science 183:1049–1052.

Worsfold, P.J. and J.R. Clinch. 1987. Spectrophotometric field monitor for water quality parameters. Anal. Chim. Acta 197:43–50.

Zhang, G. and P.K. Dasgupta. 1989. Fluorometric measurement of aqueous ammonium ion in a flow injection system. Anal. Chem. 61:408–412.

Zhang, M., Y.C. Li, and P. Stoffella. 2004. Comparison of analytical methods for organic carbon in composts, fly ash, biosolids and calcareous soils. Commun. Soil Sci. Plant Anal. (in press).

Zhao, J.Q. and M. Page. 2001. The analysis of pesticides in water with graphitized carbon SPE and HPLC. In house method, PL-AES, Florida Dept. Agric., Tallahassee, FL.

Zheng, H., W. Davison, R. Gadi, and T. Kobayashi. 1998. In situ measurement of dissolved phosphorus in natural waters using DGT. Anal. Chim. Acta 370:29–38.

Zhou, M. and D. Struve. 2002. Trace level total phosphorus determination using pre-concentration of a malachite green-phosphomolybdenum complex. Watershed 2002 Conference proceedings, Water Environmental Federation.

Zhou, M. and D. Struve. 2004a. Optimal conditions for the determination of total phosphorus concentration in water. Wat. Res. (submitted).

Zhou, M. and D. Struve. 2004b. Re-examination of optimum acid concentration of ascorbic acid method for low level phosphate analysis. Wat. Res (submitted).

Zhou, M. and D. Struve. 2004c. The effects of post-persulfate-digestion procedures on total phosphorus analysis in water. Wat. Res. 38:3893–3898.

Zhou, M. and D. Struve. 2004d. Low level phosphorus determination using single reagent malachite green method. Wat. Res. (submitted).

Zinati, G.M., Y.C. Li, and H.H. Bryan. 2001. Utilization of compost increases organic carbon and its humin, humic and fulvic acid fractions in calcareous soil. J. Compost Sci. Utilization 9:156–162.

Zui, O.V. and J.W. Birks. 2000. Trace analysis of phosphorus in water by sorption preconcentration and luminol chemiluminescence. Anal. Chem. 72:1699–1703.

15 Evaluation and Characterization of Soil Microbiological Processes

Mikael Pell and John Stenström
Swedish University of Agricultural Sciences, Uppsala, Sweden

CONTENTS

15.1 INTRODUCTION

Society today is increasingly concerned about the sustainability of soil productivity, the maintenance of biodiversity, and methods for effective soil

1-5667-0657-2/05/$0.00 + $1.50
© 2005 by CRC Press

remediation and protection. Therefore, methods to quantify the ability of the soil to function and to buffer and resist anthropogenic disturbances have come into focus. The many functions of a soil have led to an extensive discussion of the soil quality concept and how it might be defined and assessed. Two important and generally agreed-upon principles associated with soil quality, as discussed by Karlen et al. (2003), are that it is (1) determined by both inherent and dynamic properties and processes interacting with a living dynamic medium and (2) holistic, reflecting biological, chemical, and physical properties, processes, and interactions within the soil. The concept of soil fertility is often used synonymously with soil quality. However, fertility is more often related to the capacity of the soil to produce and the term as used in the literature is less applicable in natural sciences (Patzel et al., 2000).

While optimal and threshold values for many chemical and physical parameters have been identified, the biological soil component is far less specified. The reasons for this include difficulties identifying appropriate levels for biological parameters, difficulties extracting relevant information from the high natural variability in biological parameters, both between and within different ecosystems at different scales and over time, and because of the influence of external factors such as climatic conditions. These difficulties suggest soil functioning should be evaluated by integrating chemical, physical, and biological properties, where values of key parameters are used to create statistical visual descriptions by which the boundaries for functioning soils can be defined. Key microbial parameters for such an integrated approach are the activity, diversity, and size of the microbial biomass. To illustrate this, we will concentrate on the three key microbial processes—respiration, denitrification, and nitrification—focusing on arable soil. We will also briefly address some special considerations regarding soil sampling and handling of the samples for studies of soil microbiology, since the techniques used can have a large influence on the results obtained. Finally, we discuss how microbial parameters can be used in integrated evaluations of soil functioning and in assessments of their spatial variability at different scales.

15.2 BASIC SOIL MICROBIOLOGY

In order to develop a comprehensive understanding of the structure and function of soil microbial communities, the three components—activity, diversity, and biomass—should preferably be analyzed. Due to the complex nature of the soil, emphasized in Chapter 2, a vast number of methods have been developed to characterize these different properties of the microbial community. Historically, methods aiming at quantifying general or specific activities related to transformations of plant nutrients as well as methods for estimating the microbial biomass have proliferated (Torstensson, 1993a; Alef and Nannipieri, 1996; Schinner et al., 1996).

In addition, recent developments in molecular biology permit identification of specific groups of organisms, which allows us to explore the biological structure or diversity of the ecosystem (Hill et al., 2000).

15.2.1 THE ACTORS

15.2.1.1 Activity

The functioning of the biosphere is contingent on the activities of the microbial world. For instance, soil microbial activity is crucial for soil fertility. During the decomposition of organic substances, bound nutrients are liberated and transformed into inorganic forms (mineralization), making them available for plant uptake and for further transformations by other microorganisms. One such process is the oxidation of mineralized ammonium to the more available nitrate by nitrifying bacteria (nitrification). Also, soil microorganisms are important for the development and maintenance of soil structure by their production of extracellular polysaccharides (EPS), and particularly by the large production of glomalin by the vesicular-arbuscular (VA) mycorrhizal fungi. Glomalin is a glue-like, recalcitrant, iron-rich, and insoluble glycoprotein, the amount of which correlates well with soil aggregate stability (Wright and Upadhyaya, 1998).

Almost all organisms in the soil can contribute to the production of carbon dioxide. Therefore, estimates of the total soil biomass would theoretically provide information for calculations of specific respiration rates. Thus, high values could indicate an active biomass, i.e., organisms living in an environment of suitable conditions, while low values should indicate starvation or other unfavorable conditions. However, variations in the specific respiration rate could also indicate variations in the distribution between active and dormant microorganisms (Stenström et al., 2001).

15.2.1.2 Diversity

The microbial population of a soil includes members of three major domains: Bacteria, Archaea (e.g., methanogens), and Eucarya (fungi, microalgae, and protozoa). Until recently, studies on the diversity of soil microbial populations have been restricted to culturable representatives of these domains. With the use of molecular biological techniques, e.g., DNA reassociation kinetics, DNA hybridization, and cloning and sequencing ribosomal RNA genes directly from the environment, it has now become possible to explore the phylogenetic diversity of natural microbial communities (Hill et al., 2000). Such studies show an amazing microbial diversity and suggest that only a few percent of existing microbial species have been isolated and identified by culturing methods. However, these results do not yet reveal the role and quantitative importance of the different species. In addition, an increasing number of gene sequences discovered by modern molecular biological techniques cannot be given official species status, since the species concept used today is based on culture-dependent methods, for instance, availability of reference-type strains in culture collections (Schloter et al., 2000).

15.2.1.3 Biomass

Plant roots can occupy about 2–3 volume-% of the surface soil, with root exudates corresponding to about 5% of the carbon that plants photoassimilate. These exudates and the organic matter of the soil, consisting of plant, animal, and microbial residues in various stages of decay, represent the dominant source for microbial growth and activity.

The biomass in a fertile soil can exceed 20,000 kg live weight ha^{-1} and be comprised of 900 kg earthworms (> 10 mm), 20 kg mesofauna (2–10 mm; enchytraeids, microarthropods), 50 kg microfauna (< 2 mm; protozoa, nematodes), and 20,000 kg microorganisms (Lee and Pankhurst, 1992). In spite of this large number of microorganisms, microbial carbon often comprises only 2–3 weight-% of the total soil organic carbon. However, microbially immobilized carbon and nutrients have a much faster turnover time than other soil organic pools, meaning that biomass estimates can provide information on the amount of easily available plant nutrients temporarily bound in living or dead organisms.

15.2.2 SOIL AS A MICROBIAL HABITAT

Soil organisms live in the pores between and within aggregates and particles, and the spatial organization of pores of diminishing scale (see also Chapter 7) is paralleled by the presence of organisms of diminishing scale. At even smaller scales, pore necks and pore shapes that do not allow the entrance of microorganisms still allow passage and transport of organic polymers, molecules, and atoms (Figure 15.1). Microorganism-inaccessible, mostly water-filled pores can constitute a significant part of the pore volume. It has been estimated that in a clay loam with a microbial population of 10 bacteria g^{-1}, the organisms will occupy only about 0.1% of the pore volume and cover about 0.01% of the surfaces of the soil, and that at least 90% of the surfaces in soils are not accessible to organisms or enzymes (Adu and Oades, 1978). In addition, the distribution of microorganisms in the soil matrix is patchy due to spatial and temporal variations in available substrates. For example, root exudates and dead root tissue may comprise up to 30–40% of the total input of organic matter to soils, and this material is discharged directly into the rhizosphere, which represents no more than 2–3% of the total soil volume (Coleman et al., 1978). Thus, in spite of hundreds of millions of microorganisms per g of soil, most of its pores and surfaces are sterile, and substrates and other organic chemicals located in smaller pores can be inaccessible and physically protected from microbial decomposition. Consequently, in the 97–98% of the soil volume that is not directly affected by roots, microorganisms have to rely on the small amounts of organic compounds that are slowly released from more or less humified organic materials and that are transported out of protective pores, leading to starvation and dormancy as the normal physiological states for most organisms most of the time (Bakken, 1997; Stenström et al., 2001). Such a desorption-dependent mechanism is also

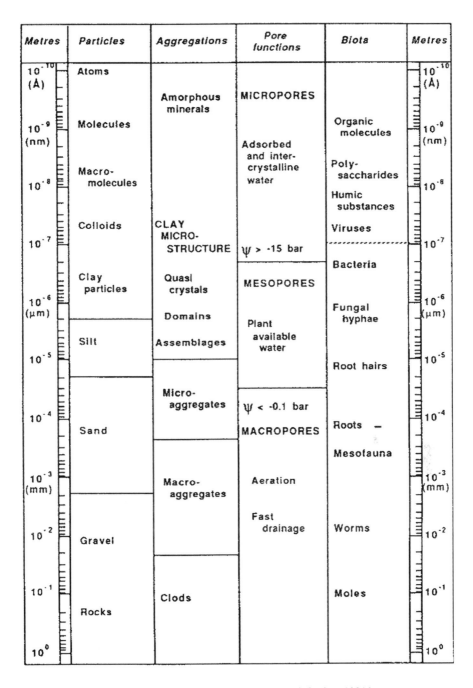

FIGURE 15.1 Scale in soil structure. (From Waters and Oades, 1991.)

important for the degradation of other organic compounds in soil (Bergström and Stenström, 1998) and can thus often constitute the rate-limiting step in, for instance, efforts to clean up contaminated soils by bioremediation.

15.3 METHODS FOR MICROBIAL SOIL CHARACTERIZATION

Three microbial soil processes — soil respiration, denitrification, and nitrification — represent key features in the turnover of carbon and nitrogen, since their result is that (1) bound or unavailable nutrients are made available to the plants and (2) nitrogen and carbon are transformed into gaseous compounds that will be emitted to the atmosphere or nitrogen ions that can leach to the groundwater, both potentially having serious environmental impact. In addition, the three processes can be used as tools to probe toxic effects of anthropogenic chemicals and be part of procedures for the assessment of soil quality. For such purposes these methods can complement each other. Soil respiration and nitrification are commonly thought to be antipodes on a scale of sensitivity — soil respiration being a basic and robust process, while nitrification is considered to be sensitive to environmental disturbances and performed by a limited group of specialists. In the following we will discuss some of the methods available for studies of these processes, with an emphasis on methods that are simple, cheap, and frequently used and that can be included in soil toxicity and soil quality assessments. The methods are also compiled in Table 15.1 according to their applicability for characterization of the activity, the diversity and the biomass of microbial communities, and whether their use is dependent on our ability to cultivate the microorganisms.

15.3.1 SAMPLING AND SOIL HANDLING

Compared to soil physical and chemical variables (see Chapters 1 and 2), microbial variables in general display higher temporal and spatial variations, depending on the physicochemical environment as well as climatic and seasonal variations in weather condition. This means that the sampling strategy is of great importance for the outcome of the results, which, however, is frequently given too little attention in soil microbiological studies.

When using a soil as a medium for laboratory tests of, e.g., chemical toxicity, large batches of soil must be collected. Sampling can be done from a limited spot as long as the soil is thoroughly mixed and sieved into a homogeneous test medium that can be well characterized chemically and physically. In cases where soil quality should be assessed, whole fields or landscape areas must be covered by the sampling procedure. If nothing is previously known about the variation, grid patterns should be the first choice (Totsche, 1996). When relevant knowledge about the spatial variation is known, stratified sampling procedures can be applied to reduce the work.

TABLE 15.1
Methods for Characterization of Microbial Communities

Microbial parameters	Culture-dependent methods	Culture-independent methods
Activity		Respiration
		O$_2$ consumption
		CO$_2$ production
		Potential denitrification activity (PDA)
		Potential ammonium oxidation (PAO)
		^{15}N-methodology
Diversity	Physiological profiles	DNA reassociation kinetics
	(e.g., BIOLOG$^®$)	PCR amplification of ribosomal
		or functional genes
		Terminal restriction fragment length
		polymorphism (T-RFLP)
		Denaturing gradient gel
		electrophoresis (DGGE)
		Temperature gradient gel
		electrophoresis (TGGE)
		Cloning and sequencing
Biomass	Viable count	Microscopic techniques
	Most probable number	Substrate-induced respiration (SIR)
	(MPN)	ATP
	Florescent polyclonal	Chloroform fumigation-extraction
	antibodies	(CFE) of biomass C, N, P, or S
		Extraction and analysis of biomarkers
		(e.g., phospholipid fatty acids, sterols,
		quinones, and lipopolysaccharides)
		Fluorescence *in situ* hybridization (FISH)

Geostatistics is a powerful method to establish sufficient sampling distances (Webster and Oliver, 2001).

Obviously, chemical testing and soil quality studies generate huge numbers of microbial analyses that cannot be processed simultaneously, and, consequently, the work must often be spread out over a period of several months and maybe a year. Therefore, the soil must be handled, transported, and stored under conditions that minimize biological changes. The time between sampling and storage should be shortened as much as possible, and transportation should be made in refrigerated/insulated containers. At the laboratory the soil should be sieved (mesh width 2–4 mm) and mixed before being divided into polyethylene plastic bags and stored. If the soils are sampled under wet conditions, it might be necessary to allow them to dry gently to about 50–60% of their water-holding capacity. Many protocols stipulate storage under a constant temperature of +4°C. However, at this low temperature microorganisms are active, resulting in mineralization of carbon and nitrogen. Under no circumstances should the soil be stored at room temperature, and probably the best method for conserving a soil is to freeze it at $-20°C$, particularly soils from regions with a temperate climate. It was found that the microbial biomass and the basal respiration, as well as the potential ammonium oxidation and potential denitrifcation rates, were not decisively affected after 13 months of storage at $-20°C$, and the effects generally were smaller than after storage at $+2°C$ (Stenberg et al., 1998a).

15.3.2 SOIL RESPIRATION, DENITRIFICATION, AND NITRIFICATION

Respiration is probably the feature most closely associated with life. It is the process where chemically bound energy in organic or inorganic compounds is converted to a form (adenosine triphosphate, ATP) that can be used for cellular work (Figure 15.2). All respiring organisms possess an electron transport chain (or respiration chain) through which electrons are transported and ATP is produced. Most respiring organisms use oxygen as the terminal electron acceptor, but alternatives like nitrate and sulfate are also widely used by microorganisms. Apart from useful energy, respiration also leads to the formation of the necessary building blocks for cell growth, reduced electron acceptors (e.g., $O_2 \rightarrow H_2O$), carbon dioxide and previously bound minerals in the substrate.

Denitrification is the anaerobic respiration process in which nitrogenous oxides, principally nitrate and nitrite, are used as terminal electron acceptors and, hence, reduced into the gaseous products nitric oxide, nitrous oxide, and dinitrogen (Figure 15.3). The process is controlled by factors such as pH, texture, organic C and mineral N supply, aeration, and water status (Aulakh et al., 1992). Normally, dinitrogen dominates the end product, but

Organic carbon

Carbon flow

ADP ATP

e^-

O_2

Electron flow

H_2O

CO_2

FIGURE 15.2 Carbon and electron flow in aerobic respiration.

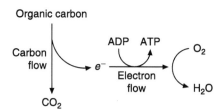

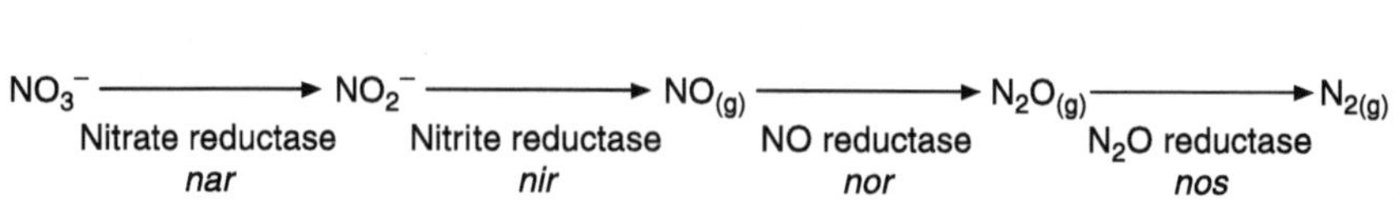

$NO_3^- \longrightarrow NO_2^- \longrightarrow NO_{(g)} \longrightarrow N_2O_{(g)} \longrightarrow N_{2(g)}$

| Nitrate reductase | Nitrite reductase | NO reductase | N$_2$O reductase |
| nar | nir | nor | nos |

FIGURE 15.3 The denitrification pathway with corresponding enzymes and genes.

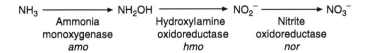

FIGURE 15.4 The autotrophic nitrification pathway with corresponding enzymes and genes.

under conditions not optimal for the organisms, nitrous oxide can constitute a considerable fraction. Nitrous oxide contributes to global warming as well as to the depletion of the ozone layer in the stratosphere (Davidson, 1991). Denitrifying capacity is represented within most taxonomical and physiological groups of bacteria (Zumft, 1992), meaning that denitrifying bacteria are probably good representatives of the bacterial community in soil. Denitrifiers are facultative anaerobes, i.e., they prefer oxygen as the terminal electron acceptor in their respiration, but upon depletion of oxygen they rapidly switch to the use of a nitrogenous oxide. Besides releasing nitrogen gases to the atmosphere, denitrification is an important nitrogen mineralization process under anaerobic conditions. Denitrifiers, being organotrophs and heterotrophs (deriving energy and carbon from organic matter, respectively), are known to use only the most readily available fraction of organic matter.

In lithotrophic nitrification, ammonia is stepwise oxidized, first via hydroxylamine to nitrite and then further to the end product nitrate (Figure 15.4). The two steps are carried out by two groups of specialists within the bacterial family Nitrobacteriaceae, consisting of five genera of ammonia-oxidizing bacteria (AOB) and four genera of nitrite-oxidizing bacteria (NOB) (Watson et al., 1989). However, the rapid developments in 16S rRNA gene analysis have led to suggestions for phylogenetic revisions of primarily the NOB group (Purkhold et al., 2000; Kowalchuk and Stephen, 2001). Through oxidation of the mineral nitrogen the bacteria derive energy (lithotrophic process) for growth, i.e., fixation of carbon dioxide into their biomass (autotrophic process). Nitrifiers are strict aerobic bacteria, i.e., they are completely dependent on oxygen in their respiration. Being both lithotrohic and autotrophic, they have complex cell machinery, leading to slow growth but also sensitivity to environmental disturbances. One consequence of disturbance on the AOB can be the formation of nitric or nitrous oxide that will be emitted to the atmosphere (Davidson, 1991).

15.3.3 ACTIVITY

The strictest way to measure aerobic soil respiration is to assay the consumption of oxygen, but it can often be more convenient to collect exhaust gas and analyze it for its content of carbon dioxide (ISO, 2002) (Chapter 13). It should, however, be noted that different volumes of oxygen are needed for the mineralization of specific amounts of various carbon sources, i.e., the respiration quotient (RQ) (the ratio of the carbon dioxide produced to the oxygen consumed in a complete combustion of the organic material) seldom has the often assumed value of one. For instance, carbohydrates are generally more oxidized, giving rise to higher RQs compared to lipids. Produced carbon dioxide can for instance be trapped in an alkaline solution of known concentration and volume (Anderson, 1982). The remaining alkali is then determined by titration with acid, and the respiration rate can be calculated. In a related method, the evolution of carbon dioxide from soil samples is automatically determined with a respirometer (Nordgren, 1988). The conductivity change in a KOH solution caused by carbon dioxide trapping is monitored. Initially, the basal respiration rate is measured, reflecting the microbial activity that is supported by the soil without addition of a substrate. The substrate-induced respiration (SIR) is then initiated by mixing glucose, nitrogen, and phosphate into the soil. The variables SIR, the specific growth rate, and the distribution between active and dormant microorganisms can then be calculated from the CO_2-production data by nonlinear regression (Stenström et al., 1998, 2001). The method is cost-effective since 96 soil samples can be measured simultaneously.

Since most denitrification products are gaseous, gas chromatography constitutes the main technique for detection and determination of denitrifying activity. Either open or closed chambers can be used to cover the soil surface from which gaseous emissions are to be determined (Aulakh et al., 1992; Smith et al., 1995) (Chapter 13). In the open system, the outside gases flow continuously through the chamber and evolved soil N gases are trapped, while in the closed system they are accumulated and sampled with a syringe. Another approach is to collect soil cores for incubation in gas-tight containers either directly in the field under ambient temperature or brought to the laboratory for incubation in a controlled environment. In both the chamber and core techniques, accumulated gas products are withdrawn for analysis. Dinitrogen will most likely dominate the end product. However, depending on the conditions, nitric oxide and nitrous oxide can be significant products. The latter gases can also be a result of nitrification activity. In addition, the atmosphere contains a background of 80% dinitrogen that will decrease the sensitivity of methods where its production is used as an endpoint. Therefore, acetylene can be injected into the soil below the field chambers or into the incubation vessels to inhibit the last step in the process, resulting in the inhibition of nitrous oxide reductase activity, and, consequently, nitrous oxide will accumulate as end product (Tiedje et al., 1989). In addition, autotrophic nitrification, and thereby production of nitrous oxide from this process, is inhibited by low concentrations of acetylene (Berg et al., 1982). A rapid method for the measurement of potential denitrification activity (PDA) has been described by Pell (1993) and Pell et al. (1996). Short-term incubations of soil slurries are made anaerobically in the presence of acetylene. The assay substrate contains optimized amounts of glucose and nitrate to saturate the dentrification enzymes. During the assay, the rate of nitrous oxide formation increases with time and the data can be fitted by nonlinear regression to a product formation equation taking exponential growth into consideration (Stenström et al., 1991; Pell et al., 1996). The equation gives the initial product formation rate (PDA) and the specific growth rate (μ_{PDA}).

An alternative approach is to use ^{15}N-based techniques (Tiedje et al., 1989; Aulakh et al., 1992). The rate of formation of ^{15}N-labeled nitrous oxide and dinitrogen after addition of a ^{15}N-enriched fertilizer source will measure the rate of denitrification. Adding a nitrogen source means that the microbial processes in the ecosystem will most likely be disturbed and, hence, obscure the interpretation of the functioning of a natural ecosystem. Sometimes estimations of nitrate or nitrite disappearance have been used to estimate denitrification, but such methods can only be regarded as indicative of the process, since nitrate assimilation or nitrate reduction to ammonia can be significant processes competing for the substrate.

Ever since nitrate formation in soil was discovered to be of microbial origin, its increase in concentration has been used to estimate nitrification activity. Nitrate can easily be extracted from soil samples before and after a period of aerobic incubation and analyzed spectrophotometrically either directly in UV light after removing nitrite and humic substances or in the red wavelength area after reducing nitrate to nitrite with cadmium followed by diazotization under acidic conditions (Keeney and Nelson, 1982). The latter method is preferred since it is very sensitive. Using the increase in nitrate or decrease in ammonium concentrations as indications of nitrification can give biased results as the assimilation of nitrate or ammonium by most soil organisms, including roots, are processes strongly competing for the substrate, but also since ammonium can be produced as a result of mineralization of organic matter. To avoid interference with competing processes, short incubations of soil slurries with and without addition of inhibitors of the ammonia monooxygenase have been adopted. Such inhibitors are, e.g., nitrapyrin, allylthiourea, and acetylene. Changes in ammonium and nitrite concentrations under such conditions provide rates of autothrophic nitrification by difference. Alternatively, chlorate can be used in short incubations of soil slurries to block the nitrite oxidation; thus, the rate of accumulation of nitrite will give the ammonia oxidation activity. A rapid and inexpensive method to assay potential ammonium oxidation rate (PAO) has been described by Belser and Mays (1980) and modified by Torstensson (1993b). In the assay, nitrite accumulation in soil slurries with optimized amounts of ammonium, chlorate, and buffer is analyzed with an automatic nitrite analyzer such as flow injection analysis (FIA). Linear regression of data is used to calculate the PAO. The method has become an ISO standard (2004).

15.3.4 DIVERSITY

Perhaps the most widely adopted biomarkers used for studies of the whole or specific subgroups of the microbial community are the universal nucleic acids RNA and DNA. Protocols for direct extraction of DNA from soil have been well established, and commercial kits are available, e.g., FastDNA® SPIN Kit for Soil, and SoilMaster™ DNA Extracton Kit. Such methods include various combinations of bead beating and treatment with detergents, enzymes, and solvents. Extracted nucleic acids provide a starting point for a wide range of techniques for estimating diversity or community structure of the soil microbial population.

One traditional method to compare similarities in nucleic acid sequences is to record the pattern of reassociation of denatured DNA from mixtures of organisms. By use of reassociation kinetics of mixtures of denatured DNA of soil microorganisms, it was found that the genotypic diversity of the major soil bacteria is very high (Torsvik et al., 1994). It could be concluded that the heterogeneity of the DNA in different soils could represent 4,000–13,000 species.

The general strategy today for genetic fingerprinting of bacterial communities in environmental samples consists of amplification of 16S rRNA genes or functional genes with the polymerase chain reaction (PCR) followed by analysis of the PCR products. One such method is to measure the difference in size of terminal restriction fragments of DNA by analysis of terminal restriction fragment length polymorphism (T-RFLP) (Clement et al., 1998; Marsh, 1999). Specific fragments of extracted DNA are marked and amplified by PCR, during which the terminal end is labeled with a fluorophore. Subsequent cleavage with selected restriction endonucleases will produce terminal fragments of different size. Fragment size determination can be made with gel or capillary electrophoresis with an automatic sequencer for determination of fluorescent molecules. The community structure can also be profiled by subjecting the PCR amplified fragments to denaturing or temperature gradient gel electrophoresis (DGGE or TGGE) analysis (Muyzer, 1999). In DGGE analysis the nucleotide bonds of the fragments are melted while traveling in a gradient of increasing concentrations of urea. Hence, the length of traveling in the gel depends on the mole percentage of guanine plus cytosine ($G + C\%$) and the distribution of these nitrogen bases within the DNA. In TGGE an increasing gradient of temperature is used to melt the DNA. Bands or patterns from the above analyses provide a fingerprint of the bacterial community structure. In addition, the PCR products can be cloned and sequenced for further analysis of the phylogenetic diversity (McCaig et al., 1999).

Bacterial ribosomal DNA contains both conserved and variable regions. The frequent sequencing of these genes in microbiology has led to a considerable database of known sequences. By choosing primer pairs targeting stable regions of the bacterial 16S rRNA genes, a genetic fingerprint of the total bacterial soil community can be obtained. Moreover, primers are available for distinguishing different larger phylogenetic groups such as the α and β proteobacteria and the Actinomycetales.

Though being a subgroup of the total bacterial community, denitrifiers have great taxonomic diversity. Therefore, ecological studies on this group normally are based on functional genes instead of ribosomal genes (Bothe et al., 2000). Genes involved in denitrification seem to be conserved in their nucleotide sequences. Most studies have focused on the diversity of the nitrite reductase, of which two types exist, a cytochrome cd_1 encoded by *nirS* and a copper-containing enzyme, encoded by *nirK*. The ecological studies made so far have revealed that bacteria containing *nirS* seem to be more widespread in natural environments. The gene encoding for nitrous oxide reductase, *nosZ* also seems to be a good candidate in ecological studies of denitrifiers. Construction and application of primer pairs against conserved regions of the above denitrification genes have been shown to be effectively amplified by PCR. DGGE analysis of denitrifiers has been proposed as an effective method to analyze their community structure of environmental samples from e.g., soil, wastewater, and compost.

The composition of the ammonia-oxidizing bacterial community has been analyzed by rRNA nucleic acid–targeted probes (Bothe et al., 2000). PCR with primer combinations targeting the 16S rRNA operon can be used to amplify 16S RNA. Profiling the bacteria with DGGE analysis in combination with sequencing of the amplified material has revealed that nitrifying bacteria are ubiquitous in soil and water environments (Hiorns et al., 1995). Moreover, based on the growing database of sequence homology, it has been suggested that the numbers of AOB genera should be reduced from five to four (Head et al., 1993).

Microbial diversity is a general term used to include the amount and distribution of species or genetic material. As discussed by Nannipieri et al. (2003), most of the molecular methods have intermediate resolution, because they allow the detection of large microbial groups rather than microbial species and are generally not quantitative. Hence, the results might be better discussed in terms of community structure.

In addition to genetic diversity, methods for the estimation of functional diversity have been developed. Such methods are based on growth or respiration responses on an array of various carbon sources consisting of amines, amino acids, carbohydrates, and carboxylic acids. In a test battery described by Degens and Harris (1997) and Degens and Vojvodić-Vuković (1999), a large number of different substrates are added to small portions of soil and incubated in gas-tight bottles. Responses to each organic compound are determined by analyzing the head-space gas for CO_2. A simple and rapid method for establishing substrate profiles is the BIOLOG® system, in which growth on up to 95 separate sole carbon sources can be determined by inoculating microbial soil suspensions into microtiter plates and quantifying color production from a redox-sensitive dye (Zak et al., 1994; Fang et al., 2001). Substrate profiles have also been used to characterize the denitrifying populations in soil (Lescure et al., 1992) and wastewater (Hallin and Pell, 1998).

15.3.5   ENUMERATION AND BIOMASS

A wide variety of methods have been developed for measuring total soil microbial biomass, e.g., substrate-induced respiration (SIR), content of ATP, and chloroform fumigation–extraction (CFE) of biomass C, N, P, or S

(Powlson, 1994). Originally, biomass calculations were based on plate counts of viable bacteria grown on various general substrates. However, this method has long been known to capture only a low fraction of the total biomass (Torsvik et al., 1994). Despite this knowledge the total viable count method probably remains the most common technique for enumeration of soil microorganisms. Estimates of the biomass with direct or indirect (after extraction from soil) observation of cells by microscopic techniques are tedious since not only numbers but also cell size and shape have to be recorded. Various stains are used for better visualization, and some stains may differentiate between living and dead cells (Bölter et al., 2002). Recent developments in image analysis in combination with new fluorescent microscopic technology such as confocal laser scanning microscopy (CLSM) have certainly led to an increase in reproducible and objective enumeration and size determinations (Bloem et al., 1995). Such techniques have confirmed earlier findings that small cells generally dominate in the soil (Bakken, 1997). Another line of methods for detection of microbes in soil is extraction followed by analysis of various biomarkers. A biomarker is any biochemical component that can be used to detect microbes. Examples of useful bio-markers are phospholipid fatty acids, sterols, quinones, and lipopolysaccharides, all of which are constituents of the envelope of cells (Alun et al., 1997). By using a relevant conversion factor, the amount of a biomarker can be used to calculate the biomass of the microbial group that contains that specific compound.

Only a few methods exist for enumerating denitrifying bacteria, whereof the most-probale-number (MPN) technique is the most widely used (Allievi et al., 1987). In this method serial dilutions are inoculated to tubes with a suitable substrate containing nitrate. Acetylene is injected to inhibit the nitrous oxide reductase. After anearobic incubation, positive tubes are identified as those having accumulated nitrous oxide in the headspace, and numbers are calculated by use of statistical tables. Another approach is based on the isolation of colonies grown on plates prepared for the total aerobe count. Randomly selected colonies are transferred into liquid medium and incubated in the presence of acetylene, after which N_2O production is determined. Multiplying the fraction of colonies thus detected as N_2O-producing by the total aerobe number yields the denitrifier number (Allievi and Møller, 1992).

One problem encountered when enumerating nitrifying bacteria seems to be that at least the AOB are extremely difficult to release by extraction from soil particles compared to most organotrophic bacteria (Aakra et al., 2000). Since nitrifying bacteria do not grow on common agar substrates, the traditional technique for enumeration has been the MPN. This technique is tedious since nitrifiers grow extremely slowly and incubations of many weeks are needed. Other techniques that have been used are direct counts in the microscope of bacteria labeled with fluorescent polyclonal antibodies (FA) (Belser and Schmidt, 1978) or by immunoblotting with monoclonal antibodies (Bartosch et al., 2002). Though problems have been encountered with specificity, the FA technique provides rapid results and is more sensitive than the MPN technique. However, FA is also a cultivation-dependent technique since pure cultures are needed to produce antibodies. Recent developments in nucleic acid–based techniques have made counting nitrifiers cultivation independent. Fluorescent oligonuclotide probes targeting the ammonia monooxygenase gene (*amo*) in AMO bacteria has been used for *in situ* hybridization (FISH) of samples from natural environments such as sewage sludge (Wagner et al., 1996). By this technique the AMO bacteria can be visualized by conventional fluorescent microscopy or CLSM.

15.3.6 CHOICE OF METHOD

In addition to high temporal and spatial variations, soil is an extremely complex medium, as discussed previously. Therefore, obviously no single variable, whether chemical, physical, or biological, will be sufficient to describe the soil quality. In addition, the biological soil functions can be divided into three levels: activity, diversity, and biomass (Figure 15.5). Temporal site-specific conditions like weather will affect the top biological level: the short-term production of enzymes and their activities. Depending on long-term environmental fluctuations in, e.g., nutrient and energy inputs or various

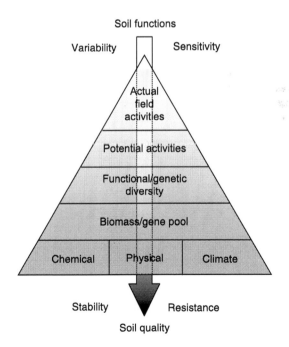

FIGURE 15.5 Levels of functions and properties of the soil ecosystem, with more variable and sensitive functions at the top of the triangle towards more stable and resistant properties at the bottom.

procedures of soil management, the next biological level, the genetic diversity and consequently the functional diversity of the microbial populations, will increase or decrease. The chemical and physical soil properties together with the climate constitute the basis for development of the fundamental biological level — the total microbial biomass with its associated gene pool. As shown in Figure 15.5, from the top of the triangle towards the bottom, the levels also reflect various degrees of variability and sensitivity, with soil quality more related to the basic and more stable functions. To get a comprehensive definition of the microbial soil component, all three properties — activity, diversity, and biomass — should be analyzed simultaneously. Such a combination will then reveal information on both the rate of a specific process as well as the status and robustness of the biomass.

Due to limited resources, studies of soil always are compromises between numbers of variables and numbers of samples that can be processed. In this situation analyses of activity should be given priority since they provide the most easily interpreted information. The next important choice to be made is whether to measure actual (field) or potential rates. Field measurements of, e.g., respiration or denitrification give information on the actual situation *in situ*. The estimated rates may be viewed as a product of intrinsic soil factors that virtually do not change, like texture and mineral composition, and soil factors changing rapidly over the year, like input of fertilizer and organic matter. The factors giving rise to the most rapid changes of microbial activities probably are precipitation and temperature. Since soil quality is a long-term concept, standardized methods for assessment of microbial activities must be looked for. Therefore, potential activities should be chosen, i.e., methods that assay enzymatic activities under standardized conditions, where the amount of enzymes instead of the actual environment becomes limiting. The amount of enzymes indicates the soil's capacity to perform a process, and, in addition, it may be to some extent viewed to mirror the microbial biomass as well as historical soil events.

Though often described in terms like handy, rapid with high throughput, and inexpensive (Marsh, 1999; Muyzer, 1999; Amann and Ludwig, 2000), nucleic acid–based methods for analysis of microbial diversity rapidly become the bottleneck in studies where high numbers of samples have to be processed. Compared to enzymatic methods for estimation of activity, nucleic acid–based methods are expensive and stand-alone they provide a limited amount of information related to, e.g., land use. In contrast, methods based on microbial utilization of arrays of various substrates for analysis of functional diversity are easy, rapid to perform and inexpensive. Using such catabolite response profiles (CRPs), Degens and Vojvodic-Vukovic (1999) were able to distinguish between land uses. In the choice between methods for estimation of biomass and activity, the latter is to be preferred since activity methods like the CRPs provide direct information on the capacity to perform specific processes and, thus, are easier to interpret.

In conclusion, enzymatic methods for measurements of potential activities, including respiration and functional diversity, should be the first choice in toxicity and soil quality estimations. Thereafter, biomass methods should be the choice, followed by methods for genetic diversity analyses. The main reason for choosing assessment of functions in the top of the triangle of Figure 15.5 is their generally better basis for interpretation and hence understanding of the soil ecosystem. However, the combination of methods should always be matched against the aim of the study, and consequently another order of priority could be the final choice.

15.4 SOME APPLICATIONS

15.4.1 TOXICITY TESTING

SIR, PDA, and PAO have all been used in assessing acute toxicity of various anthropogenic chemicals by use of growth-associated product formation kinetics. In a dose-response test, effects of 19 silver concentrations ranging from 0 to 111 ppm (dry soil) were evaluated (Johansson et al., 1998). The response by the respiring microbial community varied depending on the contact time between silver and the soil. The activity of the dormant microbial population was affected when incubated with silver for 10 days, while the activity of the active population was affected only when silver was added at the time for induction of the substrate-induced respiration. The specific growth rate of the denitrifiers was a very sensitive parameter and also indicated that part of the population was resistant to silver. The overall results indicated that silver, in concentrations known to occur in sewage sludge–amended soils, might seriously affect the microbial population. In another study, the acute toxicity effect of 54 herbicides, fungicides, and insecticides on denitrification and nitrification in soil were evaluated (Pell et al., 1998). A first screening of the pesticides at $100 \, \mu g \, g^{-1}$ dry soil revealed that 23% had a significant effect on PDA, 26% on μ_{PDA}, and 35% on PAO. In a subsequent dose-response test, the no effect concentrations (NOEC) for mancozeb were 1.0 and 0.4 $\mu g \, g^{-1}$ dry soil for PDA and PAO, respectively. The above examples of heavy metal and pesticide toxicity evaluations clearly demonstrate the advantage of applying kinetics on the test results. Growth-associated product formation kinetics allow assessment of effects on both microbial growth and their activities in the same assay. Moreover, in assessment of SIR the kinetic approach can be used to separate the observed effects on aerobic organotrophs into effects on active as well as on dormant microorganisms.

Simple tests of microbial functions are valuable and often required by authorities for decision making and legislation. In reviewing microbial toxicity tests for assessment of ecotoxicological risks of contaminants in soil, van Beelen and Doelman (1997) concluded that long-term tests generally are less sensitive than short-term tests. A compilation and evaluation of toxicity test methods for assessment of metal effects on soil microorganisms is provided by McGrath et al. (2002). However, such tests give only limited information on the functioning of the soil and its quality. In order to obtain such information, a more integrated approach is needed (Torstensson et al., 1998).

15.4.2 INTEGRATED APPROACH

Stenberg et al. (1998b) compared the variation within a single arable field with single samples from 26 different soils scattered all over Sweden. As expected, the variation in the physical, chemical, and microbiological variables analyzed displayed considerably higher coefficient of variation (CV%) among the 26 soils as compared to the single field. The CVs for PDA, PAO, and SIR, respectively, were 72, 85, and 36% as compared to 15, 62, and 18% for the single field. Despite the differences in variation of the two data sets, the functional structures (relations of physical, chemical, and microbiological variables) of the two scales seemed to have more similarities than dissimilarities, as revealed by principal component analysis (PCA). One functional group was characterized by the variables organic carbon, total nitrogen, and acidic phosphatase activity, and a second group by pH, available calcium and magnesium, and specific growth rate of denitrifiers (μ_{res}). In a third group, variables related to organotrophic microbial activity and biomass were clustered.

Stable patterns of general similarities in functional soil structures, i.e., clusters of soil variables related to each other derived from multivariate statistics, though often difficult to interpret, probably are important in the analyses of changes in soil quality. Environmentally induced changes, intentional or unintentional, probably will lead to alterations of such patterns, and they might, hence, be used as warning systems for soil quality disturbances. In addition, functional structures can be used to identify single variables that represent specific key functions. In this way variable reduction can be made without losing information. In addition to several chemical and physical variables, minimum data sets (MDS) for description of soil quality have been suggested to contain the microbial variables respiration, potential mineralizable nitrogen, and biomass (carbon and nitrogen content) (Doran and Parkin, 1994). Though manageable, three variables may be insufficient to describe the soil ecosystem. To increase the information, Stenberg (1999) proposed the additional variables PAO and PDA.

15.4.3 VARIATION

Microbial variation was studied in a single arable field with cereals and leys in the rotation, but it was not fertilized (Stenberg et al., 1998b). One of the main reasons for choosing the field was that visual inspection during the growing season suggested a uniform field. The clay content varied between 49 and 53%. The field was soil sampled in a rectangular grid of 66×200 m with grid nodes of 18×15 m, making a total of 52 samples. The mean value of the potential rate (PDA) was 19.6 with the range 13.9–25.7 ng N_2O-N g^{-1} d.s. min^{-1} and a CV of 15.4% (Figure 15.6). In contrast, when measuring actual field rates of denitrification, considerably higher variation is observed, with a CV typically being $> 100\%$ (Parkin et al., 1987; Parsons et al., 1991; Pennock et al., 1992) and sometimes exceeding $> 1000\%$ (Christensen et al., 1990). The high

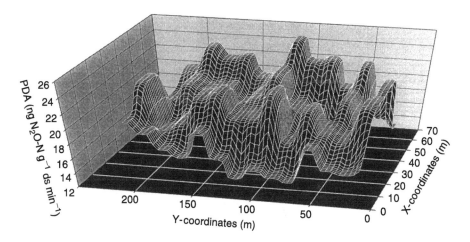

FIGURE 15.6 Interpolated mesh-plot of potential denitrification activity (PDA) in an arable field sampled in a grid pattern with nodes of 18×15 m.

variability observed is due to hot spots, i.e., patches with activities more than five standard deviations above the overall mean value (Parkin and Robinson, 1989; Christensen et al., 1990). Several reasons for transient hot-spot structures are possible. A single soil aggregate can display a high spatial gradient of carbon, nitrate, and oxygen, all factors affecting denitrification. Addition of organic matter to the surface of the aggregate stimulates respiratory activity, causing anoxia and high rates of denitrifying activity within organic hot spots (Højberg et al., 1994), but can still exhibit only low rates of denitrification due to a limiting availability of energy sources. Also, denitrification within such hot spots can rapidly become limited by diffusional supply of nitrate. Six et al. (2000) observed that increasing cultivation intensity leads to a loss of C-rich macroaggregates and an increase in C-depleted microaggregates. Vinther et al. (1999) stressed that soil in immediate contact with macropore channels could be a preferential environment for denitrification, as indicated by higher concentrations of nitrate and water-soluble carbon as well as a larger bacterial biomass. From the above it is clear that sample size will influence the result of activity measurements.

Results from studies of the effect of sample size on denitrification suggest that soil cores > 4.2 cm in diameter yielded the most reliable estimates of natural denitrification rates, and that approximately 10–15 kg of soil was necessary to obtain a representative soil mass (Parkin et al., 1987). In contrast, most soil DNA studies use rather small samples, usually < 1 g of soil. This is probably a result of the use of commercial available kits for DNA extraction. In evaluating sample sizes of 0.1–10 g, Ellingsøe and Johnsen (2002) observed the largest variation in culturable heterotrophic and genetic community structure in sample sizes < 1 g. From results of DNA extractions of sample sizes of 0.125–4 g, Ranjard et al. (2003) proposed different sampling strategies depending on the objectives. Rather large samples (≥ 1 g) should be used for global description of the genetic community structure, while a large number of smaller samples should be used for a more complete inventory of microbial diversity.

In contrast to the high variability reported for denitrification above, Ambus (1993) reported only low spatial and seasonal variation of potential denitrification enzyme activity (10–26%), suggesting a high persistence of the enzymes. This also means that potential methods are preferable to measurements of actual rates when soil quality is to be estimated.

To deal with soil variation and spatial patterns at various scales (e.g., micro, plot, field, landscape, or regional) (Parkin, 1993), an increase in the ability to find patterns can be achieved by using geostatistics (Webster and Oliver, 2001) (see also Chapter 16). The first step in geostatistics is to establish a suitable sampling pattern, usually consisting of a grid locating samples at different interdistances. Moreover, the sampling grid should cover known variations in the field such as different topographic formations. After sampling and analyses of desired variables, descriptive statistics can be used to give an overview of the results. In this step, data are tested for distribution, and if found to be non-normal, they are transformed into a normal distribution. The next step is to define a semivariogram for each soil property, i.e., the semivariance plotted versus sampling distance. The variogram provides parameters (nugget, range, and sill) that can be used to quantify the spatial variation for a range of properties by measuring the degree of correlation between sampling points at a given distance apart. The nugget corresponds to spatially uncorrelated variation due to, e.g., short-range variation not accounted for by the sampling design, whereas the sill is the point where the semivariance levels off and remains constant and, hence, equals the priori variance. The range is the distance where points are no longer spatially correlated, i.e., the sampling distance required to obtain statistically independent samples. Provided that data give a stable semivariogram, they can be used to calculate kriging maps displaying the variation of the specific variable over the field. Parkin et al. (1987) found that denitrification was characterized by large variation at distances of < 10 cm and only weak spatial dependencies at distances of 10–100 cm. In a study of two sites (250×250 m and 200×500 m, respectively), field-scale distributions and spatial trends of 28 different soil parameters were investigated (Cambardella et al., 1994). Twelve and six parameters at study site one and two, respectively, including organic carbon, total nitrogen, and pH, were strongly spatially dependent. In addition, microbial biomass was moderately spatially dependent at both sites as well as denitrification at site one. The ranges of microbial variables varied from 30 m (microbial biomass N) over 68–75 m for soil respiration and denitrification up to 270 m for fungal ergosterol. From the study it was concluded that there were similarities in the patterns of spatial variability for some of the soil parameters at both sites, which suggests that relationships derived from one set of measurements for one field may be applicable at other field sites within the same or similar landscapes. Changes in such general patterns over time may be used to trace changes in soil quality.

15.5   RECOMMENDATIONS AND FUTURE RESEARCH

15.5.1   RECOMMENDATIONS

Microbial variables chosen for assessing soil quality should ideally hold information at various levels, e.g., activity, biomass, and diversity. In addition to being representative of the microbial ecosystem, the chosen variables should also cover sensitive as well as robust and nonsensitive processes and should provide information important for the use of the soil, e.g., plant nutrient status. To become applicable tools, the methods should also be simple to perform and be inexpensive. Many of these prerequisites are fulfilled by methods to estimate activities of respiration, denitrification, and nitrification. Methods for determination of potential activities rather than actual field rates seem to be a better choice since they relate to basic microbial properties and, in addition, they are more simple to handle.

Surveys of soil quality by necessity involve large sampling programs and, hence, generate large numbers of soil samples. Therefore, it is most often better to perform many simple analyses than a few complicated ones. In combination with data from physicochemical analyses, PCA and geostatistics are able to extract a functional structure that can be used to trace changes in the soil environment. Such evaluations should preferably be done in an interdisciplinary manner where soil microbiologists cooperate with soil physicists and chemists.

15.5.2   FUTURE RESEARCH

It is a challenging task to define minimum data sets that will be both informative and cost-effective. What minimum number of variables will provide the maximum information? In soil quality test kits it is often proposed that respiration represent the microbial quality of soil (Karlen et al., 2003). However, it is not likely that one single microbial soil variable will be able to capture the complex nature of the biological component of the soil ecosystem.

Only by connecting the various scales of soil functions we can develop an extensive understanding of soil quality. How do activities at a small scale, such as activities in soil microaggregates, relate to the function of a whole field and the field, in turn, to the region? Only from such knowledge can a comprehensive picture of soil function be derived and models developed that can be understood and used in decision making.

The genetic pool in the soil may be viewed as a finite resource. Therefore, soil quality assessment must comprise tools to characterize genetic diversity. The rapid achievements in nucleic acid–based technologies have provided information on the ecological soil structure. A large number of "new" inhabitants residing in the soil has been discovered, most of which cannot be cultured. Moreover, most of these new tools provide only qualitative or semi-quantitative information. The future challenge is to make nucleic-based methods both quantitative and at the same time inexpensive (Ogram, 2000).

REFERENCES

Aakra, Å., M. Hesselsøe, and L.R. Bakken. 2000. Surface attachment of ammonia-oxidizing bacteria in soil. Microb. Ecol. 39:222–235.

Adu, J.K., and J.M. Oades. 1978. Physical factors influencing decomposition of organic materials in soil aggregates. Soil Biol. Biochem. 10:109–115.

Alef, K., and P. Nannipieri (ed.). 1996. Methods in Applied Soil Microbiology and Biochemistry. Academic Press, London.

Allievi, L., D. Catelani, A. Ferrari, and V. Treccani. 1987. A method for counting denitrifiers by N$_2$O detection. J. Microbiol. Meth. 7:67–72.

Allievi, L., and F. Møller. 1992. A method based on plate count for enumerating N$_2$O/N$_2$-producing bacteria from nitrate in the soil. J. Basic Microbiol. 32:291–298.

Alun, J., J.A.W. Morgan, and C. Winstanley. 1997. Microbial biomarkers, p. 331–352. In J.D. van Elsas, J.T. Trevors, and E.M.H. Wellington (ed.), Modern Soil Microbiology. Marcel Dekker, New York.

Amann, R., and W. Ludwig. 2000. Ribosomal RNA-targeted nucleic acid probes for studies in microbial ecology. FEMS Microbiol. Rev. 24:555–565.

Ambus, P. 1993. Control of denitrification enzyme activity in a streamside soil. FEMS Microbiol. Lett. 102:225–234.

Anderson, J.P.E. 1982. Soil respiration, p. 831–871. In A.L. Page (ed.), Methods of Soil Analysis, Part 2. 2nd ed. Agron. Monogr. 9. Amer. Soc. Agronomy and Soil Science Soc. Amer., Madison, WI.

Aulakh, M.S., J.W. Doran, and A.R. Mosier. 1992. Soil denitrification — significance, measurements, and effects of management, p. 1–57. In Advances in Soil Science, Vol. 18. Springer-Verlag, New York.

Bakken, L.R. 1997. Culturable and nonculturable bacteria in soil, p. 47–61. In J.D. van Elsas, J.T. Trevors, and E.M.H. Wellington (ed.), Modern Soil Microbiology. Marcel Dekker, New York.

Bartosch, S., C. Hartwig, E. Spieck, and E. Bock. 2002. Immunological detection of Nitrospira-like bacteria in various soils. Microbiol. Ecol. 43:26–33.

Belser, L.W., and E.L. Mays. 1980. Specific inhibition of nitrite oxidation by chlorate and its use in assessing nitrification in soil and sediments. Appl. Environ. Microbiol. 39:505–510.

Belser, L.W., and E.L. Schmidt. 1978. Serological diversity within a terrestrial ammonia-oxidizing population. Appl. Environ. Microbiol. 36:589–593.

Berg, P., L. Klemedtsson, and T. Rosswall. 1982. Inhibitory effect of low partial pressures of acetylene on nitrification. Soil Biol. Biochem. 14:301–303.

Bergström, L., and J. Stenström. 1998. Environmental fate of chemicals in soil. Ambio 27:16–23.

Bloem, J., M. Veninga, and J. Shepherd. 1995. Fully automatic determination of soil bacterium numbers, cell volumes, and frequencies of dividing cells by confocal laser scanning microscopy and image analysis. Appl. Environ. Microbiol. 61:926–936.

Bölter, M., J. Bloem, K. Meiners, and R. Möller. 2002. Enumeration and biovolume determination of microbial cells — a methodological review and recommendations for applications in ecological research. Biol. Fert. Soils 36:249–259.

Bothe, H., G. Jost, M. Schloter, B.B. Ward, and Karl-Paul Witzel. 2000. Molecular analysis of ammonia oxidation and dentitrification in natural environments. FEMS Microbiol. Rev. 24:673–690.

Cambardella, C.A., T.B. Moorman, J.M. Novak, T.B. Parkin, D.L. Karlen, R.F. Turco, and A.E. Konopka. 1994. Field-scale variability of soil properties in central Iowa soils. Soil Sci. Soc. Am. J. 58:1501–1511.

Christensen, S., S. Simkins, and J. Tiedje. 1990. Spatial variability in denitrification: dependency of activity centers on the soil environment. Soil Sci. Soc. Am. J. 54:1608–1613.

Clement, B.G., L.E. Kehl, K.L. DeBord, and C.L. Kitts. 1998. Terminal restriction fragment patterns (TRFPs), a rapid, PCR-based method for the comparison of complex bacterial communities. J. Microbiol. Meth. 31:135–142.

Coleman, D.C., C.V. Cole, H.W. Hunt, and D.A. Klein. 1978. Trophic interactions in soils as they affect energy and nutrient dynamics. I. Introduction. Microbial Ecol. 4:345–349.

Davidson, E. A. 1991. Fluxes of nitrous oxide and nitric oxide from terrestrial ecosystems, p. 219–235. In J.E. Rogers, and W.B. Whitman (ed.), Microbial Production and Consumption of Greenhouse Gases: Methane, Nitrogen Oxides, and Halomethanes. American Society for Microbiology, Washington, DC.

Degens, B.P., and J.A. Harris. 1997. Development of a physiological approach to measuring the catabolic diversity of soil microbial communities. Soil Biol. Biochem. 29:1309–1320.

Degens, B.P., and M. Vojvodić-Vuković. 1999. A sampling strategy to assess the effects of land use on microbial functional diversity in soil. Aust. J. Soil Res. 37:593–602.

Doran, J.W., and T.B. Parkin. 1994. Defining and assessing soil quality, p. 3–21. In J.W. Doran, et al. (ed.), Defining Soil Quality for a Sustainable Environment. SSSA Spec. Publ. 35, Madison, WI.

Ellingsøe, P., and K. Johnsen. 2002. Influence of soil sample sizes on the assessment of bacterial community structure. Soil Biol. Biochem. 34:1701–1707.

Fang, C., M. Radosevich, and J.J. Fuhrmann. 2001. Characterization of rhizosphere microbial community structure in five similar grass species using FAME and BIOLOG analyses. Soil Biol. Biochem. 33:679–682.

Hallin, S., and M. Pell. 1998. Metabolic properties of denitrifying bacteria adapting to methanol and ethanol in activated sludge. Water Res. 32:13–18.

Head, I.M., W.D. Hiorns, T.M. Embley, A.J. McCarthy, and J.R. Saunders. 1993. The phylogeny of autotrophic ammonia-oxidising bacteria as determined by analysis of 16S ribosomal RNA gene sequences. J. Gen. Microbiol. 139:1147–1153.

Hill, G.T., N. A. Mitkowski, L. Aldrich-Wolfe, L. R. Emele, D. D. Jurkonie, A. Ficke, S. Maldonado-Ramirez, S. T. Lynch, and E. B. Nelson. 2000. Methods for assessing the composition and diversity of soil microbial communities. Appl. Soil Ecol. 15:25–36.

Hiorns, W.D., RC Hastings, I.M. Head, A.J. McCarthy, J.R. Saunders, R.W. Pickup, and G.H. Hall. 1995. Amplification of 16S ribosomal RNA genes of autotrophic ammonia-oxidizing bacteria demonstrates the ubiquity of nitrosospiras in the environment. Microbiology 141:2793–2800.

Højberg, O., N.P. Revsbech, and J.M. Tiedje. 1994. Denitrification in soil aggregates analysed with microsensors for nitrous oxide and oxygen. Soil Sci. Soc. Am. J. 58:1691–1698.

ISO 15685. 2004. Soil quality — determination of potential nitrification — rapid test by ammonium oxidation. International Organization for Standardization, Geneva.

ISO 16072. 2002. Soil quality — laboratory methods for determination of microbial soil respiration. International Organization for Standardization, Geneva.

Johansson, M., M. Pell, and J. Stenström. 1998. Kinetics of substrate-induced respiration (SIR) and denitrification: applications to a soil amended with silver. Ambio 27:40–44.

Karlen, D.L., C.A. Ditzler, and S.S. Andrews. 2003. Soil quality: why and how? Geoderma 114:145–156.

Keeney, D.R., and D.W. Nelson. 1982. Nitrogen — inorganic forms, p. 643–698. *In* A.L. Page, et al. (ed.), Methods of Soil Analysis, Part 2, Chemical and Microbiological Properties. 2nd ed. Agron. Monogr. 9. Amer. Soc. Agronomy and Soil Science Soc. Amer. SSSA, Madison, WI.

Kowalchuk, G.A., and J.R. Stephen. 2001. Ammonium-oxidizing bacteria: a model for molecular microbial ecology. Ann. Rev. Microbiol. 55:485–529.

Lee, K.E., and C.E. Pankhurst. 1992. Soil organisms and sustainable productivity. Aust. J. Soil Res. 30:855–892.

Lescure, C., L. Menendez, R. Lensi, A. Chalamet, and A. Pidello. 1992. Effect of addition of various carbon substrates on denitrification in a vertic Mollisol. Biol. Fertil Soils 13:125–129.

Marsh, T.L. 1999. Terminal restriction fragment length polymorphism (T-RFLP): an emerging method for characterizing diversity among homologous populations of amplification products. Curr. Opin. Microbiol. 2:323–327.

McCaig, A.E., L.A. Glover, and J.I. Prosser. 1999. Molecular analysis of bacterial community structure and diversity in unimproved and improved upland grass pastures. Appl. Environ. Microbiol. 65:1721–1730.

McGrath, S.P., R.T. Checkai, J.J. Scott-Fordsmand, P.W. Glazebrook, G.I. Paton, and S. Visser. 2002. Recommendations for testing toxicity to microbes in soil, p. 17–35. *In* A. Fairbrother, et al. (ed.). Test Methods to Determine Hazards of Sparingly Soluble Metal Compounds in Soils. Society of Environmental Toxicology and Chemistry (SETAC), Pensacola, FL.

Muyzer, G. 1999. DGGE/TGGE a method for identifying genes from natural ecosystems. Curr. Opin. Microbiol. 2:317–322.

Nannipieri, P., J. Ascher, M.T. Ceccherini, L. Landi, G. Pietramellara, and G. Renella. 2003. Microbial diversity and soil functions. Eur. J. Soil Sci. 54:655–670.

Nordgren, A. 1988. Apparatus for the continuous long-term monitoring of soil respiration rate in large numbers of samples. Soil Biol. Biochem. 20:955–957.

Ogram, A. 2000. Soil molecular ecology at age 20: methodological challenges for the future. Soil Biol. Biochem. 32:1499–1504.

Parkin, T.B. 1993. Spatial variability of microbial processes in soil — a review. J. Environ. Qual. 22:409–417.

Parkin, T.B., and J.A. Robinson. 1989. Stochastic models of soil denitrification. Appl. Environ. Microbiol. 55:72–77.

Parkin, T.B., J.L. Starr, and J.J. Meisinger. 1987. Influence of sample size on measurement of soil denitrification. Soil Sci. Soc. Am. J. 51:1492–1501.

Parsons, L.L., R.E. Murray, and M.S. Smith. 1991. Soil denitrification dynamics: spatial and temporal variations of enzyme activity, populations, and nitrogen gas loss. Soil Sci. Soc. Am. J. 55:90–95.

Patzel, N., H. Sticher, and D.L. Karlen. 2000. Soil fertility — phenomenon and concept. J. Plant. Nutr. Soil Sci. 163:129–142.

Pell, M. 1993. Denitrification, a method to estimate potential denitrification rate in soil, p. 59–69. *In* L. Torstensson (ed.), Guidelines — Soil Biological Variables in Environmental Hazard Assessment. Report 4262. Swedish Environmental Protection Agency, Stockholm.

Pell, M., B. Stenberg, J. Stenström, and L. Torstensson. 1996. Potential denitrification activity assay in soil — with or without chloramphenicol? Soil Biol. Biochem. 28:393–398.

Pell, M., B. Stenberg, and L. Torstensson. 1998. Potential denitrification and nitrification tests for evaluation of pesticide effects in soil. Ambio 27:24–28.

Pennock, D.J., C. van Kessel, and R.A. Sutherland. 1992. Landscape-scale variations in denitrification. Soil Sci. Soc. Am. J. 56:770–776.

Powlson, D.S. 1994. The soil microbial biomass: before, beyond and back, p. 3–20. *In* K. Ritz, J. Dighton, and K.E. Giller (ed.), Beyond the Biomass. Wiley, Chichester.

Purkhold, U., A. Pommerening-Röser, S. Juretschko, M.C. Schmid, H.-P. Koops, and M. Wagner. 2000. Phylogeny of all recognized species of ammonia oxidizers based on comparative 16S rRNA and *amo*A sequence analysis: implications for molecular diversity surveys. Appl. Environ. Microbiol. 66:5368–5382.

Ranjard, L., D.P.H. Lejon, C. Mougel, L. Schehrer, D. Merdinoglu, and R. Chaussod. 2003. Sampling strategy in molecular microbial ecology: influence of soil sampling size on DNA fingerprinting analysis of fungal and bacterial communities. Environ. Microbiol. 5:1111–1120.

Schinner, F., R. Öhlinger, E. Kandeler, and R. Margesin (ed.). 1996. Methods in Soil Biology. Springer, Berlin.

Schloter, M., M. Lebuhn, T. Heulin, and A. Hartmann. 2000. Ecology and evolution of bacterial microdiversity. FEMS Microbiol. Rev. 24:647–660.

Six, J., K. Paustian, E.T. Elliott, and C. Combrink. 2000. Soil structure and organic matter: I. Distribution of aggregate-size classes and aggregate-associated carbon. Soil Sci. Soc. Am. J. 64:681–689.

Smith K.A., H. Clayton, I.P. McTaggart, P.E. Thomson, J.R.M. Arah, and A. Scott. 1995. The measurement of nitrous oxide emissions from soil by using chambers. Phil. Trans. R. Soc. Lond. A 351:327–338.

Stenberg, B. 1999. Monitoring soil quality of arable land: microbiological indicators. Acta Agric. Scand. Sci. Sect. B Plant Soil 49:1–24.

Stenberg, B., M. Johansson, M. Pell, K. Sjödahl-Svensson, J. Stenström and L. Torstensson. 1998a. Microbial biomass and activities in soil as affected by frozen and cold storage. Soil Biol. Biochem. 30:393–402.

Stenberg, B., M. Pell, and L. Torstensson. 1998b. Integrated evaluation of variation in biological, chemical and physical soil properties. Ambio 27:9–15.

Stenström, J., A. Hansen, and B. Svensson. 1991. Kinetics of microbial growth-associated product formation. Swedish J. Agr. Res. 21:55–62.

Stenström, J., B. Stenberg, and M. Johansson. 1998. Kinetics of substrate-induced respiration (SIR): theory. Ambio 27:35–39.

Stenström, J., K. Svensson, and M. Johansson. 2001. Reversible transition between active and dormant microbial states in soil. FEMS Microbiol. Ecol. 36:93–104.

Tiedje, J.M., S. Simkins, and P.M. Groffman. 1989. Perspectives on measurement of denitrification in the field including recommended protocols for acetylene based methods, p. 217–240. *In* M. Clarholm, and L. Bergström (ed.), Ecolgy of Arable Land Perspectives and Challanges. Kluwer Academic Publisher, Dordrecht.

Torstensson, L. (ed.) 1993a. Guidelines — Soil Biological Variables in Environmental Hazard Assessment. Report 4262. Swedish Environmental Protection Agency, Stockholm.

Torstensson, L. 1993b. Ammonium oxidation, a rapid method to estimate potential nitrification in soils, p. 40–47. *In* L. Torstensson (ed.), Guidelines — Soil Biological Variables in Environmental Hazard Assessment. Report 4262. Swedish Environmental Protection Agency, Stockholm.

Torstensson, L., M. Pell, and B. Stenberg. 1998. Need of a strategy for evaluation of arable soil quality. Ambio 27:4–8.

Torsvik, V., J. Goksoyr, F.L. Daae, R. Sorheim, J. Michalsen, and K. Salte. 1994. Use of DNA analysis to determine the diversity of microbial communities, p. 39–48. *In* K. Ritz, J. Dighton, and K.E. Giller (ed.), Beyond the Biomass. Wiley, Chichester.

Totsche, K. 1996. Quality control and quality assurance in applied soil microbiology and biochemistry. Quality — project design — spatial sampling, p. 5–24. *In* K. Alef, and P. Nannipieri (ed.), Methods in Applied Soil Microbiology and Biochemistry. Academic Press, London.

van Beelen, P., and P. Doelman. 1997. Significance and application of microbial toxicity tests in assessing ecotoxicological risks of contaminants in soil and sediment. Chemosphere 34:455–499.

Vinther, F.P., F. Eiland, A.-M. Lind, and L. Elsgaard. 1999. Microbial biomass and numbers of denitrifiers related to macropore channels in agricultural and forest soils. Soil Biol. Biochem. 31:603–611.

Wagner, M., G. Rath, H.-P. Koops, J. Flood, and R. Amann. 1996. In situ analysis of nitrifying bacteria in sewage treatment plants. Wat. Sci. Technol. 34:237–244.

Waters, A. G., and J. M. Oades. 1991. Organic matter in water-stable aggregates, p. 163–174. *In* W.S. Wilson (ed.), Advances in Soil Organic Matter Research: The Impact on Agriculture and the Environment. The Royal Society of Chemistry, Cambridge.

Watson, S.W., E. Bock, H. Harms, H.-P. Koops, and A.B. Hooper. 1989. Nitrifying bacteria, p. 1808–1834. *In* J.G. Holt, et al. (ed.), Bergey's Manual of Systematic Microbiology, Vol. 3. Williams & Wilkins, Baltimore.

Webster, R., and M.A. Oliver. 2001. Geostatistics for Environmental Scientists. J. Wiley & Sons, Chichester.

Wright, S.F., and A. Upadhyaya. 1998. A survey of soils for aggregate stability and glomalin, a glycoprotein produced by hyphae of arbuscular mycorrhizal fungi. Plant Soil 198:97–107.

Zak, J.C., M.R. Willig, D.L. Moorhead, and H.G. Wildman. 1994. Functional diversity of microbial communities: a quantitative approach. Soil Biol. Biochem. 26:1101–1108.

Zumft, W.G. 1992. The denitrifying procaryotes, p. 554–582. *In* A. Balows, et al. (ed.), The Procaryotes, Vol. 1, 2nd ed. Springer-Verlag, New York.

16 Geostatistical Procedures for Characterizing Soil Processes

*Marc Van Meirvenne, Lieven Vernaillen,
Ahmed Douaik, Niko E. C. Verhoest,
and Moira Callens*
Ghent University, Gent, Belgium

CONTENTS

1-5667-0657-2/05/$0.00 + $1.50
© 2005 by CRC Press

16.1   INTRODUCTION — WHY GEOSTATISTICS?

Soil is a spatially continuous mixture of minerals, organic matter, water, and air in a constant state of change. However, compared to other media like water and air, in soil most fluxes of matter and energy are small and slow. So it makes sense to describe and map soil properties and analyze their behavior in space and time.

But soil is difficult to observe. We need to dig or auger to investigate it, although noninvasive measurement techniques are gradually becoming available. But even then, sampling soil and analyzing it in the laboratory is still the only way to obtain hard data (i.e., data with no or a negligible uncertainty). Since we cannot observe soil continuously in three dimensions (3-D), our samples represent points. Even when we pool samples into a composite sample, we must take point samples. These point samples typically represent a fraction of $1/10^6$ to $1/10^9$ of the continuous phenomenon we want to understand and describe (Chilès and Delfinder, 1999). Characterizing a quasi-infinite population of soil "individuals," of which only a very small fraction can be observed, obviously requires statistical processing to obtain descriptive and summarizing statistics, to identify trends and relationships, and to predict its properties in space and time at different levels of scale.

The consequence is that a soil scientist must work with large uncertainties, both in attribute and in geographical space. Soil maps are among the most uncertain maps in any geographical database; moreover they are very scale dependent. Therefore, soil scientists have a close affinity with spatial statistics. It is no surprise that after its development in mining geology, soil scientists were among the first to welcome and apply geostatistics. By the end of the 1970s, shortly after the first comprehensive book on geostatisctics appeared in English (Journel and Huijbreghts, 1978), the first papers on applications of geostatistics were published in soil science journals (Campbell, 1978; Burgess and Webster, 1980). Since then many textbooks on soil statistics and geostatistical applications to soil have been published.

It will be clear that this limited ability to observe a significant proportion of the population must be offset by a sound theoretical methodology which is able to provide information on the associated uncertainty. The less observations available, the more important the applicability of the used methodology becomes. Hence it is important to understand the boundary conditions and hypotheses of the methods used, since no "one-fits-all" approach exists.

The aim of this chapter is to guide the practitioner studying soil-water-solute processes to select an appropriate geostatistical method and to set up a spatial sampling strategy. We will add a short illustration discussing alternative ways of studying the spatial distribution of soil water supported by an understanding of soil-landscape-hydrology relationships.

16.2 GEOSTATISTICS

16.2.1 THEORETICAL CONCEPTS

A number of excellent books on geostatistics is currently available. An updated list with some details can be found at http://www.ai-geostats.org/ (it provides a list with different categories of geostatistical software). Therefore, we will not extend deeply into the theoretical background of geostatistics, but some basic concepts must be understood.

Any soil property, changing continuously in space and time, can be represented as a regionalized variable $Z(\mathbf{x})$ where $\mathbf{x}$ represents a vector of spatial coordinates (and eventually the time instant of observation). This variable has a unique value at every location (e.g., soil moisture content at a particular place), but because this property is the result of the combination of complex processes acting interdependently in space and time, we are not able to predict its value in a deterministic way at every location. Therefore we lack insight and information. For this reason we turn to a stochastic approach by predicting the distribution of all possible values of this variable at $\mathbf{x}$.

In geostatistics this approach is based on the concept of a random variable. A random variable possesses a distribution function with some descriptive parameters, like the mean (a first-order moment) and other higher-order moments (e.g., variance, covariance). One selection from the distribution of the random variable at $\mathbf{x}$ is called a "realization" of the regionalized variable Z at $\mathbf{x}$.

One can extend the concept of a random variable to all locations within a study area, creating an *ensemble* of random variables. This ensemble of all random variables within a study area is called a random function (RF) or a random process or a stochastic process. By selecting a value from the distribution functions of each random variable we obtain a map of Z.

Assume that the RF Z was observed at a number of locations $\mathbf{x}_\alpha$ $\alpha = 1, \ldots, n$) and we want to predict its value at location $\mathbf{x}_0$. In classical statistics such a prediction is modeled as a combination of two components: a deterministic mean m (which can represent also a deterministic trend function) plus a random error term, or noise term, ε, which represents the spatially independent fluctuations around the mean. This error term is supposed to follow a normal distribution with zero mean and a spread around the mean represented by s^2 being a measure of the uncertainty of the prediction:

$$Z(\mathbf{x}_0) = m + \varepsilon \quad \text{with} \quad \varepsilon \in \aleph(0, s^2) \tag{16.1}$$

In the case of a regionalized variable, the hypotheses that the error term is spatially independent is often unrealistic. Therefore, in the theory of the regionalized variables the error term is split into two:

$$Z(\mathbf{x}_0) = m + \varepsilon'(\mathbf{x}) + \varepsilon'' \quad \text{with} \quad \varepsilon'' \in \aleph(0, s_\varepsilon^2) \tag{16.2}$$

where $\varepsilon'(\mathbf{x})$ is a stochastic term which represents the spatially structured component of the error. The remaining part, which can then be considered to be spatially uncorrelated, is represented by ε''. In geostatistics the aim is to characterize $\varepsilon'(\mathbf{x})$ as completely as possible, allowing us to use it to enhance the prediction of the basis of m.

The concept of a RF supposes that one is able to construct the distribution function of the random variable within the study area. But in practice only one realization of this variable at each location (the actual sample) is available. Therefore, some assumptions must be accepted in order to proceed. These are combined under the hypotheses of stationarity.

16.2.1.1 Strict Stationarity

Strict stationarity supposes that all statistical characteristics (or moments) of variable Z are constant in space, i.e., they are invariable by a spatial translation. This is a very heavy condition, which is virtually unverifiable on the basis of a unique sample per location.

16.2.1.2 Second-Order Stationarity

In practice this condition is weakened to "stationarity of the second order," i.e., only the first two order statistics remain constant after spatial translation. Stationarity of the second order assumes:

1. That the first-order statistic, being the mathematical expectation $E[Z(\mathbf{x})]$, exists and is independent from $\mathbf{x}$:

$$E[Z(\mathbf{x})] = m \qquad \forall \, \mathbf{x} \tag{16.3}$$

or, the mean is constant over the study area and does not depend on the location **x**. Therefore, it is the expected value at every location within the study area.

2. That the second-order statistic, being the autocovariance $C(\mathbf{h})$, exists and depends only on the lag **h** representing a spatial or temporal vector between observations. The autocovariance $C(\mathbf{h})$ is defined as:

$$C(\mathbf{h}) = E\big[\{Z(\mathbf{x}) - m(\mathbf{x})\} \cdot \{Z(\mathbf{x} + \mathbf{h}) - m(\mathbf{x} + \mathbf{h})\}\big] \qquad \forall\, \mathbf{x} \qquad (16.4)$$

16.2.1.3 Intrinsic Hypothesis

The autocovariance relies strongly on the assumption of a stationary mean, which is still a heavy assumption that is difficult to verify. Therefore, this condition is weakened into the "intrinsic hypothesis," which does not consider the autocovariance, but the variance between the observation points, resulting in the definition of the variogram, $\gamma(\mathbf{h})$:

$$\mathrm{Var}[Z(\mathbf{x} + \mathbf{h}) - Z(\mathbf{x})] = E\Big[\{Z(\mathbf{x} + \mathbf{h}) - Z(\mathbf{x})\}^2\Big] \equiv 2\gamma(\mathbf{h}) \qquad \forall\, \mathbf{x} \qquad (16.5)$$

16.2.2 VARIOGRAM ESTIMATION

Based on Eq. (16.5), the practical formula to calculate the variogram is:

$$\gamma(\mathbf{h}) = \frac{1}{2N(\mathbf{h})} \sum_{\alpha=1}^{N(\mathbf{h})} \{z(\mathbf{x}_\alpha + \mathbf{h}) - z(\mathbf{x}_\alpha)\}^2 \qquad (16.6)$$

with $\gamma(\mathbf{h})$ the variogram for a distance vector (lag) **h** between observations $z(\mathbf{x}_\alpha)$ and $z(\mathbf{x}_\alpha + \mathbf{h})$, and with $N(\mathbf{h})$ being the number of pairs separated by **h**.

A variogram is typically a function that increases from low values near the origin to larger values as **h** increases. Often this function stabilizes around a maximum at larger **h** values (Figure 16.1). This maximum is called the sill. The lag **h** at which the sill is reached is called the autocorrelation length, scale, or range. The range is the maximal extent of the spatial relation (correlation) between observations of the investigated variable. At distances smaller than the range, there exists a dependence between the observations that increases as the observations are situated closer to each other. At lags larger than the range, the expected difference between observations is maximal and independent from their distance.

Theoretically the variogram is zero at $\mathbf{h} = 0$, but since in practice there is always a minimal distance between the two closest observations, the

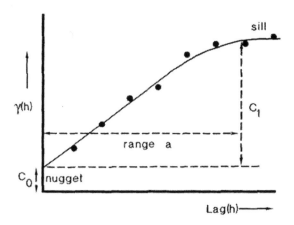

FIGURE 16.1 Typical behavior of a variogram with terminology of the three model parameters. (From Burrough and McDonnell, 1998.)

fitted model is extrapolated towards $\mathbf{h} = 0$. This can cause the curve to have a positive Y-intercept. Such an intercept is called nugget effect. This nugget effect represents the lower boundary of the structured part of the variogram. This part has been described previously as $\varepsilon'(\mathbf{x})$. The smaller the nugget effect, the bigger the proportion of $\varepsilon'(\mathbf{x})$ in s^2, i.e., the smaller the random error term ε''. All sources of random variability, like measurement or sampling errors, contribute to the nugget effect; variability at distances closer than the smallest sampling lag (microvariability) is included. Clearly it is the intention to keep this part of the total spatial variability as small as possible. This intention has implications towards both sampling strategy and analytical methodology.

Therefore, the variogram describes the average pattern of spatial variability of a regionalized variable in terms of its magnitude, scale, form, and contribution of random error. The intuitive feeling that the spatial autocorrelation between two measurements increases as they approach each other is represented by this function.

16.2.3 MODELS FOR VARIOGRAMS

The first step in fitting a theoretical model to experimental variogram values is to check its behavior near the origin. Is there a nugget effect? Do the values initially behave linearly, to level off gradually with increasing lags? Or is there a parabolic start (with a zero-gradient through the origin)? When the variogram behaves concave with an increase that is steeper than $\mathbf{h}^2$, there is a global trend (nonstationarity) in the data that must be removed first (e.g., by fitting a global trend surface). Also, the behavior at large lag values (in principle at infinity) should be investigated: if there is a stabilization, then the variogram is called "bounded." If the variogram continues to

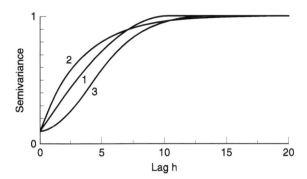

FIGURE 16.2 Behavior of three variogram models having the same nugget effect (0.1), range (10), and sill (1): 1 = spherical model, 2 = exponential model, and 3 = Gaussian model.

increase with increasing lag values ("unbounded" behavior), this indicates a nonstationarity of second order, but such a situation can still fall under the assumptions of the intrinsic hypothesis.

Not all mathematical functions are suitable to model the variogram (as a result of limitations imposed by the matrix algebra of the kriging system, where it is required that all variances obtain are strictly non-negative), but are called "permissible models." Some of these are presented in Figure 16.2.

Any linear combination of a covariance (and consequently also of a variogram) is again a covariance. Therefore, two or more variogram models can be combined, and it is not necessary that these models be the same. Such combined models are called "nested models"; one frequently used example is a double spherical model with nugget effect.

16.2.4 KRIGING INTERPOLATION

16.2.4.1 Univariate Estimation of Z

Suppose that a regionalized variable Z was sampled at n locations ($\alpha = 1, \ldots, n$) with observations $z(\mathbf{x}_\alpha)$, within a study area D. If the dimension of the support of the samples can be neglected compared to the dimension of D, we call these point observations. Suppose that one is interested to know the expected value of Z at an unsampled location $\mathbf{x}_0$, symbolized by $Z^*(\mathbf{x}_0)$. Kriging is an interpolation method developed to provide such a value.

In the most general form, the kriging algorithm can be written as a weighted linear combination of measurement points around $\mathbf{x}_0$:

$$Z^*(\mathbf{x}_0) - m(\mathbf{x}_0) = \sum_{\alpha=1}^{n(\mathbf{x}_0)} \lambda_\alpha \{ Z(\mathbf{x}_\alpha) - m(\mathbf{x}_\alpha) \} \qquad (16.7)$$

where λ_α are the weights given to the $n(\mathbf{x}_0)$ observations $z(\mathbf{x}_\alpha)$ and where $m(\mathbf{x}_0)$ and $m(\mathbf{x}_\alpha)$ represent a deterministic mean or trend (the expected values of Z) at locations $\mathbf{x}_0$ and $\mathbf{x}_\alpha$ respectively. Usually $n(\mathbf{x}_0)$ is a small subset of the total number of observations n, located within a circle centered around $\mathbf{x}_0$. Therefore, this procedure is termed a local interpolation. The aim of all interpolation methods is to find the weights λ_α.

Depending on the way the deterministic mean or trend is modeled, several kriging variants exist (Goovaerts, 1997):

1. Simple kriging requires $m(\mathbf{x}_\alpha)$ to be known and to be stationary over D
2. Ordinary kriging considers $m(\mathbf{x}_\alpha)$ to be stationary within the local neighborhood only, but its value is unknown.
3. Kriging with a trend model assumes that a smooth trend $m(\mathbf{x}_\alpha$ exists within the local neighborhood, so it is able to handle local nonstationary conditions. Therefore, it requires the variogram of the detrended data, or residual variogram.

The "workhorse" is ordinary kriging because in most practical conditions a global stationary mean is unknown. Therefore, m is eliminated from Eq. (16.7), but we have to accept one additional condition to do so: the sum of the weights must be 1. This results in the ordinary kriging system, which must be solved in order to find the λ_α weights. The ordinary kriging estimator, $Z_{OK}^*(\mathbf{x}_0)$, becomes:

$$Z_{OK}^*(\mathbf{x}_0) = \sum_{\alpha=1}^{n(\mathbf{x}_0)} \lambda_\alpha Z(\mathbf{x}_\alpha) \qquad (16.8$$

Ordinary kriging also allows one to obtain a measure of the ordinary kriging variance, $s_{OK}^2(\mathbf{x}_0)$:

$$s_{OK}^2(\mathbf{x}_0) = \sum_{\alpha=1}^{n(\mathbf{x}_0)} \{\lambda_\alpha \, \gamma(\mathbf{x}_\alpha - \mathbf{x}_0)\} + \psi \qquad (16.9$$

where ψ is a Lagrange multiplier, which is an additional unknown parameter. A schematic flowchart of ordinary kriging is presented in Figure 16.3 and of kriging with a trend model in Figure 16.4.

16.2.4.2 Multivariate Estimation of Z

16.2.4.2.1 Limited Number of Secondary Data

Sometimes it is possible to obtain observations of a second variable which are easier, quicker, or cheaper to determine than the primary variable we are interested in (e.g., soil texture determination complemented with soil moisture retained at $-1.5\,\mathrm{MPa}$). Consequently, this second variable can be

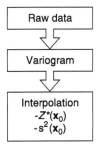

FIGURE 16.3 Schematic flowchart of ordinary kriging.

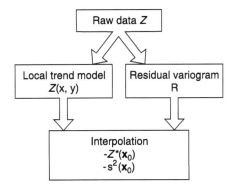

FIGURE 16.4 Schematic flowchart of kriging with a trend model.

observed at more locations than the primary one. If there is sufficient spatial correlation between both variables, then the increased spatial density of the second variable can be used to improve the mapping of the primary variable. This results in a multivariate version of kriging: cokriging.

The spatial correlation (or coregionalization) is investigated by the cross variogram:

$$\gamma_{12}(\mathbf{h}) = \frac{1}{2N(\mathbf{h})} \sum_{\alpha=1}^{N(\mathbf{h})} \{z_1(\mathbf{x}_\alpha) - z_1(\mathbf{x}_\alpha + \mathbf{h})\} \cdot \{z_2(\mathbf{x}_\alpha) - z_2(\mathbf{x}_\alpha + \mathbf{h})\} \qquad (16.10)$$

where $\gamma_{12}(\mathbf{h})$ is the cross variogram between the two variables and $N(\mathbf{h})$ is the number of pairs $[\{z_1(\mathbf{x}_\alpha), z_1(\mathbf{x}_\alpha + \mathbf{h})\}, \{z_2(\mathbf{x}_\alpha), z_2(\mathbf{x}_\alpha + \mathbf{h})\}]$ separated by lag $\mathbf{h}$. The cross variogram and the two simple variograms must be modeled jointly to form a family of simple and cross variograms to fulfill the following condition:

$$|\gamma_{12}(\mathbf{h})| \leq \sqrt{\gamma_1(\mathbf{h}) \cdot \gamma_2(\mathbf{h})} \qquad \forall \mathbf{h} \qquad (16.11)$$

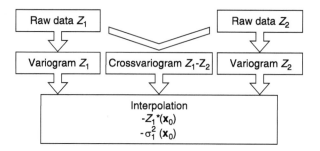

FIGURE 16.5 Schematic flowchart of ordinary cokriging.

The ordinary cokriging estimator is:

$$Z_1^*(\mathbf{x}_0) = \sum_{\alpha_1=1}^{n_1(\mathbf{x}_0)} \lambda_{1\alpha_1} Z_1(\mathbf{x}_{\alpha_1}) + \sum_{\alpha_2=1}^{n_2(\mathbf{x}_0)} \lambda_{2\alpha_2} Z_2(\mathbf{x}_{\alpha_2}) \qquad (16.12$$

with $n_1(\mathbf{x}_0)$ and $n_2(\mathbf{x}_0)$ the number of observation points of the variables Z and Z_2, respectively, used for the interpolation and with $\lambda_{1\alpha_1}$ and $\lambda_{2\alpha_2}$ the weights given to these observations. A schematic flowchart of ordinary cokriging is presented in Figure 16.5. Yates and Warrick (1987) and Van Meirvenne and Hofman (1989) provide examples of the use of cokriging.

To give more influence to the secondary variable, sometimes the two conditions imposed on the ordinary cokriging system (sum of $\lambda_{1\alpha_1} = 1$ and sum of $\lambda_{2\alpha_2} = 0$) are combined (Goovaerts, 1998; Deutsch and Journel, 1998):

$$\sum_{\alpha_1=1}^{n_1(\mathbf{x}_0)} \lambda_{1\alpha_1} + \sum_{\alpha_2=1}^{n_2(\mathbf{x}_0)} \lambda_{2\alpha_2} = 1 \qquad (16.13$$

yielding the so-called standardized ordinary cokriging because both variables first need to be standardized to a common mean first.

16.2.4.2.2 Exhaustive Secondary Data

A particular situation is created when a limited number of primary samples is complemented by a secondary variable known at every location within D Examples of such an exhaustive secondary variable are a digital elevation model, satellite images, yield maps, and soil sensors. Again, if there is a sufficiently strong (spatial) correlation between both variables, then the full coverage of the second variable can be used to improve the mapping of the primary variable. Goovaerts (2000) provides a comparison of different methods to use exhaustive secondary information.

One way of doing this is to modify the condition of a known stationary mean of simple kriging into a situation where a local mean is being estimated from the secondary variable and used as a varying local mean in simple kriging. Therefore, in Eq. (16.7), $m(\mathbf{x}_0)$ is derived as a function of the exhaustive secondary information, mostly by a linear regression. The secondary information could be both continuous or categorical. Figure 16.6 shows a schematic flowchart of the use of continuous secondary information.

The weights λ_α of simple kriging with a varying local mean are found by solving the following system:

$$\sum_{\beta=1}^{n(\mathbf{x}_0)} \lambda_\beta \, C_R(\mathbf{x}_\alpha - \mathbf{x}_\beta) = C_R(\mathbf{x}_\alpha - \mathbf{x}_0) \qquad \forall \, \alpha = 1, \dots, n(\mathbf{x}_0) \qquad (16.14)$$

with $C_R(\mathbf{h})$ the covariance of the residuals $R(\mathbf{x}) = Z(\mathbf{x}) - m(\mathbf{x})$.

Another approach is to modify kriging with a trend model to include the secondary information resulting in kriging with an external trend (or drift).

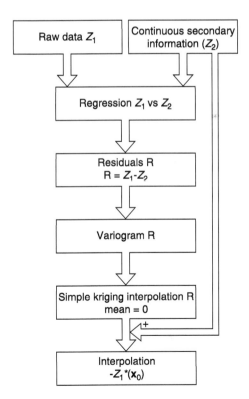

FIGURE 16.6 Schematic flowchart of simple kriging with continuous exhaustive secondary information.

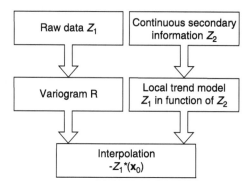

FIGURE 16.7 Schematic flowchart of kriging with an external trend.

In this method the trend is modeled as a linear function of the secondary variable $z_2(\mathbf{x}_0)$: $m(\mathbf{x}_0) = b_0 + b_1 z_2(\mathbf{x}_0)$. This function is fit within each interpolation neighborhood again, where in simple kriging with a varying local mean the relationship between the primary and secondary variables is set up only once for the entire study area. Figure 16.7 provides a schematic flowchart of kriging with an external trend.

16.2.4.3 Strongly Skewed Distributions

16.2.4.3.1 Robust Variograms

Like all second (and higher)-order statistics, the traditional variogram [Eq. (16.6)] is sensitive to outliers, even local outliers, causing the variogram to be overestimated. For this reason, a thorough exploratory data analysis, including editing and transformation of data, must proceed the variogram calculation. But even then some flexibility is recommended while fitting a variogram model.

One useful alternative is the nonergodic covariance $C_{ne}(\mathbf{h})$, which, unlike the autocovariance $C(\mathbf{h})$, does not suppose that the global mean is stationary, but takes into account a local varying mean:

$$C_{ne}(\mathbf{h}) = \frac{1}{N(\mathbf{h})} \sum_{\alpha=1}^{N(\mathbf{h})} z(\mathbf{x}_\alpha) \cdot z(\mathbf{x}_\alpha + \mathbf{h}) - m_{-\mathbf{h}} \cdot m_{+\mathbf{h}} \qquad (16.15$$

where

$$m_{-\mathbf{h}} = \frac{1}{N(\mathbf{h})} \sum_{\alpha=1}^{N(\mathbf{h})} z(\mathbf{x}_\alpha)$$

and

$$m_{+\mathbf{h}} = \frac{1}{N(\mathbf{h})} \sum_{\alpha=1}^{N(\mathbf{h})} z(\mathbf{x}_\alpha + \mathbf{h})$$

To obtain a similar representation as the variogram, the nonergodic covariance is mostly visualized as $C(0) - C_{ne}(\mathbf{h})$. The nonergodic covariance is very resistant to the combined influence of a proportional effect and a preferential sampling of larger values.

A number of alternative robust estimators of the variogram have been proposed, and Lark (2000) provides a comparison.

16.2.4.3.2 Lognormal Kriging

A common way to reduce the skew of an asymmetric distribution is to transform the data logarithmically. Usually this results in a smaller nugget effect and a more stable variogram. The next step is to use (ordinary) kriging to interpolate the transformed data, called lognormal kriging.

Suppose that $Y(\mathbf{x}) = \ln(Z(\mathbf{x}))$, then the backtransformation of the kriged values to the original units is obtained through the following formula (Webster and Oliver, 2001):

$$Z^*(\mathbf{x}_0) = \exp\left(Y^*(\mathbf{x}_0) + \sigma_{\mathrm{OK}_Y}^2(\mathbf{x}_0)/2 - \psi \right) \qquad (16.16)$$

with $\sigma_{\mathrm{OK}_Y}^2(\mathbf{x}_0)$ the ordinary kriging variance of the estimated Y. Despite the reduction of the influence of extreme values, lognormal kriging has some limitations:

It cannot be used with negative data.
For ordinary kriging estimates, there is no backtransformation of the kriging variance.
Block estimates cannot be obtained because logarithms do not average linearly.
The backtransformation is strongly influenced by the sill of the variogram, so variogram fitting has a stronger impact on the estimation than with untransformed data.

Nevertheless, lognormal kriging is frequently used with success, and Saito and Goovaerts (2000) provide an evaluation. Figure 16.8 gives a schematic flowchart of lognormal kriging.

16.2.4.4 Local Spatial Uncertainty

The previous interpolation methods focus on the estimation of the unknown value of the variable under study, but in some applications one might be

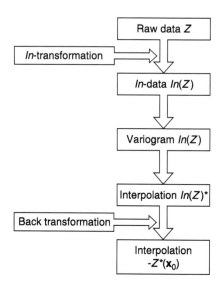

FIGURE 16.8 Schematic flowchart of lognormal kriging.

more interested in the knowledge of the local uncertainty and its influence on the evaluation of the interpolated value. One example is the evaluation of the probability to exceed a critical threshold. This uncertainty should be taken into the decision making rather than rely only on the expected value. Two approaches will be discussed: indicator kriging and Bayesian maximum entropy. In a recent study, Lark and Ferguson (2004) compared indicator kriging with disjunctive kriging and found no important differences in the performance of both methods, but they favored the latter on theoretical grounds. However, disjunctive kriging requires stronger assumptions like the condition that the data must follow a second-order stationary Gaussian diffusion process. Moreover, its mathematical background is quite complex and user-friendly software is not widely available. Therefore, most users prefer indicator geostatistics.

16.2.4.4.1 Indicator Kriging

The uncertainty about the z-value at $\mathbf{x}_0$ can be modeled through a random variable $Z(\mathbf{x}_0)$ that is characterized by its distribution function:

$$F(\mathbf{x}_0; z|(n)) = \mathrm{Prob}\{Z(\mathbf{x}_0) \leq z|(n)\} \qquad (16.17$$

The function $F(\mathbf{x}_0; z|(n))$ is referred to as a conditional cumulative distribution function (ccdf), where the notation $|(n)$ expresses the conditioning to the n data $z(\mathbf{x}_\alpha)$. The ccdf fully models the uncertainty at $\mathbf{x}_0$ since it gives the probability that the unknown is no greater than any given threshold z

It consists of estimating the value of the ccdf for a series of K threshold values z_k, discretizing the range of variation of z:

$$F(\mathbf{x}_0; z_k|(n)) = \text{Prob}\{Z(\mathbf{x}_0) \le z_k|(n)\} \qquad k = 1, 2, \ldots, K \qquad (16.18)$$

The resolution of the discrete ccdf is then increased by interpolation within each class $[z_k, z_{k+1}]$ and extrapolation beyond the two extreme threshold values z_1 and z_K.

A nonparametric estimation of ccdf values is based on the interpretation that the conditional expectation of an indicator random variable $I(\mathbf{x}_0; z_k)$ can be considered as the conditional probability in Eq. (16.18). Ccdf values can thus be estimated by interpolation of indicator transforms of data (Journel, 1983).

The indicator approach requires a preliminary coding of each observation $z(\mathbf{x}_\alpha)$ into a series of K values indicating whether the threshold z_k is exceeded or not. If the measurement errors are assumed negligible compared to the spatial variability, observations are coded into hard (0 or 1) indicator data:

$$i(x_\alpha; z_k) = \begin{cases} 1 & \text{if } z(\mathbf{x}_\alpha) \le z_k \\ 0 & \text{otherwise} \end{cases} \qquad k = 1, 2, \ldots, K \qquad (16.19)$$

At any unsampled location $\mathbf{x}_0$, each of the K ccdf values can be estimated as a linear combination of indicator transforms of neighboring observations. The ordinary indicator kriging estimator for threshold z_k is:

$$\left[F\{\mathbf{x}_0; z_k|(n)\} \right]^* = \sum_{\alpha=1}^{n} \lambda_\alpha(z_k) \, i(\mathbf{x}_\alpha; z_k) \qquad (16.20)$$

The only information required by the kriging system are K indicator variogram values for different lags, and these are derived from the variogram model $\gamma_I(\mathbf{h}; z_k)$ fitted to experimental values computed as:

$$\gamma_I(\mathbf{h}; z_k) = \frac{1}{2N(\mathbf{h})} \sum_{\alpha=1}^{N(\mathbf{h})} \{i(\mathbf{x}_\alpha; z_k) - i(\mathbf{x}_\alpha + \mathbf{h}; z_k)\}^2 \qquad (16.21)$$

At each location $\mathbf{x}_0$, the series of K ccdf values must be valued within $[0, 1]$ and be a nondecreasing function of the threshold value z_k, i.e., $[F(\mathbf{x}_0; z_k|(n))]^* \le [F(\mathbf{x}_0; z_{k'}|(n))]^* \; \forall \; z_{k'} > z_k$. These conditions are not necessarily satisfied because kriging weights can be negative and therefore the kriging estimate is a non-convex linear combination of the conditioning data. See Deutsch and Journel (1998) for more details on how to solve these problems.

Knowledge of the ccdf $F(\mathbf{x}_0; z|(n))$ at $\mathbf{x}_0$ allows one to (Goovaerts et al., 1997):

1. Assess the probability of exceeding a critical threshold z_c at $\mathbf{x}_0$:

$$\text{Prob}\{Z(\mathbf{x}_0) > z_c|(n)\} = 1 - F(\mathbf{x}_0; z_c|(n)) \qquad (16.22$$

2. Estimate the unknown value $z(\mathbf{x}_0)$. For example, using a least-squares error criterion amounts at estimating that value by the mean of the ccdf, called E-type estimate:

$$z_E^*(\mathbf{x}_0) = \int_{-\infty}^{+\infty} z \, dF(\mathbf{x}_0; z|(n)) \qquad (16.23$$

Similarly, the conditional variance of the ccdf can be calculated. Consequently, IK has been used with success in soil and environmental studies (e.g., Van Meirvenne and Goovaerts, 2001; Cattle et al., 2002).

A schematic overview of indicator kriging is provided in Figure 16.9.

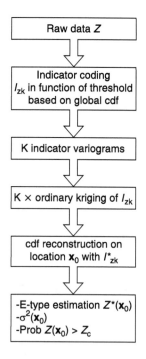

FIGURE 16.9 Schematic flowchart of indicator kriging.

16.2.4.4.2.   Bayesian Maximum Entropy

Bayesian maximum entropy (BME) (Christakos, 2000) is a recent approach developed for the spatio-temporal mapping of natural processes using uncertain information. It offers the flexibility to incorporate various sources of physical knowledge. In the BME framework the total knowledge (K) is considered to be formed from two main bases: the general knowledge (G) and the specificatory knowledge (S). The general knowledge encompasses physical laws, statistical moments of any order (including the mean and variogram or covariance functions), multipoint statistics, etc. It is said to be general because it can characterize more than one random field. The specificatory knowledge includes the data specific to a given experiment or situation. The BME has a double goal: informativeness (prior information maximization given the general knowledge) and cogency (posterior probability maximization given specificatory knowledge). The BME analysis is done in three main stages (Figure 16.10):

1. Structural or prior stage: The goal is the maximization of the information content considering only the general knowledge before any use of the data. A random field is completely defined by its multivariate probability distribution function (pdf), which forms the prior pdf. The latter should be derived by means of an estimation process that takes into consideration physical constraints under the form of prior information or knowledge. This information is measured, in the context of BME, using Shannon's entropy function, thus the E in BME. The entropy function needs to be maximized, which justifies the M in the BME acronym. At this stage we get the prior or G-based multivariate pdf.
2. Meta-prior stage, during which the specificatory knowledge is collected and organized: The available data can be divided into two main types: hard data and soft data. Hard data are considered exact and accurate measurements of the natural process, while soft data are indirect and inaccurate measures of the variable of interest. The soft data may be in the form of intervals, probability distribution functions, etc.
3. Integration or posterior stage: In this step the two knowledge bases (G and S) are integrated. The goal is the maximization of the posterior pdf given the total knowledge K. The G-based pdf is updated by considering the available data. This updating is performed using a Bayesian conditionalization based on Bayes theorem, thus the B in BME. This stage yields the K-based pdf. By substituting the expression of G-based pdf in the K-based pdf, one gets the BME or posterior pdf.

The posterior pdf, which is not limited to the Gaussian type, describes fully the random field at the estimation point. It provides a complete picture of

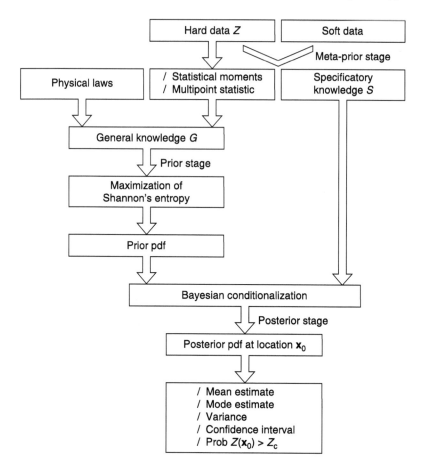

FIGURE 16.10 General flowchart for Bayesian maximum entropy.

the mapping situation as well as different estimators and their associated estimation uncertainty. Among the estimators, the mode represents the most probable situation. The mean estimate, which is in general a nonlinear function of the data, is suitable for mapping situations where one is interested in minimizing the mean-square estimation error. A measure of the uncertainty associated with the estimated values is provided by the variance of the estimation error. This is data-dependent, whereas in kriging it is data-free. In addition from the posterior pdf, one can compute directly the confidence intervals, which provide a more realistic assessment of the estimation error than the error variance.

When the general knowledge is limited to the mean and covariance functions and the specificatory knowledge is restricted to the only hard data, the BME posterior pdf is Gaussian (the mean and the mode are equal) and the BME mean estimate is equivalent to the kriging estimate. Thus, BME is a more general interpolation approach, and kriging is a special case in limiting situations.

In addition to its ability to incorporate physical laws and data with different levels of accuracy, BME does not require any assumptions about the underlying pdfs, its estimate is not limited to a linear combination of the data, multiple-point statistics and higher-order statistical moments can be considered, multipoint mapping is possible, the estimation can be extended to consider more than one variable, it offers a complete uncertainty assessment (confidence intervals and sets), and a change of support is easily handled.

Applications of BME to soil science include D'Or and Bogaert (2003) and Douaik et al. (2004).

16.2.4.5 Conditional Simulation

Conditional stochastic simulation is used to generate a number (L) of maps, each consistent with the sample histogram and/or the variogram model. This is different from kriging interpolation, which inevitably causes a smoothing effect, reducing the observed variability. Consequently a map obtained by kriging shows less variability than observed in the sample, but at every location kriging provides the "best linear unbiased" estimation, given the data.

Such a set of L realizations ("maps") can be synthesized according to the objectives of the study. Several conditional stochastic simulation algorithms are available (Deutsch and Journel, 1998), and we present here sequential Gaussian simulation that proceeds as follows (Figure 16.11) (Fagroud and Van Meirvenne, 2002):

1. Data are transformed according to $y(\mathbf{x}_i) = \phi(z(\mathbf{x}_i))$, with $z(\mathbf{x}_i)$ the original data, $\phi(\cdot)$ a transformation function and $y(\mathbf{x}_i)$ the normal scores having a standard normal (or Gaussian) histogram (Goovaerts,1997).
2. The sample variogram of the normal scores is computed and modeled.
3. L simulations are performed in normal score space as follows:

   - Define a random path visiting each of the m unsampled locations to be simulated only once.
   - At each unsampled location $\mathbf{x}_0$, estimate the parameters (mean and variance) of the Gaussian ccdf by simple kriging using the normal score variogram model and the mean value of the normal scores. The conditioning information consists of n neighboring data of both original normal score data $y(\mathbf{x}_i)$ and values $y^{(l)}(\mathbf{x}_0)$ simulated at previously estimated locations, l being the realization number ($l = 1, \dots, L$).
   - Randomly draw a simulated value $y^{(l)}(\mathbf{x}_0)$ from the ccdf, and add it to the data set.

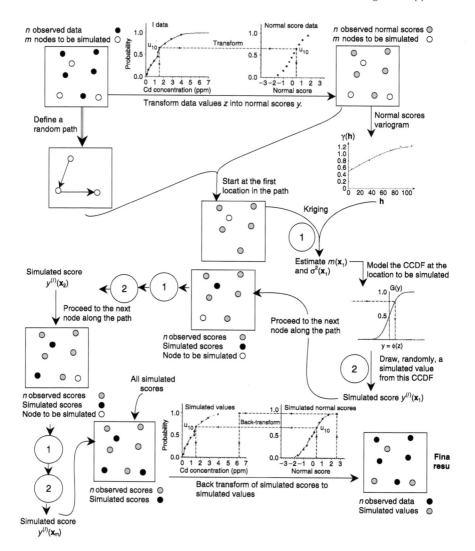

FIGURE 16.11 Flowchart of sequential Gaussian simulation. (From Fagroud and Van Meirvenne, 2002.)

- Proceed to the next location along the random path, and repeat the two previous steps.
- Loop until all m locations are simulated.
- Proceed with the next simulation by repeating the previous steps until all L realizations are available.

4. The results are finally backtransformed to the original variable space by applying the inverse of the normal score transform $\phi(\cdot)$ to the simulated y-values.

Each of these L simulations is a realization of the unknown spatial distribution of the variable Z. Differences between them provide a measure of the spatial uncertainty about Z, which can be processed into several outcomes, similar to indicator kriging or BME.

16.3 GEOSTATISTICAL SAMPLING

Designing a spatial sampling campaign must be done with care and with a specific goal in mind. Issues to consider are the sampling support, number of samples, sampling configuration, and intended methodology of data handling. An additional point to consider is whether a secondary variable can be used to support the investigation of the primary. The following is not intended to cover all aspects of these issues. It is rather an attempt to provide some general, yet practical, recommendations. More details can be found in Webster and Oliver (1990, 2001).

16.3.1 SAMPLING SUPPORT

The sampling support refers to the physical nature of the sample, as well as the procedure how to take it, e.g., in soil science the sample support is usually a soil cylinder taken by an auger down to a prescribed depth. When a variable which displays large microvariability is sampled (e.g., available nitrogen in agricultural land), it can be wise to pool a number of augerings taken within an area of a predescribed dimension (say $1\,m^2$) into a composite sample. Care should be given to follow a uniform sampling procedure so that the same support applies to all samples, avoiding a transformation to a common support. Usually we refer to samples as "point observations" despite their physical size. But "point" should be interpreted as "with a dimension that is negligible compared to the size of the study area". This interpretation also allows pooled samples to be considered as "points" as long as the size of the area within which the pooled samples are taken is small compared to the entire study area. One should also realize that the support of the samples represents the smallest unit to which one can interpolate. Maps produced by point kriging, whatever their display resolution, must be interpreted with knowledge of the sampling support to which they refer. Hence the importance to document clearly the sampling support when reporting about a spatial inventory. Downscaling is very difficult and requires information about the support area variability. Upscaling to a larger area is easier and can be done by block kriging.

16.3.2 NUMBER OF SAMPLES

The required number of samples for an experiment or survey has long been the subject of intensive research. The classical formulas to calculate the required number of samples to estimate population statistics are based on decisions about the required level of probability and which deviation of the mean is acceptable. However, these calculations are based on assumptions of normality and independence of observations, which are rarely met in spatial surveys. In situations where some degree of spatial autocorrelation is expected, it is more common to fix a number of samples, including more pragmatic considerations. For spatial surveys based on an omnidirectional variogram to interpolate in 2-D, Webster and Oliver (2001) recommend a minimum of 100 samples, but geostatistical studies based on fewer observations have been reported. Frequently the number of samples is limited by the available resources and laboratory capacity.

16.3.3 SAMPLING CONFIGURATION AND SAMPLING GOAL

To explore the nature and magnitude of the spatial variability of a target variable, the first option is to sample along a few transects oriented in different directions. But transect sampling has its limitations, since it is one-dimensional. Two-dimensional sampling configurations are more efficient because they produce many more lag combinations than 1-D configurations for the same sampling effort. Moreover, 2-D sampling configurations better represent a study area compared to a few transects, whose position is then more crucial. Three-dimensional studies (space-time or in three spatial dimensions) tend to become very data intensive with large data collection requirements, since all dimensions should have sufficient observations. Therefore, they are conducted less frequently, but with easier and quicker methods of data collection becoming more available, one can expect that such surveys will be conducted more often.

When the goal of the inventory is to produce a map, the sampling configuration must meet this objective. Therefore, a simple random (probabilistic) selection of coordinates is not recommended because such a simple random sample does not involve any spatial considerations. As a consequence, this procedure can lead to configurations containing both a spatial clustering of sampling points and consequently large unsampled areas. Simple random sampling configurations may lead to large fluctuations in the quality of the final map (expressed by the kriging variance). Therefore, simple random sampling configurations are usually limited to situations where the spatial mean (and any other statistics) of environmental variables is needed. Nevertheless, simple random selections avoid the influence of human decisions on the selection of sampling locations and have the advantage that each sample is taken with the same and known probability of selection.

Fixed grid sampling configurations avoid the clustering often found with simple random selections, but they have the disadvantage that they do not allow one to calculate the experimental variogram at lag distances smaller than the grid interval. Yet this is a critical part of the variogram for interpolation, since in a fixed grid sampling the largest distance between a unvisited location and a sampling point is equal to the grid unit distance multiplied by 0.7071. So interpolation algorithms will require information for lag intervals smaller than the grid unit distance, which are experimentally unavailable. Fixed grid sampling configurations have the advantage of being spatially unclustered, i.e., their sampling density is even. So every sample represents the same area, allowing the determination of spatially representative characteristics of the area.

The combination of both previous approaches leads to a compromise that uses the best of both: probability sampling and representative coverage. Therefore, in practice often a combination of a grid sampling and a random sampling to include clusters for short distance characterization of the variogram is used. It might be useful to check a proposed sampling configuration prior to the actual sampling in terms of the number of pairs obtained for different lag intervals. This can be done by using a dummy variable and focusing only on the number of pairs when a variogram is calculated. If insufficient pairs are found for particular lag intervals, the design of the configuration can be altered and checked again until it is found satisfactory.

When the goal of the sampling is to provide estimates of the distribution parameters of the variable, no spatial sampling scheme must be created. Then a design-based sampling, mostly a simple random sampling, is suitable. Also the number of samples can be much lower — 10–20 is sufficient to obtain reliable estimates.

16.3.4 METHOD OF DATA ANALYSIS

Some methods of data analysis, like wavelets and time series analysis, require data obtained at regular intervals (in space or time). Other methods, like geostatistics, have more gain when a range of sampling intervals is available. Therefore, the intended methods for data processing and data analysis should be taken into consideration when designing a spatial or temporal sampling scheme.

16.3.5 SECONDARY INFORMATION

In multivariate conditions it is not evident that all variables are determined at all sampling locations. Mostly a secondary variable can be selected to be easier, cheaper, or faster to sample, allowing it to be observed at many more locations than the primary variable. A special situation arises when the secondary variable is available at all locations (both the locations where the primary sample has been taken as all locations where it is being interpolated). The latter is referred to as "exhaustive secondary information" (see above). The cross-variogram required in cokriging requires locations where both the primary and the secondary variable have been observed. One way to avoid this requirement is to use the cross- and simple covariances.

16.4 CASE STUDY: EXPLORING THE SOIL MOISTURE–LANDSCAPE RELATIONSHIP

16.4.1 INTRODUCTION

Topsoil moisture is an environmentally important variable, as it influences the infiltration rate, the production of runoff, the partitioning of incoming energy into sensible and latent heat, soil erosion, etc. Because this variable often shows a large temporal and spatial variability, considerable effort is being given to techniques and methods to sample and map soil moisture in space and time (Chapter 2). The aim of this study was to characterize the spatial variability of topsoil (0–11 cm) moisture content at two different moments in time as a result of different climatological conditions and to derive conclusions with respect to soil sampling.

16.4.2 MATERIALS AND METHODS

An agricultural field of 4.2 ha, with a sandy loam texture, located within the Zwalm catchment of East Flanders, Belgium, was taken as a study area. It was located on the side of a rolling hill, with elevations ranging between 45 and 60 m above sea level (Figure 16.12).

Two sampling campaigns were conducted, one at February 25, 2003, and the second at March 7, 2003. During this period the field consisted of bare soil. Therefore, the distribution of soil moisture was only affected by topographical effects and not by evapotranspiration. At both dates the same sampling scheme was used, whereby 130 locations were analyzed for

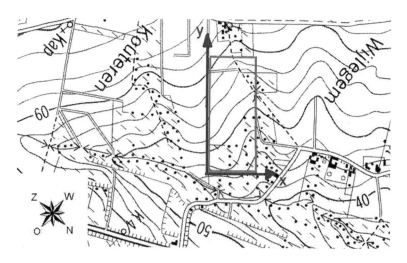

FIGURE 16.12 Study area draped on an extract of the topographic map with indications of the contour lines (contour interval = 2.5 m).

their volumetric soil moisture content using a TRIME Time Domain Reflectometer with 11 cm probes. This technique was chosen as it is a fast technique that does not need further processing in the lab such as drying of the soil samples.

Because the purpose of the sampling is to construct a reliable variogram and a representative map of the topsoil moisture distribution within the field, it is necessary to design an optimal sampling scheme. On the one hand, the sampling scheme should account for a sufficient amount of sampling points at various lag distances to be able to construct a reliable variogram. On the other hand, in order to obtain a representative map of the actual soil moisture distribution, sampling points should be spread homogeneously over the field. The latter can simply be obtained by sampling on a grid. In order to meet both conditions, a combination of systematic sampling and stratified random sampling was adopted. Systematic sampling was applied by first positioning a fixed number of grid points within the field. Then a stratified sampling was performed by subdividing the field into four equal parts within which the same number of random sampling points were chosen. Such a technique avoids obtaining large unsampled areas.

In order to reduce the variability at a very small scale, a sampling support was chosen of three observations within a circle with a diameter of 1 m. The center of the circle was assigned the average soil moisture content of the three measuments.

Because only a limited number of measurements (approx. 400) can be taken with the TRIME instrument using 1 battery, and the averaging of three measurements per sampling point, a total number of 130 sampling points was taken. This number was subdivided into 70 points on a grid of 25 m by 25 m and 60 points that were randomly positioned. Several schemes were calculated from which the most optimal scheme, containing a sufficient amount of couples of sampling points for each considered lag, was retained. This step was necessary to ensure that sufficient short-lagged sampling points were available in order to capture the short distance semivariances within the variogram. The sampling scheme finally retained is given in Figure 16.13.

During the week before the first sampling date there was almost no precipitation (1.4 mm), while during the 7 days before the second sampling it rained quite intensively (28 mm), but the 24 hours before the sampling only 0.2 mm of rain was measured.

16.4.3 Results

Figure 16.14 shows the histograms and some descriptive statistics of both data sets. Both distributions were found to be normal. On February 25 the mean moisture content was 34.7%, with a range between 27.5 and 40.4%. The coefficient of variation was 7.9%. On March 7 the soil was wetter, with

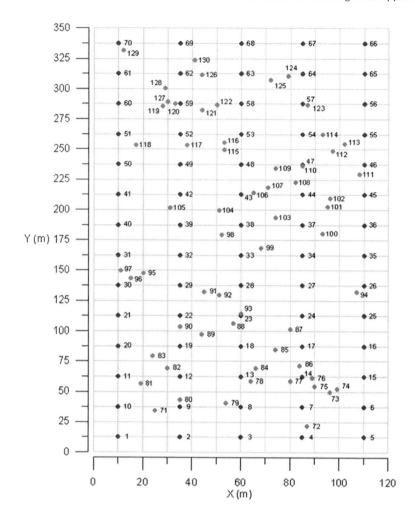

FIGURE 16.13 Sampling configuration with indication of sample ID.

an average moisture content of 38.5%, and more homogeneous, varying between 33.6 and 43.9%, and the coefficient of variation was 5.7%.

The variograms of both data sets are given in Figure 16.15. The variogram of the February 25 data shows a nested structure modeled by a double spherical model. The first structure had a range of 18 m, the second of 125 m. The variogram of the March 7 data behaved differently. Only one structure with a very small range (15.4 m) was found. Both variograms displayed a significant relative nugget effect (ratio nugget effect vs. sill): on February 25 it was 39% and on March 7, 24%. So despite the intensive sampling configuration, a large proportion of the total variance still remained at smaller scale or was due to observation errors.

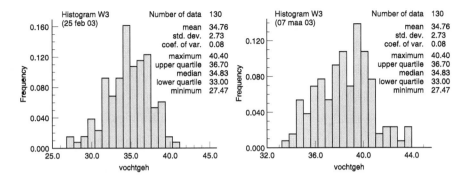

FIGURE 16.14 Histograms of volumetric moisture content of the topsoil of the study area measured on February 25 (left) and March 7, 2003 (right).

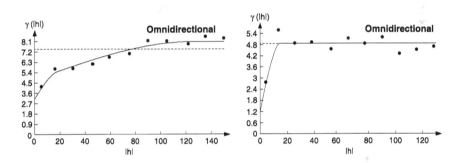

FIGURE 16.15 Variograms of the moisture content on February 25 (left) and March 7, 2003 (right).

The difference between these variograms indicate the effect of precipitation prior to the sampling dates: almost no rain before February 25 and quite intensive rain prior to March 7. A large-scale effect could be found for the February 25 data, which was completely absent on March 7, while on both dates a small scale pattern (15–18 m) was found. This resulted in the expectation that during drier periods topographic effects influence topsoil moisture, while during wet periods this influence disappears, because in drier soil moisture is redistributed in relationship with landscape position. Yet the correlation coefficient between soil moisture and elevation was very low on both dates (0.07 on February 25 and 0.13 on March 7).

On both dates soil moisture was interpolated using ordinary kriging. Both maps are given in Figure 16.16. Due to the small range on March 7, the map displayed a large amount of small scale variability, whereas on February 25 a much smoother pattern was found.

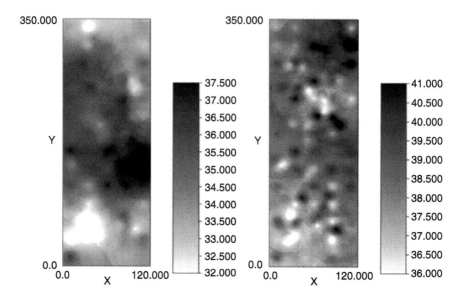

FIGURE 16.16 Topsoil volumetric moisture content interpolated with ordinary kriging and the variograms of Figure 13 on February 25 (left) and March 7, 2003 (right).

To investigate the influence of the longer-range component on February 25, we fitted a quadratic trend surface to the data (higher order terms were found to be nonsignificant). The fitted trend was (with $z =$ volumetric moisture content on February 25 and x and y the coordinates) (Figure 16.17):

$$\theta = 30.19 + 1.53 \cdot 10^{-2}x + 5.65 \cdot 10^{-2}y + 8.28 \cdot 10^{-5}x^2$$
$$- 1.31 \cdot 10^{-4}y^2 - 1.32 \cdot 10^{-4}xy$$

This trend did not correspond to the elevation as such, but indicates higher moisture conditions in the central part of the study area, which is an accumulation part of (sub)surface runoff due to the concave nature of the topography (Figure 16.12) within the study area. Removing this trend from the February 25 data, we obtained moisture residuals with an average of zero. The variogram of these residuals is shown in Figure 16.18. It was fit by a single spherical model with a range of 20.5 m and a relative nugget effect of 52%. This model is similar to the model obtained from the data of March 7 (Figure 16.15, right).

These moisture residuals were interpolated using simple kriging (with a known mean of zero), and Figure 16.19 (left) shows the result. They display a very erratic pattern, which reflects local behavior similar to the one identified on March 7 (Figure 16.16, right). When the trend was added to the map of the residuals, a similar map was obtained as on the left side of Figure 16.16 (Figure 16.19, right), indicating that in dry periods the soil

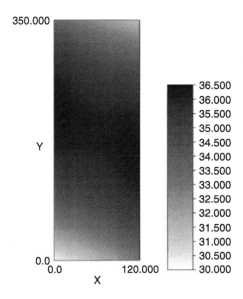

FIGURE 16.17 Quadratic trend surface fit to the February 25 data.

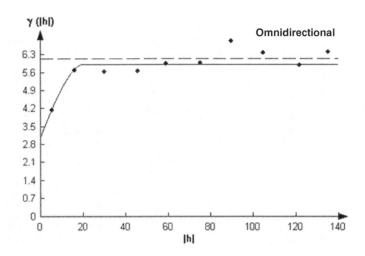

FIGURE 16.18 Variogram of the moisture residuals of March 7.

moisture content can be split into a regional trend plus a locally fluctuating component, while in wet periods the regional trend disappears and only a local pattern remains. Clearly this behavior calls for a different sampling strategy depending on the climatological conditions prevailing before the sampling campaign.

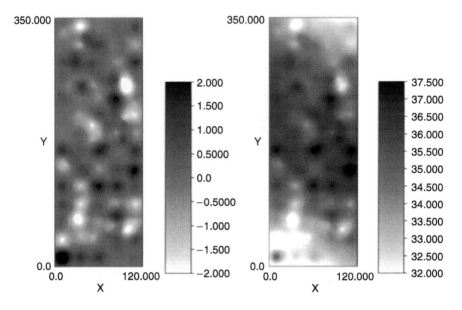

FIGURE 16.19 Moisture residuals interpolated with simple kriging (left) and residuals + trend (Figure 16.17) (right).

16.5 CONCLUSIONS

Geostatistics has been used in soil science for almost 25 years, and it has shown its value. However, despite its promise there is no solution for every situation, so there is still a need to understand the underlying hypotheses and conditions before embarking on a geostatistical analysis. This chapter has tried presented overview of the most promising methods for soil-water-solute studies, including a discussion of issues related to spatial sampling.

A practical case study has been added to illustrate the potential of a geostatistical study and to discuss alternative ways to understand the processes being studied related to soil-landscape interactions.

REFERENCES

Burgess, T.M. and Webster, R. Optimal interpolation and isarithmic mapping of soil properties. I. The semi-variogram and punctual kriging. *J. Soil Sci.* 31, 315, 1980.

Burrough, P.A. and McDonnell, R. *Principles of Geographical Information Systems* Oxford University Press, New York, 1998.

Campbell, J.B. Spatial variation of sand content and pH within single contiguous delineations of two soil mapping units. *Soil Sci. Soc. Am. J.* 42, 460, 1978.

Cattle, J.A., McBratney, A.B. and Minasny, B. Kriging method evaluation for assessing the spatial distribution of urban soil lead contamination. *J. Environ. Qual.* 31, 1576, 2002.

Chilès, J.P. and Delfinder, P. *Geostatistics.* John Wiley & Sons, New York, 1999.

Christakos, G. *Modern Spatiotemporal Geostatistics.* Oxford University Press, New York, 2000.

Deutsch, C.V. and Journel, A.G. *GSLIB Geostatistical Software Library and User's Guide.* Oxford University Press, New York, 1998.

D'Or, D. and Bogaert, P. Continuous-valued map reconstruction with the Bayesian maximum entropy. *Geoderma* 112, 169, 2003.

Douaik, A, Van Meirvenne, M, Toth, T. and Serre, M. Space-time mapping of soil salinity using probabilistic Bayesian maximum entropy. *J. Stoch. Environ. Res. Risk Ass.* 18, 219, 2004.

Fagroud, M. and Van Meirvenne, M. Accounting for soil spatial autocorrelation in the design of experimental trials on water-use efficiency. *Soil Sci. Soc. Am. J* 66, 1134, 2002.

Goovaerts, P. *Geostatistics for Natural Resources Evaluation.* Oxford University Press, New York, 1997.

Goovaerts, P. Ordinary cokriging revisited. *Math. Geol.* 30, 21, 1998.

Goovaerts, P. Geostatistical approaches for incorporating elevation into the spatial interpolation of rainfall. *J. Hydrol.* 228, 113, 2000.

Goovaerts, P., Webster, R. and Dubois, J.-P. Assessing the risk of soil contamination in the Swiss Jura using indicator geostatistics. *Envion. Ecol. Stats.* 4, 31, 1997.

Journel, A.G. Non-parametric estimation of spatial distributions. *Math. Geol.* 15, 445, 1983.

Journel, A.G. and Huijbregts, Ch.J. *Mining Geostatistics.* Academic Press Inc., New York, 1978.

Lark, R.M. A comparison of some robust estimators of the variogram for use in soil survey. *Eur. J. Soil Sci.* 51, 137, 2000.

Lark, M. and Ferguson, R.B. Mapping risk of soil nutrient deficiency or excess by disjunctive and indicator kriging. *Geoderma* 118, 39, 2004.

Saito, H. and Goovaerts, P. Geostatistical interpolation of positively skewed and censored data in a dioxin-contaminated site. *Environ. Sci. Technol.* 34, 4228, 2000.

Van Meirvenne, M. and Goovaerts, P. Evaluating the probability of exceeding a site specific soil cadmium contamination threshold. *Geoderma* 102, 75, 2001

Van Meirvenne, M. and Hofman, G. Spatial variability of soil texture in the polder area: II. Cokriging. *Pedologie* 39, 209, 1989.

Webster, R. and Oliver, M. *Statistical Methods in Soil and Land Resource Survey* Oxford University Press, New York, 1990.

Webster, R. and Oliver, M. *Geostatistics for Environmental Scientists.* John Wiley & Sons, Chichester, 2001.

Yates, S.R. and Warrick A.W. Estimating soil water content using cokriging. *Soil Sci. Soc. Am. J.* 51, 23, 1987.

17 Soil Variability Assessment with Fractal Techniques

A. N. Kravchenko
Department of Crop and Soil Sciences,
Michigan State University, East Lansing, Michigan, U.S.A.

Y. A. Pachepsky
USDA-ARS Environmental Microbial Safety Laboratory,
Beltsville, Maryland, U.S.A.

CONTENTS

17.1 INTRODUCTION

Variability is an inherent property of soils. Importance of spatial and temporal variations in soil properties has been long recognized both in collecting knowledge about soils and in applying this knowledge in soil use. The significance of soil variability has led scientists and practitioners to the realization of the need to quantify it. Statistics of soil properties have become essential components of soil descriptions. The accumulation of such statistics has eventually led to understanding that they change with scale of soil sampling or description. Much of soil data is obtained from small soil samples and cores, monoliths, or small field plots, yet the goal is to reconstruct soil properties across fields, watersheds, and landforms, or to predict physical properties of pore surfaces and structure of pore space. The representation of processes and properties at a scale different from the one at which observations and property measurements are made is a pervasive problem in soil science.

1-5667-0657-2/05/$0.00 + $1.50

The notion of scale is omnipresent in soil studies. Terms such as "research scale," "natural scale," "field scale," and "laboratory scale" are ubiquitous in the literature. Scale is a complex concept having multiple connotations. A notion of *support* is important to characterize and relate different scales. Support is the length, area, or volume for which a single value with zero variation of soil property is defined. Size of an individual soil sample and size of a discrete spatial element in a soil model are typical examples of supports. The term "resolution" is often used for supports defined in terms of lengths or areas, and the term "pixel size" is also used to define area support. An area or a volume which is sampled with given support determines the *extent* of measurements. Yet another notion, *spacing*, i.e., distance between sampling locations, is of importance in characterizing scale in research or applications. Any research into soil properties is conducted with specific support, extent, and spacing, and the triad of those three values has been suggested as a characteristic of scale (Bierkens et al., 2000). If the measured soil properties are to be used with different support, extent, or spacing, scaling becomes necessary. Scaling is used as a noun to denote a relationship between soil data at different scales or the action of relating such data on different scales.

The existence of scaling implies (1) that a property can be measured (using the same method) at several scales, (2) that the property changes as the scale changes, and (3) that the law of this change, or scaling law, remains the same across scales. Scaling law always has a range of scales within which it is applicable.

Recently, fractal geometry has become an important source of scaling laws in soil science. Fractal geometry focuses on geometric objects in which total length, area, or volume depends on the support. Such objects exhibit similar geometric shapes when observations are made at different supports. They were termed fractals by B. B. Mandelbrot (1982), who suggested that fractals rather than regular geometric shapes like segments, arcs, circles, spheres, etc., are more appropriate to approximate irregular natural shapes that have hierarchies of ever-finer detail. This approach marks the beginning of applications of fractal geometry that became very popular during the last 20 years because of its promise to relate features of natural objects observed at different scales.

Fractal models of variability can be applied to both geometric and nongeometric parameters. Consider an irregular curve $y(x)$, with its length dependent on the support. It is a purely geometric object if y is the vertical coordinate and x is the horizontal coordinate. However, this curve may represent temporal variations in soil temperature, with y denoting the temperature value and x denoting time, or variations of organic matter content along a soil transect with y denoting the organic carbon content and x being the distance along the transect. This creates an opportunity for using fractal models in soil variability studies. Further research in fractal theory led to development of multifractal models, which allow for utilizing separate scaling laws in describing variability of soil properties, hence producing more realistic characterizations of variability changes with scale.

Fractal and multifractal techniques of variability analysis are being actively developed in many fields studying various random natural surfaces and natural spatial variability (Cheng et al., 1994; Kropp et al., 1994; Goltz, 1996; Harris et al., 1996; Kropp et al., 1997; Gonçalves et al., 1998; Seuront et al., 1999; Boufadel et al., 2000; Deidda, 2000; Lammering, 2000; Lu et al., 2000; Conçalves, 1998, 2001; Finn et al., 2001; Schertzer et al., 2002; Gagnon et al., 2003). Such techniques are applied in image analysis (De Cola, 1993; Pachepsky et al., 1997; Nevado et al., 2000; Turiel and Parga, 2000; Lovejoy et al., 2001; Turiel, 2002) and plant community studies variability (Pascual et al., 1995; Drake and Weishampel, 2000). Examples and techniques for fractal/multifractal model applications in soil research are presented in this chapter.

We limit the content of this chapter to modeling soil spatial variability in the "classical" sense, i.e., variations in soil properties measured with the same support across a given area. Changes in support can also cause changes in soil properties. Readers are referred to recent books on the subject for discussions and examples of using fractals to model this type of scale-dependent variations [e.g., Pachepsky et al. (2002) and Pachepsky et al. (2003)].

We illustrate techniques and their applications using two data sets. The first data set represents a typical grid sampling example with spacing and support different from each other (Example I). The second one (Example II) represents a typical GIS (geographic information system) originated data set with the spacing and support equal to each other.

The Example I dataset contains soil test P concentration, organic matter content, and pH measured in soil samples collected from a 259 ha agricultural field located in central Illinois. Soil samples were taken on a regular 32×32 grid with the distance between the grid points of approximately 50 m. Detailed description of the data set is presented in Kravchenko et al. (1999).

The Example II dataset contains data on soil texture for the Little Washita watershed, Oklahoma, 610 km^2. Soil series occupying each individual 200×200 m pixels were identified (Ron Elliott, personal communication) (Figure 17.1a), and soil properties were extracted from the MUUF dataset (Baumer et al., 1994; Rawls et al., 2001). Contents of textural components, bulk density, and organic matter content were defined for the top horizons as average for the ranges given in MUUF in a total of 93×228 pixels. The clay content map in Figure 17.1b exemplifies the data in this dataset.

17.2 FRACTAL MODELS AND PARAMETERS OF SPATIAL VARIABILITY

Spatial variability characterizations at different scales have become widespread since geostatistical methods were introduced and adapted by soil scientists (see Chapters 16 and 18). Spatial structure of soil property variations is commonly represented using a variogram:

$$\gamma(h) = \langle |Z(x) - Z(x+h)| \rangle^2 \tag{17.1}$$

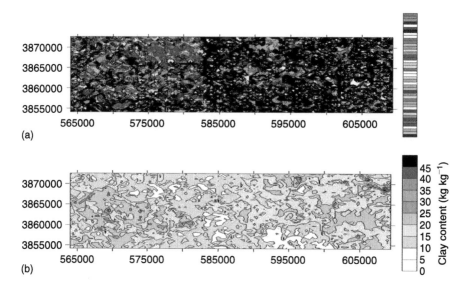

FIGURE 17.1 Map of individual soil series (a) and average clay content (b) in the Little Washita watershed, OK.

where $Z(x + h)$ and $Z(x)$ are the measurements of the studied variable collected at locations x and $x + h$ separated by distance h, and the angular brackets indicate ensemble average. The variogram gives the second statistical moment or variance for differences between measurements with spacing h. It shows how the variability changes with the change in spacing when support and extent remain the same.

17.2.1 MONOFRACTAL MODELS

Developments in spatial variability characterizations have led to a concept of self-affinity, a characteristic of data distributions along one-dimensional spatial transects or across two-dimensional sampled areas. For a transect data, self-affinity implies that in the two-dimensional xy space, $f(x, y)$ is statistically similar to $f(Rx, R^H y)$, where H is called the Hurst exponent. That is, changing the horizontal measurement scale R times has to correspond to the change in the vertical measurement scale R^H times. For a two-dimensional surface, self-affinity implies that the vertical scale, z, has to be changed R^H times and scales in horizontal directions, xy, have to be changed R times to obtain a statistically similar surface.

The Brownian motion or a random walk is a physical process that can generate self-affine lines and surfaces (Feder, 1988). A microscopic particle bombarded by random scores of molecules moves in what appears to be erratic steps in random directions. The length of the step for a particle movement, ξ, during one time interval is independent of the steps for its movement at any other time interval. The position of the particle at time t after n successive steps, where the duration of each step is equal to τ, can be determined as:

$$X(t = n\tau) = \sum_{i=1}^{n} \xi_i \qquad (17.2)$$

It can be shown (Feder, 1988) that the resulting plot or record of particle positions is a self-affine line, statistical properties of which remain the same when time, t, is scaled by an arbitrary factor R and step size, ξ, is scaled by a factor $R^{0.5}$. Hence, the Hurst exponent of the ordinary Brownian motion is equal to 0.5. An example of particle positions plotted versus time for the ordinary Brownian motion is shown on Figure 17.2a.

Characterization of spatial variability represented by self-affine lines and surfaces is related to variogram concepts. Indeed, if a spatial distribution of soil

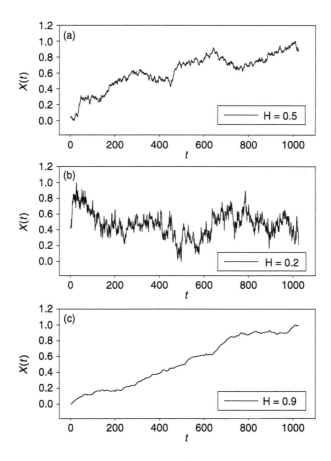

FIGURE 17.2 Example of particle positions for (a) an ordinary Brownian motion $H=0.5$), fractional Brownian motions with (b) $H=0.2$, and (c) $H=0.9$.

property values along a transect follows a Brownian motion model, the difference, $\triangle Z(h)$, between the observations separated by the distance h scales with the distance as:

$$\Delta Z(h) \propto h^H \tag{17.3}$$

where H is equal to 0.5.

The variance of these differences, i.e., variogram, is a linear function of separation distance h:

$$\gamma(h) \propto h^{2H} = h \tag{17.4}$$

where H is equal to 0.5.

Mandelbrot (1982) proposed a generalization from ordinary Brownian motion with $H=0.5$ to a fractional Brownian motion where the exponent H can assume a value in the interval $0 < H < 1$. The major difference between the ordinary Brownian motion with $H=0.5$, and its fractional counterparts with H values other than 0.5 is that the particle movement steps, ξ, are no longer independent from each other. For $H > 0.5$ the increment steps are positively correlated across infinite time t, which implies that increasing trend in the past leads to an increasing trend in the future. For $H < 0.5$, "increasing trend in the past implies a decreasing trend in the future, and a decreasing trend in the past makes an increasing trend in the future probable" (Feder, 1988). That is, high values of the Hurst exponent indicate some memory or autocorrelation in the data, while low values suggest an anti-correlation or self-correcting response. Examples of the plots for fractional Brownian motion with $H=0.2$ and $H=0.9$ are shown in Figure 17.2b and 17.2c, respectively. As shown in the figure, the higher the H value, the less erratic appear the plots. The fractal dimension of a self-affine surface is related to the Hurst exponent as:

$$D_S = 3 - H \tag{17.5}$$

The self-affine line will have a fractal dimension D_L related to the to the Hurst exponent H as:

$$D_L = 2 - H \tag{17.6}$$

Note that a line formed as a cross-section of any fractal self-affine surface and a vertical plane also displays self-affine properties. Values of D_L and D_S are interpreted similarly to H values. For example, for a set of elevation measurements along transect, any two consecutive height variations are likely to have opposite signs, i.e., negative variation follows a positive one and vice versa, when the fractal dimension D_L is larger than 1.5 at any particular scale, that is demonstrating a self-correcting response. Any two consecutive height variations are likely to have the same sign when the fractal dimension D_L is less than 1.5, indicating positive correlation.

The fractional Brownian motion model applied to spatial distribution of soil properties allows obtaining Hurst exponent, H, values based on the scaling relationship between the variogram values and separation distance h:

$$\gamma(h) \propto h^{2H} \qquad (17.7)$$

where the H value can be obtained as ½ of a slope of a log-log plot of h versus $\gamma(h)$. Examples of variograms and log-transformed variograms for the example datasets are shown in Figures 17.3 and 17.4. The figures show that variogram scale dependence [Eq. (17.7)] can hold only within a certain range of separation distances. Such behavior reflects the fact that spatial variability of soil properties as we observe it is a result of a combined effect of interacting environmental processes, many of which operate at different scales. The H value for the clay content data in Figure 17.3b equals 0.255. Being less than 0.5, it shows that any two consecutive clay content variations are likely to have opposite signs. The fractal scaling is applicable in the range of spacing from

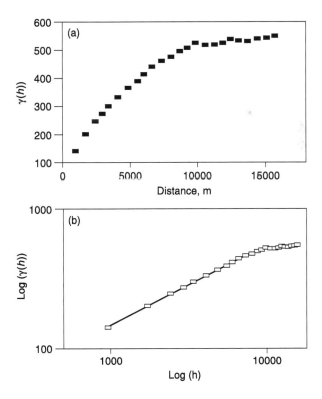

FIGURE 17.3 Variogram of clay content in the Little Washita watershed, OK, in linear (a) and double logarithmic (b) scales.

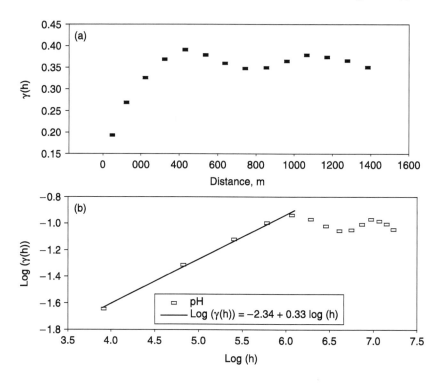

FIGURE 17.4 Variogram for soil pH in (a) linear and (b) double logarithmic scales.

900 to 8000 m. The H value for the pH data (Figure 17.4b) determined from the linear portion of the log-transformed variogram is equal to 0.165 and, similar to the clay data in the Little Washita watershed, indicates that the pH distribution in the studied field is very erratic. The region where fractal scaling is applicable extends from 50 to 400 m.

In geostatistics, the power-law relationship between the sample variogram and the separation distance, h, in Eq. (17.7) is often modeled with power-law variogram models. For example, the case of $H = 1/2$ corresponds to the linear model. However, the variogram behavior near the origin for other commonly used variogram models, such as spherical, exponential, and Gaussian, also follows the power-law relationship of Eq. (17.7). As an illustration, we plotted these three models using a zero nugget, a unit sill, and an arbitrarily selected range of 10 distance units along with Eq. (17.7) fitted to the variogram values of each model near the origin (Figure 17.5). The H values calculated for scaling at distances from 0 to 4 distance units for these variograms are 0.39, 0.95, and 0.49 for exponential, Gaussian, and spherical models, respectively. The H values reflect that short distance spatial variability patterns of the distributions described by Gaussian models are more continuous and less erratic than those of spherical and, particularly, exponential models.

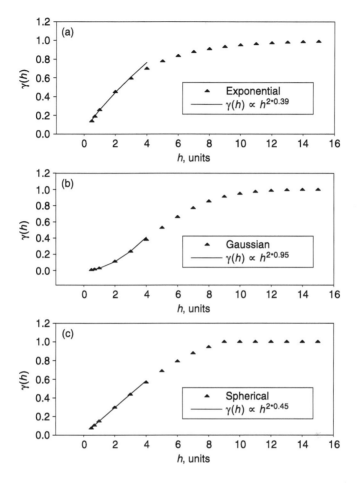

FIGURE 17.5 Exponential (a), Gaussian (b), and spherical (c) models with power-law relationship (Eq. 17.7) fitted to the variogram values near the origin. The variogram nuggets are equal to zero, sills to 1, and range to 10 distance units.

Analysis of variograms of many soil properties conducted by Burrough (1981, 1983) became the first application of fractal geometry to spatial variability assessments in soil science. It revealed that soil spatial variability exhibits scaling features that potentially can be used for its efficient characterization. Burrough found that the D_L values for soil properties' transects were much higher than those reported for distributions of geological sediments and landform data, indicating that short-distance variation constitutes a larger component of variability in soil properties than that in landscape characteristics. However, further research has shown that variations in soil properties may exhibit more complex behavior than that predicted by the self-affine model, and fractional Brownian motion is not always an adequate model for spatial variability of soil properties (Olsson et al., 2001).

17.2.2 MULTIFRACTAL MODELS

Distributions of soil properties may exhibit multifractal features. Such features can be explored using statistical moments other than the second moment (i.e., variogram). The qth order structure functions are shown to be useful for taking into account multifractal features in data spatial variability (Davis et al., 1994; Liu and Molz, 1997):

$$\langle [\Delta Z(h)]^q \rangle = \langle |Z(x) - Z(x+h)| \rangle^q \qquad (17.8$$

where q is the order of the structure function, a real number. The structure function at $q = 2$ is equivalent to the variogram, $\gamma(h)$ [Eq. (17.1)]. A variable exhibits multifractal behavior if its structure function varies with the spacing as

$$\langle [\Delta Z(h)]^q \rangle \propto h^{\zeta(q)} \qquad (17.9$$

where exponent $\zeta(q)$ describes its scaling properties.

The values of the exponent can be obtained from the measured data by plotting log of the left side of Eq. (17.9) versus log-transformed distance h. Figure 17.6 shows such plots for soil pH data from Example I. Plots with $q \neq 2$ show linear trends similar to the plot with $q = 2$, i.e., variogram (Figure 17.4b), and the linearity holds only over a certain range of distances h. Similar behavior of structure functions has been observed for topsoil concentrations of 20 chemical elements in a study by Olsson et al. (2001), as well as in studies of many other natural phenomena (Davis et al., 1994; Liu and Molz, 1997).

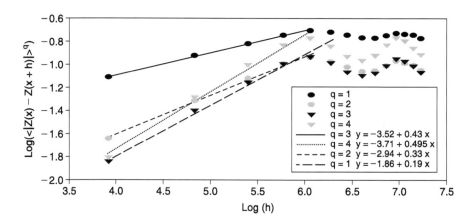

FIGURE 17.6 Structure functions for pH calculated at four q values and plotted versus separation distance h in a log-log form. The slopes of the linear portions of the plots are equal to $\zeta(q)$ values.

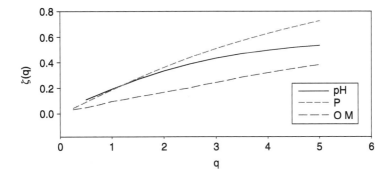

FIGURE 17.7 The $\zeta(q)$ function for pH, P, and OM plotted versus q. The $\zeta(q)$ values were obtained for q ranging from 0.25 to 5 in 0.25 intervals.

As follows from Eq. (17.9) for higher q values larger differences will dominate the left side of the equation. Hence, for large q, the exponent $\zeta(q)$ depends mostly on scaling properties of large differences. If scaling with distance occurs in a similar fashion for both average and large differences, $\zeta(q)$ function is represented by a straight line where $\zeta(q) = qH$. Such spatial distributions and processes generating them are referred to as monofractals or simple scaling and are characteristic of everywhere differentiable functions. In fact, Eq. (17.7) treats spatial distributions of soil properties as monofractals, where for $q = 2$ the exponent $\zeta(2)$ is equal to $qH = 2H$. However, for more complex distributions the $\zeta(q)$ is no longer linear but is represented by a concave curve. The curve reflects the fact that a separate fractal dimension has to be used to characterize scaling of average, large, etc., differences with distance. Such spatial distributions are, hence, called *multifractal*.

The $\zeta(q)$ versus q plots for P, OM, and pH for Example I data are shown in Figure 17.7. The plots for pH and P are nonlinear, indicating that scaling with distance in the studied field for these soil properties is multifractal. The plot for OM is linear, demonstrating that scaling for OM in the studied field is monofractal. Olsson et al. (2001) also observed both multifractal and monofractal features in scaling of the studied soil properties. In their data set, soil K, Ni, and V concentrations exhibited multifractal scaling, while monofractal scaling was observed for concentrations of several other elements, including P, Cu, and S.

17.2.3 MULTIFRACTAL SPECTRA

Besides using structure functions, multifractal random fields can be parameterized using the $f(\alpha)$ multifractal spectrum formalism (Feder, 1988; Evertsz and Mandelbrot, 1992; Baveye and Boast, 1998; Harte, 2001). The spectrum can be obtained by coarse-graining the studied field, for which the area is covered by a set of grid cells of different sizes, where each cell is characterized by a certain value or a measure of the variable of interest. The distribution of the studied variable is normalized by introducing a new variable, $\mu_i(\varepsilon)$, that describes the portion of the total mass of the studied variable contained in each map cell i of the size ε:

$$\mu_i(\varepsilon) = Z_i \bigg/ \sum_{j=1}^{N_{total}} Z_j \qquad (17.10$$

where Z_i is the data value from the map cell with size ε, and $\sum_{j=1}^{N} Z_j$ is a sum of Z_i values from all N_{total} grid cells of the studied field. An example of such coarse-graining applied to soil data is discussed in detail by Kravchenko et al. (1999).

The variable $\mu_i(\varepsilon)$ changes with the cell size and, for multifractal measures, $\mu_i(\varepsilon)$ scaling is defined as:

$$\mu_i(\varepsilon) \propto \varepsilon^{\alpha} \qquad (17.11$$

where α is called the Hölder exponent or the singularity strength. The number of cells of size ε with α values falling within α to $\alpha + d\alpha$ interval, $N_{\alpha}(\varepsilon)$, scales with the cell size as:

$$N_{\alpha}(\varepsilon) \propto \varepsilon^{-f(\alpha)} \qquad (17.12$$

where the exponent $f(\alpha)$ characterizes the scaling of the cells with a certain α A plot of $f(\alpha)$ versus α is called a multifractal spectrum. For variables displaying multifractal properties, the spectrum has a concave (bell) shape with maximum value of $f(\alpha)$ equal to the box-counting dimension of the geometric support for the variable's distribution, that is, the maximum will be equal to 1 for a one-dimensional transect, or 2 for a two-dimensional plane. In the case when the geometric support is a fractal itself, the maximum will be equal to the fractal dimension of this support. For variables displaying monofractal properties, the spectrum is a spike at $f(\alpha)$ equal to the box-counting dimension of the geometric support.

In general terms, the meaning of the α and $f(\alpha)$ values for characterizing data distributions can be inferred from Eqs. (17.11) and (17.12) as follows: larger α values correspond to the cells where $\mu_i(\varepsilon)$ values are small at small initial ε values but increase rapidly with increasing cell size. Smaller α values correspond to the cells where $\mu_i(\varepsilon)$ values are high at small initial ε values, and their increase with increasing cell size is relatively slow. If cells with very high (or very low) α values are spread relatively homogeneously through the studied field at the initial cell size, then at large cell sizes the number of cells with this α value will decrease rapidly due to averaging of the extreme data values, resulting in high $f(\alpha)$. If the cells with extreme data values are concentrated in certain locations, then at large cell sizes the number of the cells with extreme data values will still be relatively high, hence, lower $f(\alpha)$ values. The larger the deviations of the $f(\alpha)$ at high or low q from the maximum $f(\alpha$

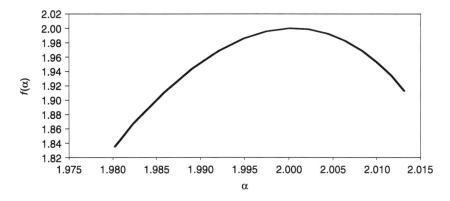

FIGURE 17.8 The multifractal spectrum $f(\alpha)/\alpha$ for soil pH. It was obtained using $\mu_i(\varepsilon$ distributions at five actual cell sizes ranging from 6.6 to 105.6 m. For the cells of the smallest size (actual cell size of 6.6 m), μ_i were obtained directly from the map. For the larger cell sizes, variable values for each cell were calculated as the sum of μ_i values of the cells included in the cell of that size.

value are, the more pronounced is the multifractality of the data, i.e., the fractal dimensions of the data with high/low values are substantially different from the dimension of the whole data set, while if $f(\alpha)$ at high or low q are similar to the maximum $f(\alpha)$ value, then the multifractal nature of the data is less pronounced. Therefore, multifractal models of variability have a built-in potential of proper simulation of rare occurrences that are known to often dominate the behavior of whole systems. An example of the multifractal spectrum for pH data from the Example I data set is shown in Figure 17.8. The spectrum is a bell-shaped curve, which indicates multifractal scaling in pH variability in the studied field. Soil clay content data in Example II did not demonstrate multifractal scaling.

Kravchenko et al. (1999) compared performances of multifractal spectra and variograms for characterizing spatial variability patterns of soil properties. They considered two data sets, which consisted of identical data values, but with certain differences in data locations. Specifically, the second data set was obtained by random relocation of the highest data values (2% of the data). The sample variograms of the studied data sets did not clearly reflect the differences in their spatial variabilities. However, the multifractal spectra showed marked dissimilarities, especially at the left side of the spectrum, which corresponds to fractal properties of rarely occurring high data values. This example demonstrates that multifractal spectra can be more effective than variograms in characterizing and comparing spatial variability features of extremely high or low data values. Similar findings were reported by Agterberg (1995), who applied multifractal analysis for characterizing giant and supergiant metal deposits.

Several methods can be used to compute multifractal spectra. One of them is the method of moments (Evertsz and Mandelbrot, 1992), which estimates α

and $f(\alpha)$ based on a partition function, $\chi_q(\varepsilon)$. The latter is calculated from the $\mu_i(\varepsilon)$ values as:

$$\chi_q(\varepsilon) = \sum_{i=1}^{N_{total}} \mu_i(\varepsilon)^q \qquad (17.13$$

For multifractal variables the partition function scales with the cell size as:

$$\chi_q(\varepsilon) \propto \varepsilon^{\tau(q)} \qquad (17.14$$

where exponent $\tau(q)$ is called a mass exponent. The mass exponent can be obtained by plotting $\log \chi_q(\varepsilon)$ versus $\log \varepsilon$ for each q value. Once $\tau(q)$ values for each q are obtained as slopes of the log-log $\chi_q(\varepsilon)$ plots, the $\alpha(q)$ and $f(\alpha(q))$ values are determined as:

$$\alpha(q) = -d\tau(q)/dq \qquad (17.15$$

$$f(\alpha(q)) = q\alpha(q) + \tau(q) \qquad (17.16$$

Another method often used for calculation of the multifractal spectrum is the method of multipliers proposed by Chhabra and Jensen (1989). Method of multipliers is based on relationships:

$$\langle \tau(q) \rangle + D = -\frac{\log\langle (\mu(\varepsilon_{min})/\mu(\varepsilon))^q \rangle}{\log(\varepsilon/\varepsilon_{min})} \qquad (17.17$$

$$\langle \alpha(q) \rangle = -\frac{\langle (\mu(\varepsilon_{min})/\mu(\varepsilon))^q \log(\mu(\varepsilon_{min})/\mu(\varepsilon)) \rangle}{\langle (\mu(\varepsilon_{min})/\mu(\varepsilon))^q \rangle \log(\varepsilon/\varepsilon_{min})} \qquad (17.18$$

where ε_{min} is the smallest cell size, and angular brackets represent ensemble average. The advantage of using the method of multipliers as compared to method of moments is that only two cell sizes are needed to calculate the multifractal spectra (Cheng, 1999). To compare, the minimum of three cell sizes is needed to obtain an estimate of $\tau(q)$ value using Eq. (17.14), while having more cell sizes is desirable for higher reliability in $\tau(q)$ estimate.

Related to $f(\alpha)$ spectrum is the co-dimension formalism developed by Schertzer and Lovejoy (1987, 1991, 1999). This formalism uses the resolution or the scale ratio $\lambda = L/\varepsilon$, where L is the largest scale of interest and ε is the cell size as defined previously. Subject of scaling here is the intensity of the studied variable, ρ_λ, related to the previously introduced multifractal measure, $\mu_i(\varepsilon)$, as $\rho_\lambda = \mu_i(\varepsilon)\varepsilon^{-D}$, where D is the Euclidian dimension of the studied field (that is, 1 for a transect, and 2 for a plane). The ρ_λ scales with resolution λ as follows:

$$\rho_\lambda \propto \lambda^\gamma \tag{17.19}$$

where the scaling exponent γ is related to the previously discussed Hölder exponent α as $\alpha = D - \gamma$.

The Probability distribution of ρ_λ varies with scale:

$$\Pr(\rho_\lambda > \lambda^\gamma) \propto \lambda^{-c(\gamma)} \tag{17.20}$$

where the co-dimension $c(\gamma)$ is related to $f(\alpha)$ as $c(\gamma) = D - f(\alpha)$. The relationships between the $f(\alpha)/\alpha$ spectra and the co-dimension formalism demonstrate that, unlike the co-dimensions, $f(\alpha)/\alpha$ values depend on the Euclidian dimension of the space where the process is observed. Hence, characterization of the variability via $f(\alpha)/\alpha$ spectra can be regarded as equivalent to enumerating the events, while the co-dimension formalism determines event frequencies (Schertzer et al., 2002). Equation (17.20) provides a complete description of how the histograms of the studied density, ρ_λ, vary with resolution, which makes it a potentially very useful tool for upscaling/downscaling in descriptions of soil properties and ecological processes.

Equivalently, scaling of the statistical moments of ρ_λ can be described as:

$$\langle \rho_\lambda^q \rangle \propto \lambda^{K(q)} \tag{17.21}$$

where $K(q)$ is the moment scaling function. The structure function exponent $\zeta(q)$ is related to $K(q)$ as $\zeta(q) = qH - K(q)$ (Schertzer and Lovejoy, 1987). The variability characterization via structure function exponents is also related to that of the multifractal $f(\alpha)/\alpha$ formalism. e.g., relationship between the structure function exponent $\zeta(q)$ and the mass exponent $\tau(q)$ [Eq. (17.14)] can be defined as $\tau(q) = (q - 1)D - qH + \zeta(q)$.

Wavelet analysis is yet another approach to analyze fractal and multifractal variability (i.e., Roux et al., 2000). For spatial data, wavelet analysis decomposes the data into components at several consecutive scales using localized filters. This decomposition retains information about both scale and spatial dependencies in the studied spatial data.

The existence of a multitude of methods for fractal/multifractal characterization of scaling properties, only few of which are briefly discussed here, warrants a word of caution regarding their application. Indeed, different methods may not necessarily render identical results when used to estimate parameters of fractal and multifractal models. One reason for this is that soils are not ideal multifractals. Other reasons include the effect of data preprocessing, such as handling missing values (Dubayah et al., 1997; Oldak et al., 2002), sensitivity of the methods to bin intervals, uncertainties in defining the linearity ranges and boundaries (Pachepsky et al., 1997), as well as the threshold levels for indicator variables (Tarquis et al., 2003).

17.3   SIMULATING SPATIAL VARIABILITY WITH FRACTAL MODELS

Simulations of spatial variability can be used for several purposes. For example, missing values in some subareas of sampling might need to be filled with the most probable values, or some parts of the area of interest might have been sampled more sparsely than others, or extrapolation may be needed both in space and across scales. Because different methods of fractal parameter estimation tend to produce different results, it is beneficial to test a candidate method using a simulated fractal dataset with known parameters. Also, synthetic datasets of spatially variable soil properties can be of use both for designing sampling strategies and for evaluating behavior of new sensors with different supports. Using multifractal characteristics in the stochastic simulation of soil properties has a potential for producing simulations that better represent features of interest (e.g., distribution properties of the high/low values) than simulations based on sample variograms or variogram models (Kravchenko, 2005). Several algorithms are available to simulate fractal and multifractal spatial fields with known spectra (Roberts and Cronin, 1996; Agterberg, 2001; Kumar, 2003).

For the Example I dataset we performed simulations of soil phosphorus and organic matter data. Here P represents an example of a multifractal scaling and OM an example of a monofractal scaling (as discussed previously for Figure 17.7). We separated the data into "model" and "test" data sets. The "model" data set included samples located on a 12×14 regular grid with 100 m distance between the grid points and an additional 20 samples randomly selected such that the distance from them to the grid points was 50 m. The remaining 836 samples constituted the "test" data set. Then we conducted simulations using the simulated annealing procedure with conditioning the simulations on the "model" data sets (Deutsch and Journel, 1998). The simulated annealing algorithm produces a new data set with the desired characteristics based on the original data. The simulated annealing procedure begins with creating an initial data set by assigning a random value at each of the grid nodes of the simulated data set. The random values are drawn from the population distribution of the soil property, which in our example was constructed based on the "model" data. Then we calculated the structure functions of the initial simulated data set using Eq. (17.8) for q values ranging from 1 to 5 in 1.0 intervals. The exponents $\zeta(q)$ were obtained for each q value as described in Eq. (17.9). The obtained set of $\zeta(q)$ was compared with the $\zeta(q)$ values desired as characteristics for the final simulated field. As the desired characteristics, we used the $\zeta(q)$ values obtained based on the whole data set, assuming them to be true structure function exponents of the studied field. After that, the initial data set was perturbed by drawing a new value for a randomly selected grid node, and the $\zeta(q)$ values for the perturbed data set were again calculated and compared with the desired ones. If the perturbed value led to a closer correspondence between the observed and desired $\zeta(q)$ values, it was retained, otherwise a new random value was drawn and the calculations and comparisons were repeated. The process continued until the $\zeta(q)$ values of the perturbed data set closely matched those of the desired spatial structure. A detailed description of the objective functions and convergence criteria of the simulated annealing procedure is provided by Deutsch and Journel (1998). To perform the simulations, we modified the program SASIM from the GSLIB package (Deutsch and Journel, 1998) by adding the structure function exponents, $\zeta(q)$, to the procedure's objective functions.

Two simulation scenarios were compared. The first scenario used histogram and the structure function with $\zeta(2)$, which is equivalent to simulating a data set with a certain variogram and to assuming that the studied variable is a monofractal; the second scenario used histogram and the structure function with $\zeta(q)$ ranging from $\zeta(1)$ to $\zeta(5)$, where the variable is assumed to be a multifractal. Based on 50 simulations, we compared the performance of the two scenarios in representation of highest/lowest values of the "test" data set. Average numbers of correctly predicted highest and lowest values obtained from 50 simulations are shown in Figure 17.9. There was no significant difference in performance of monofractal and multifractal simulations for soil organic matter. Such a result was expected since OM in this study is a

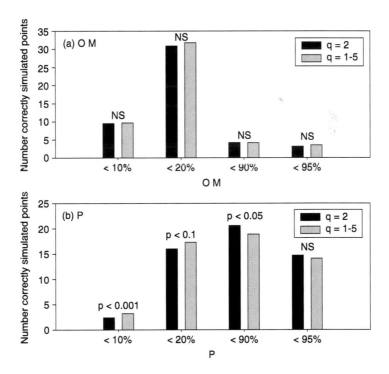

FIGURE 17.9 Numbers of correctly predicted highest and lowest observations for (a) the data set with monofractal variability (OM) and (b) the data set with multifractal variability (P). The numbers reported are averages from 50 simulations.

monofractal. However, performances of monofractal and multifractal simulations were different for P. Multifractal simulations performed better than monofractal in correct predictions of the locations with low values. Out of a total of 836 test data points, we compared, first, how many of the highest and lowest 46 observations were accurately predicted in the simulated field and, second, how many of the 80 highest and 130 lowest observations were accurately predicted. The multifractal approach produced significantly more accurate predictions of the lowest 46 observations, but significantly fewer accurate predictions of the 80 highest observations. There was no difference between the performance of the monofractal and multifractal simulations in prediction accuracy for the highest 46 values and the lowest 130 values.

17.4 SUMMARY, CRITICAL ASSESSMENT, AND FUTURE RESEARCH

Understanding soil variability becomes imperative as the management tools become more sophisticated and soil uses become more multifaceted. A correct model of variability is necessary to interpret and use results of soil sampling. Several models of the spatial variability have been proposed and are used to characterize soil structure and soil cover. Fractal models have become popular in soil studies because they are designed to mimic and parameterize irregular objects that have similar features at different scales. Such similarity is common in soil matrix, which is naturally hierarchically organized. The examples that we presented along with findings reported in the soil literature (see references in Pachepsky et al., 2000) demonstrate the applicability and usefulness of fractal and multifractal methods for characterizing spatial variability of soil properties. The applicability of fractal models to spatial variations in soil properties indicates that the natural hierarchical organization is also present in soil cover.

One important advantage of fractal models of variability is their ability to better simulate rare occurrences in soils. Rare occurrences such as large pores, preferential pathways, very high conductivities, and localized bacteria habitats, are very important because they often define soil behavior at scales coarser than observational ones. For example, macropores are rare and difficult to find when the sample size is small. However, hydraulic conductivity of a large soil monolith will be mostly defined by the macropores that occupy a tiny part of soil volume in this monolith. The presence of a rare occurrence is, in actuality, an omnipresent norm, and its proper characterization and simulation is extremely important for soil data upscaling and uncertainty characterization. One technique of upscaling soil transport properties consists in assembling of a large soil unit from many fine-scale units and using a fine-scale model to simulate behavior of and mass exchange between the fine-scale units to infer transport properties of a larger unit from such simulations (Faybishenko et al., 2003). Such an approach is feasible provided the fine-scale transport properties receive exhaustive characterization and are modeled properly. The description of rare occurrences of highly conductive regions and their connectivity is a prerequisite for proper simulation of the ensemble behavior of fine-scale units. Superior performance of multifractal models for describing rare events demonstrated in the simulation example and in the examples reported in the literature indicates that the multifractal approach has a great potential in environmental and other applications where the presence of such events is of particular interest.

Comparison of fractal model performance with that of other spatial variability models and development of criteria for such comparisons is an area where further research is much needed. The root-mean-square model error or the determination coefficients are often used as measures of model performance. These measures may suffice for investigating the performance of a single model against a single data set, yet fail to recognize that a model should not be more accurate than the data. The χ^2 test, the goodness-of-fit test (Whitmore, 1991), or the normalized root-mean-square (Loague and Green, 1991) value would account for the latter. However, when a choice of models is presented, both conceptual differences and differences in the number of adjustable parameters have to be accounted for in the performance comparison. This leads to recognition of the need for information-based measures of model performance in which the root-mean-square error is corrected to reflect the number of adjustable parameters. A wide spectrum of such measures have been suggested, criteria AIC, Cp, BIC, MDL, CAIC, MML, and SC/PMDL being examples. Other measures have been also introduced based on the predictive ability rather than on parameter estimation (Clayton et al., 1986). Obviously, research needs to be done to understand the relative value of those criteria in variability studies.

Because soils are not ideal fractals and because fractal scaling is applicable only within a range of scales, fractal models of variability techniques have their limitations. Nevertheless, such models present the opportunity to quantify the spatial variability for further comparison and use of the variability parameters, and therefore their application in soil variability studies seems to be warranted.

REFERENCES

Agterberg, F.P. 1995. Multifractal modeling of the sizes and grades of giant and supergiant deposits. Int. Geol. Rev. 37:1–8.

Agterberg, F.P. 2001. Multifractal simulation of geochemical map patterns. *In* Merriam, D.F. and J.C. Davis (eds.), Geologic Modeling and Simulation, Computer Applications in the Earth Sciences. Plenum Press, New York.

Baumer, O., P. Kenyon, and J. Bettis. 1994. MUUF v2.14 User's Manual. USDA Natural Resource Conservation Service, Lincoln, NE.

Baveye, P. and C.W. Boast. 1998. Fractal geometry, fragmentation processes and the physics of scale-invariance: an introduction. p. 1–54. *In* P. Baveye, J.-Y. Parlange and B.A. Stewart (eds.), Fractals in Soil Sciences. CRC Press, New York.

Bierkens, M.F.P., P.A. Finke, and P. de Willigen. 2000. Upscaling and Downscaling Methods for Environmental Research. Kluwer Acad. Publ., Dordrecht, The Netherlands.

Boufadel, M.C., S-L. Lu, F.J. Molz, and D. Lavallee. 2000. Multifractal scaling of the intrinsic permeability. Water Res. Res. 36:3211–3222.

Burrough, P.A. 1981. Fractal dimensions of landscapes and other environmental data. Nature 294:240–242.

Burrough, P.A. 1983. Multiscale sources of spatial variation in soil. I. The application of fractal concepts to nested levels of soil variation. J. Soil Sci. 34:577–597.

Cheng, Q. 1999. The gliding box method for multifractal modeling. Comput. Geosci. 25:1073–1079.

Cheng, Q., F.P. Agterberg, and S.B. Ballantyne 1994. The separation of geochemical anomalies from background by fractal methods. J. Geochem. Explor. 51:109–130.

Chhabra, A. and R.V. Jensen. 1989. Direct determination of the f(a) singularity spectrum. Phys. Rev. Lett. 62:1327–1330.

Clayton, M.K., S. Geisser, and D.E. Jennings. 1986. A comparison of several model selection procedures. *In* Goel, P. and A. Zellner (eds.), Bayesian Inference and Decision Techniques: Elsevier Science, New York.

Conçalves, M.A. 2001. Characterization of geochemical distributions using multifractal models. Math. Geol. 33:41–61.

Davis, A., A. Marshak, W. Wiscombe, and R. Cahalan. 1994. Multifractal characterization of nonstationaroty and intermittency in geophysical fields: observed, retrieved, or simulated. J. Geophys. Res. 99:8055–8072.

De Cola L. 1993. Multifractals in image process images. *In* Lam, N.S. and L. DeCola (eds.), Fractals in Geography. Prentice-Hall, Englewood Cliffs, NJ.

Deidda, R. 2000. Rainfall downscaling in a space-time multifractal framework. Water Res. Res. 36:1779–1794.

Deutsch, C.V. and A.G. Journel. 1998. Geostatistical Software Library and User's Guide. Oxford University Press, New York.

Drake, J.B. and J.F. Weishampel. 2000. Multifractal analysis of canopy height measures in a longleaf pine savanna. Forest Ecol. Manage. 128:121–127.

Dubayah, R., E.F. Wood, and D. Lavallee. 1997. Multiscaling analysis in distributed modeling and remote sensing: an application using soil moisture. p. 93–112. *In* Quatrocchi, D.A. and M. Goodchild (eds.), Scale in Remote Sensing and GIS. Lewis Publishers, New York.

Evertsz, C.J.G. and B.B. Mandelbrot. 1992. Multifractal measures (Appendix B). p. 922–953. *In* Peitgen. H.O., H. Jurgens, and D. Saupe (eds.), Chaos and Fractals, Springer-Verlag, New York.

Faybishenko, B, G.S. Bodvarsson, J. Hinds, and P.A. Witherspoon, 2003. Scaling and hierarchy of models for flow processes in unsaturated fractured rock. *In* Pachepsky, Y., D. Radcliffe, and M. Selim (eds.), Scaling Methods in Soil Physics. CRC Press, Boca Raton, FL.

Feder, J. 1988. Fractals. Plenum Press, New York.

Finn, D, B. Lamb, M.Y. Leclerc, S. Lovejoy, S. Pecknold, and D. Schertzer. 2001. Multifractal analysis of line-source plume concentration fluctuations in surface-layer flows. J. Appl. Meteorol. 40(2):229–245.

Gagnon, J.S., S. Lovejoy, and D. Schertzer. 2003. Multifractal surfaces and terrestrial topography. Europhys. Lett. 62(6):801–807.

Goltz, C. 1996. Multifractal and entropic properties of landslides in Japan. Geol. Rundsch. 85:71–84.

Gonçalves, M.A. 2001. Characterization of geochemical distributions using multifractal models. Math. Geol. 33:41–61.

Gonçalves, M.A., M. Vairinho, and V. Oliveira. 1998. Study of geochemical anomalies in Mombeja area using a multifractal methodology and geostatistics. p. 198. *In* Buccianti, A., G. Nardi, and R. Potenza (eds.), Proc. IAMG '98, Naples, Italy.

Harris, D., M. Menabde, A. Seed, and G. Austin. 1996. Multifractal characterization of rain fields with a strong orographic influence. J. Geophys. Res. 101: 26405–26414.

Harte, D. 2001. Multifractals: Theory and Applications. CRC Press, Boca Raton, FL.

Kravchenko A.N. 2005. Using multifractals to simulate soil property distributions. Soil Sci. Soc. Am. J. (in press).

Kravchenko, A.N., C.W. Boast, and D.G. Bullock. 1999. Multifractal analysis of the soil spatial variability. Agron. J. 91:1033–1041.

Kropp, J., A. Block, W. von Bloh, T. Klenke, and H.J. Schellnhuber. 1994. Characteristic multifractal element distribution in recent bioactive marine sediments. *In* Kruhl, J.K. (ed.), Fractals and Dynamic Systems in Geoscience. Springer, Berlin.

Kropp, J., W. von Bloh, A. Block, T. Klenke, and H.J. Schellnhuber. 1997. Multifractal characterization of microbially induced magnesian calcite formation in recent tidal flat sediments. Sediment. Geol. 109:37–51.

Kumar, P. 2003. Multiple scale conditional simulation. *In* Pachepsky, Y.A., D.L. Radcliffe, and H.M. Selim (eds.), Scaling Methods in Soil Physics, CRC Press, Boca Raton, FL.

Lammering, B. 2000. Slices of multifractal measures and applications to rainfall distributions. Fractals 8:337–348.

Loague, K.M. and Green, R.E. 1991. Statistical and graphical methods for evaluating solute transport models: overview and applications. J. Contam. Hydrol. 7:51–73.

Lovejoy, S., D. Schertzer, Y. Tessier, and H. Gaonach. 2001. Multifractals and resolution independent remote sensing algorithms: the example of ocean color. Inter. J. Remote Sens. 22(7):1191–1234.

Liu H.H. and F.J. Molz. 1997. Multifractal analyses of hydraulic conductivity distributions. Water Res. Res. 33:2483–2488.

Lu S., F.J. Molz, and D. Lavallee. 2000. Multifractal scaling of the intrinsic permeability. Water Res. Res. 36:3211–3222.

Mandelbrot, B.B. 1982. The Fractal Geometry of Nature. W.H. Freeman, New York.

Nevado, A., A. Turiel, and N. Parga. 2000. Scene dependence of the non-gaussian scaling properties of natural images. Network 11:131–152.

Oldak, A., Y.A. Pachepsky, T. Jackson, and W.J. Rawls. 2002. Statistical properties of soil moisture images revisited. J. Hydrol. 255:12–24.

Olsson, J., R. Berndtsson, A. Bahri, M. Persson, and K. Jinno. 2001. Nonlinear and scaling spatial properties of soil geochemical element contents. Water Res. Res. 37:1031–1042.

Pachepsky, Y.A., J.C. Ritchie, and D. Gimenez. 1997. Fractal modeling of airborne laser altimetry data. Remote Sens. Environ. 61:150–161.

Pachepsky, Y.A., D. Giménez, and Walter J. Rawls. 2000. Bibliography on applications of fractals in soil science. p. 273–295. *In* Pachepsky, Ya., J. Crawford, and W. Rawls (eds.), Fractals in Soil Science. Elsevier, Amsterdam.

Pachepsky, Y., J. Crawford, and W. Rawls (eds.). 2002. Fractals in Soil Science. Elsevier, New York.

Pachepsky, Y. Radcliffe, D.L., Selim H.M. (eds.). 2003. Scaling Methods in Soil Physics. CRC PRESS, Boca Raton, FL.

Pascual, M., F.A. Ascioti, and H. Caswell. 1995. Intermittency in the plankton: a multifractal analysis of zooplankton biomass variability. J. Plankton Res. 17:1209–1232.

Rawls, W.J., Y.A. Pachepsky, and M.H. Shen. 2001. Testing soil water retention estimation with the MUUF pedotransfer model using data from the southern United States. J. Hydrol. 251:177–185.

Roberts, A.J. and A. Cronin. 1996. Unbiased estimation of multi-fractal dimensions of finite data sets. Physica A 233:867–878.

Roux, S.G., A. Arnéodo, and N. Decoster. 2000. A wavelet-based method for multifractal image analysis. III. Applications to high-resolution satellite images of cloud structure. Eur. Phys. J. B. 15:765–786.

Schertzer, D. and S. Lovejoy. 1987. Physical modeling and analysis of rain and clouds by anisotropic scaling multiplicative processes. J. Geophys. Res. 92:9693–9714.

Schertzer, D. and S. Lovejoy. 1991. Non-linear Variability in Geophysics. Kluwer Academic Publishers Dordrecht, The Netherlands.

Schertzer, D. and S. Lovejoy. 1999. Multifractals and Turbulence: Fundamentals and Applications in Geophysics. World Scientific, Singapore.

Schertzer, D., S. Lovejoy, and P. Hubert. 2002. An introduction to stochastic multifractal fields. p. 106–179. In Ern A. and L. Weiping (eds.), Mathematical Problems in Environmental Science and Engineering. Higher Education Press, Beijing, PR China.

Seuront, L., F. Schmitt, Y. Lagadeuc, D. Schertzer, and S. Lovejoy. 1999. Universal multifractal analysis as a tool to characterize multiscale intermittent patterns: example of phytoplankton distribution in turbulent coastal waters. J. Plankton Res. 21:877–922.

Tarquis, A.M., D. Giménez, A. Saa, M.C. Díaz, and J.M. Gascó. 2003. Scaling and multiscaling of soil pore systems determined by image analysis. In Pachepsky, Y.A., D.L. Radcliffe, and H.M. Selim (eds.), Scaling Methods in Soil Physics. CRC Press, Boca Raton, FL.

Turiel, A. and N. Parga. 2000. Multifractal wavelet filter of natural images. Phys. Rev. Lett. 85:3325–3328.

Turiel, A. 2002. Relevance of multifractal textures in static images. Electro. Lett. Comput. Vision Image Anal. 1:35–49.

Whitmore, A.P. 1991. A method for assessing the goodness of computer simulation of soil processes. J. Soil Sci. 42:289–299.

18 Geospatial Measurements of Apparent Soil Electrical Conductivity for Characterizing Soil Spatial Variability

Dennis L. Corwin
USDA-ARS, George E. Brown, Jr. Salinity Laboratory,
Riverside, California, U.S.A.

CONTENTS

1-5667-0657-2/05/$0.00 + $1.50
© 2005 by CRC Press

18.1 INTRODUCTION

Ever since the classic paper by Nielsen et al. (1973) concerning the variability of field-measured soil water properties, the significance of within-field spatial variability of soil properties has been scientifically acknowledged and documented. Spatial variability of soil has been the focus of books (Bouma and Bregt, 1989; Mausbach and Wilding, 1991) and numerous comprehensive review articles (Warrick and Nielsen, 1980; Jury, 1985, 1986; White, 1988). The significance of soil spatial variability lies in the fact that it is a key component of any landscape-scale soil-related issue including solute transport in the vadose zone, site-specific crop management, and soil quality assessment, to mention a few.

There are a variety of methods for potentially characterizing soil spatial variability, including ground penetrating radar (GPR), aerial photography, multi- and hyperspectral imagery, time domain reflectometry (TDR), and apparent soil electrical conductivity (EC_a). However, none of these approaches has been as extensively investigated as the use of EC (Corwin and Lesch, 2005a).

Since its early agricultural use for measuring soil salinity, the application of EC_a has evolved into a widely accepted means of establishing the spatial variability of several soil physicochemical properties that influence the EC_a measurement (Corwin and Lesch, 2003, 2005a). Geospatial measurements of EC_a are well suited for characterizing spatial variability for several reasons. Geospatial measurements of EC_a are reliable, quick, and easy to take. The mobilization of EC_a measurement equipment is easy and can be accomplished at a reasonable cost. Finally, and most importantly, EC_a is influenced by a variety of soil properties for which the spatial variability of each could be potentially established. Corwin and Lesch (2005a) provide a compilation of literature pertaining to the soil physicochemical properties that are either directly or indirectly measured by EC_a.

It is the goal of this chapter to provide an overview of the characterization of soil spatial variability using EC_a-directed soil sampling for three different landscape-scale applications: (1) solute transport modeling in the vadose zone, (2) site-specific crop management, and (3) soil quality assessment. Guidelines, methodology, and strengths and limitations are presented for characterizing spatial and temporal variation in soil physicochemical properties using EC_a-directed soil sampling.

18.1.1 JUSTIFICATION FOR CHARACTERIZING SPATIAL VARIABILITY WITH GEOSPATIAL EC_a MEASUREMENTS

The prospect of feeding a projected additional 3 billion people over the next 30 years poses formidable, but not insurmountable, challenges. Feeding the ever-increasing world population will require a sustainable agricultural system that can keep pace with population growth. The concept of sustainable agriculture is predicated on maximizing crop productivity and maintaining economic stability while minimizing utilization of finite natural resources and detrimental environmental impacts of associated agrichemical pollutants. To sustain agriculture, a balance must be attained between profitability, resource utilization, crop yield, and environmental stewardship.

Conventional farming currently treats a field uniformly, ignoring the naturally inherent variability of soil and crop conditions between and within fields. However, until recently, with the introduction of global positioning systems (GPS) and yield-monitoring equipment, documentation of crop yield and soil variability at field scale was difficult to establish. Now there is well-documented evidence that spatial variability within a field is highly significant and amounts to a factor of 2–4 or more for crops (Birrel et al., 1995; Verhagen et al., 1995; Kaffka et al., 2005) and up to an order of magnitude or more for soils (Jury, 1986; Corwin et al., 2003a).

Spatial variation in crops is the result of a complex interaction of biological (e.g., pests, earthworms, microbes), edaphic (e.g., salinity, organic matter, nutrients, texture), anthropogenic (e.g., leaching efficiency, soil compaction due to farm equipment), topographic (e.g., slope, elevation), and climatic (e.g., relative humidity, temperature, rainfall) factors. To a varying extent from one field to the next, crop patterns are influenced by edaphic (i.e., soil-related) properties. Bullock and Bullock (2000) pointed out the imminent need for efficient methods to accurately measure within-field variation in soil physical and chemical properties as a key component for precision agriculture.

A fundamental component of assessing field-scale soil quality is establishing the spatial distribution of the soil properties affecting the soil's intended management goal (e.g., maximize agricultural productivity, minimize environmental impact, and/or maximize waste recycling) and its intended function (e.g., biodiversity, filtering and buffering, nutrient cycling, physical stability and structural support, resistance and resilience, and water and solute flow). It is not sufficient to take a single measurement within a field to characterize its soil quality. Rather, a sufficient number of measurements must be taken and at specific locations to representatively characterize the spatial distribution of the existing soil conditions that influence the soil's intended use. Therefore, assessing soil quality requires quantitative knowledge of each indicator property associated with a soil's quality and the spatial variability of those indicator properties.

Furthermore, spatial variability has a profound influence on solute transport, particularly of non-point source pollutants. In fact, it has become clear that the real constraint on modeling solute transport is not the detail of the model structures, but defining the characteristics of individual places (Beven, 2002). Jury (1986) provides an excellent fundamental discussion of the spatial variability of soil properties and its impact on solute transport in the vadose zone. As Jury points out, "any hope of estimating a continuous spatial pattern of chemical emissions at each point in space within a field must be abandoned due to field-scale variability of soils."

The characterization of spatial variability is without question one of the most significant areas of concern in soil science because of its broad reaching influence on all field- and landscape-scale processes. The geospatial measurement of EC_a is a sensor technology that has played, and continues to play, a major role in addressing the issue of characterizing spatial variability. Geospatial measurements of EC_a have been successfully used for (1) identifying the soil physicochemical properties influencing crop yield patterns and soil condition, (2) establishing the spatial variation of these soil properties, and (3) characterizing the spatial distribution of soil properties influencing solute transport through the vadose zone (Corwin et al., 1999, 2003a, 2003b, 2005; Kaffka et al., 2005).

18.1.2 EDAPHIC FACTORS INFLUENCING EC_a MEASUREMENTS

The earliest field applications of geophysical measurements of EC_a in soil science involved the determination of salinity within the soil profile of arid zone soils (Halvorson and Rhoades, 1976; Rhoades and Halvorson, 1977; de Jong et al., 1979; Cameron et al., 1981; Rhoades and Corwin, 1981; Corwin and Rhoades, 1982, 1984; Williams and Baker, 1982). However, it became apparent that the measurement of EC_a in the field to infer soil salinity was more complicated than initially anticipated due to the complexity of current flow pathways arising from the spatial heterogeneity of properties influencing current flow in soil.

Three pathways of current flow contribute to the EC_a of soil: (1) a liquid phase pathway via dissolved solids contained in the soil water occupying the large pores, (2) a solid-liquid phase pathway primarily via exchangeable cations associated with clay minerals, and (3) a solid pathway via soil particles that are in direct and continuous contact with one another (Rhoades et al., 1999a). Of these three pathways, the solid pathway in soil is usually negligible resulting in a dual parallel pathway system.

Rhoades et al. (1989) formulated an electrical conductance model that describes the three conductance pathways of EC_a. This model is often referred to as the dual pathway parallel conductance model:

$$EC_a = \left[\frac{(\theta_s + \theta_{ws})^2 \cdot EC_{ws} \cdot EC_s}{\theta_s \cdot EC_{ws} + \theta_{ws} \cdot EC_s} \right] + (\theta_{wc} \cdot EC_{wc}) \qquad (18.1$$

where θ_{ws} and θ_{wc} are the volumetric soil water contents in the soil-water pathway (cm^3 cm^{-3}) and in the continuous liquid pathway (cm cm^{-3}), respectively; θ_s is the volumetric content of the solid phase of soil (cm cm^{-3}); EC_{ws} and EC_{wc} are the specific electrical conductivities of the soil-water pathway (dS m^{-1}) and continuous-liquid pathway (dS m^{-1}); and EC is the electrical conductivity of the solid soil particles (dS m^{-1}). Equation (18.1) was reformulated by Rhoades et al. (1989) into Eq. (18.2):

$$EC_a = \left[\frac{(\theta_s + \theta_{ws})^2 \cdot EC_w \cdot EC_s}{(\theta_s \cdot EC_w) + (\theta_{ws} \cdot EC_s)} \right] + (\theta_w - \theta_{ws}) \cdot EC_w \qquad (18.2)$$

where $\theta_w = \theta_{ws} + \theta_{wc}$ = total volumetric water content (cm^3 cm^{-3}), and EC_w is the average electrical conductivity of the soil water assuming equilibrium (i.e., $EC_w = EC_{sw} = EC_{wc}$). The following simplifying approximations are also known:

$$\theta_w = \frac{(PW \cdot \rho_b)}{100} \qquad (18.3)$$

$$\theta_{ws} = 0.639\theta_w + 0.011 \qquad (18.4)$$

$$\theta_{ss} = \frac{\rho_b}{2.65} \qquad (18.5)$$

$$EC_{ss} = 0.019(SP) - 0.434 \qquad (18.6)$$

$$EC_w = \left[\frac{EC_e \cdot \rho_b \cdot SP}{100 \cdot \theta_w} \right] \qquad (18.7)$$

where PW is the per cent water on a gravimetric basis, ρ_b is the bulk density (Mg m^{-3}), SP is the saturation percentage, and EC_e is the electrical conductivity of the saturation extract (dS m^{-1}).

The reliability of Eqs. (18.2)–(18.7) has been evaluated by Corwin and Lesch (2003). These equations are reliable except under extremely dry soil conditions. However, Lesch and Corwin (2003) developed a means of extending equations for extremely dry soil conditions by dynamically adjusting the assumed water content function. By measuring EC_a, SP PW, and ρ_b, and using Eqs. (18.3)–(18.7), the EC_e can be estimated. The determination of EC_e is of agricultural importance because traditionally EC_e has been the standard measure of soil salinity used in all salt-tolerance plant studies. Alternatively, EC_a can be estimated by knowing EC_e SP, PW, and ρ_b.

Because of the pathways of conductance, EC_a is influenced by a complex interaction of soil properties including salinity, SP, water content, and ρ_b The SP and ρ_b are both directly influenced by clay content (or texture) and organic matter (OM). Furthermore, the exchange surfaces on clays and OM provide a solid-liquid phase pathway primarily via exchangeable cations; consequently, clay type and content (or texture), cation exchange capacity (CEC), and OM are recognized as additional factors influencing EC_a

measurements. Measurements of EC_a must be interpreted with these influencing factors in mind. Table 18.1 from Corwin and Lesch (2005a) is a compilation of work related to the influence of various edaphic properties on the EC_a measurement.

Another factor influencing EC_a is temperature. Electrolytic conductivity increases at a rate of approximately 1.9% per degree centigrade increase in temperature. Customarily, EC is expressed at a reference temperature of 25°C for purposes of comparison. The EC (i.e., EC_a, EC_e, or EC_w) measured at a particular temperature t (in degrees centigrade), EC_t, can be adjusted to a reference EC at 25°C, EC_{25}, using equations from Handbook 60 (U.S. Salinity Laboratory, 1954):

$$EC_{25} = f_t = EC_t \qquad (18.8$$

where f_t is a temperature conversion factor. Approximations for the temperature conversion factor are available in polynomial form (Stogryn, 1971; Rhoades et al., 1999b; Wraith and Or, 1999) or other equations such as Eq. (18.9) by Sheets and Hendrickx (1995):

$$f_t = 0.4470 + 1.4034e^{-t/26.815} \qquad (18.9$$

18.1.3 MOBILE EC_a MEASUREMENT EQUIPMENT

The characterization of soil spatial variability using EC_a involves the use of mobile electrical resistivity (ER) or electromagnetic induction (EMI) equipment that geo-references each EC_a measurement using a global positioning system (GPS). Mobile EC_a equipment has been developed by a variety of researchers (McNeill, 1992; Carter et al., 1993; Rhoades, 1993; Jaynes et al., 1993; Cannon et al., 1994; Kitchen et al., 1996; Freeland et al., 2002). The development of mobile EC_a measurement equipment has made it possible to produce EC_a maps with measurements taken every few meters.

Mobile EC_a measurement equipment has been developed for both ER and EMI geophysical approaches. In the case of ER, four stainless-steel electrodes are inserted into the soil generally at equal distances and connected to a resistivity meter. Current is applied to the two outer electrodes with the two inner electrodes serving as the potential electrodes. By mounting the electrodes to "fix" their spacing, considerable time for a measurement is saved. The "fixed-electrode array" has been mounted on a vehicle and coupled to a datalogger and GPS, which geo-references the EC_a measurement (Rhoades, 1992, 1993; Carter et al., 1993). Veris Technologies* (Salinas, KS; www. veristech.com) has developed a commercial mobile system for measuring EC_a using the principles of ER. In the case of EMI, an EM-38 unit*

---

*Product identification is provided solely for the benefit of the reader and does not imply the endorsement of the USDA.

TABLE 18.1
Compilation of Literature Measuring EC_a Categorized According to the Physicochemical and Soil-Related Properties Either Directly or Indirectly Measured by EC_a.

Soil property	Ref.
Directly Measured Soil Properties	
Salinity (and nutrients, e.g., NO_3^-)	Halvorson and Rhoades (1976); Rhoades et al. (1976); Rhoades and Halvorson (1977); de Jong et al. (1979); Cameron et al. (1981); Rhoades and Corwin (1981, 1990); Corwin and Rhoades (1982, 1984); Williams and Baker (1982); Greenhouse and Slaine (1983); van der Lelij (1983); Wollenhaupt et al. (1986); Williams and Hoey (1987); Corwin and Rhoades (1990); Rhoades et al. (1989, 1990, 1999a, 1999b); Slavich and Petterson (1990); Diaz and Herrero (1992); Hendrickx et al. (1992); Lesch et al. (1992, 1995a, 1995b,1998); Rhoades (1992, 1993); Cannon et al. (1994); Nettleton et al. (1994); Bennett and George (1995); Drommerhausen et al. (1995); Ranjan et al. (1995); Hanson and Kaita (1997); Johnston et al. (1997); Mankin et al. (1997); Eigenberg et al. (1998, 2002); Eigenberg and Nienaber (1998, 1999, 2001); Mankin and Karthikeyan (2002); Herrero et al. (2003); Paine (2003) ; Kaffka et al. (2005)
Water content	Fitterman and Stewart (1986); Kean et al. (1987); Kachanoski et al. (1988, 1990); Vaughan et al. (1995); Sheets and Hendrickx (1995); Hanson and Kaita (1997); Khakural et al. (1998); Morgan et al. (2000); Freeland et al. (2001); Brevik and Fenton (2002) Wilson et al. (2002); Farahani et al. (2005); Kaffka et al. (2005)
Texture-related (e.g., sand, clay, depth to claypans or sand layers)	Williams and Hoey (1987); Brus et al. (1992); Jaynes et al. (1993); Stroh et al. (1993); Sudduth and Kitchen (1993); Doolitle et al. (1994, 2002); Kitchen et al. (1996); Banton et al. (1997); Boettinger et al. (1997); Rhoades et al. (1999b); Scanlon et al. (1999); Inman et al. (2001); Triantafilis et al. (2001); Anderson- Cook et al. (2002); Brevik and Fenton (2002); Farahani et al. (2005)
Bulk density related (e.g., compaction)	Rhoades et al. (1999b); Gorucu et al. (2001)
Indirectly Measured Soil Properties	
Organic matter related (including soil organic carbon, and organic chemical plumes)	Greenhouse and Slaine (1983, 1986); Brune and Doolittle, 1990; Nyquist and Blair (1991); Jaynes (1996); Benson et al. (1997); Bowling et al. (1997); Brune et al. (1999); Nobes et al. (2000); Farahani et al. (2005)
Cation exchange capacity	McBride et al. (1990); Triantafilis et al. (2002); Farahani et al. (2005)
Leaching	Slavich and Yang (1990); Corwin et al. (1999); Rhoades et al. (1999b)
Groundwater recharge	Cook and Kilty (1992); Cook et al. (1992); Salama et al. (1994)
Herbicide partition coefficients	Jaynes et al. (1995)
Soil map unit boundaries	Fenton and Lauterbach (1999); Stroh et al. (2001)
Corn rootworm distributions	Ellsbury et al. (1999)
Soil drainage classes	Kravchenko et al. (2002)

(Geonics Ltd., Mississaugua, Ontario, Canada) has been mounted in a cylindrical nonmetallic housing in the front of a mobile spray rig that has adequate clearance to traverse fields with a crop cover (Rhoades, 1992, 1993; Carter et al., 1993). The housing can be raised and lowered to take measurements at the soil surface or at various heights above the soil or to lock into a travel position to go from one measurement site to the next. The housing can also be rotated 90° to take EMI readings at each site with the transmitter and receiver coils oriented to the soil surface in two configurations EM_h, electromagnetic induction measurement in the horizontal coil-mode configuration; EM_v, electromagnetic induction measurement in the vertical coil-mode configuration). Recently, mobile EMI equipment developed at the Salinity Laboratory was modified by the addition of a dual-dipole EM-38 unit in place of the single EM-38 unit (Corwin and Lesch, 2005a). The dual-dipole EM-38 unit permits continuous, simultaneous EC_a measurements in both the horizontal (EM_h) and vertical (EM_v) dipole configurations at time intervals of just a few seconds between readings. Other less costly mobile EMI equipment has been developed that carry the EM-38 unit on a nonmetallic cart or sled pulled by an all-terrain vehicle or tractor (Jaynes et al., 1993; Cannon et al., 1994; Kitchen et al., 1996; Freeland et al., 2002). These sleds or carts allow continuous EC_a measurements, but in only one dipole position. No commercial mobile system has been developed with EMI. The mobile "fixed-electrode array" ER and EMI equipment are both well suited for collecting detailed maps of the spatial variability at field scales and larger.

18.2 GUIDELINES FOR CONDUCTING AN EC_a-DIRECTED SOIL SAMPLING SURVEY

Because of the influence of edaphic properties on EC_a, the spatial distribution of EC_a within a field provides a potential means of mapping the spatial variability of the edaphic properties with an EC_a-directed soil sampling. Characterizing spatial variability with EC_a-directed soil sampling is based on the hypothesis that when EC_a correlates with a soil property or properties, then spatial EC_a information can be used to identify sites that reflect the range and variability of the property or properties.

In instances where EC_a correlates with a particular soil property, an EC_a-directed soil sampling approach will establish the spatial distribution of that property with an optimum number of site locations to characterize the variability and keep labor costs minimal (Corwin et al., 2003a). Also, if EC is correlated with crop yield, then an EC_a-directed soil sampling approach can be used to identify what soil properties are causing the variability in crop yield (Corwin et al., 2003b). Details for conducting a field-scale EC survey for the purpose of characterizing the spatial variability of soil properties influencing soil quality or crop yield variation can be found in Corwin and Lesch (2005b). General guidelines appear in Corwin and Lesch (2003) and Corwin et al. (2003a, 2003b).

The basic elements of a field-scale EC_a survey for characterizing spatial variability include (1) EC_a survey design, (2) geo-referenced EC_a data collection, (3) soil sample design based on geo-referenced EC_a data, (4) soil sample collection, (5) physico-chemical analysis of pertinent soil properties, (6) spatial statistical analysis, (7) determination of the dominant soil properties influencing the EC_a measurements at the study site, and (8) GIS development. The basic steps of an EC_a-directed soil sampling survey are provided in Table 18.2.

18.3 STRENGTHS AND LIMITATIONS

At present, no other single soil measurement provides a greater level of spatial information than that of geospatial measurements of EC_a. Even so, there are a variety of strengths and limitations for the use of EC_a to characterize soil spatial variability. Awareness of these strengths and weaknesses is crucial for the proper use of EC_a in characterizing spatial variability.

Apparent soil electrical conductivity is a fast, reliable, and scientifically documented approach that is reasonable in cost and can be readily mobilized for geospatial referencing when mounted on a mobile platform and coupled to a GPS. Because of these positive attributes, commercial vendors have developed mobile units using ER to measure EC_a and numerous researchers have developed EMI-based mobile units (see Sec. 18.1.3). However, ER requires direct contact between the inserted probe and the soil, which limits its use in dry or stony soils. In contrast, EMI-based units are noninvasive; consequently, contact is not an issue, so EC_a measurements of dry or stony soils can be made without difficulty. The fact that the EC_a measurement is influenced by a variety of soil physicochemical properties is both an advantage and a disadvantage to characterizing spatial variability. The advantage is that the spatial variability of each of the soil properties influencing EC_a can be potentially characterized. The disadvantage is that the relationship between EC_a and the properties influencing EC_a is complex and requires ground truth soil samples to unravel. Nevertheless, when used correctly there is no means of characterizing spatial variability that is more dependable, cost-effective, and flexible than geospatial measurements of EC_a.

Even though an EC_a survey is a quick, easy, reliable, and cost-effective means of characterizing the spatial variability of a variety of physicochemical properties, there are crucial limitations and weaknesses. A knowledge and understanding of these limitations and weaknesses is imperative for the proper use of EC_a measurements. The complex spatial heterogeneity of the soil system has subtle influences on geospatial EC_a measurements that can have significant interpretive impacts. The ability to recognize and interpret these influences can be the difference between the successful or failed application of EC_a measurements for characterizing spatial variability.

TABLE 18.2
Outline of Steps to Conduct an EC_a Field Survey.

1. Site description and EC_a survey design
   a. record site metadata
   b. define the project's/survey's objective
   c. establish site boundaries
   d. select GPS coordinate system
   e. establish EC_a measurement intensity
2. EC_a data collection with mobile GPS-based equipment
   a. geo-reference site boundaries and significant physical geographic features with GPS
   b. measure geo-referenced EC_a data at the pre-determined spatial intensity and record associated metadata
3. Soil sample design based on geo-referenced EC_a data
   a. statistically analyze EC_a data using an appropriate statistical sampling design to establish the soil sample site locations
   b. establish site locations, depth of sampling, sample depth increments, and number of cores per site
4. Soil core sampling at specified sites designated by the sample design
   a. obtain measurements of soil temperature through the profile at selected sites
   b. at randomly selected locations obtain duplicate soil cores within a 1 m distance of one another to establish local-scale variation of soil properties
   c. record soil core observations (e.g., mottling, horizonation, textural discontinuities, etc.)
5. Laboratory analysis of appropriate soil physicochemical properties defined by project objectives
6. If needed, stochastic and/or deterministic calibration of EC_a to EC_e or to other soil properties (e.g., water content and texture)
7. Spatial statistical analysis to determine the soil properties influencing EC_a and/or crop yield
   a. soil quality assessment:
      (1) perform a basic statistical analysis of physicochemical data by depth increment and by composite depth over the depth of measurement of EC_a
      (2) determine the correlation between EC_a and physicochemical soil properties by composite depth over the depth of measurement of EC_a
   b. precision agriculture applications (if EC_a correlates with crop yield, then):
      (1) perform a basic statistical analysis of physicochemical data by depth increment and by composite depths
      (2) determine the correlation between EC_a and physicochemical soil properties by depth increment and by composite depths
      (3) determine the correlation between crop yield and physicochemical soil properties by depth and by composite depths to determine depth of concern (i.e., depth with consistently highest correlation, whether positive or negative, of soil properties to yield) and the significant soil properties influencing crop yield (or crop quality)
      (4) conduct an exploratory graphical analysis to determine the relationship between the significant physicochemical properties and crop yield (or crop quality)
      (5) formulate a spatial linear regression (SLR) model that relates soil properties (independent variables) to crop yield or crop quality (dependent variable)
      (6) adjust this model for spatial auto-correlation, if necessary, using restricted maximum likelihood or some other technique
      (7) Conduct a sensitivity analysis to establish dominant soil property influencing yield or quality
8. GIS database development and graphic display of spatial distribution of soil properties

Source: Corwin and Lesch, 2005b.

First and foremost, geospatial measurements of EC_a by themselves do not directly characterize spatial variability. Actually, EC_a measurements provide limited direct information about the physicochemical properties that influence yield, affect solute transport, or determine soil quality. Rather, EC_a-survey measurements provide the spatial information necessary to direct soil sampling. It is as a cost-effective tool for directing soil sampling that EC_a-survey measurements are invaluable for characterizing spatial variability. The primary strength of geospatial EC_a measurements lies in their effectiveness as a means to direct soil sampling with a minimum number of sample sites that best characterize the spatial variability of those soil properties influencing EC_a at the site of interest.

Second, EC_a-directed soil sampling can only spatially characterize soil properties that correlate with EC_a. This correlation may be due to a direct or indirect influence on the EC_a measurement or the correlation may be a complete artifact. For example, salinity and water content will directly influence EC_a, and CEC will indirectly influence EC_a through its influence on current flow at the surface of clay particles. In many instances B and salinity distributions are similar; consequently, a correlation of B with EC_a can result. Yet, there is no cause-and-effect relationship between B and EC_a. Consequently, an understanding of the soil properties that influence EC_a and of those properties that are correlated with but may not influence EC_a at a specific site is particularly essential for temporal applications of EC_a because over time the correlation may or may not persist.

Third, as already mentioned there is a complex relationship between EC_a and those properties that influence EC_a. Apparent soil electrical conductivity is a complex measurement that requires knowledge and experience to interpret. Ground-truth soil samples are obligatory to be able to understand and interpret spatial measurements of EC_a. Without ground-truth soil samples an EC_a survey will be of minimal value. Geospatial measurements of EC_a do not supplant the need for soil sampling, but they do minimize the number of samples necessary to characterize spatial variability. Users of EC_a survey data must exercise caution and be aware of what EC_a is actually measuring at the site of interest. The only way to establish those soil properties that influence EC_a at a site is to take ground truth soil samples and establish the relationship between EC_a and the property(ies) of interest. This requires that every EC_a survey have an associated soil sampling survey based on the spatial distribution of EC_a. This generally requires a minimum of 8–16 soil core sites, where the location and number of sites are dependent on the spatial variability of EC_a. The location and number of sites is established from an intensive EC_a survey using model-based sample design software such as ESAP (Lesch et al., 2000).

Finally, the temporal stability of EC_a measurements at a site may be of potential concern due to the fact that EC_a is a product of both static and dynamic factors (these static and dynamic factors are discussed in greater detail in the subsequent section). This adds another dimension to the complexity of understanding and interpreting geospatial EC_a measurements. For this reason, greater caution must be taken to characterize spatial variability with EC_a when dynamic factors influencing EC_a are more significant than static factors. Apparent soil electrical conductivity surveys are generally conducted (1) within a set time frame to minimize the effects of dynamic properties (e.g., temperature, water content, and salinity), (2) when the soil is at or near field capacity, and (3) with regard for subtle topographic effects (e.g., bed-furrow). Protocols for conducting an EC_a survey that consider all the previously discussed limitations are presented by Corwin and Lesch (2005b, 2005c).

18.4 CHARACTERIZING SPATIAL VARIABILITY WITH EC_a-DIRECTED SOIL SAMPLING: CASE STUDIES

Measured EC_a is the product of both static and dynamic factors, which include soil salinity, clay content and mineralogy, water content, ρ_b, and temperature. Johnson et al. (2003) astutely described the observed dynamics of the general interaction of these factors. In general, the magnitude and spatial heterogeneity of EC_a in a field are dominated by one or two of these factors, which will vary from one field to the next making the interpretation of EC_a measurements highly site specific. In instances where dynamic soil properties (e.g., salinity, water content, and temperature) dominate the EC measurement, temporal changes in spatial patterns exhibit more fluidity than systems that are dominated by static factors (e.g., texture). In texture-driven systems, spatial patterns remain consistent because variations in dynamic soil properties affect only the magnitude of measured EC_a (Johnson et al., 2003). For this reason, Johnson et al. (2003) warn that EC_a maps of static-driven systems convey very different information from those of less stable dynamic-driven systems. Furthermore, the application of manure and commercial fertilizer can influence EC_a to the point where texture-dominated systems can be transformed into salt-dominated systems (Johnson et al., 2003). Although it has not been experimentally evaluated, texture-driven systems will likely be more temporally stable than salinity-driven systems.

Numerous EC_a field studies have been conducted that have revealed the site specificity and complexity of spatial EC_a measurements with respect to the particular property influencing the EC_a measurement at that study site. Table 18.1 is a compilation of various field studies and the associated dominant soil property measured. The range of factors correlated to field measurements of EC_a in Table 18.1 points to the need for ground truth soil samples associated with each EC_a survey to adequately interpret spatial EC_a data.

After the initial, largely observational work compiled in Table 18.1 involving geophysical measurements of EC_a in soil, the direction of research has gradually shifted to mapping within-field variation of EC_a to characterize the spatial distribution and variability of properties that statistically correlate with EC_a. The mapping of within-field variation of EC

to characterize the spatial distribution of properties has its roots in the early salinity mapping work by Rhoades (1992, 1993), who observed the geospatial relationship between maps of EC_a and soil salinity patterns. The earliest work in the soil science literature for the application of geospatial EC_a measurements to direct soil sampling for the purpose of characterizing the spatial variability of a soil property was by Lesch et al. (1992), who used a spatial response surface sampling (SRSS) design. The shift in the emphasis of field-related EC_a research from observation to directed-sampling design has gained momentum resulting in the accepted use of geospatial measurements of EC_a as a reliable directed-sampling tool for characterizing spatial variability at field and landscape scales (Corwin and Lesch, 2003, 2005a, 2005b).

Currently, two EC_a-directed soil sampling design approaches are used: (1) design-based sampling and (2) model-based sampling. The former consists of the use of unsupervised classification (Johnson et al., 2001), whereas the latter typically relies on optimized spatial response surface (SRS) sampling designs (Corwin and Lesch, 2005b). Throughout the statistical literature model-based designs are less common, although some statistical research has been performed in this area (Valliant et al., 2000). Nathan (1988) and Valliant et al. (2000) discuss the merits of design (probability) and model (prediction) based sampling strategies in detail. Specific model-based sampling approaches having direct application to agricultural and environmental survey work are described by McBratney and Webster (1981), Lesch et al. (1995a, 1995b) Van Groenigen et al. (1999), and Lesch (2005).

In the past the characterization of soil spatial variability using EC_a-directed soil sampling focused on three different landscape-scale applications: (1) solute transport modeling in the vadose zone (Corwin et al., 1999), (2) soil quality assessment (Johnson et al., 2001; Corwin et al., 2003a), and (3) precision agriculture (Corwin et al., 2003b). All three studies by Corwin et al. (1999, 2003a, 2003b) were conducted on irrigated, arid-zone, agricultural land located in California's Central Valley (Figure 18.1). Two of the three studies were conducted within the Broadview Water District west of Fresno, CA: (1) a landscape-scale study of salt loading through the vadose zone to tile drains from 1991 to 1996 (Corwin et al., 1999) and (2) a precision agriculture study to identify edaphic properties influencing cotton yield on an irrigated field in 1999 (Corwin et al., 2003a). The third study was a soil quality assessment study on arid zone soil conducted by Corwin et al. (2003b) at a study site located on Westlake Farm near Stratford, CA, as part of a project to assess the sustainability of drainage water reuse to mitigate drainage volumes in the San Joaquin Valley (SJV). The study by Johnson et al. (2001) was conducted on an experimental site in Colorado.

Spatial variability for each of the three studies by Corwin and colleagues was characterized following protocols and guidelines for conducting an EC_a survey to direct soil sampling that were later outlined and published by Corwin and Lesch (2003, 2005b). In each study ESAP software developed by Lesch et al. (1995a, 1995b, 2000) was used to establish the locations

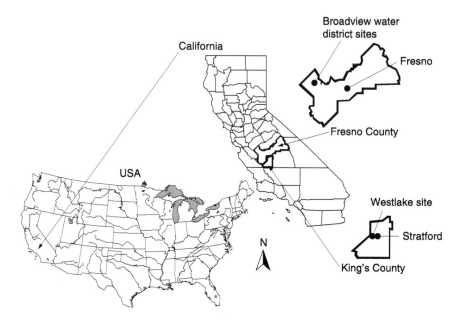

FIGURE 18.1 Map of Broadview Water District and Westlake Farm study sites located in California's San Joaquin Valley.

where soil cores were taken based on the geo-referenced EC_a data that was obtained from EC_a surveys. The ESAP software uses a spatial response surface sampling (SRSS) design, which is a model-based sampling approach. The SRSS design locates a minimum set of calibration soil sample sites based on the observed magnitudes and spatial locations of geo-referenced EC_a measurements, with the explicit goal of optimizing the estimation of a regression model by minimizing the mean-square prediction errors produced by the calibration function (Lesch et al., 2000). It is the intention of the SRSS design to characterize the variability in geo-referenced EC measurements with a minimum number of sites. These sites are the locations where soil core samples are taken and appropriate soil physicochemical properties are measured as determined by the intended application (e.g., solute transport properties, site-specific crop management properties, or soil quality properties). A detailed discussion of the SRSS design concept is found in Lesch (2005).

18.4.1 LANDSCAPE-SCALE SOLUTE TRANSPORT IN THE VADOSE ZONE

To date, the only landscape-scale study to use EC_a-directed soil sampling to characterize soil variability for use in the modeling of solute transport in the vadose zone is by Corwin et al. (1999). In a study modeling salt loading to tile drains on a 2396-ha study site in California's SJV, Corwin et al. (1999) used EC_a-directed soil sampling to define spatial domains of similar solute transport capacity in the vadose zone. These spatial domains, referred to as stream tubes, are volumes of soil that are assumed to be independent of adjacent stream tubes in the field with minimal lateral interaction (i.e., no solute exchange) so that a one-dimensional, vertical solute transport model can be applied to each stream tube without concern for lateral flow of water and transport of solute. The application of a one-dimensional solute transport model to each stream tube resulted in the prediction of salt loading for a 5-year study period.

An area of 37 contiguous quarter sections (i.e., 2396 ha) within the Broadview Water District was chosen as the experimental site to simulate salt loading to tile drains from May 1991 to May 1996 (Figure 18.2). The model that was used to simulate salt transport through the vadose zone was a one-dimensional, "tipping bucket," layer-equilibrium, functional model of solute transport (Corwin et al., 1991). The selection of a functional model, rather than a mechanistic model, to simulate salt transport through the vadose zone at landscape scale is based on organization hierarchy of spatial scales, which indicates that functional models are more appropriately applied at scales ranging from field to global (Corwin et al., 1997), whereas mechanistic models are best suited for molecular to pedon scales.

The single greatest challenge to modeling non-point source (NPS) pollutants, such as salinity, is to obtain sufficient data to characterize the temporal and spatial distribution of model parameter and variable inputs (Corwin et al., 1997). Therefore, a critical aspect of the study was to utilize a sampling strategy that would reflect the spatial heterogeneity of the physicochemical parameters and variables used in the functional solute

Broadview Water District Study Site

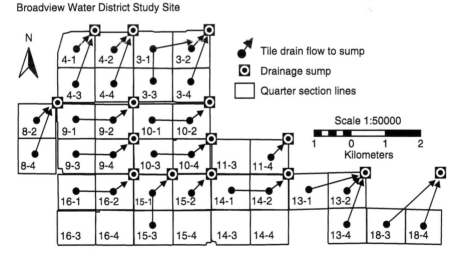

FIGURE 18.2 Broadview Water District study site showing quarter section lines, drainage sumps, and tile drain flow direction. (From Corwin et al., 1999.)

transport model. To meet this end, the statistical routine developed by Lesch et al. (1992, 1995a, 1995b) was utilized for electromagnetic induction (EMI) measurements of EC_a taken with a Geonics EM-38 to determine soil sample locations. This statistical routine selects sample sites that reflect the spatial heterogeneity exhibited for EC_a, the supposition being that the EMI measurements of EC_a are reflective of cumulative transport processes for salinity at a given location and can be used to identify spatial domains of similar transport properties for salt. Because EC_a in arid zone soils is primarily a result of salinity, but is also influenced by water content, texture, and bulk density, this supposition relies upon local-scale spatial variation in soil properties that can be characterized and upon uniform irrigation applications within a spatial domain defined as similar in its ability to transport salts through the vadose zone (i.e., stream tube).

Within the 37 quarter sections, EMI measurements (both EM_h, electro-magnetic induction measurement in the horizontal coil-mode configuration, and EM_v, electromagnetic induction measurement in the vertical coil-mode configuration) were acquired in each quarter section on a centric, systematic 8×8 grid generating 64 survey locations per quarter section (i.e., 2368 total locations for all 2396 hectares). Figure 18.3a and 18.3b show the locations of the EMI measurements for both EM_h and EM measurements, respectively. Soil spatial variability and spatial domains of solute transport were based upon the geometric mean and profile ratios calculated from the EMI measurements of EC_a. The EMI geometric mean was defined as sqrt(EM_h*EM_v). The profile ratio was defined as EM_h/EM_v.

(a) EMI survey of EM_h−1991

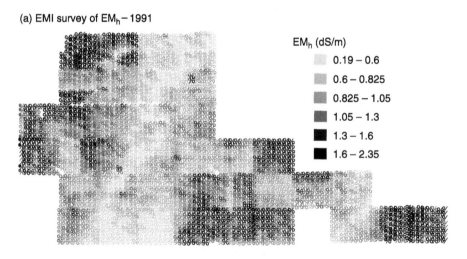

EM$_h$ (dS/m)

0.19 − 0.6
0.6 − 0.825
0.825 − 1.05
1.05 − 1.3
1.3 − 1.6
1.6 − 2.35

FIGURE 18.3 Maps of Broadview Water District showing the 1991 EC_a survey of the (a) EM_h measurements and (b) EM_v measurements, and the (c) spatial transport domains defined from the EC_a survey data. (From Corwin et al., 1999.)

(b) EMI survey of EM$_v$ – 1991

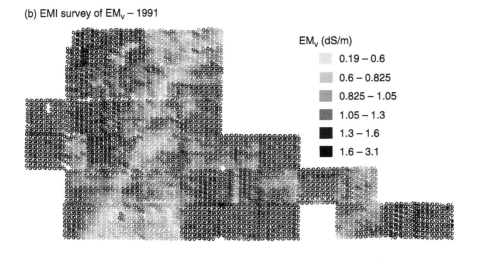

EM$_v$ (dS/m)

	0.19 – 0.6
	0.6 – 0.825
	0.825 – 1.05
	1.05 – 1.3
	1.3 – 1.6
	1.6 – 3.1

(c) Spatial transport domains

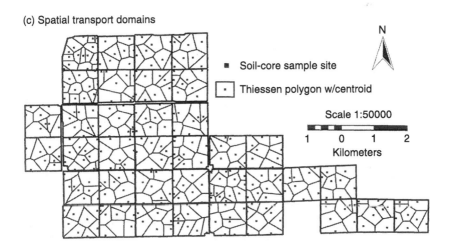

N

■ Soil-core sample site

Thiessen polygon w/centroid

Scale 1:50000

1 0 1 2
Kilometers

FIGURE 18.3 Continued

In the horizontal coil configuration, the response of the EM-38 is mainly to EC_a in the top 0.5 m of the soil profile with a volume of measurement occurring within the top 0.75 m. Sensitivity declines continuously with depth. In the vertical orientation, the response of the EM-38 is zero at the soil surface, peaks at roughly 0.4 m, then declines with increasing depth. The depth of penetration of measurement in the vertical orientation is roughly 1.5 m. Therefore, the EM_h represents a shallow measurement of EC_a while EM_v represents a deep measurement of EC_a within the root zone. Due to the response of the EM-38 in different coil orientations, the profile ratio provides an indication of the EC_a profile. Profile ratios of 1 indicate a uniform profile, profile ratios of < 1 indicate an increasing profile with depth, and profile ratios of > 1 indicate an inverted profile (i.e., conductivity decreases with depth). It is assumed that the profile ratio is analogous to the leaching fraction, which is calculated by the electrical conductivity (EC) of the soil water passing below the root zone divided by the EC of the irrigation water. As such, the profile ratio is hypothetically reflective of the hydraulic properties of the soil at that particular location. In contrast, the EMI geometric mean reflects the cumulative salinity level within the root zone.

To minimize soil sampling requirements to a realistic number of locations that could be handled with limited manpower resources, soil cores at 0.3 m increments to a depth of 1.2 m were taken at between 8 to 12 of the 64 locations within each quarter section. From the 2396 sites, a total of 315 locations was selected for soil-core sampling. The selection of the 315 soil sampling sites was based on the observed EMI field pattern utilizing the response-surface technique of Lesch et al. (1992, 1995a, 1995b). Figure 18.3c shows the location of the soil-core sample sites in relation to the quarter section boundary lines.

Thiessen polygons for each quarter section were created from the soil-core sample sites, with each soil-core site serving as the centroid of each resulting Thiessen polygon (Figure 18.3c). Each Thiessen polygon in Figure 18.3c was assumed to define a stream tube or spatial domain of solute transport properties where the variability of the properties is least (Bouma, 1990; Mayer et al., 1999), and solute transport occurs only in the vertical direction with no influence from adjacent stream tubes (Jury and Roth, 1990). In essence, the first four sample locations of each quarter section were selected so that one location satisfied each of the following four criteria: (1) high EMI geometric mean and high profile ratio, (2) high EMI geometric mean and low profile ratio, (3) low EMI geometric mean and high profile ratio, and (4) low EMI geometric mean and low profile ratio. The next four sample locations were chosen randomly within the quarter section. High and low values for the means and ratios were relative, i.e., the highs and lows were identified in each field or quarter section on a field-by-field basis.

To evaluate the validity of the stream-tube approach, measured and simulated salt loads were compared for a core area of 16 quarters where water mass balance showed no lateral flow influences. Table 18.3 shows the compared results for simulated and measured salt loads to 7 drainage sumps, which drained the 16 quarter sections as depicted in Figure 18.2. Figure 18.4 shows the spatial distribution of the salt loads after 5 years for the 16 quarter sections. Except for the drainage sump associated with quarter sections 9-3 and 9-4, the simulated salt loads show excellent agreement to within 30% of the measured salt loads. The significant difference between the measured and simulated salt loads for the 9-3/9-4 drainage sump was attributed to lateral water flows that occurred from

TABLE 18.3
Comparison of Measured and Simulated Salt Loading Amounts in the Broadview Water District, May 1991 to May 1996.

Quarter section(s) (Mg/ha)	Measured[a] (Mg/ha)	Simulated[b]
3-1, 3-2, 3-3, & 3-4	14.33	16.97
4-1 & 4-3	39.22	31.84
4-2 & 4-4	46.23	33.00
9-1 & 9-2	11.48	13.22
9-3 & 9-4	2.1	10.45
10-1 & 10-2	16.53	16.56
10-3 & 10-4	16.05	15.91

Measured at drainage sump.
Area-weighted average of 8–16 simulated Thiessen polygons within each quarter section.

Source: Corwin et al. (1999).

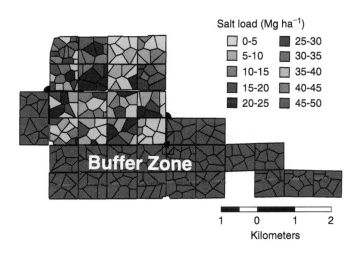

FIGURE 18.4 Map showing the spatial distribution of salt loading for the 16 quarter sections comprising the validation data set. (From Corwin et al., 1999.)

southwest to northeast from the designated buffer zone area in the south (see Figure 18.4). The general close agreement between measured and simulated results in Table 18.3 validates the delineation of stream tubes with EC_a-directed soil sampling survey data as a viable means of modeling NPS pollutant loads to tile drains or groundwater at field and landscape scales.

18.4.2 ASSESSING SOIL QUALITY AND SPATIO-TEMPORAL CHANGES IN SOIL QUALITY

The application of EC_a-directed soil sampling to characterize soil condition has been restricted to the Great Plains area and the southwestern United States. Using EC_a maps to direct soil sampling, Johnson et al. (2001) and Corwin et al. (2003a) spatially characterized the overall soil quality of physicochemical properties thought to affect yield potential.

To characterize soil quality, Johnson et al. (2001) used a stratified soil sampling design (i.e., unsupervised classification) with allocation into four geo-referenced EC_a ranges. Correlations were performed between EC and the minimum data set of physical, chemical, and biological soil attributes proposed by Doran and Parkin (1996). Their results showed a positive correlation of EC_a with percentage clay, ρ_b, pH, and $EC_{1:1}$ over a soil depth of 0–30 cm, and a negative correlation with soil moisture, total and particulate organic matter, total C and N, microbial biomass C, and microbial biomass N. Johnson et al. (2001) concluded that "EC_a classification effectively delimits distinct zones of soil condition, making it an excellent basis for soil sampling to reflect spatial variability."

Corwin et al. (2003a) characterized the soil quality of a saline-sodic soil using a SRSS design. A positive correlation (significant at the 0.01 level) was found between EC_a and the properties of volumetric water content; electrical conductivity of the saturation extract (EC_e); Cl^-, NO_3^- SO_4^-, Na^+, K^+, and Mg^{+2} in the saturation extract; SAR (sodium adsorption ratio), exchangeable sodium percentage (ESP); B; Se; and Mo. A negative correlation (significant at the 0.01 level) was found for $CaCO_3$, inorganic C, and organic C. Most of these properties are associated with soil quality for arid zone soils. The high positive and negative correlations indicated that the spatial variability of these soil properties were accurately characterized by the SRS sampling design and predictable from the EC_a survey data. However, a number of other soil properties (i.e., ρ_b; percentage clay; pH_e; SP; HCO_3^- and Ca^{+2} in the saturation extract; exchangeable Na^+, K^+, and Mg^{+2}; As; CEC; gypsum; and total N) did not correlate well with EC_a measurements. To accurately quantify these properties in this particular field, a complementary design-based sampling scheme is needed.

Neither Johnson et al. (2001) nor Corwin et al. (2003a) actually related the spatial variation in the measured soil physicochemical properties to crop yield variations. Nevertheless, both studies show the practicality and utility of using EC_a-directed soil sampling to spatially characterize soil quality for a variety of indicator properties.

Spatio-temporal variations in soil quality using EC_a-directed soil sampling have been studied by Lesch et al. (1998) and Corwin et al. (2005). In both instances, statistically significant temporal changes were identified. Corwin et al. (2005) showed that average salinity levels were reduced roughly 13% in the top 0.6 m. Sodium, B, and Mo levels were also found to be reduced from 1999 to 2002 in the top 0.6 m, with leaching the most probable cause.

In the study by Corwin et al. (2005), changes in spatial patterns were extremely complex and reflected a preferential pattern of change that was difficult to explain, except for potential microtopographic effects caused by leaching in the near surface (i.e., 0–0.3 m). Nevertheless, the ability to monitor temporal change at field scale was shown. In particular, it was clearly shown that leaching of salts had occurred for the top 0.6 m of soil. Figure 18.5 shows the initial spatial distribution of salinity in 1999, as measured by the electrical conductivity of the saturation extract (EC_e), at a saline-sodic study site where a field study was conducted to determine the sustainability of drainage water reuse on a salt-tolerant forage crop as a viable alternative to drainage water disposal on the west side of the San Joaquin Valley. Associated with the 1999 salinity levels in Figure 18.5 are the changes in salinity that have occurred up to 2002 over the 0–0.3 and 0.3–0.6 m depth increments where leaching occurred as a consequence of the application of drainage water that ranged from 5 to 10 dS m^{-1}. The preliminary 2002 results showed that salinity, sodium, B, and Mo had been leached in the top 0.6 m of soil, indicating a general improvement in soil quality (Corwin et al., 2005). Another EC_a-directed soil sampling survey was to be conducted at the end of the 5-year study.

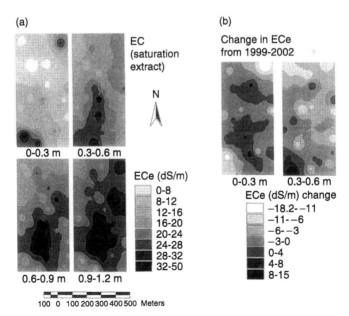

FIGURE 18.5 Maps of (a) EC_e for 1999 at depth increments of 0–0.3, 0.3–0.6, 0.6–0.9, and 0.9–1.2 m and (b) change in EC_e from 1999 to 2002 at depth increments of 0–0.3 and 0.3–0.6 m. (From Corwin et al., 2005.)

18.4.3   DELINEATING SITE-SPECIFIC MANAGEMENT UNITS FOR
PRECISON AGRICULTURE

Corwin et al. (2003b) carried the EC_a-directed soil sampling approach to
the next level by integrating crop yield to arrive at site-specific crop
management recommendations. Through spatial statistical analysis,
Corwin et al. (2003b) were able to identify those edaphic (i.e., salinity, water
content, and pH) and anthropogenic (i.e., leaching fraction) properties
(Figure 18.6) influencing the within-field spatial variation of cotton yield
(Figure 18.7a) on a 32.4-ha field in the Broadview Water District of central
California. A cotton yield response model was formulated that related
cotton yield (Y) to leaching fraction (LF), salinity (EC_e), gravimetric water
content (θ_g), and pH:

$$Y = 19.277 + 0.218(EC_e) - 0.015(EC_e)^2 -$$
$$- 4.420(LF)^2 - 1.997(\text{pH}) + 6.927(\theta_g) + \varepsilon \qquad (18.10$$

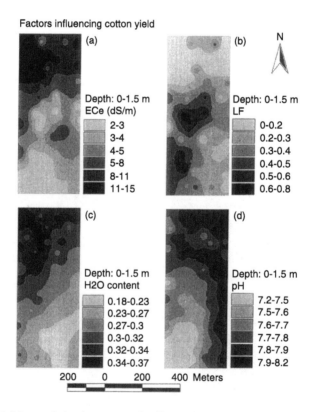

FIGURE 18.6 Maps of the four most significant factors (0–1.5 m) influencing cotton
yield: (a) EC_e (dS m^{-1}), (b) LF, (c) H$_2$O (kg kg^{-1}), and (d) pH. (From Corwin et al.,
2003b.)

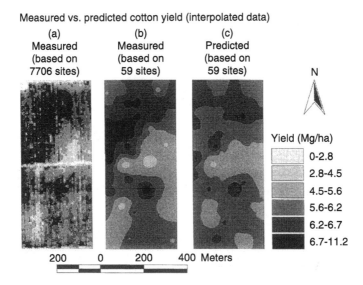

Measured vs. predicted cotton yield (interpolated data)

| (a) Measured (based on 7706 sites) | (b) Measured (based on 59 sites) | (c) Predicted (based on 59 sites) |

N

Yield (Mg/ha)

	0-2.8
	2.8-4.5
	4.5-5.6
	5.6-6.2
	6.2-6.7
	6.7-11.2

200   0   200   400 Meters

FIGURE 18.7 Comparison of (a) measured cotton yield based on 7706 yield measurements, (b) kriged data at 59 sites for measured cotton yield, and (c) kriged data at 59 sites for predicted cotton yields based on Eq. (18.10). (From Corwin et al., 2003b.)

A comparison of the measured and the simulated cotton yields at the locations where directed soil samples were taken showed close agreement, and Eq. (18.10) successfully described slightly more than 60% of the estimated spatial yield variation. A visual comparison of the measured and simulated spatial yield distributions of cotton shows a reasonably close spatial association between interpolated measured (Figure 18.7b) and predicted (Figure 18.7c) maps.

From Eq. (18.10) and scatter plots of cotton yield vs. properties, management recommendations were made that spatially prescribed what could be done to increase cotton yield at those locations with less than optimal yield. Subsequently, Corwin and Lesch (2005a) delineated site-specific management units (SSMUs), which are depicted in Figure 18.8. Highly leached zones were delineated where the LF needed to be reduced to ≤ 0.5; high-salinity areas were defined where the salinity needed to be reduced below the salinity threshold for cotton, which was established at $EC_e = 7.17$ dS m^{-1} for this field; areas of coarse texture were defined that needed more frequent irrigations; and areas were pinpointed where the pH needed to be lowered to 8 with a soil amendment such as OM. This work brought an added dimension because it delineated within-field units where associated site-specific management recommendations would optimize the yield, but it still falls short of integrating meteorological, economic, and environmental impacts on within-field crop-yield variation. Furthermore, these SSMUs have not been tested to evaluate whether their use would increase yield.

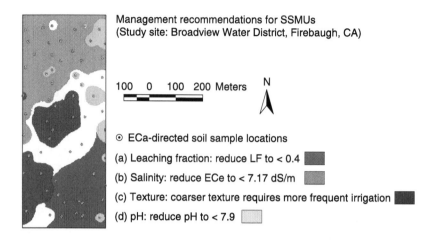

Management recommendations for SSMUs
(Study site: Broadview Water District, Firebaugh, CA)

100   0   100   200 Meters   N

⊙ ECa-directed soil sample locations
(a) Leaching fraction: reduce LF to < 0.4
(b) Salinity: reduce ECe to < 7.17 dS/m
(c) Texture: coarser texture requires more frequent irrigation
(d) pH: reduce pH to < 7.9

FIGURE 18.8 Site-specific management units for a 32.4-ha cotton field in the Broadview Water District of central California's San Joaquin Valley. Recommendations are associated with the SSMUs for (a) leaching fraction, (b) salinity, (c) texture, and (d) pH. (From Corwin and Lesch, 2005a.)

18.5   FUTURE DIRECTIONS

Because of the heterogeneous nature of soils, characterization of soil spatial variability is a fundamental component of any landscape-scale process that cannot be overlooked or superficially addressed. Significant technological advances have occurred over the past two decades, particularly in the area of sensor technology and in precisely locating geographic position with GPS. These advances have helped the progress of research in characterizing spatial variability at field and landscape scales.

Geospatial measurements of EC_a are among the ground-based sensor technologies contributing to an improved ability to characterize spatial variability. Numerous soil samples are required for representative estimates of field-scale spatial variability using traditional grid sampling, making grid sampling impractical due to labor and cost intensiveness. Soil sampling directed by geospatial EC_a measurements provides a viable alternative for characterizing spatial variability of a variety of soil-related physicochemical properties. Geospatial EC_a measurements provide a means of significantly reducing the number of soil samples needed to characterize spatial variability, provided that the target soil properties are well correlated with the conductivity survey data. The reliability of EC_a directed soil sampling for characterizing spatial variability has been shown for applications in a variety of areas including (1) solute transport in the vadose zone (Corwin et al., 1999), (2) precision agriculture (Corwin et al., 2003b), and (3) soil quality assessment (Johnson et al., 2001; Corwin et al., 2003a, 2005).

Even though geospatial measurements of EC_a provide one of the most cost-effective means of characterizing spatial variability, associated ground truth soil sampling must accompany EC_a surveys because of the complexity of the EC_a measurement and the need for ground truth samples to interpret EC_a measurements. Without associated soil samples, the interpretation of EC_a measurements is questionable and is not advised. Laboratory analyses of associated ground-truth soil samples impose the greatest cost and labor for characterizing spatial variability with geospatial EC_a measurements. Model-based sampling designs are the most efficient at reducing the number of soil sample sites to a minimum without compromising the characterization of variability, but only when a property correlates with EC_a For those instances where a property does not correlate with EC_a a random or stratified random sampling design should be used to spatially characterize the property to minimize error resulting from sampling design bias.

When geospatial measurements of EC_a are spatially correlated with georeferenced yield data, their combined use provides an excellent tool for identifying edaphic factors that influence crop yield, which can, in turn, be used to delineate site-specific management units (Corwin et al., 2003; Corwin and Lesch, 2005a). The delineation of productivity zones from geospatial measurements of EC_a provides another approach to site-specific management (Kitchen et al., 2005; Jaynes et al., 2005). Even so, an understanding of the soil-related factors influencing yield or the identification of productivity zones does not provide the whole picture for site-specific crop management because yield is influenced by a complex interaction of topographical (elevation, aspect, etc.), meteorological (humidity, temperature etc.), biological (e.g., pests), anthropogenic (management-related), and edaphic (soil-related) factors. Moreover, the precise manner in which these factors influence the dynamic process of plant growth and reproduction is not always well understood. To be able to manage within-field variation in yield, it is necessary to have an understanding within a spatial context of the relationship of all dominant factors causing the variation.

Directed soil sampling with geospatial EC_a measurements has its limitations in characterizing spatial variability because many soil properties are not measured, directly or indirectly, by EC_a; therefore, the biased sampling of a model-based sampling design approach will not be representative. Even a design-based approach (e.g., stratified random sampling) directed by EC_a may not be sufficient to characterize those properties not correlated with EC_a. In these instances, additional spatial information is needed to fill the gaps necessary to spatially characterize the variability of those properties that are not directly or indirectly measured with EC_a. The information from additional sensors is needed to either directly measure or to direct soil sampling of those soil properties not correlated with EC_a.

The integrated use of multiple remote and ground-based sensors is the future direction that research will likely take to obtain the extensive spatial data needed to characterize spatial variability. Integration of multi- and hyper-spectral imagery, TDR, GPR, aerial photography, and EC_a sensors is needed to provide the redundant and supplemental data necessary to unravel the spatial complexity of soil. Network-centric multisensor systems will likely provide the broad spectrum of overlapping and supplementary information needed to spatially characterize the compositional and structural complexity of soil.

ACKNOWLEDGMENTS

The author wishes to acknowledge the efforts and contributions of all those colleagues who collaborated with him on past projects involving the characterization of spatial variability with geospatial EC_a measurements, including Scott Lesch, Stephen Kaffka, Jim Rhoades, Jim Oster, and Pete Shouse. In particular, the author acknowledges the statistical support and collaboration provided by Scott Lesch, who has been a cornerstone to the success of this research. The author also appreciates the numerous hours of diligent technical work performed in the field and in the laboratory by several technicians whose efforts and conscientiousness were crucial to the success of the projects, including Clay Wilkinson, Nahid Vishteh, Harry Forster, Jack Jobes, JoAn Fargerlund, Derrick Lai, and Lena Ting.

REFERENCES

Anderson-Cook, C.M., Alley, M.M., Roygard, J.K.F., Khosia, R., Noble, R.B., and Doolittle, J.A., Differentiating soil types using electromagnetic conductivity and crop yield maps, *Soil Sci. Soc. Am. J.*, 66, 1562–1570, 2002.

Banton, O., Seguin, M.K., and Cimon, M.A., Mapping field-scale physical properties of soil with electrical resistivity, *Soil Sci. Soc. Am. J.*, 61(4), 1010–1017, 1997.

Bennett, D.L. and George, R.J., Using the EM38 to measure the effect of soil salinity on *Eucalyptus globulus* in south-western Australia, *Agric. Water Manage.*, 27, 69–86, 1995.

Benson, A.K., Payne, K.L., and Stubben, M.A., Mapping groundwater contamination using DC resistivity and VLF geophysical methods — case study, *Geophysics* 62(1), 80–86, 1997.

Beven, K.J., Towards an alternative blueprint for a physically-based digitally simulated hydrologic response modeling system, *Hydrol. Process.*, 16(2), 189–206, 2002.

Birrel, S.J., Borgelt, S.C., and Sudduth, K.A., Crop yield mapping: comparison of yield monitors and mapping techniques, in *Proc. 2nd International Conference on Site-specific Management for Agricultural Systems*, Robert, P.C., Rust, R.H., and Larson, W.E., Eds., ASA-CSSA-SSSA, Madison, WI, 1995, 15–32.

Boettinger, J.L., Doolittle, J.A., West, N.E., Bork, E.W., and Schupp, E.W., Nondestructive assessment of rangeland soil depth to petrocalcic horizon using electromagnetic induction, *Arid Soil Res. Rehabil.*, 11(4), 372–390, 1997.

Bouma, J., Using morphometric expressions for macropores to improve soil physical analyses of field soils, *Geoderma*, 46, 3–11, 1990.

Bouma, J., Bregt, A.K., eds., *Land Qualities in Space and Time*, Pudoc, Wageningen, The Netherlands, 1989.

Bowling, S.D., Schulte, D.D., and Woldt, W.E., A geophysical and geostatistical methodology for evaluating potential subsurface contamination from feedlot runoff retention ponds, ASAE Paper No. 972087, 1997 ASAE Winter Meetings, Dec. 1997, Chicago, IL, ASAE, St. Joseph, MI, 1997.

Brevik, E.C. and Fenton, T.E., The relative influence of soil water, clay, temperature, and carbonate minerals on soil electrical conductivity readings taken with an EM-38 along a Mollisol catena in central Iowa, *Soil Survey Horizons*, 43, 9–13, 2002.

Brune, D.E. and Doolittle, J., Locating lagoon seepage with radar and electromagnetic survey, *Environ. Geol. Water Sci.*, 16, 195–207, 1990.

Brune, D.E., Drapcho, C.M., Radcliff, D.E., Harter, T., and Zhang, R., Electromagnetic survey to rapidly assess water quality in agricultural watersheds, ASAE Paper No. 992176, ASAE, St. Joseph, MI, 1999.

Brus, D.J., Knotters, M., van Dooremolen, W.A., van Kernebeek, P., and van Seeters, R.J.M., The use of electromagnetic measurements of apparent soil electrical conductivity to predict the boulder clay depth, *Geoderma*, 55(1–2), 79–93, 1992.

Bullock, D.S. and Bullock, D.G., Economic optimality of input application rates in precision farming, *Prec. Agric.*, 2, 71–101, 2000.

Cameron, D.R., de Jong, E., Read, D.W.L., and Oosterveld, M., Mapping salinity using resistivity and electromagnetic inductive techniques, *Can. J. Soil Sci.*, 61, 67–78, 1981.

Cannon, M.E., McKenzie, R.C., and Lachapelle, G., Soil-salinity mapping with electromagnetic induction and satellite-based navigation methods, *Can. J. Soil Sci.*, 74(3), 335–343, 1994.

Carter, L.M., Rhoades, J.D., and Chesson, J.H., Mechanization of soil salinity assessment for mapping, ASAE Paper No. 931557, 1993 ASAE Winter Meetings, 12–17 Dec. 1993, Chicago, IL, ASAE, St. Joseph, MI, 1993.

Cook, P.G. and Kilty, S., A helicopter-borne electromagnetic survey to delineate groundwater recharge rates, *Water Resour. Res.*, 28 (11), 2953–2961, 1992.

Cook, P.G., Walker, G.R., Buselli, G., Potts, I., and Dodds, A.R., The application of electromagnetic techniques to groundwater recharge investigations, *J. Hydrol.* 130, 201–229, 1992.

Corwin, D.L. and Lesch, S.M., Apparent soil electrical conductivity measurements in agriculture, *Comput. Electron. Agric.*, 2005a (in press).

Corwin, D.L. and Lesch, S.M., Application of soil electrical conductivity to precision agriculture: theory, principles, and guidelines, *Agron. J.*, 95, 455–471, 2003.

Corwin, D.L. and Lesch, S.M., Characterizing soil spatial variability with apparent soil electrical conductivity: I. Survey protocols, *Comput. Electron. Agric.*, 2005b (in press).

Corwin, D.L. and Lesch, S.M., Characterizing soil spatial variability with apparent soil electrical conductivity: II. Case study, *Comput. Electron. Agric.*, 2005c (in press).

Corwin, D.L. and Rhoades, J.D., An improved technique for determining soil electrical conductivity-depth relations from above-ground electromagnetic measurements, *Soil Sci. Soc. Am. J.*, 46, 517–520, 1982.

Corwin, D.L. and Rhoades, J.D., Establishing soil electrical conductivity — depth relations from electromagnetic induction measurements, *Commun. Soil Sci. Plant Anal.*, 21(11–12), 861–901, 1990.

Corwin, D.L. and Rhoades, J.D., Measurement of inverted electrical conductivity profiles using electromagnetic induction, *Soil Sci. Soc. Am. J.*, 48, 288–291, 1984.

Corwin, D.L., Carrillo, M.L.K., Vaughan, P.J., Rhoades, J.D., and Cone, D.G., Evaluation of GIS-linked model of salt loading to groundwater, *J. Environ. Qual.*, 28, 471–480, 1999.

Corwin, D.L., Kaffka, S.R., Hopmans, J.W., Mori, Y., Lesch, S.M., and Oster, J.D., Assessment and field-scale mapping of soil quality properties of a saline-sodic soil, *Geoderma*, 114(3–4), 231–259, 2003a.

Corwin, D.L., Lesch, S.M., Oster, J.D., and Kaffka, S.R., Characterizing spatio-temporal variability with soil sampling directed by apparent soil electrical conductivity, *Geoderma*, 2005.

Corwin, D.L., Lesch, S.M., Shouse, P.J., Soppe, R., and Ayars, J.E., Identifying soil properties that influence cotton yield using soil sampling directed by apparent soil electrical conductivity, *Agron. J.*, 95(2), 352–364, 2003b.

Corwin, D.L., Waggoner, B.L., and Rhoades, J.D., A functional model of solute transport that accounts for bypass, *J. Environ. Qual.*, 20(3), 647–658, 1991.

Corwin, D.l., Vaughan, P.J., and Loague, K., Modeling nonpoint source pollutants in the vadose zone with GIS, *Environ. Sci. Technol.*, 31(8), 2157–2175, 1997.

de Jong, E., Ballantyne, A.K., Caneron, D.R., and Read, D.W., Measurement of apparent electrical conductivity of soils by an electromagnetic induction probe to aid salinity surveys, *Soil Sci. Soc. Am. J.*, 43, 810–812, 1979.

Diaz, L. and Herrero, J., Salinity estimates in irrigated soils using electromagnetic induction, *Soil Sci.*, 154, 151–157, 1992.

Doolittle, J.A., Indorante, S.J., Potter, D.K., Hefner, S.G., and McCauley, W.M., Comparing three geophysical tools for locating sand blows in alluvial soils of southeast Missouri, *J. Soil Water Conserv.*, 57 (3), 175–182, 2002.

Doolittle, J.A., Sudduth, K.A., Kitchen, N.R., and Indorante, S.J., Estimating depths to claypans using electromagnetic induction methods, *J. Soil Water Conserv.*, 49 (6), 572–575, 1994.

Doran, J.W. and Parkin, T.B., Quantitative indicators of soil quality: a minimum data set, in *Methods for Assessing Soil Quality*, Doran, J.W. and Jones, A.J., Eds., SSSA Special Publication 49, SSSA, Madison, WI, 1996, 25–38.

Drommerhausen, D.J., Radcliffe, D.E., Brune, D.E., and Gunter, H.D., Electro-magnetic conductivity surveys of dairies for groundwater nitrate, *J. Environ. Qual.*, 24, 1083–1091, 1995.

Eigenberg, R.A., Doran, J.W., Nienaber, J.A., Ferguson, R.B., and Woodbury, B.L., Electrical conductivity monitoring of soil condition and available N with animal manure and a cover crop, *Agric. Ecosyst. Environ.*, 88, 183–193, 2002.

Eigenberg, R.A., Korthals, R.L., and Neinaber, J.A., Geophysical electromagnetic survey methods applied to agricultural waste sites, *J. Environ. Qual.*, 27, 215–219, 1998.

Eigenberg, R.A. and Nienaber, J.A., Electromagnetic survey of cornfield with repeated manure applications, *J. Environ. Qual.*, 27, 1511–1515, 1998.

Eigenberg, R.A. and Nienaber, J.A., Soil conductivity map differences for monitoring temporal changes in an agronomic field, ASAE Paper No. 992176, ASAE, St. Joseph, MI, 1999.

Eigenberg, R.A. and Nienaber, J.A., Identification of nutrient distribution at abandoned livestock manure handling site using electromagnetic induction, ASAE Paper No. 012193, 2001 ASAE Annual International Meeting, 30 July–1 Aug. 2001, Sacramento, CA, ASAE St. Joseph, MI, 2001.

Ellsbury, M.M., Woodson, W.D., Malo, D.D., Clay D.E., Carlson, C.G., and Clay S.A., Spatial variability in corn rootworm distribution in relation to spatially variable soil factors and crop condition, in *Proc. 4*th *International Conference on Precision Agriculture*, Robert, P.C., Rust, R.H., and Larson, W.E., Eds., St. Paul, MN, 19–22 July 1998, ASA-CSSA-SSSA, Madison, WI, 1999, 523–533.

Farahani, H.J., Buchleiter, G.W., and Brodahl, M.K., Characterization of soil electrical conductivity variability in irrigated sandy and non-saline fields in Colorado, *Trans. ASAE*, 2005 (in press).

Fenton, T.E. and Lauterbach, M.A., Soil map unit composition and scale of mapping related to interpretations for precision soil and crop management in Iowa, in *Proc. 4*th *International Conference on Precision Agriculture*, Robert, P.C., Rust, R.H., and Larson, W.E., Eds., St. Paul, MN, 19–22 July 1998, ASA-CSSA-SSSA, Madison, WI, USA, 1999, 239–251.

Fitterman, D.V. and Stewart, M.T., Transient electromagnetic sounding for ground-water. *Geophysics*, 51, 995–1005, 1986.

Freeland, R.S., Branson, J.L., Ammons, J.T., and Leonard, L.L., Surveying perched water on anthropogenic soils using non-intrusive imagery, *Trans. ASAE*, 44, 1955–1963, 2001.

Freeland, R.S., Yoder, R.E., Ammons, J.T., and Leonard, L.L., Mobilized surveying of soil conductivity using electromagnetic induction, *Appl. Eng. Agric.*, 18(1), 121–126, 2002.

Gorucu, S., Khalilian, A., Han, Y.J., Dodd, R.B., Wolak, F.J., and Keskin, M., Variable depth tillage based on geo-referenced soil compaction data in coastal plain region of South Carolina, ASAE Paper No. 011016, 2001 ASAE Annual International Meeting, 30 July–1 Aug. 2001, Sacramento, CA, ASAE St. Joseph, MI, 2001.

Greenhouse, J.P., Slaine, D.D., The use of reconnaissance electromagnetic methods to map contaminant migration, *Ground Water Monit. Rev.*, 3(2), 47–59, 1983.

Greenhouse, J.P. and Slaine, D.D., Geophysical modelling and mapping of contaminated groundwater around three waste disposal sites in southern Ontario, *Can. Geotech. J.*, 23, 372–384, 1986.

Halvorson, A.D. and Rhoades, J.D., Field mapping soil conductivity to delineate dryland seeps with four-electrode techniques, *Soil Sci. Soc. Am. J.* 44, 571–575, 1976.

Hanson, B.R. and Kaita, K., Response of electromagnetic conductivity meter to soil salinity and soil-water content, *J. Irrig. Drain. Eng.*, 123, 141–143, 1997.

Hendrickx, J.M.H., Baerends, B., Raza, Z.I., Sadig, M., and Chaudhry, M.A., Soil salinity assessment by electromagnetic induction of irrigated land, *Soil Sci. Soc. Am. J.*, 56, 1933–1941, 1992.

Herrero, J., Ba, A.A., and Aragues, R., Soil salinity and its distribution determined by soil sampling and electromagnetic techniques, *Soil Use Manage.*, 19(2), 119–126, 2003.

Inman, D.J., Freeland, R.S., Yoder, R.E., Ammons, J.T., and Leonard, L.L., Evaluating GPR and EMI for morphological studies of loessial soils, *Soil Sci.*, 166(9), 622–630, 2001.

Jaynes, D.B., Mapping the areal distribution of soil parameters with geophysical techniques, in Applications of GIS to the Modeling of Non-point Source Pollutants in the Vadose Zone, Corwin, D.L. and Loague, K. (Eds.), SSSA Special Publication No. 48, SSSA, Madison, WI, 1996, 205–216.

Jaynes, D.B., Colvin, T.S., and Ambuel, J., Soil type and crop yield Determinations from ground conductivity surveys, ASAE Paper No. 933552, 1993 ASAE Winter Meetings, 14–17 Dec. 1993, Chicago, IL, ASAE, St. Joseph, MI, 1993.

Jaynes, D.B., Colvin, T.S., and Kaspar, T.C., Identifying potential soybean management zones from multi-year yield data, *Comput. Electron. Agric.*, 2005 (in press).

Jaynes, D.B., Novak, J.M., Moorman, T.B., and Cambardella, C.A., Estimating herbicide partition coefficients from electromagnetic induction measurements, *J. Environ. Qual.*, 24, 36–41, 1995.

Johnson, C.K., Doran, J.W., Duke, H.R., Weinhold, B.J., Eskridge, K.M., and Shanahan, J.F., Field-scale electrical conductivity mapping for delineating soil condition, *Soil Sci. Soc. Am. J.*, 65, 1829–1837, 2001.

Johnson, C.K., Doran, J.W., Eghball, B., Eigenberg, R.A., Wienhold, B.J., Woodbury, B.L, Status of soil electrical conductivity studies by central state researchers, ASAE Paper No. 032339, 2003 ASAE Annual International Meeting, 27–30 July 2003, Las Vegas, NV, ASAE, St. Joseph, MI, 2003.

Johnston, M.A., Savage, M.J., Moolman, J.H., and du Pleiss, H.M., Evaluation of calibration methods for interpreting soil salinity from electromagnetic induction measurements, *Soil Sci. Soc. Am. J.*, 61, 1627–1633, 1997.

Jury, W.A., Spatial variability of soil physical parameters in solute migration: a critical literature review, in *Electrical Power Research Institute (EPRI) Report EA-4228*, EPRI, Palo Alto, CA, 1985.

Jury, W.A., Spatial variability of soil properties, in *Vadose Zone Modeling of Organic Pollutants*, Hern, S.C. and Melancon, S.M., eds., Lewis Publishers, Chelsea, MI, 1986, 245–269.

Jury, W.A. and Roth, K., *Transfer Functions and Solute Movement Through Soil: Theory and Applications*, Birkhauser-Verlag, Basel, 1990.

Kachanoski, R.G., de Jong, E, and Van-Wesenbeeck, I.J., Field scale patterns of soil water storage from non-contacting measurements of bulk electrical conductivity, *Can. J. Soil Sci.*, 70, 537–541, 1990.

Kachanoski, R.G., Gregorich, e.g., and Van-Wesenbeeck, I.J., Estimating spatial variations of soil water content using noncontacting electromagnetic inductive methods, *Can. J. Soil Sci.*, 68, 715–722, 1988.

Kaffka, S. R., Lesch, S. M., Bali, K. M., and Corwin, D. L., Relationship of electromagnetic induction measurements, soil properties, and sugar beet yield in salt-affected fields for site-specific management, *Comput. Electron. Agric.*, 2005 (in press).

Kean, W.F., Jennings Walker, M., and Layson, H.R., Monitoring moisture migration in the vadose zone with resistivity, *Ground Water*, 25, 562–571, 1987.

Khakural, B.R., Robert, P.C., and Hugins, D.R., Use of non-contacting electromagnetic inductive method for estimating soil moisture across a landscape, *Commun. Soil Sci. Plant Anal.*, 29, 2055–2065, 1998.

Kitchen, N.R., Sudduth, K.A., and Drummond, S.T., Mapping of sand deposition from 1993 Midwest floods with electromagnetic induction measurements, *J. Soil Water Conserv.*, 51(4), 336–340, 1996.

Kitchen, N.R., Sudduth, K.A., Myers, D.B., Drummond, S.T., and Hong, S.Y., Delineating productivity zones on claypan soil fields using apparent soil electrical conductivity, *Comput. Electron. Agric.*, 2005 (in press).

Kravchenko, A.N., Bollero, G.A., Omonode, R.A., and Bullock, D.G., Quantitative mapping of soil drainage classes using topographical data and soil electrical conductivity, *Soil Sci. Soc. Am. J.*, 66, 235–243, 2002.

Lesch, S.M., Sensor-directed spatial response surface sampling designs for characterizing spatial variation in soil properties, *Comp. Electron. Agric.*, 2005 (in press).

Lesch, S.M. and Corwin, D.L., Predicting EM/soil property correlation estimates via the dual pathway parallel conductance model, *Agron. J.* 95(2), 365–379, 2003.

Lesch, S.M., Herrero, J., and Rhoades, J.D., Monitoring for temporal changes in soil salinity using electromagnetic induction techniques, 62, 232–242, 1998.

Lesch, S.M., Rhoades, J.D., and Corwin, D.L., ESAP-95 Version 2.10R: User manual and tutorial guide, Research Rpt. 146, USDA-ARS George E. Brown, Jr. Salinity Laboratory, Riverside, CA, 2000.

Lesch, S.M., Rhoades, J.D., Lund, L.J., and Corwin, D.L., Mapping soil salinity using calibrated electromagnetic measurements, *Soil Sci. Soc. Am. J.*, 56, 540–548, 1992.

Lesch, S.M., Strauss, D.J., and Rhoades, J.D., Spatial prediction of soil salinity using electromagnetic induction techniques: 1. Statistical prediction models: A comparison of multiple linear regression and cokriging, *Water Resour. Res.* 31, 373–386, 1995a.

Lesch, S.M., Strauss, D.J., and Rhoades, J.D., Spatial prediction of soil salinity using electromagnetic induction techniques: 2. An efficient spatial sampling algorithm suitable for multiple linear regression model identification and estimation, *Water Resour. Res.*, 31, 387–398, 1995b.

Mankin, K.R., Ewing, K.L., Schrock, M.D., and Kluitenberg, G.J., Field measurement and mapping of soil salinity in saline seeps, ASAE Paper No. 973145, 1997 ASAE Winter Meetings, Dec. 1997, Chicago, IL, ASAE, St. Joseph, MI, 1997.

Mankin, K.R. and Karthikeyan, R., Field assessment of saline seep remediation using electromagnetic induction, *Trans. ASAE*, 45(1), 99–107, 2002.

Mausbach, M.J., Wilding, L.P., Eds., *Spatial Variabilities of Soils and Landforms*, SSSA Special Publication 28, Soil Sci. Soc. Am., Madison, WI, 1991.

Mayer, S., Ellsworth, T.R., Corwin, D.L., and Loague, K., Identifying effective parameters for solute transport models in heterogeneous environments, in *Assessment of Non-point Source Pollution in the Vadose Zone*, Corwin, D.L., Loague, K., and Ellsworth, T.R., eds., Geophysical Monogr. 108, Am. Geophys. Union, Washington, DC, 1999, 119–133.

McBratney, A.B., Webster., R., The design of optimal sampling schemes for local estimation and mapping of regionalized variables: II. Program and examples, *Comput. Geosci.*, 7, 335–365, 1981.

McBride, R.A., Gordon, A.M., and Shrive, S.C., Estimating forest soil quality from terrain measurements of apparent electrical conductivity, *Soil Sci. Soc. Am. J.* 54, 290–293, 1990.

McNeill, J.D., Rapid, accurate mapping of soil salinity by electromagnetic ground conductivity meters, in *Advances in Measurements of Soil Physical Properties: Bringing Theory into Practice*, Topp, G.C., Reynolds, W.D., and Green, R.E., Eds., SSSA Special Publication No. 30, ASA-CSSA-SSSA, Madison, WI, 1992, 201–229.

Morgan, C.L.S., Norman, J.M., Wolkowski, R.P., Lowery, B., Morgan, G.D., and Schuler, R., Two approaches to mapping plant available water: EM-38 measurements and inverse yield modeling, in *Proceedings of the 5th International Conference on Precision Agriculture* (CD-ROM), Roberts, P.C., Rust, R.H., and Larson, W.E., eds., Minneapolis, MN 16–19 July 2000. ASA-CSSA-SSSA, Madison, WI, 2000, 14.

Nathan, G., Inference based on data from complex sample designs, in *Handbook of Statistics*, Krishnaiah, P.R. and Rao, C.R., eds., Vol. 6. Elsevier, Amsterdam, 1988, Chapter 10.

Nettleton, W.D., Bushue, L., Doolittle, J.A., Wndres, T.J., and Indorante, S.J., Sodium affected soil identification in south-central Illinois by electromagnetic induction, *Soil Sci. Soc. Am. J.*, 58, 1190–1193, 1994.

Nielsen, D.R., Biggar, J.W., and Erh, K.T., Spatial variability of field-measured soil-water properties, *Hilgardia*, 42(7), 215–259, 1973.

Nobes, D.C., Armstrong, M.J., and Close, M.E., Delineation of a landfill leachate plume and flow channels in coastal sands near Christchurch, New Zealand, using a shallow electromagnetic survey method, *Hydrogeol. J.*, 8(3), 328–336, 2000.

Nyquist, J.E. and Blair, M.S., Geophysical tracking and data logging system: description and case history, *Geophysics*, 56(7), 1114–1121, 1991.

Paine, J.G., Determining salinization extent, identifying salinity sources, and estimating chloride mass using surface, borehole, an airborne electromagnetic induction methods, *Water Resour. Res.*, 39(3), 1059, 2003.

Ranjan, R.S., Karthigesu, T., and Bulley, N.R., Evaluation of an electromagnetic method for detecting lateral seepage around manure storage lagoons. ASAE Paper No. 952440, ASAE, St. Joseph, MI, 1995.

Rhoades, J.D., Instrumental field methods of salinity appraisal, in *Advances in Measurement of Soil Physical Properties: Bring Theory into Practice*, Topp, G.C., Reynolds, W.D., and Green, R.E., eds., SSSA Special Publication No. 30, Soil Science Society of America, Madison, WI, 1992, 231–248.

Rhoades, J.D., Electrical conductivity methods for measuring and mapping soil salinity, in *Advances in Agronomy*, Sparks, D.L., ed., Vol. 49. Academic Press, San Diego, CA, 1993, 201–251.

Rhoades, J.D. and Corwin, D.L., Determining soil electrical conductivity-depth relations using an inductive electromagnetic soil conductivity meter, *Soil Sci. Soc. Am. J.*, 45, 255–260, 1981.

Rhoades, J.D. and Corwin, D.L., Soil electrical conductivity: effects of soil properties and application to soil salinity appraisal, *Commun. Soil Sci. Plant Anal.*, 21, 837–860, 1990.

Rhoades, J.D. and Halvorson, A.D., *Electrical Conductivity Methods for Detecting and Delineating Saline Seeps and Measuring Salinity in Northern Great Plains Soils*, ARS W-42. USDA-ARS Western Region, Berkeley, CA, 1977, 1–45.

Rhoades, J.D., Chanduvi, F., and Lesch, S., *Soil Salinity Assessment: Methods and Interpretation of Electrical Conductivity Measurements*, FAO Irrigation and Drainage Paper #57, Food and Agriculture Organization of the United Nations, Rome, 1999b, 1–150.

Rhoades, J.D., Corwin, D.L., and Lesch, S.M., Geospatial measurements of soil electrical conductivity to assess soil salinity and diffuse salt loading from irrigation, in *Assessment of Non-point Source Pollution in the Vadose Zone*, Corwin, D.L., Loague, K., and Ellsworth, T.R., Eds., Geophysical Monograph 108, American Geophysical Union, Washington, DC, 1999a, 197–215.

Rhoades, J.D., Manteghi, N.A., Shouse, P.J., and Alves, W.J., Soil electrical conductivity and soil salinity: New formulations and calibrations, *Soil Sci. Soc. Am. J.*, 53, 433–439, 1989.

Rhoades, J.D., Raats, P.A.C., and Prather, R.J., Effects of liquid-phase electrical conductivity, water content and surface conductivity on bulk soil electrical conductivity, *Soil Sci. Soc. Am. J.*, 40, 651–655, 1976.

Rhoades, J.D., Shouse, P.J., Alves, W.J., Manteghi, N.M., and Lesch, S.M., Determining soil salinity from soil electrical conductivity using different models and estimates, *Soil Sci. Soc. Am. J.*, 54, 46–54, 1990.

Salama, R.B., Bartle, G., Farrington, P., and Wilson, V., Basin geomorphological controls on the mechanism of recharge and discharge and its effect on salt storage and mobilization: comparative study using geophysical surveys, *J. Hydrol.*, 155 (1/2), 1–26, 1994.

Scanlon, B.R., Paine, J.G., and Goldsmith, R.S., Evaluation of electromagnetic induction as a reconnaissance technique to characterize unsaturated flow in an arid setting, *Ground Water*, 37(2), 296–304, 1999.

Sheets, K.R. and Hendrickx, J.M.H., 1995. Non-invasive soil water content measurement using electromagnetic induction, *Water Resour. Res.*, 31, 2401–2409, 1995.

Slavich, P.G. and Petterson, G.H., Estimating average rootzone salinity from electromagnetic induction (EM-38) measurements, *Aust. J. Soil Res.*, 28, 453–463, 1990.

Slavich, P.G. and Yang, J., Estimation of field-scale leaching rates from chloride mass balance and electromagnetic induction measurements, *Irrig. Sci.*, 11, 7–14, 1990.

Stogryn, A., Equations for calculating the dielectric constant of saline water, i.e., *EE Trans. Microwave Theory Technol. MIT*, 19, 733–736, 1971.

Stroh, J.C., Archer, S.R., Doolittle, J.A., and Wilding, L.P., Detection of edaphic discontinuities with ground-penetrating radar and electromagnetic induction, *Landscape Ecol.*, 16 (5), 377–390, 2001.

Stroh, J.C., Archer, S.R., Wilding, L.P., and Doolittle, J.A., Assessing the influence of subsoil heterogeneity on vegetation in the Rio Grande Plains of south Texas using electromagnetic induction and geographical information system, College Station, Texas, The Station, March 1993, 39–42, 1993.

Sudduth, K.A. and Kitchen, N.R., Electromagnetic induction sensing of claypan depth, ASAE Paper No. 931531, 1993 ASAE Winter Meetings, 12–17 Dec. 1993, Chicago, IL, ASAE, St. Joseph, MI, 1993.

Triantafilis, J., Ahmed, M.F., and Odeh, I.O.A., Application of a mobile electromagnetic sensing system (MESS) to assess cause and management of soil salinization in an irrigated cotton-growing field, *Soil Use Manage.*, 18(4), 330–339, 2002.

Triantafilis, J., Huckel, A.I., and Odeh, I.O.A., Comparison of statistical prediction methods for estimating field-scale clay content using different combinations of ancillary variables, *Soil Sci.*, 166(6), 415–427, 2001.

U.S. Salinity Laboratory Staff, Diagnosis and Improvement of Saline and Alkali Soils, USDA Handbook 60. U.S. Government Printing Office, Washington, DC, 1954, 1–160.

Valliant, R., Dorfman, A.H., and Royall, R.M., *Finite Population Sampling: A Prediction Approach*, John Wiley, New York, 2000.

van der Lelij, A., Use of an electromagnetic induction instrument (type EM38) for mapping of soil salinity, Internal Report Research Branch, Water Resources Commission, NSW, Australia, 1983.

Van Groenigen, J.W., Siderius, W., Stein, A., Constrained optimisation of soil sampling for minimisation of the kriging variance, *Geoderma*, 87, 239–259, 1999.

Vaughan, P.J., Lesch, S.M., Corwin, D.L., and Cone, D.G., Water content on soil salinity prediction: A geostatistical study using cokriging, *Soil Sci. Soc. Am. J.* 59, 1146–1156, 1995.

Verhagen, A., Booltink, H.W.G., and Bouma, J., Site-specific management: Balancing production and environmental requirements at farm level, *Agric. Syst.*, 49, 369–384, 1995.

Warrick, A.W. and Nielsen, D.R, Spatial variability of soil physical properties in the field, in *Applications of Soil Physics*, Hillel, D., ed., Academic Press, New York, 1980, 319–344.

White, I., Measurement of soil physical properties in the field, in *Flow and Transport in the Natural Environment: Advances and Applications*, Steffen, W.L. and Denmead, O.T., eds., Springer-Verlag, 1988, 59–85.

Williams, B.G. and Baker, G.C., 1982. An electromagnetic induction technique for reconnaissance surveys of soil salinity hazards, *Aust. J. Soil Res.*, 20, 107–118, 1982.

Williams, B.G. and Hoey, D., The use of electromagnetic induction to detect the spatial variability of the salt and clay contents of soils, *Aust. J. Soil Res.*, 25, 21–27, 1987.

Wilson, R.C., Freeland, R.S., Wilkerson, J.B., and Yoder, R.E., Imaging the lateral migration of subsurface moisture using electromagnetic induction, ASAE Paper No. 023070, 2002 ASAE Annual International Meeting, 28–31 July 2002, Chicago, IL, ASAE, St. Joseph, MI, 2002.

Wollenhaupt, N.C., Richardson, J.L., Foss, J.E., and Doll, E.C., A rapid method for estimating weighted soil salinity from apparent soil electrical conductivity measured with an aboveground electromagnetic induction meter, *Can. J. Soil Sci.*, 66, 315–321, 1986.

Wraith, J.M. and Or, D., Temperature effects on soil bulk dielectric permittivity measured by time domain reflectometry: experimental evidence and hypothesis development, *Water Resour. Res.*, 35, 361–369, 1999.

19 Assessment of Uncertainty Associated with the Extent of Simulation Processes from Point to Catchment. Application to 1D-Pesticide Leaching Models

Marco Trevisan
Università Cattolica del Sacro Cuore, Piacenza, Italy

Costantino Vischetti
Università Politecnica delle Marche, Ancona, Italy

CONTENTS

1-5667-0657-2/05/$0.00 + $1.50

19.1   INTRODUCTION

Mathematical models have been used to assess the fate and transport of pesticides at different scales over the past 25 years and now are increasingly used to investigate and assess virtually every type of pesticide problem (Cheng, 1990; van der Werf, 1996; Klepper et al., 1999; Bobba et al., 2000; Brawley et al., 2000; Vanclooster et al., 2000). These models are useful tools for determining pesticide concentrations in the environment and for helping in environment management. Given future input scenarios, it is claimed that the model can predict pesticide behavior. Obviously, such model prediction is uncertain because it is uncertain whether the model structure is valid, mathematical equations describing each process are correct, the model parameters are correctly chosen, and the input data are error free (Lei and Schilling, 1996).

The importance of incorporating uncertainty analysis into fate models has been emphasized by many authors (Dean et al., 1989; Tiktak, 1999; Hession et al., 2000). Ignorance of the uncertainty associated with model predictions may result in misleading interpretations when the model is compared with field measurement and used for risk assessment by the decision maker, who may draw a completely wrong conclusion from a single model prediction (Tiktak, 1999; Bobba et al., 2000; Keller et al., 2001, 2002). The inclusion of uncertainty analysis in modeling activities can be interpreted as the truthful representation of model limitation, and uncertainties must be estimated and included in modeling activities. As reported in Dubus et al. (2003), terminology related to uncertainty within the context of contaminant modeling includes variation, variability, ambiguity, heterogeneity, approximation, inexactness, vagueness, inaccuracy, subjectivity, imprecision, misclassification, misinterpretation, error, faults, mistakes, and artefacts. For us, the term uncertainty represents the combination of factors of various origins leading to a lack of confidence with regard to the description of the system under study. The terminology used encompasses both stochastic variability and incertitude (Dubus et al., 2003).

The main characteristic of pesticide fate model is that they are one-dimensional (1D) models, i.e., able to make point simulations which are accepted everywhere as representations of reality at point and/or field scale (FOCUS, 2000). Several attempts have been made to extend the possibility to simulate with pesticide fate model at the largest scale (catchments, region), but their deterministic nature clashed with a probabilistic approach to get a real understanding of the actual environmental contamination risk by pesticides.

Probabilistic approaches to environmental risk assessment for pesticides are currently receiving a vast amount of interest to account for uncertainty in exposure assessment (ECOFRAM, 1999; EUPRA, 2001). Probabilistic risk assessment (PRA) is an approach to risk assessment that integrates uncertainty considerations and probability distributions to characterize risk. In contrast to point-estimate risk assessment, the overall objective of the method is to avoid worst-case assumptions and come up with a more realistic assessment of risk. Table 19.1 shows the strengths and weaknesses of probabilistic risk assessment as proposed by the EUPRA workshop (EUPRA, 2001).

Several protocols to account for uncertainty in these approaches have been proposed (Dubus and Brown, 2002; Warren-Hicks et al., 2002; Carbone et al., 2002), and a number of articles on this topic have been published (Heuvelink, 1998; Gaunt et al., 1997; Dubus et al., 2003). In some cases software packages have been proposed to automatically incorporate uncertainty analysis into pesticide fate models (Janssen et al., 1994; Wingle et al., 1999).

The causes of model uncertainty related to the input parameters include measurement errors in parameters estimation; spatial, site-specific, and temporal natural variability; extrapolation from controlled laboratory measurement conditions to uncontrolled environmental conditions; methods to estimate the numerical values of the input parameters; use of input parameters from available data sources; or using pedo-transfer

---

TABLE 19.1

Potential Strengths and Weaknesses of PRA Compared to Point Estimate Risk Assessment Currently in Place

Potential strengths and opportunities	Potential weaknesses and threats
PRA can produce outputs which are more meaningful ecologically.	PRA techniques are more complex.
PRA techniques can quantify variability and uncertainty to some extent.	Some PRA techniques require more data.
PRA makes better use of the available data.	PRA may be difficult to communicate.
PRA techniques enable the identification of factors which most influence risk assessment results.	PRA may lead to misleading results.
PRA may provide an alternative to field testing and help to focus the testing where required.	There is no agreement on outputs to look at and on decision-making procedures.
PRA promotes better science.	PRA is difficult to validate.

function (Dean et al., 1989; Soutter et al., 1998; Bobba et al., 2000, Dubus et al., 2003).

The sources of uncertainty may be grouped into three categories: (1) errors resulting from the conceptual scheme of the world: *model error*; (2) stochasticity of the real world, (f.i. temporal and spatial variability): *natural variability*; and (3) uncertainty of the model parameters: *input parameter error* (Dean et al., 1989; Jian and Schilling, 1996; Loague and Corwin, 1996; Hession and Storm, 2000; Trevisan et al., 2001; Keller et al., 2002).

Linked to these classical sources there is the additional uncertainty of the upscaling process from point to catchments scale. This type of uncertainty derives almost completely from adapting the conceptual pesticide fate model, developed for 1D simulations, to a larger (2D and 3D) scale. The classical and additional uncertainty matched in the mapping process of pesticide behavior at regional scale should be characterized and discussed for every procedure to assess the pesticide behavior at a regional scale. To apply 1D models to a large scale, a simple and schematic method consists of two steps: (1) spatialization of a pesticide fate model through the use of informatic tools to manage spatial variable input data and to refer output data to a georeferenced system, and (2) uncertainty related to probabilistic risk assessment procedure.

19.2   SPATIALIZATION OF 1D MODELS

19.2.1   GENERAL

In this context, spatialization means transformation of pesticide fate models from 1D to 2D models, allowing simulations at catchment level. This transformation is reachable with the use of suitable informatic tools, which enable automatic simulations for the entire area. Attempts to spatialize 1D models have been made (Tiktak et al., 2002; Esposito et al., 2004), all aimed to transform 1D to 2D simulations. The approach consists of a subdivision of the entire area into a number of cells of given dimensions, each with its own characteristics concerning soil, crop, and pesticide parameters. The spatializing tools are able to solve several problems: divide the area into a grid, prepare spatial distributed input data for each cell in the grid, run multiple 1D simulations for each cell, and organize the output report following user preferences as well as generate output files compatible with geographic information systems (GIS) and other environmental software applications.

Several problems are linked to this approach, mainly due to the deterministic nature of pesticide fate models. The simulations run independently of each other, and, at the end, a series of 1D simulations have been performed without any sort of interaction between them. A mosaic output display is possible, assigning to each cell the output value found, for instance, in the central point or in any point of the cell. The uncertainty linked to this approach is clearly great and impossible to predict.

The dimension of the area is another factor that strongly influences the reliability of the results. The dimension of the side cells should not exceed a given value (100–200 m) in order to avoid assigning the same characteristics to a large area of soil that might be heterogeneous. Moreover, a large number of cells means a large number of simulations resulting in a time-consuming procedure. A reduction of simulation numbers is possible, through techniques such as unique combination and metamodel, (see below), but in this case the associated uncertainty also increases.

This approach has value as an alternative to the current pesticide registration procedure in EU, which includes the evaluation of a new substance on the basis of a number of standard scenarios that represent an approximate 80% of vulnerable locations. Different spatial patterns of pesticide leaching can be predicted with spatialized 1D models, affected by many processes. The pesticide registration procedure starts with application of a model to a single standard scenario, which should represent a realistic worst-case condition. It is not possible to find one single standard scenario that applies to the full range of registered procedures. Direct application of a spatialized 1D model is preferred, because it provides the user with frequency distributions of the leaching concentrations and gives information about areas of safe usage.

19.2.2 PROPOSED PROTOCOL

19.2.2.1 Data Collection

The data collection for model application is a complex operation that includes the recovery of information on soil, hydrology, weather conditions, and agronomic practices at the site and at agricultural loading. This operation is possible only with the help of many different authorities. Table 19.2 shows the data needed for the use of pesticide fate models at catchment scale. In the context of the spatialization of 1D pesticide fate model, two parameters exist (Table 19.3): spatially distributed and spatially constant.

19.2.2.2 Determination of Number of Simulations

The number of simulations to be performed depends on the availability of parameters and on the scale of simulation. Three possible ways could be adopted: all cells, unique combination, and meta-model. While all cells simply consists in the simulation of all cells of the grid, unique combination implies only the simulation of cells with different soil properties, climate, and/or land use. Meta-model consists of an extension of the results of a simulated cell to other cells, through a regression function that considers a correlated input parameter (e.g., organic carbon content, bulk density).

TABLE 19.2
Data Necessary for the Use of Pesticide Fate Models

Site characterization

Catchments geology/hydrology information
Agronomic practice, land use
Water table level
Soil organic matter content
Slope
Local and regional map of groundwater level, land use, slope
History of pesticide use
Weather conditions

TABLE 19.3
Spatially Distributed and Spatially Constant Simulation Parameters

Spatially constant parameters		Spatially distributed parameters			
Pesticide properties	Management	Soil	Climate	Land use	Groundwater
K_{oc}	Crop parameters	Texture	Temperature	Crop rotation	Depth
$t_{1/2}$	Pesticide use	O.C.	Rainfall		
v_p		Bulk density	ET		
Plant uptake		Hydraulic properties			

K_{oc} = soil sorption constant; $t_{1/2}$ = soil half life; vp = vapour pressure; ET = evapotranspiration. O.C. = organic carbon content.

19.2.2.2.1 All Cells

If all the cells of the grid must be simulated, at least one value for each input parameter per each cell must be available, possible only when the number of cells is limited, i.e., for simulations at field scale or small catchment scale. For each simulated cell only a value for each parameter may be used, which means that for true catchments the dimension of the cells must be different to allow the same parameter set of climatic conditions, land use, and soil heterogeneity.

19.2.2.2.2 Unique Combination Approach

Running a comprehensive model of pesticide leaching for all relevant grid cells in a catchment would require copious computation time. Hence, some authors

(Tiktak et al., 1996; Capri et al., 1999; Tiktak et al., 2002) have applied the model to unique combinations of spatially distributed model inputs. Grid cells that share the same unique combination of parameters are referred to as "megaplot."

To build unique combinations, spatially distributed parameters may be considered, including four categories: soil type, land use type, climate district, and groundwater depth (Table 19.3). A hierarchy may be followed in the model parameterization, i.e., at highest level, a distinction must be made between spatially constant parameters and spatially distributed parameters. The spatially distributed parameters are given at plot level. A relational database must be set up containing information taking from all available topographical and hydrological maps, weather stations, and national soil databases.

The overlapping of all informational levels (different classes of each spatially distributed parameter) allows the individuation of a megaplot. Only one cell per each unique combination of spatially distributed parameters is simulated, and the results are extended to all cells belonging to the same unique combination. A validation of this approach could be made comparing the results with those obtained by all simulations.

19.2.2.2.3   Meta-Model

The results obtained with the unique combination technique can be used to make regression functions between spatially distributed parameters and outputs (i.e., pesticide leached at 1 m depth) and extend the outputs to the entire area (meta-model). However, the identification of these functions is somewhat complicated by the reliability of correlation between input and output data (Capri et al., 1999). If data are not normally distributed, a transformation is needed and regression must be highly significant (high confidence limits and r^2 values). The technique can help in reducing uncertainty linked to reduction of number of simulations, but the described procedure must be strictly followed.

19.2.2.3   Mapping

The coupling between pesticide fate model and GIS can be made in different ways (see Chapter 1). The loose coupling is more reliable and easy to perform, allowing the user to manage the tool as he prefers. The linkage is established through text files. The spatial schematization results in a plot file and a plot map, which is an ASCII grid with mean resolution (200 m × 200 m). The model reads, for each individual plot, one line from the plot file (with information on soil type, weather district, etc.). Using this information, derived variables are calculated. After combining the spatially constant parameters with the spatially distributed parameters, an input file is created and the model executed. Maps of results were obtained by combining in a GIS the simulated values with the plot map.

19.2.3   Uncertainty Linked to Deterministic Simulations

19.2.3.1   General

The uncertainty linked to deterministic simulations with 1D spatialized models is high, difficult to assess, and somewhat unknown. A possible procedure to deal with this problem is proposed to better understand the problem and to provide an useful instrument that can help in EU registration procedure. The uncertainty in model structure and mathematical equations are usually referred to as conceptual errors, and the uncertainty in model output is referred to as model predictive uncertainty (Lei and Schilling, 1996).

The rainfall event was identified as the most critical parameter (Fontaine et al., 1992) in pesticide model sensitivity analysis. The weather conditions are identified as the most important parameters affecting runoff events (Wolt et al., 2002). It is evident that the variability of natural events, such as rainfall and temperature, can affect model simulations. Weather variability is at the same time temporal, during the year, and spatial, in the studied area.

The first step in the uncertainty assessment is to compile a list of the different sources of uncertainty (Dubus et al., 2003). Table 19.4 lists some of these sources. Techniques used to evaluate uncertainty in pesticide fate modeling include differential analysis, Fourier amplitude sensitivity test, Monte Carlo analysis, and fuzzy logic (Dubus et al., 2003).

These techniques have investigated the effects of input uncertainty on model predictions but not model error, parameterization problems, or

TABLE 19.4
Uncertainty Sources and Classification

	Uncertainty type
Model error	
Inability of the model to describe experimental data	B
Wrong use of model	B
Modeler subjectivity	B
Inadequacy of concepts implemented in models	B
Model calibration	B
Natural variability	
Spatial and temporal variability of environmental variables as weather, soil, crop	A
Input parameter error	
Sampling procedure and measurement error of experimental data	B
Derivation of parameters from existing data	B

modeler subjectivity. These uncertainty analysis approaches provide only an estimate of prediction error, not a full evaluation because of restrictions in the sources of uncertainty considered. Model error is quite difficult to estimate, while modeler subjectivity can be assessed by asking different individuals to simulate the same modeling situation (Brown et al., 1996; Boesten, 2000). In any case the selection and implementation of techniques designed to account for uncertainties are themselves subject to significant uncertainty (Dubus et al., 2003). It is important to differentiate in uncertainty analysis between stochastic variability (Type A uncertainty) and uncertainty due to the lack of knowledge (incertitude, or Type B uncertainty) (Dieck, 1997). Variability is an inherent and irreducible property of the scenario, while incertitude is not an inherent property and can be reduced by collecting additional data or information or performing additional analysis (Dubus et al., 2003).

19.2.3.2 Proposed Protocol

In some cases sensitivity analyses can be considered a first step of uncertainty analysis. Both one-at-a-time and Monte Carlo sensitivity analyses could be carried out. One-at-a-time sensitivity analysis consists of varying selected parameters one after the other (all other parameters being kept constant at their nominal value) and observing the influence of the changes on model predictions. In contrast, Monte Carlo sensitivity analysis involves the modification of values for all selected input parameters at the same time using Monte Carlo sampling from predefined probability density functions. There are a number of reasons why Monte Carlo approaches are often used for investigating the sensitivity of pesticide fate models. First, they allow for the simultaneous variation of the values of all the input parameters in contrast to the conceptually simpler one-at-a-time sensitivity analysis. Second, they are relatively simple to conduct when using appropriate software. Third, the use of an efficient sampling scheme (such as the Latin hypercube sampling) (McKay et al., 1979) greatly decreases the number of runs required. Fourth, Monte Carlo approaches may avoid the attribution of specific values to each parameter in a model as in the one-at-a-time sensitivity analysis. If parameters are varied within their uncertainty range, the Monte Carlo approach to sensitivity analysis can provide a simultaneous assessment of uncertainty.

Sensitivity of the model to changes in input parameters is assessed numerically for the one-at-a-time sensitivity analysis by the maximum ratio of variation of the model output and the variation of the model input. For comparison purposes, the absolute value of these ratios is taken, and the maximum absolute ratio of variation (MAROV) index for each parameter is derived as:

$$MAROV = Max(O - O_{BC})/(I - I_{BC}) \times I_{BC}/O_{BC} \quad (19.1)$$

where O is the output value, O_{BC} is the output value for the base-case scenario, I is the input value, and I_{BC} is the original input value for the base-case scenario.

The larger the MAROV for a parameter, the larger the potential influence of that parameter on model output. A MAROV of unity means that a variation in the model input by $x\%$ will result at most in the same variation $(x\%)$ in the model output.

Standardized and ranked model predictions for pesticide losses are related to standardized and ranked model inputs using multiple linear regressions:

$$Y = \sum (i = 1..k)b_i \times X_i + \varepsilon \qquad (19.2$$

where Y is a standardized model output, X_i is a standardized input parameter, b_i is the regression coefficient for each X_i, ε is the regression error, and k is the number of input parameters varied in the sensitivity analysis. The magnitude of the regression coefficients of the regression (or standardized rank regression coefficients, SRRC) allows a comparison of the relative contribution of each input parameter in the prediction of the model. Sensitivity of the model to each input parameter is thus assessed using SRRC values for this particular input parameter. The larger the SRRC for a parameter, the more influence on model predictions this parameter has.

Detailed ranking of input parameters is expected to have a number of applications in modeling procedure. First, the information can be used to guide parameterization efforts and identify those parameters whose values require the most (or the least) time and financial resources for their determination. Second, the information can assist when selecting parameters for adjustment when calibrating the model to experimental data, either manually or by inverse modeling. The third application of these results relates to probabilistic modeling. The probabilistic approach to modeling recognizes the uncertainty associated with input parameters and aims at propagating it through the modeling process to estimate the uncertainty associated with model predictions. The information on the sensitivity analysis can be combined with information on the uncertainty associated with input parameters to select those few parameters that need to be considered within a probabilistic framework.

Monte Carlo simulations (MCS) may be used to estimate the uncertainty of pesticide fate model forecasting. The use of the Monte Carlo approach requires previous setting of the probability density functions of each parameter, i.e., soil properties, pesticide properties, and crop management. The most insidious weakness of the Monte Carlo technique is a misspecification of the parameter distribution (Dubus and Brown, 2002; Warren-Hicks et al., 2002; Dubus and Jansenn, 2003). Small changes in the parameter distribution assumptions can dramatically change the shape of the resultant

Monte Carlo distribution. For example, the use of a series of independent normal distribution rather than a multivariate normal distribution (with an appropriate covariance matrix) produces large differences in resultant Monte Carlo distribution (Warren-Hicks et al., 2002), or if a normal distribution is misunderstood as a uniform one, model uncertainty may be enlarged.

So far it has not yet been clearly defined how much model iteration should be done for a successful MCS (Warren-Hicks et al., 2002). The sample size usually depends on a number of factors: the nature of the parameter being estimated, the form of underlying distribution, the variability in the observations, the degree of precision and/or accuracy desired, the level of confidence to be associated with the estimates, and the actual statistical estimator used to provide the estimate (Warren-Hicks et al., 2002). For example, probabilities of PELMO modeling results were still significantly influenced by the seed number, even for those cases in which 5000 model runs were undertaken (Dubus and Janssen, 2003). According to Parysow et al. (2000) and Dubus and Janssen (2003), when the sample size is greater than 2500 or 3000 runs, the variability of a model projection generally tends to be stable and very close to the results with 5000 or 6000 runs.

When parameter distribution was unknown, the best solution is to choose a uniform one, because it provides a reasonable alternative to guessing at the shapes of sampling distributions of few data or for a limited knowledge situation (Warren-Hicks et al., 2002). Ranges, arithmetic mean, and standard deviation are used to generate random variables of uniform, normal, or other distributions. For each variable set used for a given simulation, pseudo-random values were generated by means of a spreadsheet-based Monte Carlo approach (Janssen et al., 1994; Capri et al., 2001; Hood, 2001; Dubus et al., 2003).

Determination of uncertainty contributions may be performed using analysis of variance (ANOVA). ANOVA can be used to estimate the model error, the natural variability, and the input parameter error (Miao et al., 2004). The analysis can made using as class variables the climatic conditions and the input parameter and as dependent variables: pesticide concentration or cumulative pesticide concentration or time-weighted average pesticide concentration at different depth.

The total uncertainty (TU) was assumed to be estimated by the coefficient of variation of each output and was thus calculated as:

$$TU = \frac{\sqrt{MS_{ct}}}{\overline{X}} \times 100 \qquad (19.3)$$

where MS_{ct} is the sum of the square of the corrected total residuals, and $\overline{X}$ is the mean of model output dependent variable.

Total uncertainty can be divided into input parameter error uncertainty $IPEU$), natural variability uncertainty (NVU), and model error uncertainty

MEU). Among them, $IPEU$, NVU, and MEU were calculated as:

$$IPEU = \tau_{0.05}^{DF_r} \sqrt{\frac{MS_r}{DF_r}}$$ (19.4

where MS_r is the mean square of the input parameter variables residuals, $\tau_{0.05}$ is the confident level of 95% of Student's t-test, and DF_r is the number of degrees of freedom of the input parameter variables.

$$NVU = \tau_{0.05}^{DF_y} \sqrt{\frac{MS_y}{DF_y}}$$ (19.5

where MS_y is the mean square of the climatic conditions (i.e., years), $\tau_{0.05}$ is the confident level of 95% of Student's t-test, and DF_y is the number of degree of freedom of the climatic condition (i.e., years).

$$MEU = \tau_{0.05}^{DF_e} \sqrt{\frac{MS_e}{DF_e}}$$ (19.6

where MS_e is the mean square of the estimate error, $\tau_{0.05}$ is the confident level of 95% of Student's t-test, and DF_e is the number of degree of freedom of the estimate error.

A possible protocol for upscaling uncertainty analysis could be similar to that proposed by Thorsen et al. (2001). As the equations in pesticide fate model are basically point scale equations, a scaling procedure has to be adopted in order to apply the codes at a catchments scale (Tiktak et al., 2002; Esposito et al., 2004). Ideally, all parameters should be treated stochastically and included in uncertainty analysis, but this would result in an unrealistically high number of Monte Carlo simulations; therefore, the input uncertainty should be limited to key parameters of the model (5–10). The actual input error assessment was partly based on the analysis of available data and partly on expert judgment. Since the basis unit of calculation is a field, the variation of field-effective values was used for determining the range of the parameter probability distributions. A single realization of such a parameter can be used in the model for each grid cell. The propagation of errors in the input data to the model output can be assessed using Monte Carlo analysis. The above protocol was adopted by Thorsen et al. (2001) in the prediction of nitrate concentrations in groundwater aquifers using a spatially distributed catchments model (MIKE SHE/DAISY) (Styczen and Storm, 1993). Input data were primarily obtained from databases at a European level. The model parameters were all assessed from these data using transfer functions, and no model calibration was carried out. It appeared that the magnitude of uncertainty depends significantly on the considered temporal and spatial scale. Thus simulations of flux concentrations leaving the root zone at grid level were associated with large uncertainties, whereas uncertainty in simulated concentrations at aquifer level on a catchment scale, an interesting scale seen from a water supply and policy point of view, was much smaller so we suggest that reducing the simulated uncertainty could be accomplished by increasing of the quality of input data support by using national databases instead of the European data sets and the model calibration, as this in principle would decrease the uncertainty related to the input parameters.

19.3 EXAMPLES

19.3.1 SPATIALIZATION OF 1D MODELS

Not much work exists on this topic. Two examples of the described procedure have been performed by Tiktak et al. (2002) and Esposito et al. (2004).

Tiktak et al. (2002) proposed GeoPEARL, a spatialized model to calculate pesticide load in groundwater of The Netherlands. The model is spatialized in order to take into account the spatially distributed parameters and was used to simulate the load into the local and regional groundwater of four pesticides (atrazine, bentazone, dichloropropene, and dinoseb). The simulation was performed for 15 years + 4 warming-up years (1981–1999). Pesticides were applied annually on May 25. Atrazine, bentazone, and dinoseb were applied to the soil surface, and dichlorpropene was injected at a depth of 12.5 cm. Spatial patterns of pesticide mass flux were mapped for leachate below 1 m depth. An annual dose of $1 \, kg \, ha^{-1}$ was assumed. Frequency distribution of the concentrations as predicted by GeoPEARL are reported in Table 19.5.

Results obtained according to EU procedure for registration and properties of the Dutch standard scenario indicate a leaching of $0.03 \, \mu g \, L^-$

TABLE 19.5

Distribution of Agricultural Land over Four Leaching Concentration Classes (pesticide dosage 1 kg ha^{-1})

	Percentage of agricultural land			
C_L ($\mu g \, L^{-1}$)	Atrazine	Bentazone	Dichloropropene	Dinoseb
< 0.01	62	5	35	62
0.01–0.1	24	3	20	6
0.1–1	13	20	23	8
> 1	1	72	22	24

C_L is the 15-year average concentration in leaching water at 1 m depth.

for atrazine, $1.46\,\mu g\,L^{-1}$ for bentazone, $0.002\,\mu g\,L^{-1}$ for dichloropropene, and $0.00\,\mu g\,L^{-1}$ for dinoseb. The frequency distributions in Table 19.5 show that for dichloropropene and dinoseb an incorrect decision would have been made. It is questionable whether it is possible to find a single standard scenario that is applicable for the full range of pesticides. Authors conclude that direct application of GeoPEARL is to be preferred because it provides the user with frequency distributions of the leaching concentrations and gives information about areas of safe usage.

In the European Project PEGASE (Pesticides in European Groundwater: detailed study of representative Aquifers and Simulation of possible Evolution scenarios), a strategy to simulate the contaminant fate in unsaturated and saturated zones is under development (Mouvet et al., 2001). In this context some research groups are involved in the development of tools dedicated to the modeling of pesticide fate from topsoil compartment to and into the aquifer.

A tool has been developed to calculate the leaching potential of pesticide at catchment scale using the one-dimensional pesticide leaching model, MACRO Version 4.3 (Jarvis, 2001). It is structured in a series of MS Excel sheets in which the user can easily input the scenario data required by MACRO. The tool is able to collect the input data of the study area in a grid with homogeneous cells and prepare spatial distributed data required by MACRO. Successively, the tool can run multiple MACRO simulations for each cell of the grid and organize output reports following the user preference as well as generate output files compatible by GIS and other environmental software applications. The tool has been tested in a German catchment (Ciocanaru et al., 2002) with the simulation of behavior of water and the pesticide isoproturon, taking into account the output variability linked to space-variable input data (soil, crop, pesticide). This variability can be displayed as a map of pesticide concentration at a given depth.

The Zwischenscholle test area is a $25\,km^2$ field located in Jülich, North Rhine–Westfalia (Germany). The area is characterized by forest and agricultural land use. A grid was established over the field using cells of 200×200 meters, with a total of 14 columns and 36 rows. Soil consists of a variable horizon of Pseudogley followed by alternate layers of gravel and sand. A schematic subdivision of the area in four types of soil has been realized for simulation purpose. The main crop is sugar beet, with an annual turning of winter wheat, winter barley, and winter rye. An estimate fraction of crop distribution was used for the calculation of the random spatial distribution. This distribution was used for the modeling procedure, realizing a map of distribution for each year of simulation. A modeling exercise has been conducted for an unsaturated zone. The tool linked with the MACRO model has been used to simulate the behavior of water and isoproturon used on winter wheat at $1.75\,kg\,ha^{-1}$ of active ingredient at the Zwischenscholle site for 10 years of data and for each crop-soil combination. Simulations have been performed for the first 2 m of soil profile.

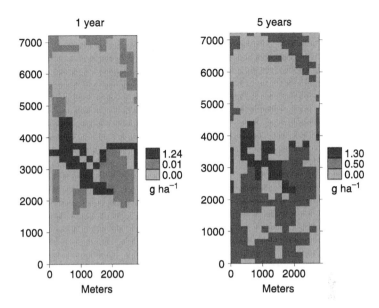

FIGURE 19.1 Concentration maps of isoproturon at 0.5 m depth of the entire area considered 1 and 5 years after treatment.

Maps of isoproturon concentration $(g\,ha^{-1})$ in liquid phase at 0.5 m depth after 1 and 5 years from starting simulation are shown in Figure 19.1. The maps of concentration obtained show that little pesticide reaches a depth of 0.5 m $(1.3\,g/ha = 0.74\%$ of applied) and suggest that the real possibility of groundwater contamination is very small in the case studied. However, depending on soil profile depth and composition, the movement of pesticide shows noticeable differences around map. In this preliminary work the tool showed good reliability in spatializing data and in managing input/output data.

19.3.2 PROBABILITY ANALYSIS OF UNCERTAINTY LINKED TO DETERMINISTIC SIMULATIONS

Dubus and Brown (2002) carried out sensitivity analyses for the preferential flow model MACRO using one-at-a-time and Monte Carlo sampling approaches. Four different scenarios were generated by simulating leaching to the depth of two hypothetical pesticides in a sandy loam and a more structured clay loam soil. Sensitivity of the model was assessed using the predictions for accumulated water percolated at a 1 m and accumulated pesticide losses in percolation. Each parameter was assigned a range of uncertainty reflecting the source of information for its derivation, the range of uncertainty associated with the attribution of values by expert judgment, and likely spatial field variability and measurement error where appropriate. The approach followed therefore differed from that where parameters are varied by a standard variation irrespective of their uncertainty. For the one-at-a-time sensitivity analysis, variation increments were broadly proportional to the variation applied (typically two 5% increments, 25% increments from 25 to 100% variation, then 100% increments for any larger variation). For the Monte Carlo approach, normal distributions were assigned to parameters for which a symmetrical variation was expected. The more uncertain parameters and those that show a large variability in the laboratory or in the field were considered to be log-normally distributed. Uniform distributions were attributed to parameters for which variation was considered to differ from the normal and log-normal distributions.

For the Monte Carlo sensitivity analysis, 250 input files were generated for each scenario using Latin hypercube sampling (LHS) from probability density functions (UNCSAM) (Janssen et al., 1994). Different seed numbers were supplied to the sampling package for each scenario. The LHS technique was used, as it provides an efficient sampling scheme that enables the number of runs to be kept to a minimum. For each scenario, input parameters and results of the 250 runs were standardized (i.e., the population mean was subtracted from the individual results and the resulting difference was divided by the standard deviation of the population) and then ranked.

Sensitivity of the model to each input parameter was thus assessed using SRRC values for this particular input parameter. The SRRC increases with the influence of a given parameter on model predictions. Results for simulated percolation were similar for the two soils. Predictions of water volumes percolated were found to be only marginally affected by changes in input parameters and the most influential parameter was the water content defining the boundary between micropores and macropores in this dual-porosity model. In contrast, predictions of pesticide losses were found to be dependent on the scenarios considered and to be significantly affected by variations in input parameters. In most scenarios, predictions for pesticide losses by MACRO were most influenced by parameters related to sorption and degradation. Under specific circumstances, pesticide losses can be largely affected by changes in hydrological properties of the soil. Since parameters were varied within ranges that approximated their uncertainty, a first-step assessment of uncertainty for the predictions of pesticide losses was possible. Large uncertainties in the predictions were reported, although these are likely to have been overestimated by considering a large number of input parameters in the exercise.

19.4   RECOMMENDATIONS AND FUTURE RESEARCH

Stochastic approaches are useful tools in uncertainty analysis, but can a 1D model accurately predict pesticide concentrations when applied at a catchment scale with input data from aggregated data sources such as national or European databases? A series of problems arises from the 1D nature of the model. Error and uncertainty at this stage derive from different sources and can be grouped as follows Gaunt et al. (1997):

Limited availability of data required to estimate and assess models: data may be collected or obtained at scales inappropriate to the model. The effects of aggregation, generalization, and derivation of collected data on the modeling process are not clarified sufficiently as a result of the diversity and complexity of models used in practice. Calibration may be required to ensure compatibility and thus to reduce error.

Error from choice of an inadequate model: Chatfield (1995) states that model uncertainty can arise from model misspecification, from nonidentification of the true model from a class of general models, and from the conflict between two or more models with different structures.

Error in estimating the model parameters.

Error propagation via linkage of submodels: any uncertainty inherent in one submodel will be transferred to the submodels that use its output.

Error in upscaling or aggregation of process-based models: models may be developed for points or small areas of land, whereas interest is in the larger scale, such as regions or catchments. Three approaches have been considered: (1) small-scale units are reused in the large-scale database, values being interpolated from relationships between variables or averaged values, weighted or otherwise substituted; (2) in a spatial sense, if stationary of covariance between areas exists, then geostatistical interpolation and extrapolation have been applied; (3) if data are strongly related to another property (such as elevation or distance to the sea), then this relation has been exploited when upscaling from a small to a larger area; regression and correlation techniques have been useful here.

Errors and uncertainties inherent in a decision model: once the probability level is determined and used, how are the model outputs to be evaluated in relation to this probability?

Some recommendations may be made to people involved in risk assessment procedure:

Uncertainty analysis could be associated with model use.

The choice of the correct model and the accuracy of input data reduce the uncertainty of output. Users must know the more sensitive input parameters for each model to reduce parameterization errors.

The use of a higher tier approach, like probabilistic risk assessment, could increase registration and management activities by clarifying the following areas of uncertainty: input parameters for modeling pesticide fate, discrepancy between exposure in the laboratory and the field, uncertainties in exposure scenarios, variability within the landscape, spatial distribution of residues, and residue dynamics (dissipation, bioaccumulation, etc.) (Mackay et al., 2002)

Maps of simulated results must have associated confidence interval of data or at least an associated map of uncertainty (see Chapter 1).

REFERENCES

Bobba, A.G., V.P. Singh, and L. Bengtsson. 2000. Application of environmental models to different hydrological systems. Ecol. Model. 125:15–49.

Boesten, J.J.T.I. 2000. Modeller subjectivity in estimating pesticide parameters for leaching models using the same laboratory dataset. Agr. Water Manage. 44:389–409.

Brawley, J.W., G. Collins, J.N. Kremer, C.H. Sham, and I. Valiela. 2000. A time-dependent model of nitrogen loading to estuaries from coastal watershed. J. Environ. Qual. 29:1448–1461.

Brown, C.D., U. Baer, P. Gunther, M. Trevisan, and A. Walker. 1996. Ring test with the models LEACHP, PRZM-2 and VARLEACH: variability between model users in prediction of pesticide leaching using a standard data set. Pestic. Sci. 47:249–258.

Capri, E., F. Ferrari, Z. Miao, and M. Trevisan. 2001. Edge field leaching study of metalaxyl-M and its main metabolite, p. 171–176. *In* 2001 BCPC Symposium Proceedings No. 78: Pesticide Behaviour in Soil and Water. The British Crop Protection Council, UK.

Capri, E., L. Padovani, and M. Trevisan. 1999. La previsione della contaminazione delle acque sotterranee da prodotti fitosanitari. Pitagora Editrice, Bologna.

Carbone, J.P., P.L. Havens, and W. Warren-Hicks. 2002. Validation of pesticide root zone models 3.12: employing uncertainty analysis. Environ. Contam. Chem. 21:1578–1590.

Chatfield, C. 1995. Model uncertainty, data mining and statistical interference (with discussion). J. Royal Statistical Soc. Series A, 419–466.

Cheng, H.H. (ed.). 1990. Pesticides in the Soil Environment: Processes, Impacts, and Modelling. Soil Science Society of America, Inc., Madison, Wisconsin.

Ciocanaru, M., R. Harms, O. Nitzsche, A. Englert, and H. Vereecken. 2002. Site Characterization of the Julicher Zwinschescholle Area for the PEGASE Project. Forschungszentrum Julich, Germany, Internal Communication.

Dean, J.D., P.S. Huyakorn, A.S. Donigian Jr., K.A. Voos, R.W. Schanz, and R.F. Carsel. 1989. Risk of unsaturated/saturated transport and transformation chemical concentrations (RUSTIC), volume I. theory and code verification, volume II. User's guide, Environmental Research Laboratory and Office of Research and Development of United States Environmental Protection Agency (EPA/600/3-89), Athens. GA.

Dieck, R.H. 1997. Measurement uncertainty models 1997. ISA Trans. 36:29–35.

Dubus, I.G., C.D. Brown, and S. Beulke. 2003. Sources of uncertainty in pesticide fate modelling. Sci. Tot. Environ. 317:53–72.

Dubus, I.G. and C.D. Brown. 2002. Sensitivity and first step uncertainty analyses for the preferential flow model MACRO. J. Environ. Qual. 31:227–240.

Dubus, I.G. and P.H.M. Janssen. 2003. Issues of replicability in Monte Carlo modeling: a case study with a pesticide lecjhing model. Environ. Toxicol. Chem. 22:3081–3087.

ECOFRAM. 1999. http://www.epa.gov/oppefed1/ecorisk/

Esposito, A., C. Vischetti, G. Errera, M. Trevisan, L. Scarponi, M. Herbst, M. Ciocanaru, and H. Vereecken. 2004. A spatializing tool to simulate pesticide fate in unsaturated zone at catchment scale. Agronomie, Special Issue "Unicum Colloquium, Crop Protection," Aix en Provence, France, May 20–24th, 2003 (in press).

EUPRA. 2001. Probabilistic risk assessment for pesticides in Europe. Implementation and research needs. Report from the European Workshop on Probabilistic Risk Assessment for the Environmental Impacts of Plant protection products, The Netherlands, June 2001.

FOCUS. 2000. FOCUS groundwater scenarios in the EU plant protection product review process. Report of the FOCUS Groundwater Scenarios Workgroup, EC Document Reference Sanco/321/2000.

Fontaine, D.D., P.L. Havens, G.E. Blau, and P.M. Tillotson. 1992. The role of sensitivity analysis in groundwater risk modeling for pesticides. Weed Technol. 6:716–724.

Gaunt, J.L., J. Riley, A. Stein, and F.W.T. Penning de Vries. 1997. Requirements for effective modelling strategies. Agr. Syst. 54(2):153–168.

Hession, W.C. and D.E. Storm. 2000. Watershed-level uncertainties: implication for phosphorus management and eutrophication. J. Environ. Qual. 29: 1172–1179.

Heuvelink, G.B.M. 1998. Uncertainty analysis in environmental modelling under a change of spatial scale. Nutr. Cycling Agroecosyst. 50:255–264.

Hood, G. 2001. Poptools version 2.2 (Build 2), CSIRO, Canberra, Australia. available at: http://www.dwe.csiro.au/vbc/poptools/index.htm

Janssen, P.H.M., P. Heuberger, and O. Klepper. 1994. UNSCAM: a tool for automating sensitivity and uncertainty of analysis. Environ. Software 9:1–11.

Jarvis, N. 2001. The MACRO Model (Version 4.3). Technical Description.SLU, Department of Soil Sciences, Box 7014, 750 07 Uppsala.

Jian, H.L. and W. Schilling. 1996. Preliminary uncertainty analysis — a prerequisite for assessing the predictive uncertainty of hydrologic models. Water Sci. Techn. 33:79–90.

Keller, A., B. von Steiger, S.E.A.T.M. van der Zee, and R. Schulin. 2001. A stochastic empirical model for regional heavy-metal balance in agro-ecosystems. J. Environ. Qual. 30:1976–1989.

Keller, A., K.C. Abbaspour, and R. Schulin. 2002. Assessment of uncertainty and risk in modeling regional heavy-metal accumulation in agricultural soils. J. Environ. Qual. 31:175–187.

Klepper, O., and H.A. den Hollander. 1999. A comparison of spatially explicit and box models for the fate of chemicals in water, air and soil in Europe. Ecol. Model. 116:183–202.

Lei J.H. and W. Schilling. 1996. Preliminary uncertainty analysis — prerequisite for assessing the predictive uncertainty of hydrologic models. Wat. Sci. Tech. 33:79–90.

Loague, K. and D.L. Corwin. 1996. Uncertainty in regional-scale assessments of non-point source pollutants. p. 131–152. In Corwin, D.L. and K. Loague (eds.), Applications of GIS to the Modeling of Non-Point Source Pollutants in the Vadose Zone, SSSA Special Publication No. 48, Soil Science Society of America, Madison, WI.

Mackay N., A. Terry, D. Arnold, and T. Pepper. 2002. Approaches and tools for higher tier assessment of environmental fate in the UK. DEFRA Contract PL0546.

McKay, M.D., W.J. Conover, and R.J. Beckman. 1979. A comparison of three methods for selecting values of input variables in the analysis of output from a computer code. Technometrics 21:239–245.

Miao, Z., M. Trevisan, E. Capri, L. Padovani, and A.A.M. Del Re. 2004. The uncertainty assessment of the model RICEWQ in northern Italy. J Environ Qual (in press).

Mouvet, C., C. Golaz, D. Thiery, N. Baran, C. Ritsema, B. Normand, C. Vischetti, and O. Nitzsche. 2001. Pesticides in European Groundwaters: detailed study of representative Aquifers and Simulation of possible Evolution scenarios (PEGASE). Proceedings of ETCA meeting, Harrogate, UK, May 21–23, 2001.

Parysow, P., G. Gertner, and J. Westervelt. 2000. Efficient approximation for building error budgets for process models. Ecol. Modell. 135:111–125.

Soutter, M. and A. Musy. 1998. Coupling 1D Monte-Carlo simulations and geostatistics to assess groundwater vulnerability to pesticide contamination on a regional scale. J. Contamin. Hydro. 32:25–39.

Styczen, M. and B. Storm. 1993. Modelling of N-movements on catchment scale — a tool for analysis and decision making. 1. Model description & 2. A case study. Fertilis. Res. 36:1–17.

Thorsen, M., J.C. Refsgaard, S. Hansen, E. Pebesma, J.B. Jensen, and S. Kleeschulte. 2001. Assessment of uncertainty in simulation of nitrate leaching to aquifers at catchment scale. J Hydrol. 242:210–227.

Tiktak, A. 1999. Modeling non-point source pollutants in soils — applications to the leaching and accumulation of pesticides and cadmium. Ph.D. thesis, University of Amsterdam, Amsterdam, the Netherlands.

Tiktak, A., D. de Nie, T. van der Linden, and R. Kruijne. 2002. Modelling the leaching and drainage of pesticides in the Netherlands: the GeoPEARL model. Agronomie 22:373–387.

Tiktak, A., A.M.A. van der linden, and R.C.M. Merkelbach. 1996. Modelling pesticide leaching at a regional scale in the Netherlands. RIVM report n° 715801008, Bilthoven, the Netherlands.

Trevisan M., R. Calandra, C. Vischetti, A. Esposito, and L. Padovani. 2001. Pesticide leaching potential in the Trasimeno lake area — assessment of uncertainty associated with the simulation process. *In* A. Walker (ed.), 2001 BCPC Symposium Proceedings No. 78: Pesticide Behaviour in Soil and Water, The British Crop Protection Council, UK.

Van der Werf, H.M.G. 1996. Assessing the impact of pesticides on the environment. Agric. Ecosyst. Environ. 60:81–96.

Vanclooster M., J.J.T.I. Boesten, M. Trevisan, C.D. Brown, E. Capri, O.M. Eklo, B. Gottesburen, V. Gouy, and A.M.A. van der Linden. 2000. A European test of pesticide leaching models: methodology and major recommendations. Agric. Water. Manage. 44:1–19.

Warren-Hicks, W., J.P. Carbone, and P.L. Havens. 2002. Using Monte Carlo techniques to judge model prediction accuracy: validation of the pesticide root zone model 3.12. Environ. Toxicol. Chem. 21:1570–1577.

Wingle, W.L., E.P. Poeter, and S.A. McKenna. 1999. UNCERT: geostatistics, uncertainty analysis and visualization software applied to groundwater flow and contaminant transport modeling. Comput. Geosci. 25:365–376.

Wolt, J., P. Singh, S. Cryer, and J. Lin. 2002. Sensitivity analysis for validating expert opinion as to ideal data set criteria for transport modeling. Environ. Toxicol. Chem. 21:1558–1565.

20 Inverse Modeling Techniques to Characterize Transport Processes in the Soil-Crop Continuum

*S. Lambot, M. Javaux, F. Hupet,
and M. Vanclooster*
Catholic University of Louvain, Louvain-la-Neuve, Belgium

CONTENTS

1-5667-0657-2/05/$0.00 + $1.50

20.1 INTRODUCTION

Simulation models are becoming readily available tools in agricultural and environmental engineering to design and evaluate management strategies which respect multiple criteria such as optimal crop production and preservation of soil and water quality. Nonetheless, the effectiveness with which these modeling tools can be adopted relies heavily on the knowledge of the system parameters, which today still constitutes an important issue for the scientific community. During the last decade, identification of system parameters from observations of the state variables of the dynamic system by means of inverse modeling procedures has gained much interest. With these methods, experimental data from a dynamic experiment are combined with a validated forward model and an appropriate optimization algorithm to estimate model parameters of the system. The inversion is generally based on the optimization of an objective function, which expresses the difference between the observed response of the system and the simulated response with the given model subject to a trial parameter set.

Progressively, inverse modeling has been introduced in soil and crop modeling as a way to remedy the pitfalls of classical parameter-identification techniques. The availability of new advanced monitoring techniques, generating detailed data of the soil-crop systems, together with fast-forward simulation models solving the many nonlinear process equations for the soil-crop continuum and advanced optimization algorithms allow the efficient inversion of the soil-crop system models, thereby retrieving the model parameters that respect the observed behavior. The success of the inverse approach, however, depends on the appropriateness of the forward model, the identifiability of the model parameters, the uniqueness and stability of the inverse solution, and the robustness of the optimization algorithm, all of which determine the accuracy with which the parameters are determined.

Compared to classical direct laboratory methods (e.g., see Klute and Dirksen, 1986), the principal advantages of the inversion procedures are that they generally require less experimental efforts and result in effective parameter estimates. Classical laboratory methods are often time-consuming because they need to reach several stages of steady-state conditions. Moreover, their utility for predicting effective behaviors is sometimes questionable (e.g., Mishra and Parker, 1989). Inversion procedures, allow much more flexibility in experimental design and yield model parameters that maximize the ability of the model to reproduce the transient event.

In the particular case of partially saturated flow and transport models, the inversion may yield simultaneous effective estimates of both the soil water retention function and the unsaturated hydraulic conductivity function. Inversion methods for estimating the flow properties of partially saturated soils were introduced in the early 1980s (Zachmann et al., 1982; Dane and Hruska, 1983), when efficient numerical solutions of the nonlinear flow governing equation became available. For instance, the one-step outflow method, first introduced by Gardner (1956), was evaluated using numerical inversion methods by Kool et al. (1985a,b) and Van Dam et al. (1994). Inverse methods for estimating the transport properties were initiated when appropriate analytical solutions of the governing transport equations (e.g., Parker and van Genuchten, 1984; Wendroth et al., 1993; Gribb, 1996; Leij et al., 1999) became available. In this area, inverse methods gained impetus when solute tracing techniques improved (e.g., Ferré and Kluitenberg, 2003) and validated numerical transport models became available. Reviews on the use of inversion methods for characterizing unsaturated flow and transport properties of soils are given in particular by Leij and van Genuchten (1999), Durner et al. (1999), and Hopmans et al. (2002). The intensive development of remote sensing and detailed monitoring of crop status increases perspectives for estimating soil-crop properties from remote sensing based on inverse modeling (Lambot et al., 2004b).

Notwithstanding the many possible applications of inverse modeling in the modeling of fate and transport processes in the soil-crop continuum and the many advantages of inverse approaches as compared to classical identification techniques, little attention is given to performance properties of the inverse modeling which determines the success or failure of an inverse approach for a given application. In this chapter we therefore review some basic principles of the inverse modeling approach and illustrate the method for identifying unsaturated flow, transport, and crop parameters of the soil-crop continuum. Particular attention is given to the properties of the forward model, the objective function, and the optimization algorithm, which determine the robustness of the inverse approach. In addition, performance evaluation techniques are presented (response surface analysis, stability analysis, uncertainty analysis), which allows one to qualify the performance of the inverse method in an objective and transparent way.

20.2   THE FORWARD MODEL

The quality of the inversion is subject to the appropriateness of the predictive forward model to describe the system of concern (Hollenbeck and Jensen, 1998; Vrugt et al., 2002, 2003). The forward (or direct) model characterizes the relationship between a dependent variable y^* and a set of explanatory variables $\mathbf{x}$:

$$y^*(\mathbf{x}) = y^*(\mathbf{x}; \mathbf{p}) \qquad (20.1)$$

with $\mathbf{p}$, a set of M adjustable parameters. Simple models can be written as straightforward analytical equations, but others may consist of a numerical resolution of a differential equation. For instance, Richards equation, often used for predicting water flow in the vadose zone, can be solved analytically or numerically following the boundary and initial conditions and the shape of hydraulic characteristic curves.

The appropriateness of a forward model to be used in an inverse procedure will depend on the existence of the solution of the model in the parameter domain, on parameter identifiability, on the sensitivity of the model to the parameters, and on the adequacy of the model to reproduce an observed system response.

20.2.1 EXISTENCE

The first condition a model must respond to is that the forward solution exists for specified model parameters, boundary conditions, and initial conditions. For instance, analytical solutions of the governing flow and transport equations only exist for a limited range of initial value and boundary value problems. For integrated soil-crop models, analytical solutions usually do not exist, and numerical resolutions are required. However, it is worth noting that numerical models may suffer from convergence problems in some parts of the parameter domain (e.g., see Lambot et al., 2002).

20.2.2 IDENTIFIABILITY, UNIQUENESS, AND SENSITIVITY

Different parameter sets must lead to different solutions; if not parameters are unidentifiable (Carrera and Neuman, 1986a,b). A theoretical illustration of identification problems is given in Figure 20.1a. The represented function is not injective, that is, the same model result can be obtained from different parameter values.

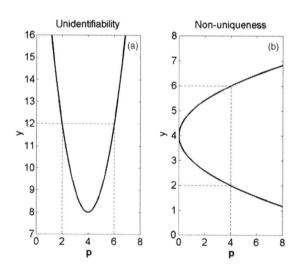

FIGURE 20.1 Unidentifiability and nonuniqueness properties of the forward model.

A second criterion to be respected in the inversion is the uniqueness criterion. The parameters and initial and boundary conditions are completely specified in the solution domain so that the problem can have a unique solution (Yeh and Šimůnek, 2002). Nonuniqueness appears when one parameter set results in more than one solution, as illustrated in Figure 20.1b (Gupta and Sorooshian, 1985; Duan et al., 1992).

Sensitivity means that the model should depend on the parameters, i.e., the derivative of the model response to the parameters should be different from zero somewhere in the parameter domain. A lack of sensitivity usually results in nonidentifiability.

Prior to using the inverse method, numerical experiments can be designed to test the uniqueness of the model. This can be performed, for instance, by visually inspecting two-dimensional response surfaces that represent the behavior of the forward model, through the definition of an objective function, as a function of parameter pairs to be optimized (e.g., Šimůnek and van Genuchten, 1996; Romano and Santini, 1999; Lambot et al., 2002). These response functions also inform the analyst as to the sensitivity and dependence of the parameters. More details on response function analysis are given in the sections below.

20.2.3 MODEL ADEQUACY

The adequacy of the forward model can never be known a priori. It can be assessed statistically on the basis of model identification criteria based on the notion of likelihood (Press et al., 1997). Moreover, adequacy and confidence in its parameters can only be evaluated relative to the uncertainty in the observations. Those topics are treated below.

20.3 OBJECTIVE FUNCTION

20.3.1 DEFINITION

The objective (or merit or goal) function is a measure of the agreement between the data and a model with a particular parameter set $\mathbf{p}$. The most used criterion is the weighted least-squares error function, defined as:

$$\Phi(\mathbf{p}) = \frac{1}{\sigma^2} \sum_{i=1}^{N} [y_i - y^*(x_i, \mathbf{p})]^2 \tag{20.2}$$

or, in matrix notation,

$$\Phi(\mathbf{p}) = (\mathbf{y} - \mathbf{y}^*)^T \mathbf{V}^{-1} (\mathbf{y} - \mathbf{y}^*) \tag{20.3}$$

where y_i is the ith element of the measured dataset $\mathbf{y}$ at x_i for $i = 1, \ldots, N$ $y^*(x_i, \mathbf{p})$ is the ith modeled answer of the vector $\mathbf{y}^*$ for x_i and parameter set $\mathbf{p}$, σ^2 is the variance, and $\mathbf{V}$ the covariance matrix of the measurement errors.

If measurement errors are independent, normally distributed, and homo-skedastic, i.e., measurement errors have the same variance, it can be shown that the likelihood function $L_y(\mathbf{p})$, which gives, as a function of the parameter vector $\mathbf{p}$, the conditional probability of observing the data $\mathbf{y}$, is related to the objective function by (see, e.g., Press et al., 1997)

$$L_\mathbf{y}(\mathbf{p}) = f \langle \mathbf{y} | \mathbf{p} \rangle \propto \exp(-0.5\Phi(\mathbf{p})) \qquad (20.4$$

where $f \langle \mathbf{y} | \mathbf{p} \rangle$ represents the probability of $\mathbf{y}$ given $\mathbf{p}$.

Therefore, minimizing Eq. (20.2) or (20.3) is equivalent to maximizing Eq. (20.4). This last expression also measures the agreement between any nominated value of $\mathbf{p}$ and the collected measurements $\mathbf{x}$ (Berger, 1985). Equation (20.2) can be extended to cases for which each data point (x_i, y_i) has a different known standard deviation σ_i (heteroskedasticity):

$$\Phi(\mathbf{p}) = \sum_{i=1}^{N} \left[\frac{y_i - y^*(x_i, \mathbf{p})}{\sigma_i} \right]^2 \equiv \chi^2 \qquad (20.5$$

which follows a chi-square distribution.

Some authors have extended the strict statistical definition of the likelihood function $L_y(\mathbf{p})$ to a "fuzzy" belief, or a probabilistic measure of how well the model conforms to the observed behavior of the system (Beven and Binley, 1992). In such an approach, it is considered that the likelihood (or objective function) increases (or decreases) monotonically with increasing model performances. The likelihood function equals zero (or is at is maximum value) for nonbehavioral models (Beven and Freer, 2001).

In practical subsurface hydrological applications, Eqs. (20.2) and (20.5) are often applied considering the error variance as being equal to 1. However, such functions do not permit to derive statistically sound indicators of parameter uncertainty (Hollenbeck and Jensen, 1998). In addition, the location of the minimum may also be affected when multi-informative objective functions are formulated.

Figure 20.2 shows the normalized breakthrough curve measured in a soil by means of a TDR probe (Javaux and Vanclooster, 2003). An analytical solution of the convection dispersion equation was fitted to this curve to obtain the soil dispersivity and velocity values. Two objective functions were considered. The first is based on Eq. (20.2) with $\sigma = 1$. A second objective function was based on the difference between the slope of the curves at each measured time. Objective functions are compared in Figure 20.3. It can be observed that neither the minimum nor the shape of the objective functions is identical, even though the optimized curves are not distinguishable (see Figure 20.2).

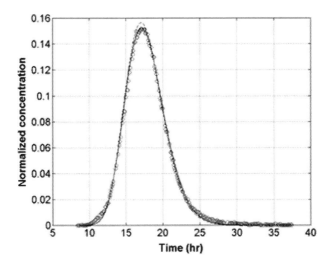

FIGURE 20.2 Normalized breakthrough curve in a soil (open circles) and optimized curve with two different objective functions.

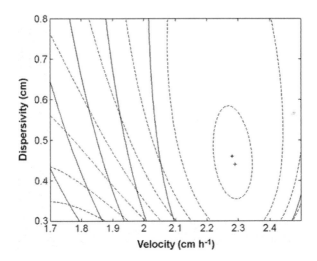

FIGURE 20.3 Comparison of the shape of two objective functions for fitting a solute breakthrough curve: least-squares error (lines) and absolute value of the slope (dashed line).

20.3.2 Multi-Informative Objective Functions

Usually objective functions are not strictly convex and include more than one global minimum: the solution is not unique. The nonuniqueness problem can be solved by including more information in the definition of the objective function. In particular, the objective function may be regularized, either by incorporating a priori knowledge on the parameter values or by considering other sources of measurement.

20.3.2.1 Use of Prior Information

If information about the distribution of the fitted parameters is available before the optimization, such information can be introduced in the objective function [Eq. (20.3)] using Bayes theorem. Bayes rules state that (Carrera and Neuman, 1986a)

$$f\langle \mathbf{p}|\mathbf{y}\rangle = \frac{f(\mathbf{p})f\langle \mathbf{y}|\mathbf{p}\rangle}{\int f(\mathbf{p})f\langle \mathbf{y}|\mathbf{p}\rangle d\mathbf{p}} = cf(\mathbf{p})L_{\mathbf{y}}(\mathbf{p}) \qquad (20.6$$

where $f\langle \mathbf{p}|\mathbf{y}\rangle$ is called the posterior distribution and is the conditional probability density function (pdf) of $\mathbf{p}$ given the sample observation $\mathbf{y}$; $f(\mathbf{p}$ is the Baysian prior distribution (i.e., the prior probability density function which represents the prior information); $f\langle \mathbf{y}|\mathbf{p}\rangle$, or $L_y(\mathbf{p})$, is the likelihood function; and c is a normalization constant that ensures that the integral of the posterior distribution is equal to 1. If $f(\mathbf{p})$ is a uniform distribution, bounded between two values, maximizing Eq. (20.6) is equivalent to minimizing Eq. (20.5) under constraints. If $f(\mathbf{p})$ is a Gaussian distribution, it can be shown that the corresponding objective function yields (Carrera and Neuman, 1986):

$$\Phi(\mathbf{p}) = (\mathbf{y} - \mathbf{y}^*)^T \mathbf{V}^{-1}(\mathbf{y} - \mathbf{y}^*) + (\mathbf{p} - \mathbf{p}^*)^T \mathbf{V}_p^{-1}(\mathbf{p} - \mathbf{p}^*) \qquad (20.7$$

where $\mathbf{V_p}$ is the covariance matrix for the parameter vector $\mathbf{p}$, and $\mathbf{p}^*$ is the prior estimate parameter vector. The first term in Eq. (20.7) penalizes for deviation of model prediction from experimental data, while the second term penalizes for deviation from model parameters from the prior estimate.

20.3.2.2 Use of Different Sources of Information

In many cases, different sources of information are available such that the observation vector $\mathbf{y}$ can be decomposed into two parts, $\mathbf{y} = \{\mathbf{y_1}, \mathbf{y_2}\}$. The different components of the observation vector can now be introduced in the objective function in different ways. The classical approach is to weigh the separated terms of the objective function, as here for two information sources:

$$\Phi(\mathbf{p}) = \Phi_1(\mathbf{p}) + W\Phi_2(\mathbf{p}) \qquad (20.8$$

where subscripts 1 and 2 refer to the two data sets $\mathbf{y_1}$ and $\mathbf{y_2}$, and W is a weight relative to the dataset magnitude ratio. In a Bayesian framework, it can be shown that the posterior pdf, given the data set $\mathbf{y}$ corresponds to:

$$f\langle\mathbf{p}|\mathbf{y_1},\mathbf{y_2}\rangle = \frac{f\langle\mathbf{p}|\mathbf{y_1}\rangle f\langle\mathbf{y_2}|\mathbf{p},\mathbf{y_1}\rangle}{f\langle\mathbf{y_2}|\mathbf{y_1}\rangle} = \frac{(f(\mathbf{p})f\langle\mathbf{y_1}|\mathbf{p}\rangle/f\langle\mathbf{y_1}\rangle)f\langle\mathbf{y_2}|\mathbf{p},\mathbf{y_1}\rangle}{f\langle\mathbf{y_2}|\mathbf{y_1}\rangle}$$

$$= cf(\mathbf{p})f\langle\mathbf{y_1},\mathbf{y_2}|\mathbf{p}\rangle \qquad (20.9)$$

If there is no prior distribution (all the elements of $\mathbf{V_p}$ equal 0), then the objective function to minimize reduces to Eq. (20.8) with

$$W = \sigma_1^2/\sigma_2^2 \qquad (20.10)$$

In soil hydrology, different sources of information are usually available and are used to improve the well-posedness of an inverse problem. For instance, Si and Kachanoski (2000) formulated the objective function using both pressure head and water content data to estimate the soil hydraulic properties during constant flux infiltration. They demonstrated by analyzing the behavior of the response surfaces the usefulness of combining the two sources of information to enable uniqueness of the inverse solution.

20.4 OPTIMIZATION ALGORITHMS

A successful inversion method also needs an efficient and robust inversion algorithm. The inherent topographical complexity of the nonlinear multi-dimensional objective functions encountered when estimating soil-crop parameters from transient experiments limits the classical gradient-based local search optimization algorithms to converge to the optimal solution (Si and Kachanoski, 2000; Abbaspour et al., 2001; Vrugt and Bouten, 2002; Lambot et al., 2002; Vrugt et al., 2003). Earlier applications of inverse methods for estimating unsaturated flow properties, for instance, emphasized the difficulty in estimating simultaneously more than two hydraulic parameters (Kool and Parker, 1988; Toorman et al., 1992; Eching and Hopmans, 1993). When gradient-based local optimization algorithms such as the traditional Levenberg-Marquardt method (Marquardt, 1963) are used, local minima may constitute traps which cause the inverse solution to be very sensitive to initial parameter guesses. To overcome this, more efficient and reliable global search optimisation algorithms have been proposed. For instance, Abbaspour et al. (1997) presented the sequential uncertainty domain parameter fitting method (SUFI) and demonstrated for different problems of increasing complexity its stability and good convergence properties. Takeshita and Kohno (1999) investigated genetic algorithms to estimate simultaneously the saturated hydraulic conductivity and parameters α and n of the van Genuchten water-retention model. Vrugt et al. (2001) combined a genetic algorithm with a local simplex optimization algorithm to infer simultaneously hydraulic and root water uptake parameters. More recently, Lambot et al. (2002, 2004) took advantage of both heuristic and stochastic global optimization methods with the global multilevel coordinate search algorithm (Huyer and Neumaier, 1999) to identify hydraulic parameters during natural infiltration events. Vrugt et al. (2003) used the shuffled complex evolution metropolis algorithm to investigate identifiability in different parametric models.

Excellent tools have been developed over the years to inverse transient flow data such as, e.g., ONESTEP (Kool et al., 1985), SFIT (Kool and Parker, 1987), RZWQM (Ahuja et al., 1999), and PEST (Doherty et al., 1995). Solute transport parameters are often obtained from column experiments assuming steady-state water flow and using parameter estimation codes such as CFITIM (van Genuchten, 1981) or CXTFIT (Toride et al., 1995) for fitting analytical solutions of the transport equation to experimental break-through curves. More recently, HYDRUS-1D (Šimůnek et al., 1998) has been developed to allow for simultaneous inversion of soil hydraulic and solute transport parameters, including simulations involving linear and nonlinear solute transport during either steady-state or transient water flow. Table 20.1 reports different optimization algorithms commonly used in the field of soil hydrology.

TABLE 20.1
Optimization Algorithms Commonly Used in Soil Hydrology[a]

Algorithm	Type	Ref
Levenberg-Marquardt	Local	Marquardt, 1963
Gauss-Newton	Local	Gill et al., 1981
Nelder-Mead simplex (downhill simplex)	Local	Nelder and Mead, 1965; Lagarias, 1998
Sequential uncertainty domain parameter fitting	Global	Abbaspour et al., 1997
Simulated annealing	Global	Ingber, 1996
Annealing-simplex	Global	Pan and Wu, 1999
Genetic	Global	Michalewicz, 1996, Vrugt et al., 2001
Shuffled complex evolution metropolis	Global	Vrugt et al., 2003
Multilevel coordinate search	Global	Huyer and Neumaier, 1999; Lambot et al., 2002

Global search algorithms are recommended to get over the complex topography of the objective functions usually encountered in soil hydrology inverse problems.

20.5 ASSESSING THE WELL-POSEDNESS OF THE INVERSE PROBLEM

A major issue with the use of inversion techniques for estimating soil and crop properties is related to the well-posedness of the inverse problem, i.e., the identifiability, uniqueness, and stability conditions must be satisfied (Carrera and Neuman, 1986b; Hopmans et al., 2002). For evaluating the well-posedness of a proposed inversion, response surface analysis, uncertainty analysis, and stability analysis can be performed.

20.5.1 RESPONSE SURFACE ANALYSIS

Response surface analysis allows one to document the problems related to nonuniqueness, model sensitivity, and parameter dependency in an objective and transparent way. Response surfaces are two-dimensional contour plots representing the objective function as a function of two parameters, while all other parameters are held constant at their true value. They represent, therefore, only cross sections of the full M-dimensional parameter space. The following numerical example illustrates how information about these properties can be derived from two-dimensional response surface analysis. In the chosen example, we assume that two soil hydraulic parameters, namely, n and α of the van Genuchten model (see Chapter 3) describing the water retention curve, and a water stress parameter, namely, the h_4 parameter in the root water uptake reduction model of Feddes (1978), are estimated by an inverse procedure using actual evapotranspiration fluxes ET_a) of a cropped soil. This could be easily conceivable at the lysimeter scale or even at the field scale if ET_a is estimated by remote sensing (see Jhorar et al., 2002). We consider here that ET_a cumulated over the considered period (28 days) constitutes the only source of information included within the objective function expressed therefore as follows:

$$\Phi(\mathbf{p}) = \frac{1}{\sigma_{ETa}^2} \sum_{i=1}^{N} \left[ETa_i - ET_a^*(x_i, \mathbf{p}) \right]^2 \qquad (20.11)$$

where $\mathbf{p}$ is the model parameter vector; ET_a and $ET_a^*(x_i, \mathbf{p})$ are, respectively, the "measured" (in our case generated with a reference run) and the simulated cumulated ET_a. To produce two-dimensional response surfaces, we first generate a reference numerical run using the SWAP model (van Dam et al., 1997) containing a soil water transport module allowing the simulation of the water transport within a medium-fine textured soil with a fully developed nongrowing maize crop (LAI = 4). Reference hydraulic parameters describing the moisture retention and the hydraulic conductivity curves of the considered soil are $\theta_s = 0.4$, $\theta_r = 0$, $\alpha = 0.004$, $n = 1.2$, $K_{sat} = 30 \text{ cm d}^{-1}$, and $\lambda = 0.5$. The bottom boundary condition is selected as free drainage at 160 cm depth (bottom of the soil profile), while upper boundary conditions are defined by crop development and driving forces. Two different climatic scenarios are considered, i.e., a very dry (ET_o Rain = 1.7) and a very wet (ET_o/Rain = 0.7) corresponding to climatic conditions encountered in Belgium between between July 6, and August 2 of 1990 and 1998, respectively. In a first step, two reference runs are launched to produce two ET_a values cumulated over the considered period (28 days). Then, numerical simulations are again performed for every combination of two parameter values (including systematically h_4 and n or α) within a predefined parameter space discretized into 50 discrete values. For each 2500 (50 × 50) simulations, values of Φ are calculated with Eq. (20.11). Note that in this step, all the other parameters of the system are kept at constant values similar to those of the reference simulation. Next, two-dimensional contour plots are drawn to show the response surface of Φ for all the different tested parameter combinations. Figure 20.4 represents response surfaces of the objective function $\Phi(\mathbf{p})$ for two different parameter planes, h_4 and n (a, b), and h_4 and α (c, d).

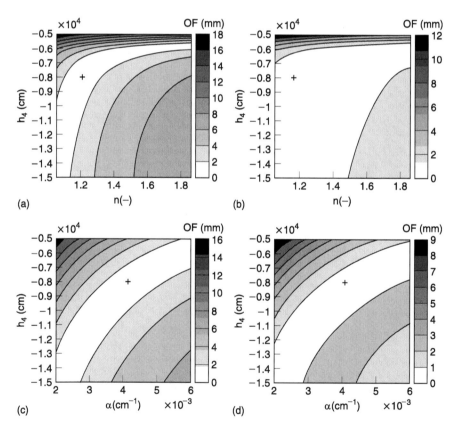

FIGURE 20.4 Two-dimensional response surfaces of the objective function in the α-n-h parameter space.

Figure 20.4a and 20.4c are given for 1990 (very dry scenario) while Figure 20.4b and 20.4d are given for 1998 (very wet scenario). Note that the plane $n - \alpha$ is not illustrated. The stars in these figures indicate the true parameter values used in the reference runs (i.e., $h_4 = -8000$ cm, $n = 1.2$ and $\alpha = 0.004$ cm^{-1}). Visual observation of Figure 20.4a–d provides very valuable information about uniqueness, sensitivity, and parameter dependence. First, all the figures show that zero or close to zero, Φ values can be found through nearly the entire considered parameter space. Very similar cumulated ET_a values can be produced by many different parameter combinations. For instance, very good fits can be produced in Figure 20.4a either with $n = 1.1$ and $h_4 = -14.000$ cm or with $n = 1.7$ and $h_4 = -6.500$ cm. These figures clearly show that the solution is nonunique and suggest that, irrespective of the optimization algorithm used, the inversion will lead to optimized parameters that can be not at all physical although producing very good fits in terms of ET_a. Figure 20.4b shows that the nonuniqueness problem is further exacerbated for the 1998 scenario compared to the 1990 scenario. Indeed, for very wet conditions ET_a fluxes are mainly controlled by the atmosphere and not by the soil-plant continuum. Visual observation of Figures 20.4a,c,d also put forward that the considered parameter pairs are positively correlated (large valley oriented from the bottom left to the upper right corners of the plots). This means that the increase of one of the two parameters can be compensated by the increase of the other one in order to obtain a similar small Φ value. In terms of sensitivity, Figure 20.4a shows that the sensitivity of ET_a to n is larger than to h_4 (at least around the true parameter values) as the topography of the response surface is steeper horizontally than vertically. This means implicitly that the confidence intervals produced on h_4 will be larger than on n (see further stability). It is certainly worth emphasizing that the different parameter planes presented above represent only cross-sections of the full tri-dimensional parameter space (i.e., h_4, n, and α). Consequently the behavior of the objective function in these parameter planes can only suggest how the objective function might behave in the three-dimensional continuum. For example, local minima of Φ could exist and not show up in the cross-sectional planes (e.g., Šimůnek and van Genuchten, 1996). Visual inspection of the response surfaces is therefore not always enough to claim that the solution is unique. It is also worth notify that uniqueness, as sensitivity and parameter dependence, can be investigated with alternative techniques. Uniqueness can be studied by launching the optimization procedure on different scenarios of synthetic data for which the true global minimum is known a priori (e.g., Lambot et al., 2002). If the optimization algorithm finds the true parameter values in each scenario, the solution is unique.

With regard to the sensitivity of the parameters, the alternative method is to compute sensitivity coefficients to directly compare the sensitivity of the model to different parameters (see Hupet et al., 2003). For parameter dependence, the alternative approach is to use classical statistical analysis, allowing one to derive correlation coefficients derived from the variance-covariance matrix (e.g., see Kool and Parker, 1988).

20.5.2 VALIDITY

It may be expected that numerically validated inversion methods, i.e., methods that have been shown to converge to unique and stable parameter solutions in numerical experiments, may fail if model errors are too large. This may become particularly important when applying inverse modeling to real-world data where system behavior may be influenced by a series of processes that are not included or well conceived within the forward model. In unsaturated flow hydrology, for instance, processes such as hysteresis, preferential flow, air entrapment, and spatio-temporal variability of the flow properties are important to describe *in situ* flow but are often poorly conceived in the forward model. In such cases, the forward model will limit the applicability of the inverse modeling scheme.

Model errors (model adequacy) can be quantified using validation criteria based on the notion of likelihood (Press et al., 1997). The goodness of fit criteria can be divided in two classes, the first taking into account the number of fitting parameters (as the Akaike criterion or the root mean-square error), the other observed and measured values (as the determination coefficient r^2, for instance) (Šimůnek and Hopmans, 2002).

20.5.3 UNCERTAINTY ANALYSIS

Uncertainties in the optimized parameter set originate from either experimental errors (or measurement errors) or model errors (including errors in the numerical resolution, among others). Both may have a systematic and a random component. However, only the statistical or random error can be assessed statistically. Moreover, before computing parameter confidence intervals, the model adequacy should be checked, so that model errors can be considered negligible.

Formulating the objective function as a maximum likelihood estimator enables the evaluation of the adequacy of a model for certain given observations (Press et al., 1997; Hollenbeck and Jensen, 1998). At the global minimum, the objective function follows a chi-square distribution, and the probability of model adequacy is then expressed as:

$$p_{adeq} = 1 - Q(\min \Phi(\mathbf{p}), N - P) \tag{20.12}$$

where $Q(\min \Phi(\mathbf{p}), N - P)$ is the chi-square cumulative density function, N is the number of observations, and P is the number of optimized parameters. The definition of the threshold probability for an adequate model is subjective, but usually $p_{adeq} > 0.5$ is used.

If the measurement errors are normally distributed and if the model is linear in its parameters or the sample size is large enough such that the uncertainties in the fitted parameters do not extend outside a region in which the model could be replaced by a suitable linearized model, then it is possible to define an approximate confidence interval (CI) with analytical formulation (Press et al., 1997). The method is based on the Cramer-Rao theorem and can be found elsewhere (Šimůnek and Hopmans, 2002). An exact method consists in computing the contour of the objective function for a given confidence level ε such that:

$$|\Phi(\mathbf{p}) - \min \Phi(\mathbf{p})| \leq \varepsilon \qquad (20.13)$$

If the objective function has been defined to be a general maximum-likelihood estimator, ε is directly related to the inverse chi-square cumulative function for a given level of confidence. Other definitions of the objective function will ask for a relative definition of ε. Bard (1974) defines it as the largest difference between risks that one is willing to consider as being insignificant. It must be noted that this method demands high computing resources but ensures a correct definition of the confidence region.

20.5.4 STABILITY ANALYSIS

The solution will be stable if it depends continuously on the measured system response so that it is not very sensitive to measurement and modeling errors, i.e., small measurement and modeling errors do not result in large changes of the optimized parameters. Measurement errors include errors in the times series of the measured system response, errors in some fixed parameters, and errors in the specification of the boundary conditions. Instability generally occurs when high measurement noise characterizes the data for the inversion and/or when the sensitivity of the parameters to the input data (i.e., the measured response of the considered system) is too low.

To illustrate stability/instability issues, the following example will be used. Assume that two different root water uptake parameters (i.e., the rooting depth, RD, and the bottom relative root length density, $BRLD$) are estimated by an inverse procedure using soil water content measured daily at different depths [each 10 cm between the soil surface and the bottom of the soil profile (160 cm depth)]. In a first step, a reference run is generated with the SWAP model (van Dam et al., 1997) for a long dry period of 28 days corresponding to actual climatic conditions in Belgium encountered between July 6 and August 2, 1999. The reference run and the subsequent numerical simulations are generated for a homogeneous soil profile, 160 cm deep with free drainage as the lower boundary condition. The hydraulic properties of the soil are similar to those used in the example presented in Section 20.5.1. The soil is assumed to be covered by a fully developed and nongrowing maize crop characterized by a LAI of 4, a crop coefficient of 1.1, and a rooting depth of 100 cm, a top relative root length density of 1, and a bottom relative root length density of 0.5. After generating the reference run producing reference time series of 420 (15 depths × 28 days) soil water content data, we investigate the stability of the solution (optimized *RD* and *BRLD* parameters) with respect to measurement errors on soil water content data. We choose five different levels of measurement errors (normally distributed with zero mean), i.e., 0.005, 0.01, 0.015, 0.02 and 0.025 in terms of standard deviation (σ), corresponding typically to soil water calibration equations of different quality. Subsequently, we add the chosen noise on the time series of the soil water content data. This is repeated 250 times for each level of error, producing 1250 different time series of soil water content data. Then, for each corrupted time series, we optimize with PEST (Doherty et al., 1995) separately *RD*, *BRLD*, and both parameters simultaneously, resulting in 250 × 3 × 5 optimized parameter sets. For each level of error and for each combination of parameters (*RD*, *BRLD*, *RD*, and *BRLD*), we obtain a distribution of the 250 optimized parameters. After visual checking for normality of the obtained distributions, confidence intervals were constructed with the true distribution of the 250 parameters. When *RD* and *BRLD* were simultaneously optimized, the correlation coefficient between the two parameters was also determined. The range of the produced confidence intervals directly constitutes a measure of the stability of the solution. Results of the stability analysis generated with the five levels of error on soil water content "measurements" are presented in Table 20.2. Means and individual parameter confidence intervals at 95% are presented separately for the optimization of *RD*, *BRLD*, and both simultaneously. For the *RD*, we observe that the range of the confidence intervals are quite small

TABLE 20.2
Impact of Measurement Errors in Soil Water Content on the Optimized RWUP[a]

| Level of error (std) | 1 parameter | | | | 2 parameters simultaneously | | | | |
| | Rooting depth | | Root length density | | Rooting depth | | Root length density | | |
	mean	CI	mean	CI	mean	CI	Mean	CI	R
0.005	99.89	96.6–103.1	0.498	0.35–0.64	99.6	87.7–111.4	0.52	0.24–0.86	−0.86
0.01	100.04	93.5–106.6	0.509	0.19–0.83	99.5	79–119.9	0.505	0.08–0.92	−0.90
0.015	99.98	89.8–110.1	0.517	0.15–0.88	—	—	—	—	—
0.02	100.05	87.4–112.6	0.5305	0.1–0.97	—	—	—	—	—
0.025	99.61	83.5–115.6	0.541	0.03–1.03	—	—	—	—	—

CI = Bounds of the 95% confidence interval, *r* = correlation coefficient. Rooting depth and root length density values are expressed in cm and cm cm^{-3}, respectively.

(6.5 < CI < 32 cm) and that the ranges increase as the magnitude of the "measurement" error increases. This clearly reveals that the instability of the solution increases with measurement errors. Confidence intervals for *BRLD* are much larger, with values covering a large proportion of the predefined parameter space already when considering the second level of error. This means that small errors on soil water content measurements have a very large effect on the stability of the optimized *BRLD*. Note that the difference in stability of identifying *RD* and *BRLD* is a direct consequence of the difference of sensitivity for these two parameters. When *RD* and *BRLD* are optimized simultaneously, results are markedly worse. Indeed, for the second error level, the confidence interval of *BRLD* (0.08–0.92) is nearly the same as the size of the initial parameter space (0–1) meaning that uncertainty on *BRLD* generated by the parameter estimation method is virtually equivalent to the original uncertainty. Given these results, optimizations for the next three error levels were not performed. Note that optimizing both parameters simultaneously substantially increases the confidence intervals as compared to the case when the parameters are optimized separately. Indeed, parameters *RD* and *BRLD* are strongly negatively correlated (see correlation coefficients in Table 20.2), meaning that a small increase of *RD* can be compensated by a small decrease of *BRLD* and can lead to very similar soil water content dynamics. The high correlation between *RD* and *BRLD* is clearly not optimal in terms of parameter estimation, as it could substantially contribute to the nonuniqueness of the solution. Note that instability issues can also be investigated with respect to errors on some fixed parameters of the system. In the above-mentioned example only root water uptake parameters were optimized, but all the other parameter values of the system (i.e., all soil hydraulic parameters) have to be specified. Specification of the fixed parameters is generally performed using measured parameter values for which measurement errors are unavoidable. Further, errors are generally introduced, as the measurement scale for the fixed parameter is very different from the scale at which the inverse procedure is performed.

REFERENCES

Abbaspour, K.C., M.T. van Genuchten, R. Schulin, and E. Schlappi, 1997. A sequential uncertainty domain inverse procedure for estimating subsurface flow and transport parameters. Water Res. Res. 33(8): 1879–1892.

Abbaspour, K.C., R. Schulin, and M.T. van Genuchten, 2001. Estimating unsaturated soil hydraulic parameters using ant colony optimisation. Adv. Water Res. 24: 827–841.

Ahuja, L.R., K.W. Rojas, J.D. Hanson, M.J. Shaffer, and L. Ma, 1999. Root Zone Water Quality Model. Modeling Management Effects on Water Quality and Crop Producion (Beta Copy). Water Resources Publications, LLC, CO.

Bard, Y., 1974. Nonlinear Parameter Estimation. Academic Press, New York. USA.

Berger, J.O., 1985. Statistical Decision Theory and Bayesian Analysis. Springer, New York.

Beven, K., and A. Binley, 1992. The future of distributed models: model calibration and uncertainty predictions. Hydr. Proc. 6: 279–298.

Beven, K., and J. Freer, 2001. Equifinality, data assimilation, and uncertainty estimation in mechanistic modelling of complex environmental systems using the GLUE methodology. J. Hydro. 249: 11–29.

Carrera, J., and S.P. Neuman, 1986a. Estimation of aquifer parameters under transient and steady state conditions: 1. Maximum likelihood method incorporating prior information. Water Res. Res. 22: 199–210.

Carrera, J., and S.P. Neuman, 1986b. Estimation of aquifer parameters under transient and steady state conditions: 2. Uniqueness, stability, and solution algorithms. Water Res. Res. 22: 211–227.

Dane, J.H., and S. Hruska, 1983. In situ determination of soil hydraulic properties during drainage. Soil Sci. Soc. Am. J. 47: 619–624.

Doherty, J., L. Brebber, and P. Whyte, 1995. PEST, Model independent parameter estimation. Australian Centre for Tropical Freshwater Research, James Cooke University, Townsville, Australia.

Duan, Q.S., S. Sorooshian, and V.K. Gupta, 1992. Effective and efficient global optimization for conceptual rainfall-runoff models. Water Resources Res. 28: 1015–1031.

Durner, W., B. Schultze, and T. Zurmul, 1999. State-of-the-art in inverse modeling of inflow/outflow experiments. In: Leij, F.J. and M.T. van Genuchten (eds.), Proceedings of the International Workshop on Characterization and Measurement of the Hydraulic Properties of Unsaturated Porous Media, Riverside, CA.

Eching, S.O., and J.W. Hopmans, 1993. Optimization of hydraulic functions from transient outflow and soil-water pressure data. Soil Sci. Soc. Am. J. 57: 1167–1175.

Feddes, R.A., P.J. Kowalik, and H. Zaradny, 1978. Simulation of field water use and crop yield. Simulation Monographs, Pudoc, Wageningen.

Ferré T.P.A., and G.J. Kluitenberg, 2003. Advances in measurement and monitoring methods: preface from the guest editors. Vadose Zone J. 2: 443.

Gardner, W.R., 1956. Calculation of capillary conductivity from pressure plate outflow data, Soil Sci. Soc. Am. Proc. 20: 317–320.

Gill, P.E., W. Murray, and M.H. Wright, 1981. Practical Optimization. Academic Press, London.

Gribb, M.M., 1996. Parameter estimation for determining hydraulic properties of a fine sand from transient flow measurements. Water Res. Res. 32: 1965–1974.

Gupta, V.K., and S. Sorooshian, 1985. Uniqueness and observability of conceptual rainfall-runoff model parameters. The percolation of process examined, Water Resources Res. 19: 269–276.

Hollenbeck, K.J., and K.H. Jensen, 1998. Maximum-likelihood estimation of unsaturated hydraulic parameters. J. Hyd. 210: 192–205.

Hopmans, J.W., J. Šimůnek, N. Romano and W. Durner, 2002. Inverse methods. In: J.H. Dane and G.C. Topp (eds.), Methods of Soil Analysis, Part 4, Physical Methods. Soil Sci. of Am. Book Series 5, Madison, WI.

Huyer, W., and A. Neumaier, 1999. Global optimization by multilevel coordinate search, J. Global Optimization 14: 331–355,.

Hupet F., S. Lambot, R.A. Feddes, J.C. van Dam, and M. Vanclooster, 2003. Estimating of root water uptake parameters by inverse modeling with soil soil water content data. Water Res. Res. 39(11), 1312, doi: 10.1029/2003 WR002046.

Ingber, L., 1996. Adaptive simulated annealing (ASA): lessons learned, Control Cybernet. 25, 33–54.

Javaux, M., and M. Vanclooster, 2003. Scale- and rate-dependent solute transport within an unsaturated sandy monolith. Soil Sci. Soc. Am. J. 67: 1334–1343.

Jhorar, R., W. Bastiaanssen, R. Feddes, and J.C. van Dam, 2002. Inversely estimating soil hydraulic functions using evapotranspiration fluxes, J. Hydrol. 258: 198–213.

Klute, A., and C. Dirksen, 1986. Methods of Soil Analysis, Part I. ASA and SSSA, Madison, WI.

Kool, J.B., J.C. Parker, and M.T. van Genuchten, 1985a. Determining soil hydraulic properties from one-step outflow experiments by parameter estimation: theory and numerical studies. Soil Sci. Soc. Am. J. 49: 1348–1354.

Kool, J.B., J.C. Parker, and M.Th. van Genuchten, 1985. ONESTEP: A nonlinear parameter estimation program for evaluating soil hydraulic properties from one-step outflow experiments. Bulletin 85-3, Virginia Polytechnic Institute and State University, Blacksburg, Virginia.

Kool, J.B., and J.C. Parker, 1987. Estimating soil hydraulic properties from transient flow experiments: SFIT User's guide. Electric Power Research Institute Report, Palo Alto, CA.

Kool, J.B., and J.C. Parker, 1988. Analysis of the inverse problem for transient unsaturated flow. Water Resour. Res. 24: 817–830.

Lagarias, J.C., J.A. Reeds, M.H. Wright, and P.E. Wright. 1998. Convergence properties of the Nelder-Mead simplex algorithm in low dimensions. SIAM J. Optim. 9: 112–147.

Lambot, S., M. Javaux, F. Hupet, and M. Vanclooster, 2002. A global multilevel coordinate search procedure for estimating the unsaturated soil hydraulic properties. Water Res. Res. 38(11), 1224, doi: 10.1029/2001WR001224.

Lambot, S., F. Hupet, M. Javaux, and M. Vanclooster, 2004a. Laboratory evaluation of a hydrodynamic inverse modeling method based on water content data. Water Res. Res. 40(3), W03506, doi: 10.1029/2003WR002641.

Lambot, S., M. Antoine, I. van den Bosch, E. Slob, and M. Vanclooster, 2004b. Electromagnetic inversion of the GPR signal and subsequent hydrodynamic inversion to reconstruct effective values of the vadose zone hydraulic properties. Vadose Zone J. (Special Issue), V04–0006 (in press).

Leij, F. J., and M.T. an Genuchten (eds.), 1999. Characterization and measurement of the hydraulic properties of unsaturated porous media. Proceedings of the International Workshop on Characterization and Measurement of the Hydraulic Properties of Unsaturated Porous Media.

Marquardt, D.W., 1963. An algorithm for least-squares estimation of nonlinear parameters. J. Soc. Indust. Appl. Math. 11: 431–441.

Michalewicz, Z., 1996. Genetic Algorithms + Data Structures = Evolution Programs, 3rd ed. Springer-Verlag, New York.

Mishra, S., and J.C. Parker, 1989. Parameter estimation for coupled unsaturated flow and transport. Water Res. Res. 25: 385–396.

Nelder, J.A., and R. Mead, 1965. A simplex method for function minimization. Comp. J. 7: 308–313.

Pan, L., and L. Wu, 1999. Inverse estimation of hydraulic parameters by using simulated annealing and downhill simplex method. In: Leij, F.J. and M.T. van Genuchten (eds.), Characterization and Measurement of the Hydraulic Properties of Unsaturated Porous Media. Proceedings of the International Workshop on Characterization and Measurement of the Hydraulic Properties of Unsaturated Porous Media, Riverside, CA.

Parker J.C., and M.Th. Van Genuchten, 1984. Determining transport parameters from laboratory and field tracer experiments. Bull 84–3. Virginia Agric. Exp. Stn., Blacksburg, VA.

Press, W.H., S.A. Teukolsky, W.T. Vetterling, and B. P. Flannery, 1997. Numerical Recipes in Fortran 77: The Art of Scientific Computing. Canbridge University Press, New York.

Romano, N., and A. Santini, 1999. Determining soil hydraulic functions from evaporation experiments by a parameter estimation approach: experimental verifications and numerical studies. Water Res. Res. 35: 3343–3359.

Si, B.C., and R.G. Kachanoski, 2000. Estimating soil hydraulic properties during constant flux infiltration: Inverse procedures. Soil Sci. Soc. Am. J. 64: 439–449.

Šimůnek, J., and Hopmans, J.N., 2002. Parameter optimization and nonlinear fitting. In: J.H. Dane and G.C. Topp (eds.), Methods of soil Analysis, part 4, Physical Methods. Soil Sci. of Am. Book Series 5, Madison, WI.

Šimůnek, J., M. Sejna, and M.T. van Genuchten. 1998. The HYDRUS-1D software package for simulating water flow and solute transport in two-dimensional variably saturated media. Version 2.0, IGWMC-TPS-70, International Ground Water Modeling Center, Colorado School of Mines, Golden, CO.

Šimůnek, J., and M.T. van Genuchten, 1996. Estimating unsaturated soil hydraulic properties from tension disc infiltrometer data by numerical inversion. Water Res. Res. 32: 2683–2696.

Takeshita, Y., and I. Kohno, 1999. Parameter estimation of unsaturated hydraulic properties from transient outflow experiments using genetic algorithms. In: Leij, F.J. and M.T. van Genuchten (eds.), Characterization and Measurement of the Hydraulic Properties of Unsaturated Porous media. Proceedings of the International Workshop on Characterization and Measurement of the Hydraulic Properties of Unsaturated Porous Media, Riverside, CA.

Toorman, A.F., P.J. Wierenga, and R.G. Hills, 1992. Parameter estimation of hydraulic properties from one-step outflow data. Water Res. Res. 28: 3021–3028.

Toride, N., F.J. Leij, and M.T. van Genuchten. 1995. The CXTFIT code for estimating transport parameters from laboratory or field tracer experiments. Version 2.0, Research Report No. 137, US Salinity Laboratory, USDA, ARS, Riverside, CA.

van Dam, J.C., J.N.M. Stricker, and P. Droogers, 1994. Inverse methods to determine soil hydraulic functions from multistep outflow experiments. Soil Sci. Soc. Am. J. 58: 647–652.

van Dam, J.C., J. Huygen, J.G. Wesseling, R.A. Feddes, P. Kabat, P.E.V. van Walsum, P. Groenendijk, and C.A. van Diepen, 1997. Theory of SWAP version 2.0. Simulation of water flow, solute transport and plant growth in the Soil-Water-Atmosphere-Plant environment, 167 pp., Report 71, Department Water Resources, Wageningen University, Technical Document 45, Alterra, Wageningen, The Netherlands.

van Genuchten, M.T., 1981. Non-equilibrium transport parameters from miscible displacement experiments. Research Report No. 119, U.S. Salinity Laboratory, USDA, ARS, Riverside, CA.

Vrugt, J.A., and W. Bouten, 2002. Validity of first order approximations to describe parameter uncertainty in soil hydrologic models. Soil Sci. Soc. Am. J. 66: 1740–1751.

Vrugt, J.A., H.V. Gupta, W. Bouten, and S. Sorooshian, 2003. A shuffled complex evolution metropolis algorithm for optimization and uncertainty assessment of hydrologic model parameters. Water Res. Res. 39, Doi: 1201 10.1029/2002WR001642.

Vrugt, J.A., W. Bouten, and A.H. Weerts, 2001. Information content of data for identifying soil hydraulic parameters from outflow experiments. Soil Sci. Soc. Am. J. 65: 19–27.

Vrugt, J.A., W. Bouten, H.V. Gupta, and S. Sorooshian, 2002. Toward improved identifiability of hydrologic model parameters: the information content of experimental data. Water Res. Res. 38, Doi: 10102g/2001 WR00–1118.

Wendroth, O., W. Ehlers, J.W. Hopmans, H. Kage, J. Halbertsma, and J.H.M. Wosten, 1993. Reevaluation of the evaporation method for determining hydraulic functions in unsaturated soils. Soil Sci. Soc. Am. J. 57: 1436–1443.

Yeh, T.C., and J. Šimůnek, 2002. Stochastic fusion of information for characterizing and monitoring the vadose zone. Vadose Zone J. 1: 207–221.

Zachmann, D.W., P.C. DuChateau, and A. Klute, 1982. Simultaneous approximation of water capacity and soil hydraulic conductivity by parameter identification. Soil Sci. 134: 157–163.

21 Computer Models for Characterizing the Fate of Chemicals in Soil: Pesticide Leaching Models and Their Practical Applications

Anna Paula Karoliina Jantunen,
Marco Trevisan and Ettore Capri
Istituto di Chimica Agraria ed Ambientale, Università
Cattolica del Sacro Cuore, Piacenza, Italy

CONTENTS

1-5667-0657-2/05/\$0.00 + \$1.50
© 2005 by CRC Press

21.1. INTRODUCTION: STATE OF THE ART ON THE USE OF PESTICIDE LEACHING AND DISSIPATION MODELS

A model may be defined as a mathematical description of a simplified conceptual representation of a certain fragment of the real world. The relationships between different processes are made explicit, translated into mathematical symbols, and inserted into algorithms that simplistically attempt to imitate reality. Once it has been established that a certain model is capable of reproducing reality with sufficient accuracy, the model may then be used to simulate reality itself at significantly lower cost and in much shorter time than experimental field trials require. In the near future, chemical-fate models are likely to have an important role in the registration and environmental risk assessment of new products.

Pesticide-fate models that predict the contamination of aquifers are an example of mathematical models that are constantly being developed and increasingly used for both scientific and normative purposes. While these models were developed with pesticide usage in agricultural contexts in mind, they can in fact be used for describing or predicting the environmental behavior of many xenobiotics in soils as long as the model sufficiently considers all the important characteristics of the chemical and the soil and the situation in question.

In order to help the reader decide whether a pesticide-fate model might serve a specific purpose, this chapter outlines the important environmental fate processes for pesticides that are spread on agricultural fields, how these processes are simulated by pesticide-fate models, and the general logic and construction of such models. We also discuss the practical aspects of choosing a suitable model for a given purpose and using it correctly, and introduce a selection of recent pesticide-fate models and their practical applications as found in literature.

21.1.1 Model Selection

No single pesticide-fate model has so far proved a satisfactory tool for all purposes (Wagenet, 1993). When such a model is needed, one has to be chosen from among a variety of available models on the basis of how they suit needs of the user. A selection of current models is introduced in Sec. 21.3.2.

Before selecting a model, the potential user should identify the purpose and objectives of the study or work, the exact objectives of the planned modeling, and the data, time, computer systems, and modeling expertise available. This helps to narrow down a model that will produce satisfactory results without demanding too much in terms of time and resources.

Various model classification schemes have been proposed to help a user select the model most suitable for a specific purpose (Addiscott and Wagenet, 1985; Wagenet, 1993; Vanclooster et al., 1994; Del Re and Trevisan, 1995). These schemes can help to answer a range of important questions about the model and thus improve the potential user's understanding of how the model functions and what it can and cannot do. Some of the issues that should be considered before applying any given model are discussed in the following subsections.

21.1.1.1 Purpose of the Model

Leaching models have been developed for a variety of different purposes — education, research, regulatory screening, etc. Specific requirements for models used for different purposes are discussed in Sec. 21.3.3. It should be carefully considered whether a model can be successfully applied for a purpose that fun0damentally differs from the one for which the model was originally designed.

The primary purpose of educational models is demonstration. While requirements of such models depend on the specific pedagogical goal, they generally do not need to produce quantitatively correct results or even address all the major chemical-fate processes.

Research models are tools for the analysis of experiment results or field data. After an agreement has been reached between data and simulation results (usually through model calibration), these models may also be used to extrapolate from data, assuming, e.g., different properties of the applied chemical or different environmental conditions. A research model must adequately address all chemical-fate processes considered significant in a given situation. Research models are frequently highly complicated and can be challenging to understand and use.

Regulatory models are needed for predicting the unknown outcome of a given pesticide application on the basis of existing information and knowledge. Predictions may need to be quantitative (e.g., concentrations in groundwater) or comparative (which of scenarios A or B is likely to lead to less serious contamination of groundwater?). Regulatory models must both have a high degree of confidence and be relatively simple, straightforward, and unambiguous to use.

21.1.1.2 Processes Considered by the Model

The reality and range of processes addressed by leaching models are discussed in Sec. 21.2.1. Obviously, the choice of a suitable model is based largely on whether or not the model can deal with the part of reality of interest. For example, a model limited to the soil column (such as VARLEACH; see Table 21.1) cannot deal with the ultimate fate of chemicals taken up by plants or removed from topsoil in runoff. Similarly, a model suitable for a given purpose adequately simulates all processes of interest or importance.

21.1.1.3 Scale

Models are designed to work at different spatial and temporal scales.

21.1.1.3.1 Temporal Scale

The lengths of both the total simulation period and the time step of the simulation may be fixed or restricted. A majority of pesticide-fate models do not limit the maximum duration of a simulation; some have a finite nonmodifiable length — in the case of VARLEACH, one crop cycle. The minimum duration of the simulation is normally one time step. A common time step in pesticide-fate models is one day, although it can sometimes be chosen by the user or fluctuate during the simulation. The time scale is of particular importance for the generation of output.

21.1.1.3.2 Spatial Scale

All models have been designed to operate at some specific spatial scale: soil column, lysimeter, plot, field, basin, district, or region. Since different processes are important at different scales, models are generally considered most useful when applied at the scale for which they were originally developed, although some recent models appear to be more versatile and may also be used at different scales. As the scale grows larger, significant variation of, e.g., the soil and the climate starts to occur inside the simulated area, but originally small-scale models can still be applied if the area is divided into smaller plots that can be considered uniform. Spatial scale also involves the extent of the simulated soil profile, which normally consists of either the root zone or the unsaturated zone (possibly together with the water table or some of the saturated zone of soil) to some specified maximum depth. The thicknesses of the computational soil layers are also important, as these have critical influence on the results of the simulation; they may interfere with dispersion estimates when layers are too thin, or flatten out distributions when they are too thick. Excessively thick layers are particularly harmful when using capacity models. Sometimes the thicknesses of layers are fixed: e.g., VARLEACH uses 1 cm layers, while in LEACHM (see Table 21.1) all layers are equally thick.

21.1.1.4 Construction of the Model

The most important questions are: what approaches have the model developers taken, how have he or she solved the basic processes, and how are these reflected in the structure of the model? The developer's vision of how reality works, and their understanding of how the basic processes operate, are reflected in the structure of the model.

On the most general level, models are either deterministic or (less commonly) stochastic: of the former kind if the answer given by the system is fixed when given a certain set of assumptions and initial conditions, of the latter if uncertainty is incorporated. Models may be further classified according to the numerical complexity of simulated processes. Mechanistic models address the highest possible number of aspects concerning each simulated process; functional models lump or ignore processes in order to limit calculation time and costs. Few models, however, are clearly and exclusively mechanistic or functional.

The general structure of leaching models is discussed in Sec. 21.3.1. Leaching models are usually built by assembling a number of subroutines, each of which simulates a basic process. A useful way to characterize a model is to identify the subroutines present and the approach chosen for simulating each process; alternative approaches are discussed in Sec. 21.2.2. When complicated equations are involved, several processes may be simulated within the same subroutine: e.g., in PRZM-2 (an older version of PRZM-3; see Table 21.1), a single routine accounts for the partition, convection, diffusion, dispersion, and degradation of pesticide in soil.

Models use mathematics to imitate reality. Leaching models are based on certain differential equations; these transport equations are generally solved with the technique of the finite differences, which approximates them to discrete ones by using finite temporal and spatial increments. This solution, however, introduces some numerical dispersion, which can lead to oscillations

TABLE 21.1

Basic Properties of Current Pesticide Leaching Models: LEACHM, MACRO, PEARL, GLEAMS, PELMO, PRZM-3, and VARLEACH

	LEACHM (3) (Leaching Estimation And CHemistry Model) Hutson and Wagenet (1992, 1995)	MACRO (5.0) Larsbo and Jarvis (2003); Stenemo and Jarvis (2003)	PEARL (1.1) (Pesticide Emission Assessment at Regional and Local scales) Tiktak et al. (2000); Leistra et al. (2001)	GLEAMS (3.0) (Groundwater Loading Effects of Agricultural Management Systems) Leonard et al. (1987); Knisel and Davis (1999)	PELMO (3.0) (PEsticide Leaching MOdel) Klein (1995)	PRZM-3 (previous versions PRZM-2, PRZM) (Pesticide Root Zone Model) Carsel et al. (1998)	VARLEACH Walker (1987); Jones and Schäfer (1995); Businelli et al. (2000)
Designed purpose	Research	Research, prediction of pesticide leaching	Pesticide regulation	Management	Management/ regulation	Management/ regulation	Research
Processes covered	Apl (ss), TrL (conv, dis, dif), TrG (conv, disp, dif), Vol, Srp (Fr), Dgr (1st, tr)^{L,m}, Plu	Apl (ss, inc), Cnp (dis, w), TrL (conv, disp, mass flow in macropores), Srp (Fr), Dgr (1st, 4 separate pools)^{L,m}, Plu, Dr	Apl (ss, cc, inc, inj, chg, res), Cnp (vol, dgr, pen, w), TrL (conv, dif, dis), TrG (dif), Vol^l, Srp (Fr), Dgr (1st, tr)^{L,m,d}, Plu, Dr, Til, effect of temperature on solubility	Apl (ss, cc, inc, inj, chg), Cnp (dgr, w), Run, Er, TrL (conv), TrG, Srp (lin)^{om}, Dgr (1st, tr)^{L,m,d}, Plu	Apl (ss, cc, inc), Cnp (dgr, w), Run, Er, TrL (conv, dif, dis), TrG (dif, dis), Vol, Srp (Fr)^{om,pH}, Dgr (1st, bd)^{L,m,d,bm}, Plu	Apl (ss, cc, inc), Cnp (w, degr, vol), Run, Er, TrL (conv, dif, dis), TrG (dif, dis), Vol^l, Srp (lin), Dgr (1st, tr, particle/ liquid/gas phase)^{t,d}, Plu, changes of bulk density	Apl (ss), TrL (conv), Srp (lin, kin)^{d,a}, Dgr (1st)^{L,m,d} effect of depth on water-holding capacity of soil
Hydrological model	Richards equation, version by Borah and Kalita (1999) also covers macropore flow	Dual-porosity: Richards equation for chromatographic flow, Darcy's law for macropore flow	Richards equation (macropore flow being implemented by FOCUS)	Capacity model, version GLEAMS-CF by Morari and Knisel (1997) covers crack flow	Capacity model (macropore flow being implemented by FOCUS)	Capacity model	Capacity model, two regions (mobile and immobile water)

Input required (for basic input, see Sec 21.3.2)							
Weather: total weekly evaporation; timing, depth and rate of precipitation events; air temperature Soil: basic composition of each segment Hydrology: fairly detailed hydr. characteristics of the soil Crop: basic development schedule, growth parameters, cover fraction at maturity Management: schedule and characterization of pesticide application and irrigation events Pesticide: basic chemical and physical qualities	Weather: detailed daily data from site Soil: very detailed characterization of composition and structure of each horizon (particles, pores) Hydrology: very detailed information on hydrological characteristics of soil (micro- and macropores), spacing of drains Crop: roots in each soil layer; basic schedule of development Pesticide: fairly detailed information about chemical and physical qualities; degradation rates separately for solid phase, liquid phase, micropores, and macropores	Weather: detailed data from site for every time step Soil: fairly detailed characterization of composition of each horizon Hydrology: fairly detailed hydr. characteristics of each soil horizon, data for chosen lower boundary condition, characterisation of drainage systems Crop: detailed data on development, rain interception and radiation extinction Management: schedule and characterization of pesticide application, irrigation, and plough events Pesticide: detailed information about all relevant chemical and physical qualities, including sorption and transformation data from site, and effect of changes in temperature, moisture and depth	Weather: daily precipitation, monthly averages of temperature, radiation, wind Soil: basic composition and structure of each horizon Hydrology: basic hydrological properties of soil Crop: basic development schedule and measurements Pesticide: basic chemical and physical properties	Weather: detailed daily data Soil: basic composition and pH of each horizon Hydrology: basic hydrological properties of each horizon Crop: basic development schedule and maximum measurements, uptake efficiency Management: schedule and characterisation of pesticide application events Site: basic topography of the field, erosion properties of soil surface Pesticide: detailed chemical and physical properties, sorption and degradation data from specific soil, effect of temperature and moisture on biodegradation	Weather: detailed daily data Soil: basic composition and structure of each horizon, albedo and reflectivity of surface Hydrology: basic hydrological properties of each horizon Crop: basic development schedule and maximum measurements, uptake efficiency, foliar extraction Management: schedule and characterisation of pesticide application and irrigation events, drainage conditions, management factor for runoff and erosion Site: basic topography of the field, erosion properties of soil surface Pesticide: detailed chemical and physical properties, sorption and degradation data from specific soil, degradation separate for different phases	Weather: basic daily data Soil: bulk density Hydrology: basic hydr. characteristics of soil Crop: none (not considered) Management: none Site: none Pesticide: basic physical qualities from site, factors for change of K_d and half-life with depth

(Continued)

TABLE 21.1
Continued

	LEACHM (3) (Leaching Estimation And CHemistry Model) Hutson and Wagenet (1992, 1995)	MACRO (5.0) Larsbo and Jarvis (2003); Stenemo and Jarvis (2003)	PEARL (1.1) (Pesticide Emission Assessment at Regional and Local scales) Tiktak et al. (2000); Leistra et al. (2001)	GLEAMS (3.0) (Groundwater Loading Effects of Agricultural Management Systems) Leonard et al. (1987); Knisel and Davis (1999)	PELMO (3.0) (PEsticide Leaching MOdel) Klein (1995)	PRZM-3 (previous versions PRZM-2, PRZM) (Pesticide Root Zone Model) Carsel et al. (1998)	VARLEACH Walker (1987); Jones and Schäfer (1995); Businelli et al. (2000)
Output given (regarding simulated chemical)	Tabular, clear, some flexibility At specified times or time intervals: mass balance for whole profile, vertical profile of chemical content Summary file: chemical fluxes at three depths and bottom of root zone, chemical content in each of three section and in root zone.	Tabular and graphic, clear, flexible, user-friendly At specified time intervals: mass balance for whole profile, fluxes, chemical content in each layer (total, micropores and macropores in liquid and solid phase)	Tabular and graphic, ersatile and flexible Summaries annually and at specified time intervals (selected variables, cumulative or not): selected mass balance terms for soil profile and crop canopy, concentrations in selected phases of soil and mass fluxes at selected depths, concentrations in drain water, percolate and groundwater Summary report for Dutch registration procedure, FOCUS summary	Tabular, flexible (restricted when more than 10 chemicals simulated) Daily/monthly/annual mass losses in runoff, erosion, percolate or total; daily pesticide residual by soil layer; daily soil concentration and concentration in water and particle phase by layer; annual max concentrations in runoff, runoff + sediment and percolation	Tabular and graphic Daily/monthly/annual balance for each layer and for the profile	Tabular; versatile; fluxes and balances, "snapshot" feature produces vertical profiles at specified times	Tabular; fixed; at specified intervals
Spatial scale	Soil column down to max. 2 m	Soil column, unsaturated zone	Soil column, unsaturated zone	Field, root zone	Field, unsaturated zone	Field, unsaturated zone	Soil column
Time step	Maximum step 0.1 d	Time step variable	User defined (generally 1 d)	1 d	1 d	1 d	0.05 d

Sensitivity (leaching)	Short-term: rate of application, K_{OC}/K_d, air entry value, bulk density of soil, water solubility (when low), dispersivity (Smith et al. 1991; Walker et al. (1995) Long-term: half-life, K_d, organic content of soil, depth of water table (Soutter and Musy, 1999)	Freundlich coefficient and exponent, degradation rate and factor for effect of depth, boundary soil water content, pore-size distribution index, bulk density; in case of structures soils, hydrological properties of the soil such as definition of the micropore and macropore regions, water content at saturation, pore-size distribution index (Dubus and Brown, 2002; Dubus et al., 2003)	Freundlich coefficient and exponent, half-life, molar activation energy of degradation, organic matter content, bulk density (Dubus and Beulke, 2001; Dubus et al., 2003)	Half-life, K_{OC}, incorporation depth (Leonard et al., 1987)	Degradation rates, factor for increase of degradation with temperature, reference soil moisture for degradation rate, exponent for effect of moisture on degradation, Freundlich coefficient and exponent, ratio of field capacity to initial soil moisture content, bulk density of soil (Dubus et al., 2003)	Application rate, bulk density of soil, K_d, degradation rates, Freundlich coefficient and exponent, effect of increase of temperature by 10°C on degradation, ratio of field capacity to initial soil moisture content (Wolt et al., 2002; Dubus et al., 2003)	Leaching depth to K_d, water solubility when low, bulk density of soil (Walker et al. 1995) Residue to parameters for correcting degradation constant K_d degradation rate (half-life), Henry constant, field capacity of soil (Del Re and Trevisan, 1993; Walker et al., 1995)
Validation status	Validation status has not been assessed or reviewed in literature	Behavior of the main bulk of chemical well simulated, but blind predictions of concentrations poor; with calibration, generally acceptable simulations of soil hydrology and pesticide fate also in preferential-flow situations; parameterisation of macropore flow difficult (Vanclooster et al., 2003b)	Behavior of the main bulk of chemical well simulated, but blind predictions of concentrations poor; with calibration (of water flow model), performance adequate when preferential flow does not occur (Vanclooster et al., 2003b)	Good conservative leaching estimates generally obtained where preferential flow not significant; persistence in surface soil easily underpredicted (Jones and Mangels, 2002); different versions widely used, although users are typically aware of the field results	Behavior of the main bulk of chemical well simulated, but blind predictions of concentrations poor; with calibration (of water flow model), performance adequate where preferential flow does not occur (new FOCUS version can also simulate pf situations) (Vanclooster et al., 2003b)	Good conservative leaching estimates generally obtained where preferential flow not significant; persistence in surface soil easily underpredicted (Jones and Mangels, 2002); different versions widely used, although users are typically aware of the field results	Validation status has not been assessed or reviewed in literature

(Continued)

TABLE 21.1
Continued

	LEACHM (3) (Leaching Estimation And CHemistry Model) Hutson and Wagenet (1992, 1995)	MACRO (5.0) Larsbo and Jarvis (2003); Stenemo and Jarvis (2003)	PEARL (1.1) (Pesticide Emission Assessment at Regional and Local scales) Tiktak et al. (2000); Leistra et al. (2001)	GLEAMS (3.0) (Groundwater Loading Effects of Agricultural Management Systems) Leonard et al. (1987); Knisel and Davis (1999)	PELMO (3.0) (PEsticide Leaching MOdel) Klein (1995)	PRZM-3 (previous versions PRZM-2, PRZM) (Pesticide Root Zone Model) Carsel et al. (1998)	VARLEACH Walker (1987); Jones and Schäfer (1995); Businelli et al. (2000)
Strengths	Ambitious and versatile description of soil hydrology; model family can simulate soil hydrology (LEACHW), pesticides in soil (LEACHP), nutrients in soil (LEACHN), or soil salinity (LEACHC); fairly widely used; numerous publications; a useful research tool	FOCUS model with graphical user interface and standard EU scenarios; ambitious mechanistic description of soil hydrology, including macropore flow; well documented; many publications; actively developed; a good research tool	FOCUS model with graphical user interface and standard EU scenarios; well documented; actively developed; up-to-date mechanistic descriptions of chemical-fate processes; above-soil processes well addressed	Versatile; widely used; well documented; numerous publications; also addresses agricultural nutrients; runoff and erosion considered; a good, simple, non data-intensive model for comparing, e.g., alternative agricultural management systems	FOCUS model with graphical user interface and standard EU scenarios; runoff and erosion considered; reasonably well documented; many publications; actively developed	FOCUS model with graphical user interface and standard EU scenarios; has a regulatory status in the United States, earlier version (PRZM-2) can be used within the PATRIOT software which supplies U.S. soil and weather scenarios; runoff and erosion considered; well documented; widely used; lots of publications; actively developed	Straightforward and valid pesticide leaching and dissipation model for many research purposes

Weaknesses						
Poorly documented; challenging to parameterize and use for nonexpert users	Data intensive and difficult to parameterize for nonexpert users; version MACRO-DB (Jarvis et al., 1997) is easier to parameterize but also less precise	New model, few publications (predecessors PESTLA and PESTRAS have been applied more); data intensive and difficult to parameterize for nonexpert users	Capacity model; process descriptions simple, no quantitative results can be expected; not actively developed any more	Capacity model	Capacity model	Poorly documented, no above-ground processes or vegetation considered; only one pesticide application allowed; not actively developed any more

Apl = Application (ss = on soil surface, cc = on crop canopy, inc = incorporation, inj = injection, chg = chemigation, res = residuals);
Cnp = canopy processes (w = washoff, dis = (lumped) dissipation, vol = volatilization, pen = penetration, dgr = degradation);
Run = runoff losses;
 = erosion losses;
TrL = transport in the liquid phase of soil, TrG = transport in the gas phase of soil (conv = convection, dif = diffusion, dis = dispersion);
 = sorption (lin = linear, Fr = Freundlich, kin = kinetic)$^{\text{effects: a = aging, t = temperature, m = moisture, om = organic matter, pH, d = depth}}$;
Dgr = degradation (f = first-order, bd = biodegradation separately considered, phas = soil phases separately considered, tr = specific transformations and fate of products can be simulated)$^{\text{effects: t = temperature, m = moisture, d = depth}}$;
 = volatilization$^{\text{effects: t = temperature}}$;
 = plant uptake;
 = lateral drainage;
 = tillage (ploughs).

of simulated pesticide concentrations along the soil profile. Various authors have studied the effects of this dispersion and its influence on the performance of the model (Chaudhari, 1971; Lantz, 1971; van Genuchten and Wierenga, 1974). The effects of numerical dispersion can be limited by introducing stability criteria in the solution of the equations, or by adopting corrective coefficients directly in the solution of the equation.

21.1.1.5 Model Inputs

Data needs vary greatly between models (see Table 21.1), although all leaching models require certain basic input (see Sec. 21.3.2). Input data may be divided into state and site parameters: the former are required to make the basic processes work, the latter define specific characteristics of the site where the model is applied. Model documentation should, but often does not, contain a list of the required input. A clear and comprehensive list may be difficult to construct, since many models contain several alternative methods for simulating certain processes as well as submodels that the user may choose to ignore, and the model itself may offer more or less universal default values for some of the required parameters. Model documentation should also state whether an approximate universal value or a literature value suffices, or whether measurements must be performed on site in order to obtain a good simulation, which evidently requires more time and effort. For instance, in order to use VARLEACH, the half-life of a given pesticide must be determined in the studied soil at different temperatures and humidities.

21.1.1.6 Model Outputs

In a majority of cases, models produce daily predictions (which can then be used to generate graphs) together with certain output parameters printed at certain times or time intervals. Different types of information are offered to the user according to the purpose of the model: pesticide mass in soil, center of mass, concentration peak, percolate mass, etc. It is, however, also advantageous to know what the model calculates but does not forward as output.

The exact content of the output presented to the user varies from model to model and may be fixed or flexible, allowing the user to choose from a selection of possibilities. It may be given in the form of tables and/or graphs, or be formatted so that graphical appliances can easily process the data. The times or intervals at which the output are printed may also be fixed or user-specified.

The most common forms of output from leaching models are the following:

Concentration: Instantaneous concentration [mg L^{-1}] of the chemical (e.g., in the liquid phase of soil at a certain depth).

Content: Instantaneous content [kg kg^{-1}] of the chemical (e.g., in the equilibrium-sorption domain of the soil at a certain depth).

Flux (or rate): Instantaneous rate [mg m^{-2} d^{-1}; mg m^{-3} d^{-1}] of a chemical-fate process, such as volatilization or degradation.

Peak: The highest concentration (at a certain depth), flux, or rate recorded during the simulation.

Cumulative mass: Total mass of chemical, e.g., leached during the simulation so far.

Balance: Sums up the fate of the chemical so far by dividing it between different phases or fates as the cumulative masses of chemical that have been, e.g., leached, transformed, taken up by plants, or lost in drainage during the simulation.

Residual: Instantaneous amount of chemical still left in the simulated soil column.

Centre of mass: Instantaneous depth of the center of mass of the pesticide remaining in the soil column.

21.1.1.7 User Requirements

Purely in terms of technical requirements, each model needs a certain amount of disk space and system resources. Since current leaching models are not major commercial products, they tend to require a specific operating system and specific accessories, which may or may not be convenient for the potential user.

The user-friendliness of a model covers all interaction between the program and the user, such as how simple the input files are to prepare, how easy the output is to read and handle and how useful it is, whether the source codes are available and documented, whether a comprehensive user manual exists, whether there is a database that contains useful literature values for important input parameters, etc. As the complexity of the structure of model subroutines increases, so do the difficulties associated with measuring, or simply finding, the data necessary to make them work. The more clearly and comprehensively the model has been documented, the more quickly as a rule the user will be able to perform useful simulations.

The complexity of the actual model may or may not be reflected in what it demands from the user. A well-designed user interface can make the most scientifically ambitious model simple and straightforward to use; in the other extremity, relatively simple models can be challenging to understand and master and slow to use if they lack proper documentation and helpful features.

21.1.1.8 Reliability

The reliability of the model in various situations can be assessed from analyses of its robustness, sensitivity, and uncertainties, which evaluate the model as a functional entity, and from validation exercises, which test the performance of the model against real data. Documentation of such exercises can usually be found in scientific literature.

Analysis of robustness, usually performed during model development, is a procedure for evaluating the stability of the model. Intrinsic control parameters are varied, and extreme values are used as input (Del Re and Trevisan, 1993). Robustness analysis firstly identifies the limits for the values of various input parameters beyond which the model will not function, pinpoints weaknesses, contradictions, and voids in the way the model describes reality. Secondly, by testing it in extreme situations, and tests the flexibility of nominally fixed model parameters. At the same time, the difficulty of the model is evaluated, and the results of robustness analysis help to develop parameterization guidance that will facilitate its use.

Sensitivity analysis addresses the effect of varying the values of input parameters, identifying and ranking the parameters that have the greatest influence on model output. This clarifies the level of accuracy required for each input parameter and the relative importance of the various basic processes simulated; it frequently turns out that some input parameters are unimportant enough to be replaced by constants, thus simplifying the use of the model.

Analysis of model uncertainty goes one step further from sensitivity analysis to evaluate the degree to which the performance of the model is influenced by unavoidable uncertainties that concern the values of input parameters. The influence of uncertainties due to current lack of knowledge or understanding of each of the basic processes is also evaluated.

The validation of a model is the final phase in the evaluation procedure. Complex environmental models cannot be truly validated, i.e., confirmed to be correct representations of reality; it is merely possible to test their performance with respect to observed data in different situations (Oreskes et al., 1994). However, the term "validation" is currently used whenever the goodness-of-fit between observed and simulated data, and thus the ability of the model to predict in this specific context, is evaluated (e.g., Garratt et al., 2002). In the simplest validation approaches, the model is used for reproducing experimental or monitoring data known to the user. In the most sophisticated validation exercises, performed, e.g., within the APECOP project in the European Union (Vanclooster et al., 2000, 2003a,b), extensive field datasets are collected and a carefully designed stepwise procedure is followed (discussed in more detail in Sec. 21.4.3).

21.1.2 CORRECT USE OF MODELS

In order to obtain useful and meaningful results, models have to be used correctly. Mathematical models are at best very simplistic approximations of reality, and simulation results should therefore never be given great confidence, particularly when significant extrapolation from known reality is involved. The highest accuracy and precision are obtained when models are used as tools for explaining experimental results. Some experimental or field data should always back up simulation results when reliable quantitative predictions are needed for, e.g., risk assessment.

The correct use of a model involves in part the selection of a model that is well suited for the purpose, i.e., simulates the desired parts of reality and covers the important processes. Once such a model has been found, correct parameterization becomes important. The results of sensitivity analyses of the model tell which parameters most strongly affect the simulation results and therefore need the most attention.

21.1.3 MODEL CALIBRATION

Models produce the most precise and reliable results when they have been calibrated to a specific, well-known situation. On the other hand, the calibrated model is rather exclusively limited to simulating these specific circumstances. When a particular situation (location, climate, chemical, crop....) is studied, and modeling is used for explaining the measured data (and possibly extrapolating from it), it usually makes sense to perform justified model calibration in order to make reality and the simulation correspond as closely as possible.

21.1.4 MODEL VALIDATION

The purpose of model validation is to prove and improve the reliability of the model in order to evoke confidence in its performance. Validation is of particular importance when the model is used for making regulatory predictions, where modeling results can have direct and significant economical as well as environmental implications. However, the credibility of simple, small-scale modeling exercises is also greatly improved if the model is known to be able to produce correct results in a similar scenario.

21.1.5 PARAMETERIZATION

Generic information, mainly the basic chemical and physical properties of the chemical, can be found from literature or databases or from the manufacturer of the chemical product. In case of generic simulations, literature and databases can also be used for obtaining the necessary information and data concerning the simulation site, soil, weather, management practices, and crops to parameterize the rest of the model. Some pesticide-fate models come with some in-built datasets or with direct access to databases that can easily provide most of the parameters needed. Care should be taken to build simulations that produce meaningful results, as it rarely makes sense to combine crops, soils, and climates that do not coincide in reality.

When a specific situation is simulated, laborious measurements and monitoring must be performed on site for best results. Laboratory studies may also be needed in order to obtain some sorption and degradation parameters, particularly the K_d of the chemical in different layers of the soil. The required weather data can often be obtained from a weather station located close to the site.

21.1.6 ASSESSING THE RELIABILITY OF MODELING RESULTS

Whenever possible, the reliability of simulation results should be assessed by comparing them to good and relevant actual data, e.g., monitoring data. Although it is important to realize that good matches between simulation results and measured data do not make the model valid, even when the exact setup where the measurements were performed is simulated, they do add to the probability that the model adequately conceptualizes the simulated reality (Oreskes et al., 1994). When simulation results notably disagree with corresponding real data in pattern or magnitude, the parameterization of the simulation should be reviewed and the choice of model reconsidered.

21.2 MODELING SOIL-PESTICIDE INTERACTIONS

21.2.1 THE ENVIRONMENTAL FATE OF PESTICIDES APPLIED ON AGRICULTURAL FIELDS

A pesticide applied on field may be sprayed on bare or crop-covered soil (pre- or postemergence), incorporated or injected directly into the soil, or applied in the irrigation water (chemigation). Some of the sprayed chemical will drift outside the target area, and if there is vegetation on the sprayed field, a part of it will settle on the canopy instead of soil surface. On crop canopy, the chemical may be volatilized or transformed, penetrate into the plant, or be washed off to the ground by rainfall or irrigation. On soil surface, the chemical may likewise be volatilized or transformed. When intensive rainfall generates runoff and erosion of soil from the field, which is particularly significant on fields that are situated on slopes, chemicals from the soil surface and topsoil are lost as solutes and with eroded soil particles (sediment). All of these processes affect the amount of pesticide that actually enters the soil system.

Pesticides enter the soil system together with infiltrating water, which is the most important medium that moves chemicals in the soil. Chemicals are constantly partitioned between the gas, liquid, and solid (particle, adsorbed) phases of the soil according to their own physicochemical properties as well as the moisture, temperature, and pH conditions and the soil type of their immediate environment. While dissolved in the liquid phase, chemicals may be transported in the soil in any direction, mainly laterally (with drainage) and downwards (with percolate as leaching), and they may eventually enter the groundwater. Chemicals can also be transported in the volatilized form in the gas phase, from where they may be lost to the atmosphere. Diffusion and dispersion also spread the chemicals in both the liquid and the gas phase. Sorption to soil particles takes several different forms: rapid and reversible equilibrium sorption, only partially reversible slow sorption, and practically irreversible very slow sorption, which operates on a time scale from weeks to years. Chemicals are degraded in soil by chemical, photochemical, and microbial processes, and important transformation products may be formed in the process. In the root zone, pesticides are taken up by plant roots both together with water and through sorption to roots surfaces and dissolution to root fats. Both sorption and degradation are affected by temperature, moisture, and pH. These concepts are addressed more thoroughly in Chapter 3.

Chemicals that leach down to the saturated zone of soil may be degraded or adsorbed in the groundwater and may also reenter the unsaturated zone through capillary rise.

21.2.2 MODELING STRATEGIES

21.2.2.1 Soil Properties

One of the most important features of soils from the modeling point of view is their more or less regular division into distinct horizontal layers. These layers result from the history of the particular soil (geology, climate, land use) and can have very different characteristics that affect the fate of chemicals. When modeling chemical fate and leaching in soils, it is customary to both distinguish and separately describe these physical layers — often called horizons — as they exist in the particular soil or soil type of interest. Horizons then define the properties of the usually much thinner computational layers, discrete volumes for each of which the model calculations are separately performed. However, horizontal heterogeneity of the soil can be remarkable even within a small area, and in order to take this properly into consideration, the model must either be stochastic (see Chapter 2) or have several spatial dimensions.

The physical properties of soil layers that immediately affect the fate of chemicals include the soil type (usually characterized by the organic matter and sand/silt/clay content of the soil), the grain-size distribution of particles, and the bulk density of the soil, which largely determine how quickly or slowly water travels through the root zone and to what extent the chemical may interact with soil particles.

The macrostructure of the soil can involve pores of different sizes that span some of the height of the soil column or even through the whole column. Pores may be left by earthworms or roots, or formed by the flow of water through the most easily penetrated sections of laterally heterogeneous soils. Especially the clayey soils may also crack in dry conditions. Pores and cracks may let water and solutes rapidly through to much deeper soil layers than would otherwise be immediately exposed to them. This preferential flow, discussed in detail in Chapter 8, can strongly increase the leaching of chemicals and cannot be accounted for if purely chromatographic flow of water through the soil column is simulated. Models may describe the development of cracks as the soil dries or simply consider the existence and extent of macropores in a particular soil as described by the user. Other types of preferential flow, e.g., fingering due to heterogeneity of the soil, can be very challenging to model in a deterministic way, and stochastic approaches that account for the variability of the soil are popular.

All soil can be divided into three phases: the liquid phase, the gas phase, and the solid (particle) phase. In chemical-fate modeling, at least the liquid and the solid phase and their volume fractions need to be simulated. While the dry bulk density of the soil does not usually change much within a relatively short time frame (although tillage can change the density of surface soil), the liquid phase of the soil is highly dynamic, and it is important to keep up with the moisture content and water flows in the soil in order to adequately describe the behavior and movement of chemicals. The flow of water in macropores needs to be described differently than the chromatographic flow.

The modeling of soil hydrology is discussed in more detail in the next section. The gas content and gas flows in the soil depend on the behavior of the liquid phase, as the sum of their volume fractions can usually be considered a constant, and it may also be important to describe these adequately if the chemical of interest is significantly volatile.

21.2.2.2 Soil Hydrology

The amount of pore water in soil changes all the time. Precipitation and irrigation (which are often simply summed up in simulations) add water to the soil from above, evaporation and transpiration by plants remove water from the soil to the atmosphere, percolation moves it downwards in the soil profile and lateral drainage removes it sideways. As water is the major transport mechanism of chemicals within and through the soil column, it is important to properly simulate water balance and movement in the soil. This is usually performed by a separate hydrology submodel that produces inputs for the actual pesticide-fate model.

The hydrology model must first define the amount of rainfall and irrigation water infiltrated by the soil. It may also quantify the fraction removed from the field as run-off when the infiltration capacity of the soil is exceeded during a rain event. The precipitation intercepted by the vegetation canopy, and its division between evaporation and flow to the soil surface, is also of interest, particularly if wash-off of chemicals from the canopy needs to be quantified. Evaporation of water from soil and transpiration by plants must be simulated and are usually calculated from the potential evapotranspiration of the soil, which can be calculated within each time step by the Penman-Monteith method or simpler methods, such as the Haude equation used in the model PELMO (see Table 21.1 and

Klein, 1995). The subject is more thoroughly addressed in Chapter 4 of this book.

In the soil, the model must describe at least chromatographic flow down through the soil column from layer to layer, subtracting water removed by plant root uptake and lateral drainage on the way. For clayey or structured soils, a description of preferential flow may also be required. At the bottom of the simulated soil column, an appropriate bottom boundary condition must be formed in order to define the fate of the remaining flow, which may simply cross the bottom boundary and continue downwards toward groundwater (free flow), enter groundwater directly in case the water table is met at the modeled bottom boundary, or meet an impenetrable or slowly penetrable layer, such as solid rock, that forces the water to either accumulate or flow laterally to surface waters.

There are currently two major techniques for modeling chromatographic flow in the soil. The simpler one, used in so-called capacity models, assumes that each computational soil layer must fill up to field capacity before any water can pass through to the layer below. This "tipping-bucket" approach is distinctly functional, but the calculations are simple, and it is often found to describe water flow through the simulated column with sufficient accuracy. A more mechanistic and considerably more demanding approach is to solve the Richards' equation (see Chapters 6 and 11), which calculates the gradient in hydraulic pressure head due to water flow for each simulated depth in soil.

Water flow in macropores is considerably faster than that in micropores and needs to be described differently, as it is primarily driven by gravity instead of the hydraulic pressure head. The initial division of the water that enters the soil system must also be defined, as well as the exchange of water between micropores and macropores. Macropore flow modeling is discussed in Chapter 8 in this book and has also been recently reviewed by Feyen et al. (1998) and Šimůnek et al. (2003).

21.2.2.3 Pesticide Properties

When the behavior of a chemical and its potential impact in the environment are assessed or modeled, three qualities are of major interest: the mobility, persistence, and toxicity of the compound. When modeling purely the fate of the chemical, however, toxicity is not of direct interest, or only comes up when the risk assessment process shifts from the characterization of exposure to the characterization of effects on biota.

The mobility of a chemical in soil is largely determined by its solubility in water and partitioning between lipids and water. The more lipophilic a chemical is, the more likely it is to become immobilized through adsorption to soil particles. K_d, the partitioning coefficient of the chemical between water and soil particles, is a soil-specific indicator of the tendency of the chemical has to adsorb to soil particles. There is usually a good correlation between such adsorption and the organic matter content of the soil, and K_{OC}, the coefficient of sorption to organic carbon, is considered a more universal property of a chemical; models may demand either the K_{OC} or a soil-specific K_d (possibly separately for different layers) of the chemical as input. Sorption is discussed in more detail in Chapter 12 of this book, and the measurement and use of soil sorption parameters has been recently reviewed by Wauchope et al. (2002). The volatility of the chemical should also be considered, as this determines its presence in the gas phase of soil and potential to escape into the atmosphere. Henry's law is commonly used for describing the partition between gas and liquid phases, usually with correction for the effect of temperature, as discussed in more detail in Chapter 3 of this book.

The persistence of the chemical in soil is expressed as experimentally determined half-lives or degradation rates of the compound in different soils. Leaching models generally accept different degradation rates for different depths in the simulated soil, and the effect of moisture and temperature can also be accounted for with experimental factors.

21.2.2.4 Pesticide-Soil Processes

The simplest approach for modeling the transport of chemicals in the liquid and gas phases is to consider only simple convection, i.e., the flow of water or gas transports the amount of chemical dissolved in it. More advanced mathematical descriptions also consider the diffusion and dispersion of the chemical. The depth, flows, and renewal rate of the local groundwater affect the potential concentrations of a leached chemical that may develop in the groundwater; such factors can be extremely complicated to model.

In the simplest approaches, sorption is described as an instantaneous and completely reversible process based on the linear adsorption coefficient of the pesticide for organic carbon and the organic content of the soil. Gradual nonequilibrium sorption can also be considered and is usually assumed to be irreversible and to stop the adsorbed chemical from being transformed. Linear or Freundlich-type sorption can be assumed for both the equilibrium and the nonequilibrium sorption domain of the soil. There can also be significant sorption to nonorganic matter in the soil, and the impact of depth, temperature, and moisture on sorption in different layers may be considered. The pH dependence of the sorption of, e.g., weak acids can also be taken into account.

The degradation of a pesticide in the soil is commonly described by first-order degradation kinetics, characterized by the observed half-life $(t_{1/2})$ of the pesticide in soil, possibly at different depths. This description is commonly assumed to cover both the chemical and the microbial transformation of the compound in the soil, although biodegradation, which is considerably more dynamic than the chemical and physical processes, may be considered separately in order to account for the dynamics of the microbe populations that metabolize the chemical (e.g., in PRZM-3). The transformation of the chemical to daughter chemicals of interest (and their subsequent environmental fate) may also be simulated. The effect of temperature, moisture, or depth in soil on the experimentally derived rate of transformation can be considered.

The most straightforward approach to describing plant uptake of pesticides is to assume passive uptake of the pesticide into plant roots with water from each layer. More complicated approaches consider the effect of the lipophilicity of the chemical on the uptake by using either the root concentration factor (RCF) or the transpiration stream concentration factor (TSCF) for correction. RCF expresses the relationship between concentrations of the chemical in the plant root and in the external solution (in this case, pore water), TSCF the relationship between the concentrations in the transpiration stream of the plant and in the external solution. Both factors can be calculated from the K_{OW} of the chemical using empirical equations. Uptake of xenobiotics by plants and modeling approaches are addressed in detail by Trapp and Mc Farlane (1995).

Pesticide losses in runoff and erosion — complicated processes of short time scale which concern a thin layer of soil at the top boundary of the soil column — can be difficult to properly integrate into the structure of a leaching model. Typical approaches assume pesticide removal from topsoil to be proportional to the volume of water runoff and mass of soil eroded from the field during a time step (or rainfall event). The volume of water runoff is commonly calculated for each storm event using the Soil Conservation Service curve number technique (SCS-CN, Mockus, 1971) based on precipitation, the properties of the soil, the type of crop, and management factors. Soil erosion is usually calculated with one of the variations of the universal soil loss equation (USLE, e.g., Williams and Berndt, 1977), considering the average annual rainfall and the erodibility and topography of the field. The depth of the affected topsoil can be defined in various ways: the simplest approaches assume homogeneous pesticide removal from a fixed layer, e.g., the top computational layer (as in PELMO), while more complicated ones decrease pesticide losses gradually with depth in soil (as in PRZM-3).

21.3 CURRENT PESTICIDE LEACHING MODELS

21.3.1 GENERAL STRUCTURE OF MATHEMATICAL PESTICIDE LEACHING MODELS

A mathematical model that needs to simulate all the fundamental fate processes in the soil column that concern a pesticide applied on the field is normally assembled in the way described below.

The typical small-scale model is one-dimensional: the locations of the simulated spatial points are defined only by their depth in soil, and the model assumes no horizontal variation of soil properties inside the field. The computed rates of change are either per unit of volume or per unit of horizontal surface area (the latter mainly for above-soil processes).

The simulated soil column is horizontally divided into computational layers, discrete volumes which are characterized according to the known properties of horizons, which in turn are thicker layers of soil that differ physically from one another. The computational layers are usually relatively thin at the top of the column and grow thicker with depth. The user may be able to define the thicknesses of the layers, or these may be fixed. The soil-fate calculations are made separately for each layer.

The model consists of at least two submodels: the soil hydrology model and the actual pesticide-fate model. The descriptions of above-soil processes often also form their own submodel which defines the actual flux of pesticide into the soil column and may also calculate pesticide losses out of the field with run-off, erosion, or drift in order to quantify surface-water loads.

The hydrological submodel forms the basis for the transport of pesticides in the soil column and produces input for the pesticide-fate part. Using weather and management data as input, it determines the amount of water that enters the soil system within a time step (as a fraction of the precipitation and/or irrigation) and leaves it at the top or at the bottom of the simulated soil column (as evaporation, transpiration, and percolation). It may describe only chromatographic water flow downwards in the soil column or also deal with preferential flow. In most models, transport within the liquid phase (convection and possibly diffusion and dispersion) is the only process that moves chemicals from one computational layer to another, but if the flows of the gas phase can be simulated, the transport of volatilized chemicals within the gas phase could also be calculated. Soil heat, which can affect the rate of various processes, is not simulated in the simplest approaches, and its actual use in calculations is often limited. Empirical or more refined heat conduction approaches may be used for calculating the temperature of the soil from air temperatures.

For each soil layer and time step, the model must calculate the water balance and solve the pesticide conservation equation for the layer in order to obtain actual predicted concentrations of the compound. This conservation equation is a differential equation that combines all the pesticide-fate processes and expresses the rate of change of the pesticide concentration in the equilibrium-sorption phase of the soil. Fluxes of water and pesticide into and out of the soil column (at the soil surface and at the bottom of the simulated column) are handled as boundary conditions at the top and bottom computational layer.

The time step of model simulations is a compromise between data requirements and accuracy needs and also affects the run-time of the simulation and the quality of output. The time scales of relevant pesticide-fate processes vary from seconds to years. The most common time step currently selected is one day, but a shorter one would actually be needed in order to use good mechanistic descriptions of several processes (spray drift, run-off, volatilization) which take place within a short time scale (minutes to hours) or have different rates during different times of day. Output from one time step act as input for the next as of chemical concentrations.

The model may either demand the parameters it needs for the simulation as direct input from the user, or include procedures for deriving these parameters from information that is more easily available for a nonexpert user, (e.g., pedotransfer functions, which derive hydrological parameters from information that can be easily obtained about the studied soil). Models are also frequently linked to databases and weather data generators in order to make simulations much faster and easier to build.

The typical user of a pesticide-fate model needs to be able to provide the model with required input (and make any necessary choices among alternative mathematical descriptions of various processes), to run simulations, and to view and handle the output data. In order to facilitate these operations, the model can be provided with a graphical user interface (GUI), which walks the user through preparing the simulation run, provides simple access to useful databases, runs the requested simulations, indicates what went wrong if the simulation cannot run properly, and allows the user to access and export the desired output needed in a useful format.

Figure 21.1 presents a generic block diagram of a physically based model of pesticide fate in soil, showing how such a model may be constructed. This computation routine is run for every soil layer during each time step of the model. Models are in practice built of blocks (represented

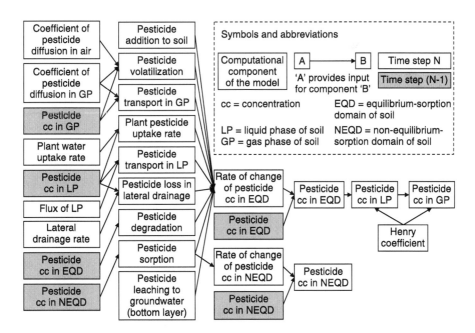

FIGURE 21.1 Block diagram of a generic physically based pesticide leaching and dissipation model. See text for discussion.

in the figure by boxes), smaller models or components that simulate some particular entity, e.g., the degradation rate of the pesticide. The output of one block act as input for another: for example, the model components that calculate the rates of pesticide volatilisation, transport in the liquid and gas phases, plant uptake, lateral drainage losses, degradation, sorption and loss to groundwater all contribute inputs to a component that calculates the rate at which the concentration of a pesticide changes in the equilibrium domain of soil. Also, the simulations of sequential time steps are linked together by calculated concentrations of the chemical in various compartments of the soil: the rates of many chemical fate processes are dependent on these concentrations, and the concentrations calculated during the previous time step ($N-1$) are inputs for the calculations of the current time step (N).

21.3.2 CURRENT LEACHING MODES

Seven recent pesticide leaching models commonly applied in different contexts — PEARL, MACRO, LEACHM, GLEAMS, PRZM-3, PELMO, and VARLEACH — are presented in Table 21.1. Since the selection of parameter values required by each model depends greatly on choices made by the user, and the models include many parameters with a default value that can be modified by users, it is not practical to make detailed lists of the required input. Instead, the general level of information required by each model concerning the site, soil, hydrology, climate, crop, management practices (irrigation, ploughing), and the chemical applied, is characterized. All models require certain basic data:

Pesticide: molecular weight, water solubility, saturated vapor pressure (or Henry constant), reference soil half-life (or degradation rate) at a specified temperature, sorption coefficient

Climate: on-site precipitation and air temperature data on site, ranging from daily (or each time step) to monthly minimum and maximum or mean

Soil: bulk density, clay/silt/sand and organic matter content of each horizon or layer, depth of the simulated soil column

Management: pesticide application rate

Start values: water content, temperature, and pesticide concentration in different soil layers or horizons at the beginning of the simulation

In order to illuminate the sensitivity of chemical-fate simulations by each model, we list input parameters that have been found to have the greatest influence on simulated leaching and residuals in the soil column. Leaching models are universally sensitive to the parameters concerning the sorption and degradation of the pesticide. It should be noted that sensitivity may depend on the length of the simulation, the values of other parameters, and other factors.

Concerning the validation status of each model, it should be remembered that each performed validation exercise only evaluates the performance of the model regarding the specific conditions, scale, chemicals, type of simulation (blind prediction, calibrated prediction, reproduction/dissection of known data), and endpoint (soil residual, loss in leachate) considered. The validation status of all of these models has not been reviewed in the literature.

21.3.3 Applications

Practical applications of pesticide-fate models can be roughly divided by type into demonstration, reproduction of obtained data, extrapolation from obtained data, and prediction. Demonstration will not be discussed here in any detail: it is usually performed for educational purposes, and the requirements of the model depend greatly on the pedagogical goal, although simple models generally suffice.

Analysis of obtained field or experimental data with the help of models can serve many purposes. Modeling can help to identify about the selection of processes that concern a chemical in a soil by offering a possible scheme of processes that explains the observations or experimental results. Models can be used for estimating the degradation rate or the distribution coefficient of a chemical in the field or in experimental conditions where several processes are simultaneously at work. Models can also help to reconstruct the past from recent data when investigating the contamination of a soil or groundwater. This type of use requires models that sufficiently cover all relevant fate processes, and it usually involves careful calibration of selected model parameters in order to obtain good agreement between data and simulation results.

Once good agreement has been reached between a dataset and a simulation, the model can — with consideration for the scope of the model and the validity of parameter sets — be used for extrapolation outside of the range of observations in order to estimate how changes in climatic conditions or the properties of the chemical or the soil affect the outcome. This is done by changing or varying the original values of model parameters, usually one or a few at a time.

When there is sufficient confidence in the validity of the model in the situation of interest, models can cautiously be used for predicting, e.g., chemical concentrations in leachate. Predictive use is common at the initial stages of the environmental risk assessment of chemicals, typically as a screening procedure in order to identify potentially problematic chemicals, and is becoming a standard part of the registration process for pesticides in both the European Union and the United States. When making predictions on the basis of model simulations, care must be taken that the limitations and uncertainties of the models are considered and well understood, the input data are of good quality, and there will not be excessive confidence in the predictions.

Modeling can be performed at different spatial scales, although consideration should be given to the original scale for which the model was designed, as this is usually the scale at which it works best. The basic processes that affect chemical fate in soil are commonly simulated at the soil column scale. Simulations on the scale of a field or a similar entity serve to assess fluxes of chemicals in agriculture or forestry, the exposure of the environment to the chemical, or the efficiency of a pesticide. Large-scale applications typically screen for potentially problematic geographic areas on a scale that may range from a municipality to a community of nations.

We will introduce by type a variety of published applications of pesticide leaching models, divided here into research, environmental management, agricultural management, large-scale, and pesticide registration applications. Three case studies of different scopes are discussed in more detail in Sec. 21.4. Whenever stated in the original paper, the authors' reasons for choosing a particular model and their views of its usefulness in the application are given; unfortunately, such considerations are rarely documented in scientific papers that involve pesticide leaching models as tools.

21.3.3.1 Research

The original purpose of most pesticide-fate models developed so far has been to help to interpret, explain, and fully utilize the results of laboratory and field studies. As methods of scientific research, these models are generally used in the calibrated form, i.e., relevant model parameters are modified in order to obtain a better correspondence between modeling results and a specific set of real data. Without calibration, models can also be used for nonquantitative comparison and ranking of leaching among different chemicals or conditions, or for studying the effect of the variation of some parameter on leaching.

Models can be used for studying the significance of a particular pesticide-fate process in specific circumstances or in a general sense. Larsson and Jarvis (2000) studied the potential influence of macropore flow on the leaching of chemicals with variable sorption and degradation properties. Using a Swedish scenario and a version of MACRO (one of the few pesticide-fate models that describe macropore flow) calibrated against extensive monitoring data from the scenario site, they carried out simulations for 60 hypothetical chemicals with variable K_{OC} values and half-lives, both considering and ignoring preferential flow.

Pesticide leaching models are often used for the estimation of degradation or sorption parameters in a field or soil column study. Krzyszowska et al. (1994) estimated the degradation rate of dicamba and picloram in an agricultural soil from Wyoming with the help of LEACHP by modifying the dispersivity and degradation rate constants until the simulated breakthrough curves matched the data measured in a soil column in the laboratory. Close et al. (1998) applied picloram, atrazine, simazine, and bromide to two different New Zealand soils and then used GLEAMS to make initial estimates of the half-lives and K_{OC}s of the pesticides in these soils by optimizing these parameters to best fit the concentrations measured in leachate at different depths; it was recognized that due to its simple description of soil and hydrology, GLEAMS was not really an adequate model for the purpose. Ares et al. (1999) observed the breakthrough of lindane to groundwater in a soil in Argentina, then used PRZM-2 and parameter optimization techniques to estimate the K_{OC} and half-life of the pesticide.

A model that can satisfactorily reproduce the observed data can be used for extrapolating from it in moderation. Hegg et al. (1988) collected field data from experimental plots in South Carolina on the leaching and degradation of aldicarb. Since PRZM had previously been shown to predict the movement of the leading edge of aldicarb in unsaturated soil fairly well, this model was used for simulating the leaching of aldicarb during the experiments, and simulations were then also performed using a 15-year local weather dataset in order to estimate the effect of variable rainfall conditions.

Models can be used, often without calibration, for studying the effect of the variation or heterogeneity of one or more factors on the outcome, e.g., pesticide leaching. Truman and Leonard (1991) used GLEAMS simulations in order to study the effect of the change of the half-life of pesticides with depth in soil, in connection with different surface and subsurface soil textures and timings of rainfall, on the potential leaching losses. Data on three soils from Georgia, a 50-year rainfall record from this region, and a small range of hypothetical pesticides with different surface and subsurface soil half-lives and K_{OC}s were used for the simulations. Since the study only involved comparison between different assumptions, GLEAMS was considered an adequate tool. Klein (1997) analyzed the influence of the variation of climatic input on the variation of pesticide concentrations in leachate with the help of PELMO and Monte Carlo simulations (combined as a stochastic model); the Monte Carlo distributions of rainfall and temperature in 13 regions of Germany were based on long-term weather data, and the leaching on three hypothetical pesticides with different leaching potentials was simulated for each region. Wu and Workman (1999) studied the effect of the spatial heterogeneity of soil on pesticide movement with the help of GLEAMS, describing some hydrological model parameters of the soil by random multivariate normal vectors. The model input was generated by field studies which represented various combinations of three different soil plots in Ohio, four different crops, and the application of three pesticides, and local weather monitoring data was used.

In a slightly different type of application, the heterogeneity or variation of an input parameter may be considered in order to demonstrate the need for, e.g., stochastic modeling or change of parameters with depth in order to obtain realistic results. Harms et al. (1996) used a modified version of PRZM for demonstrating that Monte Carlo simulations, using pesticide half-lives that are generated randomly according to the frequency distribution of values found in literature, give more realistic predictions of pesticide leaching than deterministic simulations using single-value input. Mills et al. (2001) studied the degradation rates of acetochlor in different surface soils and subsoils both in the field and in the laboratory, then used PELMO simulations for demonstrating that the inclusion of a separate realistic subsoil degradation rate in the model, instead of using the same rate of degradation for the entire soil column, can significantly improve the realism of leaching predictions.

Well-established leaching models have been used for evaluating the performance of other research methods or tools, or even for generating data in order to create a new one, although such uses require careful consideration. Loague et al. (1996) used PRZM simulations to rank the relative leaching mobilities of a variety of organic chemicals in Hawaiian soils, and compared the results with corresponding rankings performed by the simple index AF (attenuation factor), as a part of assessing the uncertainties of the model error of the index. Stewart and Loague (1999) used PRZM-2 for evaluating the performance of a so-called type transfer function model at simulating groundwater contamination by DBCP in Fresno County, California. The concentrations of DBCP in leachate at different depths were simulated for three sites representing different soils by both models, and the results were compared. Goss (1992) used over 40,000 GLEAMS runs in the development of a simple first-tier screening procedure for evaluating the potential of pesticide losses by leaching and runoff from soils, without consideration for climate or management.

Simulations are also performed in connection with laboratory and field studies, often at a small scale, in order to form a hypothesis on some practical implication of the findings. Vischetti et al. (1997) performed laboratory experiments with the herbicide rimsulfuron in an Italian topsoil under different temperatures and humidity, then used the obtained persistence data for predicting the persistence of the chemical in a typical Italian scenario with temperature and precipitation data from two very different years. Karpouzas et al. (2001), having obtained results in laboratory studies of a British agricultural soil that indicated that sufficient initial concentrations of carbofuran in subsoil activate the microflora and result in considerably greater degradation rates than in topsoil, used VARLEACH for simulating the movement of carbofuran to subsoil after a standard application in order to assess whether subsoil microflora are likely to adapt to breaking down the compound.

21.3.3.2 Environmental Management

Pesticide leaching models can simulate past, present, or future contamination of soil and groundwater following the application of a pesticide or release of a chemical to the soil, which makes them a useful tool for purposes of environmental assessment and investigation. They can also be used for comparing agricultural procedures, chemicals, etc., with regard to their environmental impact.

Leaching models are often used for evaluating the potential of groundwater contamination by pesticides that are about to be introduced, are currently in use, or have been used in the past in some agricultural area. The potential contamination can be estimated with the help of a relatively small number of scenarios which are considered to represent the conditions and pest management strategies in the area, or by organizing the relevant spatial data so that a large number of simulations can be performed and their results handled quickly and conveniently. The latter type of study is addressed separately in Sec. 21.3.3.4.

Brown et al. (1996) simulated the leaching, drainage, and runoff of alachlor with LEACHP, PRZM-2, and MACRO in order to assess its potential to contaminate Italian groundwaters, using two agricultural areas recognized as average and worst-case alachlor-use scenarios. These models were considered to cover the important pesticide fate phenomena and were validated with field and lysimeter datasets prior to the simulations. Giupponi et al. (1996) used VARLEACH for predicting the impact of different tillage systems on the degradation, sorption, and groundwater contamination of herbicides, performing at first a field study, in order to define the studied soil and measure degradation and sorption parameters for three herbicides under the different tillage practices, then simulating the leaching and residues of the chemicals. Businelli et al. (1999) used MACRO-DB for assessing the potential herbicide pollution of groundwater in the area from which the drinking water for the Italian city of Perugia is taken, considering a selection of pest-management strategies for the main crop of the area. MACRO was judged to be the best leaching model for heavy Italian soils, since it considers preferential flow. Businelli et al. (2000) used VARLEACH simulations for assessing the leaching potential of six commonly used herbicides to groundwater in two soil scenarios in northern Italy, with and without irrigation. VARLEACH was chosen for this study because it had previously been validated for two of the herbicides in a range of northern Italian soils with good results, and it had also previously been judged the most valid among four leaching models for the study area.

21.3.3.3 Farm Management

Pesticide-fate models can facilitate farm management by helping to compare the economic and environmental implications of different cropping systems or environmental policies. GLEAMS, a relatively simple model designed specifically for the purpose of comparing agricultural management systems, is frequently employed, and nutrients (mainly nitrogen and phosphorus) as well as pesticides are often considered.

Folz et al. (1993) used GLEAMS, together with EPIC and a farm-level linear programming model, Repfarm, for assessing the long-term economic and environmental trade-offs associated with different farming systems in the "corn belt" area of the United States. The role of GLEAMS was to simulate the movement and fate of agricultural chemicals in the root zone. The farming systems assessed were formed as combinations of two different sites that represent soils typical for the area, three typical crops, and three different regulatory scenarios: and the size of the farm was the one GLEAMS had previously been validated for in the area. In a later study, Folz et al. (1995) used GLEAMS similarly for predicting pesticide fate and transport as a part of a multiattribute ranking system for assessing alternative cropping systems in the Midwest, considering both profitability and environmental consequences. Seventy-two different combinations of crops or crop rotations, application practices of agrochemicals (pesticides and nutrients), tillage systems, and soil types from the area were simulated and seven different pesticides considered.

21.3.3.4 Large-Scale Vulnerability Assessment

Large-scale applications of leaching models typically screen sizeable regions for small areas potentially vulnerable to groundwater contamination. The total area may range in size from a municipality to a community of nations. Such applications generally involve, by necessity, a software system that can store, process, and illustrate large quantities of spatial data. Geographical information systems (GIS) are now common in such use.

Loague et al. (1998) used PRZM-2 for simulating the soil fate of the now banned nematocide DBCP as a part of a regional-scale groundwater vulnerability assessment in a valley in California. This study investigated whether the historical use of this chemical in agriculture following the regulations of that time can have resulted in the observed widespread DBCP contamination of groundwater in the area, or whether improper use or point sources are to blame. For the 35-year simulations of an area of $1172 \, \text{km}^2$ (in separate $1 \, \text{km}^2$ elements), the climate, land use, irrigation, and water-table depth history and soil type distribution of the study area were approximated. The vertical unsaturated fluid flow and DBCP transport were then simulated for the years 1960–1994, and the DBCP concentration profiles at the water table on the last simulation day of each year were picked for forming a water-table loading map. PRZM-2 was considered to cover all the essential fate processes of the chemical, yet not require too much data for regional simulations due to its capacity-type description of soil water balance.

Auteri et al. (1999) developed PESTIGIS, a Decision Support System (DSS) for recognizing areas at risk of pesticide pollution of soil and groundwater in Lombardy, Italy. PESTIGIS works at a regional scale in the GIS environment and utilizes a pesticide database as well as spatial information on soil, topography, land use, crop patterns, and climate in $8 \, \text{km}$ grids. The inspected chemicals are grouped according to their combined sorption and degradation properties, and the system then uses PRZM-2 for assessing the leaching of each pesticide group to groundwater in different soil-climate combinations. Soils are classified according to their vulnerability to leaching of a particular pesticide with the help of a calculated index, and the results are illustrated by GIS in the form of thematic maps. Galbiati et al. (2003) used PESTIGIS for the evaluation of soil and groundwater vulnerability to pesticide contamination in a 10,000 ha agricultural area in Lombardy. A 12-year simulation was run for five pesticides with and without irrigation.

Van Wesenbeeck and Havens (1999) used PRZM-2 to simulate expected environmental concentrations (EECs) in ground water for cloransulam-methyl, a new herbicide intended for use on U.S. soybean cultivations. The model was used within the PATRIOT software (Pesticide Assessment Tool for Rating Investigations of Transport), which integrates PRZM-2 with a large database of soil and weather scenarios in the United States. Simulations were run for 60 different soils using conservative values of soil, climate, and chemical parameters. The mobility range of the pesticide was determined for soybean-cultivation regions, the distribution of concentrations in groundwater was simulated and compared with phytotoxicology data, and the results were used together with GIS for identifying and quantifying the potential problem soils in each state.

21.3.3.5 Pesticide Registration

In the late 1980s and early 1990s, new legislature was passed in the European Union and in the United States to define new requirements for the registration of pesticides. All new active ingredients of plant protection products must fulfill these requirements in order to be allowed on the market. Existing products that the manufacturer wishes to keep on the market after a certain transition period must also be reregistered according to the new procedure, and extending the use of an existing product to new crops or application types requires extended registration.

The purpose of the current pesticide registration procedures is to ensure that the potential benefits of using an approved chemical according to instructions are greater than any adverse effects the product may have on human health, wildlife, or the environment. For existing products, observations on the field provide regulators with at least some information on the behavior of the chemical and possible effects on the environment, and the registration process may be quick and simple. When a new pesticide is introduced, however, there is usually no information available on its environmental fate or impact outside of some basic laboratory data. In order to protect consumers and the environment, new chemicals are cleared for experimental field use and for the market only through the well-grounded judgment that the intended use is probably not a substantial environmental issue.

Mathematical models are used in the registration process due to their low costs and velocity as compared to extensive field studies. Since the assessment of environmental fate needs to cover, at least in theory, the entire range of relevant combinations of crops, agricultural practices and environmental conditions, modeling can be practically the only way to come even close to meeting the requirements of registration. However, it is not possible for models to describe reality with complete accuracy, and simulation results should never be the sole basis of regulatory decisions. Instead, modeling should be considered as a tool to be used together with laboratory and field studies. Models can be used for interpreting results of these studies, recognizing areas where additional studies are needed, evaluating proposed study designs, comparing chemicals and application procedures, and generally integrating and making best possible use of the gathered data (Zubkoff, 1992). Models can also produce conservative predicted environmental concentrations (PECs) for first-tier screening purposes.

Pesticide leaching models are usually originally designed to be research tools, which means that their intended purpose is to explain real data rather than predict it. In research use, models are typically calibrated against field data, and their resulting accuracy is much higher than it can be when no comparison with real data is possible, as is usually the case in the regulatory context. It follows that these models are often capable of predicting the transport of the main bulk of a chemical downwards through the soil column with good accuracy, but predictions of specific time-and-space concentrations in soil layers or the groundwater tend to be poor. It should also be remembered that even if a model accurately describes all significant and relevant pesticide-fate processes, the quality of the output of the model can only be as good as that of the input information, and even when the same model is used to simulate the same set-up, the results can vary widely according to parameterization choices made by individual modelers. In order to minimize subjectivity and to guarantee some degree of consistency and reliability in the use of these models in the pesticide registration process, the models and the procedures of their use need to be standardized. Registration models also must be extensively validated for their intended use in order to improve general confidence in them.

The regulatory risk assessment of chemicals is based on a tiered approach. The first tier is a relatively simple and conservative screening procedure designed for recognizing — quickly and cheaply, and with a minimum of false negatives — those chemicals that may present a problem (e.g., contaminate groundwater). The usual approach for making a first-tier assessment of the leaching potential of a chemical is to use some basic information about the chemical (half-life in soil, K_{OC}, intended application rate) to perform a selection of relevant worst-case simulation runs. If the predicted groundwater concentrations systematically exceed a trigger value in these scenarios, convincing higher-tier data from lysimeter, field, or monitoring studies must be presented in order to override the modeling results and make the registration of the pesticide possible. The proposed application method and rate as well as use restrictions can also be modified in order to qualify the chemical for registration.

The environmental assessment of the current registration procedure in the United States is based on the requirements and specifications of the Federal

Insecticide, Fungicide, and Rodenticide Act (FIFRA) (1988) and the Food Quality Protection Act (FQPA) (1996). The U.S. pesticide registration authority, the Environmental Protection Agency (EPA), has developed an ecological risk assessment paradigm, which also concerns any pesticides to be registered. As a part of the aquatic exposure assessment of pesticides, FIFRA requires predicted environmental concentrations (PEC), while FQPA requires estimated drinking water concentrations (EDWC). EPA promotes the use of PRZM and other leaching models, mainly GLEAMS, for assessing the leaching potential of the chemical into groundwater. The preferred use of models is for comparing the simulated leaching of a chemical at a variety of sites to that of established pesticides for which parallel monitoring data exists. Pesticide concentrations in leachate can also be simulated, but these must be backed up with actual field data. PRZM is often used in the registration process within the PATRIOT (Pesticide Assessment Tool for Rating Investigations of Transport) shell, which contains a selection of ready-made crop, topography, and soil scenarios, automatically generates weather and irrigation data, and can quickly provide estimates of pesticide leaching under a variety of conditions. On the basis of the gathered environmental fate data on the chemical, including simulation results, EPA decides whether the chemical has the potential to contaminate groundwater and may require monitoring studies in order to make decisions about possible use restrictions.

The pesticide registration procedures in development in the European Union are discussed in Sec. 21.4.3 as a case study.

21.4 CASE STUDIES

21.4.1 Pesticides in Italian Horticulture: Potential of Groundwater Contamination and Carryover Effects

Vischetti et al. (1999) used PELMO for simulating the environmental behavior of 13 currently used herbicides, fungicides, and insecticides in three Italian soil-climate scenarios of the production of tomato and lettuce, the two most important horticultural crops in Italy. The behavior of each pesticide was compared among the different scenarios according to dissipation through various processes, potential of groundwater contamination, and possible carryover effects, which are typical for horticultural cropping systems where numerous crops are grown in sequence during the same season. PELMO was chosen for this study because it accounts for all the agricultural practices that were considered relevant, can simulate the behavior of postemergence herbicides that act on the crop canopy, and contains a simple description of water flow in soil.

Pesticide parameter values were derived from the literature, and the treatments were simulated according to current practices. The meteorological and soil data originated directly from experimental farms that represent horticulture in northern, central, and southern Italy. One-year weather datasets covering the wettest year of the previous decade at each farm were used for the simulations. Each simulation was run for one year, then continued for a year at a time in case the applied pesticide had not yet been completely dissipated; no crop or irrigation was assumed during the additional years, and the same weather dataset was used.

For each simulation, the importance of each route of dissipation of the pesticide was assessed. The simulated mean concentrations of pesticides in water leached below the simulated soil profile were compared to the legal limit of pesticide presence in drinking water, although it was realized that the leachate undergoes strong dilution once it reaches the groundwater. Simulated residuals of herbicides in the soil profile were compared to the lowest NOELs (no observable effect levels) for horticultural crops found in the literature in order to recognize situations where the residuals of the previous crop may be toxic to the different crop that immediately follows it.

21.4.2 SuSAP Decision Support System for the Region of Lombardy, Italy

The SuSAP Decision Support System (DSS) was developed by Brenna et al. (2001) as an instrument for planning sustainable use of pesticides in agriculture in the intensively cultivated area of Lombardy in northern Italy. The software combines soil and meteorological maps from the area with the major crops and their pests and a database that covers the physicochemical and ecotoxicological properties, application plan, and efficacy of 215 active substances and pesticide products that contain them. The system can be applied on the regional, local or farm level, utilizing basic soil data on either 1:250,000 or 1:50,000 scale. Twenty "macroareas" characterized by one soil and climate were defined: representative soil profiles were characterized by field and laboratory analysis (pedotransfer functions that were found to best calculate the measured data were used for producing field capacity, wilting point and bulk density in addition to the basic soil data), and weather data were collected from 125 meteorological stations and interpolated with geostatistical methods in order to produce rainfall and temperature maps for the area. PELMO (maize) or PESTLA (rice) are run for a 12-year simulation period, including a 2-year warm-up, in order to assess pesticide leaching to groundwater. On the local and regional level, the results are given in the form of GIS vulnerability maps that visualize the 80th percentiles of pesticide concentration in soil water at 1 m depth inside areas that are similar in reference to climate and soil. At farm level, PECs of pesticides in groundwater are calculated for use in the calculation of an environmental risk indicator that helps the farmer choose a suitable pesticide.

Camisa et al. (2003) used SuSAP for evaluating the potential effect on groundwater quality of adopting genetically modified glyphosate-tolerant crops of maize, soybean, and sugar beet, which require 28–43% less herbicide use than conventional crops, in the Italian province of Cremona in

Lombardy. Applications of 18 herbicides were simulated according to 12 different weed-controlling strategies (5 for maize, 3 for soybean, 4 for sugar beet), which are representative of actual practices in the area, outside of the hypothetical strategies for the glyphosate-tolerant crops.

21.4.3 FOCUS

In the European Union, the Council Directive 91/414/EEC requires the member states to jointly review all active substances of plant protection products (PPPs). When an active substance is approved for inclusion in the Annex I of this directive, any PPPs containing this chemical are then assessed and authorized nationally in each member state. The directives 91/414/EEC and 95/36/EC define this review process, which emphasizes the determination of PECs (predicted environmental concentrations) in soil, ground and surface water, and air. Mathematical models are seen as important tools for estimating PECs.

In 1993, the forum for the Co-ordination of pesticide fate models and Their Use (FOCUS), was formed for the purpose of selecting appropriate models to predict environmental concentrations of pesticides, and standardizing these models and their use, through cooperation of European regulatory authorities, industry, and researchers. FOCUS has since chosen four regulatory leaching models for predicting concentrations in groundwater: PRZM, PELMO, PEARL (formerly PESTLA), and MACRO. These models are standardized and improved for regulatory use by FOCUS working groups. The models are run within FOCUS shells, which simplify the use of the models for first-tier assessments by supplying the standard EU crop, soil, and climate scenarios also devised by FOCUS.

The nine standard EU groundwater scenarios for first-tier leaching assessments are so-called realistic worst cases (actually occurring combinations of crops, soils, and climates) designed to represent the 90th percentile of groundwater vulnerability within agricultural areas in the whole of EU. For each substance, simulations are run with at least one appropriate FOCUS leaching model for those of the standard scenarios considered relevant in the light of the intended use of the chemical. If PECs in groundwater exceed the trigger value of 0.1 μg/L for all relevant scenarios, the chemical cannot be included in Annex I unless higher-tier studies show an absence of actual danger of groundwater contamination in at least one relevant use scenario. If the PEC in groundwater is lower than 0.1 μg/L in at least one relevant standard scenario, the groundwater contamination potential of the chemical allows its inclusion in Annex I, and each member state can use the simulation results and existing higher-tier data to determine potentially vulnerable local-use situations in the course of the national authorization process.

FOCUS working groups have identified a number of potential problems with the validity of the chosen models and scenarios. The EU project APECOP (Effective Approaches for Predicting Environmental

Concentrations of Pesticides) was realized for the purpose of improving modeling concepts for PECs in groundwater and air (the latter concerning volatilization of pesticides from the soil and plants), improving the validation status of the four models, and evaluating the representativity of the current first-tier groundwater scenarios (Vanclooster et al., 2003a,b).

Within the framework of the project APECOP, suitable modules for describing macropore flow were created for the models PELMO and PEARL, considering their different hydrology submodels, and the macropore flow module of MACRO was updated (Jarvis et al., 2003; Vanclooster et al., 2003b). In order to better account for the presence of pesticides in the air, a variety of experiments and field studies were performed to define the mechanisms of pesticide volatilization from soil. A simple procedure for considering volatilization from soil was included in MACRO and the existing descriptions in PEARL and PELMO were updated to consider the increase of the sorption coefficient at low soil moisture content (van den Berg et al., 2003; Vanclooster et al., 2003b). Suitable descriptions were also included or improved in all models concerning the volatilization of pesticides from plants (Wolters et al., 2003; Vanclooster et al., 2003b).

The validation status of the FOCUS leaching models was improved with the help of a selection of old and new field datasets that were considered comprehensive and of good quality. A common agreed stepwise validation protocol was followed with each dataset and model in which the main components of the models were validated separately in order to reduce uncertainty. The hydrological component of the models was validated using measured terms of water balance, the solute transport component with the help of tracer data, and the pesticide fate and transport component using pesticide balance data. In addition, the validation progressed through an initial blind validation step (no laboratory data was made available on pesticide degradation and sorption or soil hydrology, concerning new datasets; no data was made available on hydrology, soil heat, solutes, or pesticides at the site itself for old datasets) to using calibration as a parameter-estimation technique, and to evaluating the ability of the calibrated model to extrapolate or predict (Trevisan et al., 2003; Vanclooster et al., 2003b).

Since only a limited amount of observational data can realistically exist, the real 90th percentile of leaching conditions will remain unknown, and a pragmatic technique, which uses a spatially distributed leaching model and Pan-European soil, climate, and agricultural databases, was adopted instead for estimating the 90th percentile and evaluating the representativity of the "realistic worst cases" currently selected inside each of the nine so-called FOCUS areas for regulation purposes. Simulations were carried out with PEARL for 1062 "plots," each representing a unique combination of model inputs within Europe, which constituted a spatially distributed leaching model referred to as EuroPEARL, considering two major crops (winter wheat and maize) and four hypothetical pesticides of different properties and making some simplifying assumptions. However, due to the lack of adequate soil profile data, this exercise only covered approximately

75% of the agricultural land within the EU (Tiktak et al., 2003; Vanclooster et al., 2003b). A metamodel was derived from EuroPEARL by relating the predicted annual leaching concentrations at the depth of 1 m to some basic soil and climate input parameters. In this way, the European 1:1,000,000 soil map, which covers 97% of agricultural land within the EU, could be used for leaching simulations, and the validity of the current realistic worst case scenarios could be tested statistically (Piñeros Garcet et al., 2003; Vanclooster et al., 2003b).

REFERENCES

Addiscott, T.M., and R.J. Wagenet. 1985. Concepts of solute leaching in soils: a review of new modeling approaches. J. Soil Sci. 36:411–424.

Ares, J.O., A.M. Miglierina, and R. Sanchez. 1999. Patterns of groundwater concentration and fate of lindane in an irrigated semiarid area in Argentina. Environ. Toxicol. Chem. 18:1354–1361.

Auteri, D., G. Azimonti, I. Bernandinelli, C. Catapano, L. Galbiati, V. Kambourova, T. Mammone, P. Ragni, C. Riparbelli, and M. Maroni. 1999. PESTIGIS: Decision Support System for the assessment of environmental impact of pesticides in Lombardy, Italy. p. 823–829. In A.A.M. Del Re et al. (eds.), Proc. XI Symp. Pestic. Chem., Cremona, Italy. 11–15 Sept. 1999. La Goliardica Pavese, Pavia.

Borah, M.J., and P.K. Kalita. 1999. Development and evaluation of a macropore flow component for LEACHM. Trans. ASAE 42:65–78.

Brenna, S., C. Riparbelli, M. Trevisan, E. Capri, and D. Auteri. 2001. SuSAP: un sistema di supporto alle decisioni per un uso sostenibile dei prodotti fitosanitari. (In Italian). p. 51–66. In Barberis, R., and A. Pugliese (eds.), Modellistica e qualità ambientale dei suoli. RTI CTN SSC 1/2001. Agenzia Nazionale per la Protezione dell'Ambiente (ANPA), Rome.

Brown, C.D., U. Baer, P. Gunther, M. Trevisan, and A. Walker. 1996. Ring test with the models LEACHP, PRZM-2 and VARLEACH: variability between model users in prediction of pesticide leaching using a standard data set. Pestic. Sci. 47:249–258.

Businelli, M., M. Marini, D. Businelli, and G. Gigliotti. 2000. Transport to groundwater of six commonly used herbicides: a prediction for two Italian scenarios. Pest Manag. Sci. 56:181–188.

Businelli, D., E. Tombesi, R. Calandra, and M. Trevisan. 1999. Assessment of potential groundwater pollution by winter wheat herbicides using MACRO-DB model. p. 233–244. In A.A.M. Del Re et al. (eds.), Proc. XI Symp. Pestic. Chem., Cremona, Italy. 11–15 Sept. 1999. La Goliardica Pavese, Pavia.

Camisa, M.G., G. Fontana, and B. Cambon. 2003. Using the Decision Support System SuSAP to compare potential leaching of active substances used in conventional and Roundup Ready based weed management strategies. p. 747–756. In A.A.M. Del Re et al. (eds.) Proc. XII Symp. Pestic. Chem., Piacenza, Italy. 4–6 June 2003. La Goliardica Pavese, Pavia.

Carsel, R.F., J.C. Imhoff, P.R. Hummel, J.M. Cheplick, and J.S. Donigian Jr. 1998. PRZM-3, a model for predicting pesticide and nitrogen fate in crop root and unsaturated soil zones: User's Manual for Release 3.0. U.S. Environmental Protection Agency, Athens, GA.

Chaudhari, N.M. 1971. An improved numerical technique for solving multidimensional miscible displacement equations. Soc. Pet. Eng. J. 11:277–284.

Close, M.E., L. Pang, J.P.C. Watt, and K.W. Vincent. 1998. Leaching of picloram, atrazine and simazine through two New Zealand soils. Geoderma 84:45–63.

Del Re, A.A.M., and M. Trevisan. 1993. Testing models of the unsaturated zone. p. 5–31. *In* A.A.M. Del Re et al. (eds.) Proc. IX Simp. Pestic. Chem., Piacenza, Italy. 11–13 Oct. 1993. Edizioni G. Biagini, Lucca.

Del Re, A.A.M., and M. Trevisan. 1995. Selection criteria of xenobiotic leaching models in soil. Eur. J. Agron. 4:465–472.

Dubus, I.G., and S. Beulke. 2001. Independent calibration of PEARL against lysimeter data by two modellers. Cranfield Centre for Ecochemistry research report for DEFRA PLO539. Cranfield Centre for Ecochemistry, Cranfield University, Silsoe.

Dubus, I.G., and C.D. Brown. 2002. Sensitivity and first-step uncertainty analysis for the preferential flow model MACRO. J. Environ. Qual. 31:227–240.

Dubus, I.G., C.D. Brown, and S. Beulke. 2003. Sensitivity analyses for four pesticide leaching models. Pest Manag. Sci. 59:962–982.

Feyen, J., D. Jacques, A. Timmerman, and J. Vanderborght. 1998. Modelling water flow and solute transport in heterogenous soils: a review of recent approaches. J. Agric. Eng. Res. 70:231–256.

Foltz, J.C., J.G. Lee, and M.A. Martin. 1993. Farm-level economic and environmental impacts of eastern Corn Belt cropping systems. J. Prod. Agric. 6:290–296.

Foltz, J.C., J.G. Lee, M.A. Martin, and P.V. Preckel. 1995. Multiattribute assessment of alternative cropping systems, Am. J. Agr. Econ. 77:408–420.

Galbiati, L., F. Bouraoui, C. Riparbelli, and D. Auteri. 2003. Development and application of a modelling tool to evaluate soil and groundwater vulnerability to pesticide leaching. p. 457–466. *In* A.A.M. Del Re et al. (eds.), Proc. XII Symp. Pestic. Chem., Piacenza, Italy. 4–6 June 2003. La Goliardica Pavese, Pavia.

Garratt, J.A., E. Capri, M. Trevisan, G. Errera, and R.M. Wilkins. 2002. Parameterisation, evaluation and comparison of pesticide leaching models to data from a Bologna field site, Italy. Pest Manag. Sci. 58:3–20.

Giupponi, C., G. Bonaiti, E. Capri, G. Errera, and M. Trevisan. 1996. Effects of alternative soil tillage systems on the degradation and sorption of herbicides. p. 151–161. *In* A.A.M. Del Re et al. (eds.), Proc. X Symp. Pestic. Chem., Castelnuovo Fogliani, Piacenza, Italy. 30 Sept.–2 Oct. 1996. La Goliardica Pavese, Pavia.

Goss, D.W. 1992. Screening procedure for soils and pesticides for potential water quality impacts. Weed Technol. 6:701–708.

Harms, C.T., H. Forster, and J. Hosang. 1996. Leaching of metolachlor through soils: assessing the effects of variable soil degradation and adsorption properties by means of Monte Carlo simulation. p. 171–178. *In* A.A.M. Del Re et al. (eds.), Proc. X Symp. Pestic. Chem., Castelnuovo Fogliani, Piacenza, Italy. 30 Sept–2 Oct. 1996. La Goliardica Pavese, Pavia.

Hegg, R.O., W.H. Shelley, R.L. Jones, and R.R. Romine. 1988. Movement and degradation of aldicarb residues in South Carolina loamy sand soil. Agric. Ecosyst. Environ. 20:303–315.

Hutson, J.L., and R.J. Wagenet. 1992. LEACHM (Leaching Estimation and Chemistry Model): A process-based model of water and solute movement, transformations, plant uptake and chemical reactions in the unsaturated zone, Version 3.

Department of Soil, Crop and Atmospheric Sciences, Research Series No. 92–3. Cornell University, Ithaca, NY.

Hutson, J.L., and R.J. Wagenet. 1995. An overview of LEACHM: a process based model of water and solute movement, transformations, plant uptake and chemical reactions in the unsaturated zone. p. 409–422. *In* Chemical Equilibrium and Reaction Models. SSSA Special Publication 42. Soil Science Society of America, American Society of Agronomy, Madison (WI).

Jarvis, N., J. Boesten, R. Hendriks, M. Klein, M. Larsbo, S. Roulier, F. Stenemo, and A. Tiktak. 2003. Incorporating macropore flow into FOCUS PEC models. p. 963–972. *In* A.A.M. Del Re et al. (eds.) Proc. XII Symp. Pestic. Chem., Piacenza, Italy. 4–6 June 2003. La Goliardica Pavese, Pavia.

Jarvis, N.J., J.M. Hollis, P.H. Nicholls, T. Mayer, and S.P. Evans. 1997. MACRO-DB: a decision-support tool for assessing pesticide fate and mobility in soils. Environ. Modell. Softw. 12:251–265.

Jones, R.L., and G. Mangels. 2002. Review of the validation of models used in Federal Insecticide, Fungicide, and Rodenticide Act environmental exposure assessments. Environ. Toxicol. Chem. 21:1535–1544.

Jones, R., and H. Schäfer. 1995. Assessment of various leaching models. p. 20–51. *In* FOCUS. Leaching models and EU registration. The final report of the work of the Regulatory Modelling Work group of FOCUS (FOrum for the Co-ordination of pesticide fate models and their Use). DOC.4952/VI/95.

Karpouzas, D.G., A. Walker, D.S. Drennan, and R.J. Froud-Williams. 2001. The effect of initial concentration of carbofuran on the development and stability of its enhanced biodegradation in top-soil and sub-soil. Pest Manag. Sci. 57:72–81.

Klein, M. 1995. PELMO (Pesticide Leaching Model) Version 2.01 User's Manual. Fraunhofer-Institut, Schmallenberg.

Klein, M. 1997. Statistical distribution of pesticide concentrations in leachate. Results of a Monte-Carlo analysis performed with PELMO. Chemosphere 35:379–389.

Knisel, W.G., and F.M. Davis. 1999. GLEAMS: Groundwater Loading Effects of Agricultural Management Systems, Version 3.0 (User Manual). Publication No. SEWRL-WGK/FMD-050199, revised 081500. Tifton (GA).

Krzyszowska, A.J., R.D. Allen, and G.F. Vance. 1994. Assessment of the fate of two herbicides in a Wyoming rangeland soil: column studies. J. Environ. Qual. 23:1051–1058.

Lantz, R.B. 1971. Quantitative evaluation of numerical diffusion (truncation error). Soc. Pet. Eng. J. 11:315–320.

Larsbo, M., and N. Jarvis. 2003. MACRO 5.0: A model of water flow and solute transport in macroporous soil, technical description. Swedish University of Agricultural Sciences, Sweden.

Larsson, M.H., and N.J. Jarvis. 2000. Quantifying interactions between compound properties and macropore flow effects on pesticide leaching. Pest. Manag. Sci. 56:133–141.

Leistra, M., A.M.A. van der Linden, J.J.T.I. Boesten, A. Tiktak, and F. van den Berg. 2001. PEARL model for pesticide behaviour and emissions in soil-plant systems; Descriptions of the processes in FOCUS PEARL v 1.1.1. RIVM report 711401009/Alterra-rapport 013. Alterra, Green World Research, Wageningen.

Leonard, R.A., W.G. Knisel, and D.A. Still. 1987. GLEAMS: Groundwater Loading Effects of Agricultural Management Systems. Trans. ASAE 30:1403–1418.

Loague, K., R.L. Bernknopf, R.E. Green, and T.W. Giambelluca. 1996. Uncertainty of groundwater vulnerability assessments for agricultural regions in Hawaii: review. J. Environ. Qual. 25:475–490.

Loague, K., D.A. Lloyd, A. Nguyen, S.N. Davis, and R.H. Abrams. 1998. A case study simulation of DBCP groundwater contamination in Fresno County, California. 1. Leaching through the unsaturated subsurface. J. Contam. Hydrol. 29:109–136.

Mills, M.S., I.R. Hill, A.C. Newcombe, N.D. Simmons, P.C. Vaughan, and A.A. Verity. 2001. Quantification of acetochlor degradation in the unsaturated zone using two novel *in situ* field techniques: comparisons with laboratory generated data and implications for groundwater risk assessments. Pest Manag. Sci. 57:351–359.

Mockus, V. 1971. Estimation of direct runoff from storm rainfall. p. 10.1–10.24. *In* SCS National Engineering Handbook, Section 4: Hydrology. Soil Conservation Service, U.S. Department of Agriculture, Washington (DC).

Morari, F., and W.G. Knisel. 1997. Modifications of the GLEAMS model for crack flow. Trans. ASAE 40:1337–1348.

Oreskes, N., K. Shrader-Frechette, and K. Belitz. 1994. Verification, validation, and confirmation of numerical models in the earth sciences. Science 263: 641–646.

Piñeros Garcet, J.D., M. Vanclooster, A. Tiktak, D.S. De Nie, and A. Jones. 2003. A methodology for evaluating first tier PEC groundwater scenarios supporting the prediction of environmental concentrations of pesticides at the European scale. p. 951–962. *In* A.A.M. Del Re et al. (eds.) Proc. XII Symp. Pestic. Chem., Piacenza, Italy. 4–6 June 2003. La Goliardica Pavese, Pavia.

Šimůnek, J., N.J. Jarvis, M.Th. van Genuchten, and A. Gärdenäs. 2003. Review and comparison of models for describing non-equilibrium and preferential flow and transport in the vadose zone, J. Hydrol. 272:14–35.

Smith, W.N., S.O. Prasher, and S.F. Barrington. 1991. Evaluation of PRZM and LEACHMP on intact soil columns. Trans. ASAE 34:2413–2420.

Soutter, M. and A. Musy. 1999. Global sensitivity analyses of three pesticide leaching models using a Monte Carlo approach. J. Environ. Qual. 28:1290–1297.

Stenemo, F., and N. Jarvis. 2003. Users guide to MACRO 5.0, a model of water flow and solute transport in macroporous soil. Swedish University of Agricultural Sciences, Sweden.

Stewart, I.T., and K. Loague. 1999. A type transfer function approach for regional-scale pesticide leaching assessments. J. Environ. Qual. 28:378–387.

Tiktak, A., J.D. Piñeros Garcet, D.S. De Nie, M. Vanclooster, and A. Jones. 2003. Assessment of pesticide leaching at the Pan-European level using a spatially distributed model. p. 941–950. *In* A.A.M. Del Re et al. (eds.), Proc. XII Symp. Pestic. Chem., Piacenza, Italy. 4–6 June 2003. La Goliardica Pavese, Pavia.

Tiktak, A., F. van den Berg, J.J.T.I. Boesten, D. van Kraalingen, M. Leistra, and A.M.A. van der Linden. 2000. Manual of FOCUS PEARL version 1.1.1. RIVM report 711401 008/Alterra report 28. RIVM and Alterra, The Netherlands.

Trapp, S., and J.C. Mc Farlane (eds.). 1995. Plant contamination: modeling and simulation of organic chemical processes. CRC Press. Boca Raton, FL.

Trevisan, M., L. Padovani, N. Jarvis, F. Roulier, F. Bouraoui, M. Klein, and J. Boesten. 2003. Validation status of the present PEC groundwater models. p. 933–940. *In* A.A.M. Del Re et al. (eds.), Proc. XII Symp. Pestic. Chem., Piacenza, Italy. 4–6 June 2003. La Goliardica Pavese, Pavia.

Truman, C.C., and R.A. Leonard. 1991. Effects of pesticide, soil, and rainfall characteristics on potential pesticide loss by percolation — a GLEAMS simulation. Trans. ASAE 34:2461–2468.

Vanclooster, M., A. Armstrong, F. Bouraoui, G. Bidoglio, J.J.T.I. Boesten, P. Burauel, E. Capri, D. De Nie, E. Fernandez, N. Jarvis, A. Jones, M. Klein, M. Leistra, V. Linnemann, J.D. Piñeros Garcet, J.H. Smelt, A. Tiktak, M. Trevisan, F. Van den Berg, A. Van der Linden, H. Vereecken, and A. Wolters. 2003a. Effective approaches for predicting environmental concentrations of pesticides: The APECOP project. p. 923–931. *In* A.A.M. Del Re et al. (eds.), Proc. XII Symp. Pestic. Chem., Piacenza, Italy. 4–6 June 2003. La Goliardica Pavese, Pavia.

Vanclooster, M., J.J.T.I Boesten, M. Trevisan, C.D. Brown, E. Capri, O.M. Eklo, B. Gottesbüren, V. Gouy, and A.M.A. van der Linden. 2000. Pesticide leaching modelling validation: A European experience (special issue). Agr. Water Manag. 44:1–409.

Vanclooster, M., J. Diels, D. Mallants, and J. Feyen. 1994. Characterizing the impact of uncertain soil properties on simulated pesticide leaching. p. 470–475. *In* A. Copin et al. (eds.), Proc. 5th International Workshop on Environmental Behaviour of Pesticides and Regulatory Aspects, Brussels, Belgium. 26–29 Apr. 1994. European Study Service, Rixensart.

Vanclooster, M., J.D. Piñeros Garcet, J.J.T.I. Boesten, F. Van den Berg, M. Leistra, J. Smelt, N. Jarvis, S. Roulier, P. Burauel, H. Vereecken, A. Wolters, V. Linnemann, E. Fernandez, M. Trevisan, E. Capri, L. Padovani, M. Klein, A. Tiktak, A. Van der Linden, D. De Nie, G. Bidoglio, F. Baouroui, A. Jones, and A Armstrong. 2003b. APECOP: Effective approaches for assessing the predicted environmental concentrations of pesticides — final report, Department of Environmental Sciences And Land Use Planning, Universite Catholique de Louvain, Belgium.

van den Berg, F., A. Wolters, N. Jarvis, M. Klein, J.J.T.I. Boesten, M. Leistra, V. Linnemann, J.H. Smelt, and H. Vereecken. 2003. Improvement of concepts for pesticide volatilisation from bare soil in PEARL, PELMO, and MACRO models. p. 973–983. *In* A.A.M. Del Re et al. (eds.), Proc. XII Symp. Pestic. Chem., Piacenza, Italy. 4–6 June 2003. La Goliardica Pavese, Pavia.

van Genuchten, M.T., and P.J. Wierenga. 1974. Simulation of one-dimensional solute transfer in porous media. Agr. Exp. Sta. Bull. 628. New Mexico State University, Las Cruces, New Mexico.

van Wesenbeeck, I.J., and P.L. Havens. 1999. A groundwater exposure assessment for cloransulam-methyl in the U.S. soybean market. J. Environ. Qual. 28:513–522.

Vischetti, C., A. Esposito, A. Onofri, M. Trevisan, and G. Zanin. 1999. Simulation of the behaviour of some pesticides for horticultural crops in soil using Italian scenarios. p. 491–499. *In* A.A.M. Del Re et al. (eds.), Proc. XI Symp. Pestic. Chem., Cremona, Italy. 11–15 Sept. 1999. La Goliardica Pavese, Pavia.

Vischetti, C., P. Perucci, and L. Scarponi. 1997. Rimsulfuron in soil: Effect of persistence on growth and activity of microbial biomass at varying environmental conditions. Biogeochemistry 39:165–176.

Wagenet, R.J. 1993. A review of pesticide leaching models and their application to field and laboratory data. p. 33–62. *In* A.A.M. Del Re et al. (eds.), Proc. IX Simp. Pestic. Chem., Piacenza, Italy. 11–13 Oct. 1993. Edizioni G. Biagini, Lucca.

Walker, A. 1987. Evaluation of a simulation model for prediction of herbicide movement and persistence in soil. Weed Res. 27:143–152.

Walker, A., R. Calvet, A.A.M. Del Re, W. Pestemer, and J.M. Hollis. 1995. Evaluation and improvement of mathematical models of pesticide mobility in soils and assessment of their potential to predict contamination of water systems. Final Report on European Union contract EV5V-CT92–0226. Blackwell Wissenschaft-Verlag, Berlin.

Wauchope, R.D., S. Yeh, J.B.H.J. Linders, R. Kloskowski, K. Tanaka, B. Rubin, A. Katayama, W. Kördel, Z. Gersti, M. Lane, and J.B. Unsworth. 2002. Pesticide soil sorption parameters: theory, measurement, uses, limitations and reliability. Pest Manag. Sci. 58:419–445.

Williams, J.R., and H.D. Berndt. 1977. Sediment yield prediction based on watershed hydrology. Trans. ASAE 20:1100–1104.

Wolt, J., P. Singh, S. Cryer, and J. Lin. 2002. Sensitivity analysis for validating expert opinion as to ideal data set criteria for transport modeling. Environ. Toxicol. Chem. 21:1558–1565.

Wolters, A., M. Leistra, V. Linnemann, J.H. Smelt, F. Van den Berg, M. Klein, N. Jarvis, J.J.T.I. Boesten, and H. Vereecken. 2003. Pesticide volatilisation from plants: Improvement of the PEARL, PELMO, and MACRO models. p. 985–994. *In* A.A.M. Del Re et al. (eds.), Proc. XII Symp. Pestic. Chem., Piacenza, Italy. 4–6 June 2003. La Goliardica Pavese, Pavia.

Wu, Q.J., and S.R. Workman. 1999. Stochastic simulation of pesticide transport in heterogenous unsaturated fields. J. Environ. Qual. 28:498–512.

Zubkoff, P.L. 1992. The use of runoff and surface water transport and fate models in the pesticide registration process. Weed Technol. 6:743–748.

Index

A

AA spectrometry, *see* Atomic absorption spectrometry
AC, *see* Alternating current
Accelerated solvent extraction, 446
Active sensors, 154
ADE, *see* Advection-dispersion equation
Adenosine triphosphate (ATP), 566
ADR, *see* Amplitude domain reflectometry
Advanced very high-resolution radiometry (AVHRR), 32
Advection-dispersion equation (ADE), 101, 102, 115, 374, 398, 400
Aerial photography, 28, 151, 640
Aerodynamic roughness, 481
AF, *see* Attenuation factor
Agricultural activities, movements of NPS pollutants caused by, 8
Agricultural fields, evaporation rates from, 466
Agricultural lands, salinity, 27
Agricultural productivity, pesticide use and, 436
Agriculture, precision, 660
Agrochemicals, leaching estimates, 45
Air
  -cored electromagnets, 267
  entrapment, 706
  -entry
    permeameter, modified, 211
    value, 201
  permeability, 423
  pollution fallout, 9
  sampling, 471
Akaike criterion, 706
Alkaline and Acid Persulfate Digestion Method, 535
Alternating current (AC), 329, 366
Alundum tension plate sampler, 413
American Public Health Association (APHA), 506
American Society for Testing and Materials (ASTM), 207, 507

Ammonia
  determination
    field monitoring probes, 521
    field testing kits, 521
    laboratory methods for, 517
  diffusion of, 519
  loss, 467
  open-path lasers for, 494
  -oxidizing bacteria (AOB), 567, 571
  samplers accumulating, 488
Amplitude domain reflectometry (ADR), 176, 177
Analysis of variance (ANOVA), 683
ANNs, *see* Artificial neural networks
ANOVA, *see* Analysis of variance
AOAC, *see* Association of Official Analytical Chemists–International
AOB, *see* Ammonia-oxidizing bacteria
APECOP project, 728
APHA, *see* American Public Health Association
APHA Standard Methods, 508–515, 519, 524, 526
Apparent dispersivity, 106
Apparent hysteresis effect, 108
Apparent soil bulk electrical conductivity, 322
Apparent soil electrical conductivity, 640, 650
Apparent tortuosity factor, 106
Aquifers, prevention of pollution of, 310
Arbitrary scales, 71
Artificial neural networks (ANNs), 304
ASCII grid, 679
Association of Official Analytical Chemists–International (AOAC), 507
ASTM, *see* American Society for Testing and Materials
Atmospheric turbulence theory, 111
Atomic absorption (AA) spectrometry, 539
ATP, *see* Adenosine triphosphate
Atrazine, 108, 442, 443, 449